Probability and Random Variables: Theory and Applications

Iickho Song · So Ryoung Park · Seokho Yoon

Probability and Random Variables: Theory and Applications

Iickho Song
School of Electrical Engineering
Korea Advanced Institute of Science
and Technology
Daejeon, Korea (Republic of)

So Ryoung Park
School of Information, Communications,
and Electronics Engineering
The Catholic University of Korea
Bucheon, Korea (Republic of)

Seokho Yoon
College of Information and Communication
Engineering
Sungkyunkwan University
Suwon, Korea (Republic of)

ISBN 978-3-030-97681-1   ISBN 978-3-030-97679-8 (eBook)
https://doi.org/10.1007/978-3-030-97679-8

To
our kin and academic ancestors and families
and
to
all those who appreciate and enjoy
the beauty of thinking and learning

To
Professors
Souguil J. M. Ann,
Myung Soo Cha,
Saleem A. Kassam, and
Jordan M. Stoyanov
for their invisible yet enlightening guidance

Preface

This book is a translated version, with some revisions, from *Theory of Random Variables*, originally written in Korean by the first author in the year 2020. This book is intended primarily for those who try to advance one step further beyond the basic level of knowledge and experience on probability and random variables. At the same time, this book would also be a good resource for experienced scholars to review and refine familiar concepts. For these purposes, the authors have included definitions of basic concepts in clear terms, key advanced concepts in mathematics, and diverse concepts and notions of probability and random variables with a significant number of examples and exercise problems.

The organization of this book is as follows: Chap. 1 describes the theory of sets and functions. The unit step function and impulse function, to be used frequently in the following chapters, are also discussed in detail, and the gamma function and binomial coefficients in the complex domain are introduced. In Chap. 2, the concept of sigma algebra is discussed, which is the key for defining probability logically. The notions of probability and conditional probability are then discussed, and several classes of widely used discrete and continuous probability spaces are introduced. In addition, important notions of probability mass function and probability density function are described. After discussing another important notion of cumulative distribution function, Chap. 3 is devoted to the discussion on the notions of random variables and moments, and also for the discussion on the transformations of random variables.

In Chap. 4, the concept of random variables is generalized into random vectors, also referred to as joint random variables. Transformations of random vectors are discussed in detail. The discussion on the applications of the unit step function and impulse function in random vectors in this chapter is a unique trait of this book. Chapter 5 focuses on the discussion of normal random variables and normal random vectors. The explicit formula of joint moments of normal random vectors, another uniqueness of this book, is delineated in detail. Three statistics from normal samples and three classes of impulsive distributions are also described in this chapter. In Chap. 6, the authors briefly describe the fundamental aspects of the convergence of random variables. The central limit theorem, one of the most powerful and useful results with practical applications, is among the key expositions in this chapter.

The uniqueness of this book includes, but is not limited to, interesting applications of impulse functions to random vectors, exposition of the general formula for the product moments of normal random vectors, discussion on gamma functions and binomial coefficients in the complex space, detailed procedures to the final answers for almost all results presented, and a substantially useful and extensive index for finding subjects more easily. A total of more than 320 exercise problems are included, of which a complete solution manual for all the problems is available from the authors through the publisher.

The authors feel sincerely thankful that, as is needed for the publication of any book, the publication of this book became a reality thanks to a huge amount of help from many people to the authors in a variety of ways. Unfortunately, the authors could mention only some of them explicitly: to the anonymous reviewers for constructive and helpful comments and suggestions, to Bok-Lak Choi and Seung-Ki Kim at Saengneung for allowing the use of the original Korean title, to Eva Hiarapi and Yogesh Padmanaban at Springer Nature for extensive editorial assistance, and to Amelia Youngwha Song Pegram and Yeonwha Song Wratil for improving the readability. In addition, the research grant 2018R1A2A1A05023192 from Korea Research Foundation was an essential support in successfully completing the preparation of this book.

The authors would feel rewarded if everyone who spends time and effort wisely in reading and understanding the contents of this book enjoys the pleasure of learning and advancing one step further.

Thank you!

Daejeon, Korea (Republic of)   Iickho Song
Bucheon, Korea (Republic of)   So Ryoung Park
Suwon, Korea (Republic of)   Seokho Yoon
January 2022

Contents

Chapter 1
Preliminaries

Sets and functions are key concepts that play an important role in understanding probability and random variables. In this chapter, we discuss those concepts that will be used in later chapters.

1.1 Set Theory

In this section, we introduce and review some concepts and key results in the theory of sets (Halmos 1950; Kharazishvili 2004; Shiryaev 1996; Sommerville 1958).

1.1.1 Sets

Definition 1.1.1 (*abstract space*) The collection of all entities is called an abstract space, a space, or a universal set.

Definition 1.1.2 (*element*) The smallest unit that comprises an abstract space is called an element, a point, or a component.

Definition 1.1.3 (*set*) Given an abstract space, a grouping or collection of elements of the abstract space is called a set.

An abstract space, often denoted by Ω or S, consists of elements or points, the smallest entities that we shall discuss. In the strict sense, a set is the collection of elements that can be clearly defined mathematically. For example, the collection of 'people who are taller than 1.5 m' is a set. On the other hand, the collection of 'tall people' is not a set because 'tall' is not mathematically clear. Yet, in fuzzy set theory, such a vague collection is also regarded as a set by adopting the concept of membership function.

Abstract spaces and sets are often represented with braces { } with all elements explicitly shown, e.g. $\{1, 2, 3\}$; with the property of the elements described, e.g., $\{\omega : 10 < \omega < 20\pi\}$; $\{a_i\}$; or $\{a_i\}_{i=1}^n$.

Example 1.1.1 The result of signal processing in binary digital communication can be represented by the abstract space $\Omega = \{0, 1\}$. The collection $\{A, B, \ldots, Z\}$ of capital letters of the English alphabet and the collection $S = \{(0, 0, \ldots, 0), (0, 0, \ldots, 1), \ldots, (1, 1, \ldots, 1)\}$ of binary vectors are also abstract spaces.   ◇

Example 1.1.2 In the abstract space $\Omega = \{0, 1\}$, 0 and 1 are elements. The abstract space of seven-dimensional binary vectors contains $2^7 = 128$ elements.   ◇

Example 1.1.3 The set $A = \{1, 2, 3, 4\}$ can also be depicted as, for example, $A = \{\omega : \omega$ is a natural number smaller than 5$\}$.   ◇

Definition 1.1.4 (*point set*) A set with a single point is called a point set or a singleton set.

Example 1.1.4 The sets $\{0\}$, $\{1\}$, and $\{2\}$ are point sets.   ◇

Consider an abstract space Ω and a set G of elements from Ω. When the element ω does and does not belong to G, it is denoted by

$$\omega \in G \tag{1.1.1}$$

and $\omega \notin G$, respectively. Sometimes $\omega \in G$ is expressed as $G \ni \omega$, and $\omega \notin G$ as $G \not\ni \omega$.

Example 1.1.5 For the set $A = \{0, 1\}$, we have $0 \in A$ and $2 \notin A$.   ◇

Definition 1.1.5 (*subset*) If all the elements of a set B belong to another set A, then the set B is called a subset of A, which is expressed as $B \subseteq A$ or $A \supseteq B$. When B is not a subset of A, it is expressed as $B \not\subseteq A$ or $A \not\supseteq B$.

Example 1.1.6 When $A = \{0, 1, 2, 3\}$, $B = \{0, 1\}$, and $C = \{2, 3\}$, it is clear $B \subseteq A$ and $A \supseteq C$. The set A is not a subset of B because some elements of A are not elements of B. In addition, $B \not\supseteq C$ and $B \not\subseteq C$.   ◇

Example 1.1.7 Any set is a subset of itself. In other words, $A \subseteq A$ for any set A. ◇

Definition 1.1.6 (*equality*) If all the elements of A belong to B and all the elements of B belong to A, then A and B are called equal, which is written as $A = B$.

Example 1.1.8 The set $A = \{\omega : \omega$ is a multiple of 25, larger than 15, and smaller than 99$\}$ is equal to $B = \{25, 50, 75\}$, and $C = \{1, 2, 3\}$ is equal to $D = \{3, 1, 2\}$. In other words, $A = B$ and $C = D$.   ◇

Definition 1.1.7 (*proper subset*) When $B \subseteq A$ and $B \neq A$, the set B is called a proper subset of A, which is denoted by $B \subset A$ or $A \supset B$.

Example 1.1.9 The set $B = \{0, 1\}$ is a proper set of $A = \{0, 1, 2, 3\}$; that is, $B \subset A$.
◇

In some cases, $\subseteq$ and $\subset$ are used interchangeably.

Theorem 1.1.1 *We have*

$$A = B \rightleftarrows A \subseteq B, \ B \subseteq A. \tag{1.1.2}$$

In other words, two sets A and B are equal if and only if $A \subseteq B$ and $B \subseteq A$.

As we can later see in the proof of Theorems 1.1.4 and 1.1.1 is useful especially for proving the equality of two sets.

Definition 1.1.8 (*empty set*) A set with no point is called an empty set or a null set, and is denote by $\emptyset$ or $\{\ \}$.

Note that the empty set $\emptyset = \{\ \}$ is different from the point set $\{0\}$ composed of one element 0. One interesting property of the empty set is shown in the theorem below.

Theorem 1.1.2 *An empty set is a subset of any set.*

Example 1.1.10 For the sets $A = \{0, 1, 2, 3\}$ and $B = \{1, 5\}$, we have $\emptyset \subseteq A$ and $\{\ \} \subseteq B$.
◇

Definition 1.1.9 (*finite set; infinite set*) A set with a finite or an infinite number of elements is called a finite or an infinite set, respectively.

Definition 1.1.10 (*set of natural numbers; set of integers; set of real numbers*) We will often denote the sets of natural numbers, integers, and real numbers by

$$\mathbb{J}_+ = \{1, 2, \ldots\}, \tag{1.1.3}$$
$$\mathbb{J} = \{\ldots, -1, 0, 1, \ldots\}, \tag{1.1.4}$$

and

$$\mathbb{R} = \{x : x \text{ is a real number}\}, \tag{1.1.5}$$

respectively.

Example 1.1.11 The set $\{1, 2, 3\}$ is a finite set and the null set $\{\ \} = \emptyset$ is also a finite set. The set $\{\omega : \omega \text{ is a natural number}, \ 0 < \omega < 10\}$ is a finite set and $\{\omega : \omega \text{ is a real number}, \ 0 < \omega < 10\}$ is an infinite set.
◇

Example 1.1.12 The sets $\mathbb{J}_+$, $\mathbb{J}$, and $\mathbb{R}$ are infinite sets.   ◇

Definition 1.1.11 (*interval*) An infinite set composed of all the real numbers between two distinct real numbers is called an interval or an interval set.

Let $a < b$ and $a, b \in \mathbb{R}$. Then, the sets $\{\omega : \omega \in \mathbb{R}, a \le \omega \le b\}$, $\{\omega : \omega \in \mathbb{R}, a < \omega < b\}$, $\{\omega : \omega \in \mathbb{R}, a \le \omega < b\}$, and $\{\omega : \omega \in \mathbb{R}, a < \omega \le b\}$ are denoted by $[a, b]$, (a, b), $[a, b)$, and $(a, b]$, respectively. The sets $[a, b]$ and (a, b) are called closed and open intervals, respectively, and the sets $[a, b)$ and $(a, b]$ are both called half-open and half-closed intervals.

Example 1.1.13 The set $[3, 4] = \{\omega : \omega \in \mathbb{R}, 3 \le \omega \le 4\}$ is a closed interval and the set $(2, 5) = \{\omega : \omega \in \mathbb{R}, 2 < \omega < 5\}$ is an open interval. The sets $(4, 5] = \{\omega : \omega \in \mathbb{R}, 4 < \omega \le 5\}$ and $[1, 5) = \{\omega : \omega \in \mathbb{R}, 1 \le \omega < 5\}$ are both half-closed intervals and half-open intervals.   ◇

Definition 1.1.12 (*collection of sets*) When all the elements of a 'set' are sets, the 'set' is called a set of sets, a class of sets, a collection of sets, or a family of sets.

A class, collection, and family of sets are also simply called class, collection, and family, respectively. A collection with one set is called a singleton collection. In some cases, a singleton set denotes a singleton collection similarly as a set sometimes denotes a collection.

Example 1.1.14 When $A = \{1, 2\}$, $B = \{2, 3\}$, and $C = \{\ \}$, the set $D = \{A, B, C\}$ is a collection of sets. The set $E = \{(1, 2], [3, 4)\}$ is a collection of sets.   ◇

Example 1.1.15 Assume the sets $A = \{1, 2\}$, $B = \{2, 3\}$, $C = \{4, 5\}$, and $D = \{\{1, 2\}, \{4, 5\}, 1, 2, 3\}$. Then, $A \subseteq D$, $A \in D$, $B \subseteq D$, $B \notin D$, $C \nsubseteq D$, and $C \in D$. Here, D is a set but not a collection of sets.   ◇

Example 1.1.16 The collection $A = \{\{3\}\}$ and $B = \{\{1, 2\}\}$ are singleton collections and $C = \{\{1, 2\}, \{3\}\}$ is not a singleton collection.   ◇

Definition 1.1.13 (*power set*) The class of all the subsets of a set is called the power set of the set. The power set of Ω is denoted by 2^{Ω}.

Example 1.1.17 The power set of $\Omega = \{3\}$ is $2^{\Omega} = \{\emptyset, \{3\}\}$. The power set of $\Omega = \{4, 5\}$ is $2^{\Omega} = \{\emptyset, \{4\}, \{5\}, \Omega\}$. For a set with n elements, the power set is a collection of 2^n sets.   ◇

Fig. 1.1 Set A and its
complement A^c

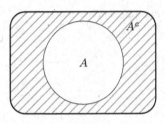

1.1.2 Set Operations

Definition 1.1.14 (*complement*) For an abstract space Ω and its subset A, the complement of A, denoted by A^c or $\overline{A}$, is defined by

$$A^c = \{\omega : \omega \notin A, \ \omega \in \Omega\}. \tag{1.1.6}$$

Figure 1.1 shows a set and its complement via a Venn diagram.

Example 1.1.18 It is easy to see that $\Omega^c = \emptyset$ and $(B^c)^c = B$ for any set B. ◇

Example 1.1.19 For the abstract space $\Omega = \{0, 1, 2, 3\}$ and $B = \{0, 1\}$, we have $B^c = \{2, 3\}$. The complement of the interval[1] $A = (-\infty, 1]$ is $A^c = (1, \infty)$. ◇

Definition 1.1.15 (*union*) The union or sum, denoted by $A \cup B$ or $A + B$, of two sets A and B is defined by

$$A \cup B = A + B$$
$$= \{\omega : \omega \in A \text{ or } \omega \in B\}. \tag{1.1.7}$$

That is, $A \cup B$ denotes the set of elements that belong to at least one of A and B.

Figure 1.2 shows the union of A and B via a Venn diagram. More generally, the union of $\{A_i\}_{i=1}^n$ is[2] denoted by

$$\bigcup_{i=1}^n A_i = A_1 \cup A_2 \cup \cdots \cup A_n. \tag{1.1.8}$$

Example 1.1.20 If $A = \{1, 2, 3\}$ and $B = \{0, 1\}$, then $A \cup B = \{0, 1, 2, 3\}$. ◇

Example 1.1.21 For any two sets A and B, we have $B \cup B = B$, $B \cup B^c = \Omega$, $B \cup \Omega = \Omega$, $B \cup \emptyset = B$, $A \subseteq (A \cup B)$, and $B \subseteq (A \cup B)$. ◇

---

[1] Because an interval assumes the set of real numbers by definition, it is not necessary to specify the abstract space when we consider an interval.

[2] We often use braces also to denote a number of items in a compact way. For example, $\{A_i\}_{i=1}^n$ here represents $A_1, A_2, \ldots, A_n$.

Fig. 1.2 Sum $A \cup B$ of A
and B

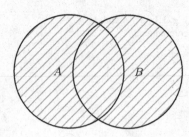

Fig. 1.3 Intersection $A \cap B$
of A and B

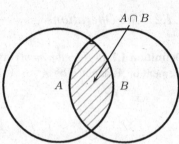

Example 1.1.22 We have $A \cup B = B$ when $A \subseteq B$, and $(A \cup B) \subseteq C$ when $A \subseteq C$
and $B \subseteq C$. In addition, for four sets A, B, C, and D, we have $(A \cup B) \subseteq (C \cup D)$
when $A \subseteq C$ and $B \subseteq D$. $\diamond$

Definition 1.1.16 (*intersection*) The intersection or product, denoted by $A \cap B$ or
AB, of two sets A and B is defined by

$$A \cap B = \{\omega : \omega \in A \text{ and } \omega \in B\}. \tag{1.1.9}$$

That is, $A \cap B$ denotes the set of elements that belong to both A and B simultaneously.

The Venn diagram for the intersection of A and B is shown in Fig. 1.3. Meanwhile,

$$\bigcap_{i=1}^{n} A_i = A_1 \cap A_2 \cap \cdots \cap A_n \tag{1.1.10}$$

denotes the intersection of $\{A_i\}_{i=1}^{n}$.

Example 1.1.23 For $A = \{1, 2, 3\}$ and $B = \{0, 1\}$, we have $A \cap B = AB = \{1\}$.
The intersection of the intervals $[1, 3)$ and $(2, 5]$ is $[1, 3) \cap (2, 5] = (2, 3)$. $\diamond$

Example 1.1.24 For any two sets A and B, we have $B \cap B = B$, $B \cap B^c = \emptyset$,
$B \cap \Omega = B$, $B \cap \emptyset = \emptyset$, $(A \cap B) \subseteq A$, and $(A \cap B) \subseteq B$. We also have $A \cap B = A$
when $A \subseteq B$. $\diamond$

Example 1.1.25 For three sets A, B, and C, we have $(A \cap B) \subseteq (A \cap C)$ when
$B \subseteq C$. $\diamond$

Fig. 1.4 Partition
$\{A_1, A_2, \ldots, A_6\}$ of S

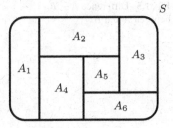

Definition 1.1.17 (*disjoint*) If A and B have no element in common, that is, if $A \cap B = AB = \emptyset$, then the sets A and B are called disjoint or mutually exclusive.

Example 1.1.26 The sets $C = \{1, 2, 3\}$ and $D = \{4, 5\}$ are mutually exclusive. The sets $A = \{1, 2, 3, 4\}$ and $B = \{4, 5, 6\}$ are not mutually exclusive because $A \cap B = \{4\} \neq \emptyset$. The intervals $[1, 3)$ and $[3, 5]$ are mutually exclusive, and $[3, 4]$ and $[4, 5]$ are not mutually exclusive. ◇

Definition 1.1.18 (*partition*) A collection of subsets of S is called a partition of S when the subsets in the collection are collectively exhaustive and every pair of subsets in the collection is disjoint. Specifically, the collection $\{A_i\}_{i=1}^n$ is a partition of S if both

$$\text{(collectively exhaustive)}: \quad \bigcup_{i=1}^{n} A_i = S \tag{1.1.11}$$

and

$$\text{(disjoint)}: \quad A_i \cap A_j = \emptyset \tag{1.1.12}$$

for all $i \neq j$ are satisfied.

The singleton collection $\{S\}$ composed only of S is not regarded as a partition of S. Figure 1.4 shows a partition $\{A_1, A_2, \ldots, A_6\}$ of S.

Example 1.1.27 When $A = \{1, 2\}$, the collection $\{\{1\}, \{2\}\}$ is a partition of A. Each of the five collections $\{\{1\}, \{2\}, \{3\}, \{4\}\}$, $\{\{1\}, \{2\}, \{3, 4\}\}$, $\{\{1\}, \{2, 3\}, \{4\}\}$, $\{\{1, 2\}, \{3\}, \{4\}\}$, and $\{\{1, 2\}, \{3, 4\}\}$ is a partition of $B = \{1, 2, 3, 4\}$ while neither $\{\{1, 2, 3\}, \{3, 4\}\}$ nor $\{\{1, 2\}, \{3\}\}$ is a partition of B. ◇

Example 1.1.28 The collection $\{A, \emptyset\}$ is a partition of A, and $\{[3, 3.3), [3.3, 3.4], (3.4, 3.6], (3.6, 4)\}$ is a partition of the interval $[3, 4)$. ◇

Example 1.1.29 For $A = \{1, 2, 3\}$, obtain all the partitions without the null set.

Solution Because the number of elements in A is three, a partition of A with non-empty sets will be that of one- and two-element sets. Thus, collections $\{\{1\}, \{2, 3\}\}$, $\{\{2\}, \{1, 3\}\}$, $\{\{3\}, \{1, 2\}\}$, and $\{\{1\}, \{2\}, \{3\}\}$ are the desired partitions. ◇

Fig. 1.5 Difference $A - B$
between A and B

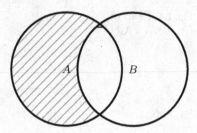

Definition 1.1.19 (*difference*) The difference $A - B$, also denoted by $A \setminus B$, is defined as

$$A - B = \{\omega : \omega \in A \text{ and } \omega \notin B\}. \tag{1.1.13}$$

Figure 1.5 shows $A - B$ via a Venn diagram. Note that we have

$$A - B = A \cap B^c$$
$$= A - AB. \tag{1.1.14}$$

Example 1.1.30 For $A = \{1, 2, 3\}$ and $B = \{0, 1\}$, we have $A - B = \{2, 3\}$ and $B - A = \{0\}$. The differences between the intervals $[1, 3)$ and $(2, 5]$ are $[1, 3) - (2, 5] = [1, 2]$ and $(2, 5] - [1, 3) = [3, 5]$. ◇

Example 1.1.31 For any set A, we have $\Omega - A = A^c$, $A - \Omega = \emptyset$, $A - A = \emptyset$, $A - \emptyset = A$, $A - A^c = A$, $(A + A) - A = \emptyset$, and $A + (A - A) = A$. ◇

Definition 1.1.20 (*symmetric difference*) The symmetric difference, denoted by $A \triangle B$, between two sets A and B is the set of elements which belong only to A or only to B.

From the definition of symmetric difference, we have

$$A \triangle B = (A - B) \cup (B - A)$$
$$= (A \cap B^c) \cup (A^c \cap B)$$
$$= (A \cup B) - (A \cap B). \tag{1.1.15}$$

Figure 1.6 shows the symmetric difference $A \triangle B$ via a Venn diagram.

Example 1.1.32 For $A = \{1, 2, 3, 4\}$ and $B = \{4, 5, 6\}$, we have $A \triangle B = \{1, 2, 3\} \cup \{5, 6\} = \{1, 2, 3, 4, 5, 6\} - \{4\} = \{1, 2, 3, 5, 6\}$. The symmetric difference between the intervals $[1, 3)$ and $(2, 5]$ is $[1, 3) \triangle (2, 5] = ([1, 3) - (2, 5]) \cup ((2, 5] - [1, 3)) = [1, 2] \cup [3, 5]$. ◇

Fig. 1.6 Symmetric difference $A \triangle B$ between A and B

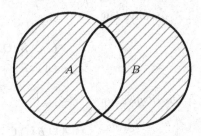

Example 1.1.33 For any set A, we have $A \triangle A = \emptyset$, $A \triangle \emptyset = \emptyset \triangle A = A$, $A \triangle \Omega = \Omega \triangle A = A^c$, and $A \triangle A^c = A^c \triangle A = \Omega$. $\diamond$

Example 1.1.34 It follows that $A - B \neq B - A$ and $A \triangle B = B \triangle A$, and that $A \triangle B = A - B$ when $B \subseteq A$. $\diamond$

Example 1.1.35 (Sveshnikov 1968) Show that every element of $A_1 \triangle A_2 \triangle \cdots \triangle A_n$ belongs to only an odd number of the sets $\{A_i\}_{i=1}^n$.

Solution Let us prove the result by mathematical induction. When $n = 1$, it is self-evident. When $n = 2$, every element of $A_1 \triangle A_2$ is an element only of A_1, or only of A_2, by definition. Next, assume that every element of $C = A_1 \triangle A_2 \triangle \cdots \triangle A_n$ belongs only to an odd number of the sets $\{A_i\}_{i=1}^n$. Then, by the definition of $\triangle$, every element of $B = C \triangle A_{n+1}$ belongs only to A_{n+1} or only to C. In other words, every element of B belongs only to A_{n+1} or only to an odd number of the sets $\{A_i\}_{i=1}^n$, concluding the proof. $\diamond$

Interestingly, the set operation similar to the addition of numbers is not the union of sets but rather the symmetric difference (Karatowski and Mostowski 1976): subtraction is the inverse operation for addition and symmetric difference is its own inverse operation while no inverse operation exists for the union of sets. More specifically, for two sets A and B, there exists only one C, which is $C = A \triangle B$, such that $A \triangle C = B$. This is clear because $A \triangle (A \triangle B) = B$ and $A \triangle C = B$.

1.1.3 Laws of Set Operations

Theorem 1.1.3 *For the operations of union and intersection, the following laws apply:*

1. *Commutative law*

$$A \cup B = B \cup A \tag{1.1.16}$$
$$A \cap B = B \cap A \tag{1.1.17}$$

2. *Associative law*

$$(A \cup B) \cup C = A \cup (B \cup C) \tag{1.1.18}$$
$$(A \cap B) \cap C = A \cap (B \cap C) \tag{1.1.19}$$

3. *Distributive law*

$$(A \cup B) \cap C = (A \cap C) \cup (B \cap C) \tag{1.1.20}$$
$$(A \cap B) \cup C = (A \cup C) \cap (B \cup C) \tag{1.1.21}$$

Note that the associative and distributive laws are for the same and different types of operations, respectively.

Example 1.1.36 Assume three sets $A = \{0, 1, 2, 6\}$, $B = \{0, 2, 3, 4\}$, and $C = \{0, 1, 3, 5\}$. It is easy to check the commutative and associative laws. Next, (1.1.20) holds true as it is clear from $(A \cup B) \cap C = \{0, 1, 2, 3, 4, 6\} \cap \{0, 1, 3, 5\} = \{0, 1, 3\}$ and $(A \cap C) \cup (B \cap C) = \{0, 1\} \cup \{0, 3\} = \{0, 1, 3\}$. In addition, (1.1.21) holds true as it is clear from $(A \cap B) \cup C = \{0, 2\} \cup \{0, 1, 3, 5\} = \{0, 1, 2, 3, 5\}$ and $(A \cup C) \cap (B \cup C) = \{0, 1, 2, 3, 5, 6\} \cap \{0, 1, 2, 3, 4, 5\} = \{0, 1, 2, 3, 5\}$. $\diamond$

Generalizing (1.1.20) and (1.1.21) for a number of sets, we have

$$B \cap \left(\bigcup_{i=1}^{n} A_i \right) = \bigcup_{i=1}^{n} (B \cap A_i) \tag{1.1.22}$$

and

$$B \cup \left(\bigcap_{i=1}^{n} A_i \right) = \bigcap_{i=1}^{n} (B \cup A_i), \tag{1.1.23}$$

respectively.

Theorem 1.1.4 *When A_1, A_2, ..., A_n are subsets of an abstract space S, we have*

$$\left(\bigcup_{i=1}^{n} A_i \right)^c = S - \left(\bigcup_{i=1}^{n} A_i \right)$$

$$= \bigcap_{i=1}^{n} (S - A_i)$$

$$= \bigcap_{i=1}^{n} A_i^c \tag{1.1.24}$$

and

$$\left(\bigcap_{i=1}^{n} A_i \right)^c = \bigcup_{i=1}^{n} A_i^c. \tag{1.1.25}$$

Proof Let us prove the theorem by using (1.1.2).

(1) Proof of (1.1.24). Let $x \in \left(\bigcup_{i=1}^{n} A_i \right)^c$. Then, $x \notin \bigcup_{i=1}^{n} A_i$, and therefore x is not an element of any A_i. This implies that x is an element of every A_i^c or, equivalently, $x \in \bigcap_{i=1}^{n} A_i^c$. Therefore, we have

$$\left(\bigcup_{i=1}^{n} A_i \right)^c \subseteq \bigcap_{i=1}^{n} A_i^c. \tag{1.1.26}$$

Next, assume $x \in \bigcap_{i=1}^{n} A_i^c$. Then, $x \in A_i^c$, and therefore x is not an element of A_i for any i: in other words, $x \notin \bigcup_{i=1}^{n} A_i$. This implies $x \in \left(\bigcup_{i=1}^{n} A_i \right)^c$. Therefore, we have

$$\bigcap_{i=1}^{n} A_i^c \subseteq \left(\bigcup_{i=1}^{n} A_i \right)^c. \tag{1.1.27}$$

From (1.1.2), (1.1.26), and (1.1.27), the result (1.1.24) is proved.

(2) Proof of (1.1.25). Replacing A_i with A_i^c in (1.1.24) and using (1.1.26), we have $\left(\bigcup_{i=1}^{n} A_i^c \right)^c = \bigcap_{i=1}^{n} \left(A_i^c \right)^c = \bigcap_{i=1}^{n} A_i$. Taking the complement completes the proof.

♠

Example 1.1.37 Consider $S = \{1, 2, 3, 4\}$ and its subsets $A_1 = \{1\}$, $A_2 = \{2, 3\}$, and $A_3 = \{1, 3, 4\}$. Then, we have $(A_1 + A_2)^c = \{1, 2, 3\}^c = \{4\}$, which is the same as $A_1^c A_2^c = \{2, 3, 4\} \cap \{1, 4\} = \{4\}$. Similarly, we have $(A_1 A_2)^c = \emptyset^c = S$, which is the same as $A_1^c + A_2^c = \{2, 3, 4\} \cup \{1, 4\} = S$. In addition, $(A_2 + A_3)^c = S^c = \emptyset$ is the same as $A_2^c A_3^c = \{1, 4\} \cap \{2\} = \emptyset$. Finally, $(A_1 A_3)^c = \{1\}^c = \{2, 3, 4\}$ is the same as $A_1^c + A_3^c = \{2, 3, 4\} \cup \{2\} = \{2, 3, 4\}$. ◇

Example 1.1.38 For three sets A, B, and C, show that $A = B - C$ when $A \subseteq B$ and $C = B - A$.

Solution First, assume $A = B$. Then, because $C = B - A = \emptyset$, we have $B - C = B - \emptyset = A$. Next, assume $A \subset B$. Then, $C = B \cap A^c$ and $C^c = (B \cap A^c)^c = B^c \cup A$ from (1.1.14) and (1.1.25), respectively. Thus, using (1.1.14) and (1.1.20), we get $B - C = B \cap C^c = B \cap (B^c \cup A) = (B \cap B^c) \cup (B \cap A) = \emptyset \cup A = A$. ◇

1.1.4   Uncountable Sets

Definition 1.1.21 (*one-to-one correspondence*) A relationship between two sets in which each element of either set is assigned with only one element of the other is called a one-to-one correspondence.

The notion of one-to-one correspondence will be redefined in Definition 1.2.9. Based on the set $\mathbb{J}_+$ of natural numbers defined in (1.1.3) and the concept of one-to-one correspondence, let us define a countable set.

Definition 1.1.22 (*countable set*) A set is called countable or denumerable if we can find a one-to-one correspondence between the set and a subset of $\mathbb{J}_+$.

The elements of a countable set can be indexed as $a_1, a_2, \ldots, a_n, \ldots$. It is easy to see that finite sets are all countable sets.

Example 1.1.39 The sets $\{1, 2, 3\}$ and $\{1, 10, 100, 1000\}$ are both countable because a one-to-one correspondence can be established between these two sets and the subsets $\{1, 2, 3\}$ and $\{1, 2, 3, 4\}$, respectively, of $\mathbb{J}_+$.   ◇

Example 1.1.40 The set $\mathbb{J}$ of integers is countable because we can establish a one-to-one correspondence as

$$
\begin{array}{cccccccccc}
0 & -1 & 1 & -2 & 2 & \cdots & -n & & n & & \cdots \\
\updownarrow & \updownarrow & \updownarrow & \updownarrow & \updownarrow & \cdots & \updownarrow & & \updownarrow & & \cdots \\
1 & 2 & 3 & 4 & 5 & \cdots & 2n & & 2n+1 & & \cdots
\end{array}
\tag{1.1.28}
$$

between $\mathbb{J}$ and $\mathbb{J}_+$. Similarly, it is easy to see that the sets $\{\omega : \omega$ is a positive even number$\}$ and $\{2, 4, \ldots, 2^n, \ldots\}$ are countable sets by noting the one-to-one correspondences $2n \leftrightarrow n$ and $2^n \leftrightarrow n$, respectively.   ◇

Theorem 1.1.5 *The set*

$$
\mathbb{Q} = \{q : q \text{ is a rational number}\}
\tag{1.1.29}
$$

of rational numbers is countable.

Proof (Method 1) A rational number can be expressed as $\left\{\frac{p}{q} : q \in \mathbb{J}_+, \ p \in \mathbb{J}\right\}$. Consider the sequence

$$
a_{ij} = \begin{cases}
\frac{0}{1}, & i = j = 0; \\
-\frac{i-j+1}{j}, & i = 1, 2, \ldots, \ j = 1, 2, \ldots, i; \\
\frac{j-i}{2i-j+1}, & i = 1, 2, \ldots, \ j = i+1, i+2, \ldots, 2i.
\end{cases}
\tag{1.1.30}
$$

In other words, consider

$$i = 0: \quad \frac{0}{1}$$

$$i = 1: \quad -\frac{1}{1}, \quad \frac{1}{1}$$

$$i = 2: \quad -\frac{2}{1}, \quad -\frac{1}{2}, \quad \frac{1}{2}, \quad \frac{2}{1}$$ (1.1.31)

$$i = 3: \quad -\frac{3}{1}, \quad -\frac{2}{2}, \quad -\frac{1}{3}, \quad \frac{1}{3}, \quad \frac{2}{2}, \quad \frac{3}{1}$$

$$\vdots$$

Reading this sequence downward from the first row and ignoring repetitions, we will have a one-to-one correspondence between the sets of rational numbers and natural numbers.

(Method 2) Assume integers $x \neq 0$ and y, and denote the rational number $\frac{y}{x}$ by the coordinates (x, y) on a two dimensional plane. Reading the integer coordinates as $(1, 0) \to (1, 1) \to (-1, 1) \to (-1, -1) \to (2, -1) \to (2, 2) \to \cdots$ while skipping a number if it had previously appeared, we have a one-to-one correspondence between $\mathbb{J}_+$ and $\mathbb{Q}$. ♠

Theorem 1.1.6 *Countable sets have the following properties:*

(1) A subset of a countable set is countable.
(2) There exists a countable subset for any infinite set.
(3) If the sets A_1, A_2, ... are all countable, then $\underset{i \in \mathbb{J}_+}{\cup} A_i = \overset{\infty}{\underset{i=1}{\cup}} A_i = \underset{n \to \infty}{\lim} \overset{n}{\underset{i=1}{\cup}} A_i$ is also countable.

Proof (1) For a finite set, it is obvious. For an infinite set, denote a countable set by $A = \{a_1, a_2, \ldots\}$. Then, a subset of A can be expressed as $B = \{a_{n_1}, a_{n_2}, \ldots\}$ and we can find a one-to-one correspondence $i \leftrightarrow a_{n_i}$ between $\mathbb{J}_+$ and B.

(2) We can choose a countable subset $\{a_1, a_2, \ldots\}$ arbitrarily from an infinite set.

(3) Consider the sequence B_1, B_2, ... defined as

$$B_1 = A_1$$
$$= \{b_{1j}\},$$ (1.1.32)
$$B_2 = A_2 - A_1$$
$$= \{b_{2j}\},$$ (1.1.33)
$$B_3 = A_3 - (A_1 \cup A_2)$$
$$= \{b_{3j}\},$$ (1.1.34)

$$\vdots$$

Clearly, $B_1, B_2, \ldots$ are mutually exclusive and $\overset{\infty}{\underset{i=1}{\cup}} A_i = \overset{\infty}{\underset{i=1}{\cup}} B_i$. Because $B_i \subseteq A_i$, the sets $B_1, B_2, \ldots$ are all countable from Property (1). Next, arrange the elements of $B_1, B_2, \ldots$ as

$$
\begin{array}{cccc}
b_{11} \to b_{12} & b_{13} \to b_{14} & \cdots \\
\swarrow \quad \nearrow & \swarrow \quad \nearrow \\
b_{21} \quad b_{22} & b_{23} \quad b_{24} & \cdots \\
\downarrow \quad \nearrow & \swarrow \quad \nearrow \\
b_{31} \quad b_{32} & b_{33} \quad b_{34} & \cdots \\
\swarrow \quad \nearrow \\
\vdots \qquad \vdots & \vdots \qquad \vdots & \ddots
\end{array}
\tag{1.1.35}
$$

and read them in the order as directed by the arrows, which represents a one-to-one correspondence between $\mathbb{J}_+$ and $\overset{\infty}{\underset{i=1}{\cup}} B_i = \overset{\infty}{\underset{i=1}{\cup}} A_i$.

♠

Property (3) of Theorem 1.1.6 also implies that a countably infinite union of countable sets is a countable set.

Example 1.1.41 (Sveshnikov 1968) Show that the Cartesian product $A_1 \times A_2 \times \cdots \times A_n = \left\{ \left(a_{1i_1}, a_{2i_2}, \ldots, a_{ni_n} \right) \right\}$ of a finite number of countable sets is countable.

Solution It suffices to show that the Cartesian product $A \times B$ is countable when A and B are countable. Denote two countable sets by $A = \{a_1, a_2, \ldots\}$ and $B = \{b_1, b_2, \ldots\}$. If we arrange the elements of the Cartesian product $A \times B$ as $(a_1, b_1), (a_1, b_2), (a_2, b_1), (a_1, b_3), (a_2, b_2), (a_3, b_1), \ldots$, then it is apparent that the Cartesian product is countable. $\diamond$

Example 1.1.42 Show that the set of finite sequences from a countable set is countable.

Solution The set B_k of finite sequences with length k from a countable set A is equivalent to the k-fold Cartesian product $A^k = \left\{ (b_1, b_2, \ldots, b_k) : b_j \in A \right\}$ of A. Then, B_k is countable from Example 1.1.41. Next, the set of finite sequences is the countable union $\overset{\infty}{\underset{k=1}{\cup}} B_k$, which is countable from (3) of Theorem 1.1.6. $\diamond$

Example 1.1.43 The set $A = \left\{ a^b : a, b \in Q_1 \right\}$ is countable, where $Q_1 = \mathbb{Q} - \{0\}$ with $\mathbb{Q}$ the set of rational numbers. $\diamond$

Example 1.1.44 The set Υ_T of infinite binary sequences with a finite period is countable. First, note that there exist[3] two sequences with period 2, six sequences with

---

[3] We assume, for example, $\cdots 1010 \cdots$ and $\cdots 0101 \cdots$ are different.

period 3, . . ., at most $2^k - 2$ sequences with period k, ... Based on this observation, we can find the one-to-one correspondence

$$\cdots 00\cdots \to 1, \qquad \cdots 11\cdots \to 2, \qquad \cdots 0101\cdots \to 3,$$
$$\cdots 1010\cdots \to 4, \quad \cdots 001001\cdots \to 5, \cdots 010010\cdots \to 6,$$
$$\cdots 100100\cdots \to 7, \cdots 011011\cdots \to 8, \cdots 101101\cdots \to 9, \cdots$$

between Υ_T and $\mathbb{J}_+$. $\diamond$

Definition 1.1.23 (*uncountable set*) When no one-to-one correspondence exists between a set and a subset of $\mathbb{J}_+$, the set is called uncountable or non-denumerable.

As it has already been mentioned, finite sets are all countable. On the other hand, some infinite sets are countable and some are uncountable.

Theorem 1.1.7 *The interval set* $[0, 1] = \mathbb{R}_{[0,1]} = \{x : 0 \leq x \leq 1\}$, *i.e., the set of real numbers in the interval* $[0, 1]$, *is uncountable.*

Proof We prove the theorem by contradiction. Letting $a_{ij} \in \{0, 1, \ldots, 9\}$, the elements of the set $\mathbb{R}_{[0,1]}$ can be expressed as $0.a_{i1}a_{i2}\cdots a_{in}\cdots$. Assume $\mathbb{R}_{[0,1]}$ is countable: in other words, assume all the elements of $\mathbb{R}_{[0,1]}$ are enumerated as

$$\alpha_1 = 0.a_{11}a_{12}\cdots a_{1n}\cdots \tag{1.1.36}$$
$$\alpha_2 = 0.a_{21}a_{22}\cdots a_{2n}\cdots \tag{1.1.37}$$
$$\vdots$$
$$\alpha_n = 0.a_{n1}a_{n2}\cdots a_{nn}\cdots \tag{1.1.38}$$
$$\vdots$$

Now, consider a number $\beta = 0.b_1 b_2 \cdots b_n \cdots$, where $b_i \neq a_{ii}$ and $b_i \in \{0, 1, \ldots, 9\}$. Then, it is clear that $\beta \in \mathbb{R}_{[0,1]}$. We also have $\beta \neq \alpha_1$ because $b_1 \neq a_{11}$, $\beta \neq \alpha_2$ because $b_2 \neq a_{22}$, $\cdots$: in short, β is not equal to any α_i. In other words, although β is an element of $\mathbb{R}_{[0,1]}$, it is not included in the enumeration, and produces a contradiction to the assumption that all the numbers in $\mathbb{R}_{[0,1]}$ have been enumerated. Therefore, $\mathbb{R}_{[0,1]}$ is uncountable.   ♠

Example 1.1.45 The set $\Upsilon = \{ = (a_1\ a_2\ \cdots) : a_i \in \{0, 1\}\}$ of one-sided infinite binary sequences is uncountable, which can be shown via a method similar to the proof of Theorem 1.1.7. Specifically, assume Υ is countable, and all the elements $_i$ of Υ are arranged as

$$_1 = (a_{11}\ a_{12}\ \cdots\ a_{1n}\ \cdots) \tag{1.1.39}$$
$$_2 = (a_{21}\ a_{22}\ \cdots\ a_{2n}\ \cdots) \tag{1.1.40}$$
$$\vdots$$

where $a_{ij} \in \{0, 1\}$. Denote the complement of a binary digit x by $\bar{x}$. Then, the sequence $(\overline{a_{11}} \ \overline{a_{22}} \cdots)$ produces a contradiction. Therefore, Υ is uncountable. ◇

Example 1.1.46 (Gelbaum and Olmsted 1964) Consider the closed interval $U = [0, 1]$. The open interval $D_{11} = \left(\frac{1}{3}, \frac{2}{3}\right)$ is removed from U in the first step, two open intervals $D_{21} = \left(\frac{1}{9}, \frac{2}{9}\right)$ and $D_{22} = \left(\frac{7}{9}, \frac{8}{9}\right)$ are removed from the remaining region $\left[0, \frac{1}{3}\right] \cup \left[\frac{2}{3}, 1\right]$ in the second step, ..., 2^{k-1} open intervals of length 3^{-k} are removed in the k-th step, ... The limit of the remaining region $\mathbb{C}$ of this procedure is called the Cantor set or Cantor ternary set. The procedure can equivalently be described as follows: Denote an open interval with the starting point

$$\zeta_{1k} = \frac{1}{3^k} + \sum_{j=0}^{k-1} \frac{2}{3^j} c_j \tag{1.1.41}$$

and ending point

$$\zeta_{2k} = \frac{2}{3^k} + \sum_{j=0}^{k-1} \frac{2}{3^j} c_j \tag{1.1.42}$$

by $A_{2c_0, 2c_1, 2c_2, \ldots, 2c_{k-1}}$, where $c_0 = 0$ and $c_j = 0$ or 1 for $j = 1, 2, \ldots, k-1$. Then, at the k-th step in the procedure of obtaining the Cantor set $\mathbb{C}$, we are removing the 2^{k-1} open intervals $\{A_{2c_0, 2c_1, 2c_2, \ldots, 2c_{k-1}}\}$ of length 3^{-k}. Specifically, we have $A_0 = \left(\frac{1}{3}, \frac{2}{3}\right)$ when $k = 1$; $A_{0,0} = \left(\frac{1}{9}, \frac{2}{9}\right)$ and $A_{0,2} = \left(\frac{7}{9}, \frac{8}{9}\right)$ when $k = 2$; $A_{0,0,0} = \left(\frac{1}{27}, \frac{2}{27}\right)$, $A_{0,0,2} = \left(\frac{7}{27}, \frac{8}{27}\right)$, $A_{0,2,0} = \left(\frac{19}{27}, \frac{20}{27}\right)$, and $A_{0,2,2} = \left(\frac{25}{27}, \frac{26}{27}\right)$ when $k = 3$; $\cdots$. ◇

The Cantor set $\mathbb{C}$ described in Example 1.1.46 has the following properties:

(1) The set $\mathbb{C}$ can be expressed as $\mathbb{C} = \bigcap\limits_{i=1}^{\infty} B_i$, where $B_1 = [0, 1]$, $B_2 = \left[0, \frac{1}{3}\right] \cup \left[\frac{2}{3}, 1\right]$, $B_3 = \left[0, \frac{1}{9}\right] \cup \left[\frac{2}{9}, \frac{3}{9}\right] \cup \left[\frac{6}{9}, \frac{7}{9}\right] \cup \left[\frac{8}{9}, 1\right]$, ...
(2) The set $\mathbb{C}$ is an uncountable and closed set.
(3) The length of the union of the open intervals removed when obtaining $\mathbb{C}$ is $\frac{1}{3} + \frac{2}{3^2} + \frac{4}{3^3} + \cdots = \frac{\frac{1}{3}}{1 - \frac{2}{3}} = 1$. Consequently, the length of $\mathbb{C}$ is 0.
(4) The set $\mathbb{C}$ is the set of ternary real numbers between 0 and 1 that can be represented without using 1. In other words, every element of $\mathbb{C}$ can be expressed as $\sum\limits_{n=1}^{\infty} \frac{x_n}{3^n}$, $x_n \in \{0, 2\}$.

In Sect. 1.3.3, the Cantor set is used as the basis for obtaining a singular function.

Example 1.1.47 (Gelbaum and Olmsted 1964) The Cantor set $\mathbb{C}$ considered in Example 1.1.46 has a length 0. A Cantor set with a length greater than 0 can be obtained similarly. For example, consider the interval $[0, 1]$ and a constant $\alpha \in (0, 1]$. In the first step, an open interval $\left(\frac{1}{2} - \frac{\alpha}{4}, \frac{1}{2} + \frac{\alpha}{4}\right)$ of length $\frac{\alpha}{2}$ is removed. In the second step, an open interval each of length $\frac{\alpha}{8}$ is removed at the center of the two

Table 1.1 Finite, infinite, countable, and uncountable sets

	Countable (enumerable, denumerable)	Uncountable (non-denumerable)
Finite	Finite Example: $\{3, 4, 5\}$	
Infinite	Countably infinite Example: $\{1, 2, \ldots\}$	Uncountably infinite breal Example: $(0, 1]$

closed intervals remaining. In the third step, an open interval each of length $\frac{\alpha}{32}$ is removed at the center of the four closed intervals remaining. .... Then, this Cantor set is a set of length $1 - \alpha$ because the sum of lengths of the regions removed is $\frac{\alpha}{2} + \frac{\alpha}{4} + \frac{\alpha}{8} + \cdots = \alpha$. A Cantor set with a non-zero length is called the Smith-Volterra-Cantor set or fat Cantor set.   ◇

Example 1.1.48 As shown in Table 1.1, the term countable set denotes a finite set or a countably infinite set, and the term infinite set denotes a countably infinite set or an uncountably infinite, simply called uncountable, set.   ◇

Definition 1.1.24 (*almost everywhere*) In real space, when the length of the union of countably many intervals is arbitrarily small, a set of points that can be contained in the union is called a set of length 0. In addition, 'at all points except for a set of length 0' is called 'almost everywhere', 'almost always', 'almost surely', 'with probability 1', 'almost certainly', or 'at almost every point'.

In the integer or discrete space (Jones 1982), 'almost everywhere' denotes 'all points except for a finite set'.

Example 1.1.49 The intervals $[1, 2)$ and $(1, 2)$ are the same almost everywhere. The sets $\{1, 2, \ldots\}$ and $\{2, 3, \ldots\}$ are the same almost everywhere.   ◇

Definition 1.1.25 (*equivalence*) If we can find a one-to-one correspondence between two sets M and N, then M and N are called equivalent, or of the same cardinality, and are denoted by $M \sim N$.

Example 1.1.50 For the two sets $A = \{4, 2, 1, 9\}$ and $B = \{8, 0, 4, 5\}$, we have $A \sim B$.   ◇

Example 1.1.51 If $A \sim B$, then $2^A \sim 2^B$.   ◇

Example 1.1.52 (Sveshnikov 1968) Show that $A \cup B \sim A$ when A is an infinite set and B is a countable set.

Solution First, arbitrarily choose a countably infinite set $A_0 = \{a_0, a_1, \ldots\}$ from A. Because $A_0 \cup B$ is also a countably infinite set, we have a one-to-one correspondence between $A_0 \cup B$ and A_0: denote the one-to-one correspondence by $g(x)$. Then,

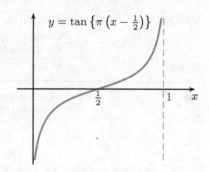

$$h(x) = \begin{cases} g(x), & x \in A_0 \cup B, \\ x, & x \notin A_0 \cup B \end{cases} \tag{1.1.43}$$

is a one-to-one correspondence between $A \cup B$ and A: that is, $A \cup B \sim A$.   ◇

Example 1.1.53 The set $\mathbb{J}_+$ of natural numbers is equivalent to the set $\mathbb{J}$ of integers. The set of irrational numbers is equivalent to the set $\mathbb{R}$ of real numbers from Exercise 1.16.   ◇

Example 1.1.54 It is interesting to note that the set of irrational numbers is not closed under certain basic operations, such as addition and multiplication, while the much smaller set of rational numbers is closed under such operations.   ◇

Example 1.1.55 The Cantor and Smith-Volterra-Cantor sets considered in Examples 1.1.46 and 1.1.47, respectively, are both equivalent to the set $\mathbb{R}$ of real numbers.   ◇

Example 1.1.56 As shown in Fig. 1.7, it is easy to see that $(0, 1) \sim \mathbb{R}$ via $y = \tan\left\{\pi\left(x - \frac{1}{2}\right)\right\}$. It is also clear that $[a, b] \sim [c, d]$ when $a < b$ and $c < d$ because a point x between a and b, and a point y between c and d, have the one-to-one correspondence $y = \frac{(d-c)}{(b-a)}(x - a) + c$.   ◇

Example 1.1.57 The set of real numbers $\mathbb{R}$ is uncountable. The intervals $[a, b]$, $[a, b)$, $(a, b]$, and (a, b) are all uncountable for any real number a and $b > a$ from Theorem 1.1.7 and Example 1.1.56.   ◇

Theorem 1.1.8 *When A is equivalent to a subset of B and B is equivalent to a subset of A, then A is equivalent to B.*

Example 1.1.58 As we have observed in Theorem 1.1.5, the set $\mathbb{J}_+$ of natural numbers is equivalent to the set $\mathbb{Q}$ of rational numbers. Similarly, the subset $\mathbb{Q}_3 = \{t : t = \frac{q}{3}, q \in \mathbb{J}\}$ of $\mathbb{Q}$ is equivalent to the set $\mathbb{J}$ of integers. Therefore, recollecting that $\mathbb{J}_+$ is a subset of $\mathbb{J}$ and $\mathbb{Q}_3$ is a subset of $\mathbb{Q}$, Theorem 1.1.8 dictates that $\mathbb{J} \sim \mathbb{Q}$.   ◇

Example 1.1.59 It is interesting to note that, although $\mathbb{J}_+$ is a proper subset of $\mathbb{J}$, which is in turn a proper subset of $\mathbb{Q}$, the three sets $\mathbb{J}_+$, $\mathbb{J}$, and $\mathbb{Q}$ are all equivalent. For finite sets, on the other hand, such an equivalence is impossible when one set is a proper subset of the other. This exemplifies that the infinite and finite spaces sometimes produce different results. $\diamond$

1.2 Functions

In this section, we will introduce and briefly review some key concepts within the theory of functions (Ito 1987; Royden 1989; Stewart 2012).

Definition 1.2.1 (*mapping*) A relation f that assigns every element of a set Ω with only one element of another set A is called a function or mapping and is often denoted by $f : \Omega \to A$.

For the function $f : \Omega \to A$, the sets Ω and A are called the domain and codomain, respectively, of f.

Example 1.2.1 Assume the domain $\Omega = [-1, 1]$ and the codomain $A = [-2, 1]$. The relation that connects all the points in $[-1, 0)$ of the domain with -1 in the codomain, and all the points in $[0, 1]$ of the domain with 1 in the codomain is a function. $\diamond$

Example 1.2.2 Assume the domain $\Omega = [-1, 1]$ and the codomain $A = [-2, 1]$. The relation that connects all the points in $[-1, 0)$ of the domain with -1 in the codomain, and all the points in $(0, 1]$ of the domain with 1 in the codomain is not a function because the point 0 in the domain is not connected with any point in the codomain. In addition, the relation that connects all the points in $[-1, 0]$ of the domain with -1 in the codomain, and all the points in $[0, 1]$ of the domain with 1 in the codomain is not a function because the point 0 in the domain is connected with more than one point in the codomain. $\diamond$

Definition 1.2.2 (*set function*) A function whose domain is a collection of sets is called a set function.

Example 1.2.3 Let the domain be the power set $2^C = \{\emptyset, \{3\}, \{4\}, \{5\}, \{3, 4\}, \{3, 5\}, \{4, 5\}, \{3, 4, 5\}\}$ of $C = \{3, 4, 5\}$. Define a function $f(B)$ for $B \in 2^C$ as the number of elements in B. Then, f is a set function, and we have $f(\{3\}) = 1$, $f(\{3, 4\}) = 2$, and $f(\{3, 4, 5\}) = 3$, for example. $\diamond$

Definition 1.2.3 (*image*) For a function $f : \Omega \to A$ and a subset G of Ω, the set

$$f(G) = \{a : a = f(\omega), \ \omega \in G\}, \tag{1.2.1}$$

which is a subset of A, is called the image of G (under f).

Fig. 1.8 Image $f(G) \subseteq A$
of $G \subseteq \Omega$ for a function
$f : \Omega \to A$

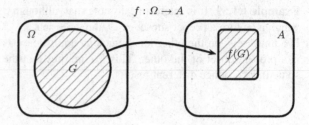

Fig. 1.9 Range $f(\Omega) \subseteq A$
of function $f : \Omega \to A$

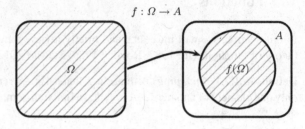

Definition 1.2.4 (*range*) For a function $f : \Omega \to A$, the image $f(\Omega)$ is called the range of the function f.

The image $f(G)$ of $G \subseteq \Omega$ and the range $f(\Omega)$ are shown in Figs. 1.8 and 1.9, respectively.

Example 1.2.4 For the domain $\Omega = [-1, 1]$ and the codomain $A = [-10, 10]$, consider the function $f(\omega) = \omega^2$. The image of the subset $G_1 = \left(-\frac{1}{2}, \frac{1}{2}\right)$ of the domain Ω is $f(G_1) = [0, 0.25)$, and the image of $G_2 = (0.1, 0.2)$ is $f(G_2) = (0.01, 0.04)$.
$\diamondsuit$

Example 1.2.5 The image of $G = \{\{3\}, \{3, 4\}\}$ in Example 1.2.3 is $f(G) = \{1, 2\}$.
$\diamondsuit$

Example 1.2.6 Consider the domain $\Omega = [-1, 1]$ and codomain $A = [-2, 1]$. Assume a function f for which all the points in $[-1, 0) \subseteq \Omega$ are mapped to $-1 \in A$ and all the points in $[0, 1] \subseteq \Omega$ are mapped to $1 \in A$. Then, the range $f(\Omega) = f([-1, 1])$ of f is $\{-1, 1\}$, which is different from the codomain A. In Example 1.2.3, the range of f is $\{0, 1, 2, 3\}$.
$\diamondsuit$

As we observed in Example 1.2.6, the range and codomain are not necessarily the same.

Definition 1.2.5 (*inverse image*) For a function $f : \Omega \to A$ and a subset H of A, the subset

$$f^{-1}(H) = \{\omega : f(\omega) \in H\}, \tag{1.2.2}$$

shown in Fig. 1.10, of Ω is called the inverse image of H (under f).

Fig. 1.10 Inverse image $f^{-1}(H) \subseteq \Omega$ of $H \subseteq A$ for a function $f : \Omega \to A$

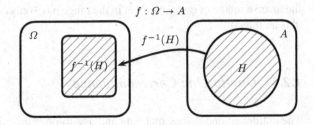

Example 1.2.7 Consider the function $f(\omega) = \omega^2$ with domain $\Omega = [-1, 1]$ and codomain $A = [-10, 10]$. The inverse image of a subset $H_1 = (-0.25, 1)$ of codomain A is $f^{-1}(H_1) = (-1, 1)$, and the inverse image of $H_2 = (-0.25, 0)$ is $f^{-1}(H_2) = f^{-1}((-0.25, 0)) = \emptyset$. ◇

Example 1.2.8 In Example 1.2.3, the inverse image of $H = \{3\}$ is $f^{-1}(H) = \{\{3, 4, 5\}\}$. ◇

Definition 1.2.6 (*surjection*) When the range and codomain of a function are the same, the function is called an onto function, a surjective function, or a surjection.

If the range and codomain of a function are not the same, that is, if the range is a proper subset of the codomain, then the function is called an into function.

Definition 1.2.7 (*injection*) When the inverse image for every element of the codomain of a function has at most one element, i.e., when the inverse image for every element of the range of a function has only one element, the function is called an injective function, a one-to-one function, a one-to-one mapping, or an injection.

In Definition 1.2.7, '... function has at most one element, i.e., ...' can be replaced with '... function is a null set, a singleton set, or a singleton collection of sets, i.e., ...', and '... has only one element, ...' with '... is a singleton set or a singleton collection of sets, ...'.

Example 1.2.9 For the domain $\Omega = [-1, 1]$ and the codomain $A = [0, 1]$, consider the function $f(\omega) = \omega^2$. Then, f is a surjective function because its range is the same as the codomain, and f is not an injective function because, for any non-zero point of the range, the inverse image has two elements. ◇

Example 1.2.10 For the domain $\Omega = [-1, 1]$ and the codomain $A = [-2, 2]$, consider the function $f(\omega) = \omega$. Then, because the range $[-1, 1]$ is not the same as the codomain, the function f is not a surjection. Because the inverse image of every element in the range is a singleton set, the function f is an injection. ◇

Example 1.2.11 For the domain $\Omega = \{\{1\}, \{2, 3\}\}$ and the codomain $A = \{3, \{4\}, \{5, 6, 7\}\}$, consider the function $f(\{1\}) = 3$, $f(\{2, 3\}) = \{4\}$. Because the range $\{3, \{4\}\}$ is not the same as the codomain, the function f is not a surjection. Because the inverse image of every element in the range have only one[4] element, the function f is an injection.   ◇

1.2.1 One-to-One Correspondence

The notions of one-to-one mapping and one-to-one correspondence defined in Definitions 1.2.7 and 1.1.21, respectively, can be alternatively defined as in the following definitions:

Definition 1.2.8 (*one-to-one mapping*) A mapping is called one-to-one if the inverse image of every singleton set in the range is a singleton set.

Definition 1.2.9 (*one-to-one correspondence*) When the inverse image of every element in the codomain is a singleton set, the function is called a one-to-one correspondence.

A one-to-one correspondence is also called a bijective, a bijective function, or a bijective mapping. A bijective function is a surjection and an injection at the same time. A one-to-one correspondence is a one-to-one mapping for which the range and codomain are the same. For a one-to-one mapping that is not a one-to-one correspondence, the range is a proper subset of the codomain.

Example 1.2.12 For the domain $\Omega = [-1, 1]$ and the codomain $A = [-1, 1]$, consider the function $f(\omega) = \omega$. Then, f is a surjective function because the range is the same as the codomain, and f is an injective function because, for every point of the range, the inverse image is a singleton set. In other words, f is a one-to-one correspondence and a bijective function.   ◇

Theorem 1.2.1 *When f is a one-to-one correspondence, we have*

$$f(A \cup B) = f(A) \cup f(B), \tag{1.2.3}$$
$$f(A \cap B) = f(A) \cap f(B), \tag{1.2.4}$$
$$f^{-1}(C \cup D) = f^{-1}(C) \cup f^{-1}(D), \tag{1.2.5}$$

and

$$f^{-1}(C \cap D) = f^{-1}(C) \cap f^{-1}(D) \tag{1.2.6}$$

for subsets A and B of the domain and subsets C and D of the range.

Proof Let us show (1.2.5) only. First, when $x \in f^{-1}(C \cup D)$, we have $f(x) \in C$ or $f(x) \in D$. Then, because $x \in f^{-1}(C)$ or $x \in f^{-1}(D)$, we have $x \in f^{-1}(C) \cup$

---

[4] The inverse image of the element {4} of the range is not {2, 3} but {{2, 3}}, which has only one element {2, 3}.

$f^{-1}(D)$. Next, when $x \in f^{-1}(C) \cup f^{-1}(D)$, we have $f(x) \in C$ or $f(x) \in D$. Thus, we have $f(x) \in C \cup D$, and it follows that $x \in f^{-1}(C \cup D)$. ♠

Theorem 1.2.1 implies that, if f is a one-to-one correspondence, not only can the images $f(A \cup B)$ and $f(A \cap B)$ be expressed in terms of the images $f(A)$ and $f(B)$, but the inverse images $f^{-1}(C \cup D)$ and $f^{-1}(C \cap D)$ can also be expressed in terms of the inverse images $f^{-1}(C)$ and $f^{-1}(D)$.

Generalizing (1.2.3)–(1.2.6), we have

$$f\left(\bigcup_{i=1}^{n} A_i\right) = \bigcup_{i=1}^{n} f(A_i), \tag{1.2.7}$$

$$f\left(\bigcap_{i=1}^{n} A_i\right) = \bigcap_{i=1}^{n} f(A_i), \tag{1.2.8}$$

$$f^{-1}\left(\bigcup_{i=1}^{m} C_i\right) = \bigcup_{i=1}^{m} f^{-1}(C_i), \tag{1.2.9}$$

and

$$f^{-1}\left(\bigcap_{i=1}^{m} C_i\right) = \bigcap_{i=1}^{m} f^{-1}(C_i) \tag{1.2.10}$$

if f is a one-to-one correspondence, where $\{A_i\}_{i=1}^{n}$ are subsets of the domain and $\{C_i\}_{i=1}^{m}$ are subsets of the range.

1.2.2 Metric Space

Definition 1.2.10 (*distance function*) A function d satisfying the three conditions below for every three points p, q, and r is called a distance function or a metric.

(1) $d(p, q) = d(q, p)$.
(2) $d(p, q) > 0$ if $p \neq q$ and $d(p, q) = 0$ if $p = q$.
(3) $d(p, q) \leq d(p, r) + d(r, q)$.

Here, $d(p, q)$ is called the distance between p and q.

Example 1.2.13 For two elements a and b in the set $\mathbb{R}$ of real numbers, assume the function $d(a, b) = |a - b|$. Then, we have $|a - b| = |b - a|$, $|a - b| > 0$ when $a \neq b$, and $|a - b| = 0$ when $a = b$. We also have $|a - c| + |c - b| \geq |a - b|$ from $(|\alpha| + |\beta|)^2 - |\alpha + \beta|^2 = 2(|\alpha||\beta| - \alpha\beta) \geq 0$ for real numbers α and β. Therefore, the function $d(a, b) = |a - b|$ is a distance function. ◇

Example 1.2.14 For two elements a and b in the set $\mathbb{R}$ of real numbers, assume the function $d(a, b) = (a - b)^2$. Then, we have $(a - b)^2 = (b - a)^2$, $(a - b)^2 > 0$ when $a \neq b$, and $(a - b)^2 = 0$ when $a = b$. Yet, because $(a - c)^2 + (c - b)^2 =$

Fig. 1.11 A function with
support $[-1, 1]$

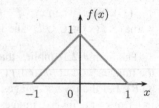

$(a - b)^2 + 2(c - a)(c - b) < (a - b)^2$ when $a < c < b$, the function $d(a, b) = (a - b)^2$ is not a distance function.   ◇

Definition 1.2.11 (*metric space; neighborhood; radius*) A set is called a metric space if a distance is defined for every two points in the set. For a metric space X with distance function d, the set of all points q such that $q \in X$ and $d(p, q) < r$ is called a neighborhood of p and denoted by $N_r(p)$, where r is called the radius of $N_r(p)$.

Example 1.2.15 For the metric space $X = \{x : -2 \le x \le 5\}$ and distance function $d(a, b) = |a - b|$, the neighborhood of 0 with radius 1 is $N_1(0) = (-1, 1)$, and $N_3(0) = [-2, 3)$ is the neighborhood of 0 with radius 3.   ◇

Definition 1.2.12 (*limit point; closure*) A point p is called a limit point of a subset E of a metric space if E contains at least one point different from p for every neighborhood of p. The union $\bar{E} = E \cup E^L$ of E and the set E^L of all the limit points of E is called the closure or enclosure of E.

Example 1.2.16 For the metric space $X = \{x : -2 \le x \le 5\}$ and distance function $d(a, b) = |a - b|$, consider a subset $Y = (-1, 2]$ of X. The set of all the limit points of Y is $[-1, 2]$ and the closure of Y is $(-1, 2] \cup [-1, 2] = [-1, 2]$.   ◇

Definition 1.2.13 (*support*) The closure of the set $\{x : f(x) \neq 0\}$ is called the support of the function $f(x)$.

Example 1.2.17 The support of the function

$$f(x) = \begin{cases} 1 - |x|, & |x| \le 1, \\ 0, & |x| > 1 \end{cases} \tag{1.2.11}$$

is $[-1, 1]$ as shown in Fig. 1.11.   ◇

Example 1.2.18 The value of the function $f(x) = \sin x$ is 0 when $x = n\pi$ for n integer. Yet, the support of $f(x)$ is the set $\mathbb{R}$ of real numbers.   ◇

1.3 Continuity of Functions

When $\{x_n\}_{n=1}^{\infty}$ is a decreasing sequence and $x_n > x$, we denote it by $x_n \downarrow x$. When $\{x_n\}$ is an increasing sequence and $x_n < x$, it is denoted by $x_n \uparrow x$. Let us also use $x^- = \lim_{\varepsilon \downarrow 0}(x - \varepsilon) = \lim_{\varepsilon \uparrow 0}(x + \varepsilon)$ and $x^+ = \lim_{\varepsilon \downarrow 0}(x + \varepsilon) = \lim_{\varepsilon \uparrow 0}(x - \varepsilon)$. In other words, x^- denotes a number smaller than, and arbitrarily close to, x; and x^+ is a number greater than, and arbitrarily close to, x. For a function f and a point x_0, when $f(x_0)$ exists, $\lim_{x \to x_0} f(x) = f(x_0^-) = f(x_0^+)$ exists, and $f(x_0) = \lim_{x \to x_0} f(x)$, the function f is called continuous at point x_0. When a function is continuous at every point in an interval, the function is called continuous on the interval. Let us now discuss the continuity of functions (Johnsonbaugh and Pfaffenberger 1981; Khaleelulla 1982; Munroe 1971; Olmsted 1961; Rudin 1976; Steen and Seebach 1970) in more detail.

1.3.1 Continuous Functions

Definition 1.3.1 (*continuous function*) If, for every positive number ϵ and every point x_0 in a region S, there exists a positive number $\delta(x_0, \epsilon)$ such that $|f(x) - f(x_0)| < \epsilon$ for all points x in S when $|x - x_0| < \delta(x_0, \epsilon)$, then the function f is called continuous on S.

In other words, for some point x_0 in S and some positive number ϵ, if there exists at least one point x in S such that $|x - x_0| < \delta(x_0, \epsilon)$ yet $|f(x) - f(x_0)| \geq \epsilon$ for every positive number $\delta(x_0, \epsilon)$, the function f is not continuous on S.

Example 1.3.1 Consider the function

$$u(x) = \begin{cases} 0, & x \leq 0, \\ 1, & x > 0. \end{cases} \tag{1.3.1}$$

Let $x_0 = 0$ and $\epsilon = 1$. For $x = \frac{\delta}{2}$ with a positive number δ, we have $|x - x_0| = \frac{\delta}{2} < \delta$, yet $|f(x) - f(x_0)| = 1 \geq \epsilon$. Thus, u is not continuous on $\mathbb{R}$. $\diamondsuit$

Theorem 1.3.1 *If f is continuous at $x_p \in E$, g is continuous at $f(x_p)$, and $h(x) = g(f(x))$ when $x \in E$, then h is continuous at x_p.*

Proof Because g is continuous at $f(x_p)$, for every $\epsilon > 0$, there exist a number η such that $|g(y) - g(f(x_p))| < \epsilon$ when $|y - f(x_p)| < \eta$ for $y \in f(E)$. In addition, because f is continuous at x_p, there exists a number δ such that $|f(x) - f(x_p)| < \eta$ when $|x - x_p| < \delta$ for $x \in E$. In other words, for every positive number ϵ, there exists a number δ such that $|h(x) - h(x_p)| = |g(y) - g(f(x_p))| < \epsilon$ when $|x - x_p| < \delta$ for $x \in E$. Therefore, h is continuous at x_p. $\spadesuit$

Definition 1.3.2 (*uniform continuity*) If, for every positive number ϵ, there exists a positive number $\delta(\epsilon)$ such that $|f(x) - f(x_0)| < \epsilon$ for all points x and x_0 in a region S when $|x - x_0| < \delta(\epsilon)$, then the function f is called uniformly continuous on S.

In other words, if there exist at least one each of x and x_0 in S for a positive number ϵ such that $|x - x_0| < \delta(\epsilon)$ yet $|f(x) - f(x_0)| \geq \epsilon$ for every positive number $\delta(\epsilon)$, then the function f is not uniformly continuous on S.

The difference between uniform continuity and continuity lies in the order of choosing the numbers x_0, δ, and ϵ. Specifically, for continuity, x_0 and ϵ are chosen first and then $\delta(x_0, \epsilon)$ is chosen, and thus $\delta(x_0, \epsilon)$ is dependent on x_0 and ϵ. On the other hand, for uniform continuity, ϵ is chosen first, $\delta(\epsilon)$ is chosen next, and then x_0 is chosen last, in which $\delta(\epsilon)$ is dependent only on ϵ and not on x or x_0. In short, the dependence of δ on x_0 is the key difference.

When a function f is uniformly continuous, we can make $f(x_1)$ arbitrarily close to $f(x_2)$ for every two points x_1 and x_2 by moving these two points together. A uniformly continuous function is always a continuous function, but a continuous function is not always uniformly continuous. In other words, uniform continuity is a stronger or more strict concept than continuity.

Example 1.3.2 For the function $f(x) = x$ in $S = \mathbb{R}$, let $\delta = \epsilon$ with $\epsilon > 0$. Then, when $|x - y| < \delta$, because $|f(x) - f(y)| = |x - y| < \delta = \epsilon$, $f(x) = x$ is uniformly continuous. As shown in Exercise 1.23, the function $f(x) = \sqrt{x}$ is uniformly continuous on the interval $(0, \infty)$. The function $f(x) = \frac{1}{x}$ is uniformly continuous for all intervals (a, ∞) with $a > 0$. On the other hand, as shown in Exercise 1.22, it is not uniformly continuous on the interval $(0, \infty)$. The function $f(x) = \tan x$ is continuous but not uniformly continuous on the interval $\left(-\frac{\pi}{2}, \frac{\pi}{2}\right)$. ◇

Example 1.3.3 In the interval $S = (0, \infty)$, consider $f(x) = x^2$. For a positive number ϵ and a point x_0 in S, let $a = x_0 + 1$ and $\delta = \min\left(1, \frac{\epsilon}{2a}\right)$. Then, for a point x in S, when $|x - x_0| < \delta$, we have $|x - x_0| < 1$ and $x < x_0 + 1 = a$ because $\delta \leq 1$. We also have $x_0 < a$. Now, we have $|x^2 - x_0^2| = (x + x_0)|x - x_0| < 2a\delta \leq 2a\frac{\epsilon}{2a} = \epsilon$ because $\delta \leq \frac{\epsilon}{2a}$, and thus f is continuous. On the other hand, let $\epsilon = 1$, assume a positive number δ, and choose $x_0 = \frac{1}{\delta}$ and $x = x_0 + \frac{\delta}{2}$. Then, we have $|x - x_0| = \frac{\delta}{2} < \delta$ but $|x^2 - x_0^2| = \left|\left(\frac{1}{\delta} + \frac{\delta}{2}\right)^2 - \frac{1}{\delta^2}\right| = 1 + \frac{\delta^2}{4} > 1 = \epsilon$, implying that f is not uniformly continuous. Note that, as shown in Exercise 1.21, function $f(x) = x^2$ is uniformly continuous on a finite interval. ◇

Theorem 1.3.2 *A function f is uniformly continuous on S if*

$$|f(x) - f(y)| \leq M|x - y| \qquad (1.3.2)$$

for a number M and every x and y in S.

In Theorem 1.3.2, the inequality (1.3.2) and the number M are called the Lipschitz inequality and Lipschitz constant, respectively.

Example 1.3.4 Consider $S = \mathbb{R}$ and $f(x) = 3x + 7$. For a positive number ϵ, let $\delta = \frac{\epsilon}{3}$. Then, we have $|f(x) - f(x_0)| = 3|x - x_0| < 3\delta = \epsilon$ when $|x - x_0| < \delta$ for every two points x and x_0 in S. Thus, f is uniformly continuous on S. ◇

As a special case of the Heine-Cantor theorem, we have the following theorem:

Theorem 1.3.3 *If a function is differentiable and has a bounded derivative, then the function is uniformly continuous.*

1.3.2 Discontinuities

Definition 1.3.3 (*type 1 discontinuity*) When the three values $f(x^+)$, $f(x^-)$, and $f(x)$ all exist and at least one of $f(x^+)$ and $f(x^-)$ is different from $f(x)$, the point x is called a type 1 discontinuity point or a jump discontinuity point, and the difference $f(x^+) - f(x^-)$ is called the jump or saltus of f at x.

Example 1.3.5 The function

$$f(x) = \begin{cases} x + 1, & x > 0, \\ 0, & x = 0, \\ x - 1, & x < 0 \end{cases} \tag{1.3.3}$$

shown in Fig. 1.12 is type 1 discontinuous at $x = 0$ and the jump is 2. ◇

Definition 1.3.4 (*type 2 discontinuity*) If at least one of $f(x^+)$ and $f(x^-)$ does not exist, then the point x is called a type 2 discontinuity point.

Example 1.3.6 The function

$$f(x) = \begin{cases} \cos \frac{1}{x}, & x \neq 0, \\ 0, & x = 0 \end{cases} \tag{1.3.4}$$

shown in Fig. 1.13 is type 2 discontinuous at $x = 0$. ◇

Fig. 1.12 An example of a type 1 discontinuous function at $x = 0$

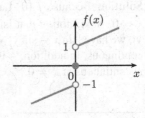

Fig. 1.13 An example of a
type 2 discontinuous
function at $x = 0$:
$f(x) = \cos \frac{1}{x}$ for $x \neq 0$ and
$f(0) = 0$

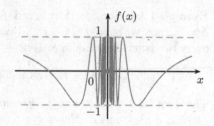

Example 1.3.7 The function

$$f(x) = \begin{cases} 1, & x = \text{rational number}, \\ 0, & x = \text{irrational number} \end{cases} \tag{1.3.5}$$

is type 2 discontinuous at any point x, and the function

$$f(x) = \begin{cases} x, & x = \text{rational number}, \\ 0, & x = \text{irrational number} \end{cases} \tag{1.3.6}$$

is type 2 discontinuous almost everywhere: that is, at all points except $x = 0$.   ◇

Example 1.3.8 The function

$$f(x) = \begin{cases} \frac{1}{q}, & x = \frac{p}{q}, p \in \mathbb{J} \text{ and } q \in \mathbb{J}_+ \text{ are coprime}, \\ 0, & x = \text{irrational number} \end{cases} \tag{1.3.7}$$

is[5] continuous almost everywhere: that is, f is continuous at all points except at rational numbers. The discontinuities are all type 2 discontinuities.   ◇

Example 1.3.9 Show that the function

$$f(x) = \begin{cases} \sin \frac{1}{x}, & x \neq 0, \\ 0, & x = 0 \end{cases} \tag{1.3.8}$$

is type 2 discontinuous at $x = 0$ and continuous at $x \neq 0$.

Solution Because $f(0^+)$ and $f(0^-)$ do not exist, $f(x)$ is type 2 discontinuous at $x = 0$. Next, noting that $|\sin x - \sin y| = 2 \left| \sin \frac{x-y}{2} \cos \frac{x+y}{2} \right| \leq 2 \left| \sin \frac{x-y}{2} \right| \leq |x - y|$, we have $|\sin x - \sin y| < \epsilon$ when $|x - y| < \delta = \epsilon$. Therefore, $\sin x$ is uniformly continuous. In addition, $\frac{1}{x}$ is continuous at $x \neq 0$. Thus, from Theorem 1.3.1, $f(x)$ is continuous at $x \neq 0$.   ◇

---

[5] This function is called Thomae's function.

1.3.3 Absolutely Continuous Functions and Singular Functions

Definition 1.3.5 (*absolute continuity*) Consider a finite collection $\{(a_k, b_k)\}_{k=1}^{n}$ of non-overlapping intervals with $a_k, b_k \in (-c, c)$ for a positive number c. If there exists a number $\delta = \delta(c, \varepsilon)$ such that

$$\sum_{k=1}^{n} |f(b_k) - f(a_k)| < \varepsilon \tag{1.3.9}$$

when $\sum_{k=1}^{n} |b_k - a_k| < \delta$ for every positive numbers c and ε, then the function f is called an absolutely continuous function.

Example 1.3.10 The functions $f_1(x) = x^2$ and $f_2(x) = \sin x$ are both absolutely continuous, and $f_3(x) = \frac{1}{x}$ is absolutely continuous for $x > 0$. ◇

If a function $f(x)$ is absolutely continuous on a finite interval (a, b), then there exists an integrable function $f'(x)$ satisfying

$$f(b) - f(a) = \int_a^b f'(x)dx, \tag{1.3.10}$$

where $-\infty < a < b < \infty$ and $f'(x)$ is the derivative of $f(x)$ at almost every point. The converse also holds true. Note that, if $f(x)$ is not absolutely continuous, the derivative does not satisfy (1.3.10) even when the derivative of $f(x)$ exists at almost every point.

Theorem 1.3.4 *If a function has a bounded derivative almost everywhere and is integrable on a finite interval, and the right and left derivatives of the function exist at the points where the derivatives do not exist, then the function is absolutely continuous.*

Definition 1.3.6 (*singular function*) A continuous, but not absolutely continuous, function is called a singular function.

In other words, a continuous function is either an absolutely continuous function or a singular function.

Example 1.3.11 Denote by D_{ij} the j-th interval that is removed at the i-th step in the procedure of obtaining the Cantor set $\mathbb{C}$ in Example 1.1.46, where $i = 1, 2, \ldots$ and $j = 1, 2, \ldots, 2^{i-1}$. Draw $2^n - 1$ line segments

$$\left\{ y = \frac{2j-1}{2^i} : x \in D_{ij}; \ j = 1, 2, \ldots, 2^{i-1}; \ i = 1, 2, \ldots, n \right\} \tag{1.3.11}$$

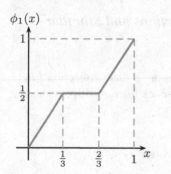 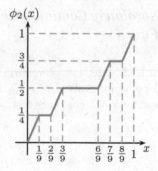

Fig. 1.14 The first two functions $\phi_1(x)$ and $\phi_2(x)$ of $\{\phi_n(x)\}_{n=1}^{\infty}$ converging to the Cantor function $\phi_C(x)$

parallel to the x-axis on an (x, y) coordinate plane. Next, draw a straight line each from the point $(0, 0)$ to the left endpoint of the nearest line segment and from the point $(1, 1)$ to the right endpoint of the nearest line segment. For every line segment, draw a straight line from the right endpoint to the left endpoint of the nearest line segment on the right-hand side. Let the function resulting from this procedure be $\phi_n(x)$. Then, $\phi_n(x)$ is continuous on the interval $(0, 1)$, and is composed of 2^n line segments of height 2^{-n} and slope $\left(\frac{3}{2}\right)^n$ connected with $2^n - 1$ horizontal line segments. Figure 1.14 shows $\phi_1(x)$ and $\phi_2(x)$. The limit

$$\phi_C(x) = \lim_{n \to \infty} \phi_n(x) \tag{1.3.12}$$

of the sequence $\{\phi_n(x)\}_{n=1}^{\infty}$ is called a Cantor function or Lebesgue function.   ◇

The Cantor function $\phi_C(x)$ described in Example 1.3.11 can be expressed as

$$\phi_C(x) = \begin{cases} 0.\frac{c_1}{2}\frac{c_2}{2}\cdots, & x \in \mathbb{C}, \\ y \text{ shown in } (1.3.11), & x \in [0, 1] - \mathbb{C} \end{cases} \tag{1.3.13}$$

when a point x in the Cantor set $\mathbb{C}$ discussed in Example 1.1.46 is written as

$$x = 0.c_1 c_2 \cdots \tag{1.3.14}$$

in a ternary number. Now, the image of $\phi_C(x)$ is a subset of $[0, 1]$. In addition, because the number

$$x = 0.(2b_1)(2b_2)\cdots \tag{1.3.15}$$

is clearly a point in $\mathbb{C}$ such that $\phi_C(x) = y$ when $y \in [0, 1]$ is expressed in a binary number as $y = 0.b_1 b_2 \cdots$, we have $[0, 1] \subseteq \phi_C(\mathbb{C})$. Therefore the range of $\phi_C(x)$ is $[0, 1]$.

Some properties of the Cantor function $\phi_C(x)$ are as follows:

(1) The Cantor function $\phi_C(x)$ is a non-decreasing function with range $[0, 1]$ and no jump discontinuity. Because there can be no discontinuity except for jump discontinuities in non-increasing and non-decreasing functions, $\phi_C(x)$ is a continuous function.

(2) Let E be the set of points represented by (1.1.41) and (1.1.42). Then, the function $\phi_C(x)$ is an increasing function at $x \in \mathbb{C} - E$ and is constant in some neighborhood of every point $x \in [0, 1] - \mathbb{C}$.

(3) As observed in Example 1.1.46, the length of $[0, 1] - \mathbb{C}$ is 1, and $\phi_C(x)$ is constant at $x \in [0, 1] - \mathbb{C}$. Therefore, the derivative of $\phi_C(x)$ is 0 almost everywhere.

Example 1.3.12 (Salem 1943) The Cantor function $\phi_C(x)$ considered in Example 1.3.11 is a non-decreasing singular function. Obtain an increasing singular function.

Solution Consider the line segment PQ connecting $P(x, y)$ and $Q(x + \Delta_x, y + \Delta_y)$ on a two-dimensional plane, where $\Delta_x > 0$ and $\Delta_y > 0$. Let the point R have the coordinate $\left(x + \frac{\Delta_x}{2}, y + \lambda_0 \Delta_y\right)$ with $0 < \lambda_0 < 1$. Denote the replacement of the line segment PQ into two line segments PR and RQ by 'transformation of PQ via $T(\lambda_0)$'. Now, starting from the line segment OA between the origin $O(0, 0)$ and the point $A(1, 1)$, consider a sequence $\{f_n(x)\}_{n=0}^{\infty}$ defined by

$f_0(x) = $ line segment OA,

$f_1(x) = $ transformation of $f_0(x)$ via $T(\lambda_0)$,

$f_2(x) = $ transformation of each of the two line segments composing $f_1(x)$
   via $T(\lambda_0)$,

$f_3(x) = $ transformation of each of the four line segments composing $f_2(x)$
   via $T(\lambda_0)$,

$\vdots$

In other words, $f_m(x)$ is increasing from $f_m(0) = 0$ to $f_m(1) = 1$ and is composed of 2^m line segments with the x coordinates of the end points $\left\{\frac{k}{2^m}\right\}_{k=1}^{2^m-1}$. Figure 1.15 shows $\{f_m(x)\}_{m=0}^{3}$ with $\lambda_0 = 0.7$.

Assume we represent the x coordinate of the end points of the line segments composing $f_m(x)$ as $x = \sum_{j=1}^{m} \frac{\theta_j}{2^j} = 0.\theta_1\theta_2 \cdots \theta_m$ in a binary number, where $\theta_j \in \{0, 1\}$. Then, the y coordinate can be written as

$$y = \sum_{k=1}^{m} \theta_k \prod_{j=1}^{k-1} \lambda_{\theta_j}, \qquad (1.3.16)$$

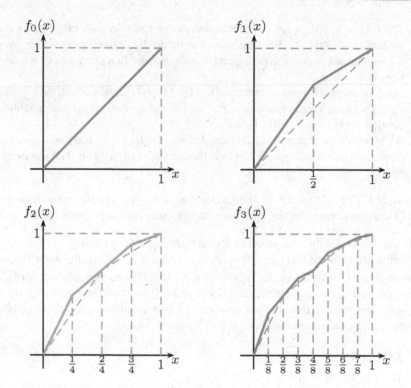

Fig. 1.15 The first four functions in the sequence $\{f_m(x)\}_{m=0}^\infty$ converging to an increasing singular function ($\lambda_0 = 0.7 = 1 - \lambda_1$)

where $\lambda_1 = 1 - \lambda_0$ and we assume $\prod_{j=1}^{k-1} \lambda_{\theta_j} = 1$ when $k = 1$. The limit $f(x) = \lim_{m \to \infty} f_m(x)$ of the sequence $\{f_m(x)\}_{m=1}^\infty$ is an increasing singular function. $\diamond$

Let us note that the convolution[6] of two absolutely continuous functions always results in an absolutely continuous function while the convolution of two singular functions may sometimes result not in a singular function but in an absolutely continuous function (Romano and Siegel 1986).

1.4   Step, Impulse, and Gamma Functions

In this section, we describe the properties of unit step, impulse (Challifour 1972; Gardner 1990; Gelfand and Moiseevich 1964; Hoskins and Pinto 2005; Kanwal 2004; Lighthill 1980), and gamma functions (Artin 1964; Carlson 1977; Zayed 1996) in detail.

---

[6] The integral $\int_{-\infty}^\infty g(x-v)f(v)dv = \int_{-\infty}^\infty g(v)f(x-v)dv$ is called the convolution of f and g, and is usually denoted by $f * g$ or $g * f$.

1.4.1 Step Function

Definition 1.4.1 (*unit step function*) The function

$$u(x) = \begin{cases} 0, & x < 0, \\ 1, & x > 0 \end{cases} \tag{1.4.1}$$

is called the unit step function, step function, or Heaviside function and is also denoted by $H(x)$.

In (1.4.1), the value $u(0)$ is not defined: usually, $u(0)$ is chosen as 0, $\frac{1}{2}$, 1, or any value u_0 between 0 and 1. Figure 1.16 shows the unit step function with $u(0) = \frac{1}{2}$. In some cases, the unit step function with value α at $x = 0$ is denoted by $u_\alpha(x)$, with $u_-(x)$, $u(x)$, and $u_+(x)$ denoting the cases of $\alpha = 0$, $\frac{1}{2}$, and 1, respectively. The unit step function can be regarded as the integral of the impulse or delta function that will be considered in Sect. 1.4.2.

The unit step function $u(x)$ with $u(0) = \frac{1}{2}$ can be represented as the limit

$$u(x) = \lim_{\alpha \to \infty} \left\{ \frac{1}{2} + \frac{1}{\pi} \tan^{-1}(\alpha x) \right\} \tag{1.4.2}$$

or

$$u(x) = \lim_{\alpha \to \infty} \frac{1}{1 + e^{-\alpha x}} \tag{1.4.3}$$

of a sequence of continuous functions. We also have

$$u(x) = \frac{1}{2} + \frac{1}{2\pi} \int_{-\infty}^{\infty} \frac{\sin(\omega x)}{\omega} d\omega. \tag{1.4.4}$$

As we have observed in (1.4.2) and (1.4.3), the unit step function can be defined alternatively by first introducing step-convergent sequence, also called the Heaviside

Fig. 1.16 Unit step function with $u(0) = \frac{1}{2}$

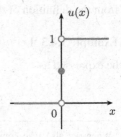

convergent sequence or Heaviside sequence. Specifically, employing the notation[7]

$$\langle a(x), b(x) \rangle = \int_{-\infty}^{\infty} a(x)b(x)dx, \tag{1.4.5}$$

a sequence $\{h_m(x)\}_{m=1}^{\infty}$ of real functions that satisfy

$$\lim_{m \to \infty} \langle f(x), h_m(x) \rangle = \int_0^{\infty} f(x)dx$$
$$= \langle f(x), u(x) \rangle \tag{1.4.6}$$

for every sufficiently smooth function $f(x)$ in the interval $-\infty < x < \infty$ is called a step-convergent sequence or a step sequence, and its limit

$$\lim_{m \to \infty} h_m(x) \tag{1.4.7}$$

is called the unit step function.

Example 1.4.1 If we let $u(0) = \frac{1}{2}$, then $u(x) = 1 - u(-x)$ and $u(a-x) = 1 - u(x-a)$. $\diamond$

Example 1.4.2 We can obtain[8]

$$u(x - |a|) = u(x + a)u(x - a)$$
$$= u\left(x^2 - a^2\right)u(x), \tag{1.4.8}$$
$$u((x-a)(x-b)) = \begin{cases} 0, \min(a,b) < x < \max(a,b), \\ 1, x < \min(a,b) \text{ or } x > \max(a,b), \end{cases} \tag{1.4.9}$$

and

$$u\left(x^2\right) = u\left(|x|\right)$$
$$= \begin{cases} u(0), x = 0, \\ 1, \quad x \neq 0 \end{cases} \tag{1.4.10}$$

from the definition of the unit step function. $\diamond$

Example 1.4.3 Let $u(0) = \frac{1}{2}$. Then, the min function $\min(t,s) = \begin{cases} t, t \leq s, \\ s, t \geq s \end{cases}$ can be expressed as

$$\min(t,s) = t\,u(s-t) + s\,u(t-s). \tag{1.4.11}$$

---

[7] When we also take complex functions into account, the notation $\langle a(x), b(x) \rangle$ is defined as $\langle a(x), b(x) \rangle = \int_{-\infty}^{\infty} a(x)b^*(x)dx$.

[8] In (1.4.8), it is implicitly assumed $u(0) = 0$ or 1.

In addition, we have

$$\min(t, s) = \int_0^t u(s - y)dy \qquad (1.4.12)$$

for $t \geq 0$ and $s \geq 0$. Similarly, the max function $\max(t, s) = \begin{cases} t, & t \geq s, \\ s, & t \leq s \end{cases}$ can be expressed as

$$\max(t, s) = t\, u(t - s) + s\, u(s - t). \qquad (1.4.13)$$

Recollecting (1.4.12), we also have $\max(t, s) = t + s - \min(t, s)$, i.e.,

$$\max(t, s) = t + s - \int_0^t u(s - y)dy \qquad (1.4.14)$$

for $t \geq 0$ and $s \geq 0$.   ◇

The unit step function is also useful in expressing piecewise continuous functions as single-line formulas.

Example 1.4.4 The function

$$F(x) = \begin{cases} x^2, & 0 < x < 1, \\ 3, & 1 < x < 2, \\ 0, & \text{otherwise} \end{cases} \qquad (1.4.15)$$

can be written as $F(x) = x^2 u(x) - (x^2 - 3)\, u(x - 1) - 3u(x - 2)$.   ◇

Example 1.4.5 Consider the function

$$h_m(x) = \begin{cases} 1, & x > \frac{1}{2m}, \\ mx + \frac{1}{2}, & -\frac{1}{2m} \leq x \leq \frac{1}{2m}, \\ 0, & x < -\frac{1}{2m} \end{cases} \qquad (1.4.16)$$

shown in Fig. 1.17. Then, we have $\lim\limits_{m \to \infty} \int_{-\infty}^{\infty} f(x)h_m(x)dx = \lim\limits_{m \to \infty} \left\{ \int_{-\frac{1}{2m}}^{\frac{1}{2m}} \left(mx + \frac{1}{2}\right) f(x)dx + \int_{\frac{1}{2m}}^{\infty} f(x)dx \right\} = \int_0^{\infty} f(x)dx = \langle f(x), u(x) \rangle$ because $\int_{-\frac{1}{2m}}^{\frac{1}{2m}} f(x)dx \leq \frac{1}{m} \max\limits_{|x| \leq \frac{1}{2m}} |f(x)| \to 0$ and $m \int_{-\frac{1}{2m}}^{\frac{1}{2m}} xf(x)dx \leq \max\limits_{|x| \leq \frac{1}{2m}} |xf(x)| \to 0$ when $m \to \infty$ and $f(0) < \infty$. In other words, the sequence $\{h_m(x)\}_{m=1}^{\infty}$ is a step-convergent sequence and its limit is $\lim\limits_{m \to \infty} h_m(x) = u(x)$.   ◇

The unit step function we have described so far is defined in the continuous space. In the discrete space, the unit step function can similarly be defined.

Fig. 1.17 A function in the step-convergent sequence $\{h_m(x)\}_{m=1}^{\infty}$

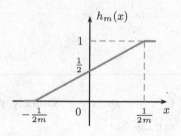

Definition 1.4.2 (*unit step function in discrete space*) The function

$$\tilde{u}(x) \; = \; \begin{cases} 1, \, x = 0, 1, \ldots, \\ 0, \, x = -1, -2, \ldots \end{cases} \tag{1.4.17}$$

is called the unit step function in discrete space.

Note that, unlike the unit step function $u(x)$ in continuous space for which the value $u(0)$ is not defined uniquely, the value $\tilde{u}(0)$ is defined uniquely as 1. In addition, for any non-zero real number a, $u(|a|x)$ is equal to $u(x)$ except possibly at $x = 0$ while $\tilde{u}(|a|x)$ and $\tilde{u}(x)$ are different[9] at infinitely many points when $|a| < 1$.

1.4.2 Impulse Function

Although an impulse function is also called a generalized function or a distribution, we will use the terms impulse function and generalized function in this book and reserve the term distribution for another concept later in Chap. 2.

An impulse function can be introduced in three ways. The first one is to define an impulse function as the symbolic derivative of the unit step function. The second way is to define an impulse function via basic properties and the third is to take the limit of an impulse-convergent sequence.

1.4.2.1 Definitions

As shown in Fig. 1.18, the ramp function

$$r_a(t) = \begin{cases} 0, \, t \le 0, \\ \frac{t}{a}, \, 0 \le t \le a, \\ 1, \, t \ge a \end{cases} \tag{1.4.18}$$

---

[9] In this case, we assume that $\tilde{u}(x)$ is defined to be 0 when x is not an integer.

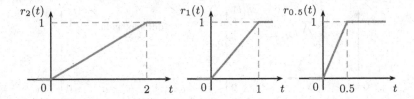

Fig. 1.18 Ramp functions $r_2(t)$, $r_1(t)$, and $r_{0.5}(t)$

is a continuous function, but not differentiable. Yet, it is differentiable everywhere except at $t = 0$ and a. Similarly, the rectangular function

$$p_a(t) = \begin{cases} 0, & t < 0 \text{ or } t > a, \\ \frac{1}{a}, & 0 < t < a \end{cases} \tag{1.4.19}$$

shown in Fig. 1.19, is not a continuous function, and therefore not differentiable. Yet, it is continuous and differentiable everywhere except at $t = 0$ and a. In addition, the rectangular function is the derivative of the ramp function almost everywhere: specifically, $p_a(t) = \frac{d}{dt} r_a(t)$ for $t \neq 0, a$. As we can observe in Fig. 1.20, the limit of the ramp function $r_a(t)$ for $a \to 0$ is the unit step function.

Consider the derivative of the unit step function $u(t)$, the limit of the ramp function $r_a(t)$. The order of operations is not always interchangeable in general: yet, we interchange the order of the derivative and limit of $r_a(t)$. Specifically, as shown in Figs. 1.21 and 1.22, the limit of the derivative of $r_a(t)$ can be regarded as the derivative of the limit of $r_a(t)$. In other words, we can imag-

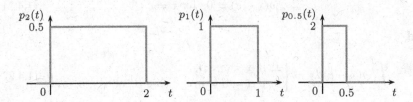

Fig. 1.19 Rectangular functions $p_2(t)$, $p_1(t)$, and $p_{0.5}(t)$

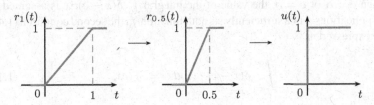

Fig. 1.20 Limit of the sequence $\{r_a(t)\}$ of ramp functions: from a ramp function to the unit step function

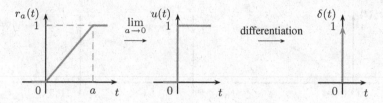

Fig. 1.21 Ramp function $\overset{\text{limit}}{\longrightarrow}$ step function $\overset{\text{differentiation}}{\longrightarrow}$ impulse function

ine $\frac{d}{dt}u(t) = \frac{d}{dt}\left\{\lim_{a\to 0} r_a(t)\right\} = \lim_{a\to 0}\left\{\frac{d}{dt}r_a(t)\right\} = \lim_{a\to 0} p_a(t)$. Based on this simple yet useful description, let us introduce the impulse function in more detail.

Definition 1.4.3 (*impulse function*) The derivative

$$\delta(x) = \frac{du(x)}{dx} \qquad (1.4.20)$$

of the unit step function $u(x)$ is called an impulse function or a generalized function.

As we have already observed, the unit step function $u(x)$ is not continuous at $x = 0$ and therefore not differentiable. This implies that (1.4.20) is technically not defined at $x = 0$: (1.4.20) can then be interpreted as "Let us define the 'symbolic' differentiation of $u(x)$ as $\delta(x)$." Clearly, the impulse function $\delta(x)$ is not defined at $x = 0$: it is often assumed that[10] $\delta(0) \to \infty$.

Let us next consider the second definition of the impulse function.

Definition 1.4.4 (*impulse function*) A function $\delta(x)$ satisfying the conditions

$$\delta(x - c) = 0 \quad \text{for } x \neq c \qquad (1.4.21)$$

and

$$\int_\alpha^\beta \delta(x - c)dx = \begin{cases} 1, & \text{if } \alpha < c < \beta, \\ 0, & \text{if } \alpha = \beta \neq c, c < \alpha < \beta, \text{ or } \alpha < \beta < c \end{cases} \qquad (1.4.22)$$

is called an impulse function.

When $c = \alpha$ or $c = \beta$, the value of the integral $\int_\alpha^\beta \delta(x - c)dx$ is assumed $\frac{1}{2}$ in some applications. For a sufficiently smooth function f, the second condition (1.4.22) can be expressed as

$$\int_{-\infty}^\infty \delta(x - c)f(x)dx = f(c), \qquad (1.4.23)$$

which is called the sifting or reproducing property of the impulse function.

---

[10] As we shall see shortly in (1.4.27), we have $\delta(0) \to -\infty$ in some cases.

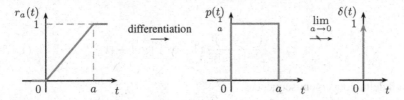

Fig. 1.22 Ramp function $\overset{\text{differentiation}}{\longrightarrow}$ rectangular function $\overset{\text{limit}}{\longrightarrow}$ impulse function

It is rather mathematically difficult to find a function satisfying the two conditions (1.4.21) and (1.4.22). For example, the integral over a non-zero interval of a function which is 0 except at one point will be zero, which contradicts the two conditions in Definition 1.4.4 if we confine ourselves to technicalities. On the other hand, some sequences of functions satisfy (1.4.23) as we can easily see[11] in

$$\lim_{m \to \infty} \int_{-\infty}^{\infty} f(x) \frac{\sin mx}{\pi x} dx = f(0), \tag{1.4.24}$$

for instance. Based on this observation, we can define the impulse function via a proper sequence of functions. Let us first introduce the concept of impulse-convergent sequence similarly as we introduced the step-convergent sequence.

Definition 1.4.5 (*impulse-convergent sequence*) When

$$\lim_{m \to \infty} \int_{-\infty}^{\infty} f(x) s_m(x) dx = f(0) \tag{1.4.25}$$

is satisfied for every sufficiently smooth function $f(x)$ over the interval $-\infty < x < \infty$, the sequence $\{s_m(x)\}_{m=1}^{\infty}$ is called an impulse sequence, an impulse-convergent sequence, or a delta-convergent sequence.

Example 1.4.6 The sequences $\left\{ \frac{\sin mx}{\pi x} \right\}_{m=1}^{\infty}$, $\left\{ \frac{1}{\pi} \frac{m}{1+m^2 x^2} \right\}_{m=1}^{\infty}$, $\left\{ \frac{m}{2} e^{-m|x|} \right\}_{m=1}^{\infty}$, $\left\{ \frac{m}{\pi(e^{mx}+e^{-mx})} \right\}_{m=1}^{\infty}$, and $\left\{ \sqrt{\frac{m}{\pi}} \exp\left(-mx^2\right) \right\}_{m=1}^{\infty}$ are all impulse-convergent sequences.

$\diamond$

Example 1.4.7 The sequences $\{s_m(x)\}_{m=1}^{\infty}$ of functions with

$$s_m(x) = m\, u \left(\frac{1}{2m} - |x| \right), \tag{1.4.26}$$

$$s_m(x) = \begin{cases} -m, & |x| < \frac{1}{2m}, \\ 2m, & \frac{1}{2m} \le |x| \le \frac{1}{m}, \\ 0, & |x| > \frac{1}{m}, \end{cases} \tag{1.4.27}$$

---

[11] Exercise 1.24 will show this result.

and

$$s_m(x) = \frac{(2m+1)!}{2^{2m+1}(m!)^2} \left(1 - x^2\right)^m u(1 - |x|) \tag{1.4.28}$$

are all delta-convergent sequences. ◇

Example 1.4.8 The sequences $\{s_m(x)\}_{m=1}^\infty$ of functions defined by

$$s_m(x) = \begin{cases} 4m^2 x + 2m, & -\frac{1}{2m} \le x \le 0, \\ -4m^2 x + 2m, & 0 < x \le \frac{1}{2m}, \\ 0, & |x| > \frac{1}{2m} \end{cases} \tag{1.4.29}$$

and

$$s_m(x) = \begin{cases} m^2 x + m, & -\frac{1}{m} \le x \le 0, \\ -m^2 x + m, & 0 \le x \le \frac{1}{m}, \\ 0, & |x| > \frac{1}{m} \end{cases} \tag{1.4.30}$$

are both delta-convergent sequences. ◇

Example 1.4.9 For a non-negative function $f(x)$ with $\int_{-\infty}^\infty f(x)dx = 1$, the sequence $\{mf(mx)\}_{m=1}^\infty$ is an impulse sequence. ◇

The impulse function can now be defined based on the impulse-convergent sequence as follows:

Definition 1.4.6 (*impulse function*) The limit

$$\lim_{m\to\infty} s_m(x) = \delta(x) \tag{1.4.31}$$

of a delta-convergent sequence $\{s_m(x)\}_{m=1}^\infty$ is called an impulse function.

Example 1.4.10 Sometimes, we have $\delta(0) \to -\infty$ as in (1.4.27). ◇

1.4.2.2 Properties

We have

$$\langle \delta(x), \phi(x) \rangle = \phi(0), \tag{1.4.32}$$
$$\langle \delta^{(k)}(x), \phi(x) \rangle = (-1)^k \langle \delta(x), \phi^{(k)}(x) \rangle, \tag{1.4.33}$$
$$\langle f(x)\delta(x), \phi(x) \rangle = \langle \delta(x), f(x)\phi(x) \rangle, \tag{1.4.34}$$

and

$$\langle \delta(ax), \phi(x) \rangle = \left\langle \delta(x), \frac{1}{|a|} \phi \left(\frac{x}{a} \right) \right\rangle \tag{1.4.35}$$

when ϕ and f are sufficiently smooth functions.

Letting $a = -1$ in (1.4.35), we get $\langle \delta(-x), \phi(x) \rangle = \langle \delta(x), \phi(-x) \rangle = \phi(0)$. Therefore

$$\delta(-x) = \delta(x) \tag{1.4.36}$$

from (1.4.32). In other words, $\delta(x)$ is an even function.

Example 1.4.11 For the minimum function $\min(t, s) = t\,u(s - t) + s\,u(t - s)$ introduced in (1.4.11), we get $\frac{\partial}{\partial t} \min(t, s) = u(s - t)$ and

$$\frac{\partial^2}{\partial s \partial t} \min(t, s) = \delta(s - t)$$
$$= \delta(t - s) \tag{1.4.37}$$

by noting that $t\delta(s - t) - s\delta(t - s) = t\delta(s - t) - t\delta(s - t) = 0$. Similarly, for $\max(t, s) = t\,u(t - s) + s\,u(s - t)$ introduced in (1.4.13), we have $\frac{\partial}{\partial t} \max(t, s) = u(t - s)$ and

$$\frac{\partial^2}{\partial s \partial t} \max(t, s) = -\delta(s - t)$$
$$= -\delta(t - s) \tag{1.4.38}$$

by noting again that $t\delta(s - t) - s\delta(t - s) = t\delta(s - t) - t\delta(s - t) = 0$. $\diamond$

Let us next introduce the concept of a test function and then consider the product of a function and the n-th order derivative $\delta^{(n)}(x)$ of the impulse function.

Definition 1.4.7 (*test function*) A real function ϕ satisfying the two conditions below is called a test function.

(1) The function $\phi(x)$ is differentiable infinitely many times at every point $x = (x_1, x_2, \ldots, x_n)$.
(2) There exists a finite number A such that $\phi(x) = 0$ for every point $x = (x_1, x_2, \ldots, x_n)$ satisfying $\sqrt{x_1^2 + x_2^2 + \cdots + x_n^2} > A$.

A function satisfying condition (1) above is often called a C^∞ function. Definition 1.4.7 allows a test function in the n-dimensional space: however, we will consider mainly one-dimensional test functions in this book.

Example 1.4.12 The function $\phi(x) = \exp\left(-\frac{a^2}{a^2 - x^2}\right) u(a - |x|)$ shown in Fig. 1.23 is a test function. $\diamond$

Fig. 1.23 A test function
$\phi(x) =$
$\exp\left(-\frac{a^2}{a^2-x^2}\right) u(a-|x|)$

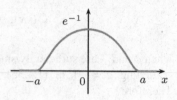

Theorem 1.4.1 *If a function f is differentiable n times consecutively, then*

$$f(x)\delta^{(n)}(x-b) = (-1)^n \sum_{k=0}^{n}(-1)^k \,_nC_k f^{(n-k)}(b)\delta^{(k)}(x-b), \qquad (1.4.39)$$

where $_nC_k$ denotes the binomial coefficient.

Proof Let $b = 0$ and assume a test function $\phi(x)$. We get $\langle f(x)\delta^{(n)}(x), \phi(x)\rangle = \int_{-\infty}^{\infty} f(x)\delta^{(n)}(x)\phi(x)dx = \int_{-\infty}^{\infty}\{f(x)\phi(x)\}\delta^{(n)}(x)dx$, i.e.,

$$\langle f(x)\delta^{(n)}(x), \phi(x)\rangle = \left[\{f(x)\phi(x)\}\,\delta^{(n-1)}(x)\right]_{-\infty}^{\infty}$$
$$- \int_{-\infty}^{\infty}\{f(x)\phi(x)\}'\delta^{(n-1)}(x)dx$$

$$\vdots$$

$$= (-1)^n \int_{-\infty}^{\infty}\{f(x)\phi(x)\}^{(n)}\delta(x)dx \qquad (1.4.40)$$

because $\phi(x) = 0$ for $x \to \pm\infty$. Now, (1.4.40) can be written as

$$\langle f(x)\delta^{(n)}(x), \phi(x)\rangle = (-1)^n \left\{ f^{(n)}(0)\langle\delta(x), \phi(x)\rangle \right.$$
$$+ (-1)^{-1}nf^{(n-1)}(0)\langle\delta^{(1)}(x), \phi(x)\rangle$$
$$+ (-1)^{-2}\frac{n(n-1)}{2!}f^{(n-2)}(0)\langle\delta^{(2)}(x), \phi(x)\rangle$$

$$\vdots$$

$$\left. + (-1)^{-n}f(0)\langle\delta^{(n)}(x), \phi(x)\rangle\right\} \qquad (1.4.41)$$

using (1.4.33) because $\{f(x)\phi(x)\}^{(n)} = \sum_{k=0}^{n} \,_nC_k f^{(n-k)}(x)\phi^{(k)}(x)$. The result (1.4.41) is the same as the symbolic expression (1.4.39).   ♠

The result (1.4.39) implies that the product of a sufficiently smooth function $f(x)$ and $\delta^{(n)}(x-b)$ can be expressed as a linear combination of $\left\{\delta^{(k)}(x-b)\right\}_{k=0}^{n}$ with the coefficient of $\delta^{(k)}(x - b)$ being the product of the number $(-1)^{n-k} \, {}_nC_k$ and[12] the value $f^{(n-k)}(b)$ of $f^{(n-k)}(x)$ at $x = b$.

Example 1.4.13 From (1.4.39), we have

$$f(x)\delta(x - b) = f(b)\delta(x - b) \tag{1.4.42}$$

when $n = 0$.   ◇

Example 1.4.14 Rewriting (1.4.39) specifically for easier reference, we have

$$f(x)\delta'(x - b) = -f'(b)\delta(x - b) + f(b)\delta'(x - b), \tag{1.4.43}$$
$$f(x)\delta''(x - b) = f''(b)\delta(x - b) - 2f'(b)\delta'(x - b) + f(b)\delta''(x - b), \tag{1.4.44}$$

and

$$f(x)\delta'''(x - b) = -f'''(b)\delta(x - b) + 3f''(b)\delta'(x - b)$$
$$- 3f'(b)\delta''(x - b) + f(b)\delta'''(x - b) \tag{1.4.45}$$

when $n = 1, 2$, and 3, respectively.   ◇

Example 1.4.15 From (1.4.43), we get $\delta'(x)\sin x = -(\cos 0)\delta(x) + (\sin 0)\delta'(x) = -\delta(x)$.   ◇

Theorem 1.4.2 *For non-negative integers m and n, we have $x^m \delta^{(n)}(x) = 0$ and $x^m \delta^{(n)}(x) = (-1)^m \frac{n!}{(n-m)!} \delta^{(n-m)}(x)$ when $m > n$ and $m \leq n$, respectively.*

Theorem 1.4.2 can be obtained directly from (1.4.39).

Theorem 1.4.3 *The impulse function $\delta(f(x))$ of a function f can be expressed as*

$$\delta(f(x)) = \sum_{m=1}^{n} \frac{\delta(x - x_m)}{|f'(x_m)|}, \tag{1.4.46}$$

where $\{x_m\}_{m=1}^{n}$ denotes the real simple zeroes of f.

Proof Assume that function f has one real simple zero x_1, and consider a sufficiently small interval $I_{x_1} = (\alpha, \beta)$ with $\alpha < x_1 < \beta$. Because x_1 is the simple zero of f, we have $f'(x_1) \neq 0$. If $f'(x_1) > 0$, then $f(x)$ increases from $f(\alpha)$ to $f(\beta)$ as x moves from α to β. Consequently, $u(f(x)) = u(x - x_1)$ and $\frac{d}{dx} u(f(x)) = \delta(x - x_1)$ on the interval I_{x_1}. On the other hand, we have $\frac{d}{dx} u(f(x)) = \frac{du(f(x))}{df(x)} \frac{df(x)}{dx} = \delta(f(x)) \frac{df(x)}{dx}\Big|_{\{x:f(x)=0\}} = \delta(f(x)) f'(x_1)$. Thus, we get

---

[12] Note that $(-1)^{n+k} = (-1)^{n-k}$.

$$\delta(f(x)) = \frac{\delta(x - x_1)}{f'(x_1)}. \tag{1.4.47}$$

Similarly, if $f'(x_1) < 0$, we have $u(f(x)) = u(x_1 - x)$ and $\delta(f(x)) = -\frac{\delta(x-x_1)}{f'(x_1)}$ on the interval I_{x_1}. In other words,

$$\delta(f(x)) = \frac{\delta(x - x_1)}{|f'(x_1)|}. \tag{1.4.48}$$

Extending (1.4.48) to all the real simple zeroes of f, we get (1.4.46). ♠

Example 1.4.16 From (1.4.46), we can get

$$\delta(ax + b) = \frac{1}{|a|}\delta\left(x + \frac{b}{a}\right) \tag{1.4.49}$$

and $\delta\left(x^3 + 3x\right) = \frac{1}{3}\delta(x)$. ◇

Example 1.4.17 Based on (1.4.46), it can be shown that $\delta((x - a)(x - b)) = \frac{1}{b-a}\{\delta(x - a) + \delta(x - b)\}$ when $b > a$, $\delta(\tan x) = \delta(x)$ when $-\frac{\pi}{2} < x < \frac{\pi}{2}$, and $\delta(\cos x) = \delta\left(x - \frac{\pi}{2}\right)$ when $0 < x < \pi$. ◇

For the function[13] $\delta'(f(x)) = \frac{d\delta(v)}{dv}\Big|_{v=f(x)}$ or

$$\delta'(f(x)) = \frac{d\delta(f(x))}{df(x)}, \tag{1.4.50}$$

we similarly have the following theorem:

Theorem 1.4.4 *We have*

$$\delta'(f(x)) = \sum_{m=1}^{n} \frac{1}{|f'(x_m)|}\left[\frac{\delta'(x - x_m)}{f'(x_m)} + \frac{f''(x_m)}{\{f'(x_m)\}^2}\delta(x - x_m)\right], \tag{1.4.51}$$

where $\{x_m\}_{m=1}^{n}$ denote the real simple zeroes of $f(x)$.

Proof We recollect that

$$\frac{\delta'(x - x_m)}{f'(x)} = \frac{f''(x_m)}{\{f'(x_m)\}^2}\delta(x - x_m) + \frac{1}{f'(x_m)}\delta'(x - x_m) \tag{1.4.52}$$

---

[13] Note that $g'(f(x)) = \left[\frac{dg(y)}{dy}\right]_{y=f(x)} \neq \frac{d}{dx}g(f(x))$. In other words, $g'(f(x))$ denotes $\left[\frac{dg(y)}{dy}\right]_{y=f(x)}$ or $\frac{dg(f)}{df}$, but not $\frac{d}{dx}g(f(x))$.

from (1.4.43) because $\left(\frac{1}{f'(x)}\right)'\Big|_{x=x_m} = -\frac{f''(x)}{\{f'(x)\}^2}\Big|_{x=x_m} = -\frac{f''(x_m)}{\{f'(x_m)\}^2}$. Recollect also that

$$\frac{d}{dx}\delta(f(x)) = \sum_{m=1}^{n} \frac{\delta'(x-x_m)}{|f'(x_m)|} \tag{1.4.53}$$

from (1.4.46). Then, because $\delta'(f) = \frac{d\delta(f)}{df} = \frac{dx}{df}\frac{d\delta(f)}{dx}$, we have

$$\delta'(f(x)) = \frac{1}{f'(x)}\sum_{m=1}^{n} \frac{\delta'(x-x_m)}{|f'(x_m)|} \tag{1.4.54}$$

from (1.4.53). Now, employing (1.4.52) into (1.4.54) results in

$$\delta'(f(x)) = \sum_{m=1}^{n} \frac{1}{|f'(x_m)|}\left[\frac{f''(x_m)}{\{f'(x_m)\}^2}\delta(x-x_m) + \frac{\delta'(x-x_m)}{f'(x_m)}\right], \tag{1.4.55}$$

completing the proof.   ♠

When the real simple zeroes of a sufficiently smooth function $f(x)$ are $\{x_m\}_{m=1}^{n}$, Theorem 1.4.3 indicates that $\delta(f(x))$ can be expressed as a linear combination of $\{\delta(x-x_m)\}_{m=1}^{n}$ and Theorem 1.4.4 similarly indicates that $\delta'(f(x))$ can be expressed as a linear combination of $\{\delta(x-x_m), \delta'(x-x_m)\}_{m=1}^{n}$.

Example 1.4.18 The function $f(x) = (x-1)(x-2)$ has two simple zeroes $x = 1$ and 2. We thus have $\delta'((x-1)(x-2)) = 2\delta(x-1) + 2\delta(x-2) - \delta'(x-1) + \delta'(x-2)$ because $f'(1) = -1$, $f''(1) = 2$, $f'(2) = 1$, and $f''(2) = 2$.   ◇

Example 1.4.19 The function $f(x) = \sinh 2x$ has one simple zero $x = 0$. Then, we get $\delta'(\sinh 2x) = \frac{1}{2}\left\{\frac{1}{2}\delta'(x) + 0\right\} = \frac{1}{4}\delta'(x)$ from $f'(0) = 2\cosh 0 = 2$ and $f''(0) = 4\sinh 0 = 0$.   ◇

1.4.3 Gamma Function

In this section, we address definitions and properties of the factorial, binomial coefficient, and gamma function (Andrews 1999; Wallis and George 2010).

1.4.3.1 Factorial and Binomial Coefficient

Definition 1.4.8 (*falling factorial; factorial*) We call

$$[m]_k = \begin{cases} 1, & k = 0, \\ m(m-1)\cdots(m-k+1), & k = 1, 2, \ldots \end{cases} \quad (1.4.56)$$

the k falling factorial of m. The number of enumeration of n distinct objects is

$$n! = [n]_n \quad (1.4.57)$$

for $n = 1, 2, \ldots$, where the symbol $!$ is called the factorial.

Consequently,

$$0! = 1. \quad (1.4.58)$$

Example 1.4.20 If we use each of the five numbers $\{1, 2, 3, 4, 5\}$ once, we can generate $5! = 5 \times 4 \times 3 \times 2 \times 1 = 120$ five-digit numbers.   ◇

Definition 1.4.9 (*permutation*) The number of ordered arrangements with k different items from n different items is

$$_n P_k = [n]_k \quad (1.4.59)$$

for $k = 0, 1, \ldots, n$, and $_n P_k$ is called the (n, k) permutation.

Example 1.4.21 If we choose two numbers from $\{2, 3, 4, 5, 6\}$ and use each of the two numbers only once, then we can make $_5 P_2 = 20$ two-digit numbers.   ◇

Theorem 1.4.5 *The number of arrangements with k different items from n different items is n^k if repetitions are allowed.*

Example 1.4.22 We can make 10^4 passwords of four digits with $\{0, 1, \ldots, 9\}$.   ◇

Definition 1.4.10 (*combination*) The number of ways to choose k different items from n items is

$$_n C_k = \frac{n!}{(n-k)!k!} \quad (1.4.60)$$

for $k = 0, 1, \ldots, n$, and $_n C_k$, written also as $\binom{n}{k}$, is called the (n, k) combination.

The symbol $_n C_k$ shown in (1.4.60) is also called the binomial coefficient, and satisfies

$$_n C_k = {}_n C_{n-k}. \quad (1.4.61)$$

From (1.4.59) and (1.4.60), we have

$$_n C_k = \frac{_n P_k}{k!}. \quad (1.4.62)$$

Example 1.4.23 We can choose two different numbers from the set $\{0, 1, \ldots, 9\}$ in $_{10}C_2 = 45$ different ways.   ◇

Theorem 1.4.6 *For n repetitions of choosing one element from $\{\omega_1, \omega_2, \ldots, \omega_m\}$, the number of results in which we have n_i of ω_i for $i = 1, 2, \ldots, m$ is*

$$
\binom{n}{n_1, n_2, \ldots, n_m} = \begin{cases} \frac{n!}{n_1! n_2! \cdots n_m!}, & \text{if } \sum_{i=1}^{m} n_i = n, \\ 0, & \text{otherwise,} \end{cases} \tag{1.4.63}
$$

where $n_i \in \{0, 1, \ldots, n\}$. The left-hand side $\binom{n}{n_1, n_2, \ldots, n_m}$ of (1.4.63) is called the multinomial coefficient.

Proof We have

$$
\binom{n}{n_1}\binom{n - n_1}{n_2} \cdots \binom{n - n_1 - n_2 - \cdots - n_{m-1}}{n_m}
$$

$$
= \frac{n!}{n_1!(n - n_1)!} \frac{(n - n_1)!}{n_2!(n - n_1 - n_2)!} \cdots \frac{(n - n_1 - n_2 - \cdots - n_{m-1})!}{n_m!}
$$

$$
= \frac{n!}{n_1! n_2! \cdots n_m!} \tag{1.4.64}
$$

because the number of desired results is the number that ω_1 occurs n_1 times among n occurrences, ω_2 occurs n_2 times among the remaining $n - n_1$ occurrences, $\cdots$, and ω_m occurs n_m times among the remaining $n - n_1 - n_2 \cdots - n_{m-1}$ occurrences.   ♠

The multinomial coefficient is clearly a generalization of the binomial coefficient.

Example 1.4.24 Let $A = \{1, 2, 3\}$, $B = \{4, 5\}$, and $C = \{6\}$ in rolling a die. When the rolling is repeated 10 times, the number of results in which A, B, and C occur four, five, and one times, respectively, is $\binom{10}{4,5,1} = 1260$.   ◇

1.4.3.2   Gamma Function

For $\alpha > 0$,

$$
\Gamma(\alpha) = \int_0^\infty x^{\alpha-1} e^{-x} dx \tag{1.4.65}
$$

is called the gamma function, which satisfies

$$
\Gamma(\alpha) = (\alpha - 1)\Gamma(\alpha - 1) \tag{1.4.66}
$$

and

$$\Gamma(n) = (n-1)! \tag{1.4.67}$$

when n is a natural number. In other words, the gamma function can be viewed as a generalization of the factorial.

Let us now consider a further generalization. When $\alpha < 0$ and $\alpha \neq -1, -2, \ldots$, we can define the gamma function as

$$\Gamma(\alpha) = \frac{\Gamma(\alpha + k + 1)}{\alpha(\alpha + 1)(\alpha + 2) \cdots (\alpha + k)}, \tag{1.4.68}$$

where k is the smallest integer such that $\alpha + k + 1 > 0$. Next, for a complex number z, let[14]

$$(z)_n = \begin{cases} 1, & n = 0, \\ z(z+1) \cdots (z+n-1), & n = 1, 2, \ldots. \end{cases} \tag{1.4.69}$$

Then, for non-negative integers α and n, we can express the factorial as

$$\begin{aligned} \alpha! &= \frac{(\alpha + n)(\alpha + n - 1) \cdots (\alpha + 1)\alpha(\alpha - 1) \cdots 1}{(\alpha + 1)(\alpha + 2) \cdots (\alpha + n)} \\ &= \frac{(\alpha + n)!}{(\alpha + 1)_n} \end{aligned} \tag{1.4.70}$$

from (1.4.67)–(1.4.69). Rewriting (1.4.70) as $\alpha! = \frac{(n+1)_\alpha n!}{(\alpha+1)_n}$ and subsequently as

$$\alpha! = \frac{n^\alpha n!}{(\alpha + 1)_n} \frac{(n+1)_\alpha}{n^\alpha}, \tag{1.4.71}$$

we have

$$\alpha! = \lim_{n \to \infty} \frac{n^\alpha n!}{(\alpha + 1)_n} \tag{1.4.72}$$

because $\lim_{n \to \infty} \frac{(n+1)_\alpha}{n^\alpha} = \left(1 + \frac{1}{n}\right)\left(1 + \frac{2}{n}\right) \cdots \left(1 + \frac{\alpha}{n}\right) = 1$. Based on (1.4.72), the gamma function for a complex number α such that $\alpha \neq 0, -1, -2, \ldots$ can be defined as

$$\Gamma(\alpha) = \lim_{n \to \infty} \frac{n^{\alpha-1} n!}{(\alpha)_n}, \tag{1.4.73}$$

which can also be written as

---

[14] Here, $(z)_n$ is called the rising factorial, ascending factorial, rising sequential product, upper factorial, Pochhammer's symbol, Pochhammer function, or Pochhammer polynomial, and is the same as Appell's symbol (z, n).

$$\Gamma(\alpha) = \lim_{n\to\infty} \frac{n^\alpha n!}{(\alpha)_{n+1}} \tag{1.4.74}$$

because $\lim_{n\to\infty} \frac{n^{\alpha-1} n!}{(\alpha)_n} = \lim_{n\to\infty} \frac{(\alpha+n)n^{\alpha-1} n!}{(\alpha)_{n+1}} = \lim_{n\to\infty} \frac{n^\alpha n!}{(\alpha)_{n+1}}\left(1 + \frac{\alpha}{n}\right)$.

Now, recollecting $\Gamma(1) = \lim_{n\to\infty} \frac{n^0 n!}{(1)_n} = \lim_{n\to\infty} \frac{n!}{n!} = 1$ and $(\alpha+1)_n = (\alpha)_n \frac{\alpha+n}{\alpha}$, we

have $\Gamma(\alpha+1) = \lim_{n\to\infty} \frac{n^\alpha n!}{(\alpha+1)_n} = \lim_{n\to\infty} \frac{\alpha n}{\alpha+n} \lim_{n\to\infty} \frac{n^{\alpha-1} n!}{(\alpha)_n}$ from (1.4.73), and therefore

$$\Gamma(\alpha+1) = \alpha\Gamma(\alpha), \tag{1.4.75}$$

which is the same as (1.4.66). Based on (1.4.75), we can obtain $\lim_{p\to0} p\Gamma(p) = \lim_{p\to0} \Gamma(p+1)$, i.e.,

$$\lim_{p\to0} p\Gamma(p) = 1. \tag{1.4.76}$$

In parallel, when $a \geq b$, we have $\lim_{n\to\infty} (cn)^{b-a} \frac{\Gamma(cn+a)}{\Gamma(cn+b)} = \lim_{n\to\infty} (cn)^{b-a}(cn + a - 1)(cn + a - 2)\cdots(cn + b) = \lim_{n\to\infty} (cn)^{b-a}(cn)^{a-b}$, i.e.,

$$\lim_{n\to\infty} (cn)^{b-a} \frac{\Gamma(cn+a)}{\Gamma(cn+b)} = 1. \tag{1.4.77}$$

We can similarly show that (1.4.77) also holds true for $a < b$.

The gamma function $\Gamma(\alpha)$ is analytic at all points except at $\alpha = 0, -1, \ldots$. In addition, noting that $\lim_{\alpha\to-k} (\alpha + k)\Gamma(\alpha) = \lim_{\alpha\to-k} \lim_{n\to\infty} \frac{(\alpha+k)n^{\alpha-1} n!}{(\alpha)_n}$ or

$$\begin{aligned}
\lim_{\alpha\to-k} (\alpha + k)\Gamma(\alpha) &= \lim_{n\to\infty} \frac{n^{-k-1} n!}{(-k)(-k+1)\ldots(-k+n-1)} \\
&= \frac{(-1)^k}{k!} \lim_{n\to\infty} \frac{n!}{n^{k+1}(n-k-1)!} \\
&= \frac{(-1)^k}{k!} \lim_{n\to\infty} \frac{(n-k)(n-k+1)\cdots n}{n^{k+1}} \\
&= \frac{(-1)^k}{k!} \lim_{n\to\infty} \left(1 - \frac{k}{n}\right)\left(1 - \frac{k-1}{n}\right)\cdots\left(1 - \frac{0}{n}\right) \\
&= \frac{(-1)^k}{k!} \tag{1.4.78}
\end{aligned}$$

for $k = 0, 1, \ldots$ because $(-k)(-k+1)\ldots(-k+n-1) = (-k)(-k+1)\ldots(-k + k - 1)(-k+k+1)\ldots(-k+n-1) = (-1)^k k!(n-k-1)!$, the residue of $\Gamma(\alpha)$ is $\frac{(-1)^{-\alpha}}{(-\alpha)!}$ at the simple pole $\alpha \in \{0, -1, -2, \ldots\}$.

As we shall show later in (1.A.31)–(1.A.39), we have (Abramowitz and Stegun 1972)

$$\Gamma(1-x)\Gamma(x) = \frac{\pi}{\sin \pi x} \qquad (1.4.79)$$

for $0 < x < 1$, which is called the Euler reflection formula. Because we have $\Gamma(\epsilon - n)\Gamma(n+1-\epsilon) = \frac{\pi}{\sin \pi\{(n+1)-\epsilon\}} = \frac{(-1)^n \pi}{\sin \pi\epsilon}$ when $x = n+1-\epsilon$ and $\Gamma(-\epsilon)\Gamma(1+\epsilon) = \frac{\pi}{\sin(\pi+\pi\epsilon)} = -\frac{\pi}{\sin \pi\epsilon}$ when $x = 1+\epsilon$ from (1.4.79), we have

$$\Gamma(\epsilon - n) = (-1)^{n-1}\frac{\Gamma(-\epsilon)\Gamma(1+\epsilon)}{\Gamma(n+1-\epsilon)}. \qquad (1.4.80)$$

Replacing x with $\frac{1}{2} + x$ and $\frac{3}{4} + x$ in (1.4.79), we have

$$\Gamma\left(\frac{1}{2}-x\right)\Gamma\left(\frac{1}{2}+x\right) = \frac{\pi}{\cos \pi x} \qquad (1.4.81)$$

and

$$\Gamma\left(\frac{1}{4}-x\right)\Gamma\left(\frac{3}{4}+x\right) = \frac{\sqrt{2}\pi}{\cos \pi x - \sin \pi x}, \qquad (1.4.82)$$

respectively.

Example 1.4.25 We get

$$\Gamma\left(\frac{1}{2}\right) = \sqrt{\pi} \qquad (1.4.83)$$

with $x = \frac{1}{2}$ in (1.4.79).   ◇

Example 1.4.26 By recollecting (1.4.75) and (1.4.83), we can obtain $\Gamma\left(\frac{1}{2}+k\right) = \left(\frac{1}{2}+k-1\right)\left(\frac{1}{2}+k-2\right)\cdots\frac{1}{2}\Gamma\left(\frac{1}{2}\right) = \frac{1\times 3\times \cdots \times (2k-1)}{2^k}\sqrt{\pi}$, i.e.,

$$\Gamma\left(\frac{1}{2}+k\right) = \frac{\Gamma(2k+1)}{2^{2k}\Gamma(k+1)}$$

$$= \frac{(2k)!}{2^{2k}k!}\sqrt{\pi} \qquad (1.4.84)$$

for $k \in \{0, 1, \ldots\}$. Similarly, we get

$$\Gamma\left(\frac{1}{2}-k\right) = (-1)^k \frac{2^{2k}k!}{(2k)!}\sqrt{\pi} \qquad (1.4.85)$$

for $k \in \{0, 1, \ldots\}$ using $\Gamma\left(\frac{1}{2}\right) = \left(\frac{1}{2} - 1\right)\left(\frac{1}{2} - 2\right) \cdots \left(\frac{1}{2} - k\right)\Gamma\left(\frac{1}{2} - k\right) =$
$(-1)^k \frac{1 \times 3 \times \cdots \times (2k-1)}{2^k}\Gamma\left(\frac{1}{2} - k\right) = (-1)^k \frac{(2k)!}{2^{2k}k!}\Gamma\left(\frac{1}{2} - k\right)$.   ◇

From (1.4.84) and (1.4.85), we have

$$\Gamma\left(\frac{1}{2} - k\right)\Gamma\left(\frac{1}{2} + k\right) = (-1)^k \pi, \qquad (1.4.86)$$

which is the same as (1.4.81) with x an integer k. We can obtain
$\Gamma\left(\frac{3}{2}\right) = \frac{1}{2}\Gamma\left(\frac{1}{2}\right) = \frac{\sqrt{\pi}}{2}$, $\Gamma\left(\frac{5}{2}\right) = \frac{3}{2}\Gamma\left(\frac{3}{2}\right) = \frac{3\sqrt{\pi}}{4}$, $\Gamma\left(\frac{7}{2}\right) = \frac{5}{2}\Gamma\left(\frac{5}{2}\right) = \frac{15\sqrt{\pi}}{8}$, $\cdots$
from (1.4.84), and $\Gamma\left(-\frac{1}{2}\right) = -2\Gamma\left(\frac{1}{2}\right) = -2\sqrt{\pi}$, $\Gamma\left(-\frac{3}{2}\right) = -\frac{2}{3}\Gamma\left(-\frac{1}{2}\right) = \frac{4\sqrt{\pi}}{3}$,
$\Gamma\left(-\frac{5}{2}\right) = -\frac{2}{5}\Gamma\left(-\frac{3}{2}\right) = -\frac{8\sqrt{\pi}}{15}$, $\cdots$ from (1.4.85). Some of such values are shown
in Table 1.2.

We can rewrite (1.4.79) as

$$\Gamma(1 - x)\Gamma(1 + x) = \frac{\pi x}{\sin \pi x} \qquad (1.4.87)$$

by recollecting $\Gamma(x + 1) = x\Gamma(x)$ shown in (1.4.75). In addition, using (1.4.79)–
(1.4.82), (1.4.86), and (1.4.87), we have[15] $\Gamma\left(\frac{1}{3}\right)\Gamma\left(\frac{2}{3}\right) = \frac{2\pi}{\sqrt{3}}$, $\Gamma\left(\frac{1}{4}\right)\Gamma\left(\frac{3}{4}\right) =$
$\sqrt{2}\pi$, and $\Gamma\left(\frac{1}{6}\right)\Gamma\left(\frac{5}{6}\right) = 2\pi$. Note also that, by letting $v = t^\beta$ when $\beta >$
0, we have $\int_0^\infty t^\alpha \exp\left(-t^\beta\right)dt = \int_0^\infty v^{\frac{\alpha}{\beta}}e^{-v}\frac{1}{\beta}v^{\frac{1}{\beta}-1}dv = \frac{1}{\beta}\int_0^\infty v^{\frac{\alpha+1}{\beta}-1}e^{-v}dv =$
$\frac{1}{\beta}\Gamma\left(\frac{\alpha+1}{\beta}\right)$. Subsequently, we have

$$\int_0^\infty t^\alpha \exp\left(-t^\beta\right)dt = \frac{1}{|\beta|}\Gamma\left(\frac{\alpha + 1}{\beta}\right) \qquad (1.4.88)$$

because $\int_0^\infty t^\alpha \exp\left(-t^\beta\right)dt = \int_\infty^0 v^{\frac{\alpha}{\beta}}e^{-v}\frac{1}{\beta}v^{\frac{1}{\beta}-1}dv = -\frac{1}{\beta}\Gamma\left(\frac{\alpha+1}{\beta}\right)$ by letting $v =$
t^β when $\beta < 0$.

When $\alpha \in \{-1, -2, \ldots\}$, we have (Artin 1964)

$$\Gamma(\alpha + 1) \rightarrow \pm\infty. \qquad (1.4.89)$$

More specifically, the value $\Gamma\left(\alpha + 1^\pm\right) = \alpha^\pm!$ can be expressed as

$$\Gamma\left(\alpha + 1^\pm\right) \rightarrow \pm(-1)^{\alpha+1}\infty. \qquad (1.4.90)$$

---

[15] Here, $\sqrt{\pi} \approx 1.7725$, $\frac{2\pi}{\sqrt{3}} \approx 3.6276$, and $\sqrt{2}\pi \approx 4.4429$. In addition, we have $\Gamma\left(\frac{1}{8}\right) \approx 7.5339$,
$\Gamma\left(\frac{1}{7}\right) \approx 6.5481$, $\Gamma\left(\frac{1}{6}\right) \approx 5.5663$, $\Gamma\left(\frac{1}{5}\right) \approx 4.5908$, $\Gamma\left(\frac{1}{4}\right) \approx 3.6256$, $\Gamma\left(\frac{1}{3}\right) \approx 2.6789$, $\Gamma\left(\frac{2}{3}\right) \approx$
1.3541, $\Gamma\left(\frac{3}{4}\right) \approx 1.2254$, $\Gamma\left(\frac{4}{5}\right) \approx 0.4022$, $\Gamma\left(\frac{5}{6}\right) \approx 0.2822$, $\Gamma\left(\frac{6}{7}\right) \approx 0.2082$, and $\Gamma\left(\frac{7}{8}\right) \approx$
0.1596.

For instance, we have $\lim_{\alpha\downarrow -1}\Gamma(\alpha+1)=+\infty$, $\lim_{\alpha\uparrow -1}\Gamma(\alpha+1)=-\infty$, $\lim_{\alpha\downarrow -2}\Gamma(\alpha+1)=-\infty$, $\lim_{\alpha\uparrow -2}\Gamma(\alpha+1)=+\infty$, $\cdots$.

Finally, when $\alpha-\beta$ is an integer, $\alpha<0$, and $\beta<0$, consider a number v for which both $\alpha-v$ and $\beta-v$ are natural numbers. Specifically, let $v=\min(\alpha,\beta)-k$ for $k=1,2,\ldots$: for instance, $v\in\{-\pi-1,-\pi-2,\ldots\}$ when $\alpha=-\pi$ and $\beta=1-\pi$, and $v\in\{-6.4,-7.4,\ldots\}$ when $\alpha=-3.4$ and $\beta=-5.4$. Rewriting $\frac{\Gamma(\alpha+1)}{\Gamma(\beta+1)}=\frac{\Gamma(\alpha+1)}{\Gamma(v)}\frac{\Gamma(v)}{\Gamma(\beta+1)}=\frac{\alpha(\alpha-1)\cdots v}{\beta(\beta-1)\cdots v}=\frac{(-1)^{\alpha-v+1}(-v)(-v-1)\cdots(-\alpha+1)(-\alpha)}{(-1)^{\beta-v+1}(-v)(-v-1)\cdots(-\beta+1)(-\beta)}=$ $(-1)^{\alpha-\beta}\frac{\Gamma(-v+1)}{\Gamma(-\alpha)}\frac{\Gamma(-\beta)}{\Gamma(-v+1)}$, we get

$$\frac{\Gamma(\alpha+1)}{\Gamma(\beta+1)}=(-1)^{\alpha-\beta}\frac{\Gamma(-\beta)}{\Gamma(-\alpha)}. \tag{1.4.91}$$

Based on (1.4.91), it is possible to obtain $\lim_{x\downarrow\beta,\,y\downarrow\alpha}\frac{\Gamma(y+1)}{\Gamma(x+1)}=\lim_{x\uparrow\beta,\,y\uparrow\alpha}\frac{\Gamma(y+1)}{\Gamma(x+1)}=$ $(-1)^{\alpha-\beta}\frac{\Gamma(-\beta)}{\Gamma(-\alpha)}$ and $\lim_{x\downarrow\beta,\,y\uparrow\alpha}\frac{\Gamma(y+1)}{\Gamma(x+1)}=\lim_{x\uparrow\beta,\,y\downarrow\alpha}\frac{\Gamma(y+1)}{\Gamma(x+1)}=(-1)^{\alpha-\beta+1}\frac{\Gamma(-\beta)}{\Gamma(-\alpha)}$: in essence, we have[16]

$$\begin{cases}\frac{\Gamma(\alpha^{\pm})}{\Gamma(\beta^{\pm})}=(-1)^{\alpha-\beta}\frac{\Gamma(-\beta+1)}{\Gamma(-\alpha+1)}, \\ \frac{\Gamma(\alpha^{\mp})}{\Gamma(\beta^{\pm})}=(-1)^{\alpha-\beta+1}\frac{\Gamma(-\beta+1)}{\Gamma(-\alpha+1)}.\end{cases} \tag{1.4.92}$$

This expression is quite useful when we deal with permutations and combinations of negative integers, as we shall see later in (1.A.3) and Table 1.4.

Example 1.4.27 We have $\frac{\Gamma(1-\pi)}{\Gamma(2-\pi)}=(-1)^{-1}\frac{\Gamma(\pi-1)}{\Gamma(\pi)}=\frac{1}{1-\pi}$ from (1.4.91). We can easily get $\frac{\Gamma(-5^{+})}{\Gamma(-4^{+})}=-\frac{\Gamma(5)}{\Gamma(6)}=-\frac{1}{5}$ from (1.4.91) or (1.4.92). $\diamond$

Example 1.4.28 It is known (Abramowitz and Stegun 1972) that $\Gamma(j)\approx -0.1550-j0.4980$, where $j=\sqrt{-1}$. Now, recollect $\Gamma(-z)=-\frac{\Gamma(1-z)}{z}$ from $\Gamma(1-z)=-z\Gamma(-z)$. Thus, using the Euler reflection formula (1.4.79), we have $\Gamma(z)\Gamma(-z)=-\frac{\Gamma(z)\Gamma(1-z)}{z}$, i.e.,

$$\Gamma(z)\Gamma(-z)=-\frac{\pi}{z\sin\pi z} \tag{1.4.93}$$

for $z\neq 0,\pm 1,\pm 2,\ldots$. Next, because $e^{\pm jx}=\cos x\pm j\sin x$, we have $\sin x=\frac{1}{2j}\left(e^{jx}-e^{-jx}\right)$, $\Gamma(\bar{z})=\overline{\Gamma(z)}$, and $\Gamma(1-j)=-j\Gamma(-j)$. Thus, we have $\Gamma(j)\Gamma(-j)=|\Gamma(j)|^{2}=-\frac{\pi}{j\sin\pi j}$, i.e.,

$$|\Gamma(j)|^{2}=\frac{2\pi}{e^{\pi}-e^{-\pi}}, \tag{1.4.94}$$

---

[16] When we use this result, we assume α^{+} and β^{+} unless specified otherwise.

Table 1.2 Values of $\Gamma(z)$ for $z = -\frac{9}{2}, -\frac{7}{2}, \ldots, \frac{9}{2}$

z	$-\frac{9}{2}$	$-\frac{7}{2}$	$-\frac{5}{2}$	$-\frac{3}{2}$	$-\frac{1}{2}$	$\frac{1}{2}$	$\frac{3}{2}$	$\frac{5}{2}$	$\frac{7}{2}$	$\frac{9}{2}$
$\Gamma(z)$	$-\frac{32\sqrt{\pi}}{945}$	$\frac{16\sqrt{\pi}}{105}$	$-\frac{8\sqrt{\pi}}{15}$	$\frac{4\sqrt{\pi}}{3}$	$-2\sqrt{\pi}$	$\sqrt{\pi}$	$\frac{\sqrt{\pi}}{2}$	$\frac{3\sqrt{\pi}}{4}$	$\frac{15\sqrt{\pi}}{8}$	$\frac{105\sqrt{\pi}}{16}$

where $\frac{2\pi}{e^\pi - e^{-\pi}} \approx 0.2720$ and $0.1550^2 + 0.4980^2 \approx 0.2720$. $\diamondsuit$

Example 1.4.29 If we consider only the region $\alpha > 0$, then the gamma function (Zhang and Jian 1996) exhibits the minimum value $\Gamma(\alpha_0) \approx 0.8856$ at $\alpha = \alpha_0 \approx 1.4616$, and is convex downward because $\Gamma''(\alpha) > 0$. $\diamondsuit$

1.4.3.3 Beta Function

The beta function is defined as[17]

$$\tilde{B}(\alpha, \beta) = \int_0^1 x^{\alpha-1}(1-x)^{\beta-1}dx \qquad (1.4.95)$$

for complex numbers α and β such that $\mathrm{Re}(\alpha) > 0$ and $\mathrm{Re}(\beta) > 0$. In this section, let us show

$$\tilde{B}(\alpha, \beta) = \frac{\Gamma(\alpha)\Gamma(\beta)}{\Gamma(\alpha+\beta)}. \qquad (1.4.96)$$

We have $\tilde{B}(\alpha, \beta+1) = \int_0^1 x^{\alpha-1}(1-x)(1-x)^{\beta-1}dx = \tilde{B}(\alpha, \beta) - \tilde{B}(\alpha+1, \beta)$ and $\tilde{B}(\alpha, \beta+1) = \frac{1}{\alpha}x^\alpha(1-x)^\beta\big|_{x=0}^1 + \frac{\beta}{\alpha}\int_0^1 x^\alpha(1-x)^{\beta-1}dx = \frac{\beta}{\alpha}\tilde{B}(\alpha+1, \beta)$. Thus,

$$\tilde{B}(\alpha, \beta) = \frac{\alpha+\beta}{\beta}\tilde{B}(\alpha, \beta+1), \qquad (1.4.97)$$

which can also be obtained as $\tilde{B}(\alpha, \beta+1) = \int_0^1 x^{\alpha+\beta-1}\left(\frac{1-x}{x}\right)^\beta dx = \frac{1}{\alpha+\beta}x^{\alpha+\beta}\left(\frac{1-x}{x}\right)^\beta\big|_{x=0}^1 + \frac{\beta}{\alpha+\beta}\int_0^1 x^{\alpha+\beta}\left(\frac{1-x}{x}\right)^{\beta-1}\frac{dx}{x^2} = \frac{\beta}{\alpha+\beta}\int_0^1 x^{\alpha-1}(1-x)^{\beta-1}dx = \frac{\beta}{\alpha+\beta}\tilde{B}(\alpha, \beta)$. Using (1.4.97) repeatedly, we get

---

[17] The right-hand side of (1.4.95) is called the Eulerian integral of the first kind.

$$\tilde{B}(\alpha, \beta) = \frac{(\alpha + \beta)(\alpha + \beta + 1)}{\beta(\beta + 1)} \tilde{B}(\alpha, \beta + 2)$$

$$= \frac{(\alpha + \beta)(\alpha + \beta + 1)(\alpha + \beta + 2)}{\beta(\beta + 1)(\beta + 2)} \tilde{B}(\alpha, \beta + 3)$$

$$\vdots$$

$$= \frac{(\alpha + \beta)_n}{(\beta)_n} \tilde{B}(\alpha, \beta + n), \tag{1.4.98}$$

which can be expressed as

$$\tilde{B}(\alpha, \beta) = \frac{(\alpha + \beta)_n}{n!} \frac{n!}{(\beta)_n} \int_0^n \left(\frac{t}{n}\right)^{\alpha - 1} \left(1 - \frac{t}{n}\right)^{\beta + n - 1} \frac{dt}{n}$$

$$= \frac{(\alpha + \beta)_n}{n^{\alpha + \beta - 1} n!} \frac{n^{\beta - 1} n!}{(\beta)_n} \int_0^n t^{\alpha - 1} \left(1 - \frac{t}{n}\right)^{\beta + n - 1} dt \tag{1.4.99}$$

after some manipulations. Now, because $\lim_{n \to \infty} \left(1 - \frac{t}{n}\right)^{\beta + n - 1} = \lim_{n \to \infty} \left(1 + \frac{-t}{n}\right)^n$ $\left(1 - \frac{t}{n}\right)^{\beta - 1} = e^{-t}$, if we let $n \to \infty$ in (1.4.99) and recollect the defining equation (1.4.73) of the gamma function, we get

$$\tilde{B}(\alpha, \beta) = \frac{\Gamma(\beta)}{\Gamma(\alpha + \beta)} \int_0^\infty t^{\alpha - 1} e^{-t} dt. \tag{1.4.100}$$

Next, from (1.4.95) with $\beta = 1$, we get $\tilde{B}(\alpha, 1) = \int_0^1 t^{\alpha - 1} dt = \frac{1}{\alpha}$. Using this result into (1.4.100) with $\beta = 1$, we get

$$\frac{1}{\alpha} = \frac{\Gamma(1)}{\Gamma(\alpha + 1)} \int_0^\infty t^{\alpha - 1} e^{-t} dt. \tag{1.4.101}$$

Therefore, recollecting $\Gamma(\alpha + 1) = \alpha \Gamma(\alpha)$ and $\Gamma(1) = 1$, we have[18]

$$\int_0^\infty t^{\alpha - 1} e^{-t} dt = \Gamma(\alpha). \tag{1.4.102}$$

From (1.4.100) and (1.4.102), we get (1.4.96). Note that (1.4.100)–(1.4.102) implicitly dictates that the defining equation (1.4.65) of the gamma function $\Gamma(\alpha)$ for $\alpha > 0$ is a special case of (1.4.73).

---

[18] The left-hand side of (1.4.102) is called the Eulerian integral of the second kind. In Exercise 1.38, we consider another way to show (1.4.96).

1.5 Limits of Sequences of Sets

In this section, the properties of infinite sequences of sets are discussed. The exposition in this section will be the basis, for instance, in discussing the σ-algebra in Sect. 2.1.2 and the continuity of probability in Appendix 2.1.

1.5.1 Upper and Lower Limits of Sequences

Let us first consider the limits of sequences of numbers before addressing the limits of sequences of sets. When $a_i \leq u$ and $a_i \geq v$ for every choice of a number a_i from the set A of real numbers, the numbers u and v are called an upper bound and a lower bound, respectively, of A.

Definition 1.5.1 (*least upper bound; greatest lower bound*) For a subset A of real numbers, the smallest among the upper bounds of A is called the least upper bound of A and is denoted by $\sup A$, and the largest among the lower bounds of A is called the greatest lower bound and is denoted by $\inf A$.

For a sequence $\{x_n\}_{n=1}^{\infty}$ of real numbers, the least upper bound and greatest lower bound are written as

$$\sup_{n \geq 1} x_n = \sup\{x_n : n \geq 1\} \tag{1.5.1}$$

and

$$\inf_{n \geq 1} x_n = \inf\{x_n : n \geq 1\}, \tag{1.5.2}$$

respectively. When there exists no upper bound and no lower bound of A, it is denoted by $\sup A \to \infty$ and $\inf A \to -\infty$, respectively.

Example 1.5.1 For the set $A = \{1, 2, 3\}$, the least upper bound is $\sup A = 3$ and the greatest lower bound is $\inf A = 1$. For the sequence $\{a_n\} = \{0, 1, 0, 1, \ldots\}$, the least upper bound is $\sup a_n = 1$ and the greatest lower bound is $\inf a_n = 0$. ◇

Definition 1.5.2 (*limit superior; limit inferior*) For a sequence $\{x_n\}_{n=1}^{\infty}$,

$$\limsup_{n \to \infty} x_n = \overline{x_n}$$
$$= \inf_{n \geq 1} \sup_{k \geq n} x_k \tag{1.5.3}$$

and

$$\liminf_{n\to\infty} x_n = \underline{x_n}$$

$$= \sup_{n\geq 1} \inf_{k\geq n} x_k \qquad (1.5.4)$$

are called the limit superior and limit inferior, respectively.

Example 1.5.2 For the sequences $x_n = \frac{1}{\sqrt{n}}\{\sqrt{4n+3} + (-1)^n\sqrt{4n-5}\}$, $y_n = \frac{5}{\sqrt{n}} + (-1)^n$, and $z_n = 3\sin\left(\frac{n\pi}{2}\right)$, we have $\overline{x_n} = 4$ and $\underline{x_n} = 0$, $\overline{y_n} = 1$ and $\underline{y_n} = -1$, and $\overline{z_n} = 3$ and $\underline{z_n} = -3$. $\diamond$

Definition 1.5.3 (*limit*) For a sequence $\{x_n\}_{n=1}^{\infty}$, $\lim_{n\to\infty} x_n = y$ and $\limsup_{n\to\infty} x_n = \liminf_{n\to\infty} x_n = y$ are the necessary and sufficient conditions of each other, where y is called the limit of the sequence $\{x_n\}_{n=1}^{\infty}$.

Example 1.5.3 For the three sequences

$$x_n = \begin{cases} \sqrt{2}, & n=1, \\ \sqrt{2+x_{n-1}}, & n=2,3,\ldots, \end{cases} \qquad (1.5.5)$$

$$y_n = \begin{cases} \sqrt{n}, & n=1,2,\ldots 9, \\ 3, & n=10,11,\ldots 100, \\ \frac{3n}{5n+4}, & n=101,102,\ldots, \end{cases} \qquad (1.5.6)$$

and

$$z_n = \frac{1}{2n}\{\sqrt{4n+3} + (-1)^n\sqrt{4n-5}\}, \qquad (1.5.7)$$

we have[19] $\overline{x_n} = \underline{x_n} = \lim_{n\to\infty} x_n = 2$, $\overline{y_n} = \underline{y_n} = \lim_{n\to\infty} y_n = \frac{3}{5}$, and $\overline{z_n} = \underline{z_n} = \lim_{n\to\infty} z_n = 0$. $\diamond$

Example 1.5.4 None of the three sequences x_n, y_n, and z_n considered in Example 1.5.2 has a limit because the limit superior is different from the limit inferior. $\diamond$

Fig. 1.24 Increasing sequence $\{B_n\}_{n=1}^{\infty}$:
$B_{n+1} \supset B_n$

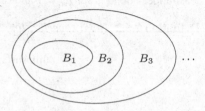

---

[19] The value $\overline{x_n} = 2$ can be obtained by solving the equation $x_n = \sqrt{2+x_n}$.

1.5.2 Limit of Monotone Sequence of Sets

Definition 1.5.4 (*increasing sequence; decreasing sequence; non-decreasing sequence; non-increasing sequence*) For a sequence $\{B_n\}_{n=1}^{\infty}$ of sets, assume that $B_{n+1} \supseteq B_n$ for every natural number n. If $B_{n+1} = B_n$ for at least one n, then the sequence is called a non-decreasing sequence; otherwise, it is called an increasing sequence. A non-increasing sequence and a decreasing sequence are defined similarly by replacing $B_{n+1} \supseteq B_n$ with $B_{n+1} \subseteq B_n$.

Example 1.5.5 The sequences $\left\{ \left[1, 2 - \frac{1}{n} \right) \right\}_{n=1}^{\infty}$ and $\{(-n, a)\}_{n=1}^{\infty}$ are increasing sequences. The sequences $\left\{ \left[1, 1 + \frac{1}{n} \right) \right\}_{n=1}^{\infty}$ and $\left\{ \left(1 - \frac{1}{n}, 1 + \frac{1}{n} \right) \right\}_{n=1}^{\infty}$ are decreasing sequences. $\diamond$

Increasing and decreasing sequences are sometimes referred to as strictly increasing and strictly decreasing sequences, respectively. A non-decreasing sequence or a non-increasing sequence is also called a monotonic sequence. Although increasing and non-decreasing sequences are slightly different from each other, they are used interchangeably when it does not cause an ambiguity. Similarly, decreasing and non-increasing sequences will be often used interchangeably. Figures 1.24 and 1.25 graphically show increasing and decreasing sequences, respectively.

Definition 1.5.5 (*limit of monotonic sequence*) We call

$$\lim_{n \to \infty} B_n = \bigcup_{i=1}^{\infty} B_i \tag{1.5.8}$$

for a non-decreasing sequence $\{B_n\}_{n=1}^{\infty}$ and

$$\lim_{n \to \infty} B_n = \bigcap_{i=1}^{\infty} B_i \tag{1.5.9}$$

for a non-increasing sequence $\{B_n\}_{n=1}^{\infty}$ the limit set or limit of $\{B_n\}_{n=1}^{\infty}$.

The limit $\lim_{n \to \infty} B_n$ denotes the set of points contained in at least one of $\{B_n\}_{n=1}^{\infty}$ and in every set of $\{B_n\}_{n=1}^{\infty}$ when $\{B_n\}_{n=1}^{\infty}$ is a non-decreasing and a non-increasing sequence, respectively.

Fig. 1.25 Decreasing sequence $\{B_n\}_{n=1}^{\infty}$: $B_{n+1} \subset B_n$

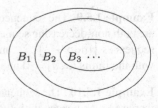

Example 1.5.6 The sequence $\left\{\left[1, 2 - \frac{1}{n}\right)\right\}_{n=1}^{\infty}$ considered in Example 1.5.5 has the limit $\lim_{n \to \infty} B_n = \bigcup_{n=1}^{\infty} \left[1, 2 - \frac{1}{n}\right) = [1, 1) \cup \left[1, \frac{3}{2}\right) \cup \cdots$ or

$$\lim_{n \to \infty} B_n = [1, 2) \tag{1.5.10}$$

because $\left\{\left[1, 2 - \frac{1}{n}\right)\right\}_{n=1}^{\infty}$ is a non-decreasing sequence. Likewise, the limit of the non-decreasing sequence $\{(-n, a)\}_{n=1}^{\infty}$ is $\lim_{n \to \infty} B_n = \bigcup_{n=1}^{\infty} (-n, a) = (-\infty, a)$. $\diamond$

Example 1.5.7 The sequence $\left\{\left[a, a + \frac{1}{n}\right)\right\}_{n=1}^{\infty}$ is a non-increasing sequence and has the limit $\lim_{n \to \infty} B_n = \bigcap_{n=1}^{\infty} \left[a, a + \frac{1}{n}\right)$ or

$$\lim_{n \to \infty} B_n = [a, a], \tag{1.5.11}$$

which is a singleton set $\{a\}$. The non-increasing sequence $\left\{\left(1 - \frac{1}{n}, 1 + \frac{1}{n}\right)\right\}_{n=1}^{\infty}$ has the limit $\lim_{n \to \infty} B_n = \bigcap_{n=1}^{\infty} \left(1 - \frac{1}{n}, 1 + \frac{1}{n}\right) = (0, 2) \cap \left(\frac{1}{2}, \frac{3}{2}\right) \cap \cdots = [1, 1]$, also a singleton set. Note that

$$\lim_{n \to \infty} \left[0, \frac{1}{n}\right) = \{0\} \tag{1.5.12}$$

is different from

$$\left[0, \lim_{n \to \infty} \frac{1}{n}\right) = \emptyset \tag{1.5.13}$$

in (1.5.11). $\diamond$

Example 1.5.8 Consider the set $S = \{x : 0 \le x \le 1\}$, sequence $\{A_i\}_{i=1}^{\infty}$ with $A_i = \left\{x : \frac{1}{i+1} < x \le 1\right\}$, and sequence $\{B_i\}_{i=1}^{\infty}$ with $B_i = \left\{x : 0 < x < \frac{1}{i}\right\}$. Then, because $\{A_i\}_{i=1}^{\infty}$ is a non-decreasing sequence, we have $\lim_{n \to \infty} A_n = \{x : \frac{1}{2} < x \le 1\} \cup \{x : \frac{1}{3} < x \le 1\} \cup \cdots = \{x : 0 < x \le 1\}$ and $S = \{0\} \cup \lim_{n \to \infty} A_n$. Similarly, because $\{B_i\}_{i=1}^{\infty}$ is a non-increasing sequence, we have $\lim_{n \to \infty} B_n = \{x : 0 < x < 1\} \cap \{x : 0 < x < \frac{1}{2}\} \cap \cdots = \{x : 0 < x \le 0\} = \emptyset$. $\diamond$

Example 1.5.9 The sequences $\left\{\left(1 + \frac{1}{n}, 2\right)\right\}_{n=1}^{\infty}$ and $\left\{\left(1 + \frac{1}{n}, 2\right]\right\}_{n=1}^{\infty}$ of interval sets are both non-decreasing sequences with the limits $(1, 2)$ and $(1, 2]$, respectively. The sequences $\left\{\left(a, a + \frac{1}{n}\right)\right\}_{n=1}^{\infty}$ and $\left\{\left[a, a + \frac{1}{n}\right)\right\}_{n=1}^{\infty}$ are both non-increasing sequences with the limits $(a, a] = \emptyset$ and $[a, a] = \{a\}$, respectively. $\diamond$

Example 1.5.10 The sequences $\left\{\left(1 - \frac{1}{n}, 2\right)\right\}_{n=1}^{\infty}$ and $\left\{\left(1 - \frac{1}{n}, 2\right]\right\}_{n=1}^{\infty}$ are both non-increasing sequences with the limits $[1, 2)$ and $[1, 2]$, respectively. The sequences

$\left\{\left(1, 2 - \frac{1}{n}\right]\right\}_{n=1}^{\infty}$ and $\left\{\left[1, 2 - \frac{1}{n}\right]\right\}_{n=1}^{\infty}$ are both non-decreasing sequences with the limits $(1, 2)$ and $[1, 2)$, respectively. $\diamond$

Example 1.5.11 The sequences $\left\{\left[1 + \frac{1}{n}, 3 - \frac{1}{n}\right]\right\}_{n=1}^{\infty}$ and $\left\{\left[1 + \frac{1}{n}, 3 - \frac{1}{n}\right)\right\}_{n=1}^{\infty}$ are both non-decreasing sequences with the common limit $(1, 3)$. Similarly, the non-decreasing sequences $\left\{\left(1 + \frac{1}{n}, 3 - \frac{1}{n}\right]\right\}_{n=1}^{\infty}$ and $\left\{\left(1 + \frac{1}{n}, 3 - \frac{1}{n}\right)\right\}_{n=1}^{\infty}$ both have the limit[20] $(1, 3)$. $\diamond$

Example 1.5.12 The four sequences $\left\{\left[1 - \frac{1}{n}, 3 + \frac{1}{n}\right]\right\}_{n=1}^{\infty}$, $\left\{\left[1 - \frac{1}{n}, 3 + \frac{1}{n}\right)\right\}_{n=1}^{\infty}$, $\left\{\left(1 - \frac{1}{n}, 3 + \frac{1}{n}\right]\right\}_{n=1}^{\infty}$, and $\left\{\left(1 - \frac{1}{n}, 3 + \frac{1}{n}\right)\right\}_{n=1}^{\infty}$ are all non-increasing sequences with the common limit[21] $[1, 3]$. $\diamond$

1.5.3 Limit of General Sequence of Sets

We have discussed the limits of monotonic sequences in Sect. 1.5.2. Let us now consider the limits of general sequences. First, note that any element in a set of an infinite sequence belongs to

(1) every set,
(2) every set except for a finite number of sets,
(3) infinitely many sets except for other infinitely many sets, or
(4) a finite number of sets.

Keeping these four cases in mind, let us define the lower bound and upper bound sets of general sequences.

Definition 1.5.6 (*lower bound set*) For a sequence of sets, the set of elements belonging to at least almost every set of the sequence is called the lower bound or lower bound set of the sequence, and is denoted by[22] $\varliminf_{n \to \infty}$ or by $\underline{\lim}_{n \to \infty}$.

Let us express the lower bound set $\varliminf_{n \to \infty} B_n = \underline{B_n}$ of the sequence $\{B_n\}_{n=1}^{\infty}$ in terms of set operations. First, note that

$$G_i = \bigcap_{k=i}^{\infty} B_k \qquad (1.5.14)$$

---

[20] In short, irrespective of the type of the parentheses, the limit is in the form of '$(a, \ldots$' for an interval when the beginning point is of the form $a + \frac{1}{n}$, and the limit is in the form of '$\ldots, b)$' for an interval when the end point is of the form $b - \frac{1}{n}$.

[21] In short, irrespective of the type of the parentheses, the limit is in the form of '$[a, \ldots$' for an interval when the beginning point is of the form $a - \frac{1}{n}$, and the limit is in the form of '$\ldots, b]$' for an interval when the end point is of the form $b + \frac{1}{n}$.

[22] The acronym inf stands for infimum or inferior.

is the set of elements belonging to B_i, B_{i+1}, ...: in other words, G_i is the set of elements belonging to all the sets of the sequence except for at most $(i-1)$ sets, possibly B_1, B_2, ..., and B_{i-1}. Specifically, G_1 is the set of elements belonging to all the sets of the sequence, G_2 is the set of elements belonging to all the sets except possibly for the first set, G_3 is the set of elements belonging to all the sets except possibly for the first and second sets, ... This implies that an element belonging to any of $\{G_i\}_{i=1}^{\infty}$ is an element belonging to almost every set of the sequence. Therefore, if we collect all the elements belonging to at least one of $\{G_i\}_{i=1}^{\infty}$, or if we take the union of $\{G_i\}_{i=1}^{\infty}$, the result would be the set of elements in every set except for a finite number of sets. In other words, the set of elements belonging to at least almost every set of the sequence $\{B_n\}_{n=1}^{\infty}$, or the lower bound set $\liminf_{n\to\infty} B_n$ of $\{B_n\}_{n=1}^{\infty}$, can be expressed as

$$\liminf_{n\to\infty} B_n = \bigcup_{i=1}^{\infty} \bigcap_{k=i}^{\infty} B_k, \tag{1.5.15}$$

which is sometimes denoted by $\{\text{eventually } B_n\}$ or $\{\text{ev. } B_n\}$.

Example 1.5.13 For the sequence

$$\{B_n\}_{n=1}^{\infty} = \{0, 1, 3\}, \{0, 2\}, \{0, 1, 2\}, \{0, 1\}, \{0, 1, 2\}, \{0, 1\}, \dots \tag{1.5.16}$$

of finite sets, obtain the lower bound set.

Solution First, 0 belongs to all sets, 1 belongs to all sets except for the second set, and 2 and 3 do not belong to infinitely many sets. Thus, the lower bound of the sequence is $\{0, 1\}$, which can be confirmed by $\liminf_{n\to\infty} B_n = \bigcup_{i=1}^{\infty} \bigcap_{k=i}^{\infty} B_k = \bigcup_{i=1}^{\infty} G_i = \{0, 1\}$ using

$$G_1 = \bigcap_{k=1}^{\infty} B_k = \{0\}, G_2 = \bigcap_{k=2}^{\infty} B_k = \{0\}, G_3 = \bigcap_{k=3}^{\infty} B_k = \{0, 1\}, G_4 = \bigcap_{k=4}^{\infty} B_k = \{0, 1\},$$

.... $\diamond$

Definition 1.5.7 (*upper bound set*) For a sequence of sets, the set of elements belonging to infinitely many sets of the sequence is called the upper bound or upper bound set of the sequence, and is denoted by[23] $\limsup_{n\to\infty}$ or by $\overline{\lim}_{n\to\infty}$.

Because an element belonging to almost every B_n^c belongs to a finite number of B_n and the converse is also true, we have $\limsup_{n\to\infty} B_n = \left(\bigcup_{i=1}^{\infty} \bigcap_{k=i}^{\infty} B_k^c \right)^c$, i.e.,

$$\limsup_{n\to\infty} B_n = \bigcap_{i=1}^{\infty} \bigcup_{k=i}^{\infty} B_k, \tag{1.5.17}$$

which is alternatively written as $\limsup_{n\to\infty} B_n = \{\text{infinitely often } B_n\}$ with 'infinitely often' also written as i.o. It is noteworthy that

---

[23] The acronym sup stands for supremum or superior.

$$\{\text{finitely often } B_n\} = \{\text{infinitely often } B_n\}^c$$

$$= \bigcup_{i=1}^{\infty} \bigcap_{k=i}^{\infty} B_k^c, \tag{1.5.18}$$

where 'finitely often' is often written as f.o.

Example 1.5.14 Obtain the upper bound set for the sequence $\{B_n\} = \{0, 1, 3\}$, $\{0, 2\}$, $\{0, 1, 2\}$, $\{0, 1\}$, $\{0, 1, 2\}$, $\{0, 1\}$, ... considered in Example 1.5.13.

Solution Because 0, 1, and 2 belong to infinitely many sets and 3 belongs to one set, the upper bound set is $\{0, 1, 2\}$. This result can be confirmed as $\lim\sup B_n =$

$\bigcap_{i=1}^{\infty} \bigcup_{k=i}^{\infty} B_k = \bigcap_{i=1}^{\infty} H_i = \{0, 1, 2\}$ by noting that $H_1 = \bigcup_{k=1}^{\infty} B_k = \{0, 1, 2, 3\}$, $H_2 = \bigcup_{k=2}^{\infty} B_k = \{0, 1, 2\}$, $H_3 = \bigcup_{k=3}^{\infty} B_k = \{0, 1, 2\}$, $H_4 = \bigcup_{k=4}^{\infty} B_k = \{0, 1, 2\}$, ... Similarly, assuming $\Omega = \{0, 1, 2, 3, 4\}$ for example, we have $B_1^c = \{2, 4\}$, $B_2^c = \{1, 3, 4\}$, $B_3^c = \{3, 4\}$, $B_4^c = \{2, 3, 4\}$, $B_5^c = \{3, 4\}$, $B_6^c = \{2, 3, 4\}$, ..., and thus $\bigcap_{k=1}^{\infty} B_k^c = \{4\}$, $\bigcap_{k=2}^{\infty} B_k^c = \{3, 4\}$, $\bigcap_{k=3}^{\infty} B_k^c = \{3, 4\}$, ... Therefore, the upper bound can be obtained also as $\lim\sup_{n\to\infty} B_n = \left(\bigcup_{i=1}^{\infty} \bigcap_{k=i}^{\infty} B_k^c \right)^c = (\{4\} \cup \{3, 4\} \cup \{3, 4\} \cup \cdots)^c = \{3, 4\}^c = \{0, 1, 2\}$. ◇

Let us note that in an infinite sequence of sets, any element belonging to almost every set belongs to infinitely many sets and thus

$$\liminf_{n\to\infty} B_n \subseteq \limsup_{n\to\infty} B_n \tag{1.5.19}$$

is always true. On the other hand, as we mentioned before, an element belonging to infinitely many sets may or may not belong to the remaining infinitely many sets: for example, we can imagine an element belonging to all the odd-numbered sets but not in any even-numbered set. Consequently, an element belonging to infinitely many sets does not necessarily belong to almost every set. In short, in some cases we have

$$\limsup_{n\to\infty} B_n \not\subseteq \liminf_{n\to\infty} B_n, \tag{1.5.20}$$

which, together with (1.5.19), confirms the intuitive observation that the upper bound is not smaller than the lower bound.

In Definition 1.5.5, we addressed the limit of monotonic sequences. Let us now extend the discussion to the limits of general sequences of sets.

Definition 1.5.8 (*convergence of sequence; limit set*) If

$$\limsup_{n\to\infty} B_n \subseteq \liminf_{n\to\infty} B_n \tag{1.5.21}$$

holds true for a sequence $\{B_n\}_{n=1}^{\infty}$ of sets, then

$$\limsup_{n\to\infty} B_n = \liminf_{n\to\infty} B_n$$

$$= B \qquad\qquad (1.5.22)$$

from (1.1.2) using (1.5.19) and (1.5.21). In such a case, the sequence $\{B_n\}_{n=1}^{\infty}$ is called to converge to B, which is denoted by $B_n \to B$ or $\lim_{n\to\infty} B_n = B$. The set B is called the limit set or limit of $\{B_n\}_{n=1}^{\infty}$.

The limit of monotonic sequences described in Definition 1.5.5 is in agreement with Definition 1.5.8: let us confirm this fact. Assume $\{B_n\}_{n=1}^{\infty}$ is a non-decreasing sequence. Because $\bigcap_{k=i}^{n} B_k = B_i$ for any i, we have $\bigcap_{k=i}^{\infty} B_k = \lim_{n\to\infty} \bigcap_{k=i}^{n} B_k = \lim_{n\to\infty} B_i = B_i$, with which we get $\liminf_{n\to\infty} B_n = \bigcup_{i=1}^{\infty} \bigcap_{k=i}^{\infty} B_k$ as

$$\liminf_{n\to\infty} B_n = \bigcup_{i=1}^{\infty} B_i$$

$$= \lim_{n\to\infty} B_n. \qquad\qquad (1.5.23)$$

We also have $\bigcup_{k=i}^{\infty} B_k = \lim_{n\to\infty} \bigcup_{k=i}^{n} B_k = \lim_{n\to\infty} B_n$ from $\bigcup_{k=i}^{n} B_k = B_n$ for any value of i. Thus, we have $\limsup_{n\to\infty} B_n = \bigcap_{i=1}^{\infty} \bigcup_{k=i}^{\infty} B_k = \bigcap_{i=1}^{\infty} \lim_{n\to\infty} B_n$, consequently resulting in

$$\limsup_{n\to\infty} B_n = \lim_{n\to\infty} B_n$$

$$= \liminf_{n\to\infty} B_n. \qquad\qquad (1.5.24)$$

Next, assume $\{B_n\}_{n=1}^{\infty}$ is a non-increasing sequence. Then, $\liminf_{n\to\infty} B_n = \bigcup_{i=1}^{\infty} \bigcap_{k=i}^{\infty} B_k = \bigcup_{i=1}^{\infty} \lim_{n\to\infty} B_n = \lim_{n\to\infty} B_n$ because $\bigcap_{k=i}^{\infty} B_k = \lim_{n\to\infty} B_n$ from $\bigcap_{k=i}^{n} B_k = B_n$ for any i. We also have $\limsup_{n\to\infty} B_n = \bigcap_{i=1}^{\infty} \bigcup_{k=i}^{\infty} B_k = \bigcap_{i=1}^{\infty} B_i = \lim_{n\to\infty} B_n$ because $\bigcup_{k=i}^{\infty} B_k = \lim_{n\to\infty} \bigcup_{k=i}^{n} B_k = B_i$ from $\bigcup_{k=i}^{n} B_k = B_i$ for any i.

Example 1.5.15 Obtain the limit of $\{B_n\}_{n=1}^{\infty} = \left\{\left(1 - \frac{1}{n}, 3 - \frac{1}{n}\right)\right\}_{n=1}^{\infty}$.

Solution First, because $B_1 = (0, 2)$, $B_2 = \left(\frac{1}{2}, \frac{5}{2}\right)$, $B_3 = \left(\frac{2}{3}, \frac{8}{3}\right)$, $B_4 = \left(\frac{3}{4}, \frac{11}{4}\right)$, $\cdots$, we have $G_1 = \bigcap_{k=1}^{\infty} B_k = [1, 2)$, $G_2 = \bigcap_{k=2}^{\infty} B_k = \left[1, \frac{5}{2}\right)$, $\cdots$ and $H_1 = \bigcup_{k=1}^{\infty} B_k = (0, 3)$, $H_2 = \bigcup_{k=2}^{\infty} B_k = \left(\frac{1}{2}, 3\right)$, $\cdots$. Therefore, the lower bound is $\liminf_{n\to\infty} B_n = \bigcup_{i=1}^{\infty} G_i = [1, 3)$, the upper bound is $\limsup_{n\to\infty} B_n = \bigcap_{i=1}^{\infty} H_i = [1, 3)$, and the limit is $\lim_{n\to\infty} B_n =$

[1, 3). We can similarly show that the limits are all[24] $[1, 3)$ for the sequences $\left\{\left(1 - \frac{1}{n}, 3 - \frac{1}{n}\right]\right\}_{n=1}^{\infty}$, $\left\{\left[1 - \frac{1}{n}, 3 - \frac{1}{n}\right)\right\}_{n=1}^{\infty}$, and $\left\{\left[1 - \frac{1}{n}, 3 - \frac{1}{n}\right]\right\}_{n=1}^{\infty}$. ◇

Example 1.5.16 Obtain the limit of the sequence $\left\{\left(1 + \frac{1}{n}, 3 + \frac{1}{n}\right)\right\}_{n=1}^{\infty}$ of intervals.

Solution First, because $B_1 = (2, 4)$, $B_2 = \left(\frac{3}{2}, \frac{7}{2}\right)$, $B_3 = \left(\frac{4}{3}, \frac{10}{3}\right)$, $B_4 = \left(\frac{5}{4}, \frac{13}{4}\right), \ldots$, we have $G_1 = \bigcap_{k=1}^{\infty} B_k = (2, 3]$, $G_2 = \bigcap_{k=2}^{\infty} B_k = \left(\frac{3}{2}, 3\right], \cdots$ and $H_1 = \bigcup_{k=1}^{\infty} B_k = (1, 4)$, $H_2 = \bigcup_{k=2}^{\infty} B_k = \left(1, \frac{7}{2}\right), \cdots$. Thus, the lower bound is $\liminf_{n \to \infty} B_n = \bigcup_{i=1}^{\infty} G_i = (1, 3]$, the upper bound is $\limsup_{n \to \infty} B_n = \bigcap_{i=1}^{\infty} H_i = (1, 3]$, and the limit set is $\lim_{n \to \infty} B_n = (1, 3]$. We can similarly show that the limits of the sequences $\left\{\left(1 + \frac{1}{n}, 3 + \frac{1}{n}\right]\right\}_{n=1}^{\infty}$, $\left\{\left[1 + \frac{1}{n}, 3 + \frac{1}{n}\right)\right\}_{n=1}^{\infty}$, and $\left\{\left[1 + \frac{1}{n}, 3 + \frac{1}{n}\right]\right\}_{n=1}^{\infty}$ are all $(1, 3]$. ◇

Appendices

Appendix 1.1 Binomial Coefficients in the Complex Space

For the factorial $n! = n(n - 1) \cdots 1$ defined in (1.4.57), the number n is a natural number. Based on the gamma function addressed in Sect. 1.4.3, we extend the factorial into the complex space, which will in turn be used in the discussion of the permutation and binomial coefficients (Riordan 1968; Tucker 2002; Vilenkin 1971) in the complex space.

(A) Factorials and Permutations in the Complex Space

Recollecting that

$$\alpha! = \Gamma(\alpha + 1) \tag{1.A.1}$$

from $\Gamma(\alpha + 1) = \alpha \Gamma(\alpha)$ shown in (1.4.75), the factorial $p!$ can be expressed as

$$p! = \begin{cases} \pm\infty, & p \in \mathbb{J}_-, \\ \Gamma(p + 1), & p \notin \mathbb{J}_- \end{cases} \tag{1.A.2}$$

for a complex number p, where $\mathbb{J}_- = \{-1, -2, \ldots\}$ denotes the set of negative integers. Therefore, $0! = \Gamma(1) = 1$ for $p = 0$.

---

[24] As it is mentioned in Examples 1.5.11 and 1.5.12, when the lower end value of an interval is in the form of $a - \frac{1}{n}$ and $a + \frac{1}{n}$, the limit is in the form of '$[a, \ldots$' and '$(a, \ldots$', respectively. In addition, when the upper end value of an interval is in the form of $b + \frac{1}{n}$ and $b - \frac{1}{n}$, the limit is in the form of '$\ldots, b]$' and '$\ldots, b)$', respectively, for both open and closed ends.

Example 1.A.1 From (1.A.2), it is easy to see that $(-2)! = \pm\infty$ and that $\left(-\frac{1}{2}\right)! = \Gamma\left(\frac{1}{2}\right) = \sqrt{\pi}$ from (1.4.83). $\diamond$

For the permutation $_nP_k = \frac{n!}{(n-k)!}$ defined in (1.4.59), it is assumed that n is a non-negative integer and $k = 0, 1, \ldots, n$. Based on (1.4.92) and (1.A.2), the permutation $_pP_q = \frac{\Gamma(p+1)}{\Gamma(p-q+1)}$ can now be generalized as

$$_pP_q = \begin{cases} \frac{\Gamma(p+1)}{\Gamma(p-q+1)}, & p \notin \mathbb{J}_- \text{ and } p - q \notin \mathbb{J}_-, \\ (-1)^q \frac{\Gamma(-p+q)}{\Gamma(-p)}, & p \in \mathbb{J}_- \text{ and } p - q \in \mathbb{J}_-, \\ 0, & p \notin \mathbb{J}_- \text{ and } p - q \in \mathbb{J}_-, \\ \pm\infty, & p \in \mathbb{J}_- \text{ and } p - q \notin \mathbb{J}_- \end{cases} \tag{1.A.3}$$

for complex numbers p and q, where the expression $(-1)^q \frac{\Gamma(-p+q)}{\Gamma(-p)}$ in the second line of the right-hand side can also be written as $(-1)^q {}_{-p+q-1}P_q$.

Example 1.A.2 For any number z, $_zP_0 = 1$ and $_zP_1 = z$. $\diamond$

Example 1.A.3 It follows that

$$_0P_z = \begin{cases} 0, & z \text{ is a natural number}, \\ \frac{1}{\Gamma(1-z)}, & \text{otherwise}, \end{cases} \tag{1.A.4}$$

$$_1P_z = \begin{cases} 0, & z = 2, 3, \ldots, \\ \frac{1}{\Gamma(2-z)}, & \text{otherwise}, \end{cases} \tag{1.A.5}$$

$$_zP_z = \begin{cases} \pm\infty, & z \in \mathbb{J}_-, \\ \Gamma(z+1), & z \notin \mathbb{J}_-; \end{cases} \tag{1.A.6}$$

and

$$\begin{aligned} _{-1}P_z &= \begin{cases} (-1)^z {}_zP_z, & z = 0, 1 \ldots, \\ \pm\infty, & \text{otherwise} \end{cases} \\ &= \begin{cases} (-1)^z \Gamma(z+1), & z = 0, 1 \ldots, \\ \pm\infty, & \text{otherwise} \end{cases} \end{aligned} \tag{1.A.7}$$

from (1.A.3). $\diamond$

Using (1.A.3), we can also get $_{-2}P_{-0.3} = \pm\infty$, $_{-0.1}P_{1.9} = 0$, $_0P_3 = \frac{\Gamma(1)}{\Gamma(-2)} = 0$, $_{\frac{1}{2}}P_3 = \frac{\Gamma\left(\frac{3}{2}\right)}{\Gamma\left(-\frac{3}{2}\right)} = \frac{3}{8}$, $_{-\frac{1}{2}}P_{0.8} = \frac{\Gamma\left(\frac{3}{2}\right)}{\Gamma(0.7)} = \frac{\sqrt{\pi}}{2\Gamma(0.7)}$, $_3P_{\frac{1}{2}} = \frac{\Gamma(4)}{\Gamma\left(\frac{7}{2}\right)} = \frac{16}{5\sqrt{\pi}}$, $_3P_{-\frac{1}{2}} = \frac{\Gamma(4)}{\Gamma\left(\frac{9}{2}\right)} = \frac{32}{35\sqrt{\pi}}$, $_3P_{-2} = \frac{\Gamma(4)}{\Gamma(6)} = \frac{1}{20}$, $_{-\frac{1}{2}}P_3 = \frac{\Gamma\left(\frac{1}{2}\right)}{\Gamma\left(-\frac{5}{2}\right)} = -\frac{8}{15}$, and $_{-2}P_3 = \frac{\Gamma(-1)}{\Gamma(-4)} = (-1)\frac{\Gamma(5)}{\Gamma(2)} = -24$. Table 1.3 shows some values of the permutation $_pP_q$.

Table 1.3 Some values of the permutation $_pP_q$ (Here, $*$ denotes $\pm\infty$)

		q								
		-2	$-\frac{3}{2}$	-1	$-\frac{1}{2}$	0	$\frac{1}{2}$	1	$\frac{3}{2}$	2
p	-2	$*$	$*$	-1	$*$	1	$*$	-2	$*$	6
	$-\frac{3}{2}$	-4	$-2\sqrt{\pi}$	-2	0	1	0	$-\frac{3}{2}$	0	$\frac{15}{4}$
	-1	$*$	$*$	$*$	$*$	1	$*$	-1	$*$	2
	$-\frac{1}{2}$	$\frac{4}{3}$	$\sqrt{\pi}$	2	$\sqrt{\pi}$	1	0	$-\frac{1}{2}$	0	$\frac{3}{4}$
	0	$\frac{1}{2}$	$\frac{4}{3\sqrt{\pi}}$	1	$\frac{2}{\sqrt{\pi}}$	1	$\frac{1}{\sqrt{\pi}}$	0	$-\frac{1}{2\sqrt{\pi}}$	0
	$\frac{1}{2}$	$\frac{4}{15}$	$\frac{\sqrt{\pi}}{4}$	$\frac{2}{3}$	$\frac{\sqrt{\pi}}{2}$	1	$\frac{\sqrt{\pi}}{2}$	$\frac{1}{2}$	0	$-\frac{1}{4}$
	1	$\frac{1}{6}$	$\frac{8}{15\sqrt{\pi}}$	$\frac{1}{2}$	$\frac{4}{3\sqrt{\pi}}$	1	$\frac{2}{\sqrt{\pi}}$	1	$\frac{1}{\sqrt{\pi}}$	0
	$\frac{3}{2}$	$\frac{4}{35}$	$\frac{\sqrt{\pi}}{8}$	$\frac{2}{5}$	$\frac{3\sqrt{\pi}}{8}$	1	$\frac{3\sqrt{\pi}}{4}$	$\frac{3}{2}$	$\frac{3\sqrt{\pi}}{4}$	$\frac{3}{4}$
	2	$\frac{1}{12}$	$\frac{32}{105\sqrt{\pi}}$	$\frac{1}{3}$	$\frac{16}{15\sqrt{\pi}}$	1	$\frac{8}{3\sqrt{\pi}}$	2	$\frac{4}{\sqrt{\pi}}$	2

(B) Binomial Coefficients in the Complex Space

For the binomial coefficient $_nC_k = \frac{n!}{(n-k)!k!}$ defined in (1.4.60), n and k are non-negative integers with $n \geq k$. Based on the gamma function described in Sect. 1.4.3, we can define the binomial coefficient in the complex space: specifically, employing (1.4.92), the binomial coefficient $_pC_q = \frac{\Gamma(p+1)}{\Gamma(p-q+1)\Gamma(q+1)}$ for p and q complex numbers can be defined as described in Table 1.4.

Example 1.A.4 When both p and $p - q$ are negative integers and q is a non-negative integer, the binomial coefficient $_pC_q = (-1)^q {}_{-p+q-1}C_q$ can be expressed also as

Table 1.4 The binomial coefficient $_pC_q = \frac{\Gamma(p+1)}{\Gamma(p-q+1)\Gamma(q+1)}$ in the complex space

Is $p \in \mathbb{J}_-$?	Is $q \in \mathbb{J}_-$?	Is $p - q \in \mathbb{J}_-$?	$_pC_q$
No	No	No	$\frac{\Gamma(p+1)}{\Gamma(p-q+1)\Gamma(q+1)}$
Yes	Yes	No	$(-1)^{p-q}\frac{\Gamma(-q)}{\Gamma(p-q+1)\Gamma(-p)}$ $= (-1)^{p-q} {}_{-q-1}C_{p-q}$
Yes	No	Yes	$(-1)^q\frac{\Gamma(-p+q)}{\Gamma(-p)\Gamma(q+1)}$ $= (-1)^q {}_{-p+q-1}C_q$
Yes	Yes	Yes	
No	Yes	No	0
No	No	Yes	
Yes	No	No	$\pm\infty$

Note. Among the three numbers p, q, and $p - q$, it is possible that only $p - q$ is not a negative integer (e.g., $p = -2$, $q = -3$, and $p - q = 1$) and only q is not a negative integer (e.g., $p = -3$, $q = 2$, and $p - q = -5$), but it is not possible that only p is not a negative integer.
In other words, when q and $p - q$ are both negative integers, p is also a negative integer.

$$_pC_q = (-1)^q \, _{-p+q-1}C_{-p-1}. \tag{1.A.8}$$

Now, when p is a negative non-integer real number and q is a non-negative integer, the binomial coefficient can be written as $_pC_q = \frac{\Gamma(p+1)}{\Gamma(p-q+1)} \frac{1}{\Gamma(q+1)} = (-1)^{p-p+q} \frac{\Gamma(-p+q)}{\Gamma(-p)} \frac{1}{\Gamma(q+1)} = (-1)^q \frac{(-p+q-1)!}{(-p-1)!q!}$ or as

$$_pC_q = (-1)^q \, _{-p+q-1}C_q \tag{1.A.9}$$

by recollecting $\frac{\Gamma(\alpha+1)}{\Gamma(\beta+1)} = (-1)^{\alpha-\beta} \frac{\Gamma(-\beta)}{\Gamma(-\alpha)}$ shown in (1.4.91) for $\alpha - \beta$ an integer, $\alpha < 0$, and $\beta < 0$. The two formulas (1.A.8) and (1.A.9) are the same as $_{-r}C_x = (-1)^x \, _{r+x-1}C_x$, which we will see in (2.5.15) for a negative real number $-r$ and a non-negative integer x. $\diamond$

Example 1.A.5 From Table 1.4, we promptly get $_zC_0 = _zC_z = 1$ and $_zC_1 = _zC_{z-1} = z$. In addition, $_0C_z = _0C_{-z}$ and $_1C_z = _1C_{1-z} = \frac{1}{\Gamma(2-z)\Gamma(1+z)}$ can be expressed as

$$
\begin{aligned}
_0C_z &= \frac{1}{\Gamma(1-z)\Gamma(1+z)} \\
&= \begin{cases} 1, & z = 0, \\ 0, & z = \pm 1, \pm 2, \ldots, \\ \frac{1}{\Gamma(1-z)\Gamma(1+z)}, & \text{otherwise} \end{cases}
\end{aligned} \tag{1.A.10}
$$

and

$$
_1C_z = \begin{cases} 1, & z = 0, 1, \\ 0, & z = -1, \pm 2, \pm 3, \ldots, \\ \frac{1}{\Gamma(2-z)\Gamma(1+z)}, & \text{otherwise,} \end{cases} \tag{1.A.11}
$$

respectively. $\diamond$

We can similarly obtain[25] $_{-3}C_{-2} = _{-3}C_{-1} = \frac{(-3)!}{(-2)!(-1)!} = 0$, $_{-3}C_2 = _{-3}C_{-5} = \frac{(-3)!}{(-5)!2!} = (-1)^2 \frac{4!}{2!2!} = _4C_2$, $_{-7}C_3 = _{-7}C_{-10} = \frac{(-7)!}{(-10)!3!} = (-1)\frac{9!}{6!3!} = _{-9}C_3$, and $_{\frac{5}{2}}C_2 = _{\frac{5}{2}}C_{\frac{1}{2}} = \frac{\Gamma(\frac{7}{2})}{\Gamma(3)\Gamma(\frac{3}{2})} = \frac{15}{8}$. Table 1.5 shows some values of the binomial coefficient.

Example 1.A.6 Obtain the series expansion of $h(z) = (1 + z)^p$ for p a real number.

Solution First, when $p \geq 0$ or when $p < 0$ and $|z| < 1$, we have $(1 + z)^p = \sum_{k=0}^{\infty} \frac{h^{(k)}(0)}{k!} z^k = \sum_{k=0}^{\infty} \frac{1}{k!} p(p-1) \cdots (p-k+1) z^k$, i.e.,

---

[25] The cases $_{-1}C_z$ and $_{-2}C_z$ are addressed in Exercise 1.39.

Table 1.5 Values of binomial coefficient $_pC_q$ (Here, $*$ denotes $\pm\infty$)

		q								
		-2	$-\frac{3}{2}$	-1	$-\frac{1}{2}$	0	$\frac{1}{2}$	1	$\frac{3}{2}$	2
p	-2	1	$*$	0	$*$	1	$*$	-2	$*$	3
	$-\frac{3}{2}$	0	1	0	0	1	0	$-\frac{3}{2}$	0	$\frac{15}{8}$
	-1	-1	$*$	1	$*$	1	$*$	-1	$*$	1
	$-\frac{1}{2}$	0	$-\frac{1}{2}$	0	1	1	0	$-\frac{1}{2}$	0	$\frac{3}{8}$
	0	0	$-\frac{2}{3\pi}$	0	$\frac{2}{\pi}$	1	$\frac{2}{\pi}$	0	$-\frac{2}{3\pi}$	0
	$\frac{1}{2}$	0	$-\frac{1}{8}$	0	$\frac{1}{2}$	1	1	$\frac{1}{2}$	0	$-\frac{1}{8}$
	1	0	$-\frac{4}{15\pi}$	0	$\frac{4}{3\pi}$	1	$\frac{4}{\pi}$	1	$\frac{4}{3\pi}$	0
	$\frac{3}{2}$	0	$-\frac{1}{16}$	0	$\frac{3}{8}$	1	$\frac{3}{2}$	$\frac{3}{2}$	1	$\frac{3}{8}$
	2	0	$-\frac{16}{105\pi}$	0	$\frac{16}{15\pi}$	1	$\frac{16}{3\pi}$	2	$\frac{16}{3\pi}$	1

$$(1+z)^p = \sum_{k=0}^{\infty} {}_pC_k z^k. \tag{1.A.12}$$

Note that (1.A.12) is the same as the binomial expansion

$$(1+z)^p = \sum_{k=0}^{p} {}_pC_k z^k \tag{1.A.13}$$

of $(1+z)^p$ because $_pC_k = 0$ for $k = p+1, p+2, \ldots$ when p is 0 or a natural number. Next, recollecting (1.A.12) and $(1+z)^p = z^p \left(1 + \frac{1}{z}\right)^p$, we get

$$(1+z)^p = \sum_{k=0}^{\infty} {}_pC_k z^{p-k} \tag{1.A.14}$$

for $p < 0$ and $|z| > 1$. Combining (1.A.12) and (1.A.14), we eventually get

$$(1+z)^p = \begin{cases} \sum_{k=0}^{\infty} {}_pC_k z^k, & \text{for } p \geq 0 \text{ or for } p < 0, |z| < 1, \\ \sum_{k=0}^{\infty} {}_pC_k z^{p-k}, & \text{for } p < 0, |z| > 1, \end{cases} \tag{1.A.15}$$

$(1+z)^p = 2^p$ for $p < 0$ and $z = 1$, and $(1+z)^p \to \infty$ for $p < 0$ and $z = -1$. Note that the term $_pC_k$ in (1.A.15) is always finite because the case of only p being a negative integer among p, k, and $p - k$ is not possible when k is an integer. $\diamond$

Example 1.A.7 Because $_{-1}C_k = (-1)^k$ for $k = 0, 1, \ldots$ as shown in (1.E.27), we get

$$\frac{1}{1+z} = \begin{cases} \sum_{k=0}^{\infty}(-1)^k z^k, & |z| < 1, \\ \sum_{k=0}^{\infty}(-1)^k z^{-1-k}, & |z| > 1 \end{cases}$$

$$= \begin{cases} 1 - z + z^2 - z^3 + \cdots, & |z| < 1, \\ \frac{1}{z} - \frac{1}{z^2} + \frac{1}{z^3} - \cdots, & |z| > 1 \end{cases} \qquad (1.A.16)$$

from (1.A.15) with $p = -1$. ◇

Example 1.A.8 Employing (1.A.15) and the result for $_{-2}C_k$ shown in (1.E.28), we get

$$\frac{1}{(1+z)^2} = \begin{cases} \sum_{k=0}^{\infty}(-1)^k(k+1)z^k, & |z| < 1, \\ \sum_{k=0}^{\infty}(-1)^k(k+1)z^{-2-k}, & |z| > 1 \end{cases}$$

$$= \begin{cases} 1 - 2z + 3z^2 - 4z^3 + \cdots, & |z| < 1, \\ \frac{1}{z^2} - \frac{2}{z^3} + \frac{3}{z^4} - \cdots, & |z| > 1. \end{cases} \qquad (1.A.17)$$

Alternatively, from $\frac{1}{(1+z)^2} = \frac{1}{1-(-2z-z^2)} = \sum_{k=0}^{\infty}\left(-2z - z^2\right)^k$ for $\left|-2z - z^2\right| < 1$, we have

$$\frac{1}{(1+z)^2} = 1 + \left(-2z - z^2\right) + \left(4z^2 + 4z^3 + z^4\right)$$
$$+ \left(-8z^3 + \cdots\right) + \cdots, \qquad (1.A.18)$$

which can be rewritten as

$$1 - 2z + \left(-z^2 + 4z^2\right) + \left(4z^3 - 8z^3\right) + \cdots$$
$$= 1 - 2z + 3z^2 - 4z^3 + \cdots \qquad (1.A.19)$$

by changing the order in the addition. The result[26] (1.A.19) is the same as (1.A.17) for $|z| < 1$.

(C) Two Equalities for Binomial Coefficients

Theorem 1.A.1 *For $\gamma \in \{0, 1, \ldots\}$ and any two numbers α and β, we have*

---

[26] In writing (1.A.19) from (1.A.18), we assume $\left\{\left|-2z - z^2\right| < 1, 0 < |\mathrm{Re}(z) + 1| \le \sqrt{2}\right\}$, a proper subset of the region $|z| < 1$. Here, $\left\{\left|-2z - z^2\right| < 1, 0 < |\mathrm{Re}(z) + 1| \le \sqrt{2}\right\}$ is the right half of the dumbbell-shaped region $\left|-2z - z^2\right| < 1$, which is a proper subset of the rectangle $\left\{|\mathrm{Im}(z)| \le \frac{1}{2}, |\mathrm{Re}(z) + 1| \le \sqrt{2}\right\}$.

$$\sum_{m=0}^{\gamma} \binom{\alpha}{\gamma-m}\binom{\beta}{m} = \binom{\alpha+\beta}{\gamma}, \qquad (1.A.20)$$

which is called Chu-Vandermonde convolution or Vandermonde convolution.

Theorem 1.A.1 is proved in Exercise 1.35. The result (1.A.20) is the same as the Hagen-Rothe identity

$$\sum_{m=0}^{\gamma} \frac{\alpha-\gamma c}{\alpha-mc}\binom{\alpha-mc}{\gamma-m}\frac{\beta}{\beta+mc}\binom{\beta+mc}{m} = \frac{\alpha+\beta-\gamma c}{\alpha+\beta}\binom{\alpha+\beta}{\gamma} \quad (1.A.21)$$

with $c = 0$ and Gauss' hypergeometric theorem

$$_2F_1(a,b;c;1) = \frac{\Gamma(c)\Gamma(c-a-b)}{\Gamma(c-a)\Gamma(c-b)}, \quad \mathrm{Re}(c) > \mathrm{Re}(a+b) \quad (1.A.22)$$

with $a = \gamma$, $b = \alpha+\beta-\gamma$, and $c = \alpha+\beta+1$. In (1.A.22),

$$_2F_1(a,b;c;z) = \sum_{n=0}^{\infty} \frac{(a)_n(b)_n}{(c)_n}\frac{z^n}{n!}, \quad \mathrm{Re}(c) > \mathrm{Re}(b) > 0 \quad (1.A.23)$$

is the hypergeometric function[27], and can be expressed also as

$$_2F_1(a,b;c;z) = \frac{1}{B(b,c-b)} \int_0^1 x^{b-1}(1-x)^{c-b-1}(1-zx)^{-a}dx \quad (1.A.24)$$

in terms of Euler's integral formula.

Example 1.A.9 In (1.A.20), assume $\alpha = 2$, $\beta = \frac{1}{2}$, and $\gamma = 2$. Then, the left-hand side is $\binom{2}{2}\binom{1/2}{0} + \binom{2}{1}\binom{1/2}{1} + \binom{2}{0}\binom{1/2}{2} = 1 + 2\frac{(\frac{1}{2})!}{(-\frac{1}{2})!1!} + \frac{(\frac{1}{2})!}{(-\frac{3}{2})!2!} = 1 + 1 + \frac{(\frac{1}{2})(-\frac{1}{2})}{2!} = \frac{15}{8}$ and the right-hand side is $\binom{5/2}{2} = \frac{(\frac{5}{2})!}{(\frac{1}{2})!2!} = \frac{15}{8}$. $\diamond$

Example 1.A.10 Consider the case $\beta = n$ for $\sum_{m=0}^{n} \binom{\alpha}{\gamma-m}\binom{\beta}{m}$, where α is not a negative integer and $n \in \{0, 1, \ldots\}$. When $\gamma = 0, 1, \ldots, n$, we have $\sum_{m=0}^{n} \binom{\alpha}{\gamma-m}\binom{\beta}{m}$

---

[27] The function $_2F_1$ is also called Gauss' hypergeometric function, and a special case of the generalized hypergeometric function

$$_pF_q(\alpha_1,\alpha_2,\ldots,\alpha_p;\beta_1,\beta_2,\ldots,\beta_q;z) = \sum_{k=0}^{\infty} \frac{(\alpha_1)_k(\alpha_2)_k\cdots(\alpha_p)_k}{(\beta_1)_k(\beta_2)_k\cdots(\beta_q)_k}\frac{z^k}{k!}.$$

Also, note that $_2F_1\left(1,1;\frac{3}{2};\frac{1}{2}\right) = \frac{\pi}{2}$.

$$= \sum_{m=0}^{\gamma} \binom{\alpha}{\gamma-m}\binom{\beta}{m} + \sum_{m=\gamma+1}^{n} \binom{\alpha}{\gamma-m}\binom{\beta}{m} = \binom{\alpha+\beta}{\gamma} \quad \text{noting} \quad \text{that} \quad \binom{\alpha}{\gamma-m} = \frac{\alpha!}{(\alpha-\gamma+m)!}$$

$\frac{1}{(\gamma-m)!} = 0$ for $m = \gamma + 1, \gamma + 2, \ldots$ due to $(\gamma - m)! = \pm\infty$. Similarly, when $\gamma = n + 1, n + 2, \ldots$, we have $\sum_{m=0}^{n} \binom{\alpha}{\gamma-m}\binom{\beta}{m} = \sum_{m=0}^{\gamma} \binom{\alpha}{\gamma-m}\binom{\beta}{m} - \sum_{m=n+1}^{\gamma} \binom{\alpha}{\gamma-m}\binom{\beta}{m} = \binom{\alpha+\beta}{\gamma}$

because $\binom{\beta}{m} = \binom{n}{m} = \frac{n!}{(n-m)!m!} = 0$ from $(n - m)! = \pm\infty$ for $m = n + 1, n + 2, \ldots$. In short, we have

$$\sum_{m=0}^{n} \binom{\alpha}{\gamma-m}\binom{n}{m} = \binom{\alpha+n}{\gamma} \tag{1.A.25}$$

for $\gamma \in \{0, 1, \ldots\}$ when $n \in \{0, 1, \ldots\}$. $\diamond$

Theorem 1.A.2 *We have*

$$\sum_{m=0}^{\gamma} [m]_\zeta \binom{\alpha}{\gamma-m}\binom{\beta}{m} = [\beta]_\zeta \binom{\alpha+\beta-\zeta}{\gamma-\zeta} \tag{1.A.26}$$

for $\zeta \in \{0, 1, \ldots, \gamma\}$ and $\gamma \in \{0, 1, \ldots\}$.

Proof (Method 1) Let us employ the mathematical induction. First, when $\zeta = 0$, (1.A.26) holds true for any values of $\gamma \in \{0, 1, \ldots\}$, α, and β from Theorem 1.A.1. Assume (1.A.26) holds true when $\zeta = \zeta_0$: in other words, for any value of $\gamma \in \{0, 1, \ldots\}$, α, and β, assume

$$\sum_{m=0}^{\gamma} [m]_{\zeta_0} \binom{\alpha}{\gamma-m}\binom{\beta}{m} = \sum_{m=\zeta_0}^{\gamma} [m]_{\zeta_0} \binom{\alpha}{\gamma-m}\binom{\beta}{m}$$

$$= [\beta]_{\zeta_0} \binom{\alpha+\beta-\zeta_0}{\gamma-\zeta_0} \tag{1.A.27}$$

holds true. Then, noting that $(m + 1)\binom{\beta}{m+1} = \beta\binom{\beta-1}{m}$ and $[m + 1]_{\zeta_0+1} = (m + 1)$ $[m]_{\zeta_0}$, we get $\sum_{m=0}^{\gamma} [m]_{\zeta_0+1}\binom{\alpha}{\gamma-m}\binom{\beta}{m} = \sum_{m=\zeta_0+1}^{\gamma} [m]_{\zeta_0+1}\binom{\alpha}{\gamma-m}\binom{\beta}{m} = \sum_{m=\zeta_0}^{\gamma-1} [m + 1]_{\zeta_0+1}$ $\binom{\alpha}{\gamma-m-1}\binom{\beta}{m+1} = \sum_{m=\zeta_0}^{\gamma-1} [m]_{\zeta_0}\binom{\alpha}{\gamma-1-m}(m + 1)\binom{\beta}{m+1}$ from (1.A.27), i.e.,

$$\sum_{m=0}^{\gamma} [m]_{\zeta_0+1} \binom{\alpha}{\gamma-m}\binom{\beta}{m} = \beta \sum_{m=\zeta_0}^{\gamma-1} [m]_{\zeta_0} \binom{\alpha}{\gamma-1-m}\binom{\beta-1}{m}$$

$$= \beta[\beta-1]_{\zeta_0} \binom{\alpha+\beta-1-\zeta_0}{\gamma-1-\zeta_0}$$

$$= [\beta]_{\zeta_0+1} \binom{\alpha+\beta-(\zeta_0+1)}{\gamma-(\zeta_0+1)}. \tag{1.A.28}$$

The result (1.A.28) implies that (1.A.26) holds true also when $\zeta = \zeta_0 + 1$ if (1.A.26) holds true when $\zeta = \zeta_0$. In short, (1.A.26) holds true for $\zeta \in \{0, 1, \dots\}$.

(Method 2) Noting (Charalambides 2002; Gould 1972) that $[m]_\zeta \binom{\beta}{m} = [m]_\zeta \frac{[\beta]_\zeta [\beta-\zeta]_{m-\zeta}}{[m]_\zeta (m-\zeta)!} = [\beta]_\zeta \binom{\beta-\zeta}{m-\zeta}$, we can rewrite (1.A.26) as $\sum_{m=\zeta}^{\gamma} \binom{\alpha}{\gamma-m}\binom{\beta-\zeta}{m-\zeta} = \binom{\alpha+\beta-\zeta}{\gamma-\zeta}$,

which is the same as the Chu-Vandermonde convolution $\sum_{k=0}^{\gamma-\zeta} \binom{\alpha}{\gamma-\zeta-k}\binom{\beta-\zeta}{k} = \binom{\alpha+\beta-\zeta}{\gamma-\zeta}$. ♠

It is noteworthy that (1.A.26) holds true also when $\zeta \in \{\gamma+1, \gamma+2, \dots\}$, in which case the value of (1.A.26) is 0.

Example 1.A.11 Assume $\alpha = 7$, $\beta = 3$, $\gamma = 6$, and $\zeta = 2$ in (1.A.26). Then, the left-hand side is $2\binom{7}{4}\binom{3}{2} + 6\binom{7}{3}\binom{3}{3} = 420$ and the right-hand side is $6\binom{8}{4} = 420$. ◇

Example 1.A.12 Assume $\alpha = -4$, $\beta = -1$, $\gamma = 3$, and $\zeta = 2$ in (1.A.26). then, the left-hand side is $2\binom{-4}{1}\binom{-1}{2} + 6\binom{-4}{0}\binom{-1}{3} = 2 \times \frac{(-4)!}{(-5)!1!}\frac{(-1)!}{(-3)!2!} + 6 \times \frac{(-4)!}{(-4)!0!}\frac{(-1)!}{(-4)!3!} = -14$ and the right-hand side is $(-1)(-2)\binom{-7}{1} = 2 \times \frac{(-7)!}{(-8)!1!} = -14$. ◇

Example 1.A.13 The identity (1.A.26) holds true also for non-integer values of α or β. For example, when $\alpha = \frac{1}{2}$, $\beta = -\frac{1}{2}$, $\gamma = 2$, and $\zeta = 1$, the left-hand side is $\binom{\frac{1}{2}}{1}\binom{-\frac{1}{2}}{1} + 2\binom{\frac{1}{2}}{0}\binom{-\frac{1}{2}}{2} = \frac{(\frac{1}{2})!}{(-\frac{1}{2})!1!}\frac{(-\frac{1}{2})!}{(-\frac{3}{2})!1!} + 2\frac{(\frac{1}{2})!}{(\frac{1}{2})!0!}\frac{(-\frac{1}{2})!}{(-\frac{5}{2})!2!} = \frac{1}{2}$ and the right-hand side is $-\frac{1}{2} \times \binom{-1}{1} = -\frac{1}{2} \times \frac{\Gamma(0)}{\Gamma(-1)} = -\frac{1}{2} \times \frac{(-1)!}{(-2)!1!} = \frac{1}{2}$. ◇

Example 1.A.14 Denoting the unit imaginary number by $j = \sqrt{-1}$, assume $\alpha = e - j$, $\beta = \pi + 2j$, $\gamma = 4$, and $\zeta = 2$ in (1.A.26). Then, the left-hand side is $0 + 0 + 2 \times \binom{\alpha}{2}\binom{\beta}{2} + 6 \times \binom{\alpha}{1}\binom{\beta}{3} + 12 \times \binom{\alpha}{0}\binom{\beta}{4} = \frac{1}{2}\alpha(\alpha-1)\beta(\beta-1) + \alpha\beta(\beta-1)(\beta-2) + \frac{1}{2}\beta(\beta-1)(\beta-2)(\beta-3) = \frac{1}{2}\beta(\beta-1)\{\alpha(\alpha-1) + 2\alpha(\beta-2) + (\beta-2)(\beta-3)\} = \frac{1}{2}\beta(\beta-1)\{\alpha^2 + \alpha(2\beta-5) + (\beta-2)(\beta-3)\} = \frac{1}{2}\beta(\beta-1)(\alpha+\beta-2)(\alpha+\beta-3)$ and the right-hand side is also $\beta(\beta-1)\binom{\alpha+\beta-2}{2} = \beta(\beta-1)\frac{(\alpha+\beta-2)!}{(\alpha+\beta-4)!2!} = \frac{1}{2}\beta(\beta-1)(\alpha+\beta-2)(\alpha+\beta-3)$. ◇

Example 1.A.15 When $\zeta = 1$ and $\zeta = 2$, (1.A.26) can be specifically written as

$$\sum_{m=0}^{\gamma} m \binom{\alpha}{\gamma - m} \binom{\beta}{m} = \beta \binom{\alpha + \beta - 1}{\gamma - 1}$$

$$= \frac{\beta \gamma}{\alpha + \beta} \binom{\alpha + \beta}{\gamma} \qquad (1.A.29)$$

and

$$\sum_{m=0}^{\gamma} m(m - 1) \binom{\alpha}{\gamma - m} \binom{\beta}{m} = \beta(\beta - 1) \binom{\alpha + \beta - 2}{\gamma - 2}$$

$$= \frac{\beta(\beta - 1)\gamma(\gamma - 1)}{(\alpha + \beta)(\alpha + \beta - 1)} \binom{\alpha + \beta}{\gamma}, \qquad (1.A.30)$$

respectively. The two results (1.A.29) and (1.A.30) will later be useful for obtaining the mean and variance of the hypergeometric distribution in Exercise 3.68.   ◇

(D) Euler Reflection Formula

We now prove the Euler reflection formula

$$\Gamma(1 - x)\Gamma(x) = \frac{\pi}{\sin \pi x} \qquad (1.A.31)$$

for $0 < x < 1$ mentioned in (1.4.79). First, if we let $x = \frac{s}{s+1}$ in the defining equation (1.4.95) of the beta function, we get

$$\tilde{B}(\alpha, \beta) = \int_0^\infty \frac{s^{\alpha - 1}}{(s + 1)^{\alpha + \beta}} ds. \qquad (1.A.32)$$

Using (1.A.32) and $\Gamma(\alpha)\Gamma(\beta) = \Gamma(\alpha + \beta)\tilde{B}(\alpha, \beta)$ from (1.4.96) will lead us to

$$\Gamma(1 - x)\Gamma(x) = \int_0^\infty \frac{s^{x-1}}{s + 1} ds \qquad (1.A.33)$$

for $\alpha = x$ and $\beta = 1 - x$. To obtain the right-hand side of (1.A.33), we consider the contour integral

$$\int_C \frac{z^{x-1}}{z - 1} dz \qquad (1.A.34)$$

in the complex space. The contour C of the integral in (1.A.34) is shown in Fig. 1.26, a counterclockwise path along the outer circle. As there exists only one pole $z = 1$

inside the contour C, we get

$$\int_C \frac{z^{x-1}}{z-1} dz = 2\pi j \qquad (1.A.35)$$

from the residue theorem $\int_C \frac{z^{x-1}}{z-1} dz = 2\pi j \operatorname*{Res}_{z=1} \frac{z^{x-1}}{z-1}$. Consider the integral along C in four segments. First, we have $z = Re^{j\theta}$ and $dz = jRe^{j\theta}d\theta$ over the segment from $z_1 = Re^{j(-\pi+\epsilon)}$ to $z_2 = Re^{j(\pi-\epsilon)}$ along the circle with radius R. Second, we have $z = re^{j(\pi-\epsilon)}$ and $dz = e^{j(\pi-\epsilon)}dr$ over the segment from $z_2 = Re^{j(\pi-\epsilon)}$ to $z_3 = pe^{j(\pi-\epsilon)}$ along the straight line toward the origin. Third, we have $z = pe^{j\theta}$ and $dz = jpe^{j\theta}d\theta$ over the segment from $z_3 = pe^{j(\pi-\epsilon)}$ to $z_4 = pe^{j(-\pi+\epsilon)}$ clockwise along the circle with radius p. Fourth, we have $z = re^{j(-\pi+\epsilon)}$ and $dz = e^{j(-\pi+\epsilon)}dr$ over the segment from $z_4 = pe^{j(-\pi+\epsilon)}$ to $z_1 = Re^{j(-\pi+\epsilon)}$ along the straight line out of the origin. Thus, we have

$$\int_C \frac{z^{x-1}}{z-1} dz = \int_{-\pi+\epsilon}^{\pi-\epsilon} \frac{\left(Re^{j\theta}\right)^{x-1} jRe^{j\theta}}{Re^{j\theta}-1} d\theta + \int_R^p \frac{\left\{re^{j(\pi-\epsilon)}\right\}^{x-1} e^{j(\pi-\epsilon)}}{re^{j(\pi-\epsilon)}-1} dr$$
$$+ \int_{\pi-\epsilon}^{-\pi+\epsilon} \frac{\left(pe^{j\theta}\right)^{x-1} jpe^{j\theta}}{pe^{j\theta}-1} d\theta + \int_p^R \frac{\left\{re^{j(-\pi+\epsilon)}\right\}^{x-1} e^{j(-\pi+\epsilon)}}{re^{j(-\pi+\epsilon)}-1} dr,$$

$$(1.A.36)$$

which can be written as

$$\int_C \frac{z^{x-1}}{z-1} dz = \int_{-\pi+\epsilon}^{\pi-\epsilon} \frac{jR^x e^{j\theta x}}{Re^{j\theta}-1} d\theta + \int_R^p \frac{r^{x-1} e^{j(\pi-\epsilon)x}}{re^{j(\pi-\epsilon)}-1} dr$$
$$+ \int_{\pi-\epsilon}^{-\pi+\epsilon} \frac{jp^x e^{j\theta x}}{pe^{j\theta}-1} d\theta + \int_p^R \frac{r^{x-1} e^{j(-\pi+\epsilon)x}}{re^{j(-\pi+\epsilon)}-1} dr \qquad (1.A.37)$$

after some steps. When $x > 0$ and $p \to 0$, the third term in the right-hand side of (1.A.37) is $\lim_{p \to 0} \int_{\pi-\epsilon}^{-\pi+\epsilon} \frac{jp^x e^{j\theta x}}{pe^{j\theta}-1} d\theta = 0$. Similarly, the first term in the right-hand side of (1.A.37) is $\left| \frac{jR^x e^{j\theta x}}{Re^{j\theta}-1} \right| = \frac{R^x}{\sqrt{R^2-2R\cos\theta+1}} \le \frac{R^x}{R-1} \to 0$ when $x < 1$ and $R \to \infty$. Therefore, (1.A.37) can be written as

$$\int_C \frac{z^{x-1}}{z-1} dz = 0 + \int_{\infty}^0 \frac{r^{x-1} e^{j\pi x}}{-r-1} dr + 0 + \int_0^{\infty} \frac{r^{x-1} e^{-j\pi x}}{-r-1} dr$$
$$= \left(e^{j\pi x} - e^{-j\pi x}\right) \int_0^{\infty} \frac{r^{x-1}}{r+1} dr$$
$$= 2j \sin \pi x \int_0^{\infty} \frac{r^{x-1}}{r+1} dr \qquad (1.A.38)$$

for $0 < x < 1$ when $R \to \infty$, $p \to 0$, and $\epsilon \to 0$. In short, we have

Fig. 1.26 The contour C of integral $\int_C \frac{z^{x-1}}{z-1} dz$, where $0 < p < 1 < R$

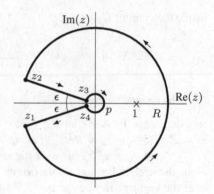

$$\int_0^\infty \frac{r^{x-1}}{r+1} dr = \frac{\pi}{\sin \pi x} \qquad (1.A.39)$$

for $0 < x < 1$ from (1.A.35) and (1.A.38) and, subsequently, we have (1.A.31) for $0 < x < 1$ from (1.A.33) and (1.A.39).

Appendix 1.2 Some Results

(A) Stepping Stone

Consider crossing a creek via $n - 1$ stepping stones with steps 0 and n denoting the two banks of the creek. Assume we can move only in one direction and skip either 0 or 1 step at each move.

(1) Obtain the number of ways we can complete the crossing in k moves.
(2) Obtain the number of ways we can complete the crossing.

Solution (1) Let $a(n, k)$ be the number of ways we can complete the crossing in k moves and let

$$n_2 = \left\lceil \frac{n}{2} \right\rceil, \qquad (1.A.40)$$

where $\lceil x \rceil$ denotes the smallest integer not smaller than x and is called the ceiling function. Denote the number of moves in which we skip 0 and 1 step by k_1 and k_2, respectively. Then, from $k_1 + k_2 = k$ and $k_1 + 2k_2 = n$, we get $k_1 = 2k - n$ and $k_2 = n - k$. The number $a(n, k)$ of ways we can complete the crossing in k moves is the same as the number $\frac{k!}{k_1! k_2!}$ of arranging k_1 of 1's and k_2 of 2's. In short, we have $a(n, k) = \frac{k!}{(2k-n)!(n-k)!}$, i.e.,

$$a(n, k) = {}_kC_{n-k} \qquad (1.A.41)$$

Table 1.6 Number $a(n, k) = {}_kC_{n-k}$ of ways we can cross a creek via $n - 1$ stepping stones in k moves when we can move only in one direction and skip either 0 or 1 step at each move

		n										
		1	2	3	4	5	6	7	8	9	10	11
k	1	1	1	0	0	0	0	0	0	0	0	0
	2	0	1	2	1	0	0	0	0	0	0	0
	3	0	0	1	3	3	1	0	0	0	0	0
	4	0	0	0	1	4	6	4	1	0	0	0
	5	0	0	0	0	1	5	10	10	5	1	0
	6	0	0	0	0	0	1	6	15	20	15	6
	7	0	0	0	0	0	0	1	7	21	35	35
	8	0	0	0	0	0	0	0	1	8	28	56
	9	0	0	0	0	0	0	0	0	1	9	36
	10	0	0	0	0	0	0	0	0	0	1	10
	11	0	0	0	0	0	0	0	0	0	0	1

for $k = n_2, n_2 + 1, \ldots, n$: some values of $a(n, k)$ are shown in Table 1.6.

(2) Denoting the number of ways we can complete the crossing by $a(n)$, which is the same as the number of arranging k of 1's and $n - k$ of 2's for $k = 1, 2, \ldots, n$, we get $a(n) = \sum_{k=1}^{n} a(n, k)$, i.e.,

$$a(n) = \sum_{k=n_2}^{n} {}_kC_{n-k}. \tag{1.A.42}$$

We also have $a(1) = 1$ and $a(2) = 2$. Therefore, $a(n)$ is the sum of the number of ways from step $n - 1$ to n and that from $n - 2$ directly to n. We thus have $a(n) = a(n - 1) + a(n - 2)$ for $n = 3, 4, \ldots$ because the number of ways to step $n - 1$ is $a(n - 1)$ and that to step $n - 2$ is $a(n - 2)$. Solving the recursion, we get [28]

$$a(n) = \frac{1}{\sqrt{5}} \left\{ \left(\frac{1 + \sqrt{5}}{2} \right)^{n+1} - \left(\frac{1 - \sqrt{5}}{2} \right)^{n+1} \right\}. \tag{1.A.43}$$

Here, as it is shown later in (1.A.58), the number $a(n)$ is an integer when $n \in \{0, 1, \ldots\}$. Table 1.7 shows $a(n)$ for $n = 1, 2, \ldots, 20$. Recollecting that ${}_kC_{n-k} = 0$ for $k = 0, 1, \ldots, n_2 - 1$, we have

---

[28] One solving method is as follows: let $a(n) = \theta^n$ in $a(n) = a(n - 1) + a(n - 2)$. Then, $\theta = \frac{1 \pm \sqrt{5}}{2}$ from $\theta^2 - \theta - 1 = 0$. Next, from $a(n) = c_1 \left(\frac{1+\sqrt{5}}{2} \right)^n + c_2 \left(\frac{1-\sqrt{5}}{2} \right)^n$ and the initial conditions $a(1) = 1$ and $a(2) = 2$, we get $c_1 = \frac{1+\sqrt{5}}{2\sqrt{5}}$ and $c_2 = -\frac{1-\sqrt{5}}{2\sqrt{5}}$.

Table 1.7 Number $a(n) = \frac{1}{\sqrt{5}} \left\{ \left(\frac{1+\sqrt{5}}{2} \right)^{n+1} - \left(\frac{1-\sqrt{5}}{2} \right)^{n+1} \right\}$ of ways we can cross a creek via $n-1$ stepping stones when we can move only in one direction and skip either 0 or 1 step at each move

n	1	2	3	4	5	6	7	8	9	10
$a(n)$	1	2	3	5	8	13	21	34	55	89
n	11	12	13	14	15	16	17	18	19	20
$a(n)$	144	233	377	610	987	1597	2584	4181	6765	10946

$$\sum_{k=0}^{n} {}_k C_{n-k} = \frac{1}{\sqrt{5}} \left\{ \left(\frac{1+\sqrt{5}}{2} \right)^{n+1} - \left(\frac{1-\sqrt{5}}{2} \right)^{n+1} \right\} \qquad (1.A.44)$$

from (1.A.42) and (1.A.43). The result (1.A.44) is the same as the well-known identity (Roberts and Tesman 2009) $F_{n+1} = \sum_{k=0}^{\lfloor \frac{n}{2} \rfloor} {}_{n-k}C_k$, where $\{F_n\}_{n=0}^{\infty}$ are Fibonacci numbers. Here, $\lfloor x \rfloor$, also expressed as $[x]$, denotes the greatest integer not larger than x and is called the floor function or Gauss function. Clearly, $\lfloor x \rfloor = \lceil x \rceil = x$ when x is an integer while $\lfloor x \rfloor = \lceil x \rceil - 1$ when x is not an integer.

If the condition 'skip either 0 or 1 step at each move' is replaced with 'skip either 0 step or 2 steps at each move', then we have

$$\sum_{m=n_3, n_3+2, \dots}^{n} {}_m C_{\frac{n-m}{2}} = a_{13} r_{13}^n + 2\mathrm{Re}\left\{ c_{13} z_{13}^n \right\} \qquad (1.A.45)$$

for $n = 0, 1, \dots$, where $n_3 = n - 2\lfloor \frac{n}{3} \rfloor$. In addition, the three numbers $r_{13} = \frac{1}{3}(1 + 2\epsilon_{13})$, $z_{13} = \frac{1}{3}\left\{ 1 - \epsilon_{13} - j\sqrt{3(\epsilon_{13}^2 - 1)} \right\}$, and z_{13}^* with the unit imaginary number $j = \sqrt{-1}$ are the solutions to the difference equation $a(n) = a(n-1) + a(n-3)$ for $n \geq 4$ with initial conditions $a(1) = 1$, $a(2) = 1$, and $a(3) = 2$. The three numbers are also the solutions to $\theta^3 - \theta^2 - 1 = 0$, and $\epsilon_{13} = \frac{1}{2}\left\{ \left(\frac{29-3\sqrt{93}}{2} \right)^{\frac{1}{3}} + \left(\frac{29+3\sqrt{93}}{2} \right)^{\frac{1}{3}} \right\} \approx 1.6984$, $a_{13} = \frac{(z_{13}-1)(z_{13}^*-1)+1}{r_{13}(r_{13}-z_{13})(r_{13}-z_{13}^*)} = \frac{|z_{13}-1|^2+1}{r_{13}|r_{13}-z_{13}|^2}$, and $c_{13} = \frac{(r_{13}-1)(z_{13}^*-1)+1}{z_{13}(z_{13}-r_{13})(z_{13}-z_{13}^*)}$.

In addition, if the condition 'skip either 0 or 1 step at each move' is replaced with 'skip either 1 step or 2 steps at each move', we can obtain

$$\sum_{m=0}^{n} {}_m C_{n-2m} = a_{23} r_{23}^n + 2\mathrm{Re}\left\{ c_{23} z_{23}^n \right\} \qquad (1.A.46)$$

for $n = 2, 3, \ldots$ by noting that $_mC_{n-2m} = \frac{m!}{(3m-n)!(n-2m)!} = 0$ when $m = 0, 1, \ldots, \lceil \frac{n}{3} \rceil - 1$ or $m = \lfloor \frac{n}{2} \rfloor + 1, \lfloor \frac{n}{2} \rfloor + 2, \ldots, n$. Here, the three numbers $r_{23} = 2\epsilon_{23}$, $z_{23} = -\epsilon_{23} - j\sqrt{3\epsilon_{23}^2 - 1}$, and z_{23}^* are the solutions to the difference equation $a(n) = a(n-2) + a(n-3)$ for $n \geq 5$ with initial conditions $a(2) = 1$, $a(3) = 1$, and $a(4) = 1$. The three numbers are also the solutions to $\theta^3 - \theta - 1 = 0$, and

$$\epsilon_{23} = \left(\frac{9-\sqrt{69}}{144}\right)^{\frac{1}{3}} + \left(\frac{9+\sqrt{69}}{144}\right)^{\frac{1}{3}} \approx 0.6624, \quad a_{23} = \frac{(z_{23}-1)(z_{23}^*-1)}{r_{23}^2(r_{23}-z_{23})(r_{23}-z_{23}^*)} = \frac{|z_{23}-1|^2}{r_{23}^2|r_{23}-z_{23}|^2}, \text{ and}$$

$$c_{23} = \frac{(r_{23}-1)(z_{23}^*-1)}{z_{23}^2(z_{23}-r_{23})(z_{23}-z_{23}^*)}.$$

The sequences of numbers calculated from (1.A.43), (1.A.45), and (1.A.46) can be written as

$$\left\{ \sum_{m=0}^{n} {_mC_{n-m}} \right\}_{n=0}^{\infty} = \{1, 1, 2, 3, 5, 8, \ldots\}, \tag{1.A.47}$$

$$\left\{ \sum_{m=n_3, n_3+2, \ldots}^{n} {_mC_{\frac{n-m}{2}}} \right\}_{n=0}^{\infty} = \{1, 1, 1, 2, 3, 4, 6, 9, \ldots\}, \tag{1.A.48}$$

and

$$\left\{ \sum_{m=0}^{n} {_mC_{n-2m}} \right\}_{n=2}^{\infty} = \{1, 1, 1, 2, 2, 3, 4, 5, 7, 9, 12, \ldots\}, \tag{1.A.49}$$

respectively.

(B) Order of Operations

When some operations such as limit, integration, and differentiation are evaluated, a change of order does not usually make a difference. However, the order is of importance in certain cases. Here, we present some examples in which a change of order yields different results.

Example 1.A.16 (Gelbaum and Olmsted 1964) For the function

$$f(x, y) = \begin{cases} y^{-2}, & 0 < x < y < 1, \\ -x^{-2}, & 0 < y < x < 1, \\ 0, & \text{otherwise}, \end{cases} \tag{1.A.50}$$

we have $\int_0^1 \int_0^1 f(x, y)dxdy = \int_0^1 dy = 1$ because $\int_0^1 f(x, y)dx = \int_0^y \frac{dx}{y^2} - \int_y^1 \frac{dx}{x^2} = \frac{1}{y} + \left(1 - \frac{1}{y}\right) = 1$. On the other hand, $\int_0^1 \int_0^1 f(x, y)dydx = \int_0^1 (-1)dx =$

-1 because $\int_0^1 f(x, y)dy = -\int_0^x \frac{dy}{x^2} + \int_x^1 \frac{dy}{y^2} = -\frac{1}{x} - \left(1 - \frac{1}{x}\right) = -1$. In other words, $\int_0^1 \int_0^1 f(x, y)dxdy \neq \int_0^1 \int_0^1 f(x, y)dydx$.   ◇

Example 1.A.17 (Gelbaum and Olmsted 1964) Consider a sequence $\{f_n(x)\}_{n=1}^\infty$ with $f_n(x) = n^2 x e^{-nx}$ on the support $[0, 1]$. Then, $\lim_{n\to\infty} \int_0^1 f_n(x)dx = \lim_{n\to\infty} \left[-(nx + 1)e^{-nx}\right]_0^1 = -\lim_{n\to\infty} \{(n + 1)e^{-n} - 1\}$, i.e.,

$$\lim_{n\to\infty} \int_0^1 f_n(x)dx = 1. \tag{1.A.51}$$

On the other hand, $\lim_{n\to\infty} f_n(x)$ is 0 for $0 \leq x \leq 1$ because $f_n(x) = 0$ for $x = 0$ and $\lim_{n\to\infty} f_n(x) = \frac{1}{x} \lim_{n\to\infty} \frac{(nx)^2}{e^{nx}} = 0$ for $0 < x \leq 1$. Thus, $\int_0^1 \lim_{n\to\infty} f_n(x)dx = \int_0^1 0 dx = 0$ and, therefore, $\int_0^1 \lim_{n\to\infty} f_n(x)dx \neq \lim_{n\to\infty} \int_0^1 f_n(x)dx$.   ◇

Example 1.A.18 (Gelbaum and Olmsted 1964) For the function

$$f(x, y) = \begin{cases} \frac{x^2 - y^2}{x^2 + y^2}, & x^2 + y^2 \neq 0, \\ 0, & x = y = 0, \end{cases} \tag{1.A.52}$$

we have $\lim_{x\to 0} \lim_{y\to 0} f(x, y) \neq \lim_{y\to 0} \lim_{x\to 0} f(x, y)$ because $\lim_{x\to 0} \lim_{y\to 0} f(x, y) = \lim_{x\to 0} \frac{x^2}{x^2} = 1$ and $\lim_{y\to 0} \lim_{x\to 0} f(x, y) = \lim_{y\to 0} \frac{-y^2}{y^2} = -1$.   ◇

Example 1.A.19 (Gelbaum and Olmsted 1964) Consider a sequence $\{f_n(x)\}_{n=1}^\infty$ of functions in which $f_n(x) = \frac{x}{1+n^2 x^2}$ for $|x| \leq 1$. Then, $\frac{d}{dx}\left\{\lim_{n\to\infty} f_n(x)\right\} = 0$, but

$$\lim_{n\to\infty}\left\{\frac{d}{dx} f_n(x)\right\} = \lim_{n\to\infty} \frac{1 - n^2 x^2}{\left(1 + n^2 x^2\right)^2}$$

$$= \begin{cases} 1, & x = 0 \\ 0, & 0 < |x| \leq 1, \end{cases} \tag{1.A.53}$$

implying that $\frac{d}{dx}\left\{\lim_{n\to\infty} f_n(x)\right\} \neq \lim_{n\to\infty}\left\{\frac{d}{dx} f_n(x)\right\}$.   ◇

Example 1.A.20 (Gelbaum and Olmsted 1964) Consider the function

$$f(x, y) = \begin{cases} \frac{x^3}{y^2} \exp\left(-\frac{x^2}{y}\right), & y > 0, \\ 0, & y = 0. \end{cases} \tag{1.A.54}$$

For any value of x, we have $\int_0^1 f(x, y)dy = x \exp\left(-x^2\right)$. If we let $g(x) = x \exp\left(-x^2\right)$, then we have $\frac{d}{dx} \int_0^1 f(x, y)dy = g'(x) = \left(1 - 2x^2\right) \exp\left(-x^2\right)$ for any value of x. On the other hand, $\frac{\partial}{\partial x} f(x, y) = 0$ for any value of y when $x = 0$ because

$$\frac{\partial}{\partial x} f(x, y) = \begin{cases} \left(\frac{3x^2}{y^2} - \frac{2x^4}{y^3}\right) \exp\left(-\frac{x^2}{y}\right), & y > 0, \\ 0, & y = 0. \end{cases} \tag{1.A.55}$$

In other words,

$$\int_0^1 \frac{\partial}{\partial x} f(x, y) \, dy = \begin{cases} \int_0^1 \left(\frac{3x^2}{y^2} - \frac{2x^4}{y^3}\right) \exp\left(-\frac{x^2}{y}\right) x \neq 0, \\ \int_0^1 0 \, dy, & x = 0 \end{cases}$$

$$= \begin{cases} (1 - 2x^2) \exp\left(-x^2\right), & x \neq 0, \\ 0, & x = 0, \end{cases} \tag{1.A.56}$$

and thus $\frac{d}{dx} \int_0^1 f(x, y) dy \neq \int_0^1 \frac{\partial}{\partial x} f(x, y) dy$.   ◇

Example 1.A.21 (Gelbaum and Olmsted 1964) Consider the Cantor function $\phi_C(x)$ discussed in Example 1.3.11 and $f(x) = 1$ for $x \in [0, 1]$. Then, both the Riemann-Stieltjes and Lebesgue-Stieltjes integrals produce $\int_0^1 f(x) d\phi_C(x) = [f(x)\phi_C(x)]_0^1 - \int_0^1 \phi_C(x) df(x) = \phi_C(1) - \phi_C(0) - 0$, i.e.,

$$\int_0^1 f(x) \, d\phi_C(x) = 1 \tag{1.A.57}$$

while the Lebesgue integral results in $\int_0^1 f(x) \phi_C'(x) dx = \int_0^1 0 dx = 0$.   ◇

(C) Sum of Powers of Two Real Numbers

Theorem 1.A.3 *If the sum $\alpha + \beta$ and product $\alpha\beta$ of two numbers α and β are both integers, then*

$$\alpha^n + \beta^n = integer \tag{1.A.58}$$

for $n \in \{0, 1, \ldots\}$.

Proof Let us prove the theorem via mathematical induction. It is clear that (1.A.58) holds true when $n = 0$ and 1. When $n = 2$, $\alpha^2 + \beta^2 = (\alpha + \beta)^2 - 2\alpha\beta$ is an integer. Assume $\alpha^n + \beta^n$ are all integers for $n = 1, 2, \ldots, k - 1$. Then,

$$\alpha^k + \beta^k = (\alpha + \beta)^k - {}_kC_1\alpha\beta\left(\alpha^{k-2} + \beta^{k-2}\right) - {}_kC_2(\alpha\beta)^2\left(\alpha^{k-4} + \beta^{k-4}\right)$$

$$- \cdots - \begin{cases} {}_kC_{\frac{k-1}{2}}(\alpha\beta)^{\frac{k-1}{2}}(\alpha + \beta), & k \text{ is odd}, \\ {}_kC_{\frac{k}{2}}(\alpha\beta)^{\frac{k}{2}}, & k \text{ is even}, \end{cases} \tag{1.A.59}$$

which implies that $\alpha^n + \beta^n$ is an integer when $n = k$ because the binomial coefficient $_kC_j$ is always an integer for $j = 0, 1, \ldots, k$ when k is a natural number. In other words, if $\alpha\beta$ and $\alpha + \beta$ are both integers, then $\alpha^n + \beta^n$ is also an integer when $n \in \{0, 1, \ldots\}$.   ♠

(D) Differences of Geometric Sequences

Theorem 1.A.4 *Consider the difference*

$$D_n = \alpha a^n - \beta b^n \tag{1.A.60}$$

of two geometric sequences, where $\alpha > 0$, $a > b > 0$, $a \neq 1$, and $b \neq 1$. Let

$$r = \frac{\ln \frac{(1-b)\beta}{(1-a)\alpha}}{\ln \frac{a}{b}}. \tag{1.A.61}$$

Then, the sequence $\{D_n\}_{n=1}^{\infty}$ has the following properties:

(1) For $0 < a < 1$, D_n is the largest at $n = r$ and $n = r + 1$ if r is an integer and at $n = \lceil r \rceil$ if r is not an integer.
(2) For $a > 1$, D_n is the smallest at $n = r$ and $n = r + 1$ if r is an integer and at $n = \lceil r \rceil$ if r is not an integer.

Proof Consider the case $0 < a < 1$. Then, from $D_{n+1} - D_n = b^n (1 - b) \beta - a^n (1 - a) \alpha$, we have $D_{n+1} > D_n$ for $n < r$, $D_{n+1} = D_n$ for $n = r$, and $D_{n+1} < D_n$ for $n > r$ and, subsequently, (1). We can similarly show (2).   ♠

Example 1.A.22 The sequence $\{\alpha a^n - \beta b^n\}_{n=1}^{\infty}$

(1) is increasing and decreasing if $a > 1$ and $0 < a < 1$, respectively, when $\frac{(1-b)\beta}{(1-a)\alpha} < \frac{a}{b}$,
(2) is first decreasing and then increasing and first increasing and then decreasing if $a > 1$ and $0 < a < 1$, respectively, when $\frac{(1-b)\beta}{(1-a)\alpha} > \frac{a}{b}$, and
(3) is increasing and decreasing if $a > 1$ and $0 < a < 1$, respectively, when $\frac{(1-b)\beta}{(1-a)\alpha} = \frac{a}{b}$ or, equivalently, when $\alpha a - \beta b = \alpha a^2 - \beta b^2$.

◇

Example 1.A.23 Assume $\alpha = \beta$. Then,

(1) $\{a^n - b^n\}_{n=1}^{\infty}$ is an increasing and decreasing sequence when $a > 1$ and $a + b < 1$, respectively,
(2) $\{a^n - b^n\}_{n=1}^{\infty}$ is a decreasing sequence and $a - b = a^2 - b^2$ when $a + b = 1$, and

(3) $\{a^n - b^n\}_{n=1}^{\infty}$ is a sequence that first increases and then decreases with the maximum at $n = \lceil r \rceil$ if r is not an integer and at $n = r$ and $n = r + 1$ if r is an integer when $a + b > 1$ and $0 < a < 1$.

$\diamond$

Example 1.A.24 Assume $\alpha = \beta = 1$, $a = 0.95$, and $b = 0.4$. Then, we have $0.95 - 4 < 0.95^2 - 4^2 < 0.95^3 - 4^3$ and $0.95^3 - 4^3 > 0.95^4 - 4^4 > \cdots$ because $\lceil r \rceil = \left\lceil \frac{\ln \frac{0.6}{0.05}}{\ln \frac{0.95}{0.4}} \right\rceil \approx \lceil 2.87 \rceil = 3$. $\diamond$

(E) Selections of Numbers with No Number Unchosen

The number of ways to select r different elements from a set of n distinct elements is $_nC_r$. Because every element will be selected as many times as any other element, each of the n elements will be selected $_nC_r \times \frac{r}{n} = {}_{n-1}C_{r-1}$ times over the $_nC_r$ selections. Each of the n elements will be included at least once if we choose appropriately

$$m_1 = \left\lceil \frac{n}{r} \right\rceil \tag{1.A.62}$$

selections among the $_nC_r$ selections. For example, assume the set $\{1, 2, 3, 4, 5\}$ and $r = 2$. Then, we have $m_1 = \left\lceil \frac{5}{2} \right\rceil = 3$, and thus, each of the five elements is included at least once in the three selections $(1, 2)$, $(3, 4)$, and $(4, 5)$.

Next, it is possible that one or more elements will not be included if we consider $_{n-1}C_r$ selections or less among the total $_nC_r$ selections. For example, for the set $\{1, 2, 3, 4, 5\}$ and $r = 2$, in some choices of $_4C_2 = 6$ selections or less such as $(1, 2)$, $(1, 3)$, $(1, 4)$, $(2, 3)$, $(2, 4)$, and $(3, 4)$ among the total of $_5C_2 = 10$ selections, the element 5 is not included.

On the other hand, each of the n elements will be included at least once in any

$$m_2 = 1 + {}_{n-1}C_r \tag{1.A.63}$$

selections. Here, we have

$$_{n-1}C_r = \left(1 - \frac{r}{n}\right) {}_nC_r$$
$$= {}_nC_r - {}_{n-1}C_{r-1}. \tag{1.A.64}$$

The identity (1.A.64) implies that the number of ways for a specific element not to be included when selecting r elements from a set of n distinct elements is the same as the following two numbers:

(1) The number of ways to select r elements from a set of $n - 1$ distinct elements.
(2) The difference between the number of ways to select r elements from a set of n distinct elements and that for a specific element to be included when selecting r elements from a set of n distinct elements.

(F) Fubini's theorem

Theorem 1.A.5 *When the function $f(x, y)$ is continuous on $A = \{(x, y) : a \leq x \leq b,\ c \leq y \leq d\}$, we have*

$$\iint_A f(x, y)dxdy = \int_c^d \int_a^b f(x, y)dxdy$$

$$= \int_a^b \int_c^d f(x, y)dydx. \qquad (1.A.65)$$

In addition, we have

$$\iint_A f(x, y)dxdy = \int_a^b \int_{g_1(x)}^{g_2(x)} f(x, y)dydx \qquad (1.A.66)$$

if $f(x, y)$ is continuous on $A = \{(x, y) : a \leq x \leq b,\ g_1(x) \leq y \leq g_2(x)\}$ and both g_1 and g_2 are continuous on $[a, b]$.

(G) Partitions of Numbers

A representation of a natural number as the sum of natural numbers is also called a partition. Denote the number of partitions for a natural number n as the sum of k natural numbers by $M(n, k)$. Then, the number $N(n)$ of partitions for a natural number n can be expressed as

$$N(n) = \sum_{k=1}^n M(n, k). \qquad (1.A.67)$$

As we can see, for example, from

$$
\begin{aligned}
&1 : \{1\}, \\
&2 : \{2\}, \{1,1\}, \\
&3 : \{3\}, \{2,1\}, \{1,1,1\}, \\
&4 : \{4\}, \{3,1\}, \{2,2\}, \{2,1,1\}, \{1,1,1,1\},
\end{aligned}
\qquad (1.A.68)
$$

we have $N(1) = M(1, 1) = 1$, $N(2) = M(2, 1) + M(2, 2) = 2$, and $N(3) = M(3, 1) + M(3, 2) + M(3, 3) = 3$. In addition, $M(4, 1) = 1$, $M(4, 2) = 2$, $M(4, 3) = 1$, and $M(4, 4) = 1$. In general, the number $M(n, k)$ satisfies

$$M(n, k) = M(n - 1, k - 1) + M(n - k, k). \qquad (1.A.69)$$

Example 1.A.25 We have $M(5, 3) = M(5 - 1, 3 - 1) + M(5 - 3, 3) = M(4, 2) + M(2, 3) = 2 + 0 = 2$ from (1.A.69). ◇

Theorem 1.A.6 *Denote the least common multiplier of k consecutive natural numbers $1, 2, \ldots, k$ by $\tilde{k}$. Let the quotient and remainder of n when divided by k be Q_k and R_k, respectively. If we write*

$$n = \tilde{k} Q_{\tilde{k}} + R_{\tilde{k}}, \qquad (1.A.70)$$

then the number $M(n, k)$ can be expressed as

$$M(n, k) = \sum_{i=0}^{k-1} c_{i,k}\left(R_{\tilde{k}}\right) Q_{\tilde{k}}^i, \quad R_{\tilde{k}} = 0, 1, \ldots, \tilde{k} - 1 \qquad (1.A.71)$$

in terms of $\tilde{k}$ polynomials of order $k - 1$ in $Q_{\tilde{k}}$, where $\left\{c_{i,k}(\cdot)\right\}_{i=0}^{k-1}$ are the coefficients of the polynomial.

Based on Theorem 1.A.6, we can obtain $M(n, 1) = 1$,

$$M(n, 2) = \begin{cases} \frac{n-1}{2}, & n \text{ is odd}, \\ \frac{n}{2}, & n \text{ is even}, \end{cases} \qquad (1.A.72)$$

$$12\, M(n, 3) = \begin{cases} n^2, & R_6 = 0; \quad n^2 - 1, \ R_6 = 1, 5; \\ n^2 - 4, \ R_6 = 2, 4; & n^2 + 3, \ R_6 = 3, \end{cases} \qquad (1.A.73)$$

and

$$144\, M(n, 4) = \begin{cases} n^3 + 3n^2, & R_{12} = 0, \\ n^3 + 3n^2 - 20, & R_{12} = 2, \\ n^3 + 3n^2 + 32, & R_{12} = 4, \\ n^3 + 3n^2 - 36, & R_{12} = 6, \\ n^3 + 3n^2 + 16, & R_{12} = 8, \\ n^3 + 3n^2 - 4, & R_{12} = 10, \\ n^3 + 3n^2 - 9n + 5, & R_{12} = 1, 7, \\ n^3 + 3n^2 - 9n - 27, & R_{12} = 3, 9, \\ n^3 + 3n^2 - 9n - 11, & R_{12} = 5, 11, \end{cases} \qquad (1.A.74)$$

for example. Table 1.8 shows the 60 polynomials of order four in Q_{60} for the representation of $M(n, 5)$.

Table 1.8 Coefficients $\{c_{j,5}(r)\}_{j=4}^{0}$ in $M(n,5) = c_{4,5}(R_{60})Q_{60}^4 + c_{3,5}(R_{60})Q_{60}^3 + c_{2,5}(R_{60})Q_{60}^2 + c_{1,5}(R_{60})Q_{60} + c_{0,5}(R_{60})$

r	$c_{4,5}(r), c_{3,5}(r), c_{2,5}(r), c_{1,5}(r), c_{0,5}(r)$	r	$c_{4,5}(r), c_{3,5}(r), c_{2,5}(r), c_{1,5}(r), c_{0,5}(r)$
0	4500, 750, 25/2, −5/2, 0	1	4500, 1050, 115/2, 1/2, 0
2	4500, 1350, 235/2, 3/2, 0	3	4500, 1650, 385/2, 17/2, 0
4	4500, 1950, 565/2, 29/2, 0	5	4500, 2250, 775/2, 55/2, 1
6	4500, 2550, 1015/2, 81/2, 1	7	4500, 2850, 1285/2, 123/2, 2
8	4500, 3150, 1585/2, 167/2, 3	9	4500, 3450, 1915/2, 229/2, 5
10	4500, 3750, 2275/2, 295/2, 7	11	4500, 4050, 2665/2, 381/2, 10
12	4500, 4350, 3085/2, 473/2, 13	13	4500, 4650, 3535/2, 587/2, 18
14	4500, 4950, 4015/2, 709/2, 23	15	4500, 5250, 4525/2, 855/2, 30
16	4500, 5550, 5065/2, 1011/2, 37	17	4500, 5850, 5635/2, 1193/2, 47
18	4500, 6150, 6235/2, 1387/2, 57	19	4500, 6450, 6865/2, 1609/2, 70
20	4500, 6750, 7525/2, 1845/2, 84	21	4500, 7050, 8215/2, 2111/2, 101
22	4500, 7350, 8935/2, 2393/2, 119	23	4500, 7650, 9685/2, 2707/2, 141
24	4500, 7950, 10465/2, 3039/2, 164	25	4500, 8250, 11275/2, 3405/2, 192
26	4500, 8550, 12115/2, 3791/2, 221	27	4500, 8850, 12985/2, 4213/2, 255
28	4500, 9150, 13885/2, 4657/2, 291	29	4500, 9450, 14815/2, 5139/2, 333
30	4500, 9750, 15775/2, 5645/2, 377	31	4500, 10050, 16765/2, 6191/2, 427
32	4500, 10350, 17785/2, 6763/2, 480	33	4500, 10650, 18835/2, 7377/2, 540
34	4500, 10950, 19915/2, 8019/2, 603	35	4500, 11250, 21025/2, 8705/2, 674
36	4500, 11550, 22165/2, 9421/2, 748	37	4500, 11850, 23335/2, 10183/2, 831
38	4500, 12150, 24535/2, 10977/2, 918	39	4500, 12450, 25765/2, 11819/2, 1014
40	4500, 12750, 27025/2, 12695/2, 1115	41	4500, 13050, 28315/2, 13621/2, 1226
42	4500, 13350, 29635/2, 14583/2, 1342	43	4500, 13650, 30985/2, 15597/2, 1469
44	4500, 13950, 32365/2, 16649/2, 1602	45	4500, 14250, 33775/2, 17755/2, 1747
46	4500, 14550, 35215/2, 18901/2, 1898	47	4500, 14850, 36685/2, 20103/2, 2062
48	4500, 15150, 38185/2, 21347/2, 2233	49	4500, 15450, 39715/2, 22649/2, 2418
50	4500, 15750, 41275/2, 23995/2, 2611	51	4500, 16050, 42865/2, 25401/2, 2818
52	4500, 16350, 44485/2, 26853/2, 3034	53	4500, 16650, 46135/2, 28367/2, 3266
54	4500, 16950, 47815/2, 29929/2, 3507	55	4500, 17250, 49525/2, 31555/2, 3765
56	4500, 17550, 51265/2, 33231/2, 4033	57	4500, 17850, 53035/2, 34973/2, 4319
58	4500, 18150, 54835/2, 36767/2, 4616	59	4500, 18450, 56665/2, 38629/2, 4932

Exercises

Exercise 1.1 Show that $B^c \subseteq A^c$ when $A \subseteq B$.

Exercise 1.2 Show that $\bigcap_{i=1}^{\infty} A_i = \left(\bigcup_{i=1}^{\infty} A_i^c\right)^c$ for a sequence $\{A_i\}_{i=1}^{\infty}$ of sets.

Exercise 1.3 Express the difference $A - B$ in terms only of intersection and symmetric difference, and the union $A \cup B$ in terms only of intersection and symmetric difference.

Exercise 1.4 Consider a sequence $\{A_i\}_{i=1}^n$ of finite sets. Show

$$|A_1 \cup A_2 \cup \cdots \cup A_n| = \sum_i |A_i| - \sum_{i<j} |A_i \cap A_j| + \sum_{i<j<k} |A_i \cap A_j \cap A_k|$$
$$- \cdots + (-1)^{n-1} |A_1 \cap A_2 \cap \cdots \cap A_n| \qquad (1.E.1)$$

and

$$|A_1 \cap A_2 \cap \cdots \cap A_n| = \sum_i |A_i| - \sum_{i<j} |A_i \cup A_j| + \sum_{i<j<k} |A_i \cup A_j \cup A_k|$$
$$- \cdots + (-1)^{n-1} |A_1 \cup A_2 \cup \cdots \cup A_n|, \qquad (1.E.2)$$

where $|A_i|$ denotes the number of elements in A_i.

Exercise 1.5 For a sequence $\{A_i\}_{i=1}^n$ of finite sets, show

$$|A_1 \triangle A_2 \triangle \cdots \triangle A_n| = \sum_i |A_i| - 2 \sum_{i<j} |A_i \cap A_j| + 4 \sum_{i<j<k} |A_i \cap A_j \cap A_k|$$
$$- \cdots + (-2)^{n-1} |A_1 \cap A_2 \cap \cdots \cap A_n|. \qquad (1.E.3)$$

(Hint. As observed in Example 1.1.35, any element in the set $A_1 \triangle A_2 \triangle \cdots \triangle A_n$ belongs to only odd number of sets among $\{A_i\}_{i=1}^n$.)

Exercise 1.6 Is the set of polynomials with integer coefficients countable?

Exercise 1.7 Is the set of algebraic numbers[29] countable?

Exercise 1.8 Show that the sets below are countable.

(1) The set of functions that map a finite subset of a countable set A onto a countable set B.
(2) The set of convergent sequences of natural numbers.

Exercise 1.9 Is the collection of all non-overlapping open intervals with real end points countable or uncountable?

Exercise 1.10 Find an injection from the set A to the set B in each of the pairs A and B below. Here, $\mathbb{J}_0 = \mathbb{J}_+ \cup \{0\}$, i.e.,

$$\mathbb{J}_0 = \{0, 1, \ldots\}. \qquad (1.E.4)$$

(1) $A = \mathbb{J}_0 \times \mathbb{J}_0$. $B = \mathbb{J}_0$.

---

[29] When $n = 1, 2, \ldots$ and $\{a_i\}_{i=0}^n$ are all integers with $a_n \neq 0$, a number z satisfying $a_n z^n + a_{n-1} z^{n-1} + \cdots + a_0 = 0$ is called an algebraic number. A number which is not an algebraic number is called a transcendental number.

(2) $A = (-\infty, \infty)$. $B = (0, 1)$.
(3) $A = \mathbb{R}$. $B =$ the set of infinite sequences of 0 and 1.
(4) $A =$ the set of infinite sequences of 0 and 1. $B = [0, 1]$.
(5) $A =$ the set of infinite sequences of natural numbers. $B =$ the set of infinite sequences of 0 and 1.
(6) $A =$ the set of infinite sequences of real numbers. $B =$ the set of infinite sequences of 0 and 1.

Exercise 1.11 Find a function from the set A to the set B in each of the pairs A and B below.

(1) $A = \mathbb{J}_0$. $B = \mathbb{J}_0 \times \mathbb{J}_0$.
(2) $A = \mathbb{J}_0$. $B = \mathbb{Q}$.
(3) $A =$ the Cantor set. $B = [0, 1]$.
(4) $A =$ the set of infinite sequences of 0 and 1. $B = [0, 1]$.

Exercise 1.12 Find a one-to-one correspondence between the set A and the set B in each of the pairs A and B below.

(1) $A = (a, b)$. $B = (c, d)$. Here, $-\infty \le a < b \le \infty$ and $-\infty \le c < d \le \infty$.
(2) $A =$ the set of infinite sequences of 0, 1, and 2. $B =$ the set of infinite sequences of 0 and 1.
(3) $A = [0, 1)$. $B = [0, 1) \times [0, 1)$.

Exercise 1.13 Assume $f_1 : A \to B$ and $f_2 : C \to D$ are both one-to-one correspondences, $A \cap C = \emptyset$, and $B \cap D = \emptyset$. Show that $f = f_1 \cup f_2$ or, equivalently, $f : A \cup C \to B \cup D$ is a one-to-one correspondence.

Exercise 1.14 Find a one-to-one correspondence[30] between $A = \mathbb{J}_0$ and $B = \mathbb{J}_0 \times \mathbb{J}_0$.

Exercise 1.15 Is the collection of intervals with rational end points in the space $\mathbb{R}$ of real numbers countable?

Exercise 1.16 Show that $U - C \sim U$ when U is an uncountable set and C is a countable set.

Exercise 1.17 Is the sum of two rational numbers a rational number? If we add one more rational number, is the result a rational number? If we add an infinite number of rational numbers, is the result a rational number? (Hint. The set of rational numbers is closed under a finite number of additions, but is not closed under an infinite number of additions.)

Exercise 1.18 Here, $\mathbb{Q}$ denotes the set of rational numbers defined in (1.1.29) and $0 < a < b < 1$.

---

[30] One of such one-to-one correspondences is a function called the Gödel pairing function.

(1) Find a rational number between a and b when $a \in \mathbb{Q}$ and $b \in \mathbb{Q}$.
(2) Find an irrational number between a and b when $a \in \mathbb{Q}$ and $b \in \mathbb{Q}^c$.
(3) Find an irrational number between a and b when $a \in \mathbb{Q}$ and $b \in \mathbb{Q}$.
(4) Find an irrational number between a and b when $a \in \mathbb{Q}^c$ and $b \in \mathbb{Q}^c$.
(5) Find a rational number between a and b when $a \in \mathbb{Q}^c$ and $b \in \mathbb{Q}^c$.
(6) Find a rational number between a and b when $a \in \mathbb{Q}$ and $b \in \mathbb{Q}^c$.

Exercise 1.19 Consider a game between two players. After a countable subset A of the interval $[0, 1]$ is determined, the two players alternately choose one number from $\{0, 1, \ldots, 9\}$. Let the numbers chosen by the first and second players be $x_0, x_1, \ldots$ and $y_0, y_1, \ldots$, respectively. When the number $0.x_1 y_1 x_2 y_2 \cdots$ belongs to A, the first player wins and otherwise the second player wins. Find a way for the second player to win.

Exercise 1.20 For a set function $f : \Omega \to \mathbb{R}$, show

$$f^{-1}\left(A^c\right) = \left(f^{-1}(A)\right)^c, \tag{1.E.5}$$
$$f^{-1}(A \cup B) = f^{-1}(A) \cup f^{-1}(B), \tag{1.E.6}$$

and

$$f^{-1}(A \cap B) = f^{-1}(A) \cap f^{-1}(B), \tag{1.E.7}$$

where $A, B \subseteq \mathbb{R}$.

Exercise 1.21 Show that the function $f(x) = x^2$ is uniformly continuous on $S = \{x : 0 < x < 4\}$.

Exercise 1.22 Show that the function $f(x) = \frac{1}{x}$ is continuous but not uniformly continuous on $S = (0, \infty)$.

Exercise 1.23 Show that the function $f(x) = \sqrt{x}$ is uniformly continuous on $S = (0, \infty)$.

Exercise 1.24 Confirm

$$\lim_{m \to \infty} \int_{-\infty}^{\infty} f(x) \frac{\sin mx}{\pi x} dx = f(0) \tag{1.E.8}$$

shown in (1.4.24). (Hint. Consider the inverse Fourier transform of product.)

Exercise 1.25 Recollect the definition of the unit step function.

(1) Express $u(ax + b)$, $u(\sin x)$, and $u(e^x - \pi)$ in other formulas.
(2) Obtain $\int_{-\infty}^{x} u(t - y)dt$.

Exercise 1.26 Obtain the Fourier transform $\mathfrak{F}\{u(x)\}$ of the unit step function $u(x)$ by following the order shown below.

(1) Let the Fourier transform of the function

$$s_\alpha(x) = \begin{cases} e^{-\alpha x}, & x > 0, \\ -e^{\alpha x}, & x < 0 \end{cases} \qquad (1.E.9)$$

with $\alpha > 0$ be $S_\alpha(\omega) = \mathfrak{F}\{s_\alpha(x)\}$. Show

$$\lim_{\alpha \to 0} S_\alpha(\omega) = \frac{2}{j\omega}. \qquad (1.E.10)$$

(2) Show that the Fourier transform of the impulse function $\delta(x)$ is $\mathfrak{F}\{\delta(x)\} = 1$. Then, show

$$\mathfrak{F}\{1\} = 2\pi\,\delta(\omega) \qquad (1.E.11)$$

using a property of Fourier transform.

(3) Noting that $\mathrm{sgn}(x) = 2u(x) - 1$ and therefore $u(x) = \frac{1}{2}\{1 + \mathrm{sgn}(x)\}$, we have $u(x) = \frac{1}{2}\left\{1 + \lim_{\alpha \to 0} s_\alpha(x)\right\}$ because $\mathrm{sgn}(x) = \lim_{\alpha \to 0} s_\alpha(x)$ from (1.E.9). Based on this result, and noting (1.E.10) and (1.E.11), obtain the Fourier transform

$$\mathfrak{F}\{u(x)\} = \pi\delta(\omega) + \frac{1}{j\omega} \qquad (1.E.12)$$

of the unit step function $u(x)$.

Exercise 1.27 Express $\delta'(x)\cos x$ in another formula.

Exercise 1.28 Calculate $\int_{-2\pi}^{2\pi} e^{\pi x}\delta\left(x^2 - \pi^2\right) dx$. When $0 \le x < 2\pi$, express $\delta(\sin x)$ in another formula.

Exercise 1.29 Calculate $\int_{-\infty}^{\infty} \delta'\left(x^2 - 3x + 2\right) dx$ and $\int_{-\infty}^{\infty} (\cos x + \sin x)\delta'\left(x^3 + x^2 + x\right) dx$.

Exercise 1.30 The multi-dimensional impulse function can be defined as

$$\delta(x_1, x_2, \ldots, x_n) = \prod_{i=1}^{n} \delta(x_i). \qquad (1.E.13)$$

Show

$$\delta(x, y) = \frac{\delta(r)}{\pi r} \qquad (1.E.14)$$

for $r = \sqrt{x^2 + y^2}$ and

$$\delta(x, y, z) = \frac{\delta(r)}{2\pi r^2} \tag{1.E.15}$$

for $r = \sqrt{x^2 + y^2 + z^2}$.

Exercise 1.31 Obtain the limits of the sequences below.

(1) $\left\{\left[1 + \frac{1}{n}, 2\right)\right\}_{n=1}^{\infty}$

(2) $\left\{\left[1 + \frac{1}{n}, 2\right]\right\}_{n=1}^{\infty}$

(3) $\left\{\left(1, 1 + \frac{1}{n}\right]\right\}_{n=1}^{\infty}$

(4) $\left\{\left[1, 1 + \frac{1}{n}\right]\right\}_{n=1}^{\infty}$

(5) $\left\{\left[1 - \frac{1}{n}, 2\right)\right\}_{n=1}^{\infty}$

(6) $\left\{\left[1 - \frac{1}{n}, 2\right]\right\}_{n=1}^{\infty}$

(7) $\left\{\left(1, 2 - \frac{1}{n}\right)\right\}_{n=1}^{\infty}$

(8) $\left\{\left[1, 2 - \frac{1}{n}\right)\right\}_{n=1}^{\infty}$

Exercise 1.32 Consider the sequence $\{f_n(x)\}_{n=1}^{\infty}$ of functions with

$$f_n(x) = \begin{cases} 2n^2 x, & 0 \le x \le \frac{1}{2n}, \\ 2n - 2n^2 x, & \frac{1}{2n} \le x \le \frac{1}{n}, \\ 0, & \frac{1}{n} \le x \le 1. \end{cases} \tag{1.E.16}$$

By obtaining $\int_0^1 \lim_{n\to\infty} f_n(x)dx$ and $\lim_{n\to\infty} \int_0^1 f_n(x)dx$, confirm that the order of integration and limit are not always interchangeable.

Exercise 1.33 For the function

$$f_n(x) = \begin{cases} 2n^3 x, & 0 \le x \le \frac{1}{2n}, \\ 2n^2 - 2n^3 x, & \frac{1}{2n} \le x \le \frac{1}{n}, \\ 0, & \frac{1}{n} \le x \le 1, \end{cases} \tag{1.E.17}$$

and a number $b \in (0, 1]$, obtain $\int_0^b \lim_{n\to\infty} f_n(x)dx$ and $\lim_{n\to\infty} \int_0^b f_n(x)dx$, which shows that the order of integration and limit are not always interchangeable.

Exercise 1.34 Obtain the number of all possible arrangements with ten distinct red balls and ten distinct black balls.

Exercise 1.35 Show[31] the identities

$$_{a-1}C_{b-1} + {}_{a-1}C_b = {}_aC_b, \tag{1.E.18}$$

$$\sum_{k=0}^{n} {}_pC_k \, {}_qC_{n-k} = {}_{p+q}C_n, \tag{1.E.19}$$

---

[31] Here, (1.E.18) is called the Pascal's identity or Pascal's rule.

and

$$\sum_{k=0}^{n} {}_kC_{j\,n-k}C_{m-j} = {}_{n+1}C_{m+1},$$ (1.E.20)

where a and b are complex numbers excluding $\{a = 0, b \neq \text{integer}\}$, p and q are complex numbers, $n \in \{0, 1, 2, \ldots\}$, and j and m are integers such that $0 \leq j \leq m \leq n$. (Hint. For (1.E.19), recollect $(1 + x)^{a+b} = (1 + x)^a (1 + x)^b$.)

Exercise 1.36 Show

$$\sum_{k=0}^{n} \binom{n}{k}\binom{n}{k} = \binom{2n}{n}$$ (1.E.21)

and

$$\sum_{k=0}^{n} \binom{k}{m} = \sum_{k=m}^{n} \binom{k}{m} = \binom{n+1}{m+1}.$$ (1.E.22)

For two integers n and q such that $n \geq q$, show

$$\sum_{k=q}^{n} \binom{n}{k}\binom{k}{q} = 2^{n-q}\binom{n}{q}.$$ (1.E.23)

Exercise 1.37 Show

$$\sum_{k_1=1}^{k_0}\sum_{k_2=1}^{k_1}\cdots\sum_{k_n=1}^{k_{n-1}}1 = \sum_{k_1=1}^{k_0}\sum_{k_2=k_1}^{k_0}\cdots\sum_{k_n=k_{n-1}}^{k_0}1 = \binom{k_0+n-1}{n}.$$ (1.E.24)

(Hint. Consider the number of combinations of choosing n from k_0 with repetitions allowed.)

Exercise 1.38 The identity

$$\tilde{B}(\alpha, \beta) = \frac{\Gamma(\beta)\Gamma(\alpha)}{\Gamma(\alpha + \beta)}$$ (1.E.25)

is shown in Appendix 1.1. Now, based on

$$\Gamma(\alpha)\Gamma(\beta) = \int_0^\infty t^{\alpha-1}e^{-t}dt \int_0^\infty t^{\beta-1}e^{-t}dt$$

$$= \int_0^\infty \int_0^\infty t^{\alpha-1}s^{\beta-1}e^{-(t+s)}dsdt,$$ (1.E.26)

confirm (1.E.25).

Exercise 1.39 Referring to Table 1.4, show

$$_{-1}C_z = \begin{cases} (-1)^z, & z = 0, 1, \ldots, \\ (-1)^{z+1}, & z = -1, -2, \ldots, \\ \pm\infty, & \text{otherwise} \end{cases} \quad (1.E.27)$$

and

$$_{-2}C_z = \begin{cases} (-1)^z(z+1), & z = -1, 0, \ldots, \\ (-1)^{-z+1}(z+1), & z = -2, -3, \ldots, \\ \pm\infty, & \text{otherwise}, \end{cases} \quad (1.E.28)$$

which imply $_{-1}C_a = {}_{-1}C_b$ for $a + b = -1$ and $_{-2}C_a = {}_{-2}C_b$ for $a + b = -2$.

Exercise 1.40 Show

(1) $\frac{1}{2}C_k = \frac{(-1)^{k-1}}{2k-1} \frac{(2k)!}{2^{2k}(k!)^2}$ for $k = 0, 1, 2, \ldots$.

(2) $_{-\frac{1}{2}}C_k = -(2k-1)\frac{1}{2}C_k = (-1)^k \frac{(2k)!}{2^{2k}(k!)^2}$ for $k = 0, 1, 2, \ldots$.

(3) $_{-2k-1}C_m = (-1)^m \frac{(2k+m)!}{(2k)!m!}$ for $k = 0, 1, 2, \ldots$ and $m = 0, 1, 2, \ldots$.

Exercise 1.41 Based on (1.A.15), obtain the values of $_pC_0 - {}_pC_1 + {}_pC_2 - {}_pC_3 + \cdots$ and $_pC_0 + {}_pC_1 + {}_pC_2 + {}_pC_3 + \cdots$ when $p > 0$. Using the results, obtain $\sum_{k=0}^{\infty} {}_pC_{2k+1}$ and $\sum_{k=0}^{\infty} {}_pC_{2k}$ when $p > 0$.

Exercise 1.42 Obtain the series expansions of $g_1(z) = (1 + z)^{\frac{1}{2}}$ and $g_2(z) = (1 + z)^{-\frac{1}{2}}$.

Exercise 1.43 For non-negative numbers α and β such that $\alpha + \beta \neq 0$, show that

$$\frac{\alpha\beta}{\alpha + \beta} \leq \min(\alpha, \beta) \leq \frac{2\alpha\beta}{\alpha + \beta} \leq \sqrt{\alpha\beta} \leq \frac{\alpha + \beta}{2} \leq \max(\alpha, \beta). \quad (1.E.29)$$

References

M. Abramowitz, I.A. Stegun (eds.), *Handbook of Mathematical Functions* (Dover, New York, 1972)

G.E. Andrews, R. Askey, R. Roy, *Special Functions* (Cambridge University, Cambridge, 1999)

E. Artin, *The Gamma Function (Translated by M Rinehart, and Winston Butler)* (Holt, New York, 1964)

B.C. Carlson, *Special Functions of Applied Mathematics* (Academic, New York, 1977)

J.L. Challifour, *Generalized Functions and Fourier Analysis: An Introduction* (W. A. Benjamin, Reading, 1972)

C.A. Charalambides, *Enumerative Combinatorics* (Chapman and Hall, New York, 2002)

W.A. Gardner, *Introduction to Random Processes with Applications to Signals and Systems*, 2nd edn. (McGraw-Hill, New York, 1990)

B.R. Gelbaum, J.M.H. Olmsted, *Counterexamples in Analysis* (Holden-Day, San Francisco, 1964)

I.M. Gelfand, I. Moiseevich, *Generalized Functions* (Academic, New York, 1964)

H.W. Gould, *Combinatorial Identities* (Morgantown Printing, Morgantown, 1972)

R.P. Grimaldi, *Discrete and Combinatorial Mathematics*, 3rd edn. (Addison-Wesley, Reading, 1994)

P.R. Halmos, *Measure Theory* (Van Nostrand Reinhold, New York, 1950)

R.F. Hoskins, J.S. Pinto, *Theories of Generalised Functions* (Horwood, Chichester, 2005)

K. Ito (ed.), *Encyclopedic Dictionary of Mathematics* (Massachusetts Institute of Technology, Cambridge, 1987)

R. Johnsonbaugh, W.E. Pfaffenberger, *Foundations of Mathematical Analysis* (Marcel Dekker, New York, 1981)

D.S. Jones, *The Theory of Generalised Functions*, 2nd edn. (Cambridge University, Cambridge, 1982)

R.P. Kanwal, *Generalized Functions: Theory and Applications* (Birkhauser, Boston, 2004)

K. Karatowski, A. Mostowski, *Set Theory* (North-Holland, Amsterdam, 1976)

S.M. Khaleelulla, *Counterexamples in Topological Vector Spaces* (Springer, Berlin, 1982)

A.B. Kharazishvili, *Nonmeasurable Sets and Functions* (Elsevier, Amsterdam, 2004)

M.J. Lighthill, *An Introduction to Fourier Analysis and Generalised Functions* (Cambridge University, Cambridge, 1980)

M.E. Munroe, *Measure and Integration*, 2nd edn. (Addison-Wesley, Reading, 1971)

I. Niven, H.S. Zuckerman, H.L. Montgomery, *An Introduction to the Theory of Numbers*, 5th edn. (Wiley, New York, 1991)

J.M.H. Olmsted, *Advanced Calculus* (Appleton-Century-Crofts, New York, 1961)

S.R. Park, J. Bae, H. Kang, I. Song, On the polynomial representation for the number of partitions with fixed length. Math. Comput. **77**(262), 1135–1151 (2008)

J. Riordan, *Combinatorial Identities* (Wily, New York, 1968)

F.S. Roberts, B. Tesman, *Applied Combinatorics*, 2nd edn. (CRC, Boca Raton, 2009)

J.P. Romano, A.F. Siegel, *Counterexamples in Probability and Statistics* (Chapman and Hall, New York, 1986)

K.H. Rosen, J.G. Michaels, J.L. Gross, J.W. Grossman, D.R. Shier, *Handbook of Discrete and Combinatorial Mathematics* (CRC, New York, 2000)

H.L. Royden, *Real Analysis*, 3rd edn. (Macmillan, New York, 1989)

W. Rudin, *Principles of Mathematical Analysis*, 3rd edn. (McGraw-Hill, New York, 1976)

R. Salem, On some singular monotonic functions which are strictly increasing. Trans. Am. Math. Soc. **53**(3), 427–439 (1943)

A.N. Shiryaev, *Probability*, 2nd edn. (Springer, New York, 1996)

N.J.A. Sloane, S. Plouffe, *Encyclopedia of Integer Sequences* (Academic, San Diego, 1995)

D.M.Y. Sommerville, *An Introduction to the Geometry of N Dimensions* (Dover, New York, 1958)

R.P. Stanley, *Enumerative Combinatorics*, Vols. 1 and 2 (Cambridge University Press, Cambridge, 1997)

L.A. Steen, J.A. Seebach Jr., *Counterexamples in Topology* (Holt, Rinehart, and Winston, New York, 1970)

J. Stewart, *Calculus: Early Transcendentals*, 7th edn. (Brooks/Coles, Belmont, 2012)

A.A. Sveshnikov (ed.), *Problems in Probability Theory, Mathematical Statistics and Theory of Random Functions* (Dover, New York, 1968)

G.B. Thomas, Jr., R.L. Finney, *Calculus and Analytic Geometry*, 9th edn. (Addison-Wesley, Reading, 1996)

J.B. Thomas, *Introduction to Probability* (Springer, New York, 1986)

A. Tucker, *Applied Combinatorics* (Wiley, New York, 2002)

N.Y. Vilenkin, *Combinatorics* (Academic, New York, 1971)

W.D. Wallis, J.C. George, *Introduction to Combinatorics* (CRC, New York, 2010)

A.I. Zayed, *Handbook of Function and Generalized Function Transformations* (CRC, Boca Raton, 1996)

S. Zhang, J. Jian, *Computation of Special Functions* (Wiley, New York, 1996)

Chapter 2
Fundamentals of Probability

Probability theory is a branch of measure theory. In measure theory and probability theory, we consider set functions of which the values are non-negative real numbers with the values called the measure and probability, respectively, of the corresponding set. In probability theory (Ross 1976, 1996), the values are in addition normalized to exist between 0 and 1: loosely speaking, the probability of a set represents the weight or size of the set. The more common concepts such as area, weight, volume, and mass are other examples of measure. As we shall see shortly, by integrating probability density or by adding probability mass, we can obtain probability. This is similar to obtaining mass by integrating the mass density or by summing point mass.

2.1 Algebra and Sigma Algebra

We first address the notions of algebra and sigma algebra (Bickel and Doksum 1977; Leon-Garcia 2008), which are the bases in defining probability.

2.1.1 Algebra

Definition 2.1.1 (*algebra*) A collection $\mathcal{A}$ of subsets of a set S satisfying the two conditions

$$\text{if } A \in \mathcal{A} \text{ and } B \in \mathcal{A}, \text{ then } A \cup B \in \mathcal{A} \tag{2.1.1}$$

© The Author(s), under exclusive license to Springer Nature Switzerland AG 2022
I. Song et al., *Probability and Random Variables: Theory and Applications*,
https://doi.org/10.1007/978-3-030-97679-8_2

and

$$\text{if } A \in \mathcal{A}, \text{ then } A^c \in \mathcal{A} \tag{2.1.2}$$

is called an algebra of S.

Example 2.1.1 The collection $\mathcal{A}_1 = \{\{1\}, \{2\}, S_1, \emptyset\}$ is an algebra of $S_1 = \{1, 2\}$, and $\mathcal{A}_2 = \{\{1\}, \{2, 3\}, S_2, \emptyset\}$ is an algebra of $S_2 = \{1, 2, 3\}$. $\diamond$

Example 2.1.2 From de Morgan's law, (2.1.1), and (2.1.2), we get $A \cap B \in \mathcal{A}$ when $A \in \mathcal{A}$ and $B \in \mathcal{A}$ for an algebra $\mathcal{A}$. Subsequently, we have

$$\bigcup_{i=1}^{n} A_i \in \mathcal{A} \tag{2.1.3}$$

and

$$\bigcap_{i=1}^{n} A_i \in \mathcal{A} \tag{2.1.4}$$

when $\{A_i\}_{i=1}^{n}$ are all the elements of the algebra $\mathcal{A}$. $\diamond$

The theorem below follows from Example 2.1.2.

Theorem 2.1.1 *An algebra is closed under a finite number of set operations.*

When $A_i \in \mathcal{A}$, we always have $A_i \cap A_i^c = \emptyset \in \mathcal{A}$ and $A_i \cup A_i^c = S \in \mathcal{A}$, expressed as a theorem below.

Theorem 2.1.2 *If $\mathcal{A}$ is an algebra of S, then $S \in \mathcal{A}$ and $\emptyset \in \mathcal{A}$.*

In other words, a collection is not an algebra of S if the collection does not include $\emptyset$ or S.

Example 2.1.3 Obtain all the algebras of $S = \{1, 2, 3\}$.

Solution The collections $\mathcal{A}_1 = \{S, \emptyset\}$, $\mathcal{A}_2 = \{S, \{1\}, \{2, 3\}, \emptyset\}$, $\mathcal{A}_3 = \{S, \{2\}, \{1, 3\}, \emptyset\}$, $\mathcal{A}_4 = \{S, \{3\}, \{1, 2\}, \emptyset\}$, and $\mathcal{A}_5 = \{S, \{1\}, \{2\}, \{3\}, \{1, 2\}, \{2, 3\}, \{1, 3\}, \emptyset\}$ are the algebras of S. $\diamond$

Example 2.1.4 Assume $\mathbb{J}_+ = \{1, 2, \ldots\}$ defined in (1.1.3), and consider the collection $\mathcal{A}_1$ of all the sets obtained from a finite number of unions of the sets $\{1\}, \{2\}, \ldots$ each containing a single natural number. Now, $\mathbb{J}_+$ is not an element of $\mathcal{A}_1$ because it is not possible to obtain $\mathbb{J}_+$ from a finite number of unions of the sets $\{1\}, \{2\}, \ldots$. Consequently, $\mathcal{A}_1$ is not an algebra of $\mathbb{J}_+$. $\diamond$

Definition 2.1.2 (*generated algebra*) For a collection C of subsets of a set, the smallest algebra to which all the element sets in C belong is called the algebra generated from C and is denoted by $\mathcal{A}(C)$.

The implication of $\mathcal{A}(C)$ being the smallest algebra is that any algebra to which all the element sets of C belong also contains all the element sets of $\mathcal{A}(C)$ as its elements.

Example 2.1.5 When $S = \{1, 2, 3\}$, the algebra generated from $C = \{\{1\}\}$ is $\mathcal{A}(C) = \mathcal{A}_1 = \{S, \{1\}, \{2, 3\}, \emptyset\}$ because $\mathcal{A}_2 = \{S, \{1\}, \{2\}, \{3\}, \{1, 2\}, \{2, 3\}, \{1, 3\}, \emptyset\}$ contains all the elements of $\mathcal{A}_1 = \{S, \{1\}, \{2, 3\}, \emptyset\}$. ◇

Example 2.1.6 For the collection $C = \{\{a\}\}$ of $S = \{a, b, c, d\}$, the algebra generated from C is $\mathcal{A}(C) = \{\emptyset, \{a\}, \{b, c, d\}, S\}$. ◇

Theorem 2.1.3 *Let $\mathcal{A}$ be an algebra of a set S, and $\{A_i\}_{i=1}^{\infty}$ be a sequence of sets in $\mathcal{A}$. Then, $\mathcal{A}$ contains a sequence $\{B_i\}_{i=1}^{\infty}$ of sets such that*

$$B_m \cap B_n = \emptyset \tag{2.1.5}$$

for $m \neq n$ and

$$\bigcup_{i=1}^{\infty} B_i = \bigcup_{i=1}^{\infty} A_i. \tag{2.1.6}$$

Proof The theorem can be proved similarly as in (1.1.32)–(1.1.34). ♠

Example 2.1.7 Assume the algebra $\mathcal{A} = \{\{1\}, \{2, 3\}, S, \emptyset\}$ of $S = \{1, 2, 3\}$. When $A_1 = \{1\}$, $A_2 = \{2, 3\}$, and $A_3 = S$, we have $B_1 = \{1\}$ and $B_2 = \{2, 3\}$. ◇

Example 2.1.8 Consider the algebra $\{\emptyset, \{a\}, \{b\}, \{a, b\}, \{c, d\}, \{b, c, d\}, \{a, c, d\}, S\}$ of $S = \{a, b, c, d\}$. For $A_1 = \{a, b\}$ and $A_2 = \{b, c, d\}$, we have $B_1 = \{a, b\}$ and $B_2 = \{c, d\}$, or $B_1 = \{a\}$ and $B_2 = \{b, c, d\}$. ◇

Example 2.1.9 Assume the algebra considered in Example 2.1.8. If $A_1 = \{a, b\}$, $A_2 = \{b, c, d\}$, and $A_3 = \{a, c, d\}$, then $B_1 = \{a\}$, $B_2 = \{b\}$, and $B_3 = \{c, d\}$. ◇

Example 2.1.10 Assume the algebra $\mathcal{A}(\{\{1\}, \{2\}, \ldots\})$ generated from $\{\{1\}, \{2\}, \ldots\}$, and let $A_i = \{2, 4, \ldots, 2i\}$. Then, we have $B_1 = \{2\}$, $B_2 = \{4\}$, $\ldots$, i.e., $B_i = \{2i\}$ for $i = 1, 2, \ldots$. It is clear that $\bigcup_{i=1}^{\infty} B_i = \bigcup_{i=1}^{\infty} A_i$. ◇

2.1.2 Sigma Algebra

In some cases, the results in finite and infinite spaces are different. For example, although the result from a finite number of set operations on the elements of an algebra is an element of the algebra, the result from an infinite number of set operations is not always an element of the algebra. This is similar to the fact that adding a finite number of rational numbers results always in a rational number while adding an infinite number of rational numbers sometimes results in an irrational number.

Example 2.1.11 Assume $S = \{1, 2, \ldots\}$, a collection C of finite subsets of S, and the algebra $A(C)$ generated from C. Then, S is an element of $A(C)$ although S can be obtained from only an infinite number of unions of the element sets in $A(C)$. On the other hand, the set $\{2, 4, \ldots\}$, a set that can also be obtained from only an infinite number of unions of the element sets in $A(C)$, is not an element of $A(C)$. In other words, while a finite number of unions of the element sets in $A(C)$ would result in an element of $A(C)$, an infinite number of unions of the element sets in $A(C)$ is not guaranteed to be an element of $A(C)$.   ◇

As it is clear from the example above, the algebra is unfortunately not closed under a countable number of set operations. We now define the notion of σ-algebra by adding one desirable property to algebra.

Definition 2.1.3 (σ-*algebra*) An algebra that is closed under a countable number of unions is called a sigma algebra or σ-algebra.

In other words, an algebra $\mathcal{F}$ is a σ-algebra if

$$\bigcup_{i=1}^{\infty} A_i \in \mathcal{F} \tag{2.1.7}$$

for all element sets A_1, A_2, $\ldots$ of $\mathcal{F}$. A sigma algebra is closed under a countable, i.e., finite and countably infinite, number of set operations while an algebra is closed under only a finite number of set operations. A sigma algebra is still an algebra, but the converse is not necessarily true. An algebra and a σ-algebra are also called an additive class of sets and a completely additive class of sets, respectively.

Example 2.1.12 For finite sets, an algebra is also a sigma algebra.   ◇

Example 2.1.13 For a σ-algebra $\mathcal{F}$ of S, we always have $S \in \mathcal{F}$ and $\emptyset \in \mathcal{F}$ from Theorem 2.1.2 because σ-algebra is an algebra.   ◇

Example 2.1.14 The collection

$$\mathcal{F} = \{\emptyset, \{a\}, \{b\}, \{c\}, \{d\}, \{a, b\}, \{a, c\}, \{a, d\}, \{b, c\}, \{b, d\}, \{c, d\},$$
$$\{a, b, c\}, \{a, b, d\}, \{a, c, d\}, \{b, c, d\}, S\} \tag{2.1.8}$$

of sets from $S = \{a, b, c, d\}$ is a σ-algebra.   ◇

When the collection of all possible outcomes is finite as in a single toss of a coin or a single rolling of a pair of dice, the limit, i.e., the infinite union in (2.1.7), does not have significant implications and an algebra is also a sigma algebra. On the other hand, when an algebra contains infinitely many element sets, the result of an infinite number of unions of the element sets of the algebra does not always belong to the algebra because an algebra is not closed under an infinite number of set operations. Such a case occurs when the collection of all possible outcomes is from, for instance, an infinite toss of a coin or an infinite rolling of a pair of dice.

Example 2.1.15 The space $\Upsilon = \{\mathbf{a} = (a_1, a_2, \ldots) : a_i \in \{0, 1\}\}$ of one-sided binary sequences is an uncountable set as discussed in Example 1.1.45. Consider the algebra

$$\mathcal{A}_\Upsilon = \mathcal{A}(G_\Upsilon) \tag{2.1.9}$$

generated from the collection $G_\Upsilon = \{\{\mathbf{a}_i\} : \mathbf{a}_i \in \Upsilon\}$ of singleton sets $\{\mathbf{a}_i\}$. Then, some useful countably infinite sets such as

$$\Upsilon_T = \{\text{periodic binary sequences}\} \tag{2.1.10}$$

described in Example 1.1.44 are not elements of the algebra $\mathcal{A}_\Upsilon$ because an infinite set cannot be obtained by a finite number of set operations on the element sets of G_Υ. $\diamond$

Example 2.1.16 Assuming $\Omega = \mathbb{J}_+$, consider the algebra $\mathcal{A}_N = \mathcal{A}(\mathcal{G})$ generated from the collection $\mathcal{G} = \{\{1\}, \{2\}, \ldots\}$. Clearly, $\mathbb{J}_+ \in \mathcal{A}_N$ and $\emptyset \in \mathcal{A}_N$. On the other hand, the set $\{2, 4, \ldots\}$ of even numbers is not an element of $\mathcal{A}_N$ because the set of even numbers cannot be obtained by a finite number of set operations on the element sets of $\mathcal{G}$, as we have already mentioned in Example 2.1.11. Therefore, $\mathcal{A}_N$ is an algebra, but is not a σ-algebra, of $\mathbb{J}_+$. $\diamond$

Example 2.1.17 Assuming $\Omega = \mathbb{Q}$, the set of rational numbers, let $\mathcal{A}_N = \mathcal{A}(\mathcal{G})$ be the algebra generated from the collection $\mathcal{G} = \{\{1\}, \{2\}, \ldots\}$ of singleton sets of natural numbers. Clearly, $\mathcal{A}_N$ is not a σ-algebra of $\mathbb{Q}$. Note that the set $\mathbb{J}_+ = \{1, 2, \ldots\}$ of natural numbers is not an element of $\mathcal{A}_N$ because $\mathbb{J}_+$ cannot be obtained by a finite number of set operations on the sets of $\mathcal{G}$. $\diamond$

Example 2.1.18 For $\Omega = [0, \infty)$, consider the collection

$$\mathcal{F}_1 = \{[a, b), [a, \infty) : 0 \leq a \leq b < \infty\} \tag{2.1.11}$$

of intervals $[a, b)$ and $[a, \infty)$ with $0 \leq a \leq b < \infty$, and the collection $\mathcal{F}_2$ obtained from a finite number of unions of the intervals in $\mathcal{F}_1$. We have $[a, a) = \emptyset \in \mathcal{F}_1$ and $[a, \infty) = \Omega \in \mathcal{F}_1$ with $a = 0$. Yet, although $[a, b) \in \mathcal{F}_1$ for $0 \leq a \leq b < \infty$, we have $[a, b)^c = [0, a) \cup [b, \infty) \notin \mathcal{F}_1$ for $0 < a < b < \infty$. Thus, $\mathcal{F}_1$ is not an algebra of Ω. On the other hand, $\mathcal{F}_2$ is an algebra[1] of Ω because a finite number of unions of the elements in $\mathcal{F}_1$ is an element of $\mathcal{F}_2$, the complement of every element in

---

[1] Here, if the condition '$0 \leq a \leq b < \infty$' is replaced with '$0 \leq a < b < \infty$', $\mathcal{F}_2$ is not an algebra because the null set is not an element of $\mathcal{F}_2$.

$\mathcal{F}_2$ is an element of $\mathcal{F}_2$, $\emptyset \in \mathcal{F}_2$, and $\Omega \in \mathcal{F}_2$. However, $\mathcal{F}_2$ is not a σ-algebra of $\Omega = [0, \infty)$ because $\overset{\infty}{\underset{n=1}{\cap}} A_n = \{0\}$, for instance, is[2] not an element of $\mathcal{F}_2$ although $A_n = \left[0, \frac{1}{n}\right) \in \mathcal{F}_2$ for $n = 1, 2, \ldots$. $\diamond$

Example 2.1.19 Assuming $\Omega = \mathbb{R}$, consider the collection $\mathcal{A}$ of results obtained from finite numbers of unions of intervals $(-\infty, a]$, $(b, c]$, and (d, ∞) with $b \leq c$. Then, $\mathcal{A}$ is an algebra of $\mathbb{R}$ but is not a σ-algebra because $\overset{\infty}{\underset{n=1}{\cap}} \left(b - \frac{1}{n}, c\right] = [b, c]$ is not an element of $\mathcal{A}$. $\diamond$

Example 2.1.20 Assume the σ-algebras $\{\mathcal{F}_i\}_{i=1}^{\infty}$ of a set Ω. Then, $\overset{\infty}{\underset{n=1}{\cap}} \mathcal{F}_n$ is a σ-algebra. However, $\overset{\infty}{\underset{n=1}{\cup}} \mathcal{F}_n$ is not always a σ-algebra. For example, for $\Omega = \{\omega_1, \omega_2, \omega_3\}$, consider the two σ-algebras $\mathcal{F}_1 = \{\emptyset, \{\omega_1\}, \{\omega_2, \omega_3\}, \Omega\}$ and $\mathcal{F}_2 = \{\emptyset, \{\omega_2\}, \{\omega_1, \omega_3\}, \Omega\}$. Then, $\mathcal{F}_1 \cap \mathcal{F}_2 = \{\emptyset, \Omega\}$ is a σ-algebra, but the collection $\mathcal{F}_1 \cup \mathcal{F}_2 = \{\emptyset, \{\omega_1\}, \{\omega_2\}, \{\omega_2, \omega_3\}, \{\omega_1, \omega_3\}, \Omega\}$ is not even an algebra. As another example, consider the sequence

$$\mathcal{F}_1 = \{\emptyset, \Omega, \{\omega_1\}, \Omega - \{\omega_1\}\}, \tag{2.1.12}$$

$$\mathcal{F}_2 = \{\emptyset, \Omega, \{\omega_2\}, \Omega - \{\omega_2\}\}, \tag{2.1.13}$$

$$\vdots$$

of sigma algebras of $\Omega = \{\omega_1, \omega_2, \ldots\}$. Then,

$$\overset{\infty}{\underset{n=1}{\cup}} \mathcal{F}_n = \{\emptyset, \Omega, \{\omega_1\}, \{\omega_2\}, \ldots, \Omega - \{\omega_1\}, \Omega - \{\omega_2\}, \ldots\} \tag{2.1.14}$$

is not an algebra. $\diamond$

Definition 2.1.4 (*generated σ-algebra*) Consider a collection $\mathcal{G}$ of subsets of Ω. The smallest σ-algebra that contains all the element sets of $\mathcal{G}$ is called the σ-algebra generated from $\mathcal{G}$ and is denoted by $\sigma(\mathcal{G})$.

The implication of the σ-algebra $\sigma(\mathcal{G})$ being the smallest σ-algebra is that any σ-algebra which contains all the elements of $\mathcal{C}$ will also contain all the elements of $\sigma(\mathcal{G})$.

Example 2.1.21 For $S = \{a, b, c, d\}$, the σ-algebra generated from $\mathcal{C} = \{\{a\}\}$ is $\sigma(\mathcal{C}) = \{\emptyset, \{a\}, \{b, c, d\}, S\}$. $\diamond$

Example 2.1.22 For the uncountable set $\Upsilon = \{\mathbf{a} = (a_1, a_2, \ldots) : a_i \in \{0, 1\}\}$ of one-sided binary sequences, consider the algebra $\mathcal{A}(G_\Upsilon)$ and σ-algebra $\sigma(G_\Upsilon)$ generated from $G_\Upsilon = \{\{\mathbf{a}_i\} : \mathbf{a}_i \in \Upsilon\}$. Then, as we have observed in Example 2.1.15, the collection

---

[2] This result is from (1.5.11).

$$\Upsilon_T = \{\text{periodic binary sequences}\} \tag{2.1.15}$$

is not included in $\mathcal{A}(G_\Upsilon)$ because all the element sets of $\mathcal{A}(G_\Upsilon)$ contain a finite number of $\mathbf{a}_i$'s while Υ_T is an infinite set. On the other hand, we have

$$\Upsilon_T \in \sigma(G_\Upsilon) \tag{2.1.16}$$

by the definition of a sigma algebra.   ◇

Based on the concept of σ-algebra, we will discuss the notion of probability space in the next section. In particular, the concept of σ-algebra plays a key role in the continuous probability space.

2.2 Probability Spaces

A probability space (Gray and Davisson 2010; Loeve 1977) is the triplet $(\Omega, \mathcal{F}, \mathsf{P})$ of an abstract space Ω, called the sample space; a sigma algebra $\mathcal{F}$, called the event space, of the sample space; and a set function P, called the probability measure, assigning a number in $[0, 1]$ to each of the element sets of the event space.

2.2.1 Sample Space

Definition 2.2.1 (*random experiment*) An experiment that can be repeated under perfect control, yet the outcome of which is not known in advance, is called a random experiment or, simply, an experiment.

Example 2.2.1 Tossing a coin is a random experiment because it is not possible to predict the exact outcome even under a perfect control of the environment.   ◇

Example 2.2.2 Making a product in a factory can be modelled as a random experiment because even the same machine would not be able to produce two same products.
                                                                              ◇

Example 2.2.3 Although the law of inheritance is known, it is not possible to know exactly, for instance, the color of eyes of a baby in advance. A probabilistic model is more appropriate.   ◇

Example 2.2.4 In any random experiment, the procedure, observation, and model should be described clearly. For example, toss a coin can be described as follows:
• Procedure. A coin will be thrown upward and fall freely down to the floor.
• Observation. When the coin stops moving, the face upward is observed.
• Model. The coin is symmetric and previous outcomes do not influence future outcomes.   ◇

Definition 2.2.2 (*sample space*) The collection of all possible outcomes of an experiment is called the sample space of the experiment.

The sample space, often denoted by S or Ω, is basically the same as the abstract space in set theory.

Definition 2.2.3 (*sample point*) An element of the sample space is called a sample point or an elementary outcome.

Example 2.2.5 In toss a coin, the sample space is $S = \{head, tail\}$ and the sample points are *head* and *tail*. In rolling a fair die, the sample space is $S = \{1, 2, 3, 4, 5, 6\}$ and the sample points are $1, 2, \ldots, 6$. In the experiment of rolling a die until a certain number appears, the sample space is $S = \{1, 2, \ldots\}$ and the sample points are $1, 2, \ldots$ when the observation is the number of rolling. $\diamond$

Example 2.2.6 In the experiment of choosing a real number between a and b randomly, the sample space is $\Omega = (a, b)$. $\diamond$

The sample spaces in Example 2.2.5 are countable sets, which are often called discrete sample spaces or discrete spaces. The sample space $\Omega = (a, b)$ considered in Example 2.2.6 is an uncountable space and is called a continuous sample space or continuous space.

A finite dimensional vector space from a discrete space is, again, a discrete space: on the other hand, it should be noted that an infinite dimensional vector space from a discrete space, which is called a sequence space, is a continuous space. A mixture of discrete and continuous spaces is called a mixed sample space or a hybrid sample space.

Let us generally denote by $\mathbb{I}$ the index set such as the set $\mathbb{R}$ of real numbers, the set

$$\mathbb{R}_0 = \{x : x \geq 0\} \tag{2.2.1}$$

of non-negative real numbers, the set

$$\mathbb{J}_k = \{0, 1, \ldots, k - 1\} \tag{2.2.2}$$

of integers from 0 to $k - 1$, the set $\mathbb{J}_+ = \{1, 2, \ldots\}$ of natural numbers, and the set $\mathbb{J} = \{\ldots, -1, 0, 1, \ldots\}$ of integers. Then, the product space $\prod_{t \in \mathbb{I}} \Omega_t$ of the sample spaces $\{\Omega_t, t \in \mathbb{I}\}$ can be described as

$$\prod_{t \in \mathbb{I}} \Omega_t = \{\text{all } \{a_t, t \in \mathbb{I}\} : a_t \in \Omega_t\}, \tag{2.2.3}$$

which can also be written as $\Omega^{\mathbb{I}}$ if it incurs no confusion or if $\Omega_t = \Omega$.

Definition 2.2.4 (*discrete combined space*) For combined random experiments on two discrete sample spaces Ω_1 of size m and Ω_2 of size n, the sample space $\Omega = \Omega_1 \times \Omega_2$ of size mn is called a discrete combined space.

Example 2.2.7 When a coin is tossed and a die is rolled at the same time, the sample space is the discrete combined space $S = \{(head, 1), (head, 2), \ldots, (head, 6), (tail, 1), (tail, 2), \ldots, (tail, 6)\}$ of size $2 \times 6 = 12$. ◇

Example 2.2.8 When two coins are tossed once or a coin is tossed twice, the sample space is the discrete combined space $S = \{(head, head), (head, tail), (tail, head), (tail, tail)\}$ of size $2 \times 2 = 4$. ◇

Example 2.2.9 Assume the sample space $\Omega = \{0, 1\}$ and let $\Omega_1 = \Omega_2 = \cdots = \Omega_k = \Omega$. Then, the space $\prod_{i=1}^{k} \Omega_i = \Omega^k = \{(a_1, a_2, \ldots, a_k) : a_i \in \{0, 1\}\}$ is the k-fold Cartesian space of Ω, a space of binary vectors of length k, and an example of a product space. ◇

Example 2.2.10 Assume the discrete space $\Omega = \{a_1, a_2, \ldots, a_m\}$. The space $\Omega^k = \{$all vectors $\boldsymbol{b} = (b_1, b_2, \ldots, b_k) : b_i \in \Omega\}$ of k-dimensional vectors from Ω, i.e., the k-fold Cartesian space of Ω, is a discrete space as we have already observed indirectly in Examples 1.1.41 and 1.1.42. On the other hand, the sequence spaces $\Omega^{\mathbb{J}} = \{$all infinite sequences $\{\ldots, c_{-1}, c_0, c_1, \ldots\} : c_i \in \Omega\}$ and $\Omega^{\mathbb{J}+} = \{$all infinite sequences $\{d_1, d_2, \ldots\} : d_i \in \Omega\}$ are continuous spaces, although they are derived from a discrete space. ◇

2.2.2 Event Space

Definition 2.2.5 (*event space; event*) A sigma algebra obtained from a sample space is called an event space, and an element of the event space is called an event.

An event in probability theory is roughly another name for a set in set theory. Nonetheless, a non-measurable set discussed in Appendix 2.4 cannot be an event and not all measurable sets are events: again, only the element sets of an event space are events.

Example 2.2.11 Consider the sample space $S = \{1, 2, 3\}$ and the event space $\mathcal{C} = \{\{1\}, \{2, 3\}, S, \emptyset\}$. Then, the subsets $\{1\}, \{2, 3\}, S$, and $\emptyset$ of S are events. However, the other subsets $\{2\}, \{3\}, \{1, 3\}$, and $\{1, 2\}$ of S are not events. ◇

As we can easily observe in Example 2.2.11, every event is a subset of the sample space, but not all the subsets of a sample space are events.

Definition 2.2.6 (*elementary event*) An event that is a singleton set is called an elementary event.

Example 2.2.12 For a coin toss, the sample space is $S = \{head, tail\}$. If we assume the event space $\mathcal{F} = \{S, \emptyset, \{head\}, \{tail\}\}$, then the sets $S, \{head\}, \{tail\}$, and $\emptyset$ are events, among which $\{head\}$ and $\{tail\}$ are elementary events. ◇

It should be noted that only a set can be an event. Therefore, no element of a sample space can be an event, and only a subset of a sample space can possibly be an event.

Example 2.2.13 For rolling a die with the sample space $S = \{1, 2, 3, 4, 5, 6\}$, the element 1, for example, of S can never be an event. The subset $\{1, 2, 3\}$ may sometimes be an event: specifically, the subset $\{1, 2, 3\}$ is and is not an event for the event spaces $\mathcal{F} = \{\{1, 2, 3\}, \{4, 5, 6\}, S, \emptyset\}$ and $\mathcal{F} = \{\{1, 2\}, \{3, 4, 5, 6\}, S, \emptyset\}$, respectively.   ◇

In addition, when a set is an event, the complement of the set is also an event even if it cannot happen.

Example 2.2.14 Consider an experiment of measuring the current through a circuit in the sample space $\mathbb{R}$. If the set $\{a : a \leq 1000 \text{ mA}\}$ is an event, then the complement $\{a : a > 1000 \text{ mA}\}$ will also be an event even if the current through the circuit cannot exceed 1000 mA.   ◇

For a sample space, several event spaces may exist. For a sample space Ω, the collection $\{\emptyset, \Omega\}$ is the smallest event space and the power set 2^{Ω}, the collection of all the subsets of Ω as described in Definition 1.1.13, is the largest event space.

Example 2.2.15 For a game of rock-paper-scissors with the sample space $\Omega = \{\text{rock, paper, scissors}\}$, the smallest event space is $\mathcal{F}_S = \{\emptyset, \Omega\}$ and the largest event space   is $\mathcal{F}_L = 2^{\Omega} = \{\emptyset, \{\text{rock}\}, \{\text{paper}\}, \{\text{scissors}\}, \{\text{rock, paper}\}, \{\text{paper, scissors}\}, \{\text{scissors, rock}\}, \Omega\}$.   ◇

Let us now briefly describe why we base the probability theory on the more restrictive σ-algebra, and not on algebra. As we have noted before, when the sample space is finite, we could also have based the probability theory on algebra because an algebra is basically the same as a σ-algebra. However, if the probability theory is based on algebra when the sample space is an infinite set, it becomes impossible to take some useful sets[3] as events and to consider the limit of events, as we can see from the examples below.

Example 2.2.16 If the event space is defined not as a σ-algebra but as an algebra, then the limit is not an element of the event space for some sequences of events. For example, even when all finite intervals (a, b) are events, no singleton set is an event because a singleton set cannot be obtained from a finite number of set operations on intervals (a, b). In a more practical scenario, even if "The voltage measured is between a and b (V)." is an event, "The voltage measured is a (V)." would not be an event if the event space were defined as an algebra.   ◇

---

[3] Among those useful sets is the set $\Upsilon_T = \{\text{all periodic binary sequences}\}$ considered in Example 2.1.15.

As we have already mentioned in Sect. 2.1.2, when the sample space is finite, the limit is not so crucial and an algebra is a σ-algebra. Consequently, the probability theory could also be based on algebra. When the sample space is an infinite set and the event space is composed of an infinite number of sets, however, an algebra is not closed under an infinite number of set operations. In such a case, the result of an infinite number of operations on sets, i.e., the limit of a sequence, is not guaranteed to be an event. In short, the fact that a σ-algebra is closed under a countable number of set operations is the reason why we adopt the more restrictive σ-algebra as an event space, and not the more general algebra.

An event space is a collection of subsets of the sample space closed under a countable number of unions. We can show, for instance via de Morgan's law, that an event space is closed also under a countable number of other set operations such as difference, complement, intersection, etc. It should be noted that an event space is closed for a countable number of set operations, but not for an uncountable number of set operations. For example, the set $\bigcup_{r=1}^{\infty} H_r$ is also an event when $\{H_r\}_{r=1}^{\infty}$ are all events, but the set

$$\bigcup_{r \in [0,1]} B_r \tag{2.2.4}$$

may or may not be an event when $\{B_r : r \in [0, 1]\}$ are all events.

Let us now discuss in some detail the condition

$$\bigcup_{i=1}^{\infty} B_i \in \mathcal{F} \tag{2.2.5}$$

shown originally in (2.1.7), where $\{B_n\}_{n=1}^{\infty}$ are all elements of the event space $\mathcal{F}$. This condition implies that the event space is closed under a countable number of union operations. Recollect that the limit $\lim_{n \to \infty} B_n$ of $\{B_n\}_{n=1}^{\infty}$ is defined as

$$\lim_{n \to \infty} B_n = \begin{cases} \bigcup_{i=1}^{\infty} B_i, & \{B_n\}_{n=1}^{\infty} \text{ is a non-decreasing sequence,} \\ \bigcap_{i=1}^{\infty} B_i, & \{B_n\}_{n=1}^{\infty} \text{ is a non-increasing sequence} \end{cases} \tag{2.2.6}$$

in (1.5.8) and (1.5.9). It is clear that a sequence $\{B_n\}_{n=1}^{\infty}$ of events in $\mathcal{F}$ will also satisfy

$$\lim_{n \to \infty} B_n \in \mathcal{F}, \quad \{B_n\}_{n=1}^{\infty} \text{ is a non-decreasing sequence} \tag{2.2.7}$$

when the sequence satisfies (2.2.5).

Example 2.2.17 Let a sequence $\{H_n\}_{n=1}^{\infty}$ of events in an event space $\mathcal{F}$ be non-decreasing. As we have seen in (1.5.8) and (2.2.6), the limit of $\{H_n\}_{n=1}^{\infty}$ can be expressed in terms of the countable union as $\lim_{n \to \infty} H_n = \bigcup_{n=1}^{\infty} H_n$. Because a countable number of unions of events results in an event, the limit $\lim_{n \to \infty} H_n$ is an event. For example, when $\left\{\left[1, 2 - \frac{1}{n}\right)\right\}_{n=1}^{\infty}$ are all events, the limit $[1, 2)$ of this non-decreasing sequence of events is an event. In addition, when finite intervals of the form (a, b) are all events, the limit $(-\infty, b)$ of $\{(-n, b)\}_{n=1}^{\infty}$ will also be an event. Similarly, assume a non-increasing sequence $\{B_n\}_{n=1}^{\infty}$ of events in $\mathcal{F}$. The limit of this sequence,

$$\lim_{n \to \infty} B_n = \bigcap_{n=1}^{\infty} B_n$$ as shown in (1.5.9) or (2.2.6), will also be an event because it is a countable intersection of events. Therefore, if finite intervals (a, b) are all events, any singleton set $\{a\}$, the limit of $\left\{\left(a - \frac{1}{n}, a + \frac{1}{n}\right)\right\}_{n=1}^{\infty}$, will also be an event. $\diamond$

Example 2.2.18 Let us show the equivalence of (2.2.5) and (2.2.7). We have already observed that (2.2.7) holds true for a collection of events satisfying (2.2.5). Let us thus show that (2.2.5) holds true for a collection of events satisfying (2.2.7). Consider a sequence $\{G_i\}_{i=1}^{\infty}$ of events chosen arbitrarily in $\mathcal{F}$ and let $H_n = \bigcup_{i=1}^{n} G_i$. Then,

$\bigcup_{n=1}^{\infty} G_n = \bigcup_{n=1}^{\infty} H_n$ and $\{H_n\}_{n=1}^{\infty}$ is a non-decreasing sequence. In addition, because

$\{H_n\}_{n=1}^{\infty}$ is a non-decreasing sequence, we have $\bigcup_{i=1}^{\infty} H_i = \lim_{n \to \infty} H_n$ from (2.2.6). There-

fore, $\bigcup_{n=1}^{\infty} G_n = \bigcup_{n=1}^{\infty} H_n = \lim_{n \to \infty} H_n \in \mathcal{F}$. In other words, for any sequence $\{G_i\}_{i=1}^{\infty}$ of events in $\mathcal{F}$ satisfying (2.2.7), we have

$$\bigcup_{n=1}^{\infty} G_n \in \mathcal{F}. \tag{2.2.8}$$

In essence, the two conditions (2.2.5) and (2.2.7) are equivalent, which implies that (2.2.7), instead of (2.2.5), can be employed as one of the requirements for a collection to be an event space. Similarly, instead of (2.2.5),

$$\lim_{n \to \infty} B_n \in \mathcal{F}, \quad \{B_n\}_{n=1}^{\infty} \text{ is a non-increasing sequence} \tag{2.2.9}$$

can be employed as one of the requirements for a collection to be an event space. $\diamond$

When the sample space is the space of real numbers, the representative of continuous spaces, we now consider a useful event space based on the notion of the smallest σ-algebra described in Definition 2.1.4.

Definition 2.2.7 (*Borel σ-algebra*) When the sample space is the real line $\mathbb{R}$, the sigma algebra

$$\mathcal{B}(\mathbb{R}) = \sigma(\text{all open intervals}) \tag{2.2.10}$$

generated from all open intervals (a, b) in $\mathbb{R}$ is called the Borel algebra, Borel sigma field, or Borel field of $\mathbb{R}$.

The members of the Borel field, i.e., the sets obtained from a countable number of set operations on open intervals, are called Borel sets.

Example 2.2.19 It is possible to see that singleton sets $\{x\} = \bigcap_{n=1}^{\infty} \left(x - \frac{1}{n}, x + \frac{1}{n}\right)$, half-open intervals $[x, y) = (x, y) \cup \{x\}$ and $(x, y] = (x, y) \cup \{y\}$, and closed intervals $[x, y] = (x, y) \cup \{x\} \cup \{y\}$ are all Borel sets after some set operations. In addition, half-open intervals $[x, +\infty) = (-\infty, x)^c$ and $(-\infty, x] = (-\infty, x) \cup \{x\}$, and open intervals $(x, \infty) = (-\infty, x]^c$ are also Borel sets. $\diamond$

The Borel σ-algebra $\mathcal{B}(\mathbb{R})$ is the most useful and widely-used σ-algebra on the set of real numbers, and contains all finite and infinite open, closed, and half-open intervals, singleton sets, and the results from set operations on these sets. On the other hand, the Borel σ-algebra $\mathcal{B}(\mathbb{R})$ is different from the collection of all subsets of $\mathbb{R}$. In other words, there exist some subsets of real numbers which are not contained in the Borel σ-algebra. One such example is the Vitali set discussed in Appendix 2.4.

When the sample space is the real line $\mathbb{R}$, we choose the Borel σ-algebra $\mathcal{B}(\mathbb{R})$ as our event space. At the same time, when a subset Ω' of real numbers is the sample space, the Borel σ-algebra

$$\mathcal{B}(\Omega') = \{G : G = H \cap \Omega', H \in \mathcal{B}(\mathbb{R})\} \tag{2.2.11}$$

of Ω' is assumed as the event space. Note that when the sample space is a discrete subset A of the set of real numbers, Borel σ-algebra $\mathcal{B}(A)$ of A is the same as the power set 2^A of A.

2.2.3 Probability Measure

We now consider the notion of probability measure, the third element of a probability space.

Definition 2.2.8 (*measurable space*) The pair $(\Omega, \mathcal{F})$ of a sample space Ω and an event space $\mathcal{F}$ is called a measurable space.

Let us again mention that when the sample space S is countable or discrete, we usually assume the power set of the sample space as the event space: in other words, the measurable space is $(S, 2^S)$. When the sample space S is uncountable or continuous, we assume the event space described by (2.2.11): in other words, the measurable space is $(S, \mathcal{B}(S))$.

Definition 2.2.9 (*probability measure*) On a measurable space $(\Omega, \mathcal{F})$, a set function P assigning a real number $\mathsf{P}(B_i)$ to each set $B_i \in \mathcal{F}$ under the constraint of the following four axioms is called a probability measure or simply probability:
Axiom 1.

$$\mathsf{P}(B_i) \geq 0. \tag{2.2.12}$$

Axiom 2.

$$P(\Omega) = 1. \tag{2.2.13}$$

Axiom 3. When a finite number of events $\{B_i\}_{i=1}^{n}$ are mutually exclusive,

$$P\left(\bigcup_{i=1}^{n} B_i\right) = \sum_{i=1}^{n} P(B_i). \tag{2.2.14}$$

Axiom 4. When a countable number of events $\{B_i\}_{i=1}^{\infty}$ are mutually exclusive,

$$P\left(\bigcup_{i=1}^{\infty} B_i\right) = \sum_{i=1}^{\infty} P(B_i). \tag{2.2.15}$$

The probability measure P is a set function assigning a value $P(G)$, called probability and also denoted by $P\{G\}$ and $\Pr\{G\}$, to an event G. A probability measure is also called a probability function, probability distribution, or distribution.

Axioms 1–4 are also intuitively appealing. The first axiom that a probability should be not smaller than 0 is in some sense chosen arbitrarily like other measures such as area, volume, and weight. The second axiom is a mathematical expression that something happens from an experiment or some outcome will result from an experiment. The third axiom is called additivity or finite additivity, and implies that the probability of the union of events with no common element is the sum of the probability of each event, which is similar to the case of adding areas of non-overlapping regions.

Axiom 4 is called the countable additivity, which is an asymptotic generalization of Axiom 3 into the limit. This axiom is the key that differentiates the modern probability theory developed by Kolmogorov from the elementary probability theory. When evaluating the probability of an event which can be expressed, for example, only by the limit of events, Axiom 4 is crucial: such an asymptotic procedure is similar to obtaining the integral as the limit of a series. It should be noted that (2.2.14) does not guarantee (2.2.15). In some cases, (2.2.14) is combined into (2.2.15) by viewing Axiom 3 as a special case of Axiom 4.

If our definition of probability is based on the space of an algebra, then we may not be able to describe, for example, some probability resulting from a countably infinite number of set operations. To guarantee the existence of the probability in such a case as well, we need sigma algebra which guarantees the result of a countably infinite number of set operations to exist within our space.

From the axioms of probability, we can obtain

$$P(\emptyset) = 0, \tag{2.2.16}$$
$$P\left(B^c\right) = 1 - P(B), \tag{2.2.17}$$

and

$$P(B) \leq 1, \qquad\qquad (2.2.18)$$

where B is an event. In addition, based on the axioms of probability and (2.2.16), the only probability measure is $P(\Omega) = 1$ and $P(\emptyset) = 0$ in the measurable space $(\Omega, \mathcal{F})$ with event space $\mathcal{F} = \{\Omega, \emptyset\}$.

Example 2.2.20 In a fair[4] coin toss, consider the sample space $\Omega = \{head, tail\}$ and event space $\mathcal{F} = 2^{\Omega}$. Then, we have[5] $P(head) = P(tail) = \frac{1}{2}$, $P(\Omega) = 1$, and $P(\emptyset) = 0$.   ◇

Example 2.2.21 Consider the sample space $\Omega = \{0, 1\}$ and event space $\mathcal{F} = \{\{0\}, \{1\}, \Omega, \emptyset\}$. Then, an example of the probability measure on this measurable space $(\Omega, \mathcal{F})$ is $P(0) = \frac{3}{10}$, $P(1) = \frac{7}{10}$, $P(\emptyset) = 0$, and $P(\Omega) = 1$.   ◇

In a discrete sample space, the power set of the sample space is assumed to be the event space and the event space may contain all the subsets of the discrete sample space. On the other hand, if we choose the event space containing all the possible subsets of a continuous sample space, not only will we be confronted with contradictions[6] but the event space will be too large to be useful. Therefore, when dealing with continuous sample spaces, we choose an appropriate event space such as the Borel sigma algebra mentioned in Definition 2.2.7 and assign a probability only to those events. In addition, in discrete sample spaces, assigning probability to a singleton set is meaningful whereas it is not useful in continuous sample spaces.

Example 2.2.22 Consider a random experiment of choosing a real number between 0 and 1 randomly. If we assign the probability to a number, then the probability of choosing any number will be zero because there are uncountably many real numbers between 0 and 1. However, it is not possible to obtain the probability of, for instance, 'the number chosen is between 0.1 and 0.5' when the probability of choosing a number is 0. In essence, in order to have a useful model in continuous sample space, we should assign probabilities to interval sets, not to singleton sets.   ◇

Definition 2.2.10 (*probability space*) The triplet $(\Omega, \mathcal{F}, P)$ of a sample space Ω, an event space $\mathcal{F}$ of Ω, and a probability measure P is called a probability space.

It is clear from Definition 2.2.10 that the event space $\mathcal{F}$ can be chosen as an algebra instead of a σ-algebra when the sample space S is finite, because an algebra is the same as a σ-algebra in finite sample spaces.

---

[4] Here, fair means '*head* and *tail* are equally likely to occur'.

[5] Because the probability measure P is a set function, $P(\{k\})$ and $P(\{head\})$, for instance, are the exact expressions. Nonetheless, the expressions $P(k)$, $P\{k\}$, $P(head)$, and $P\{head\}$ are also used.

[6] The Vitali set $\mathbb{V}_0$ discussed in Definition 2.A.12 is a subset in the space of real numbers. Denote the rational numbers in $(-1, 1)$ by $\{\alpha_i\}_{i=1}^{\infty}$ and assume the translation operation $T_t(x) = x + t$. Then, the events $\{T_{\alpha_i} \mathbb{V}_0\}_{i=1}^{\infty}$ will produce a contradiction.

2.3 Probability

We now discuss the properties of probability and alternative definitions of probability (Gut 1995; Mills 2001).

2.3.1 Properties of Probability

Assume a probability space $(\Omega, \mathcal{F}, \mathsf{P})$ and events $\{B_i \in \mathcal{F}\}_{i=1}^{\infty}$. Then, based on (2.2.12)–(2.2.15), we can obtain the following properties:

Property 1. We have

$$\mathsf{P}(B_i) \leq \mathsf{P}(B_j) \quad \text{if} \ \ B_i \subseteq B_j \tag{2.3.1}$$

and[7]

$$\mathsf{P}\left(\bigcup_{i=1}^{\infty} B_i\right) \leq \sum_{i=1}^{\infty} \mathsf{P}(B_i). \tag{2.3.2}$$

Property 2. If $\{B_i\}_{i=1}^{\infty}$ is a countable partition of the sample space Ω, then

$$\mathsf{P}(G) = \sum_{i=1}^{\infty} \mathsf{P}(G \cap B_i) \tag{2.3.3}$$

for $G \in \mathcal{F}$.

Property 3. Denoting the sum over $\binom{n}{r} = \frac{n!}{r!(n-r)!}$ ways of choosing r events from $\{B_i\}_{i=1}^{n}$ by $\sum\limits_{i_1 < i_2 < \ldots < i_r} \mathsf{P}\left(B_{i_1} B_{i_2} \cdots B_{i_r}\right)$, we have

$$\mathsf{P}\left(\bigcup_{i=1}^{n} B_i\right) = (-1)^0 \sum_{i=1}^{n} \mathsf{P}(B_i) + (-1)^1 \sum_{i_1 < i_2} \mathsf{P}\left(B_{i_1} B_{i_2}\right)$$

$$+ \cdots + (-1)^{r-1} \sum_{i_1 < i_2 < \cdots < i_r} \mathsf{P}\left(B_{i_1} B_{i_2} \cdots B_{i_r}\right)$$

$$+ \cdots + (-1)^{n-1} \mathsf{P}(B_1 B_2 \cdots B_n). \tag{2.3.4}$$

Figure 2.1 illustrates (2.3.3). We can rewrite (2.3.4) as

$$\mathsf{P}(B_1 \cup B_2) = \mathsf{P}(B_1) + \mathsf{P}(B_2) - \mathsf{P}(B_1 \cap B_2) \tag{2.3.5}$$

---

[7] Among these properties, (2.3.2) is called the Boole inequality. The Boole inequality (2.3.2) can also be written into two formulas similarly to Axioms 3 and 4 in Definition 2.2.9.

Fig. 2.1 A property $P(G) = \sum\limits_{i=1}^{\infty} P(G \cap B_i)$ of probability

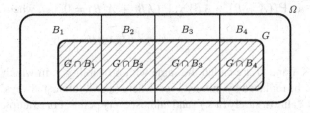

and

$$P(B_1 \cup B_2 \cup B_3) = P(B_1) + P(B_2) + P(B_3)$$
$$- P(B_1 \cap B_2) - P(B_2 \cap B_3) - P(B_3 \cap B_1)$$
$$+ P(B_1 \cap B_2 \cap B_3) \tag{2.3.6}$$

more specifically when $n = 2$ and 3, respectively.

Example 2.3.1 When $P(A \cup B) = 0.7$, $P(A) = 0.5$, and $P(B) = 0.3$ in a probability space, obtain $P(A \cap B)$.

Solution We get $P(A \cap B) = P(A) + P(B) - P(A \cup B) = 0.1$ from (2.3.5). ◇

Example 2.3.2 Consider a random experiment of toss a fair coin three times. Obtain the probability p_1 that *head* occurs at least once.

Solution There exist $2^3 = 8$ elementary events with equal probability $\frac{1}{8}$. Therefore,

$$p_1 = P(\{head, head, head\}) + P(\{head, head, tail\}) + P(\{head, tail, head\})$$
$$+ P(\{tail, head, head\}) + P(\{head, tail, tail\}) + P(\{tail, head, tail\})$$
$$+ P(\{tail, tail, head\})$$
$$= \frac{7}{8}, \tag{2.3.7}$$

which can also be obtained as $p_1 = 1 - \frac{1}{8} = \frac{7}{8}$ by noting that the event '*head* occurs at least once' is the complementary event of '*tail* occurs three times'. ◇

Note that probability being zero for an event does not necessarily mean that the event does not occur or, equivalently, that the event is the same as the 'impossible' event $\emptyset$. For example, in the space of real numbers, although $P(\{a\}) = 0$ for any value of a, the event $\{a\}$ is different from $\emptyset$. In general, A and B are called the same in probability when $P(A) = P(B)$. Similarly, when

$$P(A) = P(B)$$
$$= P(AB) \tag{2.3.8}$$

or $P((A + B) - AB) = P(AB^c + A^cB) = 0$, i.e., when

$$P(A \Delta B) = 0, \qquad (2.3.9)$$

A and B are called the same with probability 1, in which[8] 'with probability 1' can be replaced by 'almost everywhere (a.e.)', 'almost always', 'almost certainly (a.c.)', 'almost surely (a.s.)', and 'almost every point'. For example, when $P(A) = 0$, A is the same as $\emptyset$ almost surely because $P(\emptyset) = P(A) = P(A \cap \emptyset) = 0$. When $P(A) = 1$, A is the same as Ω almost surely because $P(\Omega) = P(A) = P(A \cap \Omega) = 1$. Note that A being the same as B almost surely does not necessarily mean $A = B$.

Example 2.3.3 For the sample space $\Omega = [0, 1]$, let the probability of an interval be the length of the interval. Consider the four intervals $A_1 = [0.1, 0.2]$, $A_2 = [0.1, 0.2)$, $A_3 = (0.1, 0.2]$, and $A_4 = (0.1, 0.2)$. Although $P(A_i) = P(A_j) = P(A_i A_j) = 0.1$ for any i and j, it is clear that $A_i \neq A_j$ for $i \neq j$. In other words, when $i \neq j$, A_i and A_j are the same in probability and with probability 1, but they are not the same event. In addition, A_1 and $B = [0.3, 0.4]$ are the same in probability because $P(A_1) = P(B) = 0.1$, but they are neither the same with probability 1 nor the same. $\diamond$

Theorem 2.3.1 *For events $\{A_i\}_{i=1}^n$ in a probability space, we have*[9]

$$\sum_{i=1}^n P(A_i) - \sum_{i=1}^n \sum_{k=i+1}^n P(A_i \cap A_k) \leq P\left(\bigcup_{i=1}^n A_i\right) \leq \sum_{i=1}^n P(A_i). \quad (2.3.10)$$

Proof If we let

$$B_i = \begin{cases} A_i - \bigcup_{k=i+1}^n A_k, & i = 1, 2, \ldots, n-1, \\ A_n, & i = n, \end{cases} \qquad (2.3.11)$$

then we have $P\left(\bigcup_{i=1}^n A_i\right) = P\left(\bigcup_{i=1}^n B_i\right)$ from $\bigcup_{i=1}^n B_i = \bigcup_{i=1}^n A_i$ and, subsequently,

$$P\left(\bigcup_{i=1}^n A_i\right) = \sum_{i=1}^n P(B_i) \qquad (2.3.12)$$

because $B_i \cap B_k = \emptyset$ for $i \neq k$. We also have

$$P(B_i) \leq P(A_i) \qquad (2.3.13)$$

---

[8] This has been described also in Definition 1.1.24.

[9] This formula is called the Bonferroni inequality.

from $B_i \subseteq A_i$. Combining (2.3.12) and (2.3.13), we get

$$P \left(\bigcup_{i=1}^{n} A_i \right) \leq \sum_{i=1}^{n} P(A_i). \tag{2.3.14}$$

Next, recollect that $P(C - D) = P(C) - P(CD)$ for two sets C and D from $P(C) = P(C - D) + P(CD)$ because $\{C - D, CD\}$ is a partition of C. Then, noting that $P \left(A_i \cap \left(\bigcup_{k=i+1}^{n} A_k \right) \right) = P \left(\bigcup_{k=i+1}^{n} (A_i \cap A_k) \right)$ because $A_i \cap \left(\bigcup_{k=i+1}^{n} A_k \right) = \bigcup_{k=i+1}^{n} (A_i \cap A_k)$ from (1.1.22), we have $P(B_i) = P \left(A_i - \bigcup_{k=i+1}^{n} A_k \right) = P(A_i) - P \left(A_i \cap \left(\bigcup_{k=i+1}^{n} A_k \right) \right)$, i.e.,

$$P(B_i) = P(A_i) - P \left(\bigcup_{k=i+1}^{n} (A_i \cap A_k) \right) \tag{2.3.15}$$

based on (2.3.11). We then get

$$P(A_i) - \sum_{k=i+1}^{n} P(A_i \cap A_k) \leq P(A_i) - P \left(\bigcup_{k=i+1}^{n} (A_i \cap A_k) \right)$$
$$= P(B_i) \tag{2.3.16}$$

from (2.3.15) because $P \left(\bigcup_{k=i+1}^{n} (A_i \cap A_k) \right) \leq \sum_{k=i+1}^{n} P(A_i \cap A_k)$. Now, from (2.3.12) and (2.3.16), we get

$$\sum_{i=1}^{n} P(A_i) - \sum_{i=1}^{n} \sum_{k=i+1}^{n} P(A_i \cap A_k) \leq \sum_{i=1}^{n} P(B_i)$$
$$= P \left(\bigcup_{i=1}^{n} A_i \right), \tag{2.3.17}$$

which, when combined with (2.3.14), confirms (2.3.10).   ♠

Example 2.3.4 Assume the sample space $S = \{1, 2, \ldots, 10\}$, event space 2^S, and probability measure $P(k) = \frac{k}{55}$ for $k = 1, 2, \ldots, 10$. Consider the three events $A_1 = \{1, 2, 3\}$, $A_2 = \{3, 4, 5, 6\}$, and $A_3 = \{5, 6, 7, 8\}$. First, we have $P \left(\bigcup_{i=1}^{3} A_i \right) = \frac{36}{55}$.
We also get $\sum_{i=1}^{3} P(A_i) = \frac{50}{55}$ from $P(A_1) = \frac{6}{55}$, $P(A_2) = \frac{18}{55}$, and $P(A_3) = \frac{26}{55}$.
Finally, from $P(A_1 \cap A_2) = \frac{3}{55}$, $P(A_2 \cap A_3) = \frac{11}{55}$, and $P(A_3 \cap A_1) = 0$, we have

$\sum_{i=1}^{3} P(A_i) - \sum_{i=1}^{3} \sum_{k=i+1}^{3} P(A_i \cap A_k) = \frac{50}{55} - \left(\frac{3}{55} + \frac{11}{55}\right) = \frac{36}{55}$. In short, (2.3.10) is confirmed.

$\diamond$

2.3.2 Other Definitions of Probability

Probability can be defined in several ways. In modern probability theory, probability
is defined with the four axioms by adopting the notion of σ-algebra as we have
described so far. We now introduce two other ways of defining probability.

2.3.2.1 Classical Definition

When all the outcomes from a random experiment are equally likely, the probability
of an event can be defined by the ratio of the number of desired outcomes to the total
number of outcomes. Specifically, the probability of A is given by the ratio

$$P(A) = \frac{N_A}{N}, \tag{2.3.18}$$

where N is the number[10] of all possible outcomes and N_A is the number of desired
outcomes for A. The condition of equally likely occurrence is the key in the classical
definition.

Example 2.3.5 Obtain the probability of *head* when a fair coin is tossed once.

Solution There are two equally likely possible outcomes *head* and *tail*, among which
the desired outcome is *head*. Thus, the probability of *head* is $\frac{1}{2}$. $\diamond$

Example 2.3.6 Obtain the probability P_3 that the three pieces are all longer than $\frac{1}{4}$
when a rod of length 1 is divided into three pieces by choosing two points at random.

Solution View the rod of length 1 as the interval $[0, 1]$ on the real line, and let the
coordinate of the two points of cutting be x and y with $0 < x < 1$ and $0 < y < 1$ as
shown in Fig. 2.2. The three pieces will all be longer than $\frac{1}{4}$ when $x < y$, $\frac{1}{4} < x < \frac{1}{2}$,
and $x + \frac{1}{4} < y < \frac{3}{4}$ or when $x > y$, $\frac{1}{4} < y < \frac{1}{2}$, and $y + \frac{1}{4} < x < \frac{3}{4}$. Therefore,
from

$$P_3 = \frac{\text{area of the region of desired outcomes}}{\text{area of the whole region}}, \tag{2.3.19}$$

we get $P_3 = \frac{A_1 + A_2}{A_T} = \int_{\frac{1}{4}}^{\frac{1}{2}} \int_{x+\frac{1}{4}}^{\frac{3}{4}} 1 \, dy \, dx + \int_{\frac{1}{4}}^{\frac{1}{2}} \int_{y+\frac{1}{4}}^{\frac{3}{4}} 1 \, dx \, dy$, i.e.,

---

[10] Note that the number is sometimes replaced by other quantity such as area, volume, or length as
it is shown in Example 2.3.6.

Fig. 2.2 Points of division:
$0 < x, y < 1$

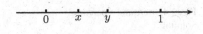

Fig. 2.3 Dividing a rod of
length 1 into three pieces all
longer than $\frac{1}{4}$

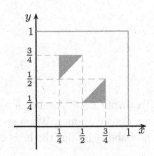

$$P_3 = \frac{1}{16} \tag{2.3.20}$$

referring to Fig. 2.3, where $A_T = 1$ denotes the area of the whole region $\{0 < x < 1, 0 < y < 1\}$, A_1 is the area of the region $\{x < y, \frac{1}{4} < x < \frac{1}{2}, x + \frac{1}{4} < y < \frac{3}{4}\}$, and A_2 is the area of the region $\{x > y, y + \frac{1}{4} < x < \frac{3}{4}, \frac{1}{4} < y < \frac{1}{2}\}$. ◇

Example 2.3.7 Obtain the probability P_T that the three pieces will form a triangle when a rod is divided into three pieces by choosing two points at random.

Solution Let the length of the rod be 1, and follow the description in Example 2.3.6. Then, the lengths of the three pieces are x, $y - x$, and $1 - y$ when $x < y$ and $y, x - y$, and $1 - x$ when $x > y$. Thus, for the three pieces to form a triangle, it is required that $\{0 < x < 1, 0 < y < 1, x < y, y > \frac{1}{2}, y < x + \frac{1}{2}, x < \frac{1}{2}\}$ or $\{0 < x < 1, 0 < y < 1, x > y, x > \frac{1}{2}, x < y + \frac{1}{2}, y < \frac{1}{2}\}$ because the sum of the lengths of two pieces should be longer than the length of the remaining piece. Consequently, from

$$P_T = \frac{\text{area of the region of desired outcomes}}{\text{area of total region}}, \tag{2.3.21}$$

we get $P_T = \frac{A_1 + A_2}{A_T}$, i.e.,

$$P_T = \frac{1}{4}, \tag{2.3.22}$$

where $A_T = 1$ is the area of the region $\{0 < x < 1, 0 < y < 1\}$, $A_1 = \frac{1}{8}$ is the area of the region $\{0 < x < 1, 0 < y < 1, x < y, y > \frac{1}{2}, y < x + \frac{1}{2}, x < \frac{1}{2}\}$, and $A_2 = \frac{1}{8}$ is the area of the region $\{0 < x < 1, 0 < y < 1, x > y, x > \frac{1}{2}, x < y + \frac{1}{2}, y < \frac{1}{2}\}$. A similar problem will be discussed in Example 3.4.2. ◇

Fig. 2.4 Bertrand's paradox.
Solution 1

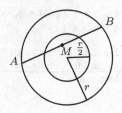

Example 2.3.8 Assume a collection of n integrated circuits (ICs), among which m are defective ones. When an IC is chosen randomly from the collection, obtain the probability α_1 that the IC is defective.

Solution The number of ways to choose one IC among n ICs is $_nC_1$ and to choose one among m defective ICs is $_mC_1$. Thus, $\alpha_1 = \frac{_mC_1}{_nC_1} = \frac{m}{n}$. $\diamond$

In obtaining the probability with the classical definition, it is important to enumerate the number of desired outcomes correctly, which is usually a problem of combinatorics. Another important consideration in the classical definition is the condition of equally likely outcomes, which will become more evident in Example 2.3.9.

Example 2.3.9 When a fair[11] die is rolled twice, obtain the probability P_7 that the sum of the two faces is 7.

Solution (Solution 1) There are 11 possible cases $\{2, 3, \ldots, 12\}$ for the sum. Therefore, one might say that $P_7 = \frac{1}{11}$.
(Solution 2) If we write the outcomes of the two rolls as $\{x, y\}$, $x \geq y$ ignoring the order, then we have 21 possible cases $\{1, 1\}, \{2, 1\}, \{2, 2\}, \ldots, \{6, 6\}$. Among the 21 cases, the three possibilities $\{4, 3\}, \{5, 2\}, \{6, 1\}$ yield a sum of 7. Based on this observation, $P_7 = \frac{3}{21} = \frac{1}{7}$.
(Solution 3) There are 36 equally likely outcomes (x, y) with $x, y = 1, 2, \ldots 6$, among which the six outcomes $(1, 6), (2, 5), (3, 4), (4, 3), (5, 2)$, and $(6, 1)$ yield a sum of 7. Therefore, $P_7 = \frac{6}{36} = \frac{1}{6}$.
Among the three solutions, neither the first nor the second is correct because the cases in these two solutions are not equally likely. For instance, in the first solution, the sum of two numbers being 2 and 3 are not equally likely. Similarly, $\{1, 1\}$ and $\{2, 1\}$ are not equally likely in the second solution. In short, Solution 3 is the correct solution. $\diamond$

In addition, especially when the number of possible outcomes is infinite, the procedure, observation, and model, mentioned in Example 2.2.4, of an experiment should be clearly depicted in advance.

Example 2.3.10 On a circle C of radius r, we draw a chord AB randomly. Obtain the probability P_B that the length l of the chord is no shorter than $\sqrt{3}r$. This problem is called the Bertrand's paradox.

---

[11] Here, 'fair' means '1, 2, 3, 4, 5, and 6 are equally likely to occur'.

Fig. 2.5 Bertrand's paradox.
Solution 2

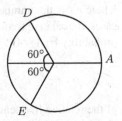

Fig. 2.6 Bertrand's paradox.
Solution 3

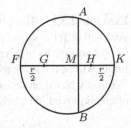

Solution (Solution 1) Assume that the center point M of the chord is chosen randomly. As shown in Fig. 2.4, $l \geq \sqrt{3}r$ is satisfied if the center point M is located in or on the circle C_1 of radius $\frac{r}{2}$ with center the same as that of C. Thus, $P_B = \frac{1}{4}\pi r^2 (\pi r^2)^{-1} = \frac{1}{4}$.

(Solution 2) Assume that the point B is selected randomly on the circle C with the point A fixed. As shown in Fig. 2.5, $l \geq \sqrt{3}r$ is satisfied when the point B is on the shorter arc DE, where D and E are the two points $\frac{2}{3}\pi r$ apart from A along C in two directions. Therefore, we have $P_B = \frac{2}{3}\pi r (2\pi r)^{-1} = \frac{1}{3}$ because the length of the shorter arc DE is $\frac{2}{3}\pi r$.

(Solution 3) Assume that the chord AB is drawn orthogonal to a diameter FK of C. As shown in Fig. 2.6, we then have $l \geq \sqrt{3}r$ if the center point M is located between the two points H and G, located $\frac{r}{2}$ apart from K toward F and from F toward K, respectively. Therefore, $P_B = \frac{r}{2r} = \frac{1}{2}$.

This example illustrates that the experiment should be described clearly to obtain the probability appropriately with the classical definition.   ◇

2.3.2.2   Definition Via Relative Frequency

Probability can also be defined in terms of the relative frequency of desired outcomes in a number of repetitions of a random experiment. Specifically, the relative frequency of a desired event A can be defined as

$$q_n(A) = \frac{n_A}{n},$$ (2.3.23)

where n_A is the number of occurrences of A and n is the number of trials. In many cases, the relative frequency $q_n(A)$ converges to a value as n becomes larger, and the probability $P(A)$ of A can be defined as the limit

$$P(A) = \lim_{n \to \infty} q_n(A) \tag{2.3.24}$$

of the relative frequency. One drawback of this definition is that the limit shown in (2.3.24) can be obtained only by an approximation in practice.

Example 2.3.11 The probability that a person of a certain age will survive for a year will differ from year to year, and thus it is difficult to use the classical definition of probability. As an alternative in such a case, we assume that the tendency in the future will be the same as that so far, and then compute the probability as the relative frequency based on the records over a long period for the same age. This method is often employed in determining an insurance premium.   ◇

2.4 Conditional Probability

Conditional probability (Helstrom 1991; Shiryaev 1996; Sveshnikov 1968; Weirich 1983) is one of the most important notions and powerful tools in probability theory. It is the probability of an event when the occurrence of another event is assumed. Conditional probability is quite useful especially when only partial information is available and, even when we can obtain some probability directly, we can often obtain the same result more easily by using conditioning in many situations.

Definition 2.4.1 (*conditional probability*) The probability of an event under the assumption that another event has occurred is called conditional probability. Specifically, the conditional probability, denoted by $P(A|B)$, of event A under the assumption that event B has occurred is defined as

$$P(A|B) = \frac{P(A \cap B)}{P(B)} \tag{2.4.1}$$

when $P(B) > 0$.

In other words, the conditional probability $P(A|B)$ of event A under the assumption that event B has occurred is the probability of A with the sample space Ω replaced by the conditioning event B. Often, the event A conditioned on B is denoted by $A|B$. From (2.4.1), we easily get

$$P(A|B) = \frac{P(A)}{P(B)} \tag{2.4.2}$$

when $A \subseteq B$ and

$$P(A|B) = 1 \qquad (2.4.3)$$

when $B \subseteq A$.

Example 2.4.1 When the conditioning event B is the sample space Ω, we have $P(A|\Omega) = P(A)$ because $A \cap B = A \cap \Omega = A$ and $P(B) = P(\Omega) = 1$. $\diamond$

Example 2.4.2 Consider the rolling of a fair die. Assume that we know the outcome is an even number. Obtain the probability that the outcome is 2.

Solution Let $A = \{2\}$ and $B = \{2, 4, 6\}$. Then, because $A \cap B = A$, $P(A \cap B) = P(A) = \frac{1}{6}$, and $P(B) = \frac{1}{2} \neq 0$, we have $P(A|B) = \frac{1}{6}\left(\frac{1}{2}\right)^{-1} = \frac{1}{3}$. Again, $P(A|B)$ is the probability of $A = \{2\}$ when $B = \{\text{an even number}\} = \{2, 4, 6\}$ is assumed as the sample space. $\diamond$

Example 2.4.3 The probability for any child to be a girl is α and is not influenced by other children. Assume that Dr. Kim has two children. Obtain the probabilities in the following two separate cases: (1) the probability p_1 that the younger child is a daughter when Dr. Kim says "The elder one is a daughter", and (2) the probability p_2 that the other child is a daughter when Dr. Kim says "One of my children is a daughter".

Solution We have $p_1 = P(D_2|D_1) = \frac{P(D_1 \cap D_2)}{P(D_1)} = \frac{\alpha^2}{\alpha} = \alpha$, where D_1 and D_2 denote the events of the first and second child being a daughter, respectively. Similarly, because $P(C_A) = P(D_1 \cap D_2) + P(D_1 \cap B_2) + P(B_1 \cap D_2) = 2\alpha - \alpha^2$, we get $p_2 = P(C_B|C_A) = \frac{P(C_A \cap C_B)}{P(C_A)} = \frac{\alpha^2}{2\alpha - \alpha^2} = \frac{\alpha}{2 - \alpha}$, where B_1 and B_2 denote the events of the first and second child being a boy, respectively, and C_A and C_B denote the events of one and the other child being a daughter, respectively. $\diamond$

2.4.1 Total Probability and Bayes' Theorems

Theorem 2.4.1 When $P(B_i| B_1 B_2 \cdots B_{i-1}) > 0$ for $i = 2, 3, \ldots, n$, the probability of the intersection $\bigcap_{i=1}^{n} B_i = \prod_{i=1}^{n} B_i$ of $\{B_i\}_{i=1}^{n}$ can be written as

$$P\left(\bigcap_{i=1}^{n} B_i\right) = P(B_1) P(B_2| B_1) \cdots P(B_n| B_1 B_2 \cdots B_{n-1}), \qquad (2.4.4)$$

which is called the multiplication theorem. Similarly, the probability of the union $\bigcup_{i=1}^{n} B_i$ can be expressed as

$$P\left(\bigcup_{i=1}^{n} B_i\right) = P(B_1) + P(B_1^c B_2) + \cdots + P(B_1^c B_2^c \cdots B_{n-1}^c B_n), \quad (2.4.5)$$

which is called the addition theorem.

Note that (2.4.5) is the same as (2.3.4). Next, (2.4.4) can be written as

$$P(B_1 \cap B_2) = P(B_1)P(B_2|B_1)$$
$$= P(B_2)P(B_1|B_2) \quad (2.4.6)$$

when $n = 2$. Now, from $1 = P(\Omega|B_2) = P(B_1^c \cup B_1|B_2) = P(B_1^c|B_2) + P(B_1|B_2)$, we have $P(B_1^c|B_2) = 1 - P(B_1|B_2)$. Using this result and (2.4.6), (2.4.5) for $n = 2$ can be written as $P(B_1 \cup B_2) = P(B_1) + P(B_1^c B_2) = P(B_1) + P(B_2)P(B_1^c|B_2) = P(B_1) + P(B_2)\{1 - P(B_1|B_2)\}$, i.e.,

$$P(B_1 \cup B_2) = P(B_1) + P(B_2) - P(B_2)P(B_1|B_2)$$
$$= P(B_1) + P(B_2) - P(B_1 \cap B_2), \quad (2.4.7)$$

which is the same as (2.3.5).

Example 2.4.4 Assume a box with three red and two green balls. We randomly choose one ball and then another without replacement. Find the probability P_{RG} that the first ball is red and the second ball is green.

Solution (Method 1) The number of all possible ways of choosing two balls is $_5P_2$, among which that of the desired outcome is $_3P_1 \times _2P_1$. Therefore, $P_{RG} = \frac{_3P_1 \times _2P_1}{_5P_2} = \frac{3}{10}$.
(Method 2) Let $A = \{$the first ball is red$\}$ and $B = \{$the second ball is green$\}$. Then, we have $P(A) = \frac{3}{5}$. In addition, $P(B|A)$ is the conditional probability that the second ball is green when the first ball is red, and is thus the probability of choosing a green ball among the remaining two red and two green balls after a red ball has been chosen: in other words, $P(B|A) = \frac{2}{4}$. Consequently, $P(AB) = P(B|A)P(A) = \frac{3}{10}$.   ◇

Consider two events A and B. The event A can be expressed as

$$A = AB \cup AB^c, \quad (2.4.8)$$

based on which we have $P(A) = P(AB) + P(AB^c) = P(A|B)P(B) + P(A|B^c)$
$P(B^c)$, i.e.,

$$P(A) = P(A|B)P(B) + P(A|B^c)\{1 - P(B)\} \quad (2.4.9)$$

from (2.2.14) because AB and AB^c are mutually exclusive. The result (2.4.9) shows that the probability of A is the weighted sum of the conditional probabilities of A when B and B^c are assumed with the weights the probabilities of the conditioning events B and B^c, respectively.

The result (2.4.9) is quite useful when a direct calculation of the probability of an event is not straightforward.

Example 2.4.5 In Box 1, we have two white and four black balls. Box 2 contains one white and one black balls. We randomly take one ball from Box 1, put it into Box 2, and then randomly take one ball from Box 2. Find the probability P_W that the ball taken from Box 2 is white.

Solution Let the events of a white ball from Box 1 and a white ball from Box 2 be C and D, respectively. Then, because $P(C) = \frac{1}{3}$, $P(D|C) = \frac{2}{3}$, $P(D|C^c) = \frac{1}{3}$, and $P(C^c) = 1 - P(C) = \frac{2}{3}$, we get $P_W = P(D) = P(D|C)P(C) + P(D|C^c) P(C^c) = \frac{4}{9}$. $\diamond$

Example 2.4.6 The two numbers of the upward faces are added after rolling a pair of fair dice. Obtain the probability α that 5 appears before 7 when we continue the rolling until the outcome is 5 or 7.

Solution Let A_n and B_n be the events that the outcome is neither 5 nor 7 and that the outcome is 5, respectively, at the n-th rolling for $n = 1, 2, \ldots$. Then, $P(A_n) = 1 - P(5) - P(7) = \frac{13}{18}$ from $P(B_n) = P(5) = \frac{4}{36} = \frac{1}{9}$ and $P(7) = \frac{6}{36} = \frac{1}{6}$. Now, $\alpha = P(B_1 \cup (A_1 B_2) \cup (A_1 A_2 B_3) \cdots) = \sum_{n=1}^{\infty} P(A_1 A_2 \ldots A_{n-1} B_n)$ from (2.4.5) because $\{A_1 A_2 \ldots A_{n-1} B_n\}_{n=1}^{\infty}$ are mutually exclusive, where we assume $A_0 B_1 = B_1$. Here,

$$P(A_1 A_2 \ldots A_{n-1} B_n) = P(B_n | A_1 A_2 \cdots A_{n-1}) P(A_1 A_2 \ldots A_{n-1})$$
$$= \frac{1}{9} \left(\frac{13}{18}\right)^{n-1}. \tag{2.4.10}$$

Therefore, $\alpha = \frac{1}{9} \sum_{n=1}^{\infty} \left(\frac{13}{18}\right)^{n-1} = \frac{2}{5}$. $\diamond$

Example 2.4.7 Example 2.4.6 can be viewed in a more intuitive way as follows: Consider two mutually exclusive events A and B from a random experiment and assume the experiments are repeated. Then, the probability that A occurs before B can be obtained as

$$\frac{\text{probability of } A}{\text{probability of } A \text{ or } B} = \frac{P(A)}{P(A) + P(B)}. \tag{2.4.11}$$

Solving Example 2.4.6 based on (2.4.11), we get $P(5 \text{ appears before } 7) = \frac{1}{9} \left(\frac{1}{9} + \frac{1}{6}\right)^{-1} = \frac{2}{5}$. $\diamond$

Let us now generalize the number of conditioning events in (2.4.9). Assume a collection $\{B_j\}_{j=1}^{n}$ of mutually exclusive events and let $A \subseteq \bigcup_{j=1}^{n} B_j$. Then, $P(A) = P\left(\bigcup_{j=1}^{n} (A \cap B_j)\right)$ can be expressed as

$$P(A) = \sum_{j=1}^{n} P\left(AB_j\right) \tag{2.4.12}$$

because $A = A \cap \left(\bigcup_{j=1}^{n} B_j\right) = \bigcup_{j=1}^{n} \left(A \cap B_j\right)$ and $\left\{A \cap B_j\right\}_{j=1}^{n}$ are all mutually exclusive. Now, recollecting that $P\left(AB_j\right) = P\left(A \,|\, B_j\right) P\left(B_j\right)$, we get the following theorem called the total probability theorem:

Theorem 2.4.2 *We have*

$$P(A) = \sum_{j=1}^{n} P\left(A \,|\, B_j\right) P\left(B_j\right) \tag{2.4.13}$$

when $\left\{B_j\right\}_{j=1}^{n}$ *is a collection of disjoint events and* $A \subseteq \bigcup_{j=1}^{n} B_j$.

Example 2.4.8 Let $A = \{1, 2, 3\}$ in the experiment of rolling a fair die. When $B_1 = \{1, 2\}$ and $B_2 = \{3, 4, 5, 6\}$, we have $P(A) = \frac{1}{2} = 1 \times \frac{1}{3} + \frac{1}{4} \times \frac{2}{3}$ because $P\left(B_1\right) = \frac{1}{3}$, $P\left(B_2\right) = \frac{2}{3}$, $P\left(A \,|\, B_1\right) = \frac{P(AB_1)}{P(B_1)} = \frac{1}{3} \times \left(\frac{1}{3}\right)^{-1} = 1$, and $P\left(A \,|\, B_2\right) = \frac{P(AB_2)}{P(B_2)} = \frac{1}{6} \times \left(\frac{2}{3}\right)^{-1} = \frac{1}{4}$. Similarly, when $B_1 = \{1\}$ and $B_2 = \{2, 3, 5\}$, we get $P(A) = \frac{1}{2} = 1 \times \frac{1}{6} + \frac{2}{3} \times \frac{1}{2}$ from $P\left(B_1\right) = \frac{1}{6}$, $P\left(B_2\right) = \frac{1}{2}$, $P\left(A \,|\, B_1\right) = \frac{P(AB_1)}{P(B_1)} = 1$, and $P\left(A \,|\, B_2\right) = \frac{P(AB_2)}{P(B_2)} = \frac{2}{3}$.   ◇

Example 2.4.9 Assume a group comprising 60% women and 40% men. Among the women, 45% play violin, and 25% of the men play violin. A person chosen randomly from the group plays violin. Find the probability that the person is a man.

Solution Denote the events of a person being a man and a woman by M and W, respectively, and playing violin by V. Then, using (2.4.1) and (2.4.13), we get $P(M|V) = \frac{P(MV)}{P(V)} = \frac{P(V|M)P(M)}{P(V|M)P(M)+P(V|W)P(W)} = \frac{10}{37}$ because $M^c = W$, $P(W) = 0.6$, $P(M) = 0.4$, $P(V|W) = 0.45$, and $P(V|M) = 0.25$.   ◇

Consider a collection $\{B_i\}_{i=1}^{n}$ of events and an event A with $P(A) > 0$. Then, we have

$$P\left(B_k \,|\, A\right) = \frac{P\left(A \,|\, B_k\right) P\left(B_k\right)}{P(A)} \tag{2.4.14}$$

because $P\left(B_k \,|\, A\right) = \frac{P(B_k A)}{P(A)}$ and $P\left(B_k A\right) = P\left(A \,|\, B_k\right) P\left(B_k\right)$ from the definition of conditional probability. Now, combining the results (2.4.13) and (2.4.14) when the events $\{B_i\}_{i=1}^{n}$ are all mutually exclusive and $A \subseteq \bigcup_{i=1}^{n} B_i$, we get the following result called the Bayes' theorem:

Theorem 2.4.3 *We have*

$$P(B_k | A) = \frac{P(A | B_k) P(B_k)}{\sum\limits_{j=1}^{n} P(A | B_j) P(B_j)}. \tag{2.4.15}$$

when $\{B_j\}_{j=1}^{n}$ *is a collection of disjoint events,* $A \subseteq \bigcup\limits_{j=1}^{n} B_j$, *and* $P(A) \neq 0$.

It should be noted in applying Theorem 2.4.3 that, when A is not a subset of $\bigcup\limits_{j=1}^{n} B_j$, using (2.4.15) to obtain $P(B_k | A)$ will yield an incorrect value. This is because $P(A) \neq \sum\limits_{j=1}^{n} P(A | B_j) P(B_j)$ when A is not a subset of $\bigcup\limits_{j=1}^{n} B_j$. To obtain $P(B_k | A)$ correctly in such a case, we must use (2.4.14), i.e., (2.4.15) with the denominator $\sum\limits_{j=1}^{n} P(A | B_j) P(B_j)$ replaced back with $P(A)$.

Example 2.4.10 In Example 2.4.5, obtain the probability that the ball drawn from Box 1 is white when the ball drawn from Box 2 is white.

Solution Using the results of Example 2.4.5 and the Bayes' theorem, we have $P(C|D) = \frac{P(CD)}{P(D)} = \frac{P(D|C)P(C)}{P(D)} = \frac{1}{2}$. $\diamond$

Example 2.4.11 For the random experiment of rolling a fair die, assume $A = \{2, 4, 5, 6\}$, $B_1 = \{1, 2\}$, and $B_2 = \{3, 4, 5\}$. Obtain $P(B_2 | A)$.

Solution We easily get $P(A) = \frac{2}{3}$. On the other hand, $\sum\limits_{j=1}^{2} P(A | B_j) P(B_j) = \frac{1}{2}$ from $P(B_1) = \frac{1}{3}$, $P(B_2) = \frac{1}{2}$, $P(A | B_1) = \frac{P(AB_1)}{P(B_1)} = \frac{1}{6} \times \left(\frac{1}{3}\right)^{-1} = \frac{1}{2}$, and $P(A | B_2) = \frac{P(AB_2)}{P(B_2)} = \frac{1}{3} \times \left(\frac{1}{2}\right)^{-1} = \frac{2}{3}$. In other words, because $A = \{2, 4, 5, 6\}$ is not a subset of $B_1 \cup B_2 = \{1, 2, 3, 4, 5\}$, we have $P(A) \neq \sum\limits_{j=1}^{2} P(A | B_j) P(B_j)$. Thus, we would get $P(B_2 | A) = \frac{P(A|B_2)P(B_2)}{\sum\limits_{j=1}^{2} P(A|B_j)P(B_j)} = \frac{2}{3}$, an incorrect answer, if we use (2.4.15) carelessly. The correct answer $P(B_2 | A) = \frac{P(A|B_2)P(B_2)}{P(A)} = \frac{1}{2}$ can be obtained by using (2.4.14) in this case. $\diamond$

Let us consider an example for the application of the Bayes' theorem.

Example 2.4.12 Assume four boxes with 2000, 500, 1000, and 1000 parts of a machine, respectively. The probability of a part being defective is 0.05, 0.4, 0.1, and 0.1, respectively, for the four boxes.

(1) When a box is chosen at random and then a part is picked randomly from the box, calculate the probability that the part is defective.

(2) Assuming the part picked is defective, calculate the probability that the part is from the second box.

(3) Assuming the part picked is defective, calculate the probability that the part is from the third box.

Solution Let A and B_i be the events that the part picked is defective and the part is from the i-th box, respectively. Then, $P(B_i) = \frac{1}{4}$ for $i = 1, 2, 3, 4$. In addition, the value of $P(A|B_2)$, for instance, is 0.4 because $P(A|B_2)$ denotes the probability that a part picked is defective when it is from the second box.

(1) Noting that $\{B_i\}_{i=1}^{4}$ are all disjoint, we get $P(A) = \sum_{i=1}^{4} P(A|B_i) P(B_i) = \frac{1}{4} \times$ $0.05 + \frac{1}{4} \times 0.4 + \frac{1}{4} \times 0.1 + \frac{1}{4} \times 0.1$, i.e.,

$$P(A) = \frac{13}{80} \qquad (2.4.16)$$

from (2.4.13).

(2) The probability to obtain is $P(B_2|A)$. We get $P(B_2|A) = \frac{P(A|B_2)P(B_2)}{\sum_{j=1}^{4}P(A|B_j)P(B_j)} =$ $\frac{0.4 \times \frac{1}{4}}{\frac{13}{80}} = \frac{8}{13}$ as shown in (2.4.15) because $P(A) = \frac{13}{80}$ from (2.4.16), $P(B_2) = \frac{1}{4}$, and $P(A|B_2) = 0.4$.

(3) Similarly, we get[12] $P(B_3|A) = \frac{0.1 \times \frac{1}{4}}{\frac{13}{80}} = \frac{2}{13}$ from $P(B_3|A) = \frac{P(A|B_3)P(B_3)}{P(A)}$, $P(B_3) = \frac{1}{4}, P(A|B_3) = 0.1$, and $P(A) = \frac{13}{80}$.

$\diamond$

If we calculate similarly the probabilities for the first and fourth boxes and then add the four values in Example 2.4.1, we will get 1: in other words, we have $\sum_{i=1}^{4} P(B_i|A) = P(\Omega|A) = 1$.

2.4.2 Independent Events

Assume two boxes. Box 1 contains one red ball and two green balls, and Box 2 contains two red balls and four green balls. If we pick a ball randomly after choosing a box with probability $P(\text{Box 1}) = p = 1 - P(\text{Box 2})$, then we have $P(\text{red ball}) = P(\text{red ball} | \text{Box 1}) P(\text{Box 1}) + P(\text{red ball} | \text{Box 2}) P(\text{Box 2}) = \frac{1}{3}p + \frac{1}{3}(1-p) = \frac{1}{3}$ and $P(\text{green ball}) = \frac{2}{3}$. Note that

$$P(\text{red ball}) = P(\text{red ball} | \text{Box 1}) = P(\text{red ball} | \text{Box 2}) \qquad (2.4.17)$$

---

[12] Here, $\frac{13}{80} = 0.1625$, $\frac{8}{13} \approx 0.6154$, and $\frac{2}{13} \approx 0.1538$.

and $P(\text{green ball}) = P(\text{green ball} \mid \text{Box 1}) = P(\text{green ball} \mid \text{Box 2})$: whichever box we choose or whatever value the probability of choosing a box is, the probability that the ball picked is red and green is $\frac{1}{3}$ and $\frac{2}{3}$, respectively. In other words, the choice of a box does not influence the probability of the color of the ball picked. On the other hand, if Box 1 contains one red ball and two green balls and Box 2 contains two red balls and one green ball, the choice of a box will influence the probability of the color of the ball picked. Such an influence is commonly represented by the notion of independence.

Definition 2.4.2 (*independence of two events*) If the probability $P(AB)$ of the intersection of two events A and B is equal to the product $P(A)P(B)$ of the probabilities of the two events, i.e., if

$$P(AB) = P(A)P(B), \tag{2.4.18}$$

then A and B are called independent (of each other) or mutually independent.

Example 2.4.13 Assume the sample space $S = \{1, 2, \ldots, 9\}$ and $P(k) = \frac{1}{9}$ for $k = 1, 2, \ldots, 9$. Consider the events $A = \{1, 2, 3\}$ and $B = \{3, 4, 5\}$. Then, $P(A) = \frac{1}{3}$, $P(B) = \frac{1}{3}$, and $P(AB) = P(3) = \frac{1}{9}$, and therefore $P(AB) = P(A)P(B)$. Thus, A and B are independent of each other. Likewise, for the sample space $S = \{1, 2, \ldots, 6\}$, the events $C = \{1, 2, 3\}$ and $D = \{3, 4\}$ are independent of each other when $P(k) = \frac{1}{6}$ for $k = 1, 2, \ldots, 6$. ◇

When one of two events has probability 1 as the sample space S or 0 as the null set $\emptyset$, the two events are independent of each other because (2.4.18) holds true.

Theorem 2.4.4 *An event with probability 1 or 0 is independent of any other event.*

Example 2.4.14 Assume the sample space $S = \{1, 2, \ldots, 5\}$ and let $P(k) = \frac{1}{5}$ for $k = 1, 2, \ldots, 5$. Then, no two sets, excluding S and $\emptyset$, are independent of each other. When $P(1) = \frac{1}{10}$, $P(2) = P(3) = P(4) = \frac{1}{5}$, and $P(5) = \frac{3}{10}$ for the sample space $S = \{1, 2, \ldots, 5\}$, the events $A = \{3, 4\}$ and $B = \{4, 5\}$ are independent because $P(A)P(B) = P(4)$ from $P(A) = \frac{2}{5}$, $P(B) = \frac{1}{2}$, and $P(4) = \frac{1}{5}$. ◇

In general, two mutually exclusive events are not independent of each other: on the other hand, we have the following theorem from Theorem 2.4.4:

Theorem 2.4.5 *If at least one event has probability 0, then two mutually exclusive events are independent of each other.*

Example 2.4.15 For the sample space $S = \{1, 2, 3\}$, let the power set $2^S = \{\emptyset, \{1\}, \{2\}, \ldots, S\}$ be the event space. Assume the probability measure $P(1) = 0$, $P(2) = \frac{1}{3}$, and $P(3) = \frac{2}{3}$. Then, the events $\{2\}$ and $\{3\}$ are mutually exclusive, but not independent of each other because $P(2)P(3) = \frac{2}{9} \neq 0 = P(\emptyset)$. On the other hand, the events $\{1\}$ and $\{2\}$ are mutually exclusive and, at the same time, independent of each other. ◇

From $P(A^c) = 1 - P(A)$ and the definition (2.4.1) of conditional probability, we can show the following theorem:

Theorem 2.4.6 *If the events A and B are independent of each other, then A and B^c are also independent of each other,* $P(A|B) = P(A)$, *and* $P(B|A) = P(B)$.

Example 2.4.16 Assume the sample space $S = \{1, 2, \ldots, 6\}$ and probability measure $P(k) = \frac{1}{6}$ for $k = 1, 2, \ldots, 6$. The events $A = \{1, 2, 3\}$ and $B = \{3, 4\}$ are independent of each other as we have already observed in Example 2.4.13. Here, $B^c = \{1, 2, 5, 6\}$ and thus $P(B^c) = \frac{2}{3}$ and $P(AB^c) = P(1, 2) = \frac{1}{3}$. In other words, A and B^c are independent of each other because $P(AB^c) = P(A)P(B^c)$. We also have $P(A|B) = \frac{1}{2} = P(A)$ and $P(B|A) = \frac{1}{3} = P(B)$.   ◇

Definition 2.4.3 (*independence of a number of events*) The events $\{A_i\}_{i=1}^{n}$ are called independent of each other if they satisfy

$$P\left(\bigcap_{i \in J} A_i\right) = \prod_{i \in J} P(A_i) \tag{2.4.19}$$

for every finite subset J of $\{1, 2, \ldots, n\}$.

Example 2.4.17 When A, B, and C are independent of each other with $P(AB) = \frac{1}{3}$, $P(BC) = \frac{1}{6}$, and $P(AC) = \frac{2}{9}$, obtain the probability of C.

Solution First, $P(A) = \frac{2}{3}$ because $P(B)P(C) = \frac{2}{27\{P(A)\}^2} = \frac{1}{6}$ from $P(B) = \frac{1}{3P(A)}$ and $P(C) = \frac{2}{9P(A)}$. Thus, $P(C) = \frac{1}{3}$ from $P(C) = \frac{2}{9P(A)}$.   ◇

A number of events $\{A_i\}_{i=1}^{n}$ with $n = 3, 4, \ldots$ may or may not be independent of each other even when A_i and A_j are independent of each other for every possible pair $\{i, j\}$. When only all pairs of two events are independent, the events $\{A_i\}_{i=1}^{n}$ with $n = 3, 4, \ldots$ are called pairwise independent.

Example 2.4.18 For the sample space $\Omega = \{1, 2, 3, 4\}$ of equally likely outcomes, consider $A_1 = \{1, 2\}$, $A_2 = \{2, 3\}$, and $A_3 = \{1, 3\}$. Then, A_1 and A_2 are independent of each other, A_2 and A_3 are independent of each other, and A_3 and A_1 are independent of each other because $P(A_1) = P(A_2) = P(A_3) = \frac{1}{2}$, $P(A_1 A_2) = P(\{2\}) = \frac{1}{4}$, $P(A_2 A_3) = P(\{3\}) = \frac{1}{4}$, and $P(A_3 A_1) = P(\{1\}) = \frac{1}{4}$. However, A_1, A_2, and A_3 are not independent of each other because $P(A_1 A_2 A_3) = P(\emptyset) = 0$ is not equal to $P(A_1)P(A_2)P(A_3) = \frac{1}{8}$.   ◇

Example 2.4.19 A malfunction of a circuit element does not influence that of another circuit element. Let the probability for a circuit element to function normally be p. Obtain the probability P_S and P_P that the circuit will function normally when n circuit elements are connected in series and in parallel, respectively.

Solution When the circuit elements are connected in series, every circuit element should function normally for the circuit to function normally. Thus, we have

$$P_S = p^n \qquad (2.4.20)$$

On the other hand, the circuit will function normally if at least one of the circuit elements functions normally. Therefore, we get

$$P_P = 1 - (1 - p)^n \qquad (2.4.21)$$

because the complement of the event that at least one of the circuit elements functions normally is the event that all elements are malfunctioning. Note that $1 - (1 - p)^n > p^n$ for $n = 1, 2, \ldots$ when $p \in (0, 1)$.   ◇

2.5   Classes of Probability Spaces

In this section, the notions of probability mass functions and probability density functions (Kim 2010), which are equivalent to the probability measure for the description of a probability space, and are more convenient tools when managing mathematical operations such as differentiation and integration, will be introduced.

2.5.1   Discrete Probability Spaces

In a discrete probability space, in which the sample space Ω is a countable set, we normally assume $\Omega = \{0, 1, \ldots\}$ or $\Omega = \{1, 2, \ldots\}$ with the event space $\mathcal{F} = 2^{\Omega}$.

Definition 2.5.1 (*probability mass function*) In a discrete probability space, a function $p(\omega)$ assigning a real number to each sample point $\omega \in \Omega$ and satisfying

$$p(\omega) \geq 0, \quad \omega \in \Omega \qquad (2.5.1)$$

and

$$\sum_{\omega \in \Omega} p(\omega) = 1 \qquad (2.5.2)$$

is called a probability mass function (pmf), a mass function, or a mass.

From (2.5.1) and (2.5.2), we have

$$p(\omega) \leq 1 \qquad (2.5.3)$$

for every $\omega \in \Omega$.

Example 2.5.1 For the sample space $\Omega = \mathbb{J}_0 = \{0, 1, \ldots\}$ and pmf

$$p(x) = \begin{cases} \frac{1}{3}, & x = 0; \quad c, \ x = 1; \\ 2c, & x = 2; \quad 0, \ \text{otherwise}, \end{cases} \tag{2.5.4}$$

determine the constant c.

Solution From (2.5.2), $\sum_{x \in \Omega} p(x) = \frac{1}{3} + 3c = 1$. Thus, $c = \frac{2}{9}$.   ◇

The probability measure P can be expressed as

$$P(F) = \sum_{\omega \in F} p(\omega) \tag{2.5.5}$$

for $F \in \mathcal{F}$ in terms of the pmf p. Conversely, the pmf p can be written as

$$p(\omega) = P(\{\omega\}) \tag{2.5.6}$$

in terms of the probability measure P.

Note that a pmf is defined for sample points and the probability measure for events. Both the probability measure P and pmf p can be used to describe the randomness of the outcomes of an experiment. Yet, the pmf is easier than the probability measure to deal with, especially when mathematical operations such as sum and difference are involved.

Some of the typical examples of pmf are provided in the examples below.

Example 2.5.2 For the sample space $\Omega = \{x_1, x_2\}$ and a number $\alpha \in (0, 1)$, the function

$$p(x) = \begin{cases} 1 - \alpha, & x = x_1, \\ \alpha, & x = x_2 \end{cases} \tag{2.5.7}$$

is called a two-point pmf.   ◇

Definition 2.5.2 (*Bernoulli trial*) An experiment with two possible outcomes, i.e., an experiment for which the sample space has two elements, is called a Bernoulli experiment or a Bernoulli trial.

Example 2.5.3 When $x_1 = 0$ and $x_2 = 1$ in the two-point pmf (2.5.7), we have

$$p(x) = \begin{cases} 1 - \alpha, & x = 0, \\ \alpha, & x = 1, \end{cases} \tag{2.5.8}$$

which is called the binary pmf or Bernoulli pmf. The binary distribution is usually denoted by $b(1, \alpha)$, where 1 signifies the number of Bernoulli trial and α represents the probability of the desired event or success.   ◇

Example 2.5.4 In the experiment of rolling a fair die, assume the events $A = \{1, 2, 3, 4\}$ and $A^c = \{5, 6\}$. Then, if we choose A as the desired event, the distribution of A is $b\left(1, \frac{2}{3}\right)$. ◇

Example 2.5.5 When the sample space is $\Omega = \mathbb{J}_n = \{0, 1, \ldots, n-1\}$, the pmf

$$p(k) = \frac{1}{n}, \quad k \in \mathbb{J}_n \tag{2.5.9}$$

is called a uniform pmf. ◇

Example 2.5.6 For the sample space $\Omega = \{1, 2, \ldots\}$ and a number $\alpha \in (0, 1)$, the pmf

$$p(k) = (1 - \alpha)^k \alpha, \quad k \in \Omega \tag{2.5.10}$$

is called a geometric pmf. The distribution represented by the geometric pmf (2.5.10) is called the geometric distribution with parameter α and denoted by $Geom(\alpha)$. ◇

When a Bernoulli trial with probability α of success is repeated until the first success, the distribution of the number of failures is $Geom(\alpha)$. In some cases, the function $p(k) = (1 - \alpha)^{k-1} \alpha$ for $k \in \{1, 2, \ldots\}$ with $\alpha \in (0, 1)$ is called the geometric pmf. In such a case, the distribution of the number of repetitions is $Geom(\alpha)$ when a Bernoulli trial with probability α of success is repeated until the first success.

Example 2.5.7 Based on the binary pmf discussed in Example 2.5.3, let us introduce the binomial pmf. Consider the sample space $\Omega = \mathbb{J}_{n+1} = \{0, 1, \ldots, n\}$ and a number $\alpha \in (0, 1)$. Then, the function

$$p(k) = {}_nC_k \alpha^k (1 - \alpha)^{n-k}, \quad k \in \mathbb{J}_{n+1} \tag{2.5.11}$$

is called a binomial pmf and the distribution is denoted by $b(n, \alpha)$. ◇

In (2.5.11), the number ${}_nC_r = \frac{n!}{(n-r)!r!}$, also denoted by $\binom{n}{r}$, is the coefficient of $x^r y^{n-r}$ in the expansion of $(x + y)^n$, and thus called the binomial coefficient as we have described in (1.4.60). Figure 2.7 shows the envelopes of binomial pmf for some values of n when $\alpha = 0.4$. The binomial pmf is discussed in more detail in Sect. 3.5.2.

Example 2.5.8 For the sample space $\Omega = \mathbb{J}_0 = \{0, 1, \ldots\}$ and a number $\lambda \in (0, \infty)$, the function

$$p(k) = e^{-\lambda} \frac{\lambda^k}{k!}, \quad k \in \mathbb{J}_0 \tag{2.5.12}$$

is called a Poisson pmf and the distribution is denoted by $P(\lambda)$. ◇

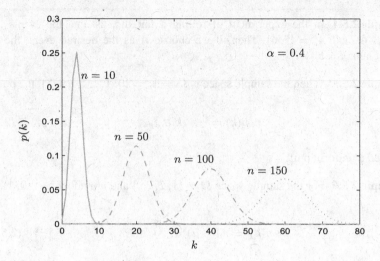

Fig. 2.7 Envelopes of binomial pmf

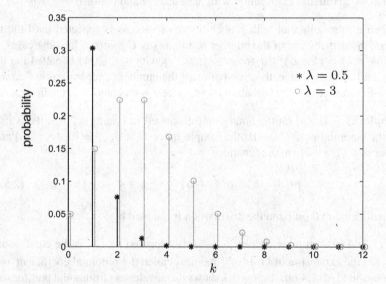

Fig. 2.8 Poisson pmf (for $\lambda = 0.5$, $p(0) = \frac{1}{\sqrt{e}} \approx 0.61$)

For the Poisson pmf (2.5.12), recollecting $\frac{p(k+1)}{p(k)} = \frac{\lambda}{k+1}$, we have $p(0) \le p(1) \le \cdots \le p(\lambda - 1) = p(\lambda) \ge p(\lambda + 1) \ge p(\lambda + 2) \ge \cdots$ when λ is an integer, and $p(0) \le p(1) \le \cdots \le p(\lfloor \lambda \rfloor - 1) \le p(\lfloor \lambda \rfloor)$ and $p(\lfloor \lambda \rfloor) \ge p(\lfloor \lambda \rfloor + 1) \ge \cdots$ when λ is not an integer, where the floor function $\lfloor x \rfloor$ is defined following (1.A.44) in Appendix 1.2. Figure 2.8 shows two examples of Poisson pmf. The Poisson pmf will be discussed in more detail in Sect. 3.5.3.

Example 2.5.9 For the sample space $\Omega = \mathbb{J}_0 = \{0, 1, \ldots\}$, $r \in (0, \infty)$, and $\alpha \in (0, 1)$, the function

$$p(x) = {}_{-r}C_x \alpha^r (\alpha - 1)^x, \quad x \in \mathbb{J}_0 \qquad (2.5.13)$$

is called a negative binomial (NB) pmf, and the distribution with the pmf (2.5.13) is denoted by $NB(r, \alpha)$. $\diamond$

When $r = 1$, the NB pmf (2.5.13) is the geometric pmf discussed in Example 2.5.6. The NB pmf with r a natural number and a real number is called the Pascal pmf and Polya pmf, respectively.

The meaning of $NB(r, \alpha)$ and the formula of the NB pmf vary depending on whether the sample space is $\{0, 1, \ldots\}$ or $\{r, r + 1, \ldots\}$, whether r represents a success or a failure, or whether α is the probability of success or failure. In (2.5.13), the parameters r and α represent the number and probability of success, respectively. When a Bernoulli trial with the probability α of success is repeated until the r-th success, the distribution of the number of repetitions is $NB(r, \alpha)$.

We clearly have $\sum_{x=0}^{\infty} p(x) = 1$ because $\sum_{x=0}^{\infty} {}_{-r}C_x (\alpha - 1)^x = (1 + \alpha - 1)^{-r} = \alpha^{-r}$ from (1.A.12) with $p = -r$ and $z = \alpha - 1$. Now, the pmf (2.5.13) can be written as $p(x) = {}_{r+x-1}C_x \alpha^r (1 - \alpha)^x$ or, equivalently, as

$$p(x) = {}_{r+x-1}C_{r-1} \alpha^r (1 - \alpha)^x, \quad x \in \mathbb{J}_0 \qquad (2.5.14)$$

using[13] ${}_{-r}C_x = \frac{1}{x!}(-r)(-r - 1) \cdots (-r - x + 1)$, i.e.,

$$ {}_{-r}C_x = (-1)^x {}_{r+x-1}C_x. \qquad (2.5.15)$$

Note that we have

$$\sum_{x=0}^{\infty} {}_{r+x-1}C_x (1 - \alpha)^x = \alpha^{-r} \qquad (2.5.16)$$

because $\sum_{x=0}^{\infty} {}_{r+x-1}C_x (1 - \alpha)^x = \sum_{x=0}^{\infty} \frac{(r+x-1)!}{(r-1)!x!} (1 - \alpha)^x = \sum_{x=0}^{\infty} {}_{r+x-1}C_{r-1} (1 - \alpha)^x$ and $\sum_{x=0}^{\infty} p(x) = 1$. Letting $x + r = y$ in (2.5.14), we get

$$p(y) = {}_{y-1}C_{r-1} \alpha^r (1 - \alpha)^{y-r}, \quad y = r, r + 1, \ldots \qquad (2.5.17)$$

when r is a natural number, which is called the NB pmf sometimes. Here, note that ${}_{x+r-1}C_x|_{x=y-r} = {}_{y-1}C_{y-r} = {}_{y-1}C_{r-1}$.

---

[13] Here, $(-r)(-r - 1) \cdots (-r - x + 1) = (-1)^x r(r + 1) \cdots (r + x - 1)$. Equation (2.5.15) can also be obtained based on Table 1.4.

2.5.2   Continuous Probability Spaces

Let us now consider the continuous probability space with the measurable space $(\Omega, \mathcal{F}) = (\mathbb{R}, \mathcal{B}(\mathbb{R}))$: in other words, the sample space Ω is the set $\mathbb{R}$ of real numbers and the event space is the Borel field $\mathcal{B}(\mathbb{R})$.

Definition 2.5.3 (*probability density function*) In a measurable space $(\mathbb{R}, \mathcal{B}(\mathbb{R}))$, a real-valued function f, with the two properties

$$f(r) \geq 0, \quad r \in \Omega \tag{2.5.18}$$

and

$$\int_{\Omega} f(r)dr = 1 \tag{2.5.19}$$

is called a probability density function (pdf), a density function, or a density.

Example 2.5.10 Determine the constant c when the pdf is $f(x) = \frac{1}{4}$, c, and 0 for $x \in [0, 1)$, $[1, 2)$, and $[0, 2)^c$, respectively.

Solution From (2.5.19), we have $\int_{-\infty}^{\infty} f(r)dr = \int_{0}^{1} \frac{1}{4}dr + \int_{1}^{2} c \, dr = \frac{1}{4} + c = 1$. Thus, $c = \frac{3}{4}$. $\diamond$

The value $f(x)$ of a pdf f does not represent the probability $\mathsf{P}(\{x\})$. Instead, the set function P defined in terms of f as

$$\mathsf{P}(F) = \int_{F} f(r)dr, \quad F \in \mathcal{B}(\mathbb{R}) \tag{2.5.20}$$

is the probability measure of the probability space on which f is defined. Note that (2.5.20) is a counterpart of (2.5.5). While we have (2.5.6), an equation describing the pmf in terms of the probability measure in the discrete probability space, we do not have its counterpart in the continuous probability space, which would describe the pdf in terms of the probability measure.

Do the integrals in (2.5.19) and (2.5.20) have any meaning? For interval events or finite unions of interval events, we can adopt the Riemann integral as in most engineering problems and calculations. On the other hand, the Riemann integral has some caveats including that the order of the limit and integral for a sequence of functions is not interchangeable. In addition, the Riemann integral is not defined in some cases. For example, when

$$f(r) = \begin{cases} 1, r \in [0, 1], \\ 0, \text{otherwise}, \end{cases} \tag{2.5.21}$$

it is not possible to obtain the Riemann integral of $f(r)$ over the set $F = \{r : r \text{ is an irrational number}, r \in [0, 1]\}$. Fortunately, such a caveat can be overcome by adopting the Lebesgue integral. Compared to the Riemann integral, the Lebesgue integral has the following three important advantages:

(1) The Lebesgue integral is defined for any Borel set.
(2) The order of the limit and integral can almost always be interchanged in the Lebesgue integral.
(3) When a function is Riemann integrable, it is also Lebesgue integrable, and the results are known to be the same.

Like the pmf, the pdf is defined on the points in the sample space, not on the events. On the other hand, unlike the pmf $p(\cdot)$ for which $p(\omega)$ directly represents the probability $P(\{\omega\})$, the value $f(x_0)$ at a point x_0 of the pdf $f(x)$ is not the probability at $x = x_0$. Instead, $f(x_0) \, dx$ represents the probability for the arbitrarily small interval $[x_0, x_0 + dx)$. While the value of a pmf cannot be larger than 1 at any point, the value of a pdf can be larger than 1 at some points. In addition, the probability of a countable event is 0 even when the value of the pdf is not 0 in the continuous space: for the pdf

$$f(x) = \begin{cases} 2, & x \in [0, 0.5], \\ 0, & \text{otherwise}, \end{cases} \tag{2.5.22}$$

we have $P(\{a\}) = 0$ for any point $a \in [0, 0.5]$. On the other hand, if we assume a very small interval around a point, the probability of that interval can be expressed as the product of the value of the pdf and the length of the interval. For example, for a pdf f with $f(3) = 4$ the probability $P([3, 3 + dx))$ of an arbitrarily small interval $[3, 3 + dx)$ near 3 is

$$f(3)dx = 4dx. \tag{2.5.23}$$

This implies that, as we can obtain the probability of an event by adding the probability mass over all points in the event in discrete probability spaces, we can obtain the probability of an event by integrating the probability density over all points in the event in continuous probability spaces.

Some of the widely-used pdf's are shown in the examples below.

Example 2.5.11 When $a < b$, the pdf

$$f(r) = \frac{1}{b - a} u(r - a)u(b - r) \tag{2.5.24}$$

shown in Fig. 2.9 is called a uniform pdf or a rectangular pdf, and its distribution is denoted by[14] $U(a, b)$.   ◇

The probability measure of $U[0, 1]$ is often called the Lebesgue measure.

---

[14] Notations $U[a, b]$, $U[a, b)$, $U(a, b]$, and $U(a, b)$ are all used interchangeably.

Fig. 2.9 The uniform pdf

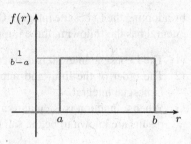

Fig. 2.10 The exponential pdf

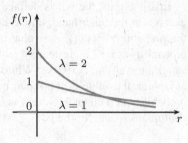

Example 2.5.12 (Romano and Siegel 1986) Countable sets are all of Lebesgue measure 0. Some uncountable sets such as the Cantor set $\mathbb{C}$ described in Example 1.1.46 are also of Lebesgue measure 0. ◇

Example 2.5.13 The pdf

$$f(r) = \lambda e^{-\lambda r} u(r) \tag{2.5.25}$$

shown in Fig. 2.10 is called an exponential pdf with $\lambda > 0$ called the rate of the pdf. The exponential pdf with $\lambda = 1$ is called the standard exponential pdf. The exponential pdf will be discussed again in Sect. 3.5.4. ◇

Example 2.5.14 The pdf

$$f(r) = \frac{\lambda}{2} e^{-\lambda|r|} \tag{2.5.26}$$

with $\lambda > 0$, shown in Fig. 2.11, is called a Laplace pdf or a double exponential pdf, and its distribution is denoted by $L(\lambda)$. ◇

Example 2.5.15 The pdf

$$f(r) = \frac{1}{\sqrt{2\pi\sigma^2}} \exp\left\{-\frac{(r-m)^2}{2\sigma^2}\right\} \tag{2.5.27}$$

shown in Fig. 2.12 is called a Gaussian pdf or a normal pdf, and its distribution is denoted by $\mathcal{N}(m, \sigma^2)$. ◇

Fig. 2.11 The Laplace pdf

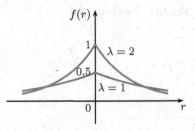

Fig. 2.12 The normal pdf

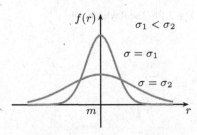

When $m = 0$ and $\sigma^2 = 1$, the normal pdf is called the standard normal pdf. The normal distribution is sometimes called the Gauss-Laplace distribution, de Moivre-Laplace distribution, or the second Laplace distribution (Lukacs 1970). The normal pdf will be addressed again in Sect. 3.5.1 and its generalizations into multidimensional spaces in Chap. 5.

Example 2.5.16 For a positive number α and a real number β, the function

$$f(r) = \frac{\alpha}{\pi} \frac{1}{(r - \beta)^2 + \alpha^2} \tag{2.5.28}$$

shown in Fig. 2.13 is called a Cauchy pdf and the distribution is denoted by $C(\beta, \alpha)$.
◇

The Cauchy pdf is also called the Lorentz pdf or Breit-Wigner pdf. We will mostly consider the case $\beta = 0$, with the notation $C(\alpha)$ in this book.

Example 2.5.17 The pdf

$$f(r) = \frac{r}{\alpha^2} \exp\left(-\frac{r^2}{2\alpha^2}\right) u(r) \tag{2.5.29}$$

shown in Fig. 2.14 is called a Rayleigh pdf.
◇

Example 2.5.18 When $f(v) = av \exp\left(-v^2\right) u(v)$ is a pdf, obtain the value of a.

Solution From $\int_{-\infty}^{\infty} f(v)dv = a \int_0^{\infty} v \exp\left(-v^2\right) dv = \frac{a}{2} = 1$, we get $a = 2$. ◇

Fig. 2.13 The Cauchy pdf

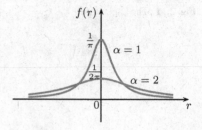

Fig. 2.14 The Rayleigh pdf

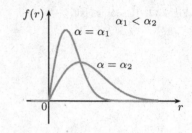

Fig. 2.15 The logistic pdf

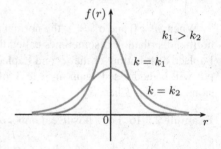

Example 2.5.19 The pdf (Balakrishnan 1992)

$$f(r) = \frac{ke^{-kr}}{\left(1 + e^{-kr}\right)^2} \tag{2.5.30}$$

shown in Fig. 2.15 is called a logistic pdf, where $k > 0$.   ◇

Example 2.5.20 The pdf

$$f(r) = \frac{1}{\beta^\alpha \Gamma(\alpha)} r^{\alpha-1} \exp\left(-\frac{r}{\beta}\right) u(r) \tag{2.5.31}$$

shown in Fig. 2.16 is called a gamma pdf and the distribution is denoted by $G(\alpha, \beta)$, where $\alpha > 0$ and $\beta > 0$. It is clear from (2.5.25) and (2.5.31) that the gamma pdf with $\alpha = 1$ is the same as an exponential pdf.   ◇

Fig. 2.16 The gamma pdf

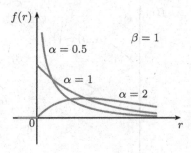

Example 2.5.21 The pdf

$$f(r) = \frac{r^{\alpha-1}(1-r)^{\beta-1}}{\tilde{B}(\alpha,\beta)} u(r)u(1-r) \tag{2.5.32}$$

shown in Fig. 2.17 is called a beta pdf and the distribution is denoted by $B(\alpha, \beta)$, where $\alpha > 0$ and $\beta > 0$. ◇

In (2.5.32), $\tilde{B}(\alpha, \beta)$ is the beta function described in (1.4.95). Unless a confusion arises regarding the beta function $\tilde{B}(\alpha, \beta)$ and the beta distribution $B(\alpha, \beta)$, we often use $B(\alpha, \beta)$ for both the beta function and beta distribution.

Table 2.1 shows some general properties of the beta pdf $f(r)$ shown in (2.5.32). When $\alpha = 1$ and $\beta > 1$, the pdf $f(r)$ is decreasing in $(0, 1)$, $f(0) = \beta$, and $f(1) = 0$. When $\alpha > 1$ and $\beta = 1$, the pdf $f(r)$ is increasing in $(0, 1)$, $f(0) = 0$, and $f(1) = \alpha$. In addition, because

$$f'(r) = r^{\alpha-2}(1-r)^{\beta-2}\{\alpha - 1 - (\alpha + \beta - 2)r\}u(r)u(1-r), \tag{2.5.33}$$

the pdf $f(r)$ increases and decreases in $\left(0, \frac{\alpha-1}{\alpha+\beta-2}\right)$ and $\left(\frac{\alpha-1}{\alpha+\beta-2}, 1\right)$, respectively, and $f(0) = f(1) = 0$ when $\alpha > 1$ and $\beta > 1$. In other words, when $\alpha > 1$ and $\beta > 1$, the pdf $f(r)$ is a unimodal function, and has its maximum at $r = \frac{\alpha-1}{\alpha+\beta-2}$ between $\frac{1}{2}$ and 1 if $\alpha > \beta$; at $r = \frac{1}{2}$ if $\alpha = \beta$; and at $r = \frac{\alpha-1}{\alpha+\beta-2}$ between 0 and $\frac{1}{2}$ if $\alpha < \beta$. The maximum point is closer to 0 when α is closer to 1 or when β is larger, and it is closer to 1 when β is closer to 1 or when α is larger. Such a property of the beta pdf can be used in the order statistics (David and Nagaraja 2003) of discrete distributions.

Example 2.5.22 The pdf of the distribution $B\left(\frac{1}{2}, \frac{1}{2}\right)$ is

$$f(r) = \frac{1}{\pi\sqrt{r(1-r)}} u(r)u(1-r). \tag{2.5.34}$$

Letting $r = \cos^2 v$, we have $\int_0^1 f(r)dr = \int_{\frac{\pi}{2}}^0 \frac{-2\cos v \sin v}{\pi \cos v \sin v} dv = 1$. The pdf (2.5.34) is also called the inverse sine pdf. ◇

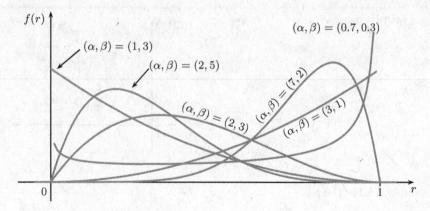

Fig. 2.17 The beta pdf

Table 2.1 Characteristics of a beta pdf $f(r), 0 < r < 1$

	$0 < \alpha < 1$	$\alpha = 1$	$\alpha > 1$
$0 < \beta < 1$	Dereasing and then increasing	Increasing function $f(0) = \beta$.	Increasing function
$\beta = 1$	Decreasing function $f(1) = \alpha$	$f(r) = 1$	Increasing function $f(0) = 0, f(1) = \alpha$
$\beta > 1$	Decreasing function	Decreasing function $f(0) = \beta, f(1) = 0$	Increasing and then decreasing $f(0) = 0, f(1) = 0$

2.5.3 Mixed Spaces

Let $\{P_i\}_{i=1}^{\infty}$ be probability measures on a common measurable space $(\Omega, \mathcal{F})$ and $\{a_i\}_{i=1}^{\infty}$ be non-negative numbers such that $\sum_{i=1}^{\infty} a_i = 1$. Then, the set function

$$P(A) = \sum_{i=1}^{\infty} a_i P_i(A) \tag{2.5.35}$$

is also a probability measure on $(\Omega, \mathcal{F})$. When some of $\{P_i\}_{i=1}^{\infty}$ are discrete while others are continuous, the probability measure (2.5.35) is called a mixed probability measure. An important example of the mixed probability measure is the sum

$$P(A) = \lambda \sum_{x \in A_d} p(x) + (1 - \lambda) \int_{x \in A_c} f(x) dx \tag{2.5.36}$$

of a continuous probability measure and a discrete probability measure, where $0 < \lambda < 1$, A_d is a discrete event, A_c is a continuous event, $A = A_d \cup A_c$, f is a pdf, and p is a pmf.

Example 2.5.23 (Thomas 1986) Consider a distribution obtained by combining a point mass of $\frac{1}{4}$ at $r = \frac{1}{2}$, a point mass of $\frac{1}{2}$ at $r = \frac{3}{4}$, and a unform density for $r \in [0, 1]$. The distribution can then be described by the pdf

$$f(r) = \frac{1}{4}\delta\left(r - \frac{1}{2}\right) + \frac{1}{2}\delta\left(r - \frac{3}{4}\right) + \frac{1}{4}, \quad r \in [0, 1]. \qquad (2.5.37)$$

This example implies that a pdf can be defined also for discrete and mixed spaces by using impulse functions. $\diamond$

Example 2.5.24 The probability space with the pmf

$$p(r) = \begin{cases} \frac{1}{2}, r = 0; & \frac{1}{3}, r = 1; \\ \frac{1}{6}, r = 2; & 0, \text{ otherwise} \end{cases} \qquad (2.5.38)$$

can also be represented by the pdf

$$f(r) = \frac{1}{2}\delta(r) + \frac{1}{3}\delta(r - 1) + \frac{1}{6}\delta(r - 2). \qquad (2.5.39)$$

Here, for example, $p(0) = \int_{0-}^{0+} f(x)dx = \frac{1}{2}$. $\diamond$

Note that what Example 2.5.24 implies is not that the pmf (2.5.38) is the same as the pdf (2.5.39), but that a discrete probability space can be expressed in terms of both a pmf and a pdf. If we have

$$\int_{a^-}^{a^+} f(x)dx = p(a) \qquad (2.5.40)$$

for an integer a, then the pdf $f(r)$ expressed in terms of impulse functions and the pmf $p(r)$ represent the same probability space. In other words, to check whether a pmf $p(r)$ and a pdf $f(r)$ represent the same probability space or not, we are required to check whether the pmf $p(r)$ and the pdf $f(r)$ satisfy (2.5.40).

Appendices

Appendix 2.1 Continuity of Probability

Theorem 2.A.1 *For a monotonic sequence* $\{B_n\}_{n=1}^{\infty}$ *of events, the probability of the limit event is equal to the limit of the probabilities of the events in the sequence. In other words,*

$$P\left(\lim_{n\to\infty} B_n\right) = \lim_{n\to\infty} P(B_n) \qquad (2.A.1)$$

holds true.

Proof First, when $\{B_n\}_{n=1}^{\infty}$ is a non-decreasing sequence, recollect that $\overset{n}{\underset{i=1}{\cup}} B_i = B_n$ and $\overset{\infty}{\underset{i=1}{\cup}} B_i = \lim_{n\to\infty} B_n$. Consider a sequence $\{F_i\}_{i=1}^{\infty}$ such that $F_1 = B_1$ and $F_n = B_n - \overset{n-1}{\underset{i=1}{\cup}} B_i = B_n \cap B_{n-1}^c$ for $n = 2, 3, \ldots$. Then, $\{F_n\}_{n=1}^{\infty}$ are all mutually exclusive, $\overset{n}{\underset{i=1}{\cup}} F_i = \overset{n}{\underset{i=1}{\cup}} B_i$ for any natural number n, and $\overset{\infty}{\underset{i=1}{\cup}} F_i = \overset{\infty}{\underset{i=1}{\cup}} B_i = \lim_{n\to\infty} B_n$. Therefore, $P\left(\lim_{n\to\infty} B_n\right) = P\left(\overset{\infty}{\underset{i=1}{\cup}} B_i\right) = P\left(\overset{\infty}{\underset{i=1}{\cup}} F_i\right) = \overset{\infty}{\underset{i=1}{\sum}} P(F_i) = \lim_{n\to\infty} \overset{n}{\underset{i=1}{\sum}} P(F_i)$ $= \lim_{n\to\infty} P\left(\overset{n}{\underset{i=1}{\cup}} F_i\right) = \lim_{n\to\infty} P\left(\overset{n}{\underset{i=1}{\cup}} B_i\right)$, i.e.,

$$P\left(\lim_{n\to\infty} B_n\right) = \lim_{n\to\infty} P(B_n) \qquad (2.A.2)$$

recollecting (2.2.15), Axiom 4 of probability, and $\overset{n}{\underset{i=1}{\cup}} B_i = B_n$.

Next, when $\{B_n\}_{n=1}^{\infty}$ is a non-increasing sequence, $\left\{B_n^c\right\}_{n=1}^{\infty}$ is a non-decreasing sequence, and thus, we have

$$P\left(\lim_{n\to\infty} B_n^c\right) = \lim_{n\to\infty} P\left(B_n^c\right) \qquad (2.A.3)$$

from (2.A.2). Noting that $\lim_{n\to\infty} B_n^c = \overset{\infty}{\underset{i=1}{\cup}} B_i^c$ because $\left\{B_n^c\right\}_{n=1}^{\infty}$ is a non-decreasing sequence and that $\overset{\infty}{\underset{i=1}{\cap}} B_i = \lim_{n\to\infty} B_n$ because $\{B_n\}_{n=1}^{\infty}$ is a non-increasing sequence, we have $\lim_{n\to\infty} B_n^c = \overset{\infty}{\underset{i=1}{\cup}} B_i^c = \left(\overset{\infty}{\underset{i=1}{\cap}} B_i\right)^c = \left(\lim_{n\to\infty} B_n\right)^c$. Thus the left-hand side of (2.A.3) can be written as $P\left(\lim_{n\to\infty} B_n^c\right) = 1 - P\left(\lim_{n\to\infty} B_n\right)$. Meanwhile, the right-hand side of (2.A.3) can easily be written as $\lim_{n\to\infty} P\left(B_n^c\right) = \lim_{n\to\infty} \{1 - P(B_n)\} = 1 - \lim_{n\to\infty} P(B_n)$. Then, (2.A.3) yields (2.A.1). ♠

The results of Theorem 2.A.1 that

$$\lim_{n\to\infty} P(B_n) = P\left(\overset{\infty}{\underset{i=1}{\cup}} B_i\right) \qquad (2.A.4)$$

for a non-decreasing sequence $\{B_n\}_{n=1}^{\infty}$ and that

$$\lim_{n\to\infty} P(B_n) = P\left(\bigcap_{i=1}^{\infty} B_i\right) \tag{2.A.5}$$

for a non-increasing sequence $\{B_n\}_{n=1}^{\infty}$ are called the continuity from below and above of probability, respectively.

Theorem 2.A.1 deals with monotonic, i.e., non-decreasing and non-increasing, sequences. The same result holds true more generally as we can see in the following theorem:

Theorem 2.A.2 *When the limit event* $\lim_{n\to\infty} B_n$ *of a sequence* $\{B_n\}_{n=1}^{\infty}$ *exists, the probability of the limit event is equal to the limit of the probabilities of the events in the sequence. In other words,*

$$P\left(\lim_{n\to\infty} B_n\right) = \lim_{n\to\infty} P(B_n) \tag{2.A.6}$$

holds true.

Proof First, recollect that, among the limit values of a sequence $\{a_n\}_{n=1}^{\infty}$ of real numbers, the largest and smallest ones are denoted by $\overline{a_n}$ and $\underline{a_n}$, respectively. When $\underline{a_n} = \overline{a_n}$, this value is called the limit of the sequence and denoted by $\lim_{n\to\infty} a_n$. Now, noting that $\left\{\bigcup_{k=n}^{\infty} B_k\right\}_{n=1}^{\infty}$ is a non-increasing sequence, we have $P\left(\limsup_{n\to\infty} B_n\right) = P\left(\bigcap_{n=1}^{\infty} \bigcup_{k=n}^{\infty} B_k\right)$, i.e.,

$$P\left(\limsup_{n\to\infty} B_n\right) = \lim_{n\to\infty} P\left(\bigcup_{k=n}^{\infty} B_k\right) \tag{2.A.7}$$

from (1.5.9), (1.5.17), and (2.A.1). In the meantime, we have

$$\overline{P(B_n)} \leq \lim_{n\to\infty} P\left(\bigcup_{k=n}^{\infty} B_k\right) \tag{2.A.8}$$

because $P(B_n) \leq P\left(\bigcup_{k=n}^{\infty} B_k\right)$ from $B_n \subseteq \bigcup_{k=n}^{\infty} B_k$. From (2.A.7) and (2.A.8), we get

$$\overline{P(B_n)} \leq P\left(\limsup_{n\to\infty} B_n\right). \tag{2.A.9}$$

Similarly, we get

$$P\left(\liminf_{n\to\infty} B_n\right) = P\left(\bigcup_{n=1}^{\infty} \bigcap_{k=n}^{\infty} B_k\right)$$

$$= \lim_{n\to\infty} P\left(\bigcap_{k=n}^{\infty} B_k\right)$$

$$\leq \underline{P(B_n)} \tag{2.A.10}$$

for the non-decreasing sequence $\left\{\bigcap_{k=n}^{\infty} B_k\right\}_{n=1}^{\infty}$. The last line

$$\lim_{n\to\infty} P\left(\bigcap_{k=n}^{\infty} B_k\right) \leq \underline{P(B_n)} \tag{2.A.11}$$

of (2.A.10) is due to $P\left(\bigcap_{k=n}^{\infty} B_k\right) \leq P(B_n)$ from $\bigcap_{k=n}^{\infty} B_k \subseteq B_n$. Now, (2.A.9) and (2.A.10) produces

$$\overline{P(B_n)} \leq P\left(\limsup_{n\to\infty} B_n\right)$$

$$= P\left(\lim_{n\to\infty} B_n\right)$$

$$= P\left(\liminf_{n\to\infty} B_n\right)$$

$$\leq \underline{P(B_n)} \tag{2.A.12}$$

if $\lim_{n\to\infty} B_n$ exists. We get the desired result

$$P\left(\lim_{n\to\infty} B_n\right) = \overline{P(B_n)}$$

$$= \underline{P(B_n)}$$

$$= \lim_{n\to\infty} P(B_n) \tag{2.A.13}$$

by combining (2.A.12) and $\overline{P(B_n)} \geq \underline{P(B_n)}$.   ♠

Definition 2.A.1 (*continuity of probability*) For the limit $\lim_{n\to\infty} B_n$ of a sequence $\{B_n\}_{n=1}^{\infty}$ of events, $\{P(B_n)\}_{n=1}^{\infty}$ converges to $P\left(\lim_{n\to\infty} B_n\right)$ as shown in (2.A.1) and (2.A.6). The relation

$$P\left(\lim_{n\to\infty} B_n\right) = \lim_{n\to\infty} P(B_n) \tag{2.A.14}$$

is called the continuity of probability.

In other words, the probability of the limit of a sequence of events is equal to the limit of the sequence of the probabilities of the events.

Appendix 2.2 Borel-Cantelli Lemma

Let us discuss the Borel-Cantelli lemma, which deals with the probability of upper bound events.

Theorem 2.A.3 *(Rohatgi and Saleh 2001) When the sum of the probabilities* $\{P(B_n)\}_{n=1}^{\infty}$ *of a sequence* $\{B_n\}_{n=1}^{\infty}$ *of events is finite, i.e., when* $\sum_{n=1}^{\infty} P(B_n) < \infty$, *the probability* $P(\overline{B_n})$ *of the upper bound of* $\{B_n\}_{n=1}^{\infty}$ *is 0.*

Proof First, from $\sum_{k=1}^{\infty} P(B_k) = \lim_{n \to \infty} \left\{ \sum_{k=1}^{n-1} P(B_k) + \sum_{k=n}^{\infty} P(B_k) \right\} = \sum_{k=1}^{\infty} P(B_k) +$ $\lim_{n \to \infty} \sum_{k=n}^{\infty} P(B_k)$, we get

$$\lim_{n \to \infty} \sum_{k=n}^{\infty} P(B_k) = 0. \tag{2.A.15}$$

Now using (2.A.7) and the Boole inequality (2.3.2), we get $P\left(\lim_{n \to \infty} \sup B_n\right) =$ $\lim_{n \to \infty} P\left(\bigcup_{k=n}^{\infty} B_k\right) \le \lim_{n \to \infty} \sum_{k=n}^{\infty} P(B_k)$, i.e.,

$$P\left(\lim_{n \to \infty} \sup B_n\right) = 0 \tag{2.A.16}$$

from (2.A.15).   ♠

Theorem 2.A.4 *When* $\{B_n\}_{n=1}^{\infty}$ *is a sequence of independent events and the sum* $\sum_{n=1}^{\infty} P(B_n)$ *is infinite, i.e.,* $\sum_{n=1}^{\infty} P(B_n) \to \infty$, *the probability* $P(\overline{B_n})$ *of the upper bound of* $\{B_n\}_{n=1}^{\infty}$ *is 1.*

Proof First, note that $P\left(\lim_{n \to \infty} \sup B_n\right) = \lim_{n \to \infty} P\left(\bigcup_{i=n}^{\infty} B_i\right)$, i.e.,

$$P\left(\lim_{n \to \infty} \sup B_n\right) = \lim_{n \to \infty} \left\{1 - P\left(\bigcap_{i=n}^{\infty} B_i^c\right)\right\} \tag{2.A.17}$$

as in the proof of Theorem 2.A.3. Next, if $\sum_{k=1}^{\infty} P(B_k) \to \infty$, then $\sum_{k=n}^{\infty} P(B_k) \to$ ∞ because $\sum_{k=1}^{n-1} P(B_k) \le n - 1$ for any number n and $\sum_{k=1}^{\infty} P(B_k) = \sum_{k=1}^{n-1} P(B_k) +$

$\sum_{k=n}^{\infty} P(B_k)$. Therefore, we get $P\left(\bigcap_{i=n}^{\infty} B_i^c\right) = \prod_{i=n}^{\infty} P(B_i^c) = \prod_{i=n}^{\infty} \{1 - P(B_i)\}$ recollecting that $\{B_i\}_{i=1}^{\infty}$ are independent of each other, and thus, $\{B_i^c\}_{i=1}^{\infty}$ are independent of each other. Finally, noting that $1 - x \le e^{-x}$ for $x \ge 0$, we get

$$P\left(\bigcap_{i=n}^{\infty} B_i^c\right) \le \prod_{i=n}^{\infty} \exp\{-P(B_i)\}$$

$$= \exp\left\{-\sum_{i=n}^{\infty} P(B_i)\right\}$$

$$= 0, \tag{2.A.18}$$

which proves the theorem when used in (2.A.17).   ♠

When $\{B_n\}_{n=1}^{\infty}$ is a sequence of independent events, the probability $P\left(\overline{B_n}\right)$ of the upper bound event of $\{B_n\}_{n=1}^{\infty}$ is either 0 or 1 from the Borel-Cantelli lemma. Borel-Cantelli lemmas will be employed when we discuss the strong law of large numbers in Sect. 6.2.2.2.

Example 2.A.1 Assume $P(X_n = 0) = \frac{1}{n^2} = 1 - P(X_n = 1)$ for a sequence $\{X_n\}_{n=1}^{\infty}$ of independent events, and let $B_n = \{X_n = 0\}$. Then, from Theorem 2.A.3, we have $P\left(\limsup_{n \to \infty} B_n\right) = P(\text{i.o. } B_n) = 0$ because $\sum_{n=1}^{\infty} P(B_n) = \frac{\pi^2}{6} < \infty$. Therefore, when n is sufficiently large, the probabilities that X_n will be 0 and 1 are 0 and 1, respectively. In other words, $\lim_{n \to \infty} X_n = 1$ almost surely.   ◇

Example 2.A.2 Assume $P(X_n = 0) = \frac{1}{n} = 1 - P(X_n = 1)$ for a sequence $\{X_n\}_{n=1}^{\infty}$ of independent events, and let $B_n = \{X_n = 0\}$. Then, from Theorem 2.A.4, we have $P(\text{i.o. } B_n) = 1$ or, equivalently, almost surely $X_n = 0$ because $\sum_{n=1}^{\infty} P(B_n) = \infty$. On the other hand, B_n^c also occurs almost surely because $\sum_{n=1}^{\infty} P(B_n^c) = \infty$. In other words, almost surely X_n is 0 infinitely many times, and at the same time, 1 infinitely many times. Consequently, the probability that $\lim_{n \to \infty} X_n$ does not exist, i.e., that X_n does not converge, is 1.   ◇

Appendix 2.3 Measures and Lebesgue Integrals

The notion of length, area, volume, and weight that we encounter in our daily lives are examples of measure. The length of a rod, the area of a house, the volume of a ball, and the weight of a package assign numbers to objects. They also assign numbers to groups of objects.

A measure is a set function assigning a number to a set. Nonetheless, not all set functions are measures. A measure should satisfy some conditions. For example, if we consider the measure of weight, the weight of a bottle filled with water is the sum of the weight of the bottle and that of the water. In other words, the measure of the union of sets is equal to the sum of the measures of the sets for mutually exclusive sets.

Definition 2.A.2 (*measure*) A non-negative additive function μ with the domain a σ-algebra is called a measure.

Here, an additive function is a function such that the value of the function for a countable union of sets is the same as the sum of the values of the function for the sets when the sets are mutually exclusive. In other words, a function μ satisfying

$$\mu \left(\bigcup_{i=1}^{\infty} A_i \right) = \sum_{i=1}^{\infty} \mu (A_i) \tag{2.A.19}$$

for countable mutually exclusive sets $\{A_i\}_{i=1}^{\infty}$ in a σ-algebra is called an additive function.

Example 2.A.3 Consider a finite set Ω and the collection $\mathcal{F} = 2^{\Omega}$. Then, the number $\mu(A)$ of elements of $A \in \mathcal{F}$ is a measure. $\diamond$

Theorem 2.A.5 *For a measure μ on a σ-algebra $\mathcal{F}$, let $\{A_n \in \mathcal{F}\}_{n=1}^{\infty}$ and $A_1 \subseteq A_2 \subseteq \cdots$. Then, $A = \bigcup_{n=1}^{\infty} A_n \in \mathcal{F}$ and* [15] $\lim_{n \to \infty} \mu (A_n) = \mu(A)$.

Proof First, because $\mathcal{F}$ is a σ-algebra, $A = \bigcup_{n=1}^{\infty} A_n$ is an element of $\mathcal{F}$. Next, let $B_1 = A_1$ and $B_n = A_n - A_{n-1}$ for $n = 2, 3, \ldots$. Then, $\{B_n\}_{n=1}^{\infty}$ are mutually exclusive, $A_n = \bigcup_{i=1}^{n} B_i$, and $A = \bigcup_{n=1}^{\infty} B_n$. Thus, $\lim_{n \to \infty} \mu (A_n) = \lim_{n \to \infty} \mu \left(\bigcup_{i=1}^{n} B_i \right) = \lim_{n \to \infty} \sum_{i=1}^{n} \mu (B_i) = \sum_{i=1}^{\infty} \mu (B_i)$ and $\mu(A) = \mu \left(\bigcup_{i=1}^{\infty} B_i \right) = \sum_{i=1}^{\infty} \mu (B_i)$ from (2.A.19) and, consequently, $\lim_{n \to \infty} \mu (A_n) = \mu(A)$. ♠

The measure for a subset A of Ω can be defined as $\mu(A) = \sum_{\omega \in A} \mu_{\omega}$ in an abstract space Ω by first choosing arbitrarily a non-negative number μ_{ω} for $\omega \in \Omega$ when Ω is a countable set.

Example 2.A.4 For an abstract space $\Omega = \{3, 4, 5\}$, let $\mu_{\omega} = 5 - \omega$ for $\omega \in \Omega$. Then, $\mu(A) = \sum_{\omega \in A} \mu_{\omega}$ is a measure. We have $\mu(\{3\}) = 2$, $\mu(\{4\}) = 1$, $\mu(\{5\}) =$

---

[15] Note that $\lim_{n \to \infty} A_n = \bigcup_{n=1}^{\infty} A_n$ and $\lim_{n \to \infty} A_n = \bigcap_{n=1}^{\infty} A_n$ when $A_1 \subseteq A_2 \subseteq \cdots$ and $A_1 \supseteq A_2 \supseteq \cdots$, respectively, as discussed in (1.5.8) and (1.5.9).

$0, \mu(\{3, 4\}) = \mu(\{3\}) + \mu(\{4\}) = 3, \mu(\{3, 5\}) = \mu(\{3\}) + \mu(\{5\}) = 2, \mu(\{4, 5\}) = \mu(\{4\}) + \mu(\{5\}) = 1$, and $\mu(\{3, 4, 5\}) = \mu(\{3\}) + \mu(\{4\}) + \mu(\{5\}) = \mu(\{3, 4\}) + \mu(\{5\}) = \mu(\{4, 5\}) + \mu(\{3\}) = \mu(\{3, 5\}) + \mu(\{4\}) = 3$.   ◇

To consider measure in uncountable sets, we introduce the notion of elementary sets, based on which the measure is defined and then extended for general sets.

Definition 2.A.3 (*rectangle*) In a Euclidean space $\mathbb{R}^p$ of p dimension, a set in the form $\{x = (x_1, x_2, \ldots, x_p) : a_i \leq x_i \leq b_i, i = 1, 2, \ldots, p\}$ with $-\infty < a_i \leq b_i \leq \infty$ is called a rectangle, an interval, or a box.

In Definition 2.A.3, $a_i \leq x_i \leq b_i$ can be replaced with $a_i < x_i < b_i$ or $a_i < x_i \leq b_i$. Note that a null set is also regarded as an interval.

Definition 2.A.4 (*elementary set*) A set is called an elementary set if it can be expressed as the union of a finite number of intervals.

Example 2.A.5 Examples of an elementary set and a non-elementary set are shown in Fig. 2.18.   ◇

Definition 2.A.5 (*outer measure; covering*) Let μ be an additive, non-negative, and finite set function defined on the collection of all elementary sets. A collection $\{A_i\}_{i=1}^{\infty}$ of elementary sets such that $E \subseteq \overset{\infty}{\underset{i=1}{\cup}} A_i$ for $E \subseteq \mathbb{R}^p$ is called a covering of E, and the lower bound

$$\mu^*(E) = \inf \sum_{i=1}^{\infty} \mu(A_i) \tag{2.A.20}$$

of $\sum_{i=1}^{\infty} \mu(A_i)$ over all the coverings of E is called the outer measure of E.

In general, we have

$$\mu^*(E) \leq \sum_{i=1}^{\infty} \mu^*(B_i) \tag{2.A.21}$$

when $E = \overset{\infty}{\underset{i=1}{\cup}} B_i$, and

$$\mu^*(E) = \mu(E) \tag{2.A.22}$$

when E is an elementary set.

Example 2.A.6 Assume the sets shown in Fig. 2.18. Let the measure of the two-dimensional interval

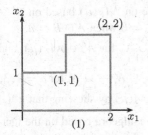 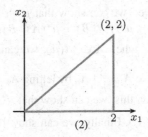

Fig. 2.18 Examples of an elementary set (1) and a non-elementary set (2) in two-dimensional space

$$A_{a,b,c,d} = \{(x_1, x_2) : a \le x_1 \le b, \, c \le x_2 \le d\} \tag{2.A.23}$$

be $\mu(A_{a,b,c,d}) = (b - a)(d - c)$. Let the set in Fig. 2.18 (1) be B_1. Then, we have $B_1 \subseteq A_{0,2,0,2}$, $B_1 \subseteq A_{0,2,0,1} \cup A_{1,2,0,2}$, and $B_1 \subseteq A_{0,2,0,1} \cup A_{1,2,1,2}$, among which the covering with the smallest measure is $\{A_{0,2,0,1}, A_{1,2,1,2}\}$. Thus, the outer measure of B_1 is $\mu^*(B_1) = 2 + 1 = 3$. Similarly, let the set in Fig. 2.18 (2) be B_2. Then, we have $B_2 \subseteq A_{0,2,0,2}$, $B_2 \subseteq A_{0,2,0,1} \cup A_{0,2,1,2}$, $B_2 \subseteq A_{0,2,0,1} \cup A_{1,2,1,2}$, ..., among which the covering with the smallest measure is $\left\{A_{\frac{2(i-1)}{n},2,\frac{2(i-1)}{n},\frac{2i}{n}}\right\}_{i=1}^{n}$ as $n \to \infty$. Thus, the outer measure of B_2 is $\mu^*(B_2) = 4 \lim_{n \to \infty} \sum_{i=1}^{n} \left(1 - \frac{i-1}{n}\right)\frac{1}{n} = 4\int_0^1 (1 - x)dx = 2$. $\qquad\qquad\qquad\diamond$

Definition 2.A.6 (*finitely μ-measurable set; μ-measurable set*) For a sequence $\{A_n\}_{n=1}^{\infty}$ of elementary sets, a set A such that

$$\lim_{n \to \infty} \mu^*(A_n \triangle A) = 0 \tag{2.A.24}$$

is called finitely μ-measurable. A set is called μ-measurable if it is obtained from a countable union of finitely μ-measurable sets.

The collections of all finitely μ-measurable sets and μ-measurable sets are denoted by $\mathcal{M}_F(\mu)$ and $\mathcal{M}(\mu)$, respectively.

Theorem 2.A.6 *The collection $\mathcal{M}(\mu)$ is a σ-algebra, and the outer measure μ^* is an additive set function on $\mathcal{M}(\mu)$.*

Proof Instead of a rigorous proof, we will simply discuss a brief outline. Assume two sequences $\{A_i\}_{i=1}^{\infty}$ and $\{B_i\}_{i=1}^{\infty}$ of elementary sets converging to A and B, respectively, when A and B are elements of $\mathcal{M}_F(\mu)$. If we let $d(A, B) = \mu^*(A \triangle B)$, then we can show that $\mathcal{M}_F(\mu)$ is an algebra by showing that $A \cup B$ and $A \cap B$ are included in $\mathcal{M}_F(\mu)$ based on $d\left(A_i \cup A_j, B_i \cup B_j\right) \le d\left(A_i, B_i\right) + d\left(A_j, B_j\right)$, $d\left(A_i \cap A_j, B_i \cap B_j\right) \le d\left(A_i, B_i\right) + d\left(A_j, B_j\right)$, and $|\mu^*(A) - \mu^*(B)| \le d(A, B)$.

Moreover, we can show that μ^* is finitely additive on $\mathcal{M}_F(\mu)$ based on $\mu^*(A) + \mu^*(B) = \mu^*(A \cup B) + \mu^*(A \cap B)$ and $\mu^*(A \cap B) = 0$ when $A \cap B = \emptyset$.

Now, when $A \in \mathcal{M}(\mu)$, we can express $A = \overset{\infty}{\underset{n=1}{\cup}} A'_n$ for $A'_n \in \mathcal{M}_F(\mu)$, and we then have $A = \overset{\infty}{\underset{n=1}{\cup}} A_n$ by letting $A_1 = A'_1$ and $A_n = \overset{n}{\underset{i=1}{\cup}} A'_i - \overset{n-1}{\underset{i=1}{\cup}} A'_i$ for $n = 2, 3, \dots$. Based on this, we can show that μ^* is additive on $\mathcal{M}(\mu)$ by showing that $\mu^*(A) = \overset{\infty}{\underset{i=1}{\sum}} \mu^*(A_i)$. Finally, we can show that $\mathcal{M}(\mu)$ is a σ-algebra based on the fact that any countable set operations on the sets in $\mathcal{M}(\mu)$ can be obtained from a countable union of $\mathcal{M}_F(\mu)$.   ♠

Based on Theorem 2.A.6, we can use μ^*, instead of μ, as the measure when we deal with μ-measurable sets. In essence, we have first defined μ for elementary sets and then extended μ into μ^*, an additive set function on the σ-algebra $\mathcal{M}(\mu)$.

Definition 2.A.7 *(Lebesgue measure)* The Lebesgue measure in the Euclidean space $\mathbb{R}^p$ is defined as

$$\mu(A) = \sum_{i=1}^{n} m(I_i) \tag{2.A.25}$$

for $A = \overset{n}{\underset{i=1}{\cup}} I_i$, where $\{I_i\}_{i=1}^{n}$ are non-overlapping intervals and

$$m(I) = \prod_{k=1}^{p} (b_k - a_k) \tag{2.A.26}$$

with $I = \{x = (x_1, x_2, \dots, x_p) : a_k \le x_k \le b_k, \ k = 1, 2, \dots, p\}$ an interval in $\mathbb{R}^p$.

Definition 2.A.7 is based on the fact that any elementary set can be obtained from a union of non-overlapping intervals $\{I_i\}_{i=1}^{n}$. An open set can be obtained from a countable union of open intervals and is a μ-measurable set. Similarly, a closed set is the complement of an open set and is also a μ-measurable set because $\mathcal{M}(\mu)$ is a σ-algebra. As discussed in Definition 2.2.7, the collection of all Borel sets is a σ-algebra and is called the Borel σ-algebra or Borel field. In addition, a μ-measurable set can always be expressed as the union of a Borel set and a set which is of measure 0 and is mutually exclusive of the Borel set. Under the Lebesgue measure, all countable sets and some[16] uncountable sets are of measure 0.

Example 2.A.7 In the one-dimensional space, the Lebesgue measure of an interval $[a, b]$ is the length $\mu([a, b]) = b - a$ of the interval. The Lebesgue measure of the set Q of rational numbers is $\mu(Q) = 0$.   ◇

---

[16] The Cantor set discussed in Example 1.1.46 is one such example.

Definition 2.A.8 (*measure space; measurable space*) In a metric space X, if there exist a σ-algebra $\mathcal{M}$ of measurable sets composed of subsets of X and a non-negative additive set function μ, then X is called a measure space. Here, if $X \in \mathcal{M}$, then $(X, \mathcal{M}, \mu)$ is called a measurable space.

Example 2.A.8 In the space $X = \mathbb{R}^p$, we have the Lebesgue measure and the collection $\mathcal{M}$ of all sets measurable by the Lebesgue measure. Then, it is easy to see that X is a measure space. $\diamond$

Example 2.A.9 In the space $X = \mathbb{J}_+$, let the number of elements in a set be the measure μ of the set and let the collection of all subsets of X be $\mathcal{M}$. Then, $(X, \mathcal{M}, \mu)$ is a measurable space. $\diamond$

Definition 2.A.9 (*measurable function*) When the set $\{x : f(x) > a\}$ is always a measurable set, a real function f defined on a measurable space is called a measurable function.

Example 2.A.10 Continuous functions in $\mathbb{R}^p$ are all measurable functions. $\diamond$

Example 2.A.11 If f is a measurable function, so is $|f|$. If f and g are both measurable functions, then $\max(f, g)$ and $\min(f, g)$ are measurable functions. $\diamond$

Example 2.A.12 If $\{f_n\}_{i=1}^{\infty}$ is a sequence of measurable functions, then $\sup f_n(x)$ and $\limsup\limits_{n \to \infty} f_n(x)$ are measurable functions. $\diamond$

Definition 2.A.10 (*simple function*) When the range of a function on a measurable space is finite, the function is called a simple function.

Example 2.A.13 When the range of a simple function f is $\{c_1, c_2, \ldots, c_n\}$, we have $f(x) = \sum\limits_{i=1}^{n} c_i K_{B_i}(x)$, where $B_i = \{x : f(x) = c_i\}$ and

$$K_E(x) = \begin{cases} 1, & x \in E, \\ 0, & x \notin E \end{cases} \tag{2.A.27}$$

is the indicator function of E. $\diamond$

Theorem 2.A.7 *There exists a sequence $\{f_n\}_{n=1}^{\infty}$ of simple functions such that $\lim\limits_{n \to \infty} f_n(x) = f(x)$ for any real function f defined on a measurable space. If f is a measurable function, then $\{f_n\}_{n=1}^{\infty}$ can be chosen as a sequence of measurable functions, and if $f \geq 0$, $\{f_n\}_{n=1}^{\infty}$ can be chosen to increase monotonically.*

Proof When $f \geq 0$, let $B_{n,i} = \left\{x : \frac{i-1}{2^n} \leq f(x) \leq \frac{i}{2^n}\right\}$ and $F_n = \{x : f(x) \geq n\}$, and then choose

$$f_n(x) = \sum_{i=1}^{n2^n} \frac{i-1}{2^n} K_{B_{n,i}}(x) + n K_{F_n}(x). \tag{2.A.28}$$

More generally, by letting $f^+ = \max(f, 0)$, $f^- = -\min(f, 0)$, and $f = f^+ - f^-$, we can prove the theorem easily. ♠

Definition 2.A.11 (*Lebesgue integral*) In a measurable space $(X, \mathcal{M}, \mu)$, let $s(x) = \sum_{i=1}^{n} c_i K_{B_i}(x)$ be a measurable function, where $c_i > 0$ and $x \in X$. In addition, let $E \in \mathcal{M}$ and $I_E(s) = \sum_{i=1}^{n} c_i \mu (E \cap B_i)$. Then, for a non-negative and measurable function f,

$$\int_E f d\mu = \sup I_E(s) \tag{2.A.29}$$

is called the Lebesgue integral.

In Definition 2.A.11, the upper bound is obtained over all measurable simple functions s such that $0 \le s \le f$. In the meantime, when the function f is not always positive, the Lebesgue integral can be defined as

$$\int_E f d\mu = \int_E f^+ d\mu - \int_E f^- d\mu \tag{2.A.30}$$

if at least one of $\int_E f^+ d\mu$ and $\int_E f^- d\mu$ is finite, where $f^+ = \max(f, 0)$ and $f^- = -\min(f, 0)$. Note that $f = f^+ - f^-$ and that f^+ and f^- are measurable functions. If both $\int_E f^+ d\mu$ and $\int_E f^- d\mu$ are finite, then $\int_E f d\mu$ is finite and the function f is called Lebesgue integrable on E for μ, which is expressed as $f \in \mathcal{L}(\mu)$ on E.

Based on mensuration by parts, the Riemann integral is the sum of products of the value of a function in an arbitrarily small interval composing the integral region and the length of the interval. On the other hand, the Lebesgue integral is the sum of products of the value of a function and the measure of the interval in the domain corresponding to an arbitrarily small interval in the range of the function. The Lebesgue integral exists not only for all Riemann integrable functions but also for other functions while the Riemann integral exists only when the function is at least piecewise continuous.

Some of the properties of the Lebesgue integral are as follows:

(1) If a function f is measurable on E and bounded and $\mu(E)$ is finite, then $f \in \mathcal{L}(\mu)$ on E.
(2) If the measure $\mu(E)$ is finite and $a \le f \le b$, then $a\mu(E) \le \int_E f d\mu \le b\mu(E)$.
(3) If $f, g \in \mathcal{L}(\mu)$ on the set E and $f(x) \le g(x)$ for $x \in E$, then $\int_E f d\mu \le \int_E g d\mu$.
(4) If $f \in \mathcal{L}(\mu)$ on the set E and c is a finite constant, then $\int_E cf d\mu \le c \int_E f d\mu$ and $cf \in \mathcal{L}(\mu)$.
(5) If $f \in \mathcal{L}(\mu)$ on the set E, then $|f| \in \mathcal{L}(\mu)$ and $\left| \int_E f d\mu \right| \le \int_E |f| d\mu$.

(6) If a function f is measurable on the set E and $\mu(E) = 0$, then $\int_E f d\mu = 0$.
(7) If a function f is Lebesgue integrable on X and $\phi(A) = \int_A f d\mu$ on $A \in \mathcal{M}$, then ϕ is additive on $\mathcal{M}$.
(8) Let $A \in \mathcal{M}$, $B \subseteq A$, and $\mu(A - B) = 0$. Then, $\int_A f d\mu = \int_B f d\mu$.
(9) Consider a sequence $\{f_n\}_{n=1}^\infty$ of measurable functions such that $\lim_{n\to\infty} f_n(x) = f(x)$ for $E \in \mathcal{M}$ and $x \in E$. If there exists a function $g \in \mathcal{L}(\mu)$ such that $|f_n(x)| \le g(x)$, then $\lim_{n\to\infty} \int_E f_n d\mu = \int_E f d\mu$.
(10) If a function f is Riemann integrable on $[a, b]$, then f is Lebesgue integrable and the Lebesgue integral with the Lebesgue measure is the same as the Riemann integral.

Appendix 2.4 Non-measurable Sets

Assume the open unit interval $J = (0, 1)$ and the set $\mathbb{Q}$ of rational numbers in the real space $\mathbb{R}$. Consider the translation operator $T_t : \mathbb{R} \to \mathbb{R}$ such that $T_t(x) = x + t$ for $x \in \mathbb{R}$. Suppose the countable set $\Gamma_t = T_t\mathbb{Q}$, i.e.,

$$\Gamma_t = \{t + q : q \in \mathbb{Q}\}. \tag{2.A.31}$$

For example, we have $\Gamma_{5445} = \{q + 5445 : q \in \mathbb{Q}\} = \mathbb{Q}$ and $\Gamma_\pi = \{q + \pi : q \in \mathbb{Q}\}$ when $t = 5445$ and $t = \pi$, respectively.

It is clear that

$$\Gamma_t \cap J \ne \emptyset \tag{2.A.32}$$

because we can always find a rational number q such that $0 < t + q < 1$ for any real number t. We have $\Gamma_t = \{t + q : q \in \mathbb{Q}\} = \{s + (t - s) + q : q \in \mathbb{Q}\} = \{s + q' : q' \in \mathbb{Q}\} = \Gamma_s$ and $\Gamma_t \cap \Gamma_s = \emptyset$ when $t - s$ is a rational number and an irrational number, respectively. Based on this observation, consider the collection

$$\mathbb{K} = \{\Gamma_t : t \in \mathbb{R}, \text{ distinct } \Gamma_t \text{ only}\} \tag{2.A.33}$$

of sets (Rao 2004). Then, we have the following facts:

(1) The collection $\mathbb{K}$ is a partition of $\mathbb{R}$.
(2) There exists only one rational number t for $\Gamma_t \in \mathbb{K}$.
(3) There exist uncountably many sets in $\mathbb{K}$.
(4) For two distinct sets Γ_t and Γ_s in $\mathbb{K}$, the number $t - s$ is not a rational number.

Definition 2.A.12 (*Vitali set*) Based on the axiom of choice[17] and (2.A.32), we can obtain an uncountable set

$$V_0 = \{x : x \in \Gamma_t \cap J, \ \Gamma_t \in \mathbb{K}\}, \tag{2.A.34}$$

where x represents a number in the interval $(0, 1)$ and an element of $\Gamma_t \in \mathbb{K}$. The set V_0 is called the Vitali set.

Note that the points in the Vitali set V_0 are all in interval $(0, 1)$ and have a one-to-one correspondence with the sets in $\mathbb{K}$. Denoting the enumeration of all the rational numbers in the interval $(-1, 1)$ by $\{\alpha_i\}_{i=1}^{\infty}$, we get the following theorem:

Theorem 2.A.8 *For the Vitali set* V_0,

$$(0, 1) \subseteq \bigcup_{i=1}^{\infty} T_{\alpha_i} V_0 \subseteq (-1, 2) \tag{2.A.35}$$

holds true.

Proof First, $-1 < \alpha_i + x < 2$ because $-1 < \alpha_i < 1$ and any point x in V_0 satisfies $0 < x < 1$. In other words, $T_{\alpha_i} x \in (-1, 2)$, and therefore

$$\bigcup_{i=1}^{\infty} T_{\alpha_i} V_0 \subseteq (-1, 2). \tag{2.A.36}$$

Next, for any point x in $(0, 1)$, $x \in \Gamma_t$ with an appropriately chosen t as we have observed in (2.A.32). Then, we have $\Gamma_t = \Gamma_x$ and $x \in \Gamma_t = \Gamma_x$ because $x - t$ is a rational number. Now, denoting a point in $\Gamma_x \cap V_0$ by y, we have $y = x + q$ because $\Gamma_x \cap V_0 \neq \emptyset$ and therefore $y - x \in \mathbb{Q}$. Here, $y - x$ is a rational number in $(-1, 1)$ because $0 < x, y < 1$ and, consequently, we can put $y - x = \alpha_i$: in other words, $y = x + \alpha_i = T_{\alpha_i} x \in T_{\alpha_i} V_0$. Thus, we have

$$(0, 1) \subseteq \bigcup_{i=1}^{\infty} T_{\alpha_i} V_0. \tag{2.A.37}$$

Subsequently, we get (2.A.35) from (2.A.36) and (2.A.37). ♠

Theorem 2.A.9 *The sets* $\{T_{\alpha_i} V_0\}_{i=1}^{\infty}$ *are all mutually exclusive: in other words,*

$$(T_{\alpha_i} V_0) \cap (T_{\alpha_j} V_0) = \emptyset \tag{2.A.38}$$

for $i \neq j$.

---

[17] The axiom of choice can be expressed as "For a non-empty set $B \subseteq A$, there exists a choice function $f : 2^A \to A$ such that $f(B) \in B$ for any set A." The axiom of choice can be phrased in various expressions, and that in Definition 2.A.12 is based on "If we assume a partition $\mathbb{P}_S$ of S composed only of non-empty sets, then there exists a set B for which the intersection with any set in $\mathbb{P}_S$ is a singleton set."

Proof We prove the theorem by contradiction. When $i \neq j$ or, equivalently, when $\alpha_i \neq \alpha_j$, assume that $(T_{\alpha_i} \mathbb{V}_0) \cap (T_{\alpha_j} \mathbb{V}_0)$ is not a null set. Letting one element of the intersection be y, we have $y = x + \alpha_i = x' + \alpha_j$ for $x, x' \in \mathbb{V}_0$. It is clear that $\Gamma_x = \Gamma_{x'}$ because $x - x' = \alpha_j - \alpha_i$ is a rational number. Thus, $x = x'$ from the definition of $\mathbb{K}$, and therefore $\alpha_i = \alpha_j$: this is contradictory to $\alpha_i \neq \alpha_j$. Consequently, $(T_{\alpha_i} \mathbb{V}_0) \cap (T_{\alpha_j} \mathbb{V}_0) = \emptyset$.   ♠

Theorem 2.A.10 *No set in* $\{T_{\alpha_i} \mathbb{V}_0\}_{i=1}^{\infty}$ *is Lebesgue measurable: in other words,* $T_{\alpha_i} \mathbb{V}_0 \notin \mathcal{M}(\mu)$ *for any i.*

Proof We prove the theorem by contradiction. Assume that the sets $\{T_{\alpha_i} \mathbb{V}_0\}_{i=1}^{\infty}$ are measurable. Then, from the translation invariance[18] of a measure, they have the same measure. Denoting the Lebesgue measure of $T_{\alpha_i} \mathbb{V}_0$ by $\mu\left(T_{\alpha_i} \mathbb{V}_0\right) = \beta$, we have

$$\mu((0, 1)) \leq \mu\left(\bigcup_{i=1}^{\infty} T_{\alpha_i} \mathbb{V}_0\right) \leq \mu((-1, 2)) \tag{2.A.39}$$

from (2.A.35). Here, $\mu((0, 1)) = 1$ and $\mu((-1, 2)) = 3$. In addition, we have $\mu\left(\bigcup_{i=1}^{\infty} T_{\alpha_i} \mathbb{V}_0\right) = \sum_{i=1}^{\infty} \mu\left(T_{\alpha_i} \mathbb{V}_0\right)$, i.e.,

$$\mu\left(\bigcup_{i=1}^{\infty} T_{\alpha_i} \mathbb{V}_0\right) = \sum_{i=1}^{\infty} \beta \tag{2.A.40}$$

because $\{T_{\alpha_i} \mathbb{V}_0\}_{i=1}^{\infty}$ is a collection of mutually exclusive sets as we have observed in (2.A.38). Combining (2.A.39) and (2.A.40) leads us to

$$1 \leq \sum_{i=1}^{\infty} \beta \leq 3, \tag{2.A.41}$$

which can be satisfied neither with $\beta = 0$ nor with $\beta \neq 0$. Consequently, no set in $\{T_{\alpha_i} \mathbb{V}_0\}_{i=1}^{\infty}$, including $\mathbb{V}_0$, is Lebesgue measurable.   ♠

Exercises

Exercise 2.1 Obtain the algebra generated from the collection $\mathcal{C} = \{\{a\}, \{b\}\}$ of the set $S = \{a, b, c, d\}$.

Exercise 2.2 Obtain the σ-algebra generated from the collection $\mathcal{C} = \{\{a\}, \{b\}\}$ of the set $S = \{a, b, c, d\}$.

---

[18] For any real number x, the measure of $A = \{a\}$ is the same as that of $A + x = \{a + x\}$.

Exercise 2.3 Obtain the sample space S in the following random experiments:

(1) An experiment measuring the lifetime of a battery.
(2) An experiment in which an integer n is selected in the interval $[0, 2]$ and then an integer m is selected in the interval $[0, n]$.
(3) An experiment of checking the color of, and the number written on, a ball selected randomly from a box containing two red, one green, and two blue balls denoted by $1, 2, \ldots, 5$, respectively.

Exercise 2.4 When $P(A) = P(B) = P(AB)$, obtain $P(AB^c + BA^c)$.

Exercise 2.5 Consider rolling a fair die. For $A = \{1\}$, $B = \{2, 4\}$, and $C = \{1, 3, 5, 6\}$, obtain $P(A \cup B)$, $P(A \cup C)$, and $P(A \cup B \cup C)$.

Exercise 2.6 Consider the events $A = (-\infty, r]$ and $B = (-\infty, s]$ with $r \leq s$ in the sample space of real numbers.

(1) Express $C = (r, s]$ in terms of A and B.
(2) Show that $B = A \cup C$ and $A \cap C = \emptyset$.

Exercise 2.7 When ten distinct red and ten distinct black balls are randomly arranged into a single line, find the probability that red and black balls are placed in an alternating fashion.

Exercise 2.8 Consider two branches between two nodes in a circuit. One of the two branches is a resistor and the other is a series connection of two resistors. Obtain the probability that the two nodes are disconnected assuming that the probability for a resistor to be disconnected is p and disconnection in a resistor is not influenced by the status of other resistors.

Exercise 2.9 Show that A^c and B are independent of each other and that A^c and B^c are independent of each other when A and B are independent of each other.

Exercise 2.10 Assume the sample space $S = \{1, 2, 3\}$ and event space $\mathcal{F} = 2^S$. Show that no two events, except S and $\emptyset$, are independent of each other for any probability measure such that $P(1) > 0$, $P(2) > 0$, and $P(3) > 0$.

Exercise 2.11 For two events A and B, show the followings:

(1) If $P(A) = 0$, then $P(AB) = 0$.
(2) If $P(A) = 1$, then $P(AB) = P(B)$.

Exercise 2.12 Among 100 lottery tickets sold each week, one is a winning ticket. When a ticket costs 10 euros and we have 500 euros, does buying 50 tickets in one week bring us a higher probability of getting the winning ticket than buying one ticket over 50 weeks?

Exercise 2.13 In rolling a fair die twice, find the probability that the sum of the two outcomes is 7 when we have 3 from the first rolling.

Exercise 2.14 When a pair of fair dice are rolled once, find $P(a - 2b < 0)$, where a and b are the face values of the two dice with $a \geq b$.

Exercise 2.15 When we choose subsets A, B, and C from $D = \{1, 2, \ldots, k\}$ randomly, find the probability that $C \cap (A - B)^c = \emptyset$.

Exercise 2.16 Denote the four vertices of a regular tetrahedron by A, B, C, and D. In each movement from one vertex to another, the probability of arriving at another vertex is $\frac{1}{3}$ for each of the three vertices. Find the probabilities $p_{n,A}$ and $p_{n,B}$ that we arrive at A and B, respectively, after n movements starting from A. Obtain the values of $p_{10,A}$ and $p_{10,B}$ when $n = 10$.

Exercise 2.17 A box contains N balls each marked with a number $1, 2, \ldots$, and N, respectively. Each of N students with identification (ID) numbers $1, 2, \ldots$, and N, respectively, chooses a ball randomly from the box. If the number marked on the ball and the ID number of the student are the same, then it is called a match.

(1) Find the probability of no match.
(2) Using conditional probability, obtain the probability in (1) again.
(3) Find the probability of k matches.

Exercise 2.18 In the interval $[0, 1]$ on a line of real numbers, two points are chosen randomly. Find the probability that the distance between the two points is shorter than $\frac{1}{2}$.

Exercise 2.19 Consider the probability space composed of the sample space $S = \{$all pairs (k, m) of natural numbers$\}$ and probability measure

$$P((k, m)) = \alpha(1 - p)^{k+m-2}, \tag{2.E.1}$$

where α is a constant and $0 < p < 1$.

(1) Determine the constant α. Then, obtain the probability $P((k, m) : k \geq m)$.
(2) Obtain the probability $P((k, m) : k + m = r)$ as a function of $r \in \{2, 3, \ldots\}$. Confirm that the result is a probability measure.
(3) Obtain the probability $P((k, m) : k$ is an odd number$)$.

Exercise 2.20 Obtain $P(A \cap B)$, $P(A | B)$, and $P(B | A)$ when $P(A) = 0.7$, $P(B) = 0.5$, and $P([A \cup B]^c) = 0.1$.

Exercise 2.21 Three people shoot at a target. Let the event of a hit by the i-th person be A_i for $i = 1, 2, 3$ and assume the three events are independent of each other. When $P(A_1) = 0.7$, $P(A_2) = 0.9$, and $P(A_3) = 0.8$, find the probability that only two people will hit the target.

Exercise 2.22 In testing circuit elements, let $A = \{$defective element$\}$ and $B = \{$element identified as defective$\}$, and $P(B | A) = p$, $P(B^c | A^c) = q$, $P(A) = r$, and $P(B) = s$. Because the test is not perfect, two types of errors could occur: a false negative, 'a defective element is identified to be fine'; or a false positive, 'a functional element is identified to be defective'. Assume that the production and testing of the elements can be adjusted such that the parameters p, q, r, and s are very close to 0 or 1.

(1) For each of the four parameters, explain whether it is more desirable to make it closer to 0 or 1.
(2) Describe the meaning of the conditional probabilities $P(B^c|A)$ and $P(B|A^c)$.
(3) Describe the meaning of the conditional probabilities $P(A^c|B)$ and $P(A|B^c)$.
(4) Given the values of the parameters p, q, r, and s, obtain the probabilities in (2) and (3).
(5) Obtain the sample space of this experiment.

Exercise 2.23 For three events A, B, and C, show the following results without using Venn diagrams:

(1) $P(A \cup B) = P(A) + P(B) - P(AB)$.
(2) $P(A \cup B \cup C) = P(A) + P(B) + P(C) + P(ABC) - P(AB) - P(AC) - P(BC)$.
(3) Union upper bound, i.e.,

$$P\left(\bigcup_{i=1}^{n} A_i\right) \leq \sum_{i=1}^{n} P(A_i). \qquad (2.E.2)$$

Exercise 2.24 For the sample space S and events E, F, and $\{B_i\}_{i=1}^{\infty}$, show that the conditional probability satisfies the axioms of probability as follows:

(1) $0 \leq P(E|F) \leq 1$.
(2) $P(S|F) = 1$.
(3) $P\left(\bigcup_{i=1}^{\infty} B_i \,\middle|\, F\right) = \sum_{i=1}^{\infty} P(B_i|F)$ when the events $\{B_i\}_{i=1}^{\infty}$ are mutually exclusive.

Exercise 2.25 Assume an event B and a collection $\{A_i\}_{i=1}^{n}$ of events, where $\{A_i\}_{i=1}^{n}$ is a partition of the sample space S.

(1) Explain whether or not A_i and A_j for $i \neq j$ are independent of each other.
(2) Obtain a partition of B using $\{A_i\}_{i=1}^{n}$.
(3) Show the total probability theorem $P(B) = \sum_{i=1}^{n} P(B|A_i) P(A_i)$.
(4) Show the Bayes' theorem $P(A_k|B) = \dfrac{P(B|A_k)P(A_k)}{\sum_{i=1}^{n} P(B|A_i)P(A_i)}$.

Exercise 2.26 Box 1 contains two red and three green balls and Box 2 contains one red and four green balls. Obtain the probability of selecting a red ball when a ball is selected from a randomly chosen box.

Exercise 2.27 Box i contains i red and $(n - i)$ green balls for $i = 1, 2, \ldots, n$. Choosing Box i with probability $\frac{2i}{n(n+1)}$, obtain the probability of selecting a red ball when a ball is selected from the box chosen.

Exercise 2.28 A group of people elects one person via rock-paper-scissors. If there is only one person who wins, then the person is chosen; otherwise, the rock-paper-scissors is repeated. Assume that the probability of rock, paper, and scissors are $\frac{1}{3}$ for every person and not affected by other people. Obtain the probability $p_{n,k}$ that n people will elect one person in k trials.

Exercise 2.29 In an election, Candidates A and B will get n and m votes, respectively. When $n > m$, find the probability that Candidate A will always have more counts than Candidate B during the ballot-counting.

Exercise 2.30 A type O cell is cultured at time 0. After one hour, the cell will become

$$
\begin{array}{lll}
\text{two type O cells,} & \text{probability} = \frac{1}{4}, & \\
\text{one type O cell, one type M cell,} & \text{probability} = \frac{2}{3}, & \text{(2.E.3)} \\
\text{two type M cells,} & \text{probability} = \frac{1}{12}. &
\end{array}
$$

A new type O cell behaves like the first type O cell and a type M cell will disappear in one hour, where a change is not influenced by any other change. Find the probability β_0 that no type M cell will appear until $n + \frac{1}{2}$ hours from the starting time.

Exercise 2.31 Find the probability of the event A that 5 or 6 appears k times when a fair die is rolled n times.

Exercise 2.32 Consider a communication channel for signals of binary digits (bits) 0 and 1. Due to the influence of noise, two types of errors can occur as shown in Fig. 2.19: specifically, 0 and 1 can be identified to be 1 and 0, respectively. Let the transmitted and received bits be X and Y, respectively. Assume *a priori* probability of $P(X = 1) = p$ for 1 and $P(X = 0) = 1 - p$ for 0, and the effect of noise on a bit is not influenced by that on other bits. Denote the probability that the received bit is i when the transmitted bit is i by $P(Y = i | X = i) = p_{ii}$ for $i = 0, 1$.

Fig. 2.19 A binary communication channel

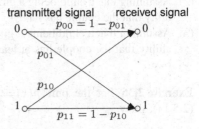

(1) Obtain the probabilities $p_{10} = P(Y = 0|X = 1)$ and $p_{01} = P(Y = 1|X = 0)$ that an error occurs when bits 1 and 0 are transmitted, respectively.

(2) Obtain the probability that an error occurs.

(3) Obtain the probabilities $P(Y = 1)$ and $P(Y = 0)$ that the received bit is identified to be 1 and 0, respectively.

(4) Obtain all *a posteriori* probabilities $P(X = j|Y = k)$ for $j = 0, 1$ and $k = 0, 1$.

(5) When $p = 0.5$, obtain $P(X = 1|Y = 0)$, $P(X = 1|Y = 1)$, $P(Y = 1)$, and $P(Y = 0)$ for a symmetric channel with $p_{00} = p_{11}$.

Exercise 2.33 Assume a pile of n integrated circuits (ICs), among which m are defective ones. When an IC is chosen randomly from the pile, the probability that the IC is defective is $\alpha_1 = \frac{m}{n}$ as shown in Example 2.3.8.

(1) Assume we pick one IC and then one more IC without replacing the first one back to the pile. Obtain the probabilities $\alpha_{1,1}$, $\alpha_{0,1}$, $\alpha_{1,0}$, and $\alpha_{0,0}$ that both are defective, the first one is not defective and the second one is defective, the first one is defective and the second one is not defective, and neither the first nor the second one is defective, respectively.

(2) Now assume we pick one IC and then one more IC after replacing the first one back to the pile. Obtain the probabilities $\alpha_{1,1}$, $\alpha_{0,1}$, $\alpha_{1,0}$, and $\alpha_{0,0}$ again.

(3) Assume we pick two ICs randomly from the pile. Obtain the probabilities β_0, β_1, and β_2 that neither is defective, one is defective and the other is not defective, and both are defective, respectively.

Exercise 2.34 Box 1 contains two old and three new erasers and Box 2 contains one old and six new erasers. We perform the experiment "choose one box randomly and pick an eraser at random" twice, during which we discard the first eraser picked.

(1) Obtain the probabilities P_2, P_1, and P_0 that both erasers are old, one is old and the other is new, and both erasers are new, respectively.

(2) When both erasers are old, obtain the probability P_3 that one is from Box 1 and the other is from Box 2.

Exercise 2.35 The probability for a couple to have k children is αp^k with $0 < p < 1$.

(1) The color of the eyes being brown for a child is of probability b and is independent of that of other children. Obtain the probability that the couple has r children with brown eyes.

(2) Assuming that a child being a girl or a boy is of probability $\frac{1}{2}$, obtain the probability that the couple has r boys.

(3) Assuming that a child being a girl or a boy is of probability $\frac{1}{2}$, obtain the probability that the couple has at least two boys when the couple has at least one boy.

Exercise 2.36 For the pmf $p(x) = {}_{r+x-1}C_{r-1}\alpha^r(1 - \alpha)^x$, $x \in \mathbb{J}_0$ introduced in (2.5.14), show

$$\lim_{r \to \infty} p(x) = \frac{\lambda^x}{x!} e^{-\lambda}, \tag{2.E.4}$$

which implies $\lim_{r \to \infty} NB\left(r, \frac{r}{r+\lambda}\right) = P(\lambda)$.

Exercise 2.37 A person plans to buy a car of price N units. The person has k units and wishes to earn the remaining from a game. In the game, the person wins and loses 1 unit when the outcome is a *head* and a *tail*, respectively, from a toss of a coin with probability p for a *head* and $q = 1 - p$ for a *tail*. Assuming $0 < k < N$ and the person continues the game until the person earns enough for the car or loses all the money, find the probability that the person loses all the money. This problem is called the gambler's ruin problem.

Exercise 2.38 A large number of bundles, each with 25 tulip bulbs, are contained in a large box. The bundles are of type R_5 and R_{15} with portions $\frac{3}{4}$ and $\frac{1}{4}$, respectively. A type R_5 bundle contains five red and twenty white bulbs and a type R_{15} bundle contains fifteen red and ten white bulbs. A bulb, chosen randomly from a bundle selected at random from the box, is planted.

(1) Obtain the probability p_1 that a red tulip blossoms.
(2) Obtain the probability p_2 that a white tulip blossoms.
(3) When a red tulip blossoms, obtain the conditional probability that the bulb is from a type R_{15} bundle.

Exercise 2.39 For a probability space with the sample space $\Omega = \mathbb{J}_0 = \{0, 1, \ldots\}$ and pmf

$$p(x) = \begin{cases} 5c^2 + c, \ x = 0; & 3 - 13c, \ x = 1; \\ c, \qquad x = 2; & 0, \qquad \text{otherwise;} \end{cases} \tag{2.E.5}$$

determine the constant c.

Exercise 2.40 Show that

$$\frac{x}{1 + x^2} \phi(x) < Q(x) < \frac{1}{2} \exp\left(-\frac{x^2}{2}\right) \tag{2.E.6}$$

for $x > 0$, where $\phi(x)$ denotes the standard normal pdf, i.e., (2.5.27) with $m = 0$ and $\sigma^2 = 1$, and

$$Q(x) = \frac{1}{\sqrt{2\pi}} \int_x^\infty \exp\left(-\frac{t^2}{2}\right) dt. \tag{2.E.7}$$

Exercise 2.41 Balls with colors $C_1, C_2, \ldots, C_n$ are contained in k boxes. Let the probability of choosing Box B_j be $P(B_j) = b_j$ and that of choosing a ball with color C_i from Box B_j be $P(C_i | B_j) = c_{ij}$, where $\left\{\sum_{i=1}^{n} c_{ij} = 1\right\}_{j=1}^{k}$ and $\sum_{j=1}^{k} b_j = 1$. A box is chosen first and then a ball is chosen from the box.

(1) Show that, if $\{c_{i1} = c_{i2} = \cdots = c_{ik}\}_{i=1}^{n}$, the color of the ball chosen is independent of the choice of a box, i.e., $\{\{P(C_i B_j) = P(C_i)P(B_j)\}_{i=1}^{n}\}_{j=1}^{k}$, for any values of $\{b_j\}_{j=1}^{k}$.

(2) When $n = 2$, $k = 3$, $b_1 = b_3 = \frac{1}{4}$, and $b_2 = \frac{1}{2}$, express the condition for $P(C_1 B_1) = P(C_1)P(B_1)$ to hold true in terms of $\{c_{11}, c_{12}, c_{13}\}$.

Exercise 2.42 Boxes 1, 2, and 3 contain four red and five green balls, one red and one green balls, and one red and two green balls, respectively. Assume that the probabilities of the event B_i of choosing Box i are $P(B_1) = P(B_3) = \frac{1}{4}$ and $P(B_2) = \frac{1}{2}$. After a box is selected, a ball is chosen randomly from the box. Denote the events that the ball is red and green by R and G, respectively.

(1) Are the events B_1 and R independent of each other? Are the events B_1 and G independent of each other?
(2) Are the events B_2 and R independent of each other? Are the events B_3 and G independent of each other?

Exercise 2.43 For the sample space $\Omega = \{1, 2, 3, 4\}$ with $\{P(i) = \frac{1}{4}\}_{i=1}^{4}$, consider $A_1 = \{1, 3, 4\}$, $A_2 = \{2, 3, 4\}$, and $A_3 = \{3\}$. Are the three events A_1, A_2, and A_3 independent of each other?

Exercise 2.44 Consider two consecutive experiments with possible outcomes A and B for the first experiment and C and D for the second experiment. When $P(AC) = \frac{1}{3}$, $P(AD) = \frac{1}{6}$, $P(BC) = \frac{1}{6}$, and $P(BD) = \frac{1}{3}$, are A and C independent of each other?

Exercise 2.45 Two people make an appointment to meet between 10 and 11 o'clock. Find the probability that they can meet assuming that each person arrives at the meeting place between 10 and 11 o'clock independently and waits only up to 10 minutes.

Exercise 2.46 Consider two children. Assume any child can be a girl or a boy equally likely. Find the probability p_1 that both are boys when the elder is a boy and the probability p_2 that both are boys when at least one is a boy.

Exercise 2.47 There are three red and two green balls in Box 1, and four red and three green balls in Box 2. A ball is randomly chosen from Box 1 and put into Box 2. Then, a ball is picked from Box 2. Find the probability that the ball picked from Box 2 is red.

Exercise 2.48 Three people A, B, and C toss a coin each. The person whose outcome is different from those of the other two wins. If the three outcomes are the same, then the toss is repeated.

(1) Show that the game is fair, i.e., the probability of winning is the same for each of the three people.

Table 2.2 Some probabilities in the game mighty

	(A) Player G_1 murmurs "Oh! I do not have the joker."	(B) Player G_1 murmurs "Oh! I have neither the *mighty* nor the joker."
(1) Probability of having the joker	0, Player G_1, $\frac{10}{43} = \frac{260}{1118}$, Players $\{G_i\}_{i=2}^5$, $\frac{3}{43} = \frac{78}{1118}$, on the table (G_6).	0, Player G_1, $\frac{10}{43} = \frac{70}{301}$, Players $\{G_i\}_{i=2}^5$, $\frac{3}{43} = \frac{21}{301}$, on the table (G_6).
(2) Probability of having the *mighty*	$\frac{5}{26} = \frac{215}{1118}$, Player G_1, $\frac{105}{559} = \frac{210}{1118}$, Players $\{G_i\}_{i=2}^5$, $\frac{63}{1118}$, on the table (G_6).	0, Player G_1, $\frac{10}{43} = \frac{70}{301}$, Players $\{G_i\}_{i=2}^5$, $\frac{3}{43} = \frac{21}{301}$, on the table (G_6).
(3) Probability of having either the *mighty* or joker	$\frac{5}{26} = \frac{215}{1118}$, Player G_1, $\frac{190}{559} = \frac{380}{1118}$, Players $\{G_i\}_{i=2}^5$, $\frac{135}{1118}$, on the table (G_6).	0, Player G_1, $\frac{110}{301}$, Players $\{G_i\}_{i=2}^5$, $\frac{40}{301}$, on the table (G_6).
(4) Probability of having at least one of the *mighty* and joker	$\frac{5}{26} = \frac{215}{1118}$, Player G_1, $\frac{425}{1118}$, Players $\{G_i\}_{i=2}^5$, $\frac{69}{559} = \frac{138}{1118}$, on the table (G_6).	0, Player G_1, $\frac{125}{301}$, Players $\{G_i\}_{i=2}^5$, $\frac{41}{301}$, on the table (G_6).
(5) Probability of having both the *mighty* and joker	0, Player G_1, $\frac{45}{1118}$, Players $\{G_i\}_{i=2}^5$, $\frac{3}{1118}$, on the table (G_6).	0, Player G_1, $\frac{15}{301}$, Players $\{G_i\}_{i=2}^5$, $\frac{1}{301}$, on the table (G_6).

(2) Find the probabilities that B wins exactly eight times and at least eight times when the coins are tossed ten times, not counting the number of no winner.

Exercise 2.49 A game called mighty can be played by three, four, or five players. When it is played with five players, 53 cards are used by adding one joker to a deck of 52 cards. Among the 53 cards, the ace of spades is called the *mighty*, except when the suit of spades[19] is declared the royal suit. In the play, ten cards are distributed to each of the five players $\{G_i\}_{i=1}^5$ and the remaining three cards are left on the table, face side down. Assume that what Player G_1 murmurs is always true and consider the two cases (A) Player G_1 murmurs "Oh! I do not have the joker." and (B) Player G_1 murmurs "Oh! I have neither the *mighty* nor the joker." For convenience, let the three cards on the table be Player G_6. Obtain the following probabilities and thereby confirm Table 2.2:

(1) Player G_i has the joker.
(2) Player G_i has the *mighty*.
(3) Player G_i has either the *mighty* or the joker.

---

[19] When the suit of spades is declared the royal suit, the ace of diamonds, not the ace of spades, becomes the *mighty*.

(4) Player G_i has at least one of the *mighty* and the joker.

(5) Player G_i has both the *mighty* and the joker.

Exercise 2.50 In a group of 30 men and 20 women, 40% of men and 60% of women play piano. When a person in the group plays piano, find the probability that the person is a man.

Exercise 2.51 The probability that a car, a truck, and a bus passes through a toll gate is 0.5, 0.3, and 0.2, respectively. Find the probability that 30 cars, 15 trucks, and 5 buses has passed when 50 automobiles have passed the toll gate.

References

N. Balakrishnan, *Handbook of the Logistic Distribution* (Marcel Dekker, New York, 1992)

P.J. Bickel, K.A. Doksum, *Mathematical Statistics* (Holden-Day, San Francisco, 1977)

H.A. David, H.N. Nagaraja, *Order Statistics*, 3rd edn. (Wiley, New York, 2003)

R.M. Gray, L.D. Davisson, *An Introduction to Statistical Signal Processing* (Cambridge University Press, Cambridge, 2010)

A. Gut, *An Intermediate Course in Probability* (Springer, New York, 1995)

C.W. Helstrom, *Probability and Stochastic Processes for Engineers*, 2nd edn. (Prentice-Hall, Englewood Cliffs, 1991)

S. Kim, *Mathematical Statistics (in Korean)* (Freedom Academy, Paju, 2010)

A. Leon-Garcia, *Probability, Statistics, and Random Processes for Electrical Engineering*, 3rd edn. (Prentice Hall, New York, 2008)

M. Loeve, *Probability Theory*, 4th edn. (Springer, New York, 1977)

E. Lukacs, *Characteristic Functions*, 2nd edn. (Griffin, London, 1970)

T.M. Mills, *Problems in Probability* (World Scientific, Singapore, 2001)

M.M. Rao, *Measure Theory and Integration*, 2nd edn. (Marcel Dekker, New York, 2004)

V.K. Rohatgi, A.KMd.E. Saleh, *An Introduction to Probability and Statistics*, 2nd edn. (Wiley, New York, 2001)

J.P. Romano, A.F. Siegel, *Counterexamples in Probability and Statistics* (Chapman and Hall, New York, 1986)

S.M. Ross, *A First Course in Probability* (Macmillan, New York, 1976)

S.M. Ross, *Stochastic Processes*, 2nd edn. (Wiley, New York, 1996)

A.N. Shiryaev, *Probability*, 2nd edn. (Springer, New York, 1996)

A.A. Sveshnikov (ed.), *Problems in Probability Theory* (Mathematical Statistics and Theory of Random Functions, Dover, New York, 1968)

J.B. Thomas, *Introduction to Probability* (Springer, New York, 1986)

P. Weirich, Conditional probabilities and probabilities given knowledge of a condition. Philos. Sci. **50**(1), 82–95 (1983)

C.K. Wong, A note on mutually independent events. Am. Stat. **26**(2), 27–28 (1972)

Chapter 3
Random Variables

Based on the description of probability in Chap. 2, let us now introduce and discuss several topics on random variables: namely, the notions of the cumulative distribution function, expected values, and moments. We will then discuss conditional distribution and describe some of the widely-used distributions.

3.1 Distributions

Let us start by introducing the notion of the random variable and its distribution (Gardner 1990; Leon-Garcia 2008; Papoulis and Pillai 2002). In describing the distributions of random variables, we adopt the notion of the cumulative distribution function, which is a useful tool in characterizing the probabilistic properties of random variables.

3.1.1 Random Variables

Generally, a random variable is a real function of which the domain is a sample space. The range of a random variable $X : \Omega \to \mathbb{R}$ on a sample space Ω is $S_X = \{x : x = X(s), \ s \in \Omega\} \subseteq \mathbb{R}$. In fact, a random variable is not a variable but a function: yet, it is customary to call it a variable. In many cases, a random variable is denoted by an upper case alphabet such as $X, Y, \ldots$.

Definition 3.1.1 (*random variable*) For a sample space Ω of the outcomes from a random experiment, a function X that assigns a real number $x = X(\omega)$ to $\omega \in \Omega$ is called a random variable.

© The Author(s), under exclusive license to Springer Nature Switzerland AG 2022
I. Song et al., *Probability and Random Variables: Theory and Applications*,
https://doi.org/10.1007/978-3-030-97679-8_3

For a more precise definition of a random variable, we need the concept of a measurable function.

Definition 3.1.2 (*measurable function*) Given a probability space $(\Omega, \mathcal{F}, \mathsf{P})$, a real-valued function g that maps the sample space Ω onto the real number $\mathbb{R}$ is called a measurable function when the condition

$$\text{if } B \in \mathcal{B}(\mathbb{R}), \text{ then } g^{-1}(B) \in \mathcal{F} \tag{3.1.1}$$

is satisfied.

Example 3.1.1 A real-valued function g for which $g^{-1}(D)$ is a Borel set for every open set D is called a Borel function, and is a measurable function.   ◇

The random variable can be redefined as follows:

Definition 3.1.3 (*random variable*) A random variable is a measurable function defined on a probability space.

In general, to show whether or not a function is a random variable is rather complicated. However,

(A) every function $g : \Omega \rightarrow \mathbb{R}$ on a probability space $(\Omega, \mathcal{F}, \mathsf{P})$ with the event space $\mathcal{F}$ being a power set is a random variable, and

(B) almost all functions such as continuous functions, polynomials, unit step function, trigonometric functions, limits of measurable functions, min and max of measurable functions, etc. that we will deal with are random variables.

Example 3.1.2 (Romano and Siegel 1986) For the sample space $\Omega = \{1, 2, 3\}$ and event space $\mathcal{F} = \{\Omega, \emptyset, \{3\}, \{1, 2\}\}$, assume the function g such that $g(1) = 1$, $g(2) = 2$, and $g(3) = 3$. Then, g is not a random variable because $g^{-1}(\{1\}) = \{1\} \notin \mathcal{F}$ although $\{1\} \in \mathcal{B}(\mathbb{R})$.   ◇

Random variables can be classified into the following three classes:

Definition 3.1.4 (*discrete random variable; continuous random variable; hybrid random variable*) A random variable is called a discrete random variable, continuous random variable, or hybrid random variable when the range is a countable set, an uncountable set, or the union of an uncountable set and a countable set, respectively.

A hybrid random variable is also called a mixed-type random variable. A discrete random variable with finite range is sometimes called a finite random variable. The probabilistic characteristics of a continuous random variable and a discrete random variable can be described by the pdf and pmf, respectively. In the meantime, based on Definitions 1.1.22 and 3.1.4, the range of a discrete random variable can be assumed as subsets of $\{0, 1, \ldots\}$ or $\{1, 2, \ldots\}$.

Example 3.1.3 When the outcome from a rolling of a fair die is n, let $X_1(n) = n$ and

$$X_2(n) = \begin{cases} 0, & n \text{ is an odd number,} \\ 1, & n \text{ is an even number.} \end{cases} \tag{3.1.2}$$

Then, X_1 and X_2 are both discrete random variables.   ◇

Example 3.1.4 The random variables L, Θ, and D defined below are all continuous random variables.

(1) When (x, y) denotes the coordinate of a randomly selected point Q inside the unit circle centered at the origin O, the length $L(Q) = \sqrt{x^2 + y^2}$ of $\overline{OQ}$. The angle $\Theta(Q) = \tan^{-1}\left(\frac{y}{x}\right)$ formed by $\overline{OQ}$ and the positive x-axis.
(2) The difference $D(r) = |r - \tilde{r}|$ between a randomly chosen real number r and its rounded integer $\tilde{r}$.   ◇

Example 3.1.5 Assume the response $g \in \{$responding, not responding, busy$\}$ in a phone call. Then, the length of a phone call is a random variable and can be expressed as

$$X(g) = t \tag{3.1.3}$$

for $t \geq 0$. Here, because $P(X = 0) > 0$ and X is continuous for $(0, \infty)$, X is a hybrid random variable.   ◇

3.1.2 Cumulative Distribution Function

Let X be a random variable defined on the probability space $(\Omega, \mathcal{F}, P)$. Denote the range of X by A and denote the inverse image of B by $X^{-1}(B)$ for $B \subseteq A$. Then, we have

$$P_X(B) = P\left(X^{-1}(B)\right), \tag{3.1.4}$$

which implies that the probability of an event is equal to the probability of the inverse image of the event. Based on (3.1.4) and the probability measure P of the original probability space $(\Omega, \mathcal{F}, P)$, we can obtain the probability measure P_X of the probability space induced by the random variable X.

Example 3.1.6 Consider a rolling of a fair die and assume $P(\omega) = \frac{1}{6}$ for $\omega \in \Omega = \{1, 2, \ldots, 6\}$. Define a random variable X by $X(\omega) = -1$ for $\omega = 1$, $X(\omega) = -2$ for $\omega = 2, 3, 4$, and $X(\omega) = -3$ for $\omega = 5, 6$. Then, we have $A = \{-3, -2, -1\}$. Logically, $X^{-1}(\{-3\}) = \{5, 6\}$, $X^{-1}(\{-2\}) = \{2, 3, 4\}$, and $X^{-1}(\{-1\}) = \{1\}$. Now, the probability measure of random variable X can be obtained as $P_X(\{-3\}) =$

$P\left(X^{-1}(\{-3\})\right) = P(\{5, 6\}) = \frac{1}{3}$, $P_X(\{-2\}) = P(\{2, 3, 4\}) = \frac{1}{2}$, and $P_X(\{-1\}) = P(\{1\}) = \frac{1}{6}$. ◇

We now describe in detail the distribution of a random variable based on (3.1.4), and then define a function with which the distribution can be managed more conveniently. Consider a random variable X defined on the probability space $(\Omega, \mathcal{F}, P)$ and the range $A \subseteq \mathbb{R}$ of X. When $B \in \mathcal{B}(A)$, the set

$$X^{-1}(B) = \{\omega : X(\omega) \in B\} \tag{3.1.5}$$

is a subset of Ω and, at the same time, an element of the event space $\mathcal{F}$ due to the definition of a random variable. Based on the set $X^{-1}(B)$ shown in (3.1.5), the distribution of the random variable X can be defined as follows:

Definition 3.1.5 (*distribution*) The set function

$$\begin{aligned} P_X(B) &= P\left(X^{-1}(B)\right) \\ &= P\left(\{\omega : X(\omega) \in B\}\right) \end{aligned} \tag{3.1.6}$$

for $B \in \mathcal{B}(A)$ represents the probability measure of X and is called the distribution of the random variable X, where A is the range of X and $\mathcal{B}(A)$ is the Borel field of A.

In essence, the distribution of X is a function representing the probabilistic characteristics of the random variable X. The probability measure P_X in (3.1.6) induces a new probability space $(A, \mathcal{B}(A), P_X)$: a consequence is that we can now deal not with the original probability space $(\Omega, \mathcal{F}, P)$ but with the equivalent probability space $(A, \mathcal{B}(A), P_X)$, where the sample points are all real numbers. Figure 3.1 shows the relationship (3.1.6).

The distribution of a random variable can be described by the pmf or pdf as we have observed in Chap. 2. First, for a discrete random variable X with range A, the pmf p_X of X can be obtained as

$$p_X(x) = P_X(\{x\}), \quad x \in A \tag{3.1.7}$$

Fig. 3.1 The distribution of random variable X

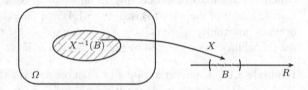

$$P_X(B) = P\left(\{\omega : X(\omega) \in B\}\right) = P\left(X^{-1}(B)\right)$$

from the distribution P_X, which in turn can be expressed as

$$P_X(B) = \sum_{x \in B} p_X(x), \quad B \in \mathcal{B}(A) \tag{3.1.8}$$

in terms of the pmf p_X. For a continuous random variable X with range A and pdf f_X, we have

$$P_X(B) = \int_B f_X(x)dx, \quad B \in \mathcal{B}(A), \tag{3.1.9}$$

which is the counterpart of (3.1.8): note that the counterpart of (3.1.7) does not exist for a continuous random variable.

Definition 3.1.6 (*cumulative distribution function*) Assume a random variable X on a sample space Ω, and let $A_x = \{s : s \in \Omega, X(s) \le x\}$ for a real number x. Then, we have $P(A_x) = P(X \le x) = P_X((-\infty, x])$ and the function

$$F_X(x) = P_X((-\infty, x]) \tag{3.1.10}$$

is called the distribution function or cumulative distribution function (cdf) of the random variable X.

The cdf $F_X(x)$ denotes the probability that X is located in the half-open interval $(-\infty, x]$. For example, $F_X(2)$ is the probability that X is in the half-open interval $(-\infty, 2]$, i.e., the probability of the event $\{-\infty < X \le 2\}$.

The pmf and cdf for a discrete random variable and the pdf and cdf for a continuous random variable can be expressed in terms of each other, as we shall see in (3.1.24), (3.1.32), and (3.1.33) later. The probabilistic characteristics of a random variable can be described by the cdf, pdf, or pmf: these three functions are all frequently indicated as the distribution function, probability distribution function, or probability function. In some cases, only the cdf is called the distribution function, and probability function in the strict sense only indicates the probability measure P as mentioned in Sect. 2.2.3. In some fields such as statistics, the name distribution function is frequently used while the name cdf is widespread in other fields including engineering.

Example 3.1.7 Let the outcome from a rolling of a fair die be X. Then, we can obtain the cdf $F_X(x) = P(X \le x)$ of X as

$$
\begin{aligned}
F_X(x) &= P(X \le x) \\
&= \begin{cases} 1, & x \ge 6, \\ \frac{i}{6}, & i \le x < i+1, \ i = 1, 2, 3, 4, 5, \\ 0, & x < 1, \end{cases}
\end{aligned} \tag{3.1.11}
$$

which is shown in Fig. 3.2.   ◇

Fig. 3.2 The cdf $F_X(x)$ of
the number X resulting from
a rolling of a fair die

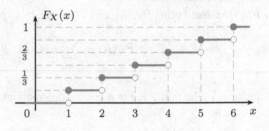

Fig. 3.3 The cdf $F_Y(x)$ of
the coordinate Y chosen
randomly in the interval
$[0, 1]$

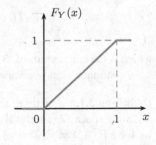

Example 3.1.8 Let the coordinate Y be a number chosen randomly in the interval
$[0, 1]$. Then, $\mathsf{P}(Y \leq x) = 1, x$, and 0 when $x \geq 1, 0 \leq x < 1$, and $x < 0$, respectively. Therefore, the cdf of Y is

$$F_Y(x) = \begin{cases} 1, & x \geq 1, \\ x, & 0 \leq x < 1, \\ 0, & x < 0. \end{cases} \tag{3.1.12}$$

Figure 3.3 shows the cdf $F_Y(x)$.   ◇

Theorem 3.1.1 *The cdf is a non-decreasing function: that is, $F(x_1) \leq F(x_2)$ when
$x_1 < x_2$ for a cdf F. In addition, we have $F(\infty) = 1$ and $F(-\infty) = 0$.*

From the definition of the cdf and probability measure, it is clear that

$$\mathsf{P}(X > x) = 1 - F_X(x) \tag{3.1.13}$$

because $\mathsf{P}(A^c) = 1 - \mathsf{P}(A)$ and that

$$\mathsf{P}(a < X \leq b) = F_X(b) - F_X(a) \tag{3.1.14}$$

for $a \leq b$. In addition, at a discontinuity point x_D of a cdf $F_X(x)$, we have $F_X(x_D) = F_X(x_D^+)$ and

$$F_X(x_D) - F_X(x_D^-) = \mathsf{P}(X = x_D) \tag{3.1.15}$$

Fig. 3.4 An example of the cdf of a hybrid random variable

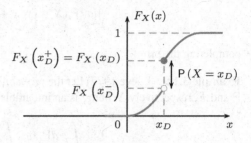

for a discrete or a hybrid random variable as shown in Fig. 3.4. On the other hand, the probability of one point is 0 for a continuous random variable: in other words, we have

$$P(X = x) = 0 \tag{3.1.16}$$

and

$$F_X(x) - F_X(x^-) = 0 \tag{3.1.17}$$

for a continuous random variable X.

Theorem 3.1.2 *The cdf is continuous from the right. That is,*

$$F_X(x) = \lim_{\epsilon \to 0} F_X(x + \epsilon), \quad \epsilon > 0 \tag{3.1.18}$$

for a cdf F_X.

Proof Consider a sequence $\{\alpha_i\}_{i=1}^{\infty}$ such that $\alpha_{i+1} \leq \alpha_i$ and $\lim_{i \to \infty} \alpha_i = 0$. Then,

$$\lim_{\varepsilon \to 0} F_X(x + \varepsilon) - F_X(x) = \lim_{i \to \infty} \{F_X(x + \alpha_i) - F_X(x)\}$$

$$= \lim_{i \to \infty} P(X \in (x, x + \alpha_i]). \tag{3.1.19}$$

Now, we have $\lim_{i \to \infty} P(X \in (x, x + \alpha_i]) = \lim_{i \to \infty} P_X((x, x + \alpha_i]) = P_X\left(\lim_{i \to \infty} \{(x, x + \alpha_i]\}\right)$ from (2.A.1) because $\{(x, x + \alpha_i]\}_{i=1}^{\infty}$ is a monotonic sequence. Subsequently, we have $P_X\left(\lim_{i \to \infty} \{(x, x + \alpha_i]\}\right) = P_X\left(\bigcap_{i=1}^{\infty} (x, x + \alpha_i]\right) = P_X(\emptyset)$ from (1.5.9) and $\bigcap_{i=1}^{\infty} (x, x + \alpha_i] = \emptyset$ as shown, for instance, in Example 1.5.9. In other words,

$$\lim_{i \to \infty} \mathsf{P}\left(X \in (x, x + \alpha_i]\right) = 0, \tag{3.1.20}$$

completing the proof.   ♠

Example 3.1.9 (Loeve 1977) Let the probability measure and corresponding cdf be P and F, respectively. When g is an integrable function,

$$\int g \, d\mathsf{P} \quad \text{or} \quad \int g \, dF \tag{3.1.21}$$

is called the Lebesgue-Stieltjes integral and is often written as, for instance,

$$\int_{[a,b)} g \, d\mathsf{P} = \int_a^b g \, dF. \tag{3.1.22}$$

When $F(x) = x$ for $x \in [0, 1]$, the measure P is called the Lebesgue measure as mentioned in Definition 2.A.7, and

$$\int_{[a,b)} g \, dx = \int_a^b g \, dx \tag{3.1.23}$$

is the Lebesgue integral. If g is continuous on $[a, b]$, then the Lebesgue-Stieltjes integral $\int_a^b g \, dF$ is the Riemann-Stieltjes integral, and the Lebesgue integral $\int_a^b g \, dx$ is the Riemann integral.   ◇

As we have already seen in Examples 3.1.7 and 3.1.8, subscripts are used to distinguish the cdf's of several random variables as in F_X and F_Y. In addition, when the cdf F_X and pdf f_X is for the random variable X with the distribution P_X, it is denoted by $X \sim \mathsf{P}_X$, $X \sim F_X$, or $X \sim f_X$. For example, $X \sim P(\lambda)$ means that the random variable X follows the Poisson distribution with parameter λ, $X \sim U[a, b)$ means that the distribution of the random variable X is the uniform distribution over $[a, b)$, and $Y \sim f_Y(t) = e^{-t}u(t)$ means that the pdf of the random variable Y is $f_Y(t) = e^{-t}u(t)$.

Theorem 3.1.3 *A cdf may have at most countably many jump discontinuities.*

Proof Assume the cdf $F(x)$ is discontinuous at x_0. Denote by D_n the set of discontinuities with the jump in the half-open interval $\left(\frac{1}{n+1}, \frac{1}{n}\right]$, where n is a natural number. Then, the number of elements in D_n is at most n, because otherwise $F(\infty) - F(-\infty) > 1$. In other words, there exists at most one discontinuity with jump between $\frac{1}{2}$ and 1, at most two discontinuities with jump between $\frac{1}{3}$ and $\frac{1}{2}$, ..., at most $n - 1$ discontinuities with jump between $\frac{1}{n}$ and $\frac{1}{n-1}$, ... Therefore the number of discontinuities is at most countable.   ♠

Theorem 3.1.3 is a special case of the more general result that a function which is continuous from the right-hand side or left-hand side at all points and a monotonic real function may have, at most, countably many jump discontinuities.

Based on the properties of the cdf, we can now redefine the continuous, discrete, and hybrid random variables as follows:

Definition 3.1.7 (*discrete random variable; continuous random variable; hybrid random variable*) A continuous, discrete, or hybrid random variable is a random variable whose cdf is a continuous, a step-like function, or a discontinuous but not a step-like function, respectively.

Here, when a function is increasing only at some points and is constant in a closed interval not containing the points, the function is called a step-like function. The cdf shown in Fig. 3.4 is an example of a hybrid random variable which is not continuous at a point x_D.

3.1.3 Probability Density Function and Probability Mass Function

In characterizing the probabilistic properties of continuous and discrete random variables, we can use a pdf and a pmf, respectively. In addition, the cdf can also be employed for the three classes of random variables: the continuous, discrete, and hybrid random variables.

Let us denote the cdf of a random variable X by F_X, the pdf by f_X when X is a continuous random variable, and the pmf by p_X when X is a discrete random variable. Then, the cdf $F_X(x) = \mathsf{P}_X((-\infty, x])$ can be expressed as

$$F_X(x) = \begin{cases} \int_{-\infty}^{x} f_X(y)dy, & \text{if } X \text{ is a continuous random variable,} \\ \sum_{y=-\infty}^{x} p_X(y), & \text{if } X \text{ is a discrete random variable.} \end{cases} \tag{3.1.24}$$

When X is a hybrid random variable, we have for $0 < \alpha < 1$

$$F_X(x) = \alpha \sum_{k=-\infty}^{x} p_X(k) + (1 - \alpha) \int_{-\infty}^{x} f_X(y)dy, \tag{3.1.25}$$

which is sufficiently general for us to deal with in this book. Note that, as described in Appendix 3.1, the most general cdf is a weighted sum of an absolutely continuous function, a discrete function, and a singular function.

The probability $\mathsf{P}_X(B) = \int_B dF_X(x)$ of an event B can be obtained as

$$P_X(B) = \begin{cases} \int_B f_X(x)dx, & \text{for a continuous random variable,} \\ \sum_{x \in B} p_X(x), & \text{for a discrete random variable.} \end{cases} \tag{3.1.26}$$

Example 3.1.10 Consider a Rayleigh random variable R. Then, from the pdf $f_R(x) = \frac{x}{\alpha^2} \exp\left(-\frac{x^2}{2\alpha^2}\right) u(x)$, the cdf $F_R(x) = \int_{-\infty}^{x} \frac{t}{\alpha^2} \exp\left(-\frac{t^2}{2\alpha^2}\right) u(t)dt$ is easily obtained as

$$F_R(x) = \left\{ 1 - \exp\left(-\frac{x^2}{2\alpha^2}\right) \right\} u(x). \tag{3.1.27}$$

When $\alpha = 1$, the probability of the event $\{1 < R < 2\}$ is $\int_1^2 f_R(t)dt = F_R(2) - F_R(1) = \sqrt{e^{-1}} - e^{-2} \approx 0.4712$. $\diamond$

Theorem 3.1.4 *The cdf F_X satisfies*

$$1 - F_X(x) = F_X(-x) \tag{3.1.28}$$

when the pdf f_X is an even function.

Proof First, $P(X > x) = \int_x^{\infty} f_X(y)dy = \int_{-x}^{-\infty} f_X(-t)(-dt) = \int_{-\infty}^{-x} f_X(t)dt = F_X(-x)$ because $f_X(x) = f_X(-x)$. Recollecting (3.1.13), we get (3.1.28). ♠

Example 3.1.11 Consider the pdf's $f_L(x) = \frac{ke^{-kx}}{(1+e^{-kx})^2}$ for $k > 0$ of the logistic distribution (Balakrishnan 1992) and $f_D(x) = \frac{\lambda}{2}e^{-\lambda|x|}$ for $\lambda > 0$ of the double exponential distribution. The cdf's of these distributions are

$$F_L(x) = \frac{1}{1 + e^{-kx}} \tag{3.1.29}$$

and

$$F_D(x) = \begin{cases} \frac{1}{2}e^{\lambda x}, & x \le 0, \\ 1 - \frac{1}{2}e^{-\lambda x}, & x \ge 0, \end{cases} \tag{3.1.30}$$

respectively, for which Theorem 3.1.4 is easily confirmed. $\diamond$

Example 3.1.12 We have the cdf

$$F_C(x) = \frac{1}{2} + \frac{1}{\pi} \tan^{-1}\left(\frac{x - \beta}{\alpha}\right) \tag{3.1.31}$$

for the Cauchy distribution with pdf $f_C(r) = \frac{\alpha}{\pi}\left\{(r-\beta)^2 + \alpha^2\right\}^{-1}$ shown in (2.5.28). ◇

From (3.1.24), we can easily see that the pdf and pmf can be obtained as

$$
\begin{aligned}
f_X(x) &= \frac{d}{dx}F_X(x) \\
&= \lim_{\varepsilon \to 0} \frac{1}{\varepsilon}P(x < X \le x+\varepsilon)
\end{aligned}
\tag{3.1.32}
$$

and

$$
p_X(x_i) = F_X(x_i) - F_X(x_{i-1})
\tag{3.1.33}
$$

from the cdf when X is a continuous random variable and a discrete random variable, respectively.

For a discrete random variable, a pmf is used normally. Yet, we can also define the pdf of a discrete random variable using the impulse function as we have observed in (2.5.37). Specifically, let the cdf and pmf of a discrete random variable X be F_X and p_X, respectively. Then, based on $F_X(x) = \sum_{x_i \le x} p_X(x_i) = \sum_i p_X(x_i)\, u(x - x_i)$, we can regard

$$
\begin{aligned}
f_X(x) &= \frac{d}{dx}\sum_i p_X(x_i)\, u(x - x_i) \\
&= \sum_i p_X(x_i)\, \delta(x - x_i)
\end{aligned}
\tag{3.1.34}
$$

as the pdf of X.

Example 3.1.13 For the pdf $f(x) = 2x$ for $x \in [0, 1]$ and 0 otherwise, sketch the cdf.

Solution Obtaining the cdf $F(x) = \int_{-\infty}^{x} f(t)dt$, we get

$$
F(x) = \begin{cases} 0, \ x < 0; & x^2, \ 0 \le x < 1; \\ 1, \ x \ge 1, \end{cases}
\tag{3.1.35}
$$

which is shown in Fig. 3.5 together with the pdf. ◇

Example 3.1.14 For the pdf

$$
f(x) = \frac{1}{2}\{u(x) - u(x - 1)\} + \frac{1}{3}\delta(x - 1) + \frac{1}{6}\delta(x - 2),
\tag{3.1.36}
$$

obtain and sketch the cdf.

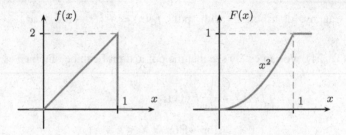

Fig. 3.5 The cdf $F(x)$ for the pdf $f(x) = 2xu(x)u(1-x)$

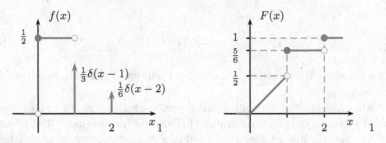

Fig. 3.6 The pdf $f(x) = \frac{1}{2}\{u(x) - u(x-1)\} + \frac{1}{3}\delta(x-1) + \frac{1}{6}\delta(x-2)$ and cdf $F(x)$

Solution First, we get the cdf $F(x) = \int_{-\infty}^{x} f(t)dt$ as

$$F(x) = \begin{cases} 0, & x < 0; & \frac{x}{2}, & 0 \le x < 1; \\ \frac{5}{6}, & 1 \le x < 2; & 1, & 2 \le x, \end{cases} \tag{3.1.37}$$

which is shown in Fig. 3.6 together with the pdf (3.1.36).   ◇

Example 3.1.15 Let X be the face of a die from a rolling. Then, the cdf of X is $F_X(x) = \frac{1}{6} \sum_{i=1}^{6} u(x-i)$, from which we get the pdf

$$f_X(x) = \frac{1}{6} \sum_{i=1}^{6} \delta(x-i) \tag{3.1.38}$$

of X by differentiation. In addition,

$$p_X(i) = \begin{cases} \frac{1}{6}, & i = 1, 2, \ldots, 6, \\ 0, & \text{otherwise} \end{cases} \tag{3.1.39}$$

is the pmf of X.   ◇

Example 3.1.16 The function (2.5.37) addressed in Example 2.5.23 is the pdf of a hybrid random variable. ◇

Example 3.1.17 A box contains G green and B blue balls. Assume we take one ball from the box n times without[1] replacement. Obtain the pmf of the number X of green balls among the n balls taken from the box.

Solution We easily get the probability of $X = k$ as

$$P(X = k) = \frac{{_B}C_{n-k}\, {_G}C_k}{{_{G+B}}C_n} \tag{3.1.40}$$

for $\{0 \le k \le G,\ 0 \le n - k \le B\}$ or, equivalently, for $\max(0, n - B) \le k \le \min(n, G)$. Thus, the pmf of X is

$$p_X(k) = \begin{cases} \dfrac{{_B}C_{n-k}\, {_G}C_k}{{_{G+B}}C_n}, & k = \check{k}, \check{k}+1, \ldots, \min(n, G), \\ 0, & \text{otherwise,} \end{cases} \tag{3.1.41}$$

where[2] $\check{k} = \max(0, n - B)$. In addition, (3.1.41) will become $p_X(k) = 1$ for $k = 0$ and 0 for $k \ne 0$ when $G = 0$, and $p_X(k) = 1$ for $k = n$ and 0 for $k \ne n$ when $B = 0$. The distribution with replacement of the balls will be addressed in Exercise 3.5. ◇

For a random variable X with pdf f_X and cdf F_X, noting that the cdf is continuous from the right-hand side, the probability of the event $\{x_1 < X \le x_2\}$ shown in (3.1.14) can be obtained as

$$P(x_1 < X \le x_2) = F_X(x_2) - F_X(x_1)$$
$$= \int_{x_1^+}^{x_2^+} f_X(x)dx. \tag{3.1.42}$$

Example 3.1.18 For the random variable Z with pdf $f_Z(z) = \frac{1}{\sqrt{2\pi}} \exp\left(-\frac{z^2}{2}\right)$, we have $P(|Z| \le 1) = \int_{-1}^{1} f_Z(z)dz \approx 0.6826$, $P(|Z| \le 2) \approx 0.9544$, and $P(|Z| \le 3) \approx 0.9974$. ◇

Using (3.1.42), the value $F(\infty) = 1$ mentioned in Theorem 3.1.1 can be confirmed as $\int_{-\infty}^{\infty} f(x)dx = P(-\infty < X \le \infty) = F(\infty) = 1$. Let us mention that although $P(x_1 \le X < x_2) = \int_{x_1^-}^{x_2^-} f_X(x)dx$, $P(x_1 \le X \le x_2) = \int_{x_1^-}^{x_2^+} f_X(x)dx$, and

---

[1] The distribution of X is a hypergeometric distribution.

[2] Here, '$\max(0, n - B) \le k \le \min(n, G)$' can be replaced with 'all integers k' by noting that $_pC_q = 0$ for $q < 0$ or $q > p$ when p is a non-negative integer and q is an integer from Table 1.4.

$P(x_1 < X < x_2) = \int_{x_1^+}^{x_2^-} f_X(x)dx$ are slightly different from (3.1.42), these four probabilities are all equal to each other unless the pdf f_X contains impulse functions at x_1 or x_2.

As it is observed, for instance, in Example 3.1.15, considering a continuous random variable with the pdf is very similar to considering a discrete random variable with the pmf. Therefore, we will henceforth focus on discussing a continuous random variable with the pdf. One final point is that

$$\lim_{x \to \pm\infty} f(x) = 0 \tag{3.1.43}$$

and

$$\lim_{x \to \pm\infty} f'(x) = 0 \tag{3.1.44}$$

hold true for all the pdf's f we will discuss in this book.

3.2 Functions of Random Variables and Their Distributions

In this section, when the cdf F_X, pdf f_X, or pmf p_X of a random variable X is known, we obtain the probability functions of a new random variable $Y = g(X)$, where g is a measurable function (Middleton 1960).

3.2.1 Cumulative Distribution Function

First, the cdf $F_Y(v) = P(Y \le v) = P(g(X) \le v)$ of $Y = g(X)$ can be obtained as

$$F_Y(v) = P(x : g(x) \le v, x \in A), \tag{3.2.1}$$

where A is the sample space of the random variable X. Using (3.2.1), the pdf or pmf of Y can be obtained subsequently: specifically, we can obtain the pdf of Y as

$$f_Y(v) = \frac{d}{dv} F_Y(v) \tag{3.2.2}$$

when Y is a continuous random variable, and the pmf of Y as

$$p_Y(v) = F_Y(v) - F_Y(v - 1) \tag{3.2.3}$$

when Y is a discrete random variable. The result (3.2.3) is for a random variable whose range is a subset of integers as described after Definition 3.1.4: more generally, we can write it as

$$p_Y(v_i) = F_Y(v_i) - F_Y(v_{i-1}) \tag{3.2.4}$$

when $A = \{v_1, v_2, \ldots\}$ instead of $A = \{0, 1, \ldots\}$.

Example 3.2.1 Obtain the cdf F_Y of $Y = aX + b$ in terms of the cdf F_X of X, where $a \neq 0$.

Solution We have the cdf $F_Y(y) = P(Y \leq y) = P(aX + b \leq y)$ as

$$
F_Y(y) = \begin{cases} P\left(X \leq \frac{y-b}{a}\right), & a > 0, \\ P\left(X \geq \frac{y-b}{a}\right), & a < 0 \end{cases}
$$
$$
= \begin{cases} F_X\left(\frac{y-b}{a}\right), & a > 0, \\ P\left(X = \frac{y-b}{a}\right) + 1 - F_X\left(\frac{y-b}{a}\right), & a < 0 \end{cases} \tag{3.2.5}
$$

by noting that the set $\{Y \leq y\}$ is equivalent to the set $\{aX + b \leq y\}$. ◇

Example 3.2.2 When the random variable X has the cdf

$$
F_X(x) = \begin{cases} 0, \; x \leq 0; & x, \, 0 \leq x \leq 1; \\ 1, \; x \geq 1; \end{cases} \tag{3.2.6}
$$

the cdf of $Y = 2X + 1$ is

$$
F_Y(y) = \begin{cases} 0, \; y \leq 1; & \frac{y-1}{2}, \, 1 \leq y \leq 3; \\ 1, \; y \geq 3; \end{cases} \tag{3.2.7}
$$

which are shown in Fig. 3.7. ◇

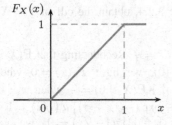 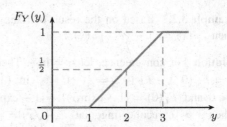

Fig. 3.7 The cdf $F_X(x)$ of X and cdf $F_Y(y)$ of $Y = 2X + 1$

Example 3.2.3 For a continuous random variable X with cdf F_X, obtain the cdf of $Y = \frac{1}{X}$.

Solution We get

$$
F_Y(y) = \mathsf{P}\left(X(yX - 1) \geq 0\right)
$$

$$
= \begin{cases} \mathsf{P}\left(X \leq 0 \text{ or } X \geq \frac{1}{y}\right), & y > 0, \\ \mathsf{P}\left(X \leq 0\right), & y = 0, \\ \mathsf{P}\left(\frac{1}{y} \leq X \leq 0\right), & y < 0 \end{cases}
$$

$$
= \begin{cases} F_X(0) + 1 - F_X\left(\frac{1}{y}\right), & y > 0, \\ F_X(0), & y = 0, \\ F_X(0) - F_X\left(\frac{1}{y}\right), & y < 0, \end{cases} \tag{3.2.8}
$$

by noting that $\left\{\frac{1}{X} \leq y\right\} = \left\{X \leq yX^2\right\} = \{(yX - 1)X \geq 0\}$.   ◇

Example 3.2.4 Obtain the cdf of $Y = aX^2$ in terms of the cdf F_X of X when $a > 0$.

Solution Because the set $\{Y \leq y\}$ is equivalent to the set $\left\{aX^2 \leq y\right\}$, the cdf of Y can be obtained as $F_Y(y) = \mathsf{P}(Y \leq y) = \mathsf{P}\left(X^2 \leq \frac{y}{a}\right)$, i.e.,

$$
F_Y(y) = \begin{cases} 0, & y < 0, \\ \mathsf{P}\left(-\sqrt{\frac{y}{a}} \leq X \leq \sqrt{\frac{y}{a}}\right), & y \geq 0, \end{cases} \tag{3.2.9}
$$

which can be rewritten as

$$
F_Y(y) = \begin{cases} 0, & y < 0, \\ F_X\left(\sqrt{\frac{y}{a}}\right) - F_X\left(-\sqrt{\frac{y}{a}}\right) + \mathsf{P}\left(X = -\sqrt{\frac{y}{a}}\right), & y \geq 0 \end{cases}
$$

$$
= \left[F_X\left(\sqrt{\frac{y}{a}}\right) - F_X\left(-\sqrt{\frac{y}{a}}\right) + \mathsf{P}\left(X = -\sqrt{\frac{y}{a}}\right)\right] u(y) \tag{3.2.10}
$$

in terms of the cdf F_X of X. In (3.2.10), it is assumed $u(0) = 1$.   ◇

Example 3.2.5 Based on the result of Example 3.2.4, obtain the cdf of $Y = X^2$ when $F_X(x) = \left(1 - \frac{2}{3}e^{-x}\right)u(x)$.

Solution For convenience, let $\alpha = \ln\frac{2}{3}$. Then, $e^\alpha = \frac{2}{3}$. Recollecting that $\mathsf{P}(X = 0) = F_X(0^+) - F_X(0^-) = \frac{1}{3} - 0 = \frac{1}{3}$ in (3.2.10), we have $F_Y(x) = 0$ when $x < 0$ and $F_Y(0) = \{1 - \exp(\alpha)\} - \{1 - \exp(\alpha)\} + \mathsf{P}(X = 0) = \frac{1}{3}$ when $x = 0$. When $x > 0$, recollecting that $F_X\left(\sqrt{x}\right) = \{1 - \exp\left(-\sqrt{x} + \alpha\right)\}u\left(\sqrt{x}\right) = 1 - \exp\left(-\sqrt{x} + \alpha\right)$ and $F_X\left(-\sqrt{x}\right) = \{1 - \exp\left(\sqrt{x} + \alpha\right)\}u\left(-\sqrt{x}\right) = 0$, we get $F_Y(x) = F_X\left(\sqrt{x}\right) - F_X\left(-\sqrt{x}\right) = 1 - \exp\left(-\sqrt{x} + \alpha\right)$. In summary,

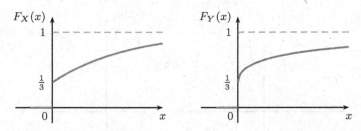

Fig. 3.8 The cdf $F_X(x) = \left(1 - \frac{2}{3}e^{-x}\right)u(x)$ and cdf $F_Y(x) = \left(1 - \frac{2}{3}e^{-\sqrt{x}}\right)u(x)$ of $Y = X^2$

$$F_Y(x) = \left(1 - \frac{2}{3}e^{-\sqrt{x}}\right)u(x), \tag{3.2.11}$$

which is shown in Fig. 3.8 together with $F_X(x)$. ◇

Example 3.2.6 Express the cdf F_Y of $Y = \sqrt{X}$ in terms of the cdf F_X of X when $P(X < 0) = 0$.

Solution We have $F_Y(y) = \begin{cases} 0, & y < 0, \\ P\left(X \le y^2\right), & y \ge 0, \end{cases}$ i.e.,

$$F_Y(y) = F_X\left(y^2\right)u(y) \tag{3.2.12}$$

from $F_Y(y) = P(Y \le y) = P\left(\sqrt{X} \le y\right)$. ◇

Example 3.2.7 Recollecting that the probability for a singleton set is 0 for a continuous random variable X, the cdf of $Y = |X|$ can be obtained as $F_Y(y) = P(Y \le y) = P(|X| \le y)$, i.e.,

$$\begin{aligned}
F_Y(y) &= \begin{cases} 0, & y < 0, \\ P(-y \le X \le y), & y \ge 0 \end{cases} \\
&= \begin{cases} 0, & y < 0, \\ F_X(y) - F_X(-y) + P(X = -y), & y \ge 0 \end{cases} \\
&= \{F_X(y) - F_X(-y)\} u(y)
\end{aligned} \tag{3.2.13}$$

in terms of the cdf F_X of X. Examples of the cdf $F_X(x)$ and $F_Y(y)$ are shown in Fig. 3.9. ◇

Example 3.2.8 When the cdf of the input X to the limiter

$$g(x) = \begin{cases} b, & x \ge b, \\ x, & -b \le x \le b, \\ -b, & x < -b \end{cases} \tag{3.2.14}$$

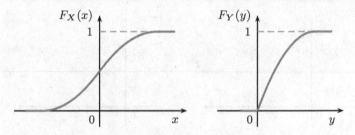

Fig. 3.9 The cdf $F_X(x)$ of X and the cdf $F_Y(y)$ of $Y = |X|$

is F_X, obtain the cdf F_Y of the output $Y = g(X)$.

Solution First, when $y < -b$ and $y \geq b$, we have $F_Y(y) = 0$ and $F_Y(y) = F_Y(b) = 1$, respectively. Next, when $-b \leq y < b$, we have $F_Y(y) = F_X(y)$ from $F_Y(y) = P(Y \leq y) = P(X \leq y)$. Thus, we eventually have

$$F_Y(y) = \begin{cases} 1, & y \geq b, \\ F_X(y), & -b \leq y < b, \\ 0, & y < -b, \end{cases} \tag{3.2.15}$$

which is continuous from the right-hand side at any point y and discontinuous at $y = \pm b$ in general.   ◇

Example 3.2.9 Obtain the cdf of $Y = g(X)$ when $X \sim U(-1, 1)$ and

$$g(x) = \begin{cases} \frac{1}{2}, & x \geq \frac{1}{2}, \\ x, & -\frac{1}{2} \leq x < \frac{1}{2}, \\ -\frac{1}{2}, & x < -\frac{1}{2}. \end{cases} \tag{3.2.16}$$

Solution The cdf of $Y = g(X)$ can be obtained as

$$F_Y(y) = \begin{cases} 1, & y \geq \frac{1}{2}, \\ \frac{1}{2}(y + 1), & -\frac{1}{2} \leq y < \frac{1}{2}, \\ 0, & y < -\frac{1}{2} \end{cases} \tag{3.2.17}$$

using (3.2.15), which is shown in Fig. 3.10.   ◇

3.2.2 Probability Density Function

Let us first introduce the following theorem which is quite useful in dealing with the differentiation of an integrated bi-variate function:

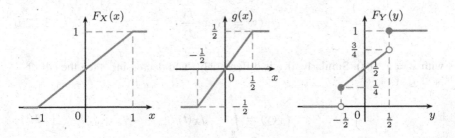

Fig. 3.10 The cdf $F_X(x)$, limiter $g(x)$, and cdf $F_Y(y)$ of $Y = g(X)$ when $X \sim U(-1, 1)$

Theorem 3.2.1 *Assume that $a(x)$ and $b(x)$ are integrable functions and that both $g(t, x)$ and $\frac{\partial}{\partial x} g(t, x)$ are continuous in x and t. Then, we have*

$$\frac{d}{dx} \int_{a(x)}^{b(x)} g(t, x) dt = g(b(x), x) \frac{db(x)}{dx} - g(a(x), x) \frac{da(x)}{dx}$$
$$+ \int_{a(x)}^{b(x)} \frac{\partial g(t, x)}{\partial x} dt, \qquad (3.2.18)$$

which is called the Leibnitz's rule.

Example 3.2.10 Assume $a(x) = x$, $b(x) = x^2$, and $g(t, x) = 2t + x$. Then, $\frac{\partial}{\partial x} \int_{a(x)}^{b(x)} g(t, x) dt = 4x^3 + 3x^2 - 4x$ from $\int_x^{x^2} (2t + x) dt = x^4 + x^3 - 2x^2$. On the other hand, $\int_{a(x)}^{b(x)} \frac{\partial g(t,x)}{\partial x} dt = x^2 - x$ from $\frac{\partial g(t,x)}{\partial x} = 1$. Therefore, $g(b(x), x) \frac{db(x)}{dx} - g(a(x), x) \frac{da(x)}{dx} + \int_{a(x)}^{b(x)} \frac{\partial g(t,x)}{\partial x} dt = 2x \left(2x^2 + x \right) - 3x + x^2 - x = 4x^3 + 3x^2 - 4x$ from $g(b(x), x) = 2x^2 + x$, $g(a(x), x) = 3x$, $\frac{da(x)}{dx} = 1$, and $\frac{db(x)}{dx} = 2x$. $\diamond$

3.2.2.1 One-to-One Transformations

We attempt to obtain the cdf $F_Y(y) = P(Y \le y)$ of $Y = g(X)$ when the pdf of X is f_X. First, if g is differentiable and increasing, then the cdf $F_Y(y) = F_X \left(g^{-1}(y) \right)$ is

$$F_Y(y) = \int_{-\infty}^{g^{-1}(y)} f_X(t) dt \qquad (3.2.19)$$

because $\{Y \le y\} = \left\{ X \le g^{-1}(y) \right\}$, where g^{-1} is the inverse of g. Thus, the pdf of $Y = g(X)$ is $f_Y(y) = \frac{d}{dy} F_Y(y) = f_X \left(g^{-1}(y) \right) \frac{dg^{-1}(y)}{dy}$, i.e.,

$$f_Y(y) = f_X(x)\frac{dx}{dy} \qquad (3.2.20)$$

with $x = g^{-1}(y)$. Similarly, if g is differentiable and decreasing, then the cdf of Y is

$$F_Y(y) = \int_{g^{-1}(y)}^{\infty} f_X(t)dt \qquad (3.2.21)$$

from $F_Y(y) = P(Y \le y) = P\left(X \ge g^{-1}(y)\right)$, and the pdf is $f_Y(y) = \frac{d}{dy}F_Y(y) = -f_X\left(g^{-1}(y)\right)\frac{dg^{-1}(y)}{dy}$, i.e.,

$$f_Y(y) = -f_X(x)\frac{dx}{dy}. \qquad (3.2.22)$$

Combining (3.2.20) and (3.2.22), we have the following theorem:

Theorem 3.2.2 *When g is a differentiable and decreasing function or a differentiable and increasing function, the pdf of $Y = g(X)$ is*

$$f_Y(y) = \left.\frac{f_X(x)}{|g'(x)|}\right|_{x=g^{-1}(y)}, \qquad (3.2.23)$$

where f_X is the pdf of X.

The result (3.2.23) can be written as $f_Y(y) = \frac{f_X(g^{-1}(y))}{|g'(g^{-1}(y))|}$, as $f_Y(y) = \left[f_X(x)\left|\frac{dx}{dy}\right|\right]_{x=g^{-1}(y)}$, or as

$$f_Y(y)|dy| = f_X(x)|dx|. \qquad (3.2.24)$$

The formula (3.2.24) represents the conservation or invariance of probability: the probability $f_X(x)|dx|$ of the region $|dx|$ of the random variable X is the same as the probability $f_Y(y)|dy|$ of the region $|dy|$ of the random variable Y when the region $|dy|$ of Y is the image of the region $|dx|$ of X under the function $Y = g(X)$.

Example 3.2.11 For a non-zero real number a, let $Y = aX + b$. Then, noting that the inverse function of $y = g(x) = ax + b$ is $x = g^{-1}(y) = \frac{y-b}{a}$ and that $\left|g'\left(g^{-1}(y)\right)\right| = |a|$, we get

$$f_{aX+b}(y) = \frac{1}{|a|}f_X\left(\frac{y-b}{a}\right) \qquad (3.2.25)$$

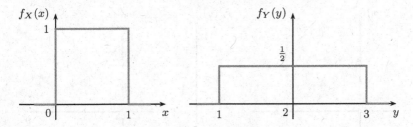

Fig. 3.11 The pdf $f_X(x)$ and pdf $f_Y(y)$ of $Y = 2X + 1$ when $X \sim U[0, 1)$

from (3.2.23). This result is the same as $f_{aX+b}(y) = \frac{d}{dy} F_{aX+b}(y)$, the derivative of the cdf (3.2.5) obtained in Example 3.2.1. Figure 3.11 shows the pdf $f_X(x)$ and pdf $f_Y(y) = \frac{1}{2}u(y + 1)u(3 - y)$ of $Y = 2X + 1$ when $X \sim U[0, 1)$. ◇

Example 3.2.12 Obtain the pdf of $Y = cX$ when $X \sim G(\alpha, \beta)$ and $c > 0$.

Solution Using (2.5.31) and (3.2.25), we get

$$f_{cX}(y) = \frac{1}{c} \frac{1}{\beta^\alpha \Gamma(\alpha)} \left(\frac{y}{c}\right)^{\alpha-1} \exp\left(-\frac{y}{c\beta}\right) u\left(\frac{y}{c}\right)$$

$$= \frac{1}{(c\beta)^\alpha \Gamma(\alpha)} y^{\alpha-1} \exp\left(-\frac{y}{c\beta}\right) u(y). \qquad (3.2.26)$$

In other words, $cX \sim G(\alpha, c\beta)$ when $X \sim G(\alpha, \beta)$ and $c > 0$. ◇

Example 3.2.13 Consider $Y = \frac{1}{X}$. Because the inverse function of $y = g(x) = \frac{1}{x}$ is $x = g^{-1}(y) = \frac{1}{y}$ and $\left|g''(g^{-1}(y))\right| = y^2$, we get

$$f_{\frac{1}{X}}(y) = \frac{1}{y^2} f_X\left(\frac{1}{y}\right) \qquad (3.2.27)$$

from (3.2.23), which can also be obtained by differentiating (3.2.8). Figure 3.12 shows the pdf $f_X(x)$ and pdf $f_Y(y)$ of $Y = \frac{1}{X}$ when $X \sim U[0, 1)$. ◇

Example 3.2.14 When $X \sim C(\alpha)$, obtain the distribution of $Y = \frac{1}{X}$.

Solution Noting that $f_X(x) = \frac{\alpha}{\pi} \frac{1}{x^2+\alpha^2}$, we get $f_Y(y) = \frac{1}{\alpha\pi} \frac{1}{y^2+\frac{1}{\alpha^2}}$ from (3.2.27). In other words, if $X \sim C(\alpha)$, then $\frac{1}{X} \sim C\left(\frac{1}{\alpha}\right)$. ◇

Example 3.2.15 Express the pdf f_Y of $Y = \sqrt{X}$ in terms of the pdf f_X of X.

Solution When $y < 0$, there is no solution to $y = \sqrt{x}$, and thus $f_Y(y) = 0$. When $y > 0$, the solution to $y = \sqrt{x}$ is $x = y^2$ and $g'(x) = \frac{1}{2\sqrt{x}}$. Therefore,

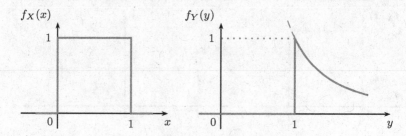

Fig. 3.12 The pdf $f_X(x)$ and pdf $f_Y(y)$ of $Y = \frac{1}{X}$ when $X \sim U[0, 1)$

$$f_{\sqrt{X}}(y) = 2yf_X\left(y^2\right)u(y), \qquad (3.2.28)$$

which is the same as $f_{\sqrt{X}}(y) = 2yf_X\left(y^2\right)u(y) + F_X(0)\delta(y)$, obtainable by differentiating $F_{\sqrt{X}}(y) = F_X\left(y^2\right)u(y)$ shown in (3.2.12), except at $y = 0$. Note that, for $\sqrt{X}$ to be meaningful, we should have $\mathsf{P}(X < 0) = 0$. Thus, when X is a continuous random variable, we have $F_X(0) = \mathsf{P}(X \le 0) = \mathsf{P}(X = 0) = 0$ and, consequently, $F_X(0)\delta(y) = 0$. We then easily obtain[3] $\int_{-\infty}^{\infty} f_{\sqrt{X}}(y)dy = \int_0^{\infty} 2yf_X\left(y^2\right)dy = \int_0^{\infty} f_X(t)dt = \int_{-\infty}^{\infty} f_X(t)dt = 1$ because $f_X(x) = 0$ for $x < 0$ from $\mathsf{P}(X < 0) = 0$.   ◇

Example 3.2.16 Obtain the pdf of $Y = \sqrt{X}$ when the pdf X is

$$f_X(x) = \begin{cases} x, & 0 \le x < 1, \\ 2 - x, & 1 \le x < 2, \\ 0, & x < 0 \text{ or } x \ge 2. \end{cases} \qquad (3.2.29)$$

Solution Noting that

$$f_X\left(y^2\right) = \begin{cases} y^2, & 0 \le y^2 < 1, \\ 2 - y^2, & 1 \le y^2 < 2, \\ 0, & y^2 < 0 \text{ or } y^2 \ge 2, \end{cases} \qquad (3.2.30)$$

we get

$$f_Y(y) = \begin{cases} 2y^3, & 0 \le y < 1, \\ 2y\left(2 - y^2\right), & 1 \le y < \sqrt{2}, \\ 0, & y < 0 \text{ or } y \ge \sqrt{2} \end{cases} \qquad (3.2.31)$$

from (3.2.28), which is shown in Fig. 3.13.   ◇

---

[3] We can equivalently obtain $\int_{-\infty}^{\infty} f_{\sqrt{X}}(y)dy = \int_0^{\infty} 2yf_X\left(y^2\right)dy + F_X(0) = \int_0^{\infty} f_X(t)dt + \int_{-\infty}^0 f_X(t)dt = \int_{-\infty}^{\infty} f_X(t)dt = 1$ using $F_X(0) = \int_{-\infty}^0 f_X(t)dt$.

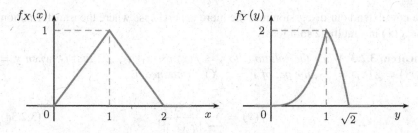

Fig. 3.13 The pdf $f_Y(y)$ of $Y = \sqrt{X}$ for the pdf $f_X(x)$ of X

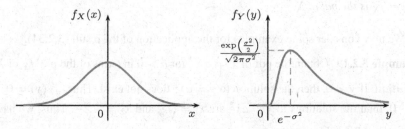

Fig. 3.14 The pdf $f_X(x)$ of $X \sim \mathcal{N}(0, \sigma^2)$ and pdf of the log-normal random variable $Y = e^X$

Example 3.2.17 Express the pdf f_Y of $Y = e^X$ in terms of the pdf f_X of X.

Solution When $y \le 0$, there is no solution to $y = e^x$, and thus $f_Y(y) = 0$. When $y > 0$, the solution to $y = e^x$ is $x = \ln y$ and $g'(x) = e^x$. We therefore get

$$f_{e^x}(y) = \frac{1}{y} f_X(\ln y) u(y), \qquad (3.2.32)$$

assuming $u(0) = 0$.　　　　　　　　　　　　　　　　　　　　　◇

Example 3.2.18 When $X \sim \mathcal{N}(m, \sigma^2)$, obtain the distribution of $Y = e^X$.

Solution Noting that $f_X(x) = \frac{1}{\sqrt{2\pi}\sigma} \exp\left\{-\frac{(x-m)^2}{2\sigma^2}\right\}$, we get

$$f_Y(y) = \frac{1}{\sqrt{2\pi\sigma^2}\, y} \exp\left\{-\frac{(\ln y - m)^2}{2\sigma^2}\right\} u(y), \qquad (3.2.33)$$

which is called the log-normal pdf. Figure 3.14 shows the pdf $f_X(x)$ of $X \sim \mathcal{N}(0, \sigma^2)$ and the pdf (3.2.33) of the log-normal random variable $Y = e^X$.　　◇

3.2.2.2　General Transformations

We have discussed the probability functions of $Y = g(X)$ in terms of those of X when the transformation g is a one-to-one correspondence, via (3.2.23) in previous section.

We now extend our discussion into the more general case where the transformation $y = g(x)$ has multiple solutions.

Theorem 3.2.3 *When the solutions to $y = g(x)$ are x_1, x_2, $\ldots$, that is, when $y = g(x_1) = g(x_2) = \cdots$, the pdf of $Y = g(X)$ is obtained as*

$$f_Y(y) = \sum_{i=1}^{\infty} \frac{f_X(x_i)}{|g'(x_i)|}, \tag{3.2.34}$$

where f_X is the pdf of X.

We now consider some examples for the application of the result (3.2.34).

Example 3.2.19 Obtain the pdf of $Y = aX^2$ for $a > 0$ in terms of the pdf f_X of X.

Solution If $y < 0$, then the solution to $y = ax^2$ does not exist. Thus, $f_Y(y) = 0$. If $y > 0$, then the solutions to $y = ax^2$ are $x_1 = \sqrt{\frac{y}{a}}$ and $x_2 = -\sqrt{\frac{y}{a}}$. Thus, we have $|g'(x_1)| = |g'(x_2)| = 2a\sqrt{\frac{y}{a}}$ from $g'(x) = 2ax$ and, subsequently,

$$f_{aX^2}(y) = \frac{1}{2\sqrt{ay}} \left\{ f_X\left(\sqrt{\frac{y}{a}}\right) + f_X\left(-\sqrt{\frac{y}{a}}\right) \right\} u(y), \tag{3.2.35}$$

which is, as expected, the same as the result obtainable by differentiating the cdf (3.2.10) of $Y = aX^2$.   ◇

Example 3.2.20 When $X \sim \mathcal{N}(0, 1)$, we can easily obtain the pdf[4] $f_Y(y) = \frac{1}{\sqrt{2\pi y}} \exp\left(-\frac{y}{2}\right) u(y)$ of $Y = X^2$ by noting that $f_X(x) = \frac{1}{\sqrt{2\pi}} \exp\left(-\frac{x^2}{2}\right)$.   ◇

Example 3.2.21 Express the pdf f_Y of $Y = |X|$ in terms of the pdf f_X of X.

Solution When $y < 0$, there is no solution to $y = |x|$, and thus $f_Y(y) = 0$. When $y > 0$, the solutions to $y = |x|$ are $x_1 = y$ and $x_2 = -y$, and $|g'(x)| = 1$. Thus, we get

$$f_Y(y) = \{f_X(y) + f_X(-y)\} u(y), \tag{3.2.36}$$

which is the same as

$$f_Y(y) = \frac{d}{dy} [\{F_X(y) - F_X(-y)\} u(y)]$$
$$= \{f_X(y) + f_X(-y)\} u(y) \tag{3.2.37}$$

---

[4] This pdf is called the central chi-square pdf with the degree of freedom of 1. The central chi-square pdf, together with the non-central chi-square pdf, is discussed in Sect. 5.4.2.

obtained by differentiating the cdf $F_Y(y)$ in (3.2.13), and then, noting that $\{F_X(y) - F_X(-y)\}\,\delta(y) = \{F_X(0) - F_X(0)\}\,\delta(y) = 0$.   ◇

Example 3.2.22 When $X \sim U[-\pi, \pi)$, obtain the pdf and cdf of $Y = a\sin(X + \theta)$, where $a > 0$ and θ are constants.

Solution First, we have $f_Y(y) = 0$ for $|y| > a$. When $|y| < a$, letting the two solutions to $y = g(x) = a\sin(x + \theta)$ in the interval $[-\pi, \pi)$ of x be x_1 and x_2, we have $f_X(x_1) = f_X(x_2) = \frac{1}{2\pi}$. Thus, recollecting that $|g'(x)| = |a\cos(x + \theta)| = \sqrt{a^2 - y^2}$, we get

$$f_Y(y) = \frac{1}{\pi\sqrt{a^2 - y^2}} u(a - |y|) \tag{3.2.38}$$

from (3.2.34). Next, let us obtain the cdf $F_Y(y)$. When $0 \le y \le a$, letting $\alpha = \sin^{-1}\frac{y}{a}$ and $0 \le \alpha < \frac{\pi}{2}$, we have $x_1 = \alpha - \theta$ and $x_2 = \pi - \alpha - \theta$ and, consequently, $F_Y(y) = \mathsf{P}(Y \le y) = \mathsf{P}(-\pi \le X \le x_1) + \mathsf{P}(x_2 \le X < \pi)$. Now, from $\mathsf{P}(-\pi \le X \le x_1) = \frac{1}{2\pi}(x_1 + \pi)$ and $\mathsf{P}(x_2 \le X < \pi) = \frac{1}{2\pi}(\pi - x_2)$, we have $F_Y(y) = \frac{1}{2\pi}(2\pi + 2\alpha - \pi)$, i.e.,

$$F_Y(y) = \frac{1}{2} + \frac{1}{\pi}\sin^{-1}\frac{y}{a}. \tag{3.2.39}$$

When $-a \le y \le 0$, letting $\beta = \sin^{-1}\frac{y}{a}$ and $-\frac{\pi}{2} \le \beta < 0$, we have $x_1 = \beta - \theta$, $x_2 = -\pi - \beta - \theta$, and $x_1 - x_2 = \pi + 2\beta$, and thus the cdf is $F_Y(y) = \mathsf{P}(Y \le y) = \mathsf{P}(x_2 \le X \le x_1) = \frac{1}{2\pi}(\pi + 2\beta)$, i.e.,

$$F_Y(y) = \frac{1}{2} + \frac{1}{\pi}\sin^{-1}\frac{y}{a}. \tag{3.2.40}$$

Combining $F_Y(y) = 0$ for $y \le -a$, $F_Y(y) = 1$ for $y \ge a$, (3.2.39), and (3.2.40), we get[5]

$$F_Y(y) = \begin{cases} 0, & y \le -a, \\ \frac{1}{2} + \frac{1}{\pi}\sin^{-1}\frac{y}{a}, & |y| \le a, \\ 1, & y \ge a. \end{cases} \tag{3.2.41}$$

The cdf (3.2.41) can of course be obtained from the pdf (3.2.38) by integration: specifically, from $F_Y(y) = \int_{-\infty}^{y} \frac{u(a-|t|)}{\pi\sqrt{a^2-t^2}}dt$, we get $F_Y(y) = 0$ when $y \le -a$, $F_Y(y) = \frac{1}{\pi}\int_{-\frac{\pi}{2}}^{\sin^{-1}\frac{y}{a}} \frac{1}{a\cos\theta} a\cos\theta d\theta = \frac{1}{\pi}\left(\sin^{-1}\frac{y}{a} + \frac{\pi}{2}\right) = \frac{1}{2} + \frac{1}{\pi}\sin^{-1}\frac{y}{a}$ when $-a \le y \le a$, and $F_Y(y) = \frac{1}{2} + \frac{1}{\pi}\sin^{-1}1 = 1$ when $y \ge a$. Figure 3.15 shows the pdf

---

[5] If $a < 0$, a will be replaced with $|a|$.

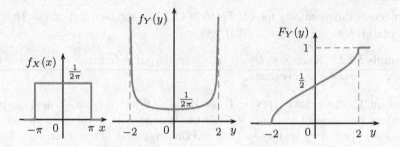

Fig. 3.15 The pdf $f_X(x)$, pdf $f_Y(y)$ of $Y = 2\sin(X + \theta)$, and cdf $F_Y(y)$ of Y when $X \sim U[-\pi, \pi)$

$f_X(x)$, pdf $f_Y(y)$, and cdf $F_Y(y)$ when $a = 2$. Exercise 3.4 discusses a slightly more general problem.   ◇

Example 3.2.23 For a continuous random variable X with cdf F_X, obtain the cdf, pdf, and pmf of $Z = \text{sgn}(X)$, where

$$
\begin{aligned}
\text{sgn}(x) &= u(x) - u(-x) \\
&= 2u(x) - 1 \\
&= \begin{cases} 1, & x > 0, \\ 0, & x = 0, \\ -1, & x < 0 \end{cases}
\end{aligned}
\tag{3.2.42}
$$

is called the sign function. First, we have the cdf $F_Z(z) = \text{P}(Z \le z) = \text{P}(\text{sgn}(X) \le z)$ as

$$
\begin{aligned}
F_Z(z) &= \begin{cases} 0, & z < -1, \\ \text{P}(X \le 0), & -1 \le z < 1, \\ 1, & z \ge 1 \end{cases} \\
&= F_X(0)u(z + 1) + \{1 - F_X(0)\}u(z - 1),
\end{aligned}
\tag{3.2.43}
$$

and thus the pdf $f_Z(z) = \frac{d}{dz} F_Z(z)$ of Z is

$$
f_Z(z) = F_X(0)\delta(z + 1) + \{1 - F_X(0)\}\delta(z - 1).
\tag{3.2.44}
$$

In addition, we also have

$$
p_Z(z) = \begin{cases} F_X(0), & z = -1, \\ 1 - F_X(0), & z = 1, \\ 0, & \text{otherwise} \end{cases}
\tag{3.2.45}
$$

as the pmf of Z.   ◇

3.2.2.3 Finding Transformations

We have so far discussed obtaining the probability functions of $Y = g(X)$ when the probability functions of X and g are given. We now briefly consider the inverse problem of finding g when the cdf's F_X and F_Y are given or, equivalently, finding the function that transforms X with cdf F_X into Y with cdf F_Y.

The problem can be solved by making use of the uniform distribution as the intermediate step: i.e.,

$$F_X(x) \rightarrow \text{ uniform distribution } \rightarrow F_Y(y). \tag{3.2.46}$$

Specifically, assume that the cdf F_X and the inverse F_Y^{-1} of the cdf F_Y are continuous and increasing. Letting

$$Z = F_X(X), \tag{3.2.47}$$

we have $X = F_X^{-1}(Z)$ and $\{F_X(X) \le z\} = \{X \le F_X^{-1}(z)\}$ because F_X is continuous and increasing. Therefore, the cdf[6] of Z is $F_Z(z) = P(Z \le z) = P(F_X(X) \le z) = P(X \le F_X^{-1}(z)) = F_X(F_X^{-1}(z)) = z$ for $0 \le z < 1$. In other words, we have

$$Z \sim U[0, 1). \tag{3.2.48}$$

Next, consider

$$V = F_Y^{-1}(Z). \tag{3.2.49}$$

Then, recollecting (3.2.48), we get the cdf $P(V \le y) = P(F_Y^{-1}(Z) \le y) = P(Z \le F_Y(y)) = F_Z(F_Y(y)) = F_Y(y)$ of V because $F_Z(x) = x$ for $x \in (0, 1)$. In other words, when $X \sim F_X$, we have $V = F_Y^{-1}(Z) = F_Y^{-1}(F_X(X)) \sim F_Y$, which is summarized as the following theorem:

Theorem 3.2.4 *The function that transforms a random variable X with cdf F_X into a random variable with cdf F_Y is $g = F_Y^{-1} \circ F_X$.*

Figure 3.16 illustrates some of the interesting results such as

$$X \sim F_X \rightarrow F_X(X) \sim U[0, 1), \tag{3.2.50}$$
$$Z \sim U[0, 1) \rightarrow F_Y^{-1}(Z) \sim F_Y, \tag{3.2.51}$$

and

$$X \sim F_X \rightarrow F_Y^{-1}(F_X(X)) \sim F_Y. \tag{3.2.52}$$

---

[6] Here, because F_X is a continuous function, $F_X(F_X^{-1}(z)) = z$ as it is discussed in (3.A.26).

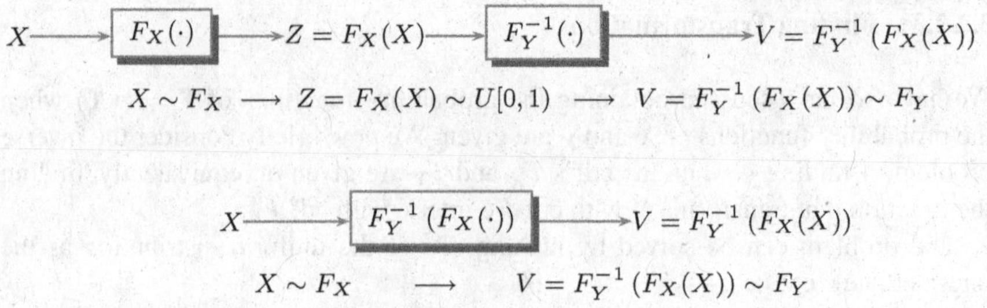

$$X \longrightarrow \boxed{F_X(\cdot)} \longrightarrow Z = F_X(X) \longrightarrow \boxed{F_Y^{-1}(\cdot)} \longrightarrow V = F_Y^{-1}(F_X(X))$$

$$X \sim F_X \longrightarrow Z = F_X(X) \sim U[0,1) \longrightarrow V = F_Y^{-1}(F_X(X)) \sim F_Y$$

$$X \longrightarrow \boxed{F_Y^{-1}(F_X(\cdot))} \longrightarrow V = F_Y^{-1}(F_X(X))$$

$$X \sim F_X \longrightarrow V = F_Y^{-1}(F_X(X)) \sim F_Y$$

Fig. 3.16 Transformation of a random variable X with cdf F_X into Y with cdf F_Y

Theorem 3.2.4 can be used in the generation of random numbers, for instance.

Example 3.2.24 From $X \sim U[0,1)$, obtain the Rayleigh random variable $Y \sim f_Y(y) = \frac{y}{\alpha^2} \exp\left(-\frac{y^2}{2\alpha^2}\right) u(y)$.

Solution Because the cdf of Y is $F_Y(y) = \left\{1 - \exp\left(-\frac{y^2}{2\alpha^2}\right)\right\} u(y)$, the function we are looking for is $g(x) = F_Y^{-1}(x) = \sqrt{-2\alpha^2 \ln(1-x)}$ as we can easily see from (3.2.51). In other words, if $X \sim U[0,1)$, then $Y = \sqrt{-2\alpha^2 \ln(1-X)}$ has the cdf $F_Y(y) = \left\{1 - \exp\left(-\frac{y^2}{2\alpha^2}\right)\right\} u(y)$. Note that, we conversely have $V = 1 - \exp\left(-\frac{y^2}{2\alpha^2}\right) \sim U(0,1)$.   ◇

Example 3.2.25 For $X \sim U(0,1)$, consider the desired pmf

$$p_Y(y_n) = \mathsf{P}(Y = y_n)$$
$$= \begin{cases} p_n, & n = 1, 2, \ldots, \\ 0, & \text{otherwise.} \end{cases} \tag{3.2.53}$$

Then, letting $p_0 = 0$, the integer Y satisfying

$$\sum_{k=0}^{Y-1} p_k < X \le \sum_{k=0}^{Y} p_k \tag{3.2.54}$$

is the random variable with the desired pmf (3.2.53).   ◇

3.3   Expected Values and Moments

The probabilistic characteristics of a random variable can be most completely described by the distribution via the cdf, pdf, or pmf of the random variable. On the other hand, the distribution is not available in some cases, and we may wish to summarize the characteristics as a few numbers in other cases.

In this section, we attempt to introduce some of the key notions for use in such cases. Among the widely employed representative values, also called central values, for describing the probabilistic characteristics of a random variable and a distribution are the mean, median, and mode (Beckenbach and Bellam 1965; Bickel and Doksum 1977; Feller 1970; Hajek 1969; McDonough and Whalen 1995).

Definition 3.3.1 (*mode*) For a random variable X with pdf f_X or pmf p_X, if

$$f_X (x_{mod}) \geq f_X(x) \text{ for } X \text{ a continuous random variable, or} \qquad (3.3.1)$$
$$p_X (x_{mod}) \geq p_X(x) \text{ for } X \text{ a discrete random variable} \qquad (3.3.2)$$

holds true for all real number x, then the value x_{mod} is called the mode of X.

The mode is the value that could happen most frequently among all the values of a random variable. In other words, the mode is the most probable value or, equivalently, the value at which the pmf or pdf of a random variable is maximum.

Definition 3.3.2 (*median*) The value α satisfying both $P(X \leq \alpha) \geq \frac{1}{2}$ and $P(X \geq \alpha) \geq \frac{1}{2}$ is called the median of the random variable X.

Roughly speaking, the median is the value at which the cumulative probability is 0.5. When the distribution is symmetric, the point of symmetry of the cdf is the median. The median is one of the quantiles of order p, or $100p$ percentile, defined as the number ξ_p satisfying $P(X \leq \xi_p) \geq p$ and $P(X \geq \xi_p) \geq 1 - p$ for $0 < p < 1$. For a random variable X with cdf F_X, we have

$$p \leq F_X (\xi_p) \leq p + P (X = \xi_p). \qquad (3.3.3)$$

Therefore, if $P (X = \xi_p) = 0$ as for a continuous random variable, the solution to $F_X(x) = p$ is ξ_p: the solution to this equation is unique when the cdf F_X is a strictly increasing function, but otherwise, there exist many solutions, each of which is the quantile of order p.

The median and mode are not unique in some cases. When there exist many medians, the middle value is regarded as the median in some cases.

Example 3.3.1 For the pmf $p_X(1) = \frac{1}{3}$, $p_X(2) = \frac{1}{2}$, and $p_X(3) = \frac{1}{6}$, because $P(X \leq 2) = \frac{1}{3} + \frac{1}{2} = \frac{5}{6} \geq \frac{1}{2}$ and $P(X \geq 2) = \frac{1}{2} + \frac{1}{6} = \frac{2}{3} \geq \frac{1}{2}$, the median[7] is 2. For the uniform distribution over the set $\{1, 2, 3, 4\}$, any real number in the interval[8] $[2, 3]$ is the median, and the mode is 1, 2, 3, or 4. ◇

---

[7] Note that if the median x_{med} is defined by $P(X \leq x_{med}) = P (X \geq x_{med})$, we do not have the median in this pmf.

[8] Note that if the median x_{med} is defined by $P(X \leq x_{med}) = P (X \geq x_{med})$, any real number in the interval $(2, 3)$ is the median.

Example 3.3.2 For the distribution $\mathcal{N}(1, 1)$, the mode is 1. When the pmf is $p_X(1)$ = $\frac{1}{3}$, $p_X(2) = \frac{1}{2}$, and $p_X(3) = \frac{1}{6}$, the mode of X is 2.   ◇

3.3.1 Expected Values

We now introduce the most widely used representative value, the expected value.

Definition 3.3.3 (*expected value*) For a random variable X with cdf F_X, the value $E\{X\} = \int_{-\infty}^{\infty} x \, dF_X(x)$, i.e.,

$$E\{X\} = \begin{cases} \int_{-\infty}^{\infty} x f_X(x) dx, & X \text{ continuous random variable}, \\ \sum_{x=-\infty}^{\infty} x p_X(x), & X \text{ discrete random variable} \end{cases} \quad (3.3.4)$$

is called the expected value or mean of X if $\int_{-\infty}^{\infty} |x| \, dF_X(x) < \infty$.

The expected value is also called the stochastic average, statistical average, or ensemble average, and $E\{X\}$ is also written as $E(X)$ or $E[X]$.

Example 3.3.3 For $X \sim U[a, b]$, we have the expected value $E\{X\} = \int_a^b \frac{x}{b-a} dx = \frac{b^2-a^2}{2(b-a)} = \frac{a+b}{2}$ of X. The mode of X is any real number between a and b, and the median is the same as the mean $\frac{a+b}{2}$.   ◇

Example 3.3.4 (Stoyanov 2013) For unimodal random variables, the median usually lies between the mode and mean: an example of exception is shown here. Assume the pdf

$$f(x) = \begin{cases} 0, & x \le 0, \\ x, & 0 < x \le c, \\ ce^{-\lambda(x-c)}, & x > c \end{cases} \quad (3.3.5)$$

of X with $c \ge 1$ and $\frac{c^2}{2} + \frac{c}{\lambda} = 1$. Then, the mean, median, and mode of X are $\mu = \frac{c^3}{3} + \frac{c^2}{\lambda} + \frac{c}{\lambda^2}$, 1, and c, respectively. If we choose $c > 1$ sufficiently close to 1, then $\lambda \approx 2$ and $\mu \approx \frac{13}{12}$, and the median is smaller than the mean and mode although $f(x)$ is unimodal.   ◇

Theorem 3.3.1 (Stoyanov 2013) *A necessary condition for the mean* $E\{X\}$ *to exist for a random variable* X *with cdf* F *is* $\lim_{x \to \infty} x\{1 - F(x)\} = 0$.

Proof Rewrite $x\{1 - F(x)\}$ as $x\{1 - F(x)\} = x\left\{\int_{-\infty}^{\infty} f(t)dt - \int_{-\infty}^{x} f(t)dt\right\} = x\int_x^{\infty} f(t)dt$. Now, letting $E\{X\} = m$, we have

$$m = \int_{-\infty}^{x} t f(t) dt + \int_{x}^{\infty} t f(t) dt$$

$$\geq \int_{-\infty}^{x} t f(t) dt + x \int_{x}^{\infty} f(t) dt \qquad (3.3.6)$$

for $x > 0$ because $\int_{x}^{\infty} t f(t) dt \geq x \int_{x}^{\infty} f(t) dt$. Here, we should have $\lim_{x \to \infty} x \int_{x}^{\infty} f(t) dt \to 0$ for (3.3.6) to hold true because $\lim_{x \to \infty} \int_{-\infty}^{x} t f(t) dt = \int_{-\infty}^{\infty} t f(t) dt = m$ when $x \to \infty$. ♠

Based on the result (3.E.2) shown in Exercise 3.1, we can show that (Rohatgi and Saleh 2001)

$$E\{X\} = \int_{0}^{\infty} P(X > x) dx - \int_{-\infty}^{0} P(X \leq x) dx \qquad (3.3.7)$$

for any continuous random variable X, dictating that a necessary and sufficient condition for $E\{|X|\} < \infty$ is that both $\int_{0}^{\infty} P(X > x) dx$ and $\int_{-\infty}^{0} P(X \leq x) dx$ converge.

3.3.2 Expected Values of Functions of Random Variables

Based on the discussions in the previous section, let us now consider the expected values of functions of a random variable. Let F_Y be the cdf of $Y = g(X)$. Then, the expected value of $Y = g(X)$ can be expressed as

$$E\{Y\} = \int_{-\infty}^{\infty} y \, dF_Y(y). \qquad (3.3.8)$$

In essence, the expected value of $Y = g(X)$ can be evaluated using (3.3.8) after we have obtained the cdf, pdf, or pmf of Y from that of X. On the other hand, the expected value of $Y = g(X)$ can be evaluated as $E\{Y\} = E\{g(X)\} = \int_{-\infty}^{\infty} g(x) dF_X(x)$, i.e.,

$$E\{Y\} = \begin{cases} \int_{-\infty}^{\infty} g(x) f_X(x) dx, & \text{continuous random variable,} \\ \\ \sum_{x=-\infty}^{\infty} g(x) p_X(x), & \text{discrete random variable.} \end{cases} \qquad (3.3.9)$$

While the first approach (3.3.8) of evaluating the expected value of $Y = g(X)$ requires that we need to first obtain the cdf, pdf, or pmf of Y from that of X, the second approach (3.3.9) does not require the cdf, pdf, or pmf of Y. In the second approach, we simply multiply the pdf $f_X(x)$ or pmf $p_X(x)$ of X with $g(x)$ and then integrate or sum without first having to obtain the cdf, pdf, or pmf of Y. In short, if there is no other reason to obtain the cdf, pdf, or pmf of $Y = g(X)$, the second approach is faster in the evaluation of the expected value of $Y = g(X)$.

Example 3.3.5 When $X \sim U[0, 1)$, obtain the expected value of $Y = X^2$.

Solution (Method 1) Based on (3.2.35), we can obtain the pdf $f_Y(y) = \frac{1}{2\sqrt{y}} \left\{ f_X \left(\sqrt{y} \right) + f_X \left(-\sqrt{y} \right) \right\} u(y) = \frac{1}{2\sqrt{y}} \{ u(y) - u(y-1) \}$ of Y. Next, using (3.3.8), we get $\mathsf{E}\{Y\} = \int_0^1 \frac{y}{2\sqrt{y}} dy = \frac{1}{2} \int_0^1 \sqrt{y} dy = \frac{1}{3}$.

(Method 2) Using (3.3.9), we can directly obtain $\mathsf{E}\{Y\} = \int_0^1 x^2 dx = \frac{1}{3}$.   ◇

From the definition of the expected value, we can deduce the following properties:

(1) When a random variable X is non-negative, i.e., when $\mathsf{P}(X \geq 0) = 1$, we have $\mathsf{E}\{X\} \geq 0$.
(2) The expected value of a constant is the constant. In other words, if $\mathsf{P}(X = c) = 1$, then $\mathsf{E}\{X\} = c$.
(3) The expected value is a linear operator: that is, we have $\mathsf{E}\left\{ \sum_{i=1}^{n} a_i g_i(X) \right\} = \sum_{i=1}^{n} a_i \mathsf{E}\{g_i(X)\}$.
(4) For any function h, we have $|\mathsf{E}\{h(X)\}| \leq \mathsf{E}\{|h(X)|\}$.
(5) If $h_1(x) \leq h_2(x)$ for every point x, then we have $\mathsf{E}\{h_1(X)\} \leq \mathsf{E}\{h_2(X)\}$.
(6) For any function h, we have $\min(h(X)) \leq \mathsf{E}\{h(X)\} \leq \max(h(X))$.

Example 3.3.6 Based on (3) above, we have $\mathsf{E}\{aX + b\} = a\mathsf{E}\{X\} + b$ when a and b are constants.   ◇

Example 3.3.7 For a continuous random variable $X \sim U(1, 9)$ and $h(x) = \frac{1}{\sqrt{x}}$, compare $h(\mathsf{E}\{X\})$, $\mathsf{E}\{h(X)\}$, $\min(h(X))$, and $\max(h(X))$.

Solution We have $h(\mathsf{E}\{X\}) = h(5) = \frac{1}{\sqrt{5}}$ from the result in Example 3.3.3 and $\mathsf{E}\{h(X)\} = \int_{-\infty}^{\infty} h(x) f_X(x) dx = \int_1^9 \frac{1}{8\sqrt{x}} dx = \frac{1}{2}$ from (3.3.9). In addition, $\min(h(X)) = \frac{1}{\sqrt{9}} = \frac{1}{3}$ and $\max(h(X)) = \frac{1}{\sqrt{1}} = 1$. Therefore, $\frac{1}{3} < \frac{1}{2} < 1$, i.e., $\min(h(X)) \leq \mathsf{E}\{h(X)\} \leq \max(h(X))$, confirming (6).   ◇

3.3.3   Moments and Variance

Definition 3.3.4 (*moment*) For a random variable X with cdf F_X, we call $m_n = \mathsf{E}\{X^n\}$, i.e.,

$$m_n = \int_{-\infty}^{\infty} x^n dF_X(x) \tag{3.3.10}$$

the n-th moment of X if $\mathsf{E}\{|X|^n\} = \int_{-\infty}^{\infty} |x|^n dF_X(x) < \infty$ for $n = 0, 1, \ldots$.

In other words, the expected value of a power of a random variable is called the moment, and the moment is one of the expected values of a function of a random variable. The n-th moment of X can specifically be written as

$$
m_n = \begin{cases} \int_{-\infty}^{\infty} x^n f_X(x)dx, & \text{continuous random variable,} \\ \sum_{x=-\infty}^{\infty} x^n p_X(x), & \text{discrete random variable.} \end{cases} \tag{3.3.11}
$$

Definition 3.3.5 (*central moment*) The expected value $\mu_n = \mathsf{E}\{(X - \mathsf{E}\{X\})^n\}$, i.e.,

$$
\mu_n = \begin{cases} \int_{-\infty}^{\infty} (x - m_1)^n f_X(x)dx, & \text{continuous random variable,} \\ \sum_{x=-\infty}^{\infty} (x - m_1)^n p_X(x), & \text{discrete random variable} \end{cases} \tag{3.3.12}
$$

is called the n-th central moment of X for $n = 0, 1, \ldots$.

From Definitions 3.3.4 and 3.3.5, it is easy to see that $m_0 = \mu_0 = 1$, $m_1 = \mathsf{E}\{X\}$, and $\mu_1 = 0$. More generally, we have

$$
\mu_n = \sum_{k=0}^{n} {}_nC_k m_k (-m_1)^{n-k} \tag{3.3.13}
$$

from $\mathsf{E}\{(X - m_1)^n\} = \mathsf{E}\left\{\sum_{k=0}^{n} {}_nC_k X^k (-m_1)^{n-k}\right\}$, and conversely,

$$
m_n = \sum_{k=0}^{n} {}_nC_k \mu_k m_1^{n-k} \tag{3.3.14}
$$

from $m_n = \mathsf{E}[\{(X - m_1) + m_1\}^n] = \mathsf{E}\left\{\sum_{k=0}^{n} {}_nC_k (X - m_1)^k m_1^{n-k}\right\}$ between the moments $\{m_n\}_{n=0}^{\infty}$ and the central moments $\{\mu_n\}_{n=0}^{\infty}$. Often, we also consider the absolute moment $\mathsf{E}\{|X|^n\}$.

Some of the moments and functions of moments are used more frequently than the others in representing the probabilistic properties of a random variable. One such important parameter is the variance.

Definition 3.3.6 (*variance; standard deviation*) The second central moment μ_2 is called the variance, and the non-negative square root of the variance is called the standard deviation.

The variance of X is often denoted by σ_X^2, $\mathsf{Var}\{X\}$, or $V\{X\}$, and can also be expressed as

$$\sigma_X^2 = \mathsf{E}\{X^2\} - \mathsf{E}^2\{X\}$$
$$= m_2 - m_1^2 \tag{3.3.15}$$

from $\mathsf{E}\{(X - \mathsf{E}\{X\})^2\} = \mathsf{E}[X^2 - 2X\mathsf{E}\{X\} + \mathsf{E}^2\{X\}] = \mathsf{E}\{X^2\} - \mathsf{E}^2\{X\}$. We also have $\mathsf{Var}\{aX\} = a^2\mathsf{Var}\{X\}$ for any real number a.

Example 3.3.8 Assume the pdf $f_X(r) = \frac{1}{b-a}\{u(r - a) - u(r - b)\}$ for a uniform random variable X. Then, we have the mean

$$\mathsf{E}\{X\} = \frac{a + b}{2} \tag{3.3.16}$$

and variance

$$\mathsf{Var}\{X\} = \frac{1}{12}(b - a)^2 \tag{3.3.17}$$

from $\mathsf{Var}\{X\} = \int_a^b \frac{x^2}{b-a} dx - \left(\frac{a+b}{2}\right)^2$.   ◇

Example 3.3.9 For the exponential random variable X with pdf $f(r) = \lambda e^{-\lambda r} u(r)$, the mean $\mathsf{E}\{X\} = \lambda \int_0^\infty r e^{-\lambda r} dr$ is

$$\mathsf{E}\{X\} = \frac{1}{\lambda} \tag{3.3.18}$$

and the variance is

$$\mathsf{Var}\{X\} = \frac{1}{\lambda^2} \tag{3.3.19}$$

from $\mathsf{Var}\{X\} = \lambda \int_0^\infty r^2 e^{-\lambda r} dr - \left(\frac{1}{\lambda}\right)^2$.   ◇

Example 3.3.10 Obtain the mean and variance for $X \sim \mathcal{N}(m, \sigma^2)$.

Solution Letting $\frac{x-m}{\sqrt{2\sigma^2}} = t$, we have the expected value $\mathsf{E}\{X\} = \int_{-\infty}^\infty \frac{\sqrt{2}\sigma t + m}{\sqrt{2\pi\sigma^2}} \exp\left(-t^2\right) \sqrt{2}\sigma dt = \frac{1}{\sqrt{\pi}} \left\{ \int_{-\infty}^\infty \sqrt{2}\sigma t \exp\left(-t^2\right) dt + m \int_{-\infty}^\infty \exp\left(-t^2\right) dt \right\} = \frac{m}{\sqrt{\pi}}\sqrt{\pi}$, i.e.,

$$\mathsf{E}\{X\} = m, \tag{3.3.20}$$

and the second moment $E\{X^2\} = \frac{1}{\sqrt{2\pi\sigma^2}} \int_{-\infty}^{\infty} \left(2\sigma^2 t^2 + 2\sqrt{2}m\sigma t + m^2\right)\sqrt{2}\sigma$
$\exp\left(-t^2\right) dt = \frac{1}{\sqrt{\pi}} \left\{ \int_{-\infty}^{\infty} 2\sigma^2 t^2 \exp\left(-t^2\right) dt + m^2\sqrt{\pi} \right\}$, i.e.,

$$E\{X^2\} = \frac{1}{\sqrt{\pi}} \left(\sigma^2\sqrt{\pi} + m^2\sqrt{\pi}\right)$$
$$= \sigma^2 + m^2. \tag{3.3.21}$$

Consequently, $\text{Var}\{X\} = E\{X^2\} - m^2 = \sigma^2$. In (3.3.20) and (3.3.21), we have used

$$\int_{-\infty}^{\infty} t^k \exp\left(-t^2\right) dt = \begin{cases} \sqrt{\pi}, & k = 0, \\ 0, & k = 1, \\ \frac{\sqrt{\pi}}{2}, & k = 2. \end{cases} \tag{3.3.22}$$

The first and third results in (3.3.22) can be shown easily by recollecting that the integration of the standard normal pdf over the entire real line is 1, i.e., $\frac{1}{\sqrt{2\pi}} \exp\left(-\frac{x^2}{2}\right) dx = 1$ with integration by parts. The second result is based on that $\int_0^{\infty} t \exp\left(-t^2\right) dt < \infty$ and that $t \exp\left(-t^2\right)$ is an odd function. ◇

Example 3.3.11 We have the mean

$$E\{X\} = np \tag{3.3.23}$$

and variance

$$\sigma_X^2 = np(1 - p) \tag{3.3.24}$$

for the binomial random variable $X \sim b(n, p)$. ◇

Example 3.3.12 We have the mean $E\{X\} = \sum_{k=1}^{\infty} k\frac{e^{-\lambda}\lambda^k}{k!} = \lambda \sum_{k=0}^{\infty} \frac{e^{-\lambda}\lambda^k}{k!}$, i.e.,

$$E\{X\} = \lambda, \tag{3.3.25}$$

second moment $E\{X^2\} = \lambda^2 + \lambda$, and variance

$$\sigma_X^2 = \lambda \tag{3.3.26}$$

for the Poisson random variable $X \sim P(\lambda)$. ◇

Example 3.3.13 Consider the Cauchy random variable with pdf $f(r) = \frac{\alpha}{\pi} \frac{1}{r^2 + \alpha^2}$. Then, because the absolute moments are $\mathsf{E}\{|X|\} = \infty$ and $\mathsf{E}\{|X|^2\} = \infty$, the mean and variance do not exist. Similarly, for a random variable with pmf

$$p(k) = \begin{cases} \frac{6}{\pi^2 k^2}, & k = 1, 2, \ldots, \\ 0, & \text{otherwise,} \end{cases} \tag{3.3.27}$$

the mean and variance do not exist.   ◇

Example 3.3.14 Consider $X \sim \mathcal{N}(0, \sigma^2)$. Recollecting $\Gamma\left(\frac{1}{2}\right) = \sqrt{\pi}$ shown in (1.4.83), we have $\int_{-\infty}^{\infty} \exp\left(-\alpha x^2\right) dx = 2 \int_0^{\infty} e^{-t} \frac{dt}{2\sqrt{\alpha t}} = \frac{1}{\sqrt{\alpha}} \Gamma\left(\frac{1}{2}\right)$, i.e.,

$$\int_{-\infty}^{\infty} \exp\left(-\alpha x^2\right) dx = \sqrt{\frac{\pi}{\alpha}}, \tag{3.3.28}$$

which can also be obtained from $\int_{-\infty}^{\infty} \frac{\sqrt{\alpha}}{\sqrt{\pi}} \exp\left(-\alpha x^2\right) dx = 1$. Differentiating (3.3.28) k times with respect to α using (3.2.18), we get

$$\int_{-\infty}^{\infty} x^{2k} \exp\left(-\alpha x^2\right) dx = \frac{(2k-1)!!}{2^k \alpha^k} \sqrt{\frac{\pi}{\alpha}} \tag{3.3.29}$$

for $k = 1, 2, \ldots$, which can also be obtained as $\int_{-\infty}^{\infty} x^{2k} \exp(-\alpha x^2) dx = 2 \int_0^{\infty} \left(\frac{t}{\alpha}\right)^k e^{-t} \frac{dt}{2\sqrt{\alpha t}} = \frac{1}{\alpha^k \sqrt{\alpha}} \Gamma\left(k + \frac{1}{2}\right) = \frac{(2k-1)!!}{2^k \alpha^k \sqrt{\alpha}} \Gamma\left(\frac{1}{2}\right)$, where

$$(2k-1)!! = (2k-1)(2k-3) \times \cdots \times 3 \times 1 \tag{3.3.30}$$

for k a natural number. Based on the symmetry of the pdf $f(x)$ of $\mathcal{N}(0, \sigma^2)$ and (3.3.29), we get

$$\mathsf{E}\{X^n\} = \begin{cases} 0, & n \text{ is odd}, \\ (n-1)!! \, \sigma^n, & n \text{ is even}. \end{cases} \tag{3.3.31}$$

We specifically have

$$\mathsf{E}\{X^4\} = 3\sigma^4 \tag{3.3.32}$$

and

$$\mathsf{E}\{X^6\} = 15\sigma^6. \tag{3.3.33}$$

In addition, when n is an even number, $\mathsf{E}\{|X|^n\} = \mathsf{E}\{X^n\}$. When n is an odd number $2k + 1$, we have $\mathsf{E}\{|X|^n\} = \int_{-\infty}^{\infty} |x|^n f_X(x) dx = \frac{\sqrt{2}}{\sqrt{\pi\sigma^2}} \int_0^{\infty} x^{2k+1} \exp$

$$\left(-\frac{x^2}{2\sigma^2}\right) dx = 2^k k! \sqrt{\frac{2}{\pi}} \sigma^{2k+1} = 2^{\frac{n-1}{2}} \Gamma\left(\frac{n+1}{2}\right) \sqrt{\frac{2}{\pi}} \sigma^n \text{ because } \int_0^\infty x^n e^{-a^2 x^2} dx = \int_0^\infty$$
$$\frac{t^{\frac{n}{2}}}{a^n} e^{-t} \frac{dt}{2a\sqrt{t}} = \frac{1}{2a^{n+1}} \int_0^\infty t^{\frac{n}{2}-\frac{1}{2}} e^{-t} dt, \text{ i.e.,}$$

$$\int_0^\infty x^n e^{-a^2 x^2} dx = \frac{1}{2a^{n+1}} \Gamma\left(\frac{n+1}{2}\right). \tag{3.3.34}$$

In summary, we have

$$E\{|X|^n\} = \begin{cases} 2^k k! \sqrt{\frac{2}{\pi}} \sigma^{2k+1}, & n = 2k+1, k = 0, 1, \ldots, \\ (n-1)!!\sigma^n, & n \text{ is even,} \end{cases} \tag{3.3.35}$$

with which

$$E\{|X|\} = \sqrt{\frac{2}{\pi}}\sigma \tag{3.3.36}$$

and

$$E\{|X|^3\} = \sqrt{\frac{8}{\pi}}\sigma^3 \tag{3.3.37}$$

can be confirmed.   ◇

Example 3.3.15 (Romano and Siegel 1986; Stoyanov 2013) When the distribution is symmetric, i.e., when the cdf F satisfies $F(-x) = 1 - F(x)$, all the odd-ordered moments are zero. On the other hand, the converse does not necessarily hold true. For example, consider the pdf

$$f_\gamma(x) = \frac{1}{24} \exp\left(-x^{\frac{1}{4}}\right) \left(1 - \gamma \sin x^{\frac{1}{4}}\right) u(x) \tag{3.3.38}$$

with $\gamma \in [-1, 1]$. Using that

$$\int x^n e^{ax} \sin bx \, dx = e^{ax} \sum_{k=1}^{n+1} \frac{(-1)^{k+1} n! \, x^{n-k+1}}{(n-k+1)!(a^2+b^2)^{\frac{k}{2}}} \sin(bx + kt) \tag{3.3.39}$$

with $t = \sin^{-1}\left(-\frac{b}{\sqrt{a^2+b^2}}\right)$, we can show that $\int_0^\infty x^k \exp\left(-x^{\frac{1}{4}}\right) \sin x^{\frac{1}{4}} dx = 0$ for $k = 0, 1, \ldots$. Thus, for any value of γ, the k-th moment is $\int_0^\infty \frac{x^k}{24} \exp\left(-x^{\frac{1}{4}}\right) dx = \frac{1}{6} \int_0^\infty v^{4k+3} e^{-v} dv = \frac{1}{6}\Gamma(4k+4)$. Now, the pdf

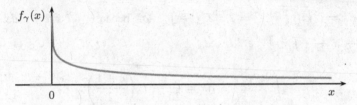

Fig. 3.17 The pdf $f_\gamma(x) = \frac{1}{24}\exp\left(-x^{\frac{1}{4}}\right)\left(1 - \gamma\sin x^{\frac{1}{4}}\right)u(x)$: when $\gamma \neq \beta$, although $f(x) = \frac{1}{2}\{f_\gamma(x)u(x) + f_\beta(-x)u(-x)\}$ is not symmetric, all odd ordered moments are 0

$$f(x) = \begin{cases} \frac{1}{2}f_\gamma(x), & x \geq 0, \\ \frac{1}{2}f_\beta(-x), & x < 0 \end{cases} \qquad (3.3.40)$$

for $\gamma \neq \beta$ is not an even function, and the $(2n+1)$-st moment of $f(x)$ is

$$\int_{-\infty}^{\infty} x^{2n+1} f(x)\, dx = \frac{1}{2}\left\{\int_0^\infty x^{2n+1} f_\gamma(x)\, dx + \int_{-\infty}^0 x^{2n+1} f_\beta(-x)\, dx\right\}$$

$$= \frac{1}{2}\left\{\int_0^\infty x^{2n+1} f_\gamma(x)\, dx - \int_0^\infty x^{2n+1} f_\beta(x)\, dx\right\}$$

$$= 0 \qquad (3.3.41)$$

because the moment of $f_\gamma(x)$ is equal to that of $f_\beta(x)$ at the same order. Specifically, when $\gamma = 1$ and $\beta = -1$, all the odd-ordered moments are 0 and the even-ordered moments are $m_{2n} = \frac{1}{6}(8n+3)!$. The pdf (3.3.38) is shown in Fig. 3.17.   ◇

3.3.4 Characteristic and Moment Generating Functions

We have discussed so far how we can obtain the moments based on the cdf, pdf, or pmf. On the other hand, we can easily obtain the moments by using the Laplace or Fourier transform as we can solve, for example, differential equations more easily by using the Laplace transform.

The set $\left\{\frac{1}{\sqrt{2\pi}}e^{j\omega x}\right\}_{\omega \in \mathbb{R}}$ of complex orthonormal basis functions has the property of $\left\langle \frac{1}{\sqrt{2\pi}}e^{j\omega x}, \frac{1}{\sqrt{2\pi}}e^{j\nu x}\right\rangle = \frac{1}{2\pi}\int_{-\infty}^\infty e^{j\omega x}e^{-j\nu x}\,dx = \frac{1}{2\pi}\int_{-\infty}^\infty e^{j(\omega-\nu)x}\,dx$, i.e.,

$$\left\langle \frac{1}{\sqrt{2\pi}}e^{j\omega x}, \frac{1}{\sqrt{2\pi}}e^{j\nu x}\right\rangle = \delta(\omega - \nu), \qquad (3.3.42)$$

which can be shown easily from, for example, (1.E.11), where $j = \sqrt{-1}$. The Fourier transform $H(\omega) = \mathfrak{F}\{h(x)\}$ of $h(x)$ and the inverse Fourier transform $h(x) = \mathfrak{F}^{-1}\{H(\omega)\}$ of $H(\omega)$ can be expressed as

$$H(\omega) = \int_{-\infty}^{\infty} h(x)e^{-j\omega x}dx \tag{3.3.43}$$

and

$$h(x) = \frac{1}{2\pi} \int_{-\infty}^{\infty} H(\omega)e^{j\omega x}d\omega, \tag{3.3.44}$$

respectively, based on the set $\left\{\frac{1}{\sqrt{2\pi}}e^{j\omega x}\right\}_{\omega \in \mathbb{R}}$ of the orthonormal basis functions.

3.3.4.1   Characteristic Functions

Definition 3.3.7 (*characteristic function*) The function

$$\varphi_X(\omega) = \begin{cases} \int_{-\infty}^{\infty} f_X(x)e^{j\omega x}dx, & \text{continuous random variable,} \\ \sum_{x=-\infty}^{\infty} p_X(x)e^{j\omega x}, & \text{discrete random variable,} \end{cases} \tag{3.3.45}$$

which is the expected value $\varphi_X(\omega) = \mathsf{E}\{e^{j\omega X}\}$, is called the characteristic function (cf) of X.

If we let $p_k = p_X(x_k) = p_X(k) = \mathsf{P}(X = x_k) = \mathsf{P}(X = k)$ for an integer k, then the cf of the discrete random variable X can be expressed as

$$\varphi_X(\omega) = \sum_{k=-\infty}^{\infty} p_k e^{jk\omega} \tag{3.3.46}$$

because we can put $x_k = k$ in discrete random variables as discussed following Definition 3.1.4.

Theorem 3.3.2 *If the cdf's of two random variables are the same, then their cf's are the same, and vice versa.*

In other words, the cf is also a function with which we can characterize the probabilistic properties of a random variable.

Example 3.3.16 For a geometric random variable with pmf $p(k) = (1 - \alpha)^k \alpha$ for $k \in \{0, 1, \ldots\}$, the cf is

$$\varphi(\omega) = \frac{\alpha}{1 - (1 - \alpha)e^{j\omega}}. \tag{3.3.47}$$

If the pmf is $p(k) = (1 - \alpha)^{k-1}\alpha$ for $k \in \{1, 2, \ldots\}$ for a geometric random variable, then the cf is $\varphi(\omega) = \frac{\alpha e^{j\omega}}{1-(1-\alpha)e^{j\omega}}$. For the NB distribution with pmf (2.5.14), the cf is

$$\varphi(\omega) = \frac{\alpha^r}{\{1 - (1 - \alpha)e^{j\omega}\}^r} \tag{3.3.48}$$

while the cf is $\varphi(\omega) = \left\{\frac{\alpha e^{j\omega}}{1-(1-\alpha)e^{j\omega}}\right\}^r$ if the NB distribution has pmf (2.5.17). ◇

Example 3.3.17 We have the cf $\varphi(\omega) = \int_{-\infty}^{\infty} \frac{1}{\sqrt{2\pi\sigma}} \exp\left\{ -\frac{1}{2\sigma^2}\left(x^2 - 2mx + m^2\right) + j\omega x\right\}dx = \exp\left(-\frac{\sigma^2\omega^2}{2} + jm\omega\right)\frac{1}{\sqrt{2\pi\sigma}}\int_{-\infty}^{\infty}\exp\left[-\frac{1}{2\sigma^2}\left\{x - \left(m + j\omega\sigma^2\right)\right\}^2\right]dx$, i.e.,

$$\varphi(\omega) = \exp\left(-\frac{\sigma^2\omega^2}{2} + jm\omega\right) \tag{3.3.49}$$

for the normal random variable with mean m and variance σ^2. ◇

3.3.4.2 Properties of Characteristic Functions

The cf $\varphi(\omega)$ has the following properties:

(1) The cf $\varphi(\omega)$ has its maximum magnitude of 1 at $\omega = 0$. In other words, $|\varphi(\omega)| \le \varphi(0) = 1$.
(2) The cf $\varphi(\omega)$ is uniformly continuous at every real number ω.
(3) The cf $\varphi(\omega)$ is positive semi-definite. In other words,

$$\sum_{l=1}^{n}\sum_{k=1}^{n} z_l z_k^* \varphi(\omega_l - \omega_k) \ge 0, \tag{3.3.50}$$

where $\{\omega_k\}_{k=1}^{n}$ are real numbers and $\{z_k\}_{k=1}^{n}$ are complex numbers.

Proof

(1) We easily have $\varphi(0) = \mathsf{E}\{1\} = 1$. Assuming the pdf f when the cf is $\varphi(\omega)$, we have $|\varphi(\omega)| \le \int_{-\infty}^{\infty} |e^{j\omega x}| f(x)dx = 1$.

(2) Consider a real number μ_0 such that $0 < \mu_0 < \frac{\varepsilon}{2}$ for a positive real number ε. Assuming a periodic function $\bar{b}(y)$ with period π and $\bar{b}(y) = |y| + \mu_0$ for $|y| \le \frac{\pi}{2}$, we have

$$|\sin y| \le \bar{b}(y). \tag{3.3.51}$$

Next, for a random variable X with pdf f_X and cf $\varphi(\omega)$, let $\mathsf{E}\{\bar{b}(X)\} = \int_{-\infty}^{\infty} \bar{b}(x) f_X(x)dx = \mu_0 + \sum_{n=-\infty}^{\infty} \int_{n\pi-\frac{\pi}{2}}^{n\pi+\frac{\pi}{2}} |x - n\pi| f_X(x)dx = \bar{\mu}$. Then, we have $0 < \mu_0 \le \bar{\mu} \le \frac{\pi}{2} + \mu_0 < \infty$ and

$$\begin{aligned}
\mathsf{E}\{\bar{b}(\nu X)\} &= \mu_0 + \sum_{n=-\infty}^{\infty} \int_{n\pi-\frac{\pi}{2}}^{n\pi+\frac{\pi}{2}} |\nu(x - n\pi)| f_X(x)dx \\
&= |\nu|(\bar{\mu} - \mu_0) + \mu_0
\end{aligned} \tag{3.3.52}$$

for a constant ν. In addition, using (3.3.51), we get

$$\begin{aligned}
\left|e^{j\alpha X} - e^{j\beta X}\right| &= 2\left|\sin\frac{(\alpha - \beta)X}{2}\right| \\
&\le 2\bar{b}\left(\frac{\alpha - \beta}{2}X\right)
\end{aligned} \tag{3.3.53}$$

from $\left|e^{j\alpha X} - e^{j\beta X}\right| = \left[\{\cos(\alpha X) - \cos(\beta X)\}^2 + \{\sin(\alpha X) - \sin(\beta X)\}^2\right]^{\frac{1}{2}} = \sqrt{2 - 2\cos\{(\alpha - \beta)X\}}$. If we let $\delta = \frac{\varepsilon - 2\mu_0}{\bar{\mu}}$, then $0 < \delta < \infty$. Therefore, from (3.3.52) and (3.3.53), we get $|\varphi(\alpha) - \varphi(\beta)| = \left|\mathsf{E}\{e^{j\alpha X} - e^{j\beta X}\}\right| \le \mathsf{E}\{\left|e^{j\alpha X} - e^{j\beta X}\right|\} \le |\alpha - \beta|(\bar{\mu} - \mu_0) + 2\mu_0 < \frac{\varepsilon - 2\mu_0}{\bar{\mu}}(\bar{\mu} - \mu_0) + 2\mu_0$, i.e.,

$$|\varphi(\alpha) - \varphi(\beta)| < \begin{cases} \frac{\varepsilon - 2\mu_0}{\bar{\mu} - \mu_0}(\bar{\mu} - \mu_0) + 2\mu_0, & \text{if } \bar{\mu} - \mu_0 \ne 0, \\ 2\mu_0, & \text{if } \bar{\mu} - \mu_0 = 0 \end{cases}$$
$$\le \varepsilon \tag{3.3.54}$$

when $|\alpha - \beta| < \delta$. Thus, for any $\varepsilon > 0$, we have $|\varphi(\alpha) - \varphi(\beta)| < \varepsilon$ if $|\alpha - \beta| < \delta = \frac{\varepsilon - 2\mu_0}{\bar{\mu}}$, implying that $\varphi(\omega)$ is uniformly continuous at every real number ω.

(3) For a random variable with cf $\varphi(\omega)$ and pdf f, we have

$$\sum_{l=1}^{n}\sum_{k=1}^{n} z_l z_k^* \varphi(\omega_l - \omega_k) = \mathsf{E}\left\{\left|\sum_{l=1}^{n} z_l e^{j\omega_l X}\right|^2\right\}$$
$$\ge 0 \tag{3.3.55}$$

because $\displaystyle\sum_{l=1}^{n}\sum_{k=1}^{n} z_l z_k^* \varphi(\omega_l - \omega_k) = \sum_{l=1}^{n}\sum_{k=1}^{n} z_l z_k^* \int_{-\infty}^{\infty} e^{j(\omega_l - \omega_k)x} f(x)dx = \int_{-\infty}^{\infty}$

$\displaystyle\sum_{l=1}^{n} z_l e^{j\omega_l x} \sum_{k=1}^{n} z_k^* e^{-j\omega_k x} f(x)dx.$ ♠

Theorem 3.3.3 *The cf of* $Y = aX + b$ *is*

$$\varphi_Y(\omega) = e^{jb\omega} \varphi_X(a\omega) \tag{3.3.56}$$

when φ_X *is the cf of* X *and* $a, b \in \mathbb{R}$.

Example 3.3.18 We have obtained the cf $\varphi_X(\omega) = \exp\left(-\frac{\sigma^2 \omega^2}{2} + jm\omega\right)$ of $X \sim \mathcal{N}(m, \sigma^2)$ in Example 3.3.17. Based on this result and (3.3.56), we can obtain the cf of $Y = \frac{1}{\sigma}(X - m)$ as $\varphi_Y(\omega) = \exp\left\{j\omega\left(-\frac{m}{\sigma}\right)\right\} \exp\left\{-\frac{\sigma^2}{2}\left(\frac{\omega^2}{\sigma^2}\right) + jm\left(\frac{\omega}{\sigma}\right)\right\}$, i.e.,

$$\varphi_Y(\omega) = \exp\left(-\frac{\omega^2}{2}\right). \tag{3.3.57}$$

This result implies that if $X \sim \mathcal{N}(m, \sigma^2)$, then $\frac{1}{\sigma}(X - m) \sim \mathcal{N}(0, 1)$. ◇

3.3.4.3　Moment Generating Functions

Definition 3.3.8 (*moment generating function*) The function

$$M_X(t) = \int_{-\infty}^{\infty} e^{tx} dF_X(x), \tag{3.3.58}$$

which is the expected value $M_X(t) = \mathsf{E}\left\{e^{tX}\right\}$, is called the moment generating function (mgf) of X.

Denoting the Laplace transform of the pdf f of X by $\tilde{M}(t) = \mathcal{L}(f)$, the mgf of X is $\tilde{M}(-t)$. The mgf $M(t)$ and the cf $\varphi(\omega)$ of a random variable is related as

$$\varphi(\omega) = M(t)\Big|_{t=j\omega}, \tag{3.3.59}$$

implying that the cf and mgf are basically the same in the sense that, by taking the inverse transform of the cf or mgf, we can obtain the cdf. Normally, the cf is guaranteed its convergence whereas the convergence region of the mgf should be considered for the inverse transform, and for some distributions the mgf does not exist.

Based on the discussion in Sect. 3.2 and Definitions 3.3.7 and 3.3.8, the cf of $Y = g(X)$ can be obtained as $\varphi_Y(\omega) = \mathsf{E}\left\{e^{j\omega Y}\right\} = \int_{-\infty}^{\infty} e^{j\omega y} dF_Y(y)$, i.e.,

$$\varphi_Y(\omega) = E\left\{e^{j\omega g(X)}\right\}$$

$$= \int_{-\infty}^{\infty} e^{j\omega g(x)} dF_X(x). \tag{3.3.60}$$

We can subsequently obtain the mgf of $Y = g(X)$ as

$$M_Y(t) = \int_{-\infty}^{\infty} e^{tg(x)} dF_X(x) \tag{3.3.61}$$

by replacing $j\omega$ with t in (3.3.60).

3.3.4.4 Characteristic and Cumulative Distribution Functions

It is easy to see that

$$\varphi(\omega) = \mathfrak{F}\{f(x)\}_{\omega \to -\omega} \tag{3.3.62}$$

from the definition of the cf. In other words, the cf is the complex conjugate of the Fourier transform of the pdf. Hence, we can obtain the pdf from the cf as

$$f(x) = \mathfrak{F}^{-1}\{\varphi(-\omega)\}$$

$$= \frac{1}{2\pi} \int_{-\infty}^{\infty} \varphi(\omega) e^{-j\omega x} d\omega \tag{3.3.63}$$

from the property of the Fourier transform.

Now, let us express the cdf $F(x) = \int_{-\infty}^{x} f(t)dt$ as the convolution

$$F(x) = f(x) * u(x) \tag{3.3.64}$$

of the pdf $f(x)$ and the unit step function $u(x)$. The Fourier transform of the convolution of two functions is the product of the Fourier transforms of the two functions. Noting that the Fourier transform of the unit step function $u(x)$ is

$$\mathfrak{F}\{u(x)\} = \pi\delta(\omega) + \frac{1}{j\omega} \tag{3.3.65}$$

as discussed in Exercise 1.26, the Fourier transform of the cdf $F(x)$ is $\mathfrak{F}\{F(x)\} = \mathfrak{F}\{f(x)\}\mathfrak{F}\{u(x)\} = \varphi(-\omega)\left\{\pi\delta(\omega) + \frac{1}{j\omega}\right\}$, i.e.,

$$\mathfrak{F}\{F(x)\} = \pi\varphi(0)\delta(\omega) + \frac{\varphi(-\omega)}{j\omega}. \tag{3.3.66}$$

Inverse transforming (3.3.66), the cdf $F(x) = \frac{1}{2\pi} \int_{-\infty}^{\infty} \left\{ \pi\varphi(0)\delta(\omega) + \frac{\varphi(-\omega)}{j\omega} \right\}$
$\exp{(j\omega x)}\, d\omega$ can be expressed as (Papoulis 1962)

$$F(x) = \frac{\varphi(0)}{2} + \frac{1}{2\pi j} \int_{-\infty}^{\infty} \frac{\varphi(-\omega)}{\omega} \exp{(j\omega x)}\, d\omega$$

$$= \frac{\varphi(0)}{2} + \frac{j}{2\pi} \int_{-\infty}^{\infty} \frac{\varphi(\omega)}{\omega} \exp{(-j\omega x)}\, d\omega. \tag{3.3.67}$$

in terms of the cf $\varphi(\omega)$.

3.3.4.5 Cumulants

Expanding the natural logarithm $\psi(\omega) = \ln \varphi(\omega) = \ln \left\{ 1 + \sum_{s=1}^{\infty} (j\omega)^s \frac{m_s}{s!} \right\}$ of the cf
$\varphi(\omega)$ in the power series of $j\omega$ near $\omega = 0$, we get

$$\psi(\omega) = \left\{ \sum_{s=1}^{\infty} (j\omega)^s \frac{m_s}{s!} \right\} - \frac{1}{2}\left\{ \sum_{s=1}^{\infty} (j\omega)^s \frac{m_s}{s!} \right\}^2 + \frac{1}{3}\left\{ \sum_{s=1}^{\infty} (j\omega)^s \frac{m_s}{s!} \right\}^3 + \cdots$$

$$= m_1 \frac{j\omega}{1!} + \left(m_2 - m_1^2 \right)\frac{(j\omega)^2}{2!} + \left(m_3 - 3m_1 m_2 + 2m_1^3 \right)\frac{(j\omega)^3}{3!} + \cdots$$

$$= \sum_{n=1}^{\infty} k_n \frac{(j\omega)^n}{n!}, \tag{3.3.68}$$

based on which the cumulant is defined as follows:

Definition 3.3.9 (*cumulant*) The parameter k_n in (3.3.68) can be expressed as

$$k_n = \left. \frac{\partial^n}{\partial (j\omega)^n} \psi(\omega) \right|_{\omega=0} \tag{3.3.69}$$

and is called the n-th cumulant.

Example 3.3.19 The first, second, and third cumulants are the same as the mean
$k_1 = m_1$, the variance $k_2 = m_2 - m_1^2 = \sigma^2$, and the third central moment $k_3 = m_3 -$
$3m_2 m_1 + 2m_1^3 = \mu_3$, respectively. In addition, the fourth cumulant is $k_4 = m_4 -$
$4m_3 m_1 - 3m_2^2 + 12m_2 m_1^2 - 6m_1^4 = \mu_4 - 3\left(m_2 - m_1^2 \right)^2 = \mu_4 - 3\sigma^4$. ◇

Definition 3.3.10 (*coefficient of variation; skewness; kurtosis*) Let the mean, variance, n-th central moment, and n-th cumulant be m, σ^2, μ_n, and k_n, respectively.
Then, $v_1 = \frac{\sigma}{m}$, $v_2 = \frac{\mu_3}{\sigma^3} = \frac{k_3}{\sqrt{k_2^3}}$, and $v_3 = \frac{\mu_4}{\sigma^4} = 3 + \frac{k_4}{k_2^2}$ are called the coefficient of
variation, skewness, and kurtosis, respectively.

Fig. 3.18 The skewness v_2 and symmetry of pdf

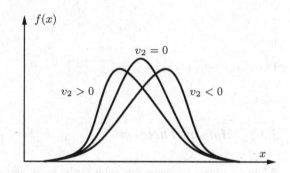

In characterizing the probabilistic properties of a random variable, we can first consider the expected value, and then the variance. The coefficient of variation, skewness, and kurtosis can then be considered in the characterization. These three parameters represent deviations of a distribution from the normal distribution. The coefficient of variation is a measure of dispersion normalized by the mean. The skewness represents the degree of asymmetry: specifically, the distribution is symmetric about the mean when $v_2 = 0$, the mean is greater than the mode and median (called right-skewed or positively-skewed) when $v_2 > 0$, and the mean is smaller than the mode and median (called left-skewed or negatively-skewed) when $v_2 < 0$. Figure 3.18 shows an example. Skewness is frequently used along with kurtosis to better judge the likelihood of events falling in the tails of a probability distribution.

Example 3.3.20 The skewness of the geometric distribution with parameter α is $v_2 = \frac{2-\alpha}{\sqrt{1-\alpha}}$, and that of $NB(r, \alpha)$ is $v_2 = \frac{2-\alpha}{\sqrt{(1-\alpha)r}}$. ◇

Example 3.3.21 Assume the pdf $f_X(x) = \lambda \exp(-\lambda x) u(x)$ of X. Then, as we have observed in Example 3.3.9, we have $\mathsf{E}\{X\} = m = \frac{1}{\lambda}$, $\mu_2 = \mathsf{Var}\{X\} = \sigma^2 = \frac{1}{\lambda^2}$, and $\mu_3 = \int_0^\infty \left(x - \frac{1}{\lambda}\right)^3 \lambda e^{-\lambda x} dx = -\left[\left\{\frac{6}{\lambda^3} + \frac{6}{\lambda^2}\left(x - \frac{1}{\lambda}\right) + \frac{3}{\lambda}\left(x - \frac{1}{\lambda}\right)^2 + \left(x - \frac{1}{\lambda}\right)^3\right\} e^{-\lambda x}\right]_0^\infty = \frac{2}{\lambda^3}$. Therefore, the coefficient of variation is $v_1 = \frac{\sigma}{m} = \left(\frac{1}{\lambda}\right)\left(\frac{1}{\lambda}\right)^{-1} = 1$ and the skewness is $v_2 = \frac{\mu_3}{\sigma^3} = \left(\frac{2}{\lambda^3}\right)\left(\frac{1}{\lambda^3}\right)^{-1} = 2$. ◇

The kurtosis v_3 represents how sharp the peak is when compared to the normal distribution: when $v_3 = 3$, the sharpness of the peak of the distribution is the same as that of the normal distribution, when $v_3 < 3$, the distribution is less sharp than the normal distribution, called platykurtic or mild peak, and when $v_3 > 3$, the distribution is sharper than the normal distribution, called leptokurtic or sharp peak.

Example 3.3.22 When the pdf is $f_X(x) = \lambda \exp(-\lambda x) u(x)$ for X, $\sigma = \frac{1}{\lambda}$ as we have observed in Example 3.3.21. In addition, $\mu_4 = \int_0^\infty \left(x - \frac{1}{\lambda}\right)^4 \lambda e^{-\lambda x} dx = -\left[\left\{\left(x - \frac{1}{\lambda}\right)^4 + \frac{4}{\lambda}\left(x - \frac{1}{\lambda}\right)^3 + \frac{12}{\lambda^2}\left(x - \frac{1}{\lambda}\right)^2 + \frac{24}{\lambda^3}\left(x - \frac{1}{\lambda}\right) + \frac{24}{\lambda^4}\right\} e^{-\lambda x}\right]_0^\infty$, i.e.,

$$\mu_4 = \frac{9}{\lambda^4}. \tag{3.3.70}$$

Thus, the kurtosis is $v_3 = \frac{\mu_4}{\sigma^4} = \left(\frac{9}{\lambda^4}\right)\left(\frac{1}{\lambda^4}\right)^{-1} = 9$.   ◇

3.3.5 Moment Theorem

When we obtain the moments such as the mean and variance, we need to evaluate one integral for each of the moments. While the number of integration is the same as the number of moments that we want to obtain based on the definition of moments, we can first obtain the cf or mgf by one integration and then obtain the moments by differentiation if we use the moment theorem: note that differentiation is easier in general to evaluate than integration.

Theorem 3.3.4 *The k-th moment of X can be obtained as*

$$m_k = j^{-k} \frac{\partial^k}{\partial \omega^k} \varphi_X(\omega) \Big|_{\omega=0}$$
$$= j^{-k} \varphi_X^{(k)}(0) \tag{3.3.71}$$

or

$$m_k = M_X^{(k)}(0) \tag{3.3.72}$$

if the cf $\varphi_X(\omega)$ or the mgf $M_X(t)$ of X is differentiable k times at 0.

Proof First, if we evaluate $\mathsf{E}\left\{X^k\right\} = \int_{-\infty}^{\infty} x^k f_X(x)dx = \int_{-\infty}^{\infty} f_X(x) \frac{1}{j^k} \frac{\partial^k}{\partial \omega^k} e^{j\omega x}\Big|_{\omega=0} dx = j^{-k} \frac{\partial^k}{\partial \omega^k} \int_{-\infty}^{\infty} f_X(x) e^{j\omega x} dx\Big|_{\omega=0}$ recollecting $j^{-k} \frac{\partial^k}{\partial \omega^k} e^{j\omega x}\Big|_{\omega=0} = x^k$, we get

$$\mathsf{E}\{X^k\} = j^{-k} \frac{\partial^k}{\partial \omega^k} \varphi_X(\omega) \Big|_{\omega=0}. \tag{3.3.73}$$

Similarly, by differentiating the mgf $M_X(t)$ k times, we get $M_X^{(k)}(t) = \mathsf{E}\left\{X^k e^{tX}\right\}$ and, subsequently, the desired result.   ♠

Theorem 3.3.4 is referred to as the moment theorem.

Example 3.3.23 For $X \sim \mathcal{N}\left(m, \sigma^2\right)$, we have $\varphi_X(\omega) = \exp\left(-\frac{\omega^2 \sigma^2}{2} + jm\omega\right)$ as observed in Example 3.3.17. Thus, $\mathsf{E}\{X\} = j^{-1}\varphi_X'(0) = m$, $\mathsf{E}\left\{X^2\right\} = j^{-2}\varphi_X''(0) = m^2 + \sigma^2$, and $\mathsf{Var}\{X\} = \sigma^2$.   ◇

Example 3.3.24 For a random variable X with pdf $f_X(x) = \lambda \exp(-\lambda x) u(x)$, we have the mgf $M_X(t) = \int_0^\infty \lambda \exp(-\lambda x) \exp(tx) dx = \frac{\lambda}{\lambda - t}$ for[9] $t < \lambda$. Thus, we get $M_X'(0) = \frac{\lambda}{(\lambda - t)^2}\big|_{t=0} = \frac{1}{\lambda}$ and $M_X''(0) = \frac{2(\lambda - t)\lambda}{(\lambda - t)^4}\big|_{t=0} = \frac{2}{\lambda^2}$. Based on these two results, we obtain $E\{X\} = \frac{1}{\lambda}$ and $Var\{X\} = \frac{1}{\lambda^2}$, which are the same as those obtained directly from the definition of moments in Example 3.3.9. $\diamond$

In evaluating the moments of discrete random variables via the moment theorem, it is often convenient to let $z = e^{j\omega}$ and $s = e^t$ when using the cf and mgf, respectively.

Example 3.3.25 For $K \sim b(n, p)$, the cf is $\varphi_K(\omega) = \sum_{k=0}^n {}_nC_k p^k (1-p)^{n-k} e^{jk\omega}$, i.e.,

$$\varphi_K(\omega) = \left\{ pe^{jw} + (1-p) \right\}^n. \tag{3.3.74}$$

Now, letting $e^{j\omega} = z$ and writing the cf as $\gamma_K(z) = \varphi_K(\omega)\big|_{e^{j\omega}=z}$, we get

$$\gamma_K(z) = (pz + 1 - p)^n. \tag{3.3.75}$$

We then have

$$\gamma_K^{(i)}(1) = E\{K(K-1)\cdots(K-i+1)\} \tag{3.3.76}$$

because $\frac{\partial^i}{\partial z^i}\gamma_K(z) = \frac{\partial^i}{\partial z^i}E\{z^K\} = E\{K(K-1)\cdots(K-i+1)z^{K-i}\}$. Therefore, $\gamma_K(1) = 1$, $\gamma_K'(1) = E\{K\} = np$, and $\gamma_K''(1) = E\{K^2\} - E\{K\} = n(n-1)p^2$. From these results, we have $E\{K\} = np$ and $Var\{K\} = np(1-p)$. $\diamond$

Example 3.3.26 Consider $K \sim P(\lambda)$. Then, $\gamma_K(z) = e^{-\lambda} \sum_{k=0}^\infty \frac{\lambda^k z^k}{k!} = e^{\lambda(z-1)}$ and $\varphi_K(\omega) = \exp\left\{ \lambda\left(e^{j\omega} - 1\right) \right\}$ from $P(K = k) = e^{-\lambda}\frac{\lambda^k}{k!}$ for $k = 0, 1, 2, \ldots$. In other words, $\gamma_K'(1) = E\{K\} = \lambda$ and $\gamma_K''(1) = E\{K^2\} - E\{K\} = \lambda^2$. Therefore, $E\{K\} = \lambda$ and $Var\{K\} = \lambda$. Meanwhile, the mgf of K is $G_K(s) = M_K(t)\big|_{s=e^t} = \sum_{k=0}^\infty s^k \frac{\lambda^k}{k!} e^{-\lambda}$, i.e.,

$$G_K(s) = e^{\lambda(s-1)}, \tag{3.3.77}$$

with which we can also obtain the mean, variance, . . ., etc. $\diamond$

---

[9] Unless stated otherwise, an appropriate region of convergence is assumed when we consider the mgf.

The moment theorem also implies the following: similar to how a function can be expressed in terms of the coefficients of the Taylor series or Fourier series, the moments or central moments are the coefficients with which the pdf can be expressed. Specifically, if we express the mgf $M_X(t)$ of a random variable X in a series expansion, we have

$$M_X(t) = \sum_{n=0}^{\infty} \frac{E\{X^n\}}{n!} t^n. \tag{3.3.78}$$

Now, when the coefficients $\left\{ \frac{E\{X^n\}}{n!} \right\}_{n=0}^{\infty}$ of two distributions are the same, i.e., when the moments are the same, the two distributions will be the same. Based on this observation, by comparing the first few coefficients such as the mean and second moment, we can investigate how similar a distribution is to another.

3.4 Conditional Distributions

In this section, we discuss the conditional distribution (Park et al. 2017; Sveshnikov 1968) of a random variable under conditions given in the form of an event.

3.4.1 Conditional Probability Functions

Definition 3.4.1 (*conditional pmf*) When the occurrence of an event A with $P(A) > 0$ is assumed, the function $p_{X|A}(x) = P(X = x|A)$, i.e.,

$$p_{X|A}(x) = \frac{P(X = x, A)}{P(A)} \tag{3.4.1}$$

is called the conditional pmf of the discrete random variable X.

Example 3.4.1 Let X be the face number from rolling a fair die. When we know that the number is an odd number, the pmf of X is

$$p_{X|A}(x) = \begin{cases} \frac{1}{3}, & x = 1, 3, 5, \\ 0, & \text{otherwise} \end{cases} \tag{3.4.2}$$

because $P(A) = \frac{1}{2}$ for the event $A = \{$the number is an odd number$\}$. $\qquad \diamond$

Definition 3.4.2 (*conditional cdf; conditional pdf*) When the occurrence of an event A with $P(A) > 0$ is assumed, the function

$$F_{X|A}(x) = P(X \leq x|A)$$
$$= \frac{P(X \leq x, A)}{P(A)} \qquad (3.4.3)$$

is called the conditional cdf of X, and the function

$$f_{X|A}(x) = \frac{d}{dx} F_{X|A}(x) \qquad (3.4.4)$$

is called the conditional pdf of X.

Here, $F_{X|A}(x|A)$ or $F_X(x|A)$ is sometimes used to denote $F_{X|A}(x)$. Similarly, $f_{X|A}(x)$ is also written as $f_{X|A}(x|A)$ or $f_X(x|A)$. Because the conditional cdf is a cdf, we have $F_{X|A}(\infty) = 1$, $F_{X|A}(-\infty) = 0$, and $F_{X|A}(x_1) \leq F_{X|A}(x_2)$ for $x_1 < x_2$.

Recollecting $P(A|B) = \frac{P(B|A)P(A)}{P(B)}$ shown in (2.4.1) with the conditioning event $B = \{x_1 < X \leq x_2\}$, we have $P(A|x_1 < X \leq x_2) = \frac{P(x_1 < X \leq x_2|A)}{P(x_1 < X \leq x_2)} P(A)$, i.e.,

$$P(A|x_1 < X \leq x_2) = \frac{F_{X|A}(x_2) - F_{X|A}(x_1)}{F_X(x_2) - F_X(x_1)} P(A). \qquad (3.4.5)$$

Let $x_1 = x$ and $x_2 = x + \Delta x$ in (3.4.5). Then, we get $\lim\limits_{\Delta x \to 0} P(A|B) = P(A|X = x) = \lim\limits_{\Delta x \to 0} \frac{F_{X|A}(x+\Delta x) - F_{X|A}(x)}{F_X(x+\Delta x) - F_X(x)} P(A)$, i.e.,

$$P(A|X = x) = \frac{f_{X|A}(x)}{f_X(x)} P(A), \qquad (3.4.6)$$

because $\lim\limits_{\Delta x \to 0} \frac{F_{X|A}(x+\Delta x) - F_{X|A}(x)}{F_X(x+\Delta x) - F_X(x)}$ can be written as $\lim\limits_{\Delta x \to 0} \frac{F_{X|A}(x+\Delta x) - F_{X|A}(x)}{\Delta x} \frac{\Delta x}{F_X(x+\Delta x) - F_X(x)}$. The result (3.4.6) can be expressed as

$$f_{X|A}(x)P(A) = P(A|X = x) f_X(x). \qquad (3.4.7)$$

By integrating (3.4.7) and noting that $\int_{-\infty}^{\infty} f_{X|A}(x)dx = 1$, we get the following theorem:

Theorem 3.4.1 *We have*

$$P(A) = \int_{-\infty}^{\infty} P(A|X = x) f_X(x)dx, \qquad (3.4.8)$$

which is called the total probability theorem for continuous random variables. Similarly,

$$P(A) = \sum_{x=-\infty}^{\infty} P(A|X = x)p_X(x) \qquad (3.4.9)$$

is called the total probability theorem for discrete random variables.

Example 3.4.2 Consider a rod with thickness 0. Cut the rod into two parts. Choose one of the two parts at random and cut it into two. Find the probability P_{T2} that the three parts obtained in this way can make a triangle.

Solution Let the length of the rod be 1. As in Examples 2.3.6 and 2.3.7, let the point of the first cutting be X. Then, the pdf of X is $f_X(v) = u(v)u(1 - v)$. Call the interval $[0, X]$ the left piece and the interval $[X, 1]$ the right piece on the real line. When $X = t$, we get

$$P_{T2} = \int_{-\infty}^{\infty} P(\text{triangle with the three pieces} | X = t) \, f_X(t) \, dt$$

$$= \int_{0}^{1} P(\text{triangle with the three pieces} | X = t) \, dt \qquad (3.4.10)$$

based on (3.4.8). We can make a triangle when the sum of the lengths of two pieces is larger than the length of the third piece. When $0 < t < \frac{1}{2}$, we should choose the right piece and the second cutting should be placed[10] in $\left(\frac{1}{2}, t + \frac{1}{2}\right)$. Thus, we have   P (triangle with the three pieces| $X = t$) = P(choose the right piece)P (the second cutting is in $\left(\frac{1}{2}, t + \frac{1}{2}\right)$| choose the right piece), i.e.,

$$P(\text{triangle with the three pieces} | X = t) = \frac{1}{2} \frac{\text{length of } \left(\frac{1}{2}, t + \frac{1}{2}\right)}{\text{length of the right piece}}$$

$$= \frac{1}{2} \frac{t}{1 - t}. \qquad (3.4.11)$$

Similarly, when $\frac{1}{2} < t < 1$, we should choose the left piece and the second cutting should be placed[11] in $\left(t - \frac{1}{2}, \frac{1}{2}\right)$. Thus, P (triangle with the three pieces| $X = t$) = P(choose the left piece)P$\left(\text{the second cutting is in } \left(t - \frac{1}{2}, \frac{1}{2}\right)\right|$ choose the left piece$)$, i.e.,

$$P(\text{triangle with the three pieces} | X = t) = \frac{1}{2} \frac{1 - t}{t}. \qquad (3.4.12)$$

---

[10] Denoting the location of the second cutting by $y \in (t, 1)$, the lengths of the three pieces are t, $y - t$, and $1 - y$, resulting in the condition $\frac{1}{2} < y < t + \frac{1}{2}$ of y to make a triangle.
[11] Denoting the location of the second cutting by $y \in (1, t)$, the lengths of the three pieces are y, $t - y$, and $1 - y$, resulting in the condition $t - \frac{1}{2} < y < \frac{1}{2}$ of y to make a triangle.

Using (3.4.11) and (3.4.12) in (3.4.10), we get $P_{T2} = \int_0^{\frac{1}{2}} \frac{1}{2}\frac{t}{1-t}dt + \int_{\frac{1}{2}}^1 \frac{1}{2}\frac{1-t}{t}dt = \left[-\frac{1}{2}(t + \ln|1-t|)\right]_{t=0}^{\frac{1}{2}} + \left[\frac{1}{2}(-t + \ln|t|)\right]_{t=\frac{1}{2}}^1 = \ln 2 - \frac{1}{2} \approx 0.1931$.

Meanwhile, considering that a triangle cannot be made if the shorter piece is chosen among the first two pieces, assume that we choose the longer of the two pieces and then cut it into two. Then, we have the probability

$$2P_{T2} = 2\ln 2 - 1$$
$$\approx 0.3863 \tag{3.4.13}$$

of making a triangle, which is higher than the probability $\frac{1}{4}$ obtained in Example 2.3.7. This is a consequence that we have used the information "We should choose the longer piece among the first two pieces." ◇

The following theorem similar to Theorem 3.4.1 can be obtained from the total probability theorem $P(A) = \sum_{i=1}^n P(A|B_i)P(B_i)$ shown in (2.4.13) with $A = \{X \leq x\}$:

Theorem 3.4.2 *When the collection* $\{B_i\}_{i=1}^n$ *is a partition of the range of X, the cdf, the pdf for a continuous random variable X, and the pmf for a discrete random variable X can be expressed as*

$$F_X(x) = \sum_{i=1}^n F_{X|B_i}(x)P(B_i), \tag{3.4.14}$$

$$f_X(x) = \sum_{i=1}^n f_{X|B_i}(x)P(B_i), \tag{3.4.15}$$

and

$$p_X(x) = \sum_{i=1}^n p_{X|B_i}(x)P(B_i), \tag{3.4.16}$$

respectively.

Example 3.4.3 Consider a communication system transmitting bits of 0 and 1, and let H_0 and H_1 be the events that a bit of 0 and 1 is sent, respectively. Assume $P(H_0) = \frac{1}{4}$, $P(H_1) = \frac{3}{4}$, and the conditional pdf's of X are

$$f_{X|H_0}(x) = \frac{2}{3}u\left(\frac{3}{4} - |x|\right) \tag{3.4.17}$$

and

$$f_{X|H_1}(x) = \frac{2}{3}u\left(x - \frac{1}{4}\right)u\left(\frac{7}{4} - x\right).$$ (3.4.18)

Then, we have

$$f_X(x) = \frac{1}{4}f_{X|H_0}(x) + \frac{3}{4}f_{X|H_1}(x)$$

$$= \begin{cases} \frac{1}{6}, & -\frac{3}{4} \le x < \frac{1}{4}; \ \frac{2}{3}, \ \frac{1}{4} \le x < \frac{3}{4}; \\ \frac{1}{2}, & \frac{3}{4} \le x < \frac{7}{4}; \quad 0, \ \text{otherwise} \end{cases}$$ (3.4.19)

as the pdf of X.   ◇

Theorem 3.4.3 *Using (3.4.8) in (3.4.7), we get*

$$f_{X|A}(x) = \frac{P(A|X = x)}{P(A)}f_X(x)$$

$$= \frac{P(A|X = x)f_X(x)}{\int_{-\infty}^{\infty} P(A|X = x)f_X(x)dx},$$ (3.4.20)

another expression of the Bayes' theorem.

By integrating the conditional pdf (3.4.20), the conditional cdf discussed in (3.4.3) can be obtained as

$$F_{X|A}(x) = \int_{-\infty}^{x} f_{X|A}(t)dt$$

$$= \frac{\int_{-\infty}^{x} P(A|X = t)f_X(t)dt}{\int_{-\infty}^{\infty} P(A|X = t)f_X(t)dt}.$$ (3.4.21)

From (3.4.3) and (3.4.21), we also get

$$P(A, X \le x) = \int_{-\infty}^{x} P(A|X = t)f_X(t)dt.$$ (3.4.22)

Example 3.4.4 Express the conditional cdf $F_{X|X \le a}(x)$ and the conditional pdf $f_{X|X \le a}(x)$ in terms of the pdf f_X and the cdf F_X of a continuous random variable X.

Solution First, the conditional cdf $F_{X|X \le a}(x) = P(X \le x|X \le a)$ can be written as

$$F_{X|X \le a}(x) = \frac{P(X \le x, X \le a)}{P(X \le a)}.$$ (3.4.23)

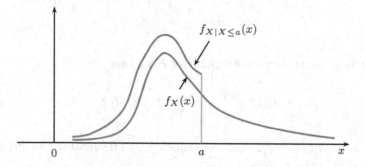

Fig. 3.19 The pdf $f_X(x)$ and conditional pdf $f_{X|X \leq a}(x)$

Here, we have

$$F_{X|X \leq a}(x) = \frac{P(X \leq x)}{P(X \leq a)}$$

$$= \frac{F_X(x)}{F_X(a)} \qquad (3.4.24)$$

when $x \leq a$ because $P(X \leq x, X \leq a) = P(X \leq x)$, and

$$F_{X|X \leq a}(x) = 1 \qquad (3.4.25)$$

when $x > a$ because $P(X \leq x, X \leq a) = P(X \leq a)$. From (3.4.24) and (3.4.25), we finally have

$$f_{X|X \leq a}(x) = \frac{f_X(x)}{F_X(a)} u(a - x), \qquad (3.4.26)$$

which is shown in Fig. 3.19. ◇

Example 3.4.5 Let F be the cdf of the time X of a failure for a system: i.e., $F(t) = P(X \leq t)$ is the probability that the system fails before time t, and $1 - F(t) = P(X > t)$ is the probability that the system does not fail before time t. We also define the conditional rate of failure $\beta(t)$ via

$$\beta(t)dt = P(\text{when the system does not fail before time } t,$$
$$\text{the system fails in the interval } (t, t + dt)). \qquad (3.4.27)$$

Letting $A = \{X > t\}$ in (3.4.3) and differentiating the result, we get the conditional pdf $f_{X|X>t}(x) = \frac{F'(x)}{1-F(t)} = \frac{f(x)}{\int_t^\infty f(x)dx}$ of X. Using this result, the conditional rate of failure $\beta(t) = f_{X|X>t}(t)$ can be expressed as

$$\beta(t) = \frac{F'(t)}{1 - F(t)}. \tag{3.4.28}$$

Solving the differential equation (3.4.28) for F, we have

$$F(x) = 1 - \exp\left(-\int_0^x \beta(t)dt\right) \tag{3.4.29}$$

from $-\ln(1 - F(x)) = \int_0^x \beta(t)dt$. Subsequently, by differentiating (3.4.29), we get
$f(x) = \beta(x)\exp\left(-\int_0^x \beta(t)dt\right)$.   ◇

3.4.2 Expected Values Conditional on Event

Definition 3.4.3 (*conditional expected value*) The expectation

$$E\{X|A\} = \begin{cases} \int_{-\infty}^{\infty} xf_{X|A}(x)dx, & \text{continuous random variable,} \\ \\ \sum_{x=-\infty}^{\infty} xp_{X|A}(x), & \text{discrete random variable} \end{cases} \tag{3.4.30}$$

is the conditional expected value or conditional mean of X when the event A is assumed.

Example 3.4.6 When the pdf of X is f_X, obtain the conditional mean of X under the assumption $A = \{X \le a\}$.

Solution Using the conditional pdf (3.4.26), we can obtain the conditional mean
$E\{X|A\} = \int_{-\infty}^{\infty} xf_{X|A}(x)dx = \int_{-\infty}^{\infty} xf_{X|X\le a}(x)dx$ as

$$E\{X|A\} = \frac{1}{F_X(a)} \int_{-\infty}^{a} xf_X(x)dx \tag{3.4.31}$$

from (3.4.30). Here, $\lim_{a\to\infty} E\{X|A\} = \frac{1}{F_X(\infty)} \int_{-\infty}^{\infty} xf_X(x)dx$. Thus, (3.4.31) is in agreement with that the mean $E\{X\}$ can be written as $E\{X|\Omega\}$, i.e.,

$$E\{X\} = \frac{1}{F_X(\infty)} \int_{-\infty}^{\infty} xf_X(x)dx \tag{3.4.32}$$

because $F_X(\infty) = 1$.   ◇

Definition 3.4.3 can be extended to the conditional expected value $E\{g(X)|A\} = \int_{-\infty}^{\infty} g(x)dF_{X|A}(x)$ of $Y = g(X)$ as

$$E\{g(X)|A\} = \begin{cases} \int_{-\infty}^{\infty} g(x) f_{X|A}(x)dx, & \text{continuous random variable,} \\ \sum_{x=-\infty}^{\infty} g(x) p_{X|A}(x), & \text{discrete random variable} \end{cases} \quad (3.4.33)$$

when the event A is assumed.

3.4.3 Evaluation of Expected Values via Conditioning

We have observed in Sect. 2.4 that the probability of an event can often be obtained quite easily by first obtaining the conditional probability under an appropriate condition. We will now similarly see that obtaining the conditional expected value first will be quite useful when we try to obtain the expected value. Evaluation of the expected values via conditioning will be discussed again in Sect. 4.4.3.

Theorem 3.4.4 *The expected value* $E\{X\}$ *of X can be expressed as*

$$E\{X\} = \sum_{k=1}^{n} E\{X|A_k\} P(A_k), \quad (3.4.34)$$

where the collection $\{A_1, A_2, \ldots, A_n\}$ *is a partition of the range of X.*

Proof We show the theorem for discrete random variables only. From (2.4.13), we easily get $E\{X\} = \sum_{x=-\infty}^{\infty} x p_X(x) = \sum_{x=-\infty}^{\infty} \sum_{k=1}^{n} x p_{X|A_k}(x) P(A_k) = \sum_{k=1}^{n} \sum_{x=-\infty}^{\infty} x p_{X|A_k}(x) P(A_k) = \sum_{k=1}^{n} E\{X|A_k\} P(A_k).$ ♠

Example 3.4.7 There are a red balls and b green balls in a box. Pick one ball at random from the box: if it is red, we put it back into the box; and if it is green, we discard it and put one red ball from another source into the box. Let X_n be the number of red balls in the box and $E\{X_n\} = M_n$ be the expected value of X_n after repeating this experiment n times. Show that

$$M_n = \left(1 - \frac{1}{a+b}\right) M_{n-1} + 1 \quad (3.4.35)$$

and, based on this result, confirm

$$M_n = a + b - b\left(1 - \frac{1}{a+b}\right)^n. \quad (3.4.36)$$

Obtain the probability P_n that the ball picked at the n-th trial is red.

Solution We have

$$M_n = \mathrm{E}\{X_n | n\text{-th ball is red}\}\,\mathrm{P}(n\text{-th ball is red})$$
$$+ \mathrm{E}\{X_n | n\text{-th ball is green}\}\,\mathrm{P}(n\text{-th ball is green}), \qquad (3.4.37)$$

which can be rewritten as $M_n = M_{n-1}\frac{M_{n-1}}{a+b} + (M_{n-1} + 1)\left(1 - \frac{M_{n-1}}{a+b}\right) = M_{n-1}\left(1 - \frac{1}{a+b}\right) + 1$ because

$$\mathrm{E}\{X_n | n\text{-th ball is red}\} = M_{n-1}, \qquad (3.4.38)$$
$$\mathrm{P}(n\text{-th ball is red}) = \frac{M_{n-1}}{a+b}, \qquad (3.4.39)$$
$$\mathrm{E}\{X_n | n\text{-th ball is green}\} = M_{n-1} + 1, \qquad (3.4.40)$$

and

$$\mathrm{P}(n\text{-th ball is green}) = 1 - \frac{M_{n-1}}{a+b}. \qquad (3.4.41)$$

Letting $\mu = 1 - \frac{1}{a+b}$, we get $M_n = a\mu^n + \frac{1-\mu^n}{1-\mu} = a+b - b\left(1 - \frac{1}{a+b}\right)^n$ from $M_n = \mu M_{n-1} + 1 = \mu(\mu M_{n-2} + 1) + 1 = \mu^2 M_{n-2} + 1 + \mu = \cdots = \mu^n M_0 + 1 + \mu + \cdots + \mu^{n-1}$ and $M_0 = a$. We also have $P_n = \frac{M_{n-1}}{a+b} = 1 - \frac{b}{a+b}\left(1 - \frac{1}{a+b}\right)^{n-1}$ for $n = 1, 2, \ldots$. $\diamond$

3.5 Classes of Random Variables

In this section, we discuss four classes of widely-used random variables (Hahn and Shapiro 1967; Johnson and Kotz 1970; Kassam 1988; Song et al. 2002; Thomas 1986; Zwillinger and Kokoska 1999) in detail. We start with the normal random variables, followed by the binomial, Poisson, and exponential random variables. The normal distributions are again discussed extensively in Chap. 5.

3.5.1 Normal Random Variables

Definition 3.5.1 (*normal random variable*) A random variable with the pdf

$$f(x) = \frac{1}{\sqrt{2\pi\sigma^2}} \exp\left\{-\frac{(x-m)^2}{2\sigma^2}\right\} \qquad (3.5.1)$$

is called the normal random variable and its distribution is denoted by $\mathcal{N}\left(m, \sigma^2\right)$.

The normal distribution is the most important and widely-used distribution as we will see from the central limit theorem, Theorem 6.2.12. We have already mentioned in Example 2.5.15 that the distribution $\mathcal{N}(0, 1)$ is called the standard normal distribution.

Now, consider the standard normal pdf

$$\phi(x) = \frac{1}{\sqrt{2\pi}} \exp\left(-\frac{x^2}{2}\right) \tag{3.5.2}$$

and its integral, the standard normal cdf

$$\Phi(x) = \frac{1}{\sqrt{2\pi}} \int_{-\infty}^{x} \exp\left(-\frac{t^2}{2}\right) dt. \tag{3.5.3}$$

Then, from the symmetry of $\phi(x)$, we get

$$\Phi(-x) = 1 - \Phi(x) \tag{3.5.4}$$

as we observed in (3.1.28). In addition, the cdf of $\mathcal{N}(m, \sigma^2)$ can be expressed as

$$F(x) = \Phi\left(\frac{x - m}{\sigma}\right). \tag{3.5.5}$$

We have already seen in (3.3.49) that the cf and mgf of $\mathcal{N}(m, \sigma^2)$ are

$$\varphi(\omega) = \exp\left(jm\omega - \frac{\sigma^2\omega^2}{2}\right) \tag{3.5.6}$$

and

$$M(t) = \exp\left(mt + \frac{\sigma^2 t^2}{2}\right), \tag{3.5.7}$$

respectively.

The tail probability of the normal distribution is used quite frequently in many areas such as statistics, communications, and signal processing. Let us briefly discuss an approximation of the tail probability of the normal distribution. First, the error function $\Theta(x) = \text{erf}(x) = \frac{2}{\sqrt{\pi}} \int_0^x \exp\left(-t^2\right) dt$ can be expressed as

$$\text{erf}(x) = 2\Phi\left(\sqrt{2}x\right) - 1 \tag{3.5.8}$$

in terms of the standard normal cdf $\Phi(x)$. For the tail integral

$$Q(x) = \frac{1}{\sqrt{2\pi}} \int_x^\infty \exp\left(-\frac{t^2}{2}\right) dt \qquad (3.5.9)$$

of the standard normal pdf, also called the complementary standard normal cdf, we
have

$$\frac{x}{1+x^2}\phi(x) < Q(x) < \frac{\phi(x)}{x} \qquad (3.5.10)$$

for $x > 0$. Assume we approximate $Q(x)$ as $Q_a(x) = \frac{1}{(1-a)x+a\sqrt{x^2+b}}\phi(x)$ with
$Q_a(0) = Q(0) = \frac{1}{2}$, in which case $a\sqrt{2\pi b} = 2$ should be satisfied. When we consider only the case $x > 0$, it is known (Börjesson and Sundberg Mar. 1979) that
$Q_a(x)$ is the optimum upper bound on $Q(x)$ when $a \approx 0.344$ and $b \approx 5.334$ and
optimum lower bound when $a = \frac{1}{\pi}$ and $b = 2\pi$. In addition, when $a \approx 0.339$ and
$b \approx 5.510$, $Q_a(x)$ minimizes max $\left|\frac{Q_a(x)-Q(x)}{Q(x)}\right|$.

3.5.2 Binomial Random Variables

Definition 3.5.2 (*binomial random variable*) The number of occurrences of an event
in a repetition of n Bernoulli trials with distribution $b(1, p)$ is a binomial random
variable with distribution $b(n, p)$.

If we let K be the number of occurrences of a desired event A with probability
$P(A) = p$ in n Bernoulli trials, then we have the pmf $p_K(k) = P_n(k)$ of $K \sim b(n, p)$,
where

$$P_n(k) = P(\text{desired event occurs } k \text{ times among } n \text{ trials})$$
$$= {}_nC_k p^k q^{n-k} \qquad (3.5.11)$$

for $k = 0, 1, \ldots, n$, with $q = 1 - p = P(A^c)$. We also have the cdf

$$F(x) = \sum_{k=0}^{\lfloor x \rfloor} {}_nC_k p^k q^{n-k} \qquad (3.5.12)$$

for $\lfloor x \rfloor \le x < \lfloor x \rfloor + 1$ of $b(n, p)$, and the probability

$$P(x \le K \le y) = \sum_{k=\lceil x \rceil}^{\lfloor y \rfloor} P_n(k) \qquad (3.5.13)$$

of the event $\{x \le K \le y\}$ for $K \sim b(n, p)$. The pdf of $b(n, p)$ can be written as
$$f(x) = \sum_{k=0}^{n} {}_nC_k p^k q^{n-k} \delta(x - k).$$

Example 3.5.1 Let X be the number of 2's when we roll a fair die five times. Then, $X \sim b\left(5, \frac{1}{6}\right)$ and $P_5(k) = {}_5C_k \left(\frac{1}{6}\right)^k \left(\frac{5}{6}\right)^{5-k}$ for $k = 0, 1, \ldots, 5$.   ◇

Example 3.5.2 A fair die is rolled seven times. Let Y be the number of even numbers. Then, $Y \sim b\left(7, \frac{1}{2}\right)$ and $P_7(k) = \binom{7}{k}\left(\frac{1}{2}\right)^k \left(\frac{1}{2}\right)^{7-k} = \frac{1}{128}\binom{7}{k}$ for $k = 0, 1, \ldots, 7$.   ◇

The sequence $\{P_n(k)\}_{k=0}^{n}$ increases until $k = (n+1)p - 1$ and $k = (n+1)p$ when $(n+1)p$ is an integer and until $k = [(n+1)p]$ when $(n+1)p$ is not an integer, and then decreases.

Example 3.5.3 In $b\left(3, \frac{1}{4}\right)$, $P_n(k)$ is maximum at $k = (n+1)p - 1 = 0$ and $k = 1$. Specifically, $P_3(0) = P_3(1) = \frac{27}{64}$, $P_3(2) = \frac{9}{64}$, and $P_3(3) = \frac{1}{64}$.   ◇

In evaluating $P_n(k) = {}_nC_k p^k (1-p)^{n-k}$, we need to calculate ${}_nC_k$, which becomes rather difficult when n is large and k is near $\frac{n}{2}$. We now discuss some methods to alleviate this problem by considering the asymptotic approximations of $P_n(k)$ as $n \to \infty$.

Definition 3.5.3 (*small o*) When

$$\lim_{x \to \infty} \frac{f(x)}{g(x)} = 0, \tag{3.5.14}$$

the function $f(x)$ is of lower order than $g(x)$ for $x \to \infty$, and is denoted by $f = o(g)$.

Definition 3.5.3 implies that, when $f = o(g)$ for $x \to \infty$, $f(x)$ increases slower than $g(x)$ as $x \to \infty$.

Definition 3.5.4 (*big O*) Suppose that $f(x) > 0$ and $g(x) > 0$ for a sufficiently large number x. When there exists a natural number M satisfying

$$\frac{f(x)}{g(x)} \le M \tag{3.5.15}$$

for a sufficiently large number x, $f(x)$ is said to be of, at most, the order of $g(x)$ for $x \to \infty$, and is denoted by $f = O(g)$.

From Definitions 3.5.3 and 3.5.4, when $f(x)$ and $g(x)$ are both positive for a sufficiently large x, we have $f = O(g)$ if $f = o(g)$.

Example 3.5.4 We have $\ln x = o(x)$ for $x \to \infty$ from $\lim_{x \to \infty} \frac{\ln x}{x} = 0$ for the two functions $f(x) = \ln x$ and $g(x) = x$, and $x^2 = o\left(x^3 + 1\right)$ for $x \to \infty$ from $\lim_{x \to \infty} \frac{x^2}{x^3+1} = 0$.   ◇

Example 3.5.5 We have $x + \sin x = O(x)$ for $x \to \infty$ because $\frac{x+\sin x}{x} \le 2$ when x is sufficiently large. We also have $e^x + x^2 = O(e^x)$ for $x \to \infty$ because $\frac{e^x+x^2}{e^x} \to 1$ when $x \to \infty$, and $x = o(e^x)$ for $x \to \infty$ because $\frac{x}{e^x} \to 0$ when $x \to \infty$. $\diamondsuit$

Example 3.5.6 (Khuri 2003) We have $\cos x = O(1)$ and $\sin x = O(|x|)$ for any real number x, and $x = O(x^2)$ and $x^2 + 2x = O(x^2)$ for a large number x. $\diamondsuit$

Theorem 3.5.1 *When $npq \gg 1$, we have the approximation*

$$_nC_k p^k (1-p)^{n-k} \approx \frac{1}{\sqrt{2\pi npq}} \exp\left\{ -\frac{(k-np)^2}{2npq} \right\} \qquad (3.5.16)$$

for $k = np \pm O\left(\sqrt{npq}\right)$, that is, for $k \in \left(np - a\sqrt{npq}, np + a\sqrt{npq}\right)$ with some $a > 0$.

Proof The theorem is proved in Appendix 3.2. $\spadesuit$

Theorem 3.5.1 is called the de Moivre-Laplace theorem or the Gaussian approximation of binomial distribution.

Example 3.5.7 Consider a random variable $K \sim b\left(10, \frac{1}{2}\right)$. Then, we have $P(K = 5) = P_{10}(5) = {}_{10}C_5 \left(\frac{1}{2}\right)^{10} = \frac{63}{256} \approx 0.246$. The Gaussian approximation produces $P(K = 5) \approx \frac{e^0}{\sqrt{2\pi \times 10 \times \frac{1}{2} \times \frac{1}{2}}} \approx 0.252$. $\diamondsuit$

Example 3.5.8 The distribution of even numbers from 1000 rollings of a fair die is $b\left(1000, \frac{1}{2}\right)$. Thus, the Gaussian approximation of the probability that we have 500 times of even numbers is $P_{1000}(500) \approx \left(\sqrt{2\pi \times 1000 \times \frac{1}{2} \times \frac{1}{2}}\right)^{-1} \approx 0.0252$. Similarly, $P_{1000}(510) \approx \left(\sqrt{500\pi}\right)^{-1} e^{-\frac{100}{500}} \approx 0.0207$ from (3.5.16). $\diamondsuit$

Let us try to approximate $P(k_1 \le k \le k_2) = \sum_{k=k_1}^{k_2} {}_nC_k p^k (1-p)^{n-k}$ by making use of the steps in the proof of Theorem 3.5.1 shown in Appendix 3.2. First, when $npq \gg 1$, we have $\sum_{k=k_1}^{k_2} {}_nC_k p^k (1-p)^{n-k} \approx \frac{1}{\sqrt{2\pi npq}} \sum_{k=k_1}^{k_2} \exp\left\{ -\frac{(k-np)^2}{2npq} \right\} = \frac{1}{\sqrt{2\pi npq}} \sum_{k=k_1}^{k_2} \exp\left\{ -\frac{(k-np)^2}{2npq} \right\} \{(k + \frac{1}{2}) - (k - \frac{1}{2})\}$, i.e.,

$$\sum_{k=k_1}^{k_2} {}_nC_k p^k (1-p)^{n-k} \approx \frac{1}{\sqrt{2\pi npq}} \int_{k_1}^{k_2} \exp\left\{ -\frac{(x - np)^2}{2npq} \right\} dx$$

$$= \Phi\left(\frac{k_2 - np}{\sqrt{npq}}\right) - \Phi\left(\frac{k_1 - np}{\sqrt{npq}}\right) \qquad (3.5.17)$$

for $k_2 - k_1 = O\left(\sqrt{npq}\right)$. The integral in (3.5.17) implies that the approximation error will be small when $|k_1 - k_2| \gg 1$, but it could be large otherwise. To reduce such an error, we often use the approximation

$$\sum_{k=k_1}^{k_2} {}_nC_k p^k(1-p)^{n-k} \approx \Phi\left(\frac{k_2 - np + \frac{1}{2}}{\sqrt{npq}}\right) - \Phi\left(\frac{k_1 - np - \frac{1}{2}}{\sqrt{npq}}\right), \quad (3.5.18)$$

which is called the continuity correction and is considered also in (6.2.66), Example 6.2.26, and Exercise 6.10.

Example 3.5.9 In Example 3.5.8, the probability of 500, 501, or 502 times of even numbers is $P_{1000}(500) + P_{1000}(501) + P_{1000}(502) \approx 0.0754$. With the two approximations above, we have $\Phi\left(\frac{2}{\sqrt{250}}\right) - \Phi\left(\frac{0}{\sqrt{250}}\right) \approx 0.0503$ from (3.5.17) and $\Phi\left(\frac{2.5}{\sqrt{250}}\right) - \Phi\left(\frac{-0.5}{\sqrt{250}}\right) \approx 0.0754$ from (3.5.18).   ◇

Theorem 3.5.2 *If $np \to \lambda$ when $n \to \infty$ and $p \to 0$, then*

$$ {}_nC_k p^k(1-p)^{n-k} \to \frac{\lambda^k}{k!}e^{-\lambda} \quad (3.5.19)$$

for $k = O(np)$.

Proof A proof is given in Appendix 3.2.   ♠

Theorem 3.5.2 is called the Poisson limit theorem or the Poisson approximation of binomial distribution.

Example 3.5.10 For $b\left(1000, 10^{-3}\right)$, we have $P_{1000}(0) = 0.999^{1000} \approx 0.3677$, for which the Poisson approximation provides $\exp(-np) = \exp(-1) \approx 0.3679$.   ◇

Example 3.5.11 In Example 3.5.7, we have observed that $P(K = 5) = 0.246$ with its Gaussian approximation 0.252 for $K \sim b\left(10, \frac{1}{2}\right)$. From the Poisson approximation, we can get $\frac{5^5}{5!}e^{-5} \approx 0.1755$. In Example 3.5.10, when $K \sim b\left(1000, 10^{-3}\right)$, the Poisson approximation for $P(K = 0) \approx 0.3677$ is 0.3679. With the Gaussian approximation, we would get $\frac{1}{\sqrt{1.998\pi}}\exp\left(-\frac{1}{1.998}\right) \approx 0.2420$ from the normal distribution $\mathcal{N}(1, 0.999)$.   ◇

As we can see from Examples 3.5.10 and 3.5.11, the Poisson approximation is more accurate than the Gaussian approximation when p is close to 0, and vice versa.

3.5.3 Poisson Random Variables

Definition 3.5.5 (*Poisson random variable*) A random variable with the pmf

$$p_k = e^{-\lambda}\frac{\lambda^k}{k!} \tag{3.5.20}$$

for $k = 0, 1, \ldots$ is called the Poisson random variable. The distribution is denoted by $P(\lambda)$ and $\lambda > 0$ is called the Poisson rate or Poisson parameter.

The Poisson pmf $\{p_k\}_{k=0}^{\infty}$ is a sequence which first increases and then decreases or is a decreasing sequence: it is maximum at $k = 0$ when $\lambda < 1$; at $k = \lambda$ and $\lambda - 1$ when $\lambda \geq 1$ is an integer; and at $k = [\lambda]$ when λ is a non-integer not smaller than 1. We have the cdf

$$F_K(x) = e^{-\lambda}\sum_{k=0}^{\lfloor x \rfloor}\frac{\lambda^k}{k!} \tag{3.5.21}$$

of the Poisson random variable $K \sim P(\lambda)$.

Assume that we choose n points randomly in the interval $\left[-\frac{T}{2}, \frac{T}{2}\right]$. Then,

$$P(k \text{ points exists in the interval } (t_1, t_2)) = {}_nC_k\left(\frac{t_a}{T}\right)^k\left(1 - \frac{t_a}{T}\right)^{n-k}$$

$$= {}_nC_k p^k q^{n-k}, \tag{3.5.22}$$

where $-\frac{T}{2} \leq t_1 < t_2 \leq \frac{T}{2}$, $p = \frac{t_2-t_1}{T} = \frac{t_a}{T}$, and $q = 1 - p$. Now, even when an interval D_a of length t_a does not overlap with another interval D_b of length t_b, the two events $A = \{k_a \text{ points in } D_a\}$ and $B = \{k_b \text{ points in } D_b\}$ are not independent of each other if the interval $\left[-\frac{T}{2}, \frac{T}{2}\right]$ is finite because

$$P(AB) = \frac{n!}{k_a!k_b!(n - k_a - k_b)!}\left(\frac{t_a}{T}\right)^{k_a}\left(\frac{t_b}{T}\right)^{k_b}\left(1 - \frac{t_a}{T} - \frac{t_b}{T}\right)^{n-k_a-k_b}$$

$$\neq P(A)P(B). \tag{3.5.23}$$

Now, we let $\lambda = \frac{n}{T}$, we have $p = \frac{t_a}{T} \to 0$ and $np = \lambda t_a$ when $n \to \infty$ and $T \to \infty$. Thus, recollecting the Poisson limit theorem (3.5.19), we get $P(k_a \text{ points in } D_a) \to e^{-\lambda t_a}\frac{(\lambda t_a)^k}{k!}$, $\left(1 - \frac{t_a}{T} - \frac{t_b}{T}\right)^{n-k_a-k_b} \to e^{-n\left(\frac{t_a}{T} + \frac{t_b}{T}\right)} = \exp\{-\lambda(t_a + t_b)\}$, and $\frac{n!}{k_a!k_b!(n-k_a-k_b)!} \to \frac{n^{k_a+k_b}}{k_a!k_b!}$. Therefore, $P(AB) = e^{-\lambda t_a}\frac{(\lambda t_a)^{k_a}}{k_a!}e^{-\lambda t_b}\frac{(\lambda t_b)^{k_b}}{k_b!}$, i.e.,

$$P(AB) = P(A)P(B) \tag{3.5.24}$$

implying that the two events $\{k_a$ points in $D_a\}$ and $\{k_b$ points in $D_b\}$ are independent of each other. The set of infinitely many points described above is called the random Poisson points or Poisson points as defined below.

Definition 3.5.6 (*random Poisson points*) A collection of points satisfying the two properties below is called random Poisson points, or simply Poisson points, with parameter λ.

(1) $\mathsf{P}\,(k$ points in an inteval of length $t) = e^{-\lambda t}\frac{(\lambda t)^k}{k!}$.
(2) If two intervals D_a and D_b are non-overlapping, then the events $\{k_a$ points in $D_a\}$ and $\{k_b$ points in $D_b\}$ are independent of each other.

The parameter λ in Definition 3.5.6 represents the average number of points in a unit interval.

Example 3.5.12 Assume a set of Poisson points with parameter λ. Find the probability $\mathsf{P}(A|C)$ of $A = \{k_a$ points in $D_a = (t_1, t_2)\}$ when there are k_c points in $D_c = (t_1, t_3)$, where $t_1 \leq t_2 \leq t_3$.

Solution Let $B = \{k_b$ points in $D_b\}$ and $C = \{k_c$ points in $D_c\}$, where $D_b = (t_2, t_3)$ and $k_b = k_c - k_a$. Then, because $AC = \{k_a$ points in D_a, k_c points in $D_c\}$ $= \{k_a$ points in D_a, k_b points in $D_b\} = AB$, and D_a and D_b are non-overlapping, we get $\mathsf{P}(A|C) = \frac{\mathsf{P}(AC)}{\mathsf{P}(C)} = \frac{\mathsf{P}(AB)}{\mathsf{P}(C)}$, i.e.,

$$\mathsf{P}(A|C) = \frac{\mathsf{P}(A)\mathsf{P}(B)}{\mathsf{P}(C)}. \tag{3.5.25}$$

We thus finally get $\mathsf{P}(A|C) = \left\{ e^{-\lambda t_a}\frac{(\lambda t_a)^{k_a}}{k_a!} e^{-\lambda t_b}\frac{(\lambda t_b)^{k_b}}{k_b!} \right\} \left\{ e^{-\lambda t_c}\frac{(\lambda t_c)^{k_c}}{k_c!} \right\}^{-1}$, i.e.,

$$\mathsf{P}(A|C) = \frac{k_c!}{k_a!k_b!} \left(\frac{t_a}{t_c}\right)^{k_a} \left(\frac{t_b}{t_c}\right)^{k_b}, \tag{3.5.26}$$

where $t_a = t_2 - t_1$, $t_b = t_3 - t_2$, and $t_c = t_3 - t_1$.   ◇

Example 3.5.13 Assume a set of Poisson points. Let X be the distance from a fixed point t_0 to the nearest point to the right-hand direction. Then, we have the cdf $F_X(x) = \mathsf{P}(X \leq x) = \mathsf{P}$ (at least one point exists in $(t_0, t_0 + x))$ can be obtained as

$$F_X(x) = 1 - \mathsf{P}\,(\text{no point in } (t_0, t_0 + x))$$
$$= 1 - e^{-\lambda x} \tag{3.5.27}$$

for $x \geq 0$. Thus, we have the pdf $f_X(x) = \lambda e^{-\lambda x} u(x)$ and X is an exponential random variable.   ◇

Example 3.5.14 Consider a constant α and a set of Poisson points with parameter λ. Then, for the number N of Poisson points in the interval $(0, \alpha)$, we have $\mathsf{P}(N = k) = e^{-\lambda \alpha}\frac{(\lambda \alpha)^k}{k!}$. In other words, $N \sim P(\lambda \alpha)$.   ◇

3.5.4 Exponential Random Variables

Definition 3.5.7 (*exponential random variable*) A random variable with the pdf

$$f(x) = \lambda e^{-\lambda x} u(x) \tag{3.5.28}$$

is called an exponential random variable, where the parameter $\lambda > 0$ is called the exponential rate or rate.

When X is an exponential random variable with rate $\lambda > 0$, the mgf $M(t) = \int_0^\infty e^{tx} \lambda e^{-\lambda x} dx$ is

$$M(t) = \frac{\lambda}{\lambda - t}, \quad t < \lambda, \tag{3.5.29}$$

with which we can easily obtain the expected value $\mathsf{E}\{X\} = \frac{1}{\lambda}$, the second moment $\mathsf{E}\{X^2\} = \frac{2\lambda}{(\lambda-t)^3}\Big|_{t=0} = \frac{2}{\lambda^2}$, and the variance $\mathsf{Var}\{X\} = \frac{2}{\lambda^2} - \frac{1}{\lambda^2} = \frac{1}{\lambda^2}$. In addition, from the cdf

$$F(x) = \left(1 - e^{-\lambda x}\right) u(x) \tag{3.5.30}$$

of X, we get

$$\mathsf{P}(X > s + t \mid X > t) = \mathsf{P}(X > s) \tag{3.5.31}$$

for $s, t \geq 0$ because $\mathsf{P}(X > s + t \mid X > t) = \frac{\mathsf{P}(X > s+t)}{\mathsf{P}(X > t)} = \frac{1 - F(s+t)}{1 - F(t)} = e^{-\lambda s}$. The property expressed by (3.5.31) is called the memoryless property of the exponential distribution.

Example 3.5.15 Assume that the lifetime of an electric bulb follows an exponential distribution. The result (3.5.31) implies that, if the electric bulb is on at some moment, the distribution of the remaining lifetime of the bulb is the same as that of the original lifetime: the remaining lifetime of the bulb at any instant follows the same distribution as a new bulb. In other words, for a bulb that is on at time t, the probability that the bulb will be on at $t + s$ is simply the probability that the bulb will be on for s time units. This can be exemplified as follows. When a person finds a bulb is lit in a place and the person does not know from when the bulb has been lit, how long does the person expect the bulb will be on? Surprisingly, if the lifetime of the bulb is an exponential random variable, the remaining lifetime is the same as a new bulb. In a slightly different way, this can be described as 'the past does not influence the future', which is called the Markov property.   ◇

Rewriting (3.5.31), we get

$$P(X > s + t) = P(X > s)P(X > t), \qquad (3.5.32)$$

which is satisfied by only exponential distribution (Komjath and Totik 2006) among continuous[12] distributions. This can be proved as follows: Assume a function $g(\cdot)$ satisfies[13]

$$g(s + t) = g(s)g(t). \qquad (3.5.33)$$

Then, $g\left(\frac{1}{n}\right) = g^{\frac{1}{n}}(1)$ and $g\left(\frac{m}{n}\right) = g\left(\frac{1}{n} + \frac{1}{n} + \cdots + \frac{1}{n}\right) = g^m\left(\frac{1}{n}\right)$ because $g(1) = g\left(\frac{1}{n} + \frac{1}{n} + \cdots + \frac{1}{n}\right) = g^n\left(\frac{1}{n}\right)$ for any choice of natural numbers m and n. Thus, $g\left(\frac{m}{n}\right) = g^{\frac{m}{n}}(1)$, and we can write as $g(x) = g^x(1)$ when $g(\cdot)$ is continuous from right-hand side for $x \geq 0$. Now, because $g(1) = g^2\left(\frac{1}{2}\right) \geq 0$, we have $g(x) = e^{-\lambda x}$, where $\lambda = -\log\{g(1)\}$. From this result and (3.5.30), we have the conclusion.

Example 3.5.16 The time required to finish a transaction in a bank is an exponential random variable with a mean of 10 minutes. Assume that every customer will leave the bank immediately after finishing the transaction. Find the probability P_1 for a customer to wait more than 15 minutes to finish the transaction after the arrival at the bank. Also find the probability P_2 that a customer will still be waiting at 10:15 when the customer arrived at the bank at 10:00 and has been waiting for 10 min.

Solution Let X be the waiting time for a customer at the bank to finish the transaction, and denote by F the cdf of X. Then, the waiting time is the same as the time required to finish the transaction, and X is an exponential random variable with rate $\lambda = \frac{1}{10}$. Thus, we get $P_1 = P(X > 15) = 1 - P(X \leq 15) = 1 - F(15) = e^{-15\lambda} = e^{-\frac{3}{2}} \approx 0.2231$. Next, because an exponential random variable is not influenced by the past, the probability P_2 is the same as the probability that a customer will wait more than 5 minutes: in other words, $P_2 = P(X > 5) = e^{-5\lambda} = e^{-\frac{1}{2}} \approx 0.6065$. ◇

Example 3.5.17 The lifetime of an electric bulb is an exponential random variable with a mean of 10 hrs. A person returns to a room and finds that the bulb is lit. The person needs five hours to complete a task in the room. Find the probability P_5 that the person can complete the task before the light gets off. Discuss what happens if the lifetime is not an exponential random variable.

Solution Let X be the lifetime of the electric bulb. Then, because $\lambda = \frac{1}{10}$, we have $P_5 = P(X > 5) = 1 - F(5) = e^{-5\lambda} = e^{-\frac{1}{2}} \approx 0.6065$. Next, assume that the lifetime is not an exponential random variable. Then, denoting by t the time that the bulb was lit on, we have $P_5 = P(X > t + 5 | X > t)$, i.e.,

---

[12] Only the geometric distribution satisfies (3.5.32) among discrete distributions.

[13] A similar relationship $g(s + t) = g(s) + g(t)$ is called the Cauchy equation.

$$P_5 = \frac{1 - F(t + 5)}{1 - F(t)},\tag{3.5.34}$$

where F is the cdf of the lifetime of the bulb. Thus, the probability will be available only if we know how long the bulb has been lit on at t.   ◇

In discussing the memoryless property of the exponential random variable, we often employ the failure rate function, also called the hazard rate function or conditional rate of failure. The failure rate function is the probability that an object that has been operated for t time units will become inoperable within the next dt time units, i.e., the probability that the object will become inoperable after operating t time units. As described in (3.4.28), the failure rate function $\beta(t)$ can be expressed as

$$\beta(t) = \frac{f(t)}{1 - F(t)}\tag{3.5.35}$$

for a random variable with pdf f and cdf F. The function $\beta(t)$ is the conditional rate of failure for an object to become inoperable after being operated for t time units.

Let us now discuss the failure rate function $\beta(t)$ for an exponential random variable. The rate of failure of an object that has operated for t time units is the same as that of a new object because an exponential random variable is not influenced by the past. In other words, as we can observe from $\beta(t) = \frac{\lambda e^{-\lambda t}}{e^{-\lambda t}} = \lambda$, the failure rate function for an exponential random variable is the rate of the exponential random variable, a constant independent of t. The rate is the inverse of the mean and represents how many events occurs on the average over a unit time interval: for example, when the time interval between occurrences of an event is an exponential random variable with mean $\frac{1}{10}$, the rate $\lambda = 10$ tells us that the event occurs ten times on the average in a unit time.

Example 3.5.18 (Yates and Goodman 1999) Consider an exponential random variable X with parameter λ. Show that $K = \lceil X \rceil$ is a geometric random variable with parameter $p = 1 - e^{-\lambda}$.

Solution The pmf of K is $p_K(k) = \mathsf{P}(k - 1 < X \le k) = F_X(k) - F_X(k - 1) = \exp\{-\lambda(k - 1)\} - \exp(-\lambda k) = \{\exp(-\lambda)\}^{k-1}\{1 - \exp(-\lambda)\} = p(1 - p)^{k-1}$, which implies that $K = \lceil X \rceil$ is a geometric random variable with parameter $p = 1 - e^{-\lambda}$. Note that, although $\lfloor x \rfloor + 1 \ne \lceil x \rceil$ in general for a real number x, the distribution of $\lfloor X \rfloor + 1$ here is the same as that of $\lceil X \rceil$ because X is a continuous random variable: in other words, $\lfloor X \rfloor + 1$ is equal to $\lceil X \rceil$ in distribution, which is often denoted by $\lfloor X \rfloor + 1 \stackrel{d}{=} \lceil X \rceil$.   ◇

Appendices

Appendix 3.1 Cumulative Distribution Functions and Their Inverse Functions

(A) Cumulative Distribution Functions

We now discuss the cdf in the context of function theory (Gelbaum and Olmsted 1964).

Definition 3.A.1 (*cdf*) A real function $F(x)$ possessing all of the three following properties is a cdf:

(1) The function $F(x)$ is non-decreasing: $F(x + h) \geq F(x)$ for $h > 0$.
(2) The function $F(x)$ is continuous from the right-hand side: $F(x^+) = F(x)$.
(3) The function $F(x)$ has the limits $\lim_{x \to -\infty} F(x) = 0$ and $\lim_{x \to \infty} F(x) = 1$.

A cdf is a finite and monotonic function. A point x such that $F(x + \varepsilon) - F(x - \varepsilon) > 0$ for every positive number ε, $F(x) = F(x^-)$, and $F(x^+) = F(x) \neq F(x^-)$ is called an increasing point, a continuous point, and a discontinuity, respectively, of $F(x)$. Here, as we have already seen in Definition 1.3.3, $p_x = F(x^+) - F(x^-) = F(x) - F(x^-)$ is the jump of $F(x)$ at x. A cdf may have only type 1 discontinuity, i.e., jump discontinuity, with every jump between 0 and 1.

Example 3.A.1 Consider the function g shown in Fig. 3.20. Here, y_1 is the local minimum of $y = g(x)$, and x_1 and x_{11} are the two solutions to $y_1 = g(x)$. In addition, y_2 is the local maximum of $y = g(x)$, and x_2 and x_{22} are the solutions to $y_2 = g(x)$. Let $x_3 < x_4 < x_5$ be the X coordinates of the crossing points of the straight line $Y = y$ and the function $Y = g(X)$ for $y_1 < y < y_2$. Then, $x_{11} < x_3 < x_2 < x_4 < x_1 < x_5 < x_{22}$ and $y = g(x_3) = g(x_4) = g(x_5)$ for $y_1 < y < y_2$. Obtain the cdf F_Y of $Y = g(X)$ in terms of the cdf F_X of X, and discuss if the cdf F_Y is a continuous function.

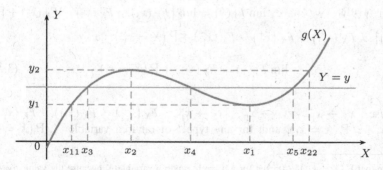

Fig. 3.20 The function $Y = g(X)$

Solution When $y > y_2$ or $y < y_1$, we get

$$F_Y(y) = F_X\left(g^{-1}(y)\right) \tag{3.A.1}$$

from Fig. 3.20 because $P(Y \leq y) = P(g(X) \leq y) = P\left(X \leq g^{-1}(y)\right)$. When $y = y_1$, because $\{Y \leq y_1\} = \{g(X) \leq y_1\} = \{X \leq x_{11}\} + \{X = x_1\}$, we get

$$F_Y(y) = F_X(x_{11}) + P(X = x_1). \tag{3.A.2}$$

In addition, when $y_1 < y < y_2$, the cdf is

$$F_Y(y) = F_X(x_3) + F_X(x_5) - F_X(x_4) + P(X = x_4) \tag{3.A.3}$$

because $\{g(X) \leq y\} = \{X \leq x_3\} + \{x_4 \leq X \leq x_5\} = \{X \leq x_3\} + \{x_4 < X \leq x_5\} + \{X = x_4\}$. Finally, from $\{g(X) \leq y_2\} = \{X \leq x_{22}\}$ we get

$$F_Y(y) = F_X(x_{22}) \tag{3.A.4}$$

when $y = y_2$. Combining the results (3.A.1)–(3.A.4), we have the cdf of Y as

$$F_Y(y) = \begin{cases} F_X\left(g^{-1}(y)\right), & y < y_1 \text{ or } y > y_2, \\ F_X(x_{11}) + P(X = x_1), & y = y_1, \\ F_X(x_3) + F_X(x_5) - F_X(x_4) \\ \quad + P(X = x_4), & y_1 < y < y_2, \\ F_X(x_{22}), & y = y_2. \end{cases} \tag{3.A.5}$$

Let us next discuss the continuity of the cdf (3.A.5). When $y \uparrow y_1$, because $g^{-1}(y) \to x_{11}^-$, we get $\lim_{y \uparrow y_1} F_Y(y) = \lim_{y \uparrow y_1} F_X\left(g^{-1}(y)\right)$, i.e.,

$$\lim_{y \uparrow y_1} F_Y(y) = F_X\left(x_{11}^-\right). \tag{3.A.6}$$

When $y \downarrow y_1$, we have $\lim_{y \downarrow y_1} F_Y(y) = \lim_{y \downarrow y_1} \{F_X(x_3) + F_X(x_5) - F_X(x_4) + P(X = x_4)\} = F_X\left(x_{11}^+\right) + F_X\left(x_1^+\right) - F_X\left(x_1^-\right) + P\left(X = x_1^-\right)$, i.e.,

$$\lim_{y \downarrow y_1} F_Y(y) = F_X(x_{11}) + P(X = x_1) \tag{3.A.7}$$

because $x_3 \to x_{11}^+$, $x_4 \to x_1^-$, $x_5 \to x_1^+$, $F_X\left(x_{11}^+\right) = F_X(x_{11})$, $F_X\left(x_1^+\right) - F_X\left(x_1^-\right) = P(X = x_1)$, and, for any type[14] of random variable X, $P\left(X = x_1^-\right)$

---

[14] Note that $P\left(X = k^-\right) = 0$ even for a discrete random variable X because the value $p_X(k) = P(X = k)$ is 0 when k is not an integer for a pmf $p_X(k)$.

$= 0$. Thus, from the second line of (3.A.5), (3.A.6), and (3.A.7), the continuity of the cdf $F_Y(y)$ at $y = y_1$ can be summarized as follows: The cdf $F_Y(y)$ is (A) continuous from the right-hand side at $y = y_1$ and (B) continuous from the left-hand side at $y = y_1$ only if $F_X(x_{11}) + P(X = x_1) - F_X(x_{11}^-) \rightleftharpoons P(X = x_1) + P(X = x_{11})$ is 0 or, equivalently, only if $P(X = x_1) = P(X = x_{11}) = 0$.

Next, when $y \uparrow y_2$, recollecting that $x_3 \to x_2^-$, $x_4 \to x_2^+$, $x_5 \to x_{22}^-$, $F_X(x_2^+) - F_X(x_2^-) = P(X = x_2)$, and $P(X = x_2^+) = 0$, we get $\lim_{y \uparrow y_2} F_Y(y) = \lim_{y \uparrow y_2} \{F_X(x_3) + F_X(x_5) - F_X(x_4) + P(X = x_4)\} = F_X(x_2^-) + F_X(x_{22}^-) - F_X(x_2^+) + P(X = x_2^+)$, i.e.,

$$\lim_{y \uparrow y_2} F_Y(y) = F_X(x_{22}^-) - P(X = x_2). \tag{3.A.8}$$

In addition,

$$\lim_{y \downarrow y_2} F_Y(y) = \lim_{y \downarrow y_2} F_X(g^{-1}(y))$$
$$= F_X(x_{22}) \tag{3.A.9}$$

because $F_X(x_{22}^+) = F_X(x_{22})$ and $g^{-1}(y) \to x_{22}^+$ for $y \downarrow y_2$. Thus, from the fourth case of (3.A.5), (3.A.8), and (3.A.9), the continuity of the cdf $F_Y(y)$ at $y = y_2$ can be summarized as follows: The cdf $F_Y(y)$ is (A) continuous from the right-hand side at $y = y_2$ and (B) continuous from the left-hand side at $y = y_2$ only if $F_X(x_{22}) - F_X(x_{22}^-) + P(X = x_2) = P(X = x_2) + P(X = x_{22})$ is 0 or, equivalently, only if $P(X = x_2) = P(X = x_{22}) = 0$. Exercises 3.56 and 3.57 also deal with the continuity of the cdf. $\diamond$

Let $\{x_\nu\}$ be the set of discontinuities of the cdf $F(x)$, and $p_{x_\nu} = F(x_\nu^+) - F(x_\nu^-)$ be the jump of $F(x)$ at $x = x_\nu$. Denote by $\Psi(x)$ the sum of jumps of $F(x)$ at discontinuities not larger than x, i.e.,

$$\Psi(x) = \sum_{x_\nu \le x} p_{x_\nu}. \tag{3.A.10}$$

The function $\Psi(x)$ is increasing only at $\{x_\nu\}$ and is constant in a closed interval not containing x_ν: thus, it is a step-like function described following Definition 3.1.7.

If we now let

$$\psi(x) = F(x) - \Psi(x), \tag{3.A.11}$$

then $\psi(x)$ is a continuous function while $\Psi(x)$ is continuous only from the right-hand side. In addition, $\Psi(x)$ and $\psi(x)$ are both non-decreasing and satisfy $\Psi(-\infty) = \psi(-\infty) = 0$, $\Psi(+\infty) = a_1 \le 1$, and $\psi(+\infty) = b \le 1$. Here, the functions $F_d(x) = \frac{1}{a_1}\Psi(x)$ and $F_c(x) = \frac{1}{b}\psi(x)$ are both cdf, where $F_d(x)$ is a step-like function and $F_c(x)$ is a continuous function. Rewriting (3.A.11), we get

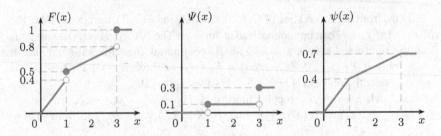

Fig. 3.21 The decomposition of cdf $F(x) = \Psi(x) + \psi(x)$ with a step-like function $\Psi(x)$ and a continuous function $\psi(x)$

$$F(x) = \Psi(x) + \psi(x), \tag{3.A.12}$$

i.e.,

$$F(x) = a_1 F_d(x) + b F_c(x). \tag{3.A.13}$$

Figure 3.21 shows an example of the cdf $F(x) = \Psi(x) + \psi(x)$ with a step-like function $\Psi(x)$ and a continuous function $\psi(x)$.

The decomposition (3.A.13) is unique because the decomposition shown in (3.A.12) is unique. Let us prove this result. Assume that two decompositions of $F(x) = \Psi(x) + \psi(x) = \Psi_1(x) + \psi_1(x)$ are possible. Rewriting this equation, we get $\Psi(x) - \Psi_1(x) = \psi_1(x) - \psi(x)$. The left-hand side is a step-like function because it is the difference of two step-like functions and, similarly, the right-hand side is a continuous function. Therefore, both sides should be 0. In other words, $\Psi_1(x) = \Psi(x)$ and $\psi_1(x) = \psi(x)$. By the discussion so far, we have shown the following theorem:

Theorem 3.A.1 *Any cdf $F(x)$ can be decomposed as*

$$F(x) = a_1 F_d(x) + b F_c(x) \tag{3.A.14}$$

into a step-like cdf $F_d(x)$ and a continuous cdf $F_c(x)$, where $0 \le a_1 \le 1$, $0 \le b \le 1$, and $a_1 + b = 1$.

In Theorem 3.A.1, $F_d(x)$ and $F_c(x)$ are called the discontinuous or discrete part and continuous part, respectively, of $F(x)$. Now, from Theorem 3.1.3, we see that there exists a countable set D such that $\int_D dF_d(x) = 1$, where the integral is the Lebesgue-Stieltjes integral. The function $F_c(x)$ is continuous but not differentiable at all points. Yet, any cdf is differentiable at almost every point and thus, based on the Lebesgue decomposition theorem, the continuous part $F_c(x)$ in (3.A.14) can be decomposed into two continuous functions $F_{ac}(x)$ and $F_s(x)$ as

$$F_c(x) = b_1 F_{ac}(x) + b_2 F_s(x), \tag{3.A.15}$$

where $b_1 \geq 0$, $b_2 \geq 0$, and $b_1 + b_2 = 1$. The function $F_{ac}(x)$ is the integral of the derivative of $F_{ac}(x)$: in other words, $F_{ac}(x) = \int_{-\infty}^{x} F'_{ac}(y)dy$, indicating that $F_{ac}(x)$ is an absolutely continuous function and that $\int_N dF_{ac}(x) = 0$ for a set N of Lebesgue measure 0. The function $F_s(x)$ is a continuous function but the derivative is 0 at almost every point. In addition, for a suitably chosen set N of Lebesgue measure 0, we have $\int_N dF_s(x) = 1$, indicating that $F_s(x)$ is a singular function. Combining (3.A.14) and (3.A.15), we have the following theorem (Lukacs 1970):

Theorem 3.A.2 *Any cdf $F(x)$ can be decomposed into a step-like function $F_d(x)$, an absolutely continuous function $F_{ac}(x)$, and a singular function $F_s(x)$ as*

$$F(x) = a_1 F_d(x) + a_2 F_{ac}(x) + a_3 F_s(x), \tag{3.A.16}$$

where $a_1 \geq 0$, $a_2 \geq 0$, $a_3 \geq 0$, and $a_1 + a_2 + a_3 = 1$.

In (3.A.16), when one of the three coefficients a_1, a_2, and a_3 is 1, the cdf is called pure: when $a_1 = 1$, $a_2 = 1$, or $a_3 = 1$, the cdf is a discrete cdf, an absolutely continuous cdf, or a singular cdf, respectively. Almost all cdf's are practically discrete cdf's or absolutely continuous cdf's. Because a singular cdf is only theoretically meaningful with very rare practical applications, an absolutely continuous cdf is considered to be a continuous cdf and no singular cdf is considered in statistics. In this book, a continuous cdf indicates an absolutely continuous cdf unless specified otherwise.

When the intervals between discontinuity points are all the same for a distribution, the discrete distribution is called a lattice distribution, the discontinuities are called lattice points, and the interval between adjacent two lattice points is called the span.

Example 3.A.2 Let us consider an example of a singular cdf. Assume the closed interval [0, 1] and the ternary expression

$$x = \sum_{i=1}^{\infty} \frac{a_i(x)}{3^i}$$
$$= 0.a_1(x)a_2(x)a_3(x)\cdots \tag{3.A.17}$$

of a point x in the Cantor set $\mathbb{C}$ discussed in Example 1.1.46, where $a_i(x) \in \{0, 1\}$. Now, let $n(x)$ be the location of the first 1 in the ternary expression (3.A.17) of x with $n(x) = \infty$ if no 1 appears eventually. Define a cdf as (Romano and Siegel 1986)

$$F(x) = \begin{cases} 0, & x < 0, \\ g(x), & x \in ([0, 1] - \mathbb{C}), \\ \sum_j \frac{c_j}{2^j}, & x = \sum_j \frac{2c_j}{3^j} \in \mathbb{C}, \\ 1, & x \geq 1, \end{cases} \tag{3.A.18}$$

i.e., as

$$F(x) = \begin{cases} 0, & x < 0, \\ \phi_C(x), & 0 \le x \le 1, \\ 1, & x \ge 1 \end{cases}$$

$$= \begin{cases} 0, & x \le 0, \\ \frac{1}{2^{1+n(x)}} + \sum_{i=1}^{n(x)} \frac{a_i(x)}{2^{i+1}}, & 0 \le x \le 1, \\ 1, & x \ge 1 \end{cases} \qquad (3.A.19)$$

based on the Cantor function $\phi_C(x)$ discussed in Example 1.3.11, where

$$g(x) = \begin{cases} \frac{1}{2}, & \frac{1}{3} < x < \frac{2}{3}, \\ \frac{1}{2^k} + \sum_{j=1}^{k-1} \frac{c_j}{2^j}, & x \in A_{2c_1,2c_2,\dots,2c_{k-1}} \end{cases} \qquad (3.A.20)$$

with the open interval $A_{2c_1,2c_2,\dots,2c_{k-1}}$ defined by (1.1.41) and (1.1.42). Then, it is easy to see that $F(x)$ is a continuous cdf. In addition, as we have observed in Example 1.3.11, the derivative of $F(x)$ is 0 at almost every point in $([0, 1] - \mathbb{C})$ and thus is not a pdf. In other words, $F(x)$ is not an absolutely continuous cdf but is a singular cdf.

Some specific values of the cdf (3.A.19) are as follows: We have $F\left(\frac{1}{9}\right) = \frac{1}{2^{1+2}} + \sum_{i=1}^{2} \frac{a_i\left(\frac{1}{9}\right)}{2^{i+1}} = \frac{1}{8} + \frac{1}{8} = \frac{1}{4}$ from $\frac{1}{9} = 0.01_3$ and $n\left(\frac{1}{9}\right) = 2$, we have $F\left(\frac{2}{9}\right) = \frac{1}{2^\infty} + \sum_{i=1}^{\infty} \frac{a_i\left(\frac{2}{9}\right)}{2^{i+1}} = \frac{1}{4}$ from $\frac{2}{9} = 0.02_3$ and $n\left(\frac{2}{9}\right) = \infty$, and we have $F\left(\frac{1}{3}\right) = \frac{1}{2^{1+1}} + \sum_{i=1}^{1} \frac{a_i\left(\frac{1}{3}\right)}{2^{i+1}} = \frac{1}{2}$ from $\frac{1}{3} = 0.1_3$ and $n\left(\frac{1}{3}\right) = 1$. Similarly, from $\frac{2}{3} = 0.2_3$ and $n\left(\frac{2}{3}\right) = \infty$, we have $F\left(\frac{2}{3}\right) = 0 + \frac{2}{4} = \frac{1}{2}$; from $0 = 0.0_3$ and $n(0) = \infty$, we have $F(0) = 0$; and from $1 = 0.22\cdots_3$ and $n(1) = \infty$, we have $F(1) = 0 + \sum_{i=1}^{\infty} \frac{2}{2^{i+1}} = 1$. $\diamond$

(B) Inverse Cumulative Distribution Functions

The inverse cdf F^{-1}, the inverse function of the cdf F, can be defined specifically as

$$F^{-1}(u) = \inf\{x : F(x) \ge u\} \qquad (3.A.21)$$

for $0 < u < 1$. Because F is a non-decreasing function, we have $\{x : F(x) \ge y_2\} \subseteq \{x : F(x) \ge y_1\}$ when $y_1 < y_2$ and, consequently, $\inf\{x : F(x) \ge y_1\} \le \inf\{x : F(x) \ge y_2\}$. Thus,

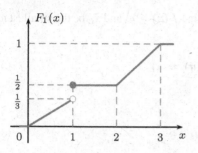
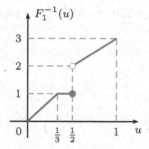

Fig. 3.22 The cdf F_1 and the inverse cdf F_1^{-1}

$$F^{-1}(y_1) \leq F^{-1}(y_2) \qquad (3.A.22)$$

for $y_1 < y_2$. In other words, like a cdf, an inverse cdf is a non-decreasing function.

Example 3.A.3 Consider the cdf

$$F_1(x) = \begin{cases} 0, \ x \leq 0; & \frac{x}{3}, \qquad 0 \leq x < 1; \\ \frac{1}{2}, \ 1 \leq x \leq 2; & \frac{1}{2}(x-1), \ 2 \leq x \leq 3; \\ 1, \ x \geq 3. \end{cases} \qquad (3.A.23)$$

Then, the inverse cdf is

$$F_1^{-1}(u) = \begin{cases} 3u, & 0 < u \leq \frac{1}{3}; \quad 1, \ \frac{1}{3} \leq u \leq \frac{1}{2}; \\ 2u + 1, \ \frac{1}{2} < u < 1. \end{cases} \qquad (3.A.24)$$

For example, we have $F_1^{-1}\left(\frac{1}{3}\right) = \inf\left\{x : F_1(x) \geq \frac{1}{3}\right\} = \inf\{x : x \geq 1\} = 1$ and $F_1^{-1}\left(\frac{1}{2}\right) = \inf\left\{x : F_1(x) \geq \frac{1}{2}\right\} = \inf\{x : x \geq 1\} = 1$. Note that, unlike a cdf, an inverse cdf is continuous from the left-hand side. Figure 3.22 shows the cdf F_1 and the inverse cdf F_1^{-1}. $\diamond$

Theorem 3.A.3 (Hajek et al. 1999) *Let F and F^{-1} be a cdf and its inverse, respectively. Then,*

$$F\left(F^{-1}(u)\right) \geq u \qquad (3.A.25)$$

for $0 < u < 1$: in addition,

$$F\left(F^{-1}(u)\right) = u \qquad (3.A.26)$$

if F is continuous.

Proof Let S_u be the set[15] of all x such that $F(x) \geq u$, and x_L be the smallest number in S_u. Then,

$$F^{-1}(u) = x_L \qquad (3.A.27)$$

and

$$F(x_L) \geq u \qquad (3.A.28)$$

because $F(x) \geq u$ for any point $x \in S_u$. In addition, because S_u is the set of all x such that $F(x) \geq u$, we have $F(x) < u$ for any point $x \in S_u^c$. Now, x_L is the smallest number in S_u and thus $x_L - \epsilon \in S_u^c$ because $x_L - \epsilon < x_L$ when $\epsilon > 0$. Consequently, we have

$$F(x_L - \epsilon) < u \qquad (3.A.29)$$

when $\epsilon > 0$. Using (3.A.27) in (3.A.28), we get $F\left(F^{-1}(u)\right) \geq u$. Next, recollecting (3.A.27), and combining (3.A.28) and (3.A.29), we get $F\left(F^{-1}(u) - \epsilon\right) < u \leq F\left(F^{-1}(u)\right)$. In other words, u is a number between $F\left(F^{-1}(u) - \epsilon\right)$ and $F\left(F^{-1}(u)\right)$. Now, $u = F\left(F^{-1}(u)\right)$ because $\lim_{\epsilon \to 0} F\left(F^{-1}(u) - \epsilon\right) = F\left(F^{-1}(u)\right)$ if F is a continuous function. ♠

Example 3.A.4 For the cdf

$$F_2(x) = \begin{cases} 0, \ x < 0; & \frac{x}{4}, \qquad 0 \leq x < 1; \\ \frac{1}{2}, \ 1 \leq x < 2; & \frac{1}{4}(x+1), \ 2 \leq x < 3; \\ 1, \ x \geq 3 \end{cases} \qquad (3.A.30)$$

and the inverse cdf

$$F_2^{-1}(u) = \begin{cases} 4u, \ 0 < u \leq \frac{1}{4}; & 1, \qquad \frac{1}{4} \leq u \leq \frac{1}{2}; \\ 2, \ \frac{1}{2} < u \leq \frac{3}{4}; & 4u - 1, \ \frac{3}{4} \leq u < 1, \end{cases} \qquad (3.A.31)$$

we have $F_2\left(F_2^{-1}\left(\frac{1}{4}\right)\right) = F_2(1) = \frac{1}{2} \geq \frac{1}{4}$, $F_2\left(F_2^{-1}\left(\frac{1}{2}\right)\right) = F_2(1) = \frac{1}{2} \geq \frac{1}{2}$, $F_2^{-1}(F_2(1)) = F_2^{-1}\left(\frac{1}{2}\right) = 1 \leq 1$, and $F_2^{-1}\left(F_2\left(\frac{3}{2}\right)\right) = F_2^{-1}\left(\frac{1}{2}\right) = 1 \leq \frac{3}{2}$. Figure 3.23 shows the cdf F_2 and the inverse cdf F_2^{-1}. ◇

Note that even if F is a continuous function, we have not $F^{-1}(F(x)) = x$ but

$$F^{-1}(F(x)) \leq x. \qquad (3.A.32)$$

It is also noteworthy that

---

[15] Here, because a cdf is continuous from the right-hand side, S_u is either in the form $S_u = [x_L, \infty)$ or in the form $S_u = \{x_L, x_L + 1, \ldots\}$.

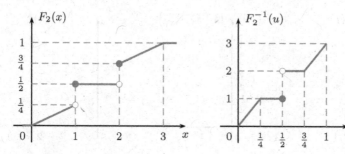

Fig. 3.23 The cdf F_2 and the inverse cdf F_2^{-1}

$$P\left(F^{-1}(F(X)) \neq X\right) = 0 \qquad (3.A.33)$$

when $X \sim F$.

Example 3.A.5 Consider the cdf $F_1(x)$ and the inverse cdf $F_1^{-1}(u)$ discussed in Example 3.A.3. Then, we have

$$
F_1\left(F_1^{-1}(u)\right) = \begin{cases}
0, & F_1^{-1}(u) \leq 0, \\
\frac{1}{3}F_1^{-1}(u), & 0 \leq F_1^{-1}(u) < 1, \\
\frac{1}{2}, & 1 \leq F_1^{-1}(u) \leq 2, \\
\frac{1}{2}\left\{F_1^{-1}(u) - 1\right\}, & 2 \leq F_1^{-1}(u) \leq 3, \\
1, & F_1^{-1}(u) \geq 3
\end{cases}
$$
$$
= \begin{cases}
u, \, 0 < u < \frac{1}{3}; & \frac{1}{2}, \frac{1}{3} \leq u \leq \frac{1}{2}; \\
u, \, \frac{1}{2} < u < 1
\end{cases} \qquad (3.A.34)
$$

and

$$
F_1^{-1}\left(F_1(x)\right) = \begin{cases}
3F_1(x), & 0 < F_1(x) \leq \frac{1}{3}; \quad 1, \frac{1}{3} \leq F_1(x) \leq \frac{1}{2}; \\
2F_1(x) + 1, \, \frac{1}{2} < F_1(x) < 1
\end{cases}
$$
$$
= \begin{cases}
x, \, 0 < x < 1; \quad 1, \, 1 \leq x \leq 2; \\
x, \, 2 < x < 3,
\end{cases} \qquad (3.A.35)
$$

which are shown in Fig. 3.24. The results (3.A.34) and (3.A.35) clearly confirm $F_1\left(F_1^{-1}(u)\right) \geq u$ shown in (3.A.26) and $F_1^{-1}\left(F_1(x)\right) \leq x$ shown in (3.A.32). ◇

The results of Example 3.A.5 and Exercises 3.79 and 3.80 imply the following: In general, even when a cdf is a continuous function, if it is constant over an interval, its inverse is discontinuous. Specifically, if $F(a) = F\left(b^-\right) = \alpha$ for $a < b$ or, equivalently, if $F(x) = \alpha$ over $a \leq x < b$, the inverse cdf $F^{-1}(u)$ is discontinuous at $u = \alpha$ and $F^{-1}(\alpha) = a$. On the other hand, even if a cdf is not a continuous function, its inverse is a continuous function when the cdf is not constant over an interval. Figure 3.25 shows an example of a discontinuous cdf with a continuous inverse.

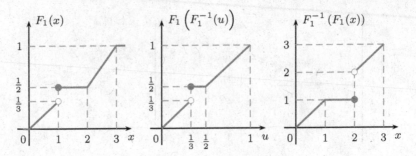

Fig. 3.24 The results $F_1\left(F_1^{-1}(u)\right)$ and $F_1^{-1}(F_1(x))$ from the cdf F_1

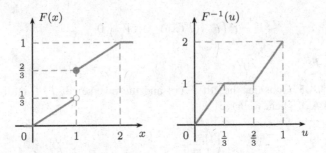

Fig. 3.25 A discontinuous cdf F with a continuous inverse F^{-1}

Theorem 3.A.4 (Hajek et al. 1999) *If a cdf F is continuous, its inverse F^{-1} is strictly increasing.*

Proof When F is continuous, assume $F^{-1}(y_1) = F^{-1}(y_2)$ for $y_1 < y_2$. Then, from (3.A.26), we have $y_1 = F\left(F^{-1}(y_1)\right) = F\left(F^{-1}(y_2)\right) = y_2$, which is a contradiction to $y_1 < y_2$. In other words,

$$F^{-1}(y_1) \neq F^{-1}(y_2) \tag{3.A.36}$$

when $y_1 < y_2$. From (3.A.22) and (3.A.36), we have $F^{-1}(y_1) < F^{-1}(y_2)$ for $y_1 < y_2$ when the cdf F is continuous.   ♠

Theorem 3.A.5 (Hajek et al. 1999) *When the pdf f is continuous, we have*

$$\frac{d}{du}F^{-1}(u) = \frac{1}{f\left(F^{-1}(u)\right)} \tag{3.A.37}$$

if $f\left(F^{-1}(u)\right) \neq 0$, where F is the cdf.

Proof When f is continuous, F is also continuous. Letting $F^{-1}(u) = v$, we get $\frac{d}{du}F^{-1}(u) = \frac{1}{f(v)}\frac{dv}{dv}$, i.e.,

$$\frac{d}{du}F^{-1}(u) = \frac{1}{f\left(F^{-1}(u)\right)} \tag{3.A.38}$$

because $F(v) = F\left(F^{-1}(u)\right) = u$ and $du = f(v)dv$ from (3.A.26). ♠

Appendix 3.2 Proofs of Theorems

(A) Proof of Theorem 3.5.1

Letting $k - np = l$, rewrite $P_n(k) = {}_nC_k p^k(1-p)^{n-k}$ as

$$\begin{aligned}
{}_nC_k p^k(1-p)^{n-k} &= \frac{n!}{(np+l)!(nq-l)!}p^{np+l}q^{nq-l} \\
&= \frac{n!p^{np}q^{nq}}{(np)!(nq)!} \times \frac{(np)!(nq)!}{(np+l)!(nq-l)!}\frac{p^l}{q^l}. \tag{3.A.39}
\end{aligned}$$

First, using the Stirling approximation $n! \approx \sqrt{2\pi n}\left(\frac{n}{e}\right)^n$, which can be obtained from $\sqrt{2\pi n}\left(\frac{n}{e}\right)^n < n! < \sqrt{2\pi n}\left(1+\frac{1}{4n}\right)\left(\frac{n}{e}\right)^n$, the first part of the right-hand side of (3.A.39) becomes

$$\begin{aligned}
\frac{n!p^{np}q^{nq}}{(np)!(nq)!} &\approx \frac{\sqrt{2\pi n}\,n^n e^{-n}}{\sqrt{2\pi np}(np)^{np}e^{-np}\sqrt{2\pi nq}(nq)^{nq}e^{-nq}}p^{np}q^{nq} \\
&= \frac{\sqrt{2\pi n}}{\sqrt{2\pi np}\sqrt{2\pi nq}} \times \frac{n^n\,p^{np}q^{nq}}{(np)^{np}(nq)^{nq}} \times \frac{e^{-n}}{e^{-np}e^{-nq}} \\
&= \frac{1}{\sqrt{2\pi npq}}. \tag{3.A.40}
\end{aligned}$$

The second part of the right-hand side of (3.A.39) can be rewritten as

$$\frac{(np)!(nq)!}{(np+l)!(nq-l)!}\frac{p^l}{q^l} = \frac{(np)!(npq)^l}{(np+l)!q^l} \times \frac{(nq)!p^l}{(nq-l)!(npq)^l}. \tag{3.A.41}$$

Letting $t_j = \frac{j}{npq}$, and using $e^x \approx 1+x$ for $x \approx 0$, the first part of the right-hand side of (3.A.41) becomes $\frac{(np)!(npq)^l}{(np+l)!q^l} = \frac{(np)^l}{(np+l)(np+l-1)\cdots(np+1)} = \left\{\prod_{j=1}^{l}\left(1+\frac{j}{np}\right)\right\}^{-1} =$

$$\left\{ \prod_{j=1}^{l} (1 + qt_j) \right\}^{-1} \approx \left(\prod_{j=1}^{l} e^{qt_j} \right)^{-1}$$ and the second part of the right-hand side

of (3.A.41) can be rewritten as $\dfrac{(nq)! p^l}{(nq-l)!(npq)^l} = \dfrac{\prod_{j=1}^{l}(nq+1-j)}{(nq)^l} = \prod_{j=1}^{l} \left(\dfrac{nq+1}{nq} - \dfrac{j}{nq} \right) \approx$

$\prod_{j=1}^{l} \left(1 - \dfrac{j}{nq} \right) = \prod_{j=1}^{l} \left(1 - pt_j \right) \approx \prod_{j=1}^{l} e^{-pt_j}$. Employing these two results in (3.A.41),

we get $\dfrac{(np)!(nq)!}{(np+l)!(nq-l)!} \dfrac{p^l}{q^l} \approx \prod_{j=1}^{l} \dfrac{e^{-pt_j}}{e^{qt_j}} = \prod_{j=1}^{l} e^{-t_j} = \exp\left\{ -\dfrac{l(l+1)}{2npq} \right\}$, i.e.,

$$\frac{(np)!(nq)!}{(np+l)!(nq-l)!} \frac{p^l}{q^l} \approx \exp\left(-\frac{l^2}{2npq} \right). \tag{3.A.42}$$

Now, recollecting $l = k - np$, we get the desired result (3.5.16) from (3.A.39), (3.A.40), and (3.A.42).

(B) Proof of Theorem 3.5.2

For $k = 0$, we have

$$\lim_{n \to \infty} P_n(0) = e^{-\lambda} \tag{3.A.43}$$

because $\lim_{n \to \infty} (1 - p)^n = \lim_{n \to \infty} \left(1 - \dfrac{\lambda}{n} \right)^n = e^{-\lambda}$ when $np \to \lambda$. Next, for $k = 1, 2, \ldots, n$, we have

$$\begin{aligned}
P_n(k) &= \frac{n(n-1)\cdots(n-k+1)}{k!} p^k (1-p)^{n-k} \\
&= \frac{(np)^k}{k!} (1-p)^n \frac{\left(1 - \frac{1}{n}\right)\left(1 - \frac{2}{n}\right)\cdots\left(1 - \frac{k-1}{n}\right)}{(1-p)^k}.
\end{aligned} \tag{3.A.44}$$

Now, $1 - p \approx e^{-p}$ when p is small, and

$$\frac{\left(1 - \frac{1}{n}\right)\left(1 - \frac{2}{n}\right)\cdots\left(1 - \frac{k-1}{n}\right)}{(1-p)^k} \to \frac{1 - \frac{1}{n}}{1 - \frac{\lambda}{n}} \frac{1 - \frac{2}{n}}{1 - \frac{\lambda}{n}} \cdots \frac{1 - \frac{k-1}{n}}{1 - \frac{\lambda}{n}} \frac{1}{1 - \frac{\lambda}{n}}$$

$$\to 1 \tag{3.A.45}$$

when $k = 1, 2, \ldots$ is fixed, $p \to \dfrac{\lambda}{n}$, and $n \to \infty$. Thus, letting $np \to \lambda$, we get $P_n(k) \to \left(e^{-p} \right)^n \dfrac{\lambda^k}{k!} = e^{-\lambda} \dfrac{\lambda^k}{k!}$: from this result and (3.A.43), we have (3.5.19).

Appendix 3.3 Distributions and Moment Generating Functions

(A) Discrete distributions: pmf $p(k)$ and mgf $M(t)$

Bernoulli distribution: $\alpha \in (0, 1)$

$$p(k) = \begin{cases} 1 - \alpha, & k = 0, \\ \alpha, & k = 1 \end{cases} \tag{3.A.46}$$

$$M(t) = 1 - \alpha + \alpha e^t \tag{3.A.47}$$

Binomial distribution: $n = 1, 2, \ldots, \alpha \in (0, 1)$

$$p(k) = {}_nC_k \alpha^k (1 - \alpha)^{n-k}, \ k = 0, 1, \ldots \tag{3.A.48}$$

$$M(t) = \left(1 - \alpha + \alpha e^t\right)^n \tag{3.A.49}$$

Geometric distribution 1: $\alpha \in (0, 1)$

$$p(k) = \alpha(1 - \alpha)^k, \ k = 0, 1, \ldots \tag{3.A.50}$$

$$M(t) = \frac{\alpha}{1 - (1 - \alpha)e^t} \tag{3.A.51}$$

Geometric distribution 2: $\alpha \in (0, 1)$

$$p(k) = \alpha(1 - \alpha)^{k-1}, \ k = 1, 2, \ldots \tag{3.A.52}$$

$$M(t) = \frac{\alpha e^t}{1 - (1 - \alpha)e^t} \tag{3.A.53}$$

Negative binomial distribution: $r > 0, \ \alpha \in (0, 1)$

$$p(k) = {}_{-r}C_k \alpha^r (\alpha - 1)^k, \ k = 0, 1, \ldots \tag{3.A.54}$$

$$M(t) = \left\{\frac{\alpha}{1 - (1 - \alpha)e^t}\right\}^r \tag{3.A.55}$$

Pascal distribution: $r > 0, \ \alpha \in (0, 1)$

$$p(k) = {}_{k-1}C_{r-1} \alpha^r (1 - \alpha)^{k-r}, \ k = r, r + 1, \ldots \tag{3.A.56}$$

$$M(t) = \left\{\frac{\alpha e^t}{1 - (1 - \alpha)e^t}\right\}^r \tag{3.A.57}$$

Poisson distribution: $\lambda > 0$

$$p(k) = \frac{\lambda^k}{k!}e^{-\lambda}, \; k = 0, 1, \ldots \tag{3.A.58}$$

$$M(t) = \exp\left\{-\lambda(1 - e^t)\right\} \tag{3.A.59}$$

Uniform distribution

$$p(k) = \frac{1}{n}, \; k = 0, 1, \ldots, n - 1 \tag{3.A.60}$$

$$M(t) = \frac{1 - e^{nt}}{n(1 - e^t)} \tag{3.A.61}$$

(B) Continuous distributions: pdf $f(x)$ and mgf $M(t)$

Cauchy distribution: $\alpha > 0$ **(cf $\varphi(\omega)$ instead of mgf)**

$$f(x) = \frac{\alpha}{\pi} \frac{1}{(x - \beta)^2 + \alpha^2} \tag{3.A.62}$$

$$\varphi(\omega) = \exp\left(j\beta\omega - \alpha|\omega|\right) \tag{3.A.63}$$

Central chi-square distribution: $n = 1, 2, \ldots$

$$f(x) = \frac{x^{\frac{n}{2}-1}}{\Gamma\left(\frac{n}{2}\right)2^{\frac{n}{2}}} \exp\left(-\frac{x}{2}\right) u(x) \tag{3.A.64}$$

$$M(t) = (1 - 2t)^{-\frac{n}{2}} \tag{3.A.65}$$

Exponential distribution: $\lambda > 0$

$$f(x) = \lambda e^{-\lambda x} u(x) \tag{3.A.66}$$

$$M(t) = \frac{\lambda}{\lambda - t} \tag{3.A.67}$$

Gamma distribution: $\alpha > 0, \beta > 0$

$$f(x) = \frac{x^{\alpha-1}}{\Gamma(\alpha)\beta^{\alpha}} \exp\left(-\frac{x}{\beta}\right) u(x) \qquad (3.A.68)$$

$$M(t) = \frac{1}{(1-\beta t)^{\alpha}} \qquad (3.A.69)$$

Laplace (double exponential) distribution: $\lambda > 0$

$$f(x) = \frac{\lambda}{2} e^{-\lambda|x|} \qquad (3.A.70)$$

$$M(t) = \frac{\lambda^2}{\lambda^2 - t^2} \qquad (3.A.71)$$

Normal distribution

$$f(x) = \frac{1}{\sqrt{2\pi\sigma^2}} \exp\left\{-\frac{(x-m)^2}{2\sigma^2}\right\} \qquad (3.A.72)$$

$$M(t) = \exp\left(mt + \frac{\sigma^2 t^2}{2}\right) \qquad (3.A.73)$$

Rayleigh distribution[16]**:** $\alpha > 0$

$$f(x) = \frac{x}{\alpha^2} \exp\left(-\frac{x^2}{2\alpha^2}\right) u(x) \qquad (3.A.74)$$

$$M(t) = 1 + \sqrt{2\pi}\,\alpha t \, \exp\left(\frac{\alpha^2 t^2}{2}\right) \Phi(\alpha t) \qquad (3.A.75)$$

Uniform distribution: $b > a$

$$f(x) = \frac{1}{b-a} u(x-a)u(b-x) \qquad (3.A.76)$$

$$M(t) = \frac{e^{bt} - e^{at}}{(b-a)t} \qquad (3.A.77)$$

---

[16] In (3.A.75), the function $\Phi(\cdot)$ is the standard normal cdf defined in (3.5.3).

Exercises

Exercise 3.1 Show that

$$\lim_{x \to \infty} x F(-x) = 0 \tag{3.E.1}$$

for an absolutely continuous cdf F. Using this result, show that

$$\mathsf{E}\{X\} = \int_0^\infty \{1 - F_X(x)\} dx - \int_{-\infty}^0 F_X(x) dx \tag{3.E.2}$$

for a random variable X with a continuous and absolutely integrable pdf.

Exercise 3.2 Express the cdf of

$$g(X) = \begin{cases} X - c, & X > c, \\ 0, & -c \le X \le c, \\ X + c, & X < -c \end{cases} \tag{3.E.3}$$

in terms of the cdf F_X of a continuous random variable X, where $c > 0$.

Exercise 3.3 Express the cdf of

$$g(X) = \begin{cases} X + c, & X \ge 0, \\ X - c, & X < 0 \end{cases} \tag{3.E.4}$$

in terms of the cdf F_X of X, where $c > 0$.

Exercise 3.4 Express the pdf f_Y of $Y = a \sin(X + \theta)$ in terms of the pdf of X, where $a > 0$ and θ are constants.

Exercise 3.5 Obtain the pmf of X in Example 3.1.17 assuming that each ball taken is replaced into the box before the following trial.

Exercise 3.6 Obtain the pdf and cdf of $Y = X^2 + 1$ when $X \sim U[-1, 2)$.

Exercise 3.7 Obtain the cdf of $Y = X^3 - 3X$ when the pmf of X is $p_X(k) = \frac{1}{7}$ for $k \in \{0, \pm 1, \pm 2, \pm 3\}$.

Exercise 3.8 Obtain the expected value $\mathsf{E}\{X^{-1}\}$ when the pdf of X is $f_X(r) = \frac{1}{2^{\frac{n}{2}} \Gamma(\frac{n}{2})} r^{\frac{n}{2}-1} \exp\left(-\frac{r}{2}\right) u(r)$.

Exercise 3.9 Express the pdf of the output $Y = X u(X)$ of a half-wave rectifier in terms of the pdf f_X of X.

Exercise 3.10 Obtain the pdf of $Y = \left(X - \frac{1}{\theta}\right)^2$ when the pdf of X is $f_X(x) = \theta e^{-\theta x} u(x)$.

Exercise 3.11 Let the pdf and cdf of a continuous random variable X be f_X and F_X, respectively. Obtain the conditional cdf $F_{X|b<X\leq a}(x)$ and the conditional pdf $f_{X|b<X\leq a}(x)$ in terms of f_X and F_X, where $a > b$.

Exercise 3.12 For $X \sim U[0, 1)$, obtain the conditional mean $\mathsf{E}\{X|X > a\}$ and conditional variance $\mathsf{Var}\{X|X > a\} = \mathsf{E}\left\{\left[X - \mathsf{E}\{X|X > a\}\right]^2 \middle| X > a\right\}$ when $0 < a < 1$. Obtain the limits of the conditional mean and conditional variance when $a \to 1$.

Exercise 3.13 Obtain the probability $\mathsf{P}(950 \leq R < 1050)$ when the resistance R of a resistor has the uniform distribution $U(900, 1100)$.

Exercise 3.14 The cost of being early and late by s minutes for an appointment is cs and ks, respectively. Denoting by f_X the pdf of the time X taken to arrive at the location of the appointment, find the time of departure for the minimum cost.

Exercise 3.15 Let ω be the outcome of a random experiment of taking one ball from a box containing one each of red, green, and blue balls. Obtain $\mathsf{P}(X \leq \alpha)$, $\mathsf{P}(X \leq 0)$, and $\mathsf{P}(2 \leq X < 4)$, where

$$X(\omega) = \begin{cases} \pi, & \omega = \text{green ball or blue ball}, \\ 0, & \omega = \text{red ball} \end{cases} \qquad (3.E.5)$$

with α a real number.

Exercise 3.16 For $V \sim U[-1, 1]$, obtain $\mathsf{P}(V > 0)$, $\mathsf{P}\left(|V| < \frac{1}{3}\right)$, $\mathsf{P}\left(|V| \geq \frac{3}{4}\right)$, and $\mathsf{P}\left(\frac{1}{3} < V < \frac{1}{2}\right)$.

Exercise 3.17 In successive tosses of a fair coin, let the number of tosses until the first *head* be K. For the two events $A = \{K > 5\}$ and $B = \{K > 10\}$, obtain the probabilities of A, B, B^c, $A \cap B$, and $A \cup B$.

Exercise 3.18 Data is transmitted via a sequence of N bits through two independent channels C_A and C_B. Due to channel noise during the transmission, w_A and w_B bits are in error among the sequences A and B of N bits received through channels C_A and C_B, respectively. Assume that the noise on a bit does not influence that on others.

(1) Obtain the probability $\mathsf{P}(D = d)$ that the number D of error bits common to A and B is d.
(2) Assume the sequence of N bits is reconstructed by selecting each bit from A with probability p or from B with probability $1 - p$. Obtain the probability $\mathsf{P}(K = k)$ that the number K of error bits is k in the reconstructed sequence of N bits.
(3) When $N = 3$ and $w_A = w_B = 1$, obtain $\mathsf{P}(D = d)$ and $\mathsf{P}(K = k)$.

Exercise 3.19 Assume the pdf $f(x) = \frac{c}{x^3}u(x-1)$ of X.

(1) Determine the constant c.
(2) Obtain the mean $E\{X\}$ of X.
(3) Show that the variance of X does not exist.

Exercise 3.20 Assume the cdf

$$F(x) = \begin{cases} 0, & x < 0; & \frac{1}{4}(x^2+1), \ 0 \le x < 1; \\ \frac{1}{4}(x+2), \ 1 \le x < 2; & 1, & x \ge 2 \end{cases} \qquad (3.E.6)$$

of X.

(1) Obtain $P(0 < X < 1)$ and $P(1 \le X < 1.5)$.
(2) Obtain the mean $\mu = E\{X\}$ and variance $\sigma^2 = Var\{X\}$ of X.

Exercise 3.21 For $Y \sim U[0, 1)$ and $a < b$, consider $W = a + (b-a)Y$.

(1) Obtain the cdf of W.
(2) Obtain the distribution of W.

Exercise 3.22 Obtain the pdf of $Y = \frac{X}{1+X}$ when $X \sim U[0, 1)$.

Exercise 3.23 Express the pdf f_Y of $Y = \frac{X}{1-X}$ in terms of the pdf f_X of X, and obtain f_Y when $X \sim U[0, 1)$.

Exercise 3.24 Consider $Y = \frac{X}{1+X}$ and $Z = \frac{Y}{1-Y}$. Then, for the pdf's f_Z and f_X of Z and X, respectively, it should hold true that $f_Z(z) = f_X(z)$ because $Z = \frac{Y}{1-Y} = \frac{\frac{X}{1+X}}{1-\frac{X}{1+X}} = X$. Confirm this fact from the results of Exercise 3.22 and 3.23.

Exercise 3.25 Obtain the pdf of $Z = \frac{X}{1-X}$ when the pdf of X is $f_X(y) = \frac{1}{(1-y)^2}\left\{u(y) - u\left(y - \frac{1}{2}\right)\right\}$.

Exercise 3.26 When $X \sim U[0, 1)$, obtain the pdf of $Y = \frac{1+X}{1-X}$ and the pdf of $Z = \frac{X-1}{X+1}$.

Exercise 3.27 Assuming the pdf

$$f_X(x) = \begin{cases} 1+x, & -1 \le x < 0, \\ 1-x, & 0 \le x < 1, \\ 0, & x < -1 \text{ or } x \ge 1 \end{cases} \qquad (3.E.7)$$

of X, obtain the pdf of $Y = |X|$.

Exercise 3.28 Consider $Y = a\cos(X + \theta)$ for a random variable X, where $a > 0$ and θ are constants.

(1) Obtain the pdf f_Y of Y when $X \sim U[-\pi, \pi)$.

(2) Obtain the pdf f_Y of Y when $X \sim U\left(-\frac{\pi}{2}, \frac{\pi}{2}\right)$, $a = 1$, and $\theta = 0$.

(3) Obtain the cdf F_Y of Y when $X \sim U\left(0, \frac{3}{2}\pi\right)$.

Exercise 3.29 Consider $Y = \tan X$ for a random variable X.

(1) Express the pdf f_Y of Y in terms of the pdf f_X of X.

(2) Obtain the pdf f_Y of Y when $X \sim U\left(-\frac{\pi}{2}, \frac{\pi}{2}\right)$.

(3) Obtain the pdf f_Y of Y when $X \sim U(0, 2\pi)$.

Exercise 3.30 Find a function g for $Y = g(X)$ with which we can obtain the exponential random variable $Y \sim f_Y(y) = \lambda e^{-\lambda y} u(y)$ from the uniform random variable $X \sim U[0, 1)$.

Exercise 3.31 For a random variable X with pmf $p_X(v) = \frac{1}{6}$ for $v = 1, 2, \ldots, 6$, obtain the expected value, mode, and median of X.

Exercise 3.32 Assume $p_X(k) = \frac{1}{7}$ for $k \in \{1, 2, 3, 5, 15, 25, 50\}$. Show that the value of c that minimizes $\mathsf{E}\{|X - c|\}$ is 5. Compare this value with the value of b that minimizes $\mathsf{E}\left\{(X - b)^2\right\}$.

Exercise 3.33 Obtain the mean and variance of a random variable X with pdf $f(x) = \frac{\lambda}{2}e^{-\lambda|x|}$, where $\lambda > 0$.

Exercise 3.34 For a random variable X with pdf $f(r) = \frac{r^{\alpha-1}(1-r)^{\beta-1}}{B(\alpha,\beta)}u(r)u(1 - r)$, show that the k-th moment is

$$\mathsf{E}\left\{X^k\right\} = \frac{\Gamma(\alpha + k)\Gamma(\alpha + \beta)}{\Gamma(\alpha + \beta + k)\Gamma(\alpha)} \tag{3.E.8}$$

and obtain the mean and variance of X.

Exercise 3.35 Show that $\mathsf{E}\{X\} = R'(0)$ and $\mathrm{Var}(X) = R''(0)$, where $R(t) = \ln M(t)$ and $M(t)$ is the mgf of X.

Exercise 3.36 Obtain the pdf's f_Y, f_Z, and f_W of $Y = X - 2$, $Z = 2Y$, and $W = Z + 1$, respectively, when the pdf of X is $f_X(x) = u(x) - u(x - 1)$.

Exercise 3.37 Obtain the pmf's p_Y, p_Z, and p_W of $Y = X + 2$, $Z = X - 2$, and $W = \frac{X-2}{X+2}$, respectively, when the pmf of X is $p_X(k) = \frac{1}{4}$ for $k = 1, 2, 3, 4$.

Exercise 3.38 For a random variable X such that $\mathsf{P}(0 \le X \le a) = 1$, show the following:

(1) $\mathsf{E}\{X^2\} \le a\mathsf{E}\{X\}$. Specify when the equality holds true.

(2) $\mathrm{Var}\{X\} \le \mathsf{E}\{X\}(a - \mathsf{E}\{X\})$.

(3) $\mathrm{Var}\{X\} \le \frac{a^2}{4}$. Specify when the equality holds true.

Exercise 3.39 The value of a random variable X is not less than 0.

(1) When X can assume values in $\{0, 1, 2, \ldots\}$, show that the expected value $E\{X\} = \sum_{n=0}^{\infty} nP(X = n)$ is

$$E\{X\} = \sum_{n=0}^{\infty} P(X > n)$$

$$= \sum_{n=1}^{\infty} P(X \geq n). \qquad (3.E.9)$$

(2) Express $E\left\{X_c^-\right\}$ in terms of $F_X(x) = P(X \leq x)$, where $X_c^- = \min(X, c)$ for a constant c.

(3) Express $E\left\{X_c^+\right\}$ in terms of $F_X(x)$, where $X_c^+ = \max(X, c)$ for a constant c.

Exercise 3.40 Assume the cf $\varphi_X(\omega) = \exp\left(\mu\left[\exp\left\{\lambda\left(e^{j\omega} - 1\right)\right\} - 1\right]\right)$ of X, where $\lambda > 0$ and $\mu > 0$.

(1) Obtain $E\{X\}$ and $\text{Var}\{X\}$.

(2) Show that $P(X = 0) = \exp\left\{-\mu\left(1 - e^{-\lambda}\right)\right\}$.

Exercise 3.41 Let N be the number of hydrogen molecules in a sphere of radius r with volume $V = \frac{4\pi}{3}r^3$. Assuming that N has the Poisson pmf $p_N(n) = \frac{1}{n!}e^{-\rho V}(\rho V)^n$ for $n = 0, 1, 2, \ldots$, obtain the pdf of the distance X from the center of the sphere to the closest hydrogen molecule. (Hint. Try to express $P(X > x)$ via p_N.)

Exercise 3.42 Determine the constant A when

$$f(x) = \begin{cases} Ax, & 0 \leq x \leq 4, \\ A(8 - x), & 4 \leq x \leq 8, \\ 0, & \text{otherwise} \end{cases} \qquad (3.E.10)$$

is the pdf of X. Sketch the cdf and pdf, and obtain $P(X \leq 6)$.

Exercise 3.43 The median α of a continuous random variable X can be defined via $\int_{-\infty}^{\alpha} f_X(x)dx = \int_{\alpha}^{\infty} f_X(x)dx = \frac{1}{2}$. Show that the value of b minimizing $E\{|X - b|\}$ is α.

Exercise 3.44 When F is the cdf of continuous random variable X, obtain the expected value $E\{F(X)\}$ of $F(X)$.

Exercise 3.45 Obtain the mgf and cf of a negative exponential random variable X with pdf $f_X(x) = e^x u(-x)$. Using the mgf, obtain the first four moments. Compare the results with those obtained directly.

Exercise 3.46 Show that $f_X(x) = \frac{1}{\cosh(\pi x)}$ can be a pdf. Obtain the corresponding mgf.

Exercise 3.47 When $f(x) = \frac{\alpha}{\cosh^n(\beta x)}$ is a pdf, determine the value of α in terms of β. Note that this pdf is the same as the logistic pdf (2.5.30) when $n = 2$ and $\beta = \frac{k}{2}$.

Exercise 3.48 Show that the mgf is as shown in (3.A.75) for the Rayleigh random variable with pdf $f(x) = \frac{x}{\alpha^2} \exp\left(-\frac{x^2}{2\alpha^2}\right) u(x)$.

Exercise 3.49 For $X \sim b(n, p)$ with $q = 1 - p$, show

$$P(X = \text{an even number}) = \frac{1}{2}\left\{1 + (q - p)^n\right\} \qquad (3.\text{E}.11)$$

and the recurrence formula

$$P(X = k + 1) = \frac{(n - k)p}{(k + 1)q} P(X = k) \qquad (3.\text{E}.12)$$

for $k = 0, 1, \ldots, n - 1$.

Exercise 3.50 When $X \sim P(\lambda)$, show

$$P(X = \text{an even number}) = \frac{1}{2}\left(1 + e^{-2\lambda}\right) \qquad (3.\text{E}.13)$$

and the recurrence formula

$$P(X = k + 1) = \frac{\lambda}{k + 1} P(X = k) \qquad (3.\text{E}.14)$$

for $k = 0, 1, 2, \ldots$.

Exercise 3.51 When the pdf of X is

$$f_X(x) = \begin{cases} \frac{1}{2}, & -1 < x \le 0, \\ \frac{1}{4}(2 - x), & 0 < x \le 2, \\ 0, & \text{otherwise,} \end{cases} \qquad (3.\text{E}.15)$$

obtain the pdf f_Y of $Y = |X|$.

Exercise 3.52 Find a cdf that has infinitely many jumps in the finite interval (a, b).

Exercise 3.53 Assume that

$$F_X(x) = \begin{cases} 0, & x < -2; \\ ax + b, & -1 \le x < 3; \end{cases} \quad \begin{cases} \frac{1}{3}(x + 2), & -2 \le x < -1; \\ 1, & x \ge 3 \end{cases} \qquad (3.\text{E}.16)$$

is the cdf of X.

(1) Obtain the condition that the two constants a and b should satisfy and sketch the region of the condition on a plane with the a-b coordinates.
(2) When $a = \frac{1}{8}$ and $b = \frac{5}{8}$, obtain the cdf of $Y = X^2$, $P(Y = 1)$, and $P(Y = 4)$.

Exercise 3.54 Obtain the cdf of $Y = X^2$ for the cdf

$$F_X(x) = \begin{cases} 0, & x < -1; & \frac{1}{2}(x+1), & -1 \le x < 0; \\ \frac{1}{8}(x+4), & 0 \le x < 4; & 1, & x \ge 4 \end{cases} \quad (3.\text{E}.17)$$

of X.

Exercise 3.55 Show that

$$F_Y(y) = \begin{cases} F_X(\alpha), & y < -2 \text{ or } y > 2, \\ F_X(-2) + P(X = 1); & y = -2, \\ F_X(\beta_3) - F_X(\beta_2) & \\ \quad + F_X(\beta_1) + P(X = \beta_2), & -2 < y < 2, \\ F_X(2), & y = 2 \end{cases} \quad (3.\text{E}.18)$$

is the cdf of $Y = X^3 - 3X$ expressed in terms of the cdf F_X of X. Here, α is the only real root of the equation $y = x^3 - 3x$ when $y > 2$ or $y < -2$, and $\beta_1 < \beta_2 < \beta_3$ are the three real roots of the equation when $-2 < y < 2$ with α, β_1, β_2, and β_3 being all functions of y.

Exercise 3.56 Discuss the continuity of the cdf $F_Y(y)$ shown in (3.E.18) when X is a continuous random variable.

Exercise 3.57 For a random variable X with pmf

$$p_X(k) = P(X = k), \quad k \in \{\cdots, -1, 0, 2, \ldots\}, \quad (3.\text{E}.19)$$

obtain the cdf F_Y of $Y = X^3 - 3X$ and discuss its continuity.

Exercise 3.58 Obtain the cdf of $Y = X^2$ for the cdf

$$F_X(x) = \begin{cases} 0, & x < -1, \\ -\frac{1}{2}\left(x^2 - 1\right), & -1 \le x < 0, \\ \frac{1}{2}\left(x^2 + 1\right), & 0 \le x < 1, \\ 1, & x \ge 1 \end{cases} \quad (3.\text{E}.20)$$

of X.

Exercise 3.59 Assume the cf $\varphi(\omega) = \frac{1}{2\pi} \int_0^{2\pi} \exp\left\{-\frac{\omega^2}{2}\alpha(\theta)\right\} d\theta$ of a random variable.

(1) Obtain the cf's by completing the integral when $\alpha(\theta) = \frac{1}{2}$ and when $\alpha(\theta) = \cos^2 \theta$.

(2) When $\alpha(\theta) = \frac{1}{2}$, obtain the pdf f_1.
(3) Show that

$$f_2(x) = \frac{1}{\sqrt{2\pi^3}} \exp\left(-\frac{x^2}{4}\right) K_0\left(\frac{x^2}{4}\right) \qquad (3.E.21)$$

is the pdf when $\alpha(\theta) = \cos^2\theta$. Here,

$$K_0(x) = \int_0^\infty \frac{\cos(xt)}{\sqrt{1+t^2}}\, dt\, u(x) \qquad (3.E.22)$$

is the zeroth order modified Bessel function of the second type.
(4) Obtain the mean and variance for f_1 and those for f_2.
(5) Show that the pdf f_2 is heavier-tailed than f_1: that is, $\frac{f_2(x)}{f_1(x)} > 1$ when x is sufficiently large.

Exercise 3.60 From (3.3.31), we have

$$\mathsf{E}\{X^n\} = \begin{cases} 0, & n \text{ is odd}, \\ 1 \times 3 \times \cdots \times (n-1), & n \text{ is even} \end{cases} \qquad (3.E.23)$$

when $X \sim \mathcal{N}(0,1)$. Show that

$$\mathsf{E}\{Y^p\} = m^p \sum_{n=0}^{\lfloor \frac{p}{2} \rfloor} \frac{p!}{2^n n!(p-2n)!} \left(\frac{\sigma}{m}\right)^{2n} \qquad (3.E.24)$$

for $p = 0, 1, \ldots$ when $Y \sim \mathcal{N}(m, \sigma^2)$. The result (3.E.24) implies that we have $\mathsf{E}\{Y^0\} = 1$, $\mathsf{E}\{Y^1\} = m$, $\mathsf{E}\{Y^2\} = m^2 + \sigma^2$, $\mathsf{E}\{Y^3\} = m^3 + 3m\sigma^2$, $\mathsf{E}\{Y^4\} = m^4 + 6m^2\sigma^2 + 3\sigma^4$, $\mathsf{E}\{Y^5\} = m^5 + 10m^3\sigma^2 + 15m\sigma^4, \ldots$ when $Y \sim \mathcal{N}(m, \sigma^2)$. (Hint. Note that $Y = \sigma X + m$.)

Exercise 3.61 Show that

$$\mathsf{E}\{X^n\} = \begin{cases} 1 \times 3 \times \cdots \times n\alpha^n \sqrt{\frac{\pi}{2}}, & n = 2k+1, \\ 2^k k! \alpha^{2k}, & n = 2k \end{cases} \qquad (3.E.25)$$

and obtain the mean and variance for a Rayleigh random variable X with pdf $f_X(x) = \frac{x}{\alpha^2} \exp\left(-\frac{x^2}{2\alpha^2}\right) u(x)$.

Exercise 3.62 Show that

$$\mathsf{E}\{X^n\} = \begin{cases} 1 \times 3 \times \cdots \times (n+1)\alpha^n, & n = 2k, \\ 2^k k! \alpha^{2k-1} \sqrt{\frac{2}{\pi}}, & n = 2k-1 \end{cases} \qquad (3.E.26)$$

for a random variable X with pdf $f_X(x) = \frac{\sqrt{2}x^2}{\alpha^3\sqrt{\pi}} \exp\left(-\frac{x^2}{2\alpha^2}\right) u(x)$.

Exercise 3.63 For a random variable with pdf $f_X(x) = \frac{1}{2}u(x)u(\pi - x)\sin x$, obtain the mean and second moment.

Exercise 3.64 Show that the mean is $E\{X\} = \frac{1-\alpha}{\alpha}r$ and variance is $Var\{X\} = \frac{1-\alpha}{\alpha^2}r$ for the NB random variable X with the pmf $p(x) = {}_{-r}C_x\alpha^r(\alpha - 1)^x$, $x \in \mathbb{J}_0$ shown in (2.5.13). When the pmf of Y is $p_Y(y) = {}_{y-1}C_{r-1}\alpha^r(1 - \alpha)^{y-r}$ for $y = r, r + 1, \ldots$ as shown in (2.5.17), show that

$$E\{Y\} = \frac{r}{\alpha} \tag{3.E.27}$$

and $Var\{Y\} = \frac{1-\alpha}{\alpha^2}$.

Exercise 3.65 Consider the absolute value $Y = |X|$ for a continuous random variable X with pdf f_X. If we consider the half mean[17]

$$m_X^{\pm} = \int_{-\infty}^{\infty} x f_X(x) u(\pm x) dx, \tag{3.E.28}$$

then the mean of X can be expressed as $m_X = m_X^+ + m_X^-$. Show that the mean of $Y = |X|$ can be expressed as

$$E\{|X|\} = m_X^+ - m_X^-, \tag{3.E.29}$$

and obtain the variance of $Y = |X|$ in terms of the variance and half means of X.

Exercise 3.66 For a continuous random variable X with cdf F_X, show that

$$E\{Z\} = 1 - 2F_X(0) \tag{3.E.30}$$

and

$$Var\{Z\} = 4F_X(0)\{1 - F_X(0)\} \tag{3.E.31}$$

are the mean and variance, respectively, of $Z = sgn(X)$.

Exercise 3.67 For a random variable X with pmf $p(k) = (1 - \alpha)^k\alpha$ for $k \in \{0, 1, \ldots\}$, obtain the mean and variance. For a random variable X with pmf $p(k) = (1 - \alpha)^{k-1}\alpha$ for $k \in \{1, 2, \ldots\}$, obtain the mean and variance.

---

[17] More generally, $\int_0^{\infty} x^m f_X(x)dx$ for $m = 1, 2, \ldots$ are called the half moments, incomplete moments, or partial moments.

Exercise 3.68 Consider a hypergeometric random variable X with pmf

$$p_X(x) = \frac{1}{\binom{\alpha+\beta}{\gamma}} \binom{\alpha}{\gamma-x}\binom{\beta}{x}, \qquad (3.E.32)$$

where α, β, and γ are natural numbers such that $\alpha + \beta \geq \gamma$. Note that $\min(\beta, \gamma) \geq \max(0, \gamma - \alpha)$ and that the pmf (3.E.32) is zero when $x \notin \{\max(0, \gamma - \alpha), \max(0, \gamma - \alpha) + 1, \ldots, \min(\beta, \gamma)\}$. Obtain the mean and variance of X. Show that the ν-th moment of X is

$$\mathsf{E}\{X^\nu\} = \sum_{k=0}^{\nu} \begin{Bmatrix} \nu \\ k \end{Bmatrix} [\beta]_k \frac{\binom{\alpha+\beta-k}{\gamma-k}}{\binom{\alpha+\beta}{\gamma}}. \qquad (3.E.33)$$

Here,

$$\begin{Bmatrix} \nu \\ k \end{Bmatrix} = \frac{1}{k!} \sum_{i=0}^{k} (-1)^i \binom{k}{i} (k-i)^\nu \qquad (3.E.34)$$

is the Stirling number of the second kind, and $\alpha + \beta$, β, and γ represent the size of the group, number of 'successes', and number of trials, respectively.

Exercise 3.69 For a random variable X with pdf $f(x) = \frac{ke^{-kx}}{\left(1+e^{-kx}\right)^2}$, show that $\mathsf{E}\{X^2\} = \frac{\pi^2}{3k^2}$, $\mathsf{E}\{X^4\} = \frac{7\pi^4}{15k^4}$, and $m_L^+ = \frac{\ln 2}{k} = -m_L^-$, where the half means m_L^+ and m_L^- are defined in (3.E.28).

Exercise 3.70 Obtain the mgf, expected value, and variance of a random variable Y with pdf $f_Y(x) = \frac{\lambda^n}{(n-1)!} x^{n-1} e^{-\lambda x} u(x)$.

Exercise 3.71 A coin with probability p of *head* is tossed twice in one trial. Define X_n as

$$X_n = \begin{cases} 1, & \text{if the outcome is } head \text{ and then } tail, \\ -1, & \text{if the outcome is } tail \text{ and then } head, \\ 0, & \text{if the two outcomes are the same} \end{cases} \qquad (3.E.35)$$

based on the two outcomes from the n-th trial. Obtain the cdf and mean of $Y = \min\{n : n \geq 1, X_n = 1 \text{ or } -1\}$.

Exercise 3.72 Assume a cdf F such that $F(x) = 0$ for $x < 0$, $F(x) < 1$ for $0 \leq x < \infty$, and

$$\frac{1 - F(x+y)}{1 - F(y)} = 1 - F(x) \qquad (3.E.36)$$

for $0 \leq x < \infty$ and $0 \leq y < \infty$. Show that there exists a positive number β satisfying $1 - F(x) = \exp\left(-\frac{x}{\beta}\right) u(x)$.

Exercise 3.73 For a geometric random variable X, show

$$P(X > m + n | X > m) = P(X > n). \tag{3.E.37}$$

Exercise 3.74 In the distribution $b\left(10, \frac{1}{3}\right)$, at which value of k is $P_{10}(k)$ the largest? At which value of k is $P_{11}(k)$ the largest in the distribution $b\left(11, \frac{1}{2}\right)$?

Exercise 3.75 The probability of a side effect from a flu shot is 0.005. When 1000 people get the flu shot, obtain the following probabilities and their approximate values:

(1) The probability P_{01} that at most one person experiences the side effect.
(2) The probability P_{456} that four, five, or six persons experience the side effect.

Exercise 3.76 Show that the skewness and kurtosis[18] of $b(n, p)$ are $\frac{1-2p}{\sqrt{np(1-p)}}$ and $\frac{3(n-2)}{n} + \frac{1}{np(1-p)}$, respectively, based on Definition 3.3.10.

Exercise 3.77 Consider a Poisson random variable X with rate λ.

(1) Show that $\mathsf{E}\left\{X^3\right\} = \lambda + 3\lambda^2 + \lambda^3$, $\mathsf{E}\left\{X^4\right\} = \lambda + 7\lambda^2 + 6\lambda^3 + \lambda^4$, $\mu_2 = \mu_3 = \lambda$, and $\mu_4 = \lambda + 3\lambda^2$.
(2) Obtain the coefficient of variation.
(3) Obtain the skewness and kurtosis, and compare them with those of normal distribution.

Exercise 3.78 When X is an exponential random variable with parameter λ, show that $Y = \sqrt{2\sigma^2 \lambda X}$ is a Rayleigh random variable with parameter σ.

Exercise 3.79 Consider the continuous cdf

$$F_3(x) = \begin{cases} 0, \ x \leq 0; & \frac{x}{2}, \quad 0 \leq x \leq 1; \\ \frac{1}{2}, \ 1 \leq x \leq 2; & \frac{1}{2}(x-1), \ 2 \leq x \leq 3; \\ 1, \ x \geq 3. \end{cases} \tag{3.E.38}$$

Confirm that $F_3\left(F_3^{-1}(u)\right) = u$, $F_3^{-1}(F_3(x)) \leq x$, and $\mathsf{P}\left(F_3^{-1}(F_3(X)) \neq X\right) = 0$ when $X \sim F_3$. Sketch $F_3(x)$, $F_3^{-1}(u)$ and $F_3^{-1}(F_3(x))$.

Exercise 3.80 Consider the continuous cdf

---

[18] Here, when $n \to \infty$, $p \to 0$, and $np \to \lambda$, the skewness is $\frac{1}{\sqrt{\lambda}}$ and the kurtosis is $3 + \frac{1}{\lambda}$. In addition, when $\sigma^2 = np(1-p) \to \infty$, the skewness is $\frac{1}{\sigma} \to 0$ and the kurtosis is $3 + \frac{1}{\sigma^2} \to 3$.

$$F_4(x) = \begin{cases} 0, \ x \leq 0; & \frac{2x}{3}, \ 0 \leq x \leq 1; \\ \frac{2}{3}, \ 1 \leq x \leq 2; & \frac{x}{3}, \ 2 \leq x \leq 3; \\ 1, \ x \geq 3. \end{cases} \qquad (3.E.39)$$

Confirm that $F_4\left(F_4^{-1}(u)\right) = u$, $F_4^{-1}\left(F_4(x)\right) \leq x$, and $\mathsf{P}\left(F_4^{-1}\left(F_4(X)\right) \neq X\right) = 0$ when $X \sim F_4$. Sketch $F_4(x)$, $F_4^{-1}(u)$, and $F_4^{-1}\left(F_4(x)\right)$.

Exercise 3.81 When the pdf of X is

$$f_X(x) = \frac{1}{2}u(x)u(1 - x) + \frac{1}{4}u(x - 1)u(3 - x)$$

$$= \begin{cases} \frac{1}{2}, \ 0 \leq x < 1, \\ \frac{1}{4}, \ 1 \leq x < 3, \\ 0, \ \text{otherwise}, \end{cases} \qquad (3.E.40)$$

obtain and sketch the pdf of $Y = \sqrt{X}$.

References

N. Balakrishnan, *Handbook of the Logistic Distribution* (Marcel Dekker, New York, 1992)

E.F. Beckenbach, R. Bellam, *Inequalities* (Springer, Berlin, 1965)

P.J. Bickel, K.A. Doksum, *Mathematical Statistics* (Holden-Day, San Francisco, 1977)

P.O. Börjesson, C.-E.W. Sundberg, Simple approximations of the error function $Q(x)$ for communications applications. IEEE Trans. Commun. **27**(3), 639–643 (Mar. 1979)

W. Feller, *An Introduction to Probability Theory and Its Applications*, 3rd edn., revised printing (Wiley, New York, 1970)

W.A. Gardner, *Introduction to Random Processes with Applications to Signals and Systems*, 2nd edn. (McGraw-Hill, New York, 1990)

B.R. Gelbaum, J.M.H. Olmsted, *Counterexamples in Analysis* (Holden-Day, San Francisco, 1964)

G.J. Hahn, S.S. Shapiro, *Statistical Models in Engineering* (Wiley, New York, 1967)

J. Hajek, *Nonparametric Statistics* (Holden-Day, San Francisco, 1969)

J. Hajek, Z. Sidak, P.K. Sen, *Theory of Rank Tests*, 2nd edn. (Academic, New York, 1999)

N.L. Johnson, S. Kotz, *Distributions in Statistics: Continuous Univariate Distributions*, vol. I, II (Wiley, New York, 1970)

S.A. Kassam, *Signal Detection in Non-Gaussian Noise* (Springer, New York, 1988)

A.I. Khuri, *Advanced Calculus with Applications in Statistics* (Wiley, New York, 2003)

P. Komjath, V. Totik, *Problems and Theorems in Classical Set Theory* (Springer, New York, 2006)

A. Leon-Garcia, *Probability, Statistics, and Random Processes for Electrical Engineering*, 3rd edn. (Prentice Hall, New York, 2008)

M. Loeve, *Probability Theory*, 4th edn. (Springer, New York, 1977)

E. Lukacs, *Characteristic Functions*, 2nd edn. (Griffin, London, 1970)

R.N. McDonough, A.D. Whalen, *Detection of Signals in Noise*, 2nd edn. (Academic, New York, 1995)

D. Middleton, *An Introduction to Statistical Communication Theory* (McGraw-Hill, New York, 1960)

A. Papoulis, *The Fourier Integral and Its Applications* (McGraw-Hill, New York, 1962)

A. Papoulis, S.U. Pillai, *Probability, Random Variables, and Stochastic Processes*, 4th edn. (McGraw-Hill, New York, 2002)

S.R. Park, Y.H. Kim, S.C. Kim, I. Song, *Fundamentals of Random Variables and Statistics (in Korean)* (Freedom Academy, Paju, 2017)

V.K. Rohatgi, A.KMd.E. Saleh, *An Introduction to Probability and Statistics*, 2nd edn. (Wiley, New York, 2001)

J.P. Romano, A.F. Siegel, *Counterexamples in Probability and Statistics* (Chapman and Hall, New York, 1986)

I. Song, J. Bae, S.Y. Kim, *Advanced Theory of Signal Detection* (Springer, Berlin, 2002)

J.M. Stoyanov, *Counterexamples in Probability*, 3rd edn. (Dover, New York, 2013)

A.A. Sveshnikov (ed.), *Problems in Probability Theory, Mathematical Statistics and Theory of Random Functions* (Dover, New York, 1968)

J.B. Thomas, *Introduction to Probability* (Springer, New York, 1986)

R.D. Yates, D.J. Goodman, *Probability and Stochastic Processes* (Wiley, New York, 1999)

D. Zwillinger, S. Kokoska, *CRC Standard Probability and Statistics Tables and Formulae* (CRC, New York, 1999)

Chapter 4
Random Vectors

We consider the concept and applications of random vectors in this chapter. In describing the probabilistic properties of a random vector, we need to specify not only the probabilistic properties of each of the element random variables, but also the relationships among random variables.

4.1 Distributions of Random Vectors

In this section, we concentrate on the distributions (Abramowitz and Stegun 1972; Kassam 1988; Mardia 1970; Song et al. 2002; Stuart and Ord 1987) of random vectors. The notion of joint probability functions, i.e., joint cdf, joint pdf, and joint pmf, are discussed to completely characterize a random vector.

4.1.1 Random Vectors

Definition 4.1.1 (random vector; continuous random vector; discrete random vector) A vector consisting of a number of random variables is called a random vector, multi-dimensional random vector, multi-variate random variables, or joint random variables. If the components of a random vector are all continuous random variables or all discrete random variables, then the random vector is called a continuous random vector or a discrete random vector, respectively.

Conversely, a random variable is a one-dimensional random vector, and a random process, also called a stochastic process, is a random vector of infinite dimension.

I. Song et al., *Probability and Random Variables: Theory and Applications*,
https://doi.org/10.1007/978-3-030-97679-8_4

Often, the terms random variable, random vector, and random process are used interchangeably.

When the size of a random vector is n, it is called an n-dimensional, n-variate,
or n-variable random vector. When some of the components of a random vector are
discrete random variables and some are continuous random variables, the random
vector is called a mixed-type or hybrid random vector. In this book, we mostly
consider only continuous and discrete random vectors.

Definition 4.1.2 (*joint cdf*) For a random vector $X = (X_1, X_2, \ldots, X_n)$, the function

$$F_X(x) = P(X_1 \leq x_1, X_2 \leq x_2, \ldots, X_n \leq x_n), \tag{4.1.1}$$

describing the probabilistic characteristics of X via the probability of the joint event
$\overset{n}{\underset{i=1}{\cap}} \{X_i \leq x_i\}$, is called the joint cdf of X, where $x = (x_1, x_2, \ldots, x_n)$.

The joint cdf is often simply called the cdf.

Example 4.1.1 Letting X_1 and X_2 be the first and second numbers on the face of a
fair die from two rollings, $X = (X_1, X_2)$ is a bi-variate discrete random vector. ◇

Example 4.1.2 (Thomas 1986) A fair coin is tossed three times. Let $X = (X_1, X_2, X_3)$, where X_i denotes the outcome from the i-th toss with 1 and 0 representing the *head* and *tail*, respectively. Then, the value of the discrete random vector
X is one of $(0, 0, 0)$, $(0, 0, 1)$, $(0, 1, 0)$, $(1, 0, 0)$, $(1, 1, 0)$, $(1, 0, 1)$, $(0, 1, 1)$, and
$(1, 1, 1)$. The joint cdf $F_X(x) = P\left(\overset{3}{\underset{i=1}{\cap}} \{X_i \leq x_i\}\right)$ of X is

$$F_X(x) = \begin{cases} 0, & \{x_1 < 0\}, \{x_2 < 0\}, \text{ or } \{x_3 < 0\}; \\ \frac{1}{8}, & \{0 \leq x_1 < 1, 0 \leq x_2 < 1, 0 \leq x_3 < 1\}; \\ \frac{1}{4}, & \{0 \leq x_1 < 1, 0 \leq x_2 < 1, x_3 \geq 1\}, \\ & \{0 \leq x_1 < 1, x_2 \geq 1, 0 \leq x_3 < 1\}, \text{ or} \\ & \{x_1 \geq 1, 0 \leq x_2 < 1, 0 \leq x_3 < 1\}; \\ \frac{1}{2}, & \{0 \leq x_1 < 1, x_2 \geq 1, x_3 \geq 1\}, \\ & \{x_1 \geq 1, 0 \leq x_2 < 1, x_3 \geq 1\}, \text{ or} \\ & \{x_1 \geq 1, x_2 \geq 1, 0 \leq x_3 < 1\}; \\ 1, & \{x_1 \geq 1, x_2 \geq 1, x_3 \geq 1\}; \end{cases} \tag{4.1.2}$$

where $x = (x_1, x_2, x_3)$.   ◇

Often, but not necessarily, the probability functions of a subvector of a random
vector are indicated with the term 'marginal' in front of the probability functions.
For example, for the random vector (X, Y), the cdf, pdf, or pmf of X is called the
marginal cdf, marginal pdf, or marginal pmf of X, respectively.

Theorem 4.1.1 *For a random vector* X, *the cdf of* X_i *can be obtained as*

$$F_{X_i}(x_i) = \lim_{\substack{x_j \to \infty \\ j \neq i}} F_X(x) \tag{4.1.3}$$

from the joint cdf $F_X(x)$ *of* X.

Proof We prove the theorem in the bi-variate case because the proofs in other cases are similar. Assume a sequence $\{y_n\}_{n=1}^{\infty}$ increasing to infinity. Then, the events $\{X_1 \leq x, X_2 \leq y_n\}_{n=1}^{\infty}$ are increasing sets, and thus $\lim_{n \to \infty} \{X_1 \leq x, X_2 \leq y_n\} = \bigcup_{n=1}^{\infty} \{X_1 \leq x, X_2 \leq y_n\} = \{X_1 \leq x, X_2 \leq \infty\} = \{X_1 \leq x\}$. This result implies $\lim_{n \to \infty} P(X_1 \leq x, X_2 \leq y_n) = P(X_1 \leq x)$ from the continuity of probability. In other words, we have

$$F_{X_1}(x) = \lim_{y \to \infty} F_{X_1, X_2}(x, y), \tag{4.1.4}$$

which completes the proof. ♠

In general, for a subvector $X_s = \left(X_{s_1}, X_{s_2}, \ldots, X_{s_m} \right)$ of X, we have

$$F_{X_s}(x_s) = \lim_{\substack{x_j \to \infty \\ j \notin I_s}} F_X(x), \tag{4.1.5}$$

where $I_s = \{s_1, s_2, \ldots, s_m\}$ and $x_s = \left(x_{s_1}, x_{s_2}, \ldots, x_{s_m} \right)$.

Definition 4.1.3 (*joint pdf*) For a measurable space $(\Omega, \mathcal{F}) = \left(\mathbb{R}^k, \mathcal{B}\left(\mathbb{R}^k \right) \right)$, a real function f is called a (k-dimensional) joint pdf if it satisfies

$$f(x) \geq 0, \quad x \in \mathbb{R}^k \tag{4.1.6}$$

and

$$\int_{\mathbb{R}^k} f(x) dx = 1, \tag{4.1.7}$$

where $dx = dx_1 dx_2 \cdots dx_k$.

Often, a joint pdf is called simply a pdf if it does not incur any confusion. Consider the set function

$$P(G) = \int_{x \in G} f(x) \, dx \tag{4.1.8}$$

for an event $G \in \mathcal{B}\left(\mathbb{R}^k\right)$. The set function P shown in (4.1.8) is a probability measure and is an extension of (2.5.20), a relationship between the pdf and probability measure in the one-dimensional space.

Example 4.1.3 When $\{f_i\}_{i=1}^k$ are one dimensional pdf's, a pdf $f(x)$ that can be expressed as $f(x) = \prod_{i=1}^k f_i(x_i)$ is called a product pdf.   ◇

Theorem 4.1.2 *For a random vector* $X = (X_1, X_2, \ldots, X_n)$ *with joint cdf* $F_X(x)$, *the joint pdf can be obtained as*

$$f_X(x) = \frac{\partial^n}{\partial x} F_X(x), \qquad (4.1.9)$$

where $\partial x = \partial x_1 \partial x_2 \cdots \partial x_n$.

Conversely, based on Definition 4.1.2 and Theorem 4.1.2, the joint cdf can be obtained as

$$F_X(x) = \int_{-\infty}^{x_n} \int_{-\infty}^{x_{n-1}} \cdots \int_{-\infty}^{x_1} f_X(t)\, dt \qquad (4.1.10)$$

from the joint pdf, where $t = (t_1, t_2, \ldots, t_n)$ and $dt = dt_1 dt_2 \cdots dt_n$. The joint cdf F_X and joint pdf f_X characterize the probabilistic properties of X.

The marginal pdf $f_{X_i}(x_i) = \frac{d}{dx_i} F_{X_i}(x_i)$ of X_i can be obtained as

$$f_{X_i}(x_i) = \frac{d}{dx_i} \int_{-\infty}^{\infty} \int_{-\infty}^{\infty} \cdots \int_{-\infty}^{\infty} \int_{-\infty}^{x_i} \int_{-\infty}^{\infty} \int_{-\infty}^{\infty} \cdots \int_{-\infty}^{\infty} f_X(t)\, dt$$

$$= \int_{-\infty}^{\infty} \int_{-\infty}^{\infty} \cdots \int_{-\infty}^{\infty} f_X(x) dx_a dx_b \qquad (4.1.11)$$

by differentiating the marginal cdf $F_{X_i}(x_i)$ of X_i after using (4.1.10) in (4.1.3), where $dx_a = dx_1 dx_2 \cdots dx_{i-1}$ and $dx_b = dx_{i+1} dx_{i+2} \cdots dx_n$. More generally, for a subvector $X_s = \left(X_{s_1}, X_{s_2}, \ldots, X_{s_m}\right)$ of X, we have

$$f_{X_s}(x_s) = \frac{\partial^m}{\partial x_s} \left\{ \lim_{\substack{x_j \to \infty \\ j \notin I_s}} F_X(x) \right\} \qquad (4.1.12)$$

from (4.1.5), where $I_s = \{s_1, s_2, \ldots, s_m\}$ and $\partial x_s = \partial x_{s_1} \partial x_{s_2} \cdots \partial x_{s_m}$.

Definition 4.1.4 (*joint pmf*) A real-valued function p is called a k-dimensional joint pmf if it satisfies

$$p(x) \geq 0 \qquad (4.1.13)$$

and

$$\sum_{x \in \Omega^k} p(x) = 1 \qquad (4.1.14)$$

for all $x = (x_1, x_2, \ldots, x_k) \in \Omega^k$ on a measurable space $(\Omega^k, \mathcal{F}^k)$, where Ω and $\mathcal{F}$ are a discrete sample space and the corresponding event space, respectively.

Consider the set function

$$\mathsf{P}(G) = \sum_{x \in G} p(x) \qquad (4.1.15)$$

for an event $G \in \mathcal{F}^k$. The set function P shown in (4.1.15) is a probability measure, and is an extension of (2.5.5), a relationship between the pmf and probability measure in the one-dimensional space.

Example 4.1.4 Consider the sample space $\Omega^k = \mathbb{J}_{n+1}^k = \{0, 1, \ldots, n\}^k$ and k numbers $\{\alpha_j \in (0, 1)\}_{j=1}^k$ such that $\sum_{j=1}^k \alpha_j = 1$. Then,

$$p(x) = \begin{cases} \binom{n}{x_1, x_2, \ldots, x_k} \alpha_1^{x_1} \alpha_2^{x_2} \cdots \alpha_k^{x_k}, & \sum_{i=1}^k x_i = n, \ x_i \in \mathbb{J}_{n+1}, \\ 0, & \text{otherwise} \end{cases} \qquad (4.1.16)$$

is called the multinomial[1] pmf, where

$$\binom{n}{n_1, n_2, \ldots, n_k} = \begin{cases} \frac{n!}{n_1! n_2! \cdots n_k!}, & \sum_{i=1}^k n_i = n, \\ 0, & \text{otherwise} \end{cases} \qquad (4.1.17)$$

is the multinomial coefficient discussed in (1.4.63). The multinomial pmf is a generalization of the binomial pmf, and denotes the binomial pmf when $k = 2$. ◇

Example 4.1.5 Consider the face number from a rolling of a fair die, and let $A = \{1, 2, 3\}$, $B = \{4, 5\}$, and $C = \{6\}$ as in Example 1.4.24. The probability of the occurrence of four times of A, five times of B, and one time of C is $\binom{10}{4,5,1} \left(\frac{1}{2}\right)^4 \left(\frac{1}{3}\right)^5 \left(\frac{1}{6}\right)^1 = \frac{35}{648} \approx 0.054$ from ten rollings. ◇

When $X = (X_1, X_2, \ldots, X_n)$ is an n-dimensional discrete random vector, the joint cdf can be expressed as (4.1.1) and the joint pmf can be written as

$$p_X(x) = \mathsf{P}(X = x). \qquad (4.1.18)$$

---

[1] The multinomial pmf is discussed also in Appendix 4.1.

The joint cdf (4.1.1) can be expressed as

$$F_X(x) = \sum_{\{t \le x\}} p_X(t) \tag{4.1.19}$$

in terms of the joint pmf (4.1.18), where $\{t \le x\}$ denotes $\{t_1 \le x_1, t_2 \le x_2, \ldots, t_n \le x_n\}$.

4.1.2 Bi-variate Random Vectors

We now consider two-dimensional random vectors in detail because they are used more frequently than, and provide insights on, higher dimensional random vectors.

Let $F_{X,Y}(x, y) = P(X \le x, Y \le y)$ and $f_{X,Y}$ be the joint cdf and pdf, respectively, of a two-dimensional random vector (X, Y). Then, the joint pdf can be written as

$$f_{X,Y}(x, y) = \frac{\partial^2}{\partial x \partial y} F_{X,Y}(x, y) \tag{4.1.20}$$

in terms of the joint cdf $F_{X,Y}$, and the joint cdf can be expressed as

$$F_{X,Y}(x, y) = \int_{-\infty}^{y} \int_{-\infty}^{x} f_{X,Y}(u, v) du dv \tag{4.1.21}$$

in terms of the joint pdf $f_{X,Y}$. In addition, we can obtain the (marginal) cdf $F_X(x) = P(X \le x)$ as

$$F_X(x) = \lim_{y \to \infty} F_{X,Y}(x, y) \tag{4.1.22}$$

from the joint cdf $F_{X,Y}$, and the (marginal) pdf $f_X(x) = \frac{\partial}{\partial x} F_X(x)$ as

$$f_X(x) = \int_{-\infty}^{\infty} f_{X,Y}(x, y) dy \tag{4.1.23}$$

from the joint pdf $f_{X,Y}$. In (4.1.22) and (4.1.23), interchanging the two random variables X and Y, we have the cdf $F_Y(y) = P(Y \le y)$ of Y as

$$F_Y(y) = \lim_{x \to \infty} F_{X,Y}(x, y) \tag{4.1.24}$$

and the pdf $f_Y(y) = \int_{-\infty}^{\infty} f_{X,Y}(x, y) dx$ of Y from $f_Y(y) = \frac{\partial}{\partial y} F_Y(y)$.

The two-dimensional cdf $F_{X,Y}(x, y)$ has the following properties:

$$F_{X,Y}(x + h, y + k) + F_{X,Y}(x, y) \geq F_{X,Y}(x + h, y)$$
$$+ F_{X,Y}(x, y + k). \qquad (4.1.25)$$
$$F_{X,Y}(x + h, y) \geq F_{X,Y}(x, y). \qquad (4.1.26)$$
$$F_{X,Y}(x, y + k) \geq F_{X,Y}(x, y). \qquad (4.1.27)$$
$$F_{X,Y}(-\infty, y) = 0. \qquad (4.1.28)$$
$$F_{X,Y}(x, -\infty) = 0. \qquad (4.1.29)$$
$$F_{X,Y}(\infty, \infty) = 1. \qquad (4.1.30)$$

Here, h and k are non-negative constants. Properties (4.1.25)–(4.1.27) can be obtained from

$$P(a < X \leq b, c < Y \leq d) = F_{X,Y}(b, d) - F_{X,Y}(a, d) - F_{X,Y}(b, c)$$
$$+ F_{X,Y}(a, c)$$
$$\geq 0 \qquad (4.1.31)$$

for $b \geq a$ and $d \geq c$, and imply that a joint cdf is a non-decreasing function. In addition, (4.1.28) and (4.1.29) imply $P(\emptyset) = 0$ described by (2.2.16), and (4.1.30) implies $P(\Omega) = 1$ discussed in (2.2.13).

Example 4.1.6 (Thomas 1986) Consider the function

$$F(x, y) = \begin{cases} 1, & x \geq 0, \ y \geq 0, \ x + y \geq 1, \\ 0, & \text{otherwise.} \end{cases} \qquad (4.1.32)$$

The function $F(x, y)$ is continuous from the right and non-decreasing. The values of $F(x, y)$ for $x, y \to \pm\infty$ satisfy the property of a cdf. However, $F(x, y)$ is not a cdf because it does not satisfy (4.1.25) or, equivalently, (4.1.31): for example, we have $F(b, d) - F(a, d) - F(b, c) + F(a, c) = F(1, 1) - F\left(\frac{1}{3}, 1\right) - F\left(1, \frac{1}{3}\right) + F\left(\frac{1}{3}, \frac{1}{3}\right) = 1 - 1 - 1 + 0 = -1$ when $a = c = \frac{1}{3}$ and $b = d = 1$. $\diamond$

Because a joint cdf $F(x, y)$ is a non-decreasing function, its derivative, the joint pdf $f(x, y)$ satisfies $f(x, y) \geq 0$ as described also in (4.1.6). In addition, from (4.1.21) and $F(\infty, \infty) = 1$ observed in (4.1.30), or as mentioned in (4.1.7), we have $\int_{-\infty}^{\infty} \int_{-\infty}^{\infty} f(x, y) dx dy = 1$.

For a discrete random vector (X, Y), similar results can be obtained. First, the joint pmf $p_{X,Y}(x, y) = P(X = x, Y = y)$, satisfying

$$\sum_{x=-\infty}^{\infty} \sum_{y=-\infty}^{\infty} p_{X,Y}(x, y) = 1, \qquad (4.1.33)$$

denotes the probability of the intersection $\{X = x, Y = y\}$ of the two events $\{X = x\}$ and $\{Y = y\}$. The joint cdf can be expressed as

$$F_{X,Y}(x, y) = \sum_{u \leq x, v \leq y} p_{X,Y}(u, v) \tag{4.1.34}$$

in terms of the joint pmf and, conversely, the joint pmf $p_{X,Y}$ can be expressed as

$$p_{X,Y}(x, y) = F_{X,Y}(x, y) + F_{X,Y}(x - 1, y - 1) \\ - F_{X,Y}(x - 1, y) - F_{X,Y}(x, y - 1) \tag{4.1.35}$$

in terms of the joint cdf $F_{X,Y}$ when the support of $p_{X,Y}(x, y)$ is a subset of the integer lattice points $\{(x, y) : x, y \in \mathbb{J}\}$ in the two-dimensional space. In addition, from the joint pmf $p_{X,Y}$, the pmf $p_X(x) = \mathsf{P}(X = x)$ of X can be obtained as

$$p_X(x) = \sum_{y=-\infty}^{\infty} p_{X,Y}(x, y) \tag{4.1.36}$$

and the pmf of Y as $p_Y(y) = \mathsf{P}(Y = y) = \sum_{x=-\infty}^{\infty} p_{X,Y}(x, y)$.

Example 4.1.7 When the joint pmf of (X, Y) is

$$p_{X,Y}(u, v) = \begin{cases} \frac{1}{3}, & (u, v) = (1, 1), (2, 4), (3, 9), \\ 0, & \text{otherwise}, \end{cases} \tag{4.1.37}$$

obtain the pmf of X.

Solution The pmf $p_X(x) = \sum_v p_{X,Y}(x, v) = p_{X,Y}(x, 1) + p_{X,Y}(x, 4) + p_{X,Y}(x, 9)$ of X can be obtained as

$$p_X(x) = \begin{cases} p_{X,Y}(1, 1), \ x = 1; & p_{X,Y}(2, 4), \ x = 2; \\ p_{X,Y}(3, 9), \ x = 3; & 0, \quad\quad\quad \text{otherwise} \end{cases}$$
$$= \begin{cases} \frac{1}{3}, \ x = 1, 2, 3, \\ 0, \ \text{otherwise} \end{cases} \tag{4.1.38}$$

using (4.1.36).   ◇

The probability that the random vector (X, Y) will have a value in a region $\mathcal{D}$ is obtained as

$$\int\!\!\int_{\mathcal{D}} dF_{X,Y}(x, y) = \begin{cases} \int\!\!\int_{\mathcal{D}} f_{X,Y}(x, y) dx dy, & \text{continuous random vector}, \\ \sum\sum_{(x,y)\in\mathcal{D}} p_{X,Y}(x, y), & \text{discrete random vector}. \end{cases} \tag{4.1.39}$$

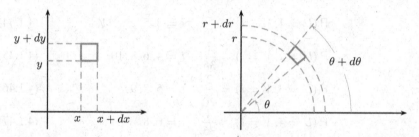

Fig. 4.1 The differential areas $dx\,dy$ and $dr\,rd\theta$ in the perpendicular and polar coordinate systems, respectively

Example 4.1.8 For a random vector (X, Y) with the joint pdf $f_{X,Y}(x, y) = \frac{1}{2\pi\sigma^2} \exp\left(-\frac{x^2+y^2}{2\sigma^2}\right)$, obtain the probability $P\left(X^2 + Y^2 \leq a^2\right)$ that (X, Y) will be on or inside the circle of radius a and center at the origin.

Solution From (4.1.39), we get

$$P\left(X^2 + Y^2 \leq a^2\right) = \iint\limits_{x^2+y^2\leq a^2} \frac{1}{2\pi\sigma^2} \exp\left(-\frac{x^2+y^2}{2\sigma^2}\right) dxdy. \qquad (4.1.40)$$

We then have $P\left(X^2 + Y^2 \leq a^2\right) = \frac{1}{2\pi\sigma^2} \int_0^a \int_0^{2\pi} \exp\left(-\frac{r^2}{2\sigma^2}\right) rd\theta dr = -\exp\left(-\frac{r^2}{2\sigma^2}\right)\Big|_{r=0}^a$, i.e.,

$$P\left(X^2 + Y^2 \leq a^2\right) = 1 - \exp\left(-\frac{a^2}{2\sigma^2}\right) \qquad (4.1.41)$$

by changing the perpendicular coordinate system (x, y) into the polar coordinate system (r, θ) as shown in Fig. 4.1. ◇

Example 4.1.9 The probability of a discrete random vector is sometimes called a point mass. Let X be the face value from a rolling of a fair die and $Y = 2X$. Then, the joint pmf $p_{ij} = p_{X,Y}(i, j) = P(X = i, Y = j)$ of (X, Y) is

$$p_{ij} = \begin{cases} \frac{1}{6}, & i = 1, 2, \ldots, 6, \ j = 2i, \\ 0, & \text{otherwise.} \end{cases} \qquad (4.1.42)$$

Next, let $U = |\text{face number of the first die} - \text{face number of the second die}|$ and $V = (\text{face number of the first die} + \text{face number of the second die})$ from a rolling of a pair of dice. Then, we have

$$P(U = 0, V = i) = \frac{1}{36}, \quad i = 2, 4, \ldots, 12, \qquad (4.1.43)$$

$$P(U = 1, V = i) = \frac{2}{36}, \quad i = 3, 5, \ldots, 11, \tag{4.1.44}$$

$$P(U = 2, V = i) = \frac{2}{36}, \quad i = 4, 6, 8, 10, \tag{4.1.45}$$

$$P(U = 3, V = i) = \frac{2}{36}, \quad i = 5, 7, 9, \tag{4.1.46}$$

$$P(U = 4, V = i) = \frac{2}{36}, \quad i = 6, 8, \tag{4.1.47}$$

$$P(U = 5, V = 7) = \frac{2}{36} \tag{4.1.48}$$

as the point mass of (U, V).   ◇

Example 4.1.10 The probability of a two dimensional hybrid random vector of which one element is a discrete random variable and the other is a continuous random variable is called a line mass. For example, when X is the face number from a rolling of a die and Y is a real number chosen randomly in the interval $(0, 1)$, $P(X = x, y_1 \le Y \le y_2)$ is a line mass.   ◇

Example 4.1.11 Assume that we repeatedly roll a fair die until the number of even-numbered outcomes is 10. Let N and X_i denote the numbers of rolls and outcome i, respectively, when the rolling ends. Obtain the pmf of X_2 and the pmf of X_1.

Solution (1) Let us obtain the pmf of X_2 in two ways.

(Method 1) The pmf $p_2(k) = P(X_2 = k) = \sum_{j=10}^{\infty} P(X_2 = k, N = j)$ of X_2 can be obtained[2] as

$$
\begin{aligned}
p_2(k) =\ & \sum_{j=10}^{\infty} \Big[P(2 \text{ at the } j\text{-th rolling; among the remaining } j - 1 \text{ rollings,} \\
& k - 1 \text{ times of } 2,\ 10 - k \text{ times of } 4 \text{ or } 6, \text{ and } j - 10 \text{ times of } 1, 3, \text{ or } 5) \\
& + P(4 \text{ or } 6 \text{ at the } j\text{-th rolling; among the remaining } j - 1 \text{ rollings,} \\
& k \text{ times of } 2,\ 9 - k \text{ times of } 4 \text{ or } 6, \text{ and } j - 10 \text{ times of } 1, 3, \text{ or } 5)\Big] \\
=\ & \sum_{j=10}^{\infty} \left\{ \frac{1}{6} \frac{(j-1)!}{(k-1)!(10-k)!(j-10)!} \left(\frac{1}{6}\right)^{k-1} \left(\frac{1}{3}\right)^{10-k} \left(\frac{1}{2}\right)^{j-10} \right. \\
& \left. + \frac{1}{3} \frac{(j-1)!}{k!(9-k)!(j-10)!} \left(\frac{1}{6}\right)^{k} \left(\frac{1}{3}\right)^{9-k} \left(\frac{1}{2}\right)^{j-10} \right\} \\
=\ & \left\{ \frac{1}{(k-1)!(10-k)!} + \frac{1}{k!(9-k)!} \right\} \frac{1}{6^k} \frac{1}{3^{10-k}} \\
& \times \sum_{j=10}^{\infty} \frac{(j-1)!}{(j-10)!} \frac{1}{2^{j-10}}
\end{aligned}
\tag{4.1.49}
$$

----

[2] Here, the events '2 occurs $k - 1$ times' for $k = 0$ and '4 or 6 occurs $9 - k$ times' for $k = 10$ are both empty events and thus have probability 0, which can be confirmed from $(-1)! \to \pm\infty$.

for $k = 0, 1, \ldots, 10$. We subsequently have

$$p_2(k) = {}_{10}C_k \left(\frac{1}{3}\right)^k \left(\frac{2}{3}\right)^{10-k} \tag{4.1.50}$$

for $k = 0, 1, \ldots, 10$ because $\displaystyle\sum_{j=10}^{\infty} \frac{(j-1)!}{(j-10)!} \frac{1}{2^{j-10}} = \sum_{q=0}^{\infty} \frac{(q+9)!}{q!9!} \frac{9!}{2^q} = 2^{10}9!$ from

$\displaystyle\sum_{x=0}^{\infty} {}_{r+x-1}C_x (1-\alpha)^x = \alpha^{-r}$ shown in (2.5.16). Thus, $X_2 \sim b\left(10, \frac{1}{3}\right)$.

(Method 2) Among the 10 times of the occurrences of even numbers, 2 can occur $0, 1, \ldots, 10$ times and the probability of 2 among $\{2, 4, 6\}$ is $\frac{1}{3}$. Thus, the probability that 2 will occur k times is

$$p_2(k) = {}_{10}C_k \left(\frac{1}{3}\right)^k \left(\frac{2}{3}\right)^{10-k} \tag{4.1.51}$$

for $k = 0, 1, \ldots, 10$.

(2) Similarly, recollecting $\displaystyle\sum_{x=0}^{\infty} {}_{r+x-1}C_x(1-\alpha)^x = \alpha^{-r}$ shown in (2.5.16), we can

obtain the pmf $p_1(k) = \mathsf{P}(X_1 = k) = \displaystyle\sum_{j=10+k}^{\infty} \mathsf{P}(X_1 = k, N = j)$ of X_1 as

$p_1(k) = \displaystyle\sum_{j=10+k}^{\infty} \mathsf{P}$(an even number at the j-th rolling; among the remaining

$\qquad\qquad j - 1$ rollings, k times of 1, 9 times of even numbers, and

$\qquad\qquad j - k - 10$ times of 3 or 5)

$$= \sum_{j=10+k}^{\infty} \frac{1}{2} \frac{(j-1)!}{(j-k-10)!k!9!} \left(\frac{1}{6}\right)^k \left(\frac{1}{2}\right)^9 \left(\frac{1}{3}\right)^{j-k-10} \tag{4.1.52}$$

for $k = 0, 1, \ldots$, which can be rewritten as

$$p_1(k) = {}_{k+9}C_9 \left(\frac{1}{4}\right)^k \left(\frac{3}{4}\right)^{10} \tag{4.1.53}$$

from $\displaystyle\sum_{j=10+k}^{\infty} \frac{1}{2} \frac{(j-1)!}{(j-k-10)!k!9!} \left(\frac{1}{6}\right)^k \left(\frac{1}{2}\right)^9 \left(\frac{1}{3}\right)^{j-k-10} = \sum_{i=0}^{\infty} \frac{(i+k+9)!}{i!k!9!} \left(\frac{1}{6}\right)^k \left(\frac{1}{2}\right)^{10} \left(\frac{1}{3}\right)^i =$

$\displaystyle\sum_{i=0}^{\infty} \frac{(i+k+9)!}{i!(k+9)!} \left(\frac{1}{3}\right)^i \frac{(k+9)!}{k!9!} \left(\frac{1}{6}\right)^k \left(\frac{1}{2}\right)^{10} = \left(\frac{2}{3}\right)^{-k-10} \frac{(k+9)!}{k!9!} \left(\frac{1}{6}\right)^k \left(\frac{1}{2}\right)^{10} = \frac{(k+9)!}{k!9!} \left(\frac{1}{4}\right)^k \left(\frac{3}{4}\right)^{10}$.

Thus, $X_1 \sim NB\left(10, \frac{3}{4}\right)$ with the pmf (2.5.14). $\diamond$

4.1.3 Independent Random Vectors

Definition 4.1.5 (*independent random vector*) When the joint probability function of a random vector is the product of the marginal probability functions of the element random variables, the random vector is called an independent random vector.

In other words, if the joint cdf F_X, joint pdf f_X, or joint pmf p_X of a random vector $X = (X_1, X_2, \ldots, X_n)$ can be expressed as

$$F_X(x) = \prod_{i=1}^{n} F_{X_i}(x_i), \tag{4.1.54}$$

$$f_X(x) = \prod_{i=1}^{n} f_{X_i}(x_i), \quad \text{(continuous random vector)}, \tag{4.1.55}$$

or

$$p_X(x) = \prod_{i=1}^{n} p_{X_i}(x_i), \quad \text{(discrete random vector)}, \tag{4.1.56}$$

respectively, for every real vector x, the random vector X is an independent random vector. The random variables of an independent random vector are all independent of each other (Burdick 1992; Davenport 1970; Dawid 1979; Geisser and Mantel 1962; Gray and Davisson 2010; Papoulis and Pillai 2002; Wang 1979).

Example 4.1.12 Assume the joint pmf

$$p_{X,Y}(u, v) = \begin{cases} \frac{1}{3}, & \text{for } (u, v) = (1, 1), (2, 4), (3, 9), \\ 0, & \text{otherwise} \end{cases} \tag{4.1.57}$$

of (X, Y) discussed in Example 4.1.7. Then, using (4.1.36), we can obtain the pmf $p_Y(w) = \sum_u p_{X,Y}(u, w) = p_{X,Y}(1, w) + p_{X,Y}(2, w) + p_{X,Y}(3, w)$ of Y as

$$
\begin{aligned}
p_Y(w) &= \begin{cases} p_{X,Y}(1, 1), & w = 1; \quad p_{X,Y}(2, 4), \ w = 4; \\ p_{X,Y}(3, 9), & w = 9; \quad 0, \qquad\qquad \text{otherwise} \end{cases} \\
&= \begin{cases} \frac{1}{3}, & w = 1, 4, \text{ or } 9, \\ 0, & \text{otherwise.} \end{cases}
\end{aligned} \tag{4.1.58}
$$

It is clear that $p_{X,Y}(u, w) \neq p_X(u) p_Y(w)$ from (4.1.38) and (4.1.58). Thus, (X, Y) is not an independent random vector. ◇

Example 4.1.13 A needle of length $2a$ is tossed at random on a plane ruled with infinitely many parallel lines of an infinite length, where the distance between adja-

Fig. 4.2 Buffon's needle when $a \le b$

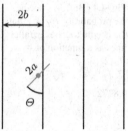

Fig. 4.3 Buffon's needle when $a \ge b$, where $\theta_T = \sin^{-1}\frac{b}{a}$

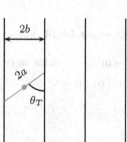

cent lines is $2b$ as shown in Figs. 4.2 and 4.3. Assuming that the thickness of the needle is negligible, find the probability P_B that the needle touches one or more of the parallel lines when $a \le b$.

Solution Let us denote by X the distance from the center of the needle to the nearest parallel line and by Θ the smaller angle that the needle makes with the lines. Then, $X \sim U[0, b)$, $\Theta \sim U[0, \frac{\pi}{2})$, and X and Θ are independent. Thus, the joint pdf $f_{X,\Theta}(x, \theta) = f_X(x) f_\Theta(\theta)$ of (X, Θ) can be expressed as

$$f_{X,\Theta}(x, \theta) = \frac{2}{\pi b}, \qquad 0 \le x < b, \, 0 \le \theta < \frac{\pi}{2}. \qquad (4.1.59)$$

Now, when $a \le b$, recollecting that $\{(x, \theta) : x \le a \sin \theta\} = \{(x, \theta) : 0 \le \theta < \frac{\pi}{2}, 0 \le x \le a \sin \theta\}$, we get[3] $P_B = P((X, \Theta) : X < a \sin \Theta) = \frac{2}{\pi b} \int_0^{\frac{\pi}{2}} a \sin \theta d\theta$ as

$$P_B = \frac{2a}{\pi b}. \qquad (4.1.60)$$

The probability (4.1.60) is proportional to the length of the needle and inversely proportional to the interval of the parallel lines, which is also appealing intuitively. When $a \to 0$ or $b \to \infty$, we have $P_B \to 0$. ◇

---

[3] Considering $\{(x, \theta) : x < a \sin \theta\} = \{(x, \theta) : 0 \le x \le a, \sin^{-1}\frac{x}{a} \le \theta < \frac{\pi}{2}\}$, the result (4.1.60) can be obtained also as $P_B = \frac{2}{\pi b} \int_0^a \int_{\sin^{-1}\frac{x}{a}}^{\frac{\pi}{2}} d\theta dx = \frac{2}{\pi b} \int_0^a \left(\frac{\pi}{2} - \sin^{-1}\frac{x}{a}\right) dx = \frac{a}{b} - \frac{2}{\pi b} \int_0^a \sin^{-1}\frac{x}{a} dx = \frac{a}{b} - \frac{2a}{\pi b} \int_0^{\frac{\pi}{2}} t \cos t dt = \frac{a}{b} - \frac{2a}{\pi b}(t \sin t + \cos t)\big|_{t=0}^{\frac{\pi}{2}} = \frac{2a}{\pi b}$.

Fig. 4.4 Buffon's needle: the probability $P_B\left(\frac{a}{b}\right)$ that the needle touches parallel lines as a function of $\frac{a}{b}$. $P_B(2) = \frac{2}{3\pi}\left(6+\pi-3\sqrt{3}\right) \approx 0.8372$

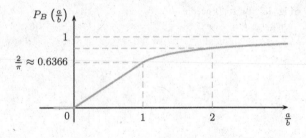

Example 4.1.14 In Example 4.1.13, find the probability when $a \geq b$.

Solution With the interval of integration $\left\{(x,\theta): \sin^{-1}\frac{b}{a} \leq \theta < \frac{\pi}{2}, 0 \leq x \leq b\right\} \cup \left\{(x,\theta): 0 \leq \theta \leq \sin^{-1}\frac{b}{a}, 0 \leq x \leq a\sin\theta\right\}$ in mind, $P_B = \frac{2}{\pi b}\left(\int_{\sin^{-1}\frac{b}{a}}^{\frac{\pi}{2}}\int_0^b dx\, d\theta + \int_0^{\sin^{-1}\frac{b}{a}}\int_0^{a\sin\theta}dx\, d\theta\right) = \frac{2}{\pi}\left(\frac{\pi}{2}-\sin^{-1}\frac{b}{a}\right) + \frac{2a}{\pi b}\int_0^{\sin^{-1}\frac{b}{a}}\sin\theta\, d\theta$, i.e.,

$$P_B = 1 + \frac{2}{\pi b}\left(a - \sqrt{a^2-b^2} - b\sin^{-1}\frac{b}{a}\right). \tag{4.1.61}$$

The same result can be obtained also as $P_B = \frac{2}{\pi b}\int_0^b \int_{\sin^{-1}\frac{x}{a}}^{\frac{\pi}{2}} d\theta dx = \frac{2}{\pi b}\int_0^b \left(\frac{\pi}{2}-\sin^{-1}\frac{x}{a}\right)dx = 1 - \frac{2a}{\pi b}\int_0^{\sin^{-1}\frac{b}{a}} t\cos t\, dt = 1 - \frac{2a}{\pi b}\left(t\sin t + \cos t\right)\big|_{t=0}^{\sin^{-1}\frac{b}{a}} = 1 - \frac{2}{\pi}\sin^{-1}\frac{b}{a} + \frac{2a}{\pi b} - \frac{2\sqrt{a^2-b^2}}{\pi b}$ with the interval of integration $\left\{(x,\theta): 0 \leq x \leq b, \sin^{-1}\frac{x}{a} \leq \theta < \frac{\pi}{2}\right\}$. ◇

It is easy to see that $P_B \to \frac{2}{\pi}$ when $a \to b$ from (4.1.60) and (4.1.61) and that $P_B \to 1$ when b is finite and $a \to \infty$ from (4.1.61). Figure 4.4 shows the probability P_B as a function of $\frac{a}{b}$.

Definition 4.1.6 (*independent and identically distributed random vectors*) An independent random vector is called an independent and identically distributed (i.i.d.) random vector when the marginal distributions of all the element random variables are identical.

For an i.i.d. random vector $X = (X_1, X_2, \ldots, X_n)$, we have the cdf

$$F_X(x) = \prod_{i=1}^n F_X(x_i), \tag{4.1.62}$$

the pdf

$$f_X(x) = \prod_{i=1}^n f_X(x_i) \tag{4.1.63}$$

when X is a continuous random vector, or the pmf

$$p_X(x) = \prod_{i=1}^{n} p_X(x_i) \tag{4.1.64}$$

when X is a discrete random vector for every real-valued vector x.

Example 4.1.15 Let $X = (X_1, X_2)$ be the pair of face numbers from a rolling of a pair of fair dice. Then, X is an i.i.d. random vector. Here, we have $p_X(x_1, x_2) = p_{X_1}(x_1) p_{X_2}(x_2)$ for every value of x_1 and x_2, and $p_{X_1}(x_i) = p_{X_2}(x_i) = p_X(x_i)$ for every value of x_i. In addition, $p_X(x_i) = \frac{1}{6}$ for $x_i = 1, 2, \ldots, 6$ and 0 otherwise. $\diamond$

Definition 4.1.7 (*two random vectors independent of each other*) Consider two random vectors $X = (X_1, X_2, \ldots, X_n)$ and $Y = (Y_1, Y_2, \ldots, Y_m)$. When the joint cdf F_{X^d, Y^d} of two subvectors $X^d = (X_{t_1}, X_{t_2}, \ldots, X_{t_k})$ and $Y^d = (Y_{s_1}, Y_{s_2}, \ldots, Y_{s_l})$ satisfies

$$F_{X^d, Y^d}(x^d, y^d) = F_{X^d}(x^d) F_{Y^d}(y^d) \tag{4.1.65}$$

for all natural numbers $k = 1, 2, \ldots, n$, $l = 1, 2, \ldots, m$, $t_1, t_2, \ldots, t_k$, $s_1, s_2, \ldots$, and s_l, the two random vectors X and Y are called independent of each other, where $x^d = (x_1, x_2, \ldots, x_k)$ and $y^d = (y_1, y_2, \ldots, y_l)$.

If the random vectors $X = (X_1, X_2, \ldots, X_n)$ and $Y = (Y_1, Y_2, \ldots, Y_m)$ are independent of each other, then

$$f_{X^d, Y^d}(x^d, y^d) = f_{X^d}(x^d) f_{Y^d}(y^d) \tag{4.1.66}$$

when X and Y are continuous random vectors and

$$p_{X^d, Y^d}(x^d, y^d) = p_{X^d}(x^d) p_{Y^d}(y^d) \tag{4.1.67}$$

when X and Y are discrete random vectors, for all subvectors X^d of X and Y^d of Y.

We can similarly define the independence among several random vectors by generalizing Definition 4.1.7. Note that the independence between X and Y has nothing to do with if X or Y is an independent random vector. In other words, even when X and Y are independent of each other, X or Y may or may not be independent random vectors, and even when X and Y are both independent random vectors, X and Y may or may not be independent of each other.

Theorem 4.1.3 *Assume the two random vectors $X_1 = (X_1, X_2, \ldots, X_m)$ and $X_2 = (X_{m+1}, X_{m+2}, \ldots, X_{m+n})$ are independent of each other, and the elements of the two vectors $g(\cdot) = (g_1(\cdot), g_2(\cdot), \ldots, g_m(\cdot))$ and $h(\cdot) = (h_1(\cdot), h_2(\cdot), \ldots, h_n(\cdot))$ are all continuous functions. Then, the two random vectors $Y_1 = g(X_1)$ and $Y_2 = h(X_2)$ are independent of each other.*

Proof For convenience, we show the result for the simplest case of $m = n = 1$. Let A_{y_1} and B_{y_2} be the inverse images of $\{Y_1 \leq y_1\}$ for $Y_1 = g(X_1)$ and $\{Y_2 \leq y_2\}$ for $Y_2 = h(X_2)$, respectively. Then, we have the joint cdf $F_{Y_1, Y_2}(y_1, y_2) = \mathsf{P}\left(X_1 \in A_{y_1}, X_2 \in B_{y_2}\right) = \mathsf{P}\left(X_1 \in A_{y_1}\right)\mathsf{P}\left(X_2 \in B_{y_2}\right) = F_{Y_1}(y_1) F_{Y_2}(y_2)$ because $\mathsf{P}(Y_1 \leq y_1, Y_2 \leq y_2) = \mathsf{P}\left(X_1 \in A_{y_1}, X_2 \in B_{y_2}\right)$.   ♠

Example 4.1.16 (Stoyanov 2013) We can easily show that, if $g(X)$ and $h(Y)$ are independent of each other and g and h are both one-to-one correspondences, then X and Y are also independent of each other. On the other hand, the converse of Theorem 4.1.3 does not hold true in general. In other words, when g or h is a continuous function but is not a one-to-one correspondence, X and Y may not be independent of each other even when $g(X)$ and $h(Y)$ are independent of each other. Let us consider one such example of discrete random variables.

Assume the joint pmf

$$p_{X,Y}(-1, -1) = \frac{3}{32}, \quad p_{X,Y}(0, -1) = \frac{5}{32}, \quad p_{X,Y}(1, -1) = \frac{3}{32},$$

$$p_{X,Y}(-1, 0) = \frac{5}{32}, \quad p_{X,Y}(0, 0) = \frac{8}{32}, \quad p_{X,Y}(1, 0) = \frac{3}{32}, \qquad (4.1.68)$$

$$p_{X,Y}(-1, 1) = \frac{1}{32}, \quad p_{X,Y}(0, 1) = \frac{3}{32}, \quad p_{X,Y}(1, 1) = \frac{1}{32}$$

of (X, Y). Then, $p_X(-1) = p_{X,Y}(-1, -1) + p_{X,Y}(-1, 0) + p_{X,Y}(-1, 1) = \frac{9}{32}$, and similarly $p_X(0) = \frac{1}{2}$ and $p_X(1) = \frac{7}{32}$. In addition, $p_Y(-1) = p_{X,Y}(-1, -1) + p_{X,Y}(0, -1) + p_{X,Y}(1, -1) = \frac{11}{32}$, $p_Y(0) = \frac{1}{2}$, and $p_Y(1) = \frac{5}{32}$. Thus, $p_{X,Y}(-1, -1) = \frac{3}{32} \neq \frac{99}{32^2} = p_X(-1)p_Y(-1)$ and X and Y are not independent of each other.

Now, consider $g(X) = X^2$ and $h(Y) = Y^2$. Then, $p_{X^2,Y^2}(0, 0) = p_{X,Y}(0, 0) = \frac{1}{4}$, $p_{X^2,Y^2}(0, 1) = p_{X,Y}(0, -1) + p_{X,Y}(0, 1) = \frac{1}{4}$, $p_{X^2,Y^2}(1, 0) = p_{X,Y}(-1, 0) + p_{X,Y}(1, 0) = \frac{1}{4}$, and $p_{X^2,Y^2}(1, 1) = p_{X,Y}(-1, -1) + p_{X,Y}(-1, 1) + p_{X,Y}(1, -1) + p_{X,Y}(1, 1) = \frac{1}{4}$. In other words, $p_{X^2}(0) = p_{X^2,Y^2}(0, 0) + p_{X^2,Y^2}(0, 1) = \frac{1}{2}$. Similarly, we get $p_{X^2}(1) = \frac{1}{2}$, $p_{Y^2}(0) = \frac{1}{2}$, and $p_{Y^2}(1) = \frac{1}{2}$. Thus,

$$p_{X^2,Y^2}(i, j) = p_{X^2}(i)p_{Y^2}(j) \qquad (4.1.69)$$

for $i, j = 0, 1$, implying that X^2 and Y^2 are independent of each other.   ◇

4.2 Distributions of Functions of Random Vectors

In this section, we will focus on the distributions of functions (Horn and Johnson 1985; Leon-Garcia 2008) of random vectors in terms of the cdf, pdf, and pmf.

4.2.1　Joint Probability Density Function

For the transformation $Y = g(X)$, if the region $|dx|$ of X is mapped to the region $|dy|$ of Y, we have

$$f_Y(y)|dy| = f_X(x)|dx|. \tag{4.2.1}$$

In other words, the probability $f_X(x)|dx|$ that X is in $|dx|$ will be the same as the probability $f_Y(y)|dy|$ that Y is in $|dy|$, a type of conservation laws.

4.2.1.1　General Formula

Based on (4.2.1), the pdf f_Y for the function $Y = (Y_1, Y_2, \ldots, Y_m) = g(X)$ of $X = (X_1, X_2, \ldots, X_n)$ for $m = n$ can be obtained as described in the following theorem:

Theorem 4.2.1 *Denote the Jacobian[4] of $g(x) = (g_1(x), g_2(x), \ldots, g_n(x))$ by $J(g(x)) = \left|\frac{\partial}{\partial x} g(x)\right|$, i.e.,*

$$J(g(x)) = \begin{vmatrix} \frac{\partial g_1}{\partial x_1} & \frac{\partial g_1}{\partial x_2} & \cdots & \frac{\partial g_1}{\partial x_n} \\ \frac{\partial g_2}{\partial x_1} & \frac{\partial g_2}{\partial x_2} & \cdots & \frac{\partial g_2}{\partial x_n} \\ \vdots & \vdots & \ddots & \vdots \\ \frac{\partial g_n}{\partial x_1} & \frac{\partial g_n}{\partial x_2} & \cdots & \frac{\partial g_n}{\partial x_n} \end{vmatrix}. \tag{4.2.2}$$

Then, the joint pdf of $Y = (Y_1, Y_2, \ldots, Y_n) = g(X) = (g_1(X), g_2(X), \ldots, g_n(X))$ for a random vector $X = (X_1, X_2, \ldots, X_n)$ can be obtained as

$$f_Y(y) = \sum_{j=1}^{Q} \frac{f_X(x)}{|J(g(x))|} \Bigg|_{x=x^j(y)}, \tag{4.2.3}$$

where $\{x^j(y)\}_{j=1}^{Q} = \left\{\left(x_1^j(y), x_2^j(y), \ldots, x_n^j(y)\right)\right\}_{j=1}^{Q}$ are the solutions to the simultaneous equations $g_1(x) = y_1, \ g_2(x) = y_2, \ \ldots, \ g_n(x) = y_n$, i.e., $\{x_i\}_{i=1}^{n}$ expressed in terms of $\{y_i\}_{i=1}^{n}$.

---

[4] The Jacobian is also referred to as the transformation Jacobian or Jacobian determinant. In addition, from the property of determinant, we also have $J(g(x)) = \begin{vmatrix} \frac{\partial g_1}{\partial x_1} & \frac{\partial g_2}{\partial x_1} & \cdots & \frac{\partial g_n}{\partial x_1} \\ \frac{\partial g_1}{\partial x_2} & \frac{\partial g_2}{\partial x_2} & \cdots & \frac{\partial g_n}{\partial x_2} \\ \vdots & \vdots & \ddots & \vdots \\ \frac{\partial g_1}{\partial x_n} & \frac{\partial g_2}{\partial x_n} & \cdots & \frac{\partial g_n}{\partial x_n} \end{vmatrix}.$

Example 4.2.1 When the joint pdf of $X = (X_1, X_2)$ is

$$f_X(x_1, x_2) = u(x_1) u(1 - x_1) u(x_2) u(1 - x_2), \qquad (4.2.4)$$

obtain the joint pdf f_Y of $Y = (Y_1, Y_2) = (3X_1 - X_2, -X_1 + X_2)$. Next, based on the joint pdf f_Y, obtain the pdf's of $Y_1 = 3X_1 - X_2$ and $Y_2 = -X_1 + X_2$.

Solution The Jacobian of the transformation $g(x_1, x_2) = (3x_1 - x_2, -x_1 + x_2)$ is

$$J(g(x_1, x_2)) = \left| \frac{\partial g(x_1, x_2)}{\partial(x_1, x_2)} \right| = \begin{vmatrix} 3 & -1 \\ -1 & 1 \end{vmatrix} = 2.$$ In addition, solving the simultaneous

equations $y_1 = 3x_1 - x_2$ and $y_2 = -x_1 + x_2$, we get $x_1 = \frac{y_1}{2} + \frac{y_2}{2}$ and $x_2 = \frac{y_1}{2} + \frac{3y_2}{2}$ and, consequently, we have $Q = 1$. Therefore, the joint pdf $f_Y(y_1, y_2)$ of Y can be obtained as

$$
\begin{aligned}
f_Y(y_1, y_2) &= \left. \frac{1}{|2|} f_X(x_1, x_2) \right|_{(x_1, x_2) = \left(\frac{y_1}{2} + \frac{y_2}{2}, \frac{y_1}{2} + \frac{3y_2}{2} \right)} \\
&= \frac{1}{2} u\left(\frac{y_1}{2} + \frac{y_2}{2} \right) u\left(1 - \frac{y_1}{2} - \frac{y_2}{2} \right) u\left(\frac{y_1}{2} + \frac{3y_2}{2} \right) \\
&\quad \times u\left(1 - \frac{y_1}{2} - \frac{3y_2}{2} \right) \\
&= \frac{1}{2} u(y_1 + y_2) u(2 - y_1 - y_2) u(y_1 + 3y_2) u(2 - y_1 - 3y_2). \quad (4.2.5)
\end{aligned}
$$

Figure 4.5 shows the support V_X of the pdf $f_X(x_1, x_2)$ and also the support

$$V_Y = \{(y_1, y_2) : 0 < y_1 + y_2 < 2, \ 0 < y_1 + 3y_2 < 2\} \qquad (4.2.6)$$

of the pdf $f_Y(y_1, y_2)$. Here, $f_Y(y_1, y_2) = \frac{1}{2}$ for $(y_1, y_2) \in V_Y$ and $f_Y(y_1, y_2) = 0$ otherwise. It is also easy to see that the area of V_Y is 2. Now, keeping in mind the support V_Y of the joint pdf $f_Y(y_1, y_2)$ shown in Fig. 4.5 for integration, we can obtain the marginal pdf

$$
\begin{aligned}
f_{Y_1}(y_1) &= \int_{-\infty}^{\infty} f_Y(y_1, y_2) \, dy_2 \\
&= \begin{cases} \frac{1}{2} \int_{-y_1}^{-\frac{1}{3}(y_1 - 2)} dy_2 = \frac{1}{3}(y_1 + 1), & -1 < y_1 \leq 0, \\ \frac{1}{2} \int_{-\frac{y_1}{3}}^{-\frac{1}{3}(y_1 - 2)} dy_2 = \frac{1}{3}, & 0 < y_1 \leq 2, \\ \frac{1}{2} \int_{-\frac{y_1}{3}}^{2 - y_1} dy_2 = 1 - \frac{y_1}{3}, & 2 < y_1 \leq 3, \\ 0, & \text{otherwise} \end{cases} \qquad (4.2.7)
\end{aligned}
$$

of Y_1 and the marginal pdf

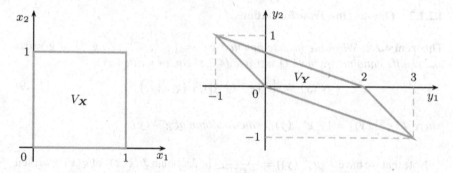

Fig. 4.5 Supports V_X and V_Y of the joint pdf's $f_X(x_1, x_2)$ and $f_Y(y_1, y_2)$, respectively, when the transformation is $(Y_1, Y_2) = (3X_1 - X_2, -X_1 + X_2)$

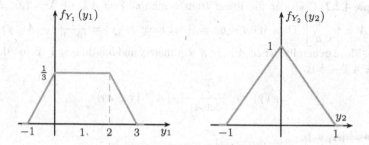

Fig. 4.6 The pdf's $f_{Y_1}(y_1)$ of $Y_1 = 3X_1 - X_2$ and $f_{Y_2}(y_2)$ of $Y_2 = -X_1 + X_2$ when the joint pdf of $X = (X_1, X_2)$ is $f_X(x_1, x_2) = u(x_1) u(1 - x_1) u(x_2) u(1 - x_2)$

$$f_{Y_2}(y_2) = \begin{cases} \frac{1}{2} \int_{-3y_2}^{2-y_2} dy_1 = 1 + y_2, & -1 < y_2 \le 0, \\ \frac{1}{2} \int_{-y_2}^{-3y_2+2} dy_1 = 1 - y_2, & 0 < y_2 \le 1, \\ 0, & \text{otherwise} \end{cases} \qquad (4.2.8)$$

of Y_2 by integrating the joint pdf $f_Y(y_1, y_2)$. Figure 4.6 shows the pdf's $f_{Y_1}(y_1)$ and $f_{Y_2}(y_2)$. ◇

When $m < n$, the joint pdf of Y can be obtained from Theorem 4.2.1 by employing auxiliary variables: details will be discussed in Sect. 4.2.2. It is not possible to obtain the joint pdf of $g(X)$ for $m > n$ except in very special cases as discussed in Sect. 4.5. When $J(g(x)) = 0$ and $m = n$, we cannot use Theorem 4.2.1 in obtaining the joint pdf of $g(X)$ because the denominator in (4.2.3) is 0: in Sect. 4.5, we briefly discuss with some examples on how we can deal with such cases.

4.2.1.2 One-to-One Transformations

Theorem 4.2.2 *When the inverse function* $g^{-1} = \left(g_1^{-1}, g_2^{-1}, \ldots, g_n^{-1}\right)$ *of* g *exists and is differentiable, we have* $Q = 1$ *and (4.2.3) can be written as*

$$f_Y(y) = \left| J\left(g^{-1}(y)\right)\right| f_X\left(g^{-1}(y)\right), \tag{4.2.9}$$

where $J\left(g^{-1}(y)\right) = \left|\frac{\partial}{\partial y}g^{-1}(y)\right|$ *is the Jacobian of* $g^{-1}(y)$.

Note that we have $J\left(g^{-1}(y)\right) = \frac{1}{J(g(x))}$. The Jacobian $J(g(x))$ of $g(x)$ is written also as $J(y)$ or $\left|\frac{\partial y}{\partial x}\right|$, and the Jacobian $J\left(g^{-1}(y)\right)$ is expressed also as $J(x)$ or $\left|\frac{\partial x}{\partial y}\right|$.

Example 4.2.2 Consider the linear transformation $Y = AX$ of $X = (X_1, X_2)^T$, where $A = \begin{bmatrix} a & b \\ c & d \end{bmatrix}$. Then, if $ad - bc \neq 0$, we have $f_Y(y) = \frac{1}{|\det A|} f_X\left(A^{-1}y\right)$ from (4.2.9). More generally, when A is an $n \times n$ matrix and b is an $n \times 1$ vector, the pdf of $Y = AX + b$ is

$$f_Y(y) = \frac{1}{|\det A|} f_X\left(A^{-1}(y - b)\right) \tag{4.2.10}$$

if $\det A = |A| \neq 0$.   ◇

Example 4.2.3 When $X \sim G(\alpha_1, \beta)$ and $Y \sim G(\alpha_2, \beta)$ are independent of each other, show that $Z = X + Y$ and $S = \frac{X}{X+Y}$ are independent of each other, and obtain the pdf of S.

Solution Expressing X and Y as $X = ZS$ and $Y = Z - ZS$ in terms of Z and S, we have the Jacobian $J\left(g^{-1}(z, s)\right) = \left|\frac{\partial}{\partial(z,s)}g^{-1}(z, s)\right| = \begin{vmatrix} s & z \\ 1-s & -z \end{vmatrix} = -z$ of the inverse transformation $(X, Y) = g^{-1}(Z, S) = (ZS, Z - ZS)$. Thus, the joint pdf $f_{Z,S}(z, s) = |z| f_{X,Y}(x, y)\big|_{x=zs,\, y=z-zs}$ of (Z, S) is

$$f_{Z,S}(z, s) = \frac{|z|(zs)^{\alpha_1 - 1}(z - zs)^{\alpha_2 - 1}}{\beta^{\alpha_1 + \alpha_2} \Gamma(\alpha_1) \Gamma(\alpha_2)} \exp\left(-\frac{zs}{\beta}\right) \exp\left(-\frac{z - zs}{\beta}\right)$$

$$\times u(zs) u(z - zs)$$

$$= \left\{ \frac{z^{\alpha_1 + \alpha_2 - 1}}{\beta^{\alpha_1 + \alpha_2} \Gamma(\alpha_1 + \alpha_2)} \exp\left(-\frac{z}{\beta}\right) u(z) \right\}$$

$$\times \left\{ \frac{\Gamma(\alpha_1 + \alpha_2)}{\Gamma(\alpha_1) \Gamma(\alpha_2)} s^{\alpha_1 - 1}(1 - s)^{\alpha_2 - 1} u(1 - s) u(s) \right\}. \tag{4.2.11}$$

This result dictates that $Z = X + Y$ and $S = \frac{X}{X+Y}$ are independent of each other, the pdf of Z is $f_Z(z) = \frac{z^{\alpha_1 + \alpha_2 - 1}}{\beta^{\alpha_1 + \alpha_2}} \frac{1}{\Gamma(\alpha_1 + \alpha_2)} \exp\left(-\frac{z}{\beta}\right) u(z)$, and the pdf of S is $f_S(s) = \frac{\Gamma(\alpha_1 + \alpha_2)}{\Gamma(\alpha_1) \Gamma(\alpha_2)} s^{\alpha_1 - 1}(1 - s)^{\alpha_2 - 1} u(1 - s) u(s)$. Note that $S = \frac{X}{X+Y} \sim B(\alpha_1, \alpha_2)$.   ◇

4.2.2   Joint Probability Density Function: Method of Auxiliary Variables

When $m < n$ for a random vector $X = (X_1, X_2, \ldots, X_n)$ and its function $Y = g(X) = (Y_1, Y_2, \ldots, Y_m)$, we first choose auxiliary variables as $Y_{m+1} = X_{m+1}$, $Y_{m+2} = X_{m+2}$, $\ldots$, $Y_n = X_n$. We next obtain the pdf f_{Y^a} of $Y^a = (Y_1, Y_2, \ldots, Y_m, Y_{m+1}, Y_{m+2}, \ldots, Y_n)$ using Theorem 4.2.1. Then, we obtain the pdf of Y as

$$f_Y(y) = \int_{-\infty}^{\infty} \int_{-\infty}^{\infty} \cdots \int_{-\infty}^{\infty} f_{Y^a}(y^a)\, dy_{m+1}\, dy_{m+2} \cdots dy_n \qquad (4.2.12)$$

by integrating f_{Y^a} over the auxiliary variables, where $y = (y_1, y_2, \ldots, y_m)$ and $y^a = (y_1, y_2, \ldots, y_n)$.

Example 4.2.4 Obtain the pdf of $Y = X_1 + X_2$ for $X = (X_1, X_2)$ with the joint pdf f_X.

Solution Let $Y_1 = X_1 + X_2$, $Y_2 = X_2$, $Y^a = [Y_1\ Y_2]^T$, and $X = [X_1\ X_2]^T$. Then, we have $Y^a = AX$, where $A = \begin{bmatrix} 1 & 1 \\ 0 & 1 \end{bmatrix}$. Because $A^{-1} = \begin{bmatrix} 1 & -1 \\ 0 & 1 \end{bmatrix}$, we get $f_{Y^a}(y^a) = \frac{1}{|\det(A)|} f_X(A^{-1}y^a) = f_{X_1, X_2}(y_1 - y_2, y_2)$ based on Theorem 4.2.1. Integrating this result, we have

$$f_Y(y) = \int_{-\infty}^{\infty} f_{X_1, X_2}(y - y_2, y_2)\, dy_2. \qquad (4.2.13)$$

This example is discussed again in Example 4.2.13.   ◇

When X_1 and X_2 are independent of each other, we have

$$f_Y(y) = \int_{-\infty}^{\infty} f_{X_1}(y - x) f_{X_2}(x)\, dx$$
$$= \int_{-\infty}^{\infty} f_{X_1}(x) f_{X_2}(y - x)\, dx \qquad (4.2.14)$$

from (4.2.13). In other words, the pdf of the sum of two random variables independent of each other is the convolution

$$f_{X+Y} = f_X * f_Y \qquad (4.2.15)$$

of the pdf's of the two random variables.

Example 4.2.5 (Thomas 1986) Obtain the pdf of X_1, the pdf of X_2, and the pdf of $Y_1 = X_1 + X_2$ for a random vector (X_1, X_2) with the joint pdf $f_{X_1, X_2}(x_1, x_2) = 24x_1(1 - x_2) u(x_1) u(x_2 - x_1) u(1 - x_2)$.

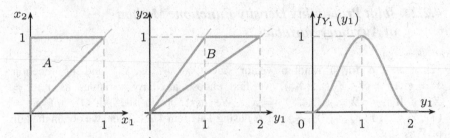

Fig. 4.7 The region $A = \{(x_1, x_2) : u(x_1) u(x_2 - x_1) u(1 - x_2) > 0\} = \{(x_1, x_2) : 0 \le x_1 \le x_2, 0 \le x_2 \le 1\}$ and the corresponding region $B = \{(y_1, y_2) : u(y_1 - y_2) u(2y_2 - y_1) u(1 - y_2) > 0\} = \{(y_1, y_2) : y_2 \le y_1 \le 2y_2, 0 \le y_2 \le 1\}$ of the transformation $(y_1, y_2) = (x_1 + x_2, x_2)$. The pdf $f_{Y_1}(y_1)$ of $Y_1 = X_1 + X_2$ when the joint pdf of X_1 and X_2 is $f_{X_1, X_2}(x_1, x_2) = 24x_1(1 - x_2) u(x_1) u(x_2 - x_1) u(1 - x_2)$

Solution First, the support of $f_{X_1, X_2}(x_1, x_2)$ is the region A shown in Fig. 4.7. Recollecting that $u(x_2 - x_1) u(1 - x_2)$ is non-zero only on $\{(x_1, x_2) : x_1 < x_2 < 1\} = \{x_1 : x_1 < 1\} \cap \{x_2 : x_1 < x_2 < 1\}$, the pdf of X_1 can be obtained as[5]

$$f_{X_1}(x_1) = 12x_1(1 - x_1)^2 u(x_1) u(1 - x_1), \qquad (4.2.16)$$

for which $\int_{-\infty}^{\infty} f_{X_1}(x_1) dx_1 = \int_0^1 12x_1(1 - x_1)^2 dx_1 = 12\left(\frac{1}{2} - \frac{2}{3} + \frac{1}{4}\right) = 1$. Next, recollecting that $u(x_1) u(x_2 - x_1)$ is non-zero only on $\{(x_1, x_2) : 0 < x_1 < x_2\} = \{x_1 : 0 < x_1 < x_2\} \cap \{x_2 : x_2 > 0\}$, we can obtain[6] the pdf of X_2 as

$$f_{X_2}(x_2) = \int_{-\infty}^{\infty} 24x_1(1 - x_2) u(x_1) u(x_2 - x_1) dx_1 u(1 - x_2)$$
$$= 12x_2^2(1 - x_2) u(x_2) u(1 - x_2), \qquad (4.2.17)$$

for which we clearly have $\int_{-\infty}^{\infty} f_{X_2}(x_2) dx_2 = \int_0^1 12\left(x_2^2 - x_2^3\right) dx_2 = 1$.

Next, let us obtain the pdf of $Y_1 = X_1 + X_2$. Choosing the auxiliary variable as $Y_2 = X_2$, we have $X_1 = Y_1 - Y_2$, $X_2 = Y_2$, and $(X_1, X_2) = \boldsymbol{g}^{-1}(Y_1, Y_2) = (Y_1 - Y_2, Y_2)$. The Jacobian of the inverse transformation is $\left| \frac{\partial \boldsymbol{g}^{-1}(y_1, y_2)}{\partial (y_1, y_2)} \right| = \begin{vmatrix} 1 & -1 \\ 0 & 1 \end{vmatrix} = 1$. Thus, the joint pdf of (Y_1, Y_2) is $f_{Y_1, Y_2}(y_1, y_2) = f_{X_1, X_2}(x_1, x_2)\big|_{x_1 = y_1 - y_2, x_2 = y_2}$, i.e.,

$$f_{Y_1, Y_2}(y_1, y_2) = 24(y_1 - y_2)(1 - y_2) u(y_1 - y_2) u(2y_2 - y_1), \quad (4.2.18)$$

---

[5] More specifically, we have $f_{X_1}(x_1) = \int_{-\infty}^{\infty} f_{X_1, X_2}(x_1, x_2) dx_2 = \int_{-\infty}^{\infty} 24x_1(1 - x_2) u(x_2 - x_1) u(1 - x_2) dx_2 u(x_1) = \int_{x_1}^1 24x_1(1 - x_2) dx_2 u(x_1) u(1 - x_1) = 12x_1(1 - x_1)^2 u(x_1) u(1 - x_1)$.

[6] More specifically, we have $f_{X_2}(x_2) = \int_{-\infty}^{\infty} f_{X_1, X_2}(x_1, x_2) dx_1 = \int_{-\infty}^{\infty} 24x_1(1 - x_2) u(x_1) u(x_2 - x_1) dx_1 u(1 - x_2) = \int_0^{x_2} 24x_1(1 - x_2) dx_1 u(x_2) u(1 - x_2) = 12x_2^2(1 - x_2) u(x_2) u(1 - x_2)$.

of which the support is the region B shown in Fig. 4.7. Now, we can obtain the pdf $f_{Y_1}(y_1) = \int_{-\infty}^{\infty} f_{Y_1,Y_2}(y_1, y_2)\,dy_2$ of $Y_1 = X_1 + X_2$ as $f_{Y_1}(y_1) = \int_{\frac{1}{2}y_1}^{y_1} 24(y_1 - y_2)(1 - y_2)\,dy_2$ for $0 \le y_1 \le 1$ and $f_{Y_1}(y_1) = \int_{\frac{1}{2}y_1}^{1} 24(y_1 - y_2)(1 - y_2)\,dy_2$ for $1 \le y_1 \le 2$, i.e.,

$$f_{Y_1}(y_1) = \begin{cases} -2y_1^3 + 3y_1^2, & 0 \le y_1 \le 1, \\ 2y_1^3 - 9y_1^2 + 12y_1 - 4, & 1 \le y_1 \le 2, \\ 0, & \text{otherwise} \end{cases} \qquad (4.2.19)$$

based[7] on (4.2.18). Here, we have $\int_{-\infty}^{\infty} f_{Y_1}(v)dv = \int_0^1 \left(-2v^3 + 3v^2\right) dv + \int_1^2 \left(2v^3 - 9v^2 + 12v - 4\right) dv = \left[-\frac{1}{2}v^4 + v^3\right]_0^1 + \left[2v^4 - 3v^3 + 6v^2 - 4v\right]_1^2 = 1$. The pdf $f_{Y_1}(y_1)$ is also shown in Fig. 4.7. $\diamond$

The distribution of differences can be obtained similarly.

Example 4.2.6 Obtain the pdf of $Y = X_1 - X_2$ when the joint pdf of $X = (X_1, X_2)$ is f_X.

Solution Following steps similar to those for (4.2.13), we get $f_{X_1-X_2,X_2}(y_1, y_2) = f_{X_1,X_2}(y_1 + y_2, y_2)$. We then have

$$f_{X_1-X_2}(y) = \int_{-\infty}^{\infty} f_{X_1,X_2}(y + y_2, y_2)\,dy_2 \qquad (4.2.20)$$

from integration. $\diamond$

Example 4.2.7 When the joint pdf of X and Y is $f_{X,Y}(x, y) = \frac{1}{4}u(x)u(2 - x)u(y)u(2 - y)$, obtain the pdf of $V = X - Y$.

Solution We first have

$$f_{X,Y}(v + y, y) = \frac{1}{4}u(v + y)u(2 - v - y)u(y)u(2 - y). \qquad (4.2.21)$$

Then, from (4.2.20), we get

$$f_V(v) = \begin{cases} \int_{-v}^{2} \frac{1}{4}dy, & -2 < v < 0, \\ \int_0^{-v+2} \frac{1}{4}dy, & 0 < v < 2, \\ 0, & \text{otherwise,} \end{cases} \qquad (4.2.22)$$

---

[7] Because $u(y_1 - y_2)u(2y_2 - y_1)$ is non-zero only when $\{(y_1, y_2) : y_2 < y_1 < 2y_2\} = \{y_1 : y_2 < y_1 < 2y_2\} \cap \{y_2 : y_2 > 0\}$, we have $f_{Y_2}(y_2) = \int_{-\infty}^{\infty} 24(y_1 - y_2)(1 - y_2)\,u(y_1 - y_2)u(2y_2 - y_1)u(1 - y_2)\,dy_1 = (1 - y_2)\int_{y_2}^{2y_2} 24(y_1 - y_2)u(1 - y_2)u(y_2)\,dy_1 = 24(1 - y_2)\left[\frac{1}{2}y_1^2 - y_2 y_1\right]_{y_2}^{2y_2} u(1 - y_2)u(y_2) = 12(1 - y_2)y_2^2 u(1 - y_2)u(y_2)$, which is the same as (4.2.17): note that we have chosen $Y_2 = X_2$.

Fig. 4.8 The region of integration in obtaining the pdf of $V = X - Y$ when the joint pdf of (X, Y) is $f_{X,Y}(x, y) = \frac{1}{4}u(x)u(2-x)u(y)u(2-y)$

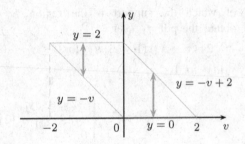

i.e.,

$$f_V(v) = \begin{cases} \frac{v+2}{4}, & -2 < v < 0, \\ -\frac{v-2}{4}, & 0 < v < 2, \\ 0, & \text{otherwise}, \end{cases} \qquad (4.2.23)$$

for which Fig. 4.8 can be used to identify the upper and lower limits of the integration.

◇

Example 4.2.8 Obtain the pdf of $Z = XY$ for a random vector (X, Y) with the joint pdf $f_{X,Y}$.

Solution Let $W = X$. Then, $X = W$, $Y = \frac{Z}{W}$, and the Jacobian of the inverse trans-

formation $(X, Y) = \mathbf{g}^{-1}(Z, W) = \left(W, \frac{Z}{W}\right)$ is $\left| \frac{\partial}{\partial(z,w)} \mathbf{g}^{-1}(z, w) \right| = \begin{vmatrix} \frac{\partial x}{\partial z} & \frac{\partial y}{\partial z} \\ \frac{\partial x}{\partial w} & \frac{\partial y}{\partial w} \end{vmatrix} = -\frac{1}{w}$.

Thus, we have $f_{Z,W}(z, w) = \frac{1}{|w|} f_{X,Y}(w, \frac{z}{w})$. We can then obtain the pdf $f_Z(z) = \int_{-\infty}^{\infty} f_{Z,W}(z, w) dw$ of $Z = XY$ as

$$f_Z(z) = \int_{-\infty}^{\infty} f_{X,Y}\left(x, \frac{z}{x}\right) \frac{1}{|x|} dx$$

$$= \int_{-\infty}^{\infty} f_{X,Y}\left(\frac{z}{y}, y\right) \frac{1}{|y|} dy \qquad (4.2.24)$$

after integration.

◇

Example 4.2.9 When X_1 and X_2 are independent of each other with the identical pdf $f(x) = u(x)u(1 - x)$, obtain the pdf of $Z = X_1 X_2$.

Solution The joint pdf of (X_1, X_2) is $f_{X_1,X_2}(x_1, x_2) = f_{X_1}(x_1) f_{X_2}(x_2) = u(x_1) u(1 - x_1) u(x_2) u(1 - x_2)$ because X_1 and X_2 are independent of each other. Using (4.2.24), we then get

$$f_Z(z) = \int_{-\infty}^{\infty} u\left(\frac{z}{y}\right) u\left(1 - \frac{z}{y}\right) u(y) u(1 - y) \frac{1}{|y|} dy. \qquad (4.2.25)$$

Fig. 4.9 The pdf
$f_Z(z) = (-\ln z)\, u(z)$
$u(1 - z)$ of $Z = X_1 X_2$ when
X_1 and X_2 are i.i.d. with the
marginal pdf
$f(x) = u(x)u(1 - x)$

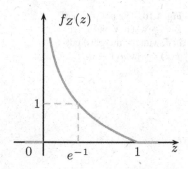

Now, noting that the integral of (4.2.25) is non-zero only when $\{(y, z) : 0 < y < 1, 0 < \frac{z}{y} < 1\} = \{(y, z) : 0 < z < y < 1\} = \{(y, z) : z < y < 1, 0 < z < 1\}$, we have $f_Z(z) = \int_z^1 \frac{1}{y} dy\, u(z) u(1 - z)$, i.e.,

$$f_Z(z) = (-\ln z)\, u(z) u(1 - z), \tag{4.2.26}$$

which is shown in Fig. 4.9. ◇

Example 4.2.10 Obtain the pdf of $Z = \frac{X}{Y}$ when the joint pdf of X and Y is $f_{X,Y}$.

Solution Let $V = Y$. Then, $X = ZV, Y = V$, and the Jacobian of the inverse transformation $(X, Y) = g^{-1}(Z, V) = (ZV, V)$ is $\left| \frac{\partial(zv,v)}{\partial(z,v)} \right| = \begin{vmatrix} v & 0 \\ z & 1 \end{vmatrix} = v$. Thus, we have $f_{Z,V}(z, v) = |v| f_{X,Y}(zv, v)$. Finally, we get the pdf $f_Z(z) = \int_{-\infty}^{\infty} f_{Z,V}(z, v) dv$ of $Z = \frac{X}{Y}$ as

$$f_Z(z) = \int_{-\infty}^{\infty} |y| f_{X,Y}(zy, y) dy \tag{4.2.27}$$

from integration. ◇

Example 4.2.11 Obtain the pdf of $Z = \frac{X}{Y}$ when X and Y are i.i.d. with the marginal pdf $f(x) = u(x)u(1 - x)$.

Solution We have $f_Z(z) = \int_{-\infty}^{\infty} |y| u(zy) u(y) u(1 - zy) u(1 - y) dy$ from (4.2.27). Next, noting that $\{(y, z) : zy > 0, y > 0, 1 - zy > 0, 1 - y > 0\}$ is the same as $\{(y, z) : z > 0, 0 < y < \min\left(1, \frac{1}{z}\right)\}$, the pdf of $Z = \frac{X}{Y}$ can be obtained as $f_Z(z) = \int_0^{\min(1, \frac{1}{z})} y\, dy\, u(z)$, i.e.,

$$f_Z(z) = \begin{cases} 0, & z < 0, \\ \frac{1}{2}, & 0 < z \le 1, \\ \frac{1}{2z^2}, & z \ge 1. \end{cases} \tag{4.2.28}$$

Fig. 4.10 The pdf $f_Z(z)$ of $Z = \frac{X}{Y}$ when X and Y are i.i.d. with the marginal pdf $f(x) = u(x)u(1-x)$

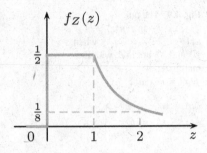

Note that the value $f_Z(0)$ is not, and does not need to be, specified. Figure 4.10 shows the pdf (4.2.28) of $Z = \frac{X}{Y}$. ◇

4.2.3 Joint Cumulative Distribution Function

Theorem 4.2.1 is useful when we obtain the joint pdf f_Y of $Y = g(X)$ directly from the joint pdf f_X of X. In some cases, it is more convenient and easier to obtain the joint pdf f_Y after first obtaining the joint cdf $F_Y(y) = \mathsf{P}(Y \le y) = \mathsf{P}(g(X) \le y)$ as

$$F_Y(y) = \mathsf{P}(X \in A_y). \tag{4.2.29}$$

Here, $Y \le y$ denotes $\{Y_1 \le y_1, Y_2 \le y_2, \ldots, Y_n \le y_n\}$, and A_y denotes the inverse image of $Y \le y$, i.e., the region of X such that $\{Y \le y\} = \{X \in A_y\}$. For example, when $n = 1$, if $g(x)$ is non-decreasing at every point x and has an inverse function, we have $\{X \in A_y\} = \{X \le g^{-1}(y)\}$ as we observed in Chap. 3, and we get $F_Y(y) = F_X(g^{-1}(y))$.

Example 4.2.12 When the joint pdf of (X, Y) is

$$f_{X,Y}(x, y) = \frac{1}{2\pi\sigma^2} \exp\left(-\frac{x^2 + y^2}{2\sigma^2}\right), \tag{4.2.30}$$

obtain the joint pdf of $Z = \sqrt{X^2 + Y^2}$ and $W = \frac{Y}{X}$.

Solution The joint cdf $F_{Z,W}(z, w) = \mathsf{P}(Z \le z, W \le w)$ of (Z, W) is

$$F_{Z,W}(z, w) = \mathsf{P}\left(\sqrt{X^2 + Y^2} \le z, \frac{Y}{X} \le w\right)$$

$$= \iint\limits_{D_{zw}} \frac{1}{2\pi\sigma^2} \exp\left(-\frac{x^2 + y^2}{2\sigma^2}\right) dx dy \tag{4.2.31}$$

for $z \geq 0$, where D_{zw} is the union of the two fan shapes $\{(x, y) : x^2 + y^2 \leq z^2, x > 0, y \leq wx\}$ and $\{(x, y) : x^2 + y^2 \leq z^2, x < 0, y \geq wx\}$ when $w > 0$ and also when $w < 0$. Changing the integration in the perpendicular coordinate system into that in the polar coordinate system as indicated in Fig. 4.1 and noting the symmetry of $f_{X,Y}$, we get $F_{Z,W}(z, w) = 2 \int_{\theta=-\frac{\pi}{2}}^{\theta_w} \int_{r=0}^{z} \frac{1}{2\pi\sigma^2} \exp\left(-\frac{r^2}{2\sigma^2}\right) r dr d\theta = \frac{1}{\pi\sigma^2}\left(\theta_w + \frac{\pi}{2}\right)\left[-\sigma^2 \exp\left(-\frac{r^2}{2\sigma^2}\right)\right]_{r=0}^{z}$, i.e.,

$$F_{Z,W}(z, w) = \frac{1}{2\pi}\left(\pi + 2\tan^{-1} w\right)\left\{1 - \exp\left(-\frac{z^2}{2\sigma^2}\right)\right\}, \qquad (4.2.32)$$

where $\theta_w = \tan^{-1} w \in \left(-\frac{\pi}{2}, \frac{\pi}{2}\right)$. ◇

Recollect that we obtained the total probability theorems (2.4.13), (3.4.8), and (3.4.9) based on $P(A|B)P(B) = P(AB)$ derived from (2.4.1). Now, extending the results into the multi-dimensional space, we similarly[8] have

$$P(A) = \begin{cases} \int_{\text{all } x} P(A|X = x) f_X(x) dx, & \text{continuous random vector } X, \\ \sum_{\text{all } x} P(A|X = x) p_X(x), & \text{discrete random vector} X, \end{cases} \qquad (4.2.33)$$

which are useful in obtaining the cdf, pdf, and pmf in some cases.

Example 4.2.13 Obtain the pdf of $Y = X_1 + X_2$ when $X = (X_1, X_2)$ has the joint pdf f_X.

Solution This problem has already been discussed in Example 4.2.4 based on the pdf. We now consider the problem based on the cdf.
(Method 1) Recollecting (4.2.33), the cdf $F_Y(y) = P(X_1 + X_2 \leq y)$ of Y can be expressed as

$$F_Y(y) = \int_{-\infty}^{\infty} \int_{-\infty}^{\infty} P(X_1 + X_2 \leq y | X_1 = x_1, X_2 = x_2) f_X(x_1, x_2) dx_1 dx_2. \qquad (4.2.34)$$

Here, $\{X_1 + X_2 \leq y | X_1 = x_1, X_2 = x_2\}$ does and does not hold true when $x_1 + x_2 \leq y$ and $x_1 + x_2 > y$, respectively. Thus, we have

$$P(X_1 + X_2 \leq y | X_1 = x_1, X_2 = x_2) = \begin{cases} 1, & x_1 + x_2 \leq y, \\ 0, & x_1 + x_2 > y \end{cases} \qquad (4.2.35)$$

and the cdf of Y can be expressed as

---

[8] Conditional distribution in random vectors will be discussed in Sect. 4.4 in more detail.

Fig. 4.11 The region $A = \{(X_1, X_2) : X_1 + X_2 \leq y\}$ and the interval $(-\infty, y - x_2)$ of integration for the value x_1 of X_1 when the value of X_2 is x_2

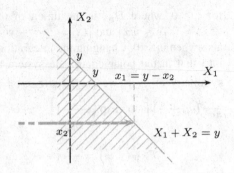

$$F_Y(y) = \int\!\!\int_{x_1 + x_2 \leq y} f_X(x_1, x_2)\, dx_1 dx_2$$

$$= \int_{-\infty}^{\infty} \int_{-\infty}^{y - x_2} f_X(x_1, x_2)\, dx_1 dx_2. \tag{4.2.36}$$

Then, the pdf $f_Y(y) = \frac{\partial}{\partial y} F_Y(y)$ of Y is

$$f_Y(y) = \int_{-\infty}^{\infty} f_X(y - x_2, x_2)\, dx_2, \tag{4.2.37}$$

which can also be expressed as $f_Y(y) = \int_{-\infty}^{\infty} f_X(x_1, y - x_1)\, dx_1$. In obtaining (4.2.37), we used $\frac{\partial}{\partial y} \int_{-\infty}^{y-x_2} f_X(x_1, x_2)\, dx_1 = f_X(y - x_2, x_2)$ from Leibnitz's rule (3.2.18).

(Method 2) Referring to the region $A = \{(X_1, X_2) : X_1 + X_2 \leq y\}$ shown in Fig. 4.11, the value x_1 of X_1 runs from $-\infty$ to $y - x_2$ when the value of x_2 of X_2 runs from $-\infty$ to ∞. Thus we have[9] $F_Y(y) = \mathsf{P}(Y \leq y) = \mathsf{P}(X_1 + X_2 \leq y) = \int\!\!\int_A f_X(x_1, x_2)\, dx_1 dx_2$, i.e.,

$$F_Y(y) = \int_{x_2 = -\infty}^{\infty} \int_{x_1 = -\infty}^{y - x_2} f_X(x_1, x_2)\, dx_1 dx_2, \tag{4.2.38}$$

and subsequently (4.2.37).   ◇

Example 4.2.14 Obtain the pdf of $Z = X + Y$ when X with the pdf $f_X(x) = \alpha e^{-\alpha x} u(x)$ and Y with the pdf $f_Y(y) = \beta e^{-\beta y} u(y)$ are independent of each other.

Solution We first obtain the pdf of Z directly. The joint pdf of X and Y is $f_{X,Y}(x, y) = f_X(x) f_Y(y) = \alpha \beta e^{-\alpha x} e^{-\beta y} u(x) u(y)$. Recollecting that $u(y) u(z - y)$

---

[9] If the order of integration is interchanged, then $\int_{x_2 = -\infty}^{\infty} \int_{x_1 = -\infty}^{y - x_2} f_X(x_1, x_2)\, dx_1 dx_2$ will become $\int_{x_1 = -\infty}^{\infty} \int_{x_2 = -\infty}^{y - x_1} f_X(x_1, x_2)\, dx_2 dx_1$.

is non-zero only when $0 < y < z$, the pdf of $Z = X + Y$ can be obtained as
$f_Z(z) = \int_{-\infty}^{\infty} \alpha\beta e^{-\alpha(z-y)} e^{-\beta y} u(y) u(z-y) dy = \alpha\beta e^{-\alpha z} \int_0^z e^{(\alpha-\beta)y} dy u(z)$, i.e.,

$$f_Z(z) = \begin{cases} \frac{\alpha\beta}{\beta-\alpha} \left(e^{-\alpha z} - e^{-\beta z} \right) u(z), & \beta \neq \alpha, \\ \alpha^2 z e^{-\alpha z} u(z), & \beta = \alpha \end{cases} \tag{4.2.39}$$

from (4.2.37).

Next, the cdf of Z can be expressed as

$$F_Z(z) = \alpha\beta \int_{-\infty}^{\infty} \int_{-\infty}^{z-y} e^{-\alpha x} e^{-\beta y} u(x) u(y) \, dx dy \tag{4.2.40}$$

based on (4.2.38). Here, (4.2.40) is non-zero only when $\{x > 0, y > 0, z - y > 0\}$
due to $u(x)u(y)$. With this fact in mind and by noting that $\{y > 0, z - y > 0\} = \{z > y > 0\}$, we can rewrite (4.2.40) as

$$\begin{aligned} F_Z(z) &= \alpha\beta \int_0^z \int_0^{z-y} e^{-\alpha x} e^{-\beta y} dx dy \, u(z) \\ &= \beta \int_0^z \left\{ 1 - e^{-\alpha(z-y)} \right\} e^{-\beta y} dy \, u(z) \\ &= \begin{cases} \left\{ 1 - \frac{1}{\beta-\alpha} \left(\beta e^{-\alpha z} - \alpha e^{-\beta z} \right) \right\} u(z), & \beta \neq \alpha, \\ \left\{ 1 - (1 + \alpha z) e^{-\alpha z} \right\} u(z), & \beta = \alpha. \end{cases} \end{aligned} \tag{4.2.41}$$

By differentiating this cdf, we can obtain the pdf (4.2.39). ◇

Example 4.2.15 For a continuous random vector (X, Y), let $Z = \max(X, Y)$ and
$W = \min(X, Y)$. Referring to Fig. 4.12, we first have $F_Z(z) = P(\max(X, Y) \leq z) = P(X \leq z, Y \leq z)$, i.e.,

$$F_Z(z) = F_{X,Y}(z, z). \tag{4.2.42}$$

Fig. 4.12 The region
$\{(X, Y) : \max(X, Y) \leq z\}$

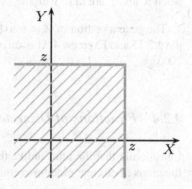

Fig. 4.13 The region
$\{(X, Y) : \min(X, Y) \le w\}$

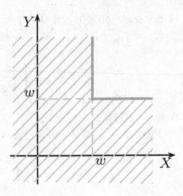

Next, when $A = \{X \le w\}$ and $B = \{Y \le w\}$, we have

$$P(X > w, Y > w) = 1 - F_X(w) - F_Y(w) + F_{X,Y}(w, w) \qquad (4.2.43)$$

from $\quad P(X > w, Y > w) = P(A^c \cap B^c) = P((A \cup B)^c) = 1 - P(A \cup B) =$
$1 - P(A) - P(B) + P(A \cap B)$, $\quad P(A) = P(X \le w) = F_X(w)$, $\quad P(B) = P(Y \le$
$w) = F_Y(w)$, and $P(A \cap B) = P(X \le w, Y \le w) = F_{X,Y}(w, w)$. Therefore, we
get the cdf $F_W(w) = P(W \le w) = 1 - P(W > w) = 1 - P(\min(X, Y) > w) =$
$1 - P(X > w, Y > w)$ of W as

$$F_W(w) = F_X(w) + F_Y(w) - F_{X,Y}(w, w), \qquad (4.2.44)$$

which can also be obtained intuitively from Fig. 4.13. Note that the pdf $f_Z(z) =$
$\frac{d}{dz} F_{X,Y}(z, z)$ of $Z = \max(X, Y)$ becomes

$$f_Z(z) = 2F(z)f(z) \qquad (4.2.45)$$

and the pdf $f_W(w) = \frac{d}{dw} \{F_X(w) + F_Y(w) - F_{X,Y}(w, w)\}$ of $W = \min(X, Y)$
becomes

$$f_W(w) = 2\{1 - F(w)\}f(w) \qquad (4.2.46)$$

when X and Y are i.i.d. with the marginal cdf F and marginal pdf f.   ◇

The generalization of $Z = \max(X, Y)$ and $W = \min(X, Y)$ discussed in Example 4.2.15 and Exercise 4.31 is referred to as the order statistic (David and Nagaraja 2003).

4.2.4   Functions of Discrete Random Vectors

Considering that the pmf, unlike the pdf, represents a probability, we now discuss functions of discrete random vectors.

Example 4.2.16 (Rohatgi and Saleh 2001) Obtain the pmf of $Z = X + Y$ and the pmf of $W = X - Y$ when $X \sim b(n, p)$ and $Y \sim b(n, p)$ are independent of each other.

Solution First, the pmf $P(Z = z) = \sum_{k=0}^{n} P(X = k, Y = z - k)$ of $Z = X + Y$

can be obtained as $P(Z = z) = \sum_{k=0}^{n} {}_nC_k p^k (1 - p)^{n-k} \, {}_nC_{z-k} p^{z-k} (1 - p)^{n-z+k} =$

$\sum_{k=0}^{n} {}_nC_k \, {}_nC_{z-k} p^z (1 - p)^{2n-z}$, i.e.,

$$P(Z = z) = {}_{2n}C_z \, p^z (1 - p)^{2n-z} \tag{4.2.47}$$

for $z = 0, 1, \ldots, 2n$, where we have used $\sum_{k=0}^{n} {}_nC_k \, {}_nC_{z-k} = {}_{2n}C_z$ based on (1.A.25).

Next, the pmf $P(W = w) = \sum_{k=0}^{n} P(X = k + w, Y = k) = \sum_{k=0}^{n} {}_nC_{k+w} \, {}_nC_k p^{2k+w} (1 - p)^{2n-2k-w}$ of $W = X - Y$ can be obtained as

$$P(W = w) = \left(\frac{p}{1 - p} \right)^w \sum_{k=0}^{n} {}_nC_{k+w} \, {}_nC_k p^{2k} (1 - p)^{2n-2k} \tag{4.2.48}$$

for $w = -n, -n + 1, \ldots, n$. $\diamond$

Example 4.2.17 Assume that X and Y are i.i.d. with the marginal pmf $p(x) = (1 - \alpha)\alpha^{x-1}\tilde{u}(x - 1)$, where $0 < \alpha < 1$ and $\tilde{u}(x)$ is the discrete space unit step function defined in (1.4.17). Obtain the joint pmf of $(X + Y, X)$, and based on the result, obtain the pmf of X and the pmf of $X + Y$.

Solution First we have $p_{X+Y,X}(v, x) = P(X + Y = v, X = x) = P(X = x, Y = v - x)$, i.e.,

$$p_{X+Y,X}(v, x) = (1 - \alpha)^2 \alpha^{v-2} \tilde{u}(x - 1)\tilde{u}(v - x - 1). \tag{4.2.49}$$

Thus, we have $p_{X+Y}(v) = \sum_{x=-\infty}^{\infty} p_{X+Y,X}(v, x)$, i.e.,

$$p_{X+Y}(v) = (1 - \alpha)^2 \alpha^{v-2} \sum_{x=-\infty}^{\infty} \tilde{u}(x - 1)\tilde{u}(v - x - 1) \tag{4.2.50}$$

from (4.2.49). Now noting that $\tilde{u}(x - 1)\tilde{u}(v - x - 1) = 1$ for $\{x - 1 \geq 0, v - x - 1 \geq 0\}$ and 0 otherwise and that[10] $\{x : x - 1 \geq 0, v - x - 1 \geq 0\} = \{x : 1 \leq x \leq v - 1, v \geq 2\}$, we have $p_{X+Y}(v) = (1 - \alpha)^2 \alpha^{v-2} \sum_{x=1}^{v-1} \tilde{u}(v - 2)$, i.e.,

---

[10] Here, $v - x - 1 \geq 0$, for example, can more specifically be written as $v - x - 1 = 0, 1, \ldots$.

$$p_{X+Y}(v) = (1 - \alpha)^2 (v - 1)\alpha^{v-2}\tilde{u}(v - 2). \tag{4.2.51}$$

Next, the pmf of X can be obtained as $p_X(x) = \sum_{v=-\infty}^{\infty} p_{X+Y,X}(v, x)$, i.e.,

$$p_X(x) = (1 - \alpha)^2 \alpha^{-2} \sum_{v=-\infty}^{\infty} \alpha^v \tilde{u}(x - 1)\tilde{u}(v - x - 1) \tag{4.2.52}$$

from (4.2.49), which can be rewritten as $p_X(x) = (1 - \alpha)^2 \alpha^{-2} \sum_{v=x+1}^{\infty} \alpha^v \tilde{u}(x - 1)$, i.e.,

$$p_X(x) = (1 - \alpha)\alpha^{x-1}\tilde{u}(x - 1) \tag{4.2.53}$$

by noting that $\{v : x - 1 \geq 0, v - x - 1 \geq 0\} = \{v : v \geq x + 1, x \geq 1\}$ and $\tilde{u}(x - 1)\tilde{u}(v - x - 1) = 1$ for $\{x - 1 \geq 0, v - x - 1 \geq 0\}$ and 0 otherwise. ◇

4.3 Expected Values and Joint Moments

For random vectors, we will describe here the basic properties (Balakrishnan 1992; Kendall and Stuart 1979; Samorodnitsky and Taqqu 1994) of expected values. New notions will also be defined and explored.

4.3.1 Expected Values

The expected values for random vectors can be described as, for example,

$$E\{g(X)\} = \int g(x)dF_X(x) \tag{4.3.1}$$

by extending the notion of the expected values discussed in Chap. 3 into multiple dimensions. Because the expectation is a linear operator, we have

$$E\left\{\sum_{i=1}^{n} a_i g_i(X_i)\right\} = \sum_{i=1}^{n} a_i E\{g_i(X_i)\} \tag{4.3.2}$$

for an n-dimensional random vector X when $\{g_i\}_{i=1}^{n}$ are all measurable functions. In addition,

$$E\left\{\prod_{i=1}^{n} g_i(X_i)\right\} = \prod_{i=1}^{n} E\{g_i(X_i)\} \tag{4.3.3}$$

when X is an independent random vector.

Example 4.3.1 Assume that we repeatedly roll a fair die until the number of even-numbered outcomes is 10. Let N denote the number of rolls and X_i denote the number of outcome i when the repetition ends. Obtain the pmf of N, expected value of N, expected value of X_1, and expected value of X_2.

Solution First, the pmf of N can be obtained as $P(N = k) = P(A_k \cap B) = P(A_k| B) P(B) = P(A_k) P(B) = \left(\frac{1}{2}\right) {}_{k-1}C_9 \left(\frac{1}{2}\right)^9 \left(\frac{1}{2}\right)^{k-1-9}$, i.e.,

$$P(N = k) = {}_{k-1}C_9 \left(\frac{1}{2}\right)^k \tag{4.3.4}$$

for $k = 10, 11, \ldots$, where $A_k = \{$an even number at the k-th rolling$\}$ and $B = \{9$ times of even numbers until $(k-1)$-st rolling$\}$. Using $\sum_{x=0}^{\infty} {}_{r+x-1}C_x (1-\alpha)^x = \alpha^{-r}$ shown in (2.5.16) and noting that $k {}_{k-1}C_9 = k \frac{(k-1)!}{(k-10)!9!} = 10 \frac{k!}{(k-10)!10!} = 10 {}_kC_{10}$, we get $E\{N\} = \sum_{k=10}^{\infty} {}_{k-1}C_9 \left(\frac{1}{2}\right)^k k = 10 \sum_{k=10}^{\infty} {}_kC_{10} \left(\frac{1}{2}\right)^k = \frac{10}{2^{10}} \sum_{j=0}^{\infty} {}_{j+10}C_{10} \left(\frac{1}{2}\right)^j = \frac{10}{2^{10}} \left(\frac{1}{2}\right)^{-11} = 20$. This result can also be obtained from the formula (3.E.27) of the mean of the NB distribution with the pmf (2.5.17) by using $(r, p) = \left(10, \frac{1}{2}\right)$. Subsequently, until the end, even numbers will occur 10 times, among which 2, 4, and 6 will occur equally likely. Thus, $E\{X_2\} = \frac{10}{3}$. Next, from $N = \sum_{i=1}^{6} X_i$, we get $E\{N\} = \sum_{i=1}^{6} E\{X_i\}$. Here, because $E\{X_2\} = E\{X_4\} = E\{X_6\} = \frac{10}{3}$ and $E\{X_1\} = E\{X_3\} = E\{X_5\}$, the expected value[11] of X_1 is $E\{X_1\} = \left(20 - 3 \times \frac{10}{3}\right) \frac{1}{3} = \frac{10}{3}$. ◇

4.3.2 Joint Moments

We now generalize the concept of moments discussed in Chap. 3 for random vectors. The moments for bi-variate random vectors will first be considered and then those for higher dimensions will be discussed.

---

[11] The expected values of X_1 and X_2 can of course be obtained with the pmf's of X_1 and X_2 obtained already in Example 4.1.11.

4.3.2.1   Bi-variate Random Vectors

Definition 4.3.1 (*joint moment; joint central moment*) The expected value

$$m_{jk} = \mathsf{E}\left\{X^j Y^k\right\} \tag{4.3.5}$$

is termed the (j, k)-th joint moment or product moment of X and Y, and

$$\mu_{jk} = \mathsf{E}\left\{(X - m_X)^j \, (Y - m_Y)^k\right\} \tag{4.3.6}$$

is termed the (j, k)-th joint central moment or product central moment of X and Y, for $j, k = 0, 1, \ldots$, where m_X and m_Y are the means of X and Y, respectively.

It is easy to see that $m_{00} = \mu_{00} = 1$, $m_{10} = m_X = \mathsf{E}\{X\}$, $m_{01} = m_Y = \mathsf{E}\{Y\}$, $m_{20} = \mathsf{E}\left\{X^2\right\}$, $m_{02} = \mathsf{E}\left\{Y^2\right\}$, $\mu_{10} = \mu_{01} = 0$, $\mu_{20} = \sigma_X^2$ is the variance of X, and $\mu_{02} = \sigma_Y^2$ is the variance of Y.

Example 4.3.2 The expected value $\mathsf{E}\left\{X_1 X_2^3\right\}$ is the $(1, 3)$-rd joint moment of $X = (X_1, X_2)$.   ◇

Definition 4.3.2 (*correlation; covariance*) The $(1, 1)$-st joint moment m_{11} and the $(1, 1)$-st joint central moment μ_{11} are termed the correlation and covariance, respectively, of the two random variables. The ratio of the covariance to the product of the standard deviations of two random variables is termed the correlation coefficient.

The correlation $m_{11} = \mathsf{E}\{XY\}$ is often denoted[12] by R_{XY}, and the covariance $\mu_{11} = \mathsf{E}\{(X - m_X)(Y - m_Y)\} = \mathsf{E}\{XY\} - m_X m_Y$ by K_{XY}, $\mathsf{Cov}(X, Y)$, or C_{XY}. Specifically, we have

$$K_{XY} = R_{XY} - m_X m_Y \tag{4.3.7}$$

for the covariance, and

$$\rho_{XY} = \frac{K_{XY}}{\sqrt{\sigma_X^2 \sigma_Y^2}} \tag{4.3.8}$$

for the correlation coefficient.

Definition 4.3.3 (*orthogonal; uncorrelated*) When the correlation is 0 or, equivalently, when the mean of the product is 0, the two random variables are called orthogonal. When the mean of the product is the same as the product of the means or, equivalently, when the covariance or correlation coefficient is 0, the two random variables are called uncorrelated.

---

[12] When there is more than one subscript, we need commas in some cases: for example, the joint pdf $f_{X,Y}$ of (X, Y) should be differentiated from the pdf f_{XY} of the product XY. In other cases, we do not need to use commas: for instance, $R_{XY}, \mu_{jk}, K_{XY}, \ldots$ denote relations among two or more random variables and thus is expressed without any comma.

In other words, when $R_{XY} = \mathsf{E}\{XY\} = 0$, X and Y are orthogonal. When $\rho_{XY} = 0$, $K_{XY} = \mathsf{Cov}(X, Y) = 0$, or $\mathsf{E}\{XY\} = \mathsf{E}\{X\}\mathsf{E}\{Y\}$, X and Y are uncorrelated.

Theorem 4.3.1 *If two random variables are independent of each other, then they are uncorrelated, but the converse is not necessarily true.*

In other words, there exist some uncorrelated random variables that are not independent of each other. In addition, when two random variables are independent and at least one of them has mean 0, the two random variables are orthogonal.

Theorem 4.3.2 *The absolute value of a correlation coefficient is no larger than 1.*

Proof From the Cauchy-Schwarz inequality $\mathsf{E}^2\{XY\} \le \mathsf{E}\left\{X^2\right\}\mathsf{E}\left\{Y^2\right\}$ shown in (6.A.26), we get

$$\mathsf{E}^2\{(X - m_X)(Y - m_Y)\} \le \mathsf{E}\left\{(X - m_X)^2\right\}\mathsf{E}\left\{(Y - m_Y)^2\right\}, \qquad (4.3.9)$$

which implies $K_{XY}^2 \le \sigma_X^2 \sigma_Y^2$. Thus, $\rho_{XY}^2 \le 1$ and $|\rho_{XY}| \le 1$. ♠

Example 4.3.3 When the two random variables X and Y are related by $Y - m_Y = c(X - m_X)$ or $Y = cX + d$, we have $|\rho_{XY}| = 1$. ◇

4.3.2.2 Multi-dimensional Random Vectors

Let $\mathsf{E}\{X\} = m_X = (m_1, m_2, \ldots, m_n)^T$ be the mean vector of $X = (X_1, X_2, \ldots, X_n)^T$. In subsequent discussions, especially when we discuss joint moments of random vectors, we will often assume the random vectors are complex. The discussion on complex random vectors is almost the same as that on real random vectors on which we have so far focused.

Definition 4.3.4 *(correlation matrix; covariance matrix)* The matrix

$$R_X = \mathsf{E}\left\{X X^H\right\} \qquad (4.3.10)$$

is termed the correlation matrix and the matrix

$$K_X = R_X - m_X m_X^H \qquad (4.3.11)$$

is termed the covariance matrix or variance-covariance matrix of X, where the superscript H denotes the complex conjugate transpose, also called the Hermitian transpose or Hermitian conjugate.

The correlation matrix $R_X = [R_{ij}]$ is of size $n \times n$: the (i, j)-th element of R_X is the correlation $R_{ij} = R_{X_iX_j} = \mathsf{E}\left\{X_i X_j^*\right\}$ between X_i and X_j when $i \neq j$ and the second absolute moment $R_{ii} = \mathsf{E}\left\{|X_i|^2\right\}$ when $i = j$. The covariance matrix $K_X = [K_{ij}]$ is also an $n \times n$ matrix: the (i, j)-th element of K_X is the covariance $K_{ij} = K_{X_iX_j} = \mathsf{E}\left\{(X_i - m_i)\left(X_j - m_j\right)^*\right\}$ of X_i and X_j when $i \neq j$ and the variance $K_{ii} = \mathsf{Var}(X_i)$ of X_i when $i = j$.

Example 4.3.4 For a random vector $X = (X_1, X_2, \ldots, X_n)^T$ and an $n \times n$ linear transformation matrix $L = [L_{ij}]$, consider the random vector

$$Y = L\,X, \tag{4.3.12}$$

where $Y = (Y_1, Y_2, \ldots, Y_n)^T$. Then, letting $m_X = \left(m_{X_1}, m_{X_2}, \ldots, m_{X_n}\right)^T$ be the mean vector of X, the mean vector $m_Y = \mathsf{E}\{Y\} = \left(m_{Y_1}, m_{Y_2}, \ldots, m_{Y_n}\right)^T = \mathsf{E}\{LX\} = L\mathsf{E}\{X\}$ of Y can be obtained as

$$m_Y = L\,m_X. \tag{4.3.13}$$

Similarly, denoting by R_X and K_X the correlation and covariance matrices, respectively, of X, the correlation matrix $R_Y = \mathsf{E}\left\{Y\,Y^H\right\} = \mathsf{E}\left\{L\,X\,(L\,X)^H\right\} = L\,\mathsf{E}\left\{X\,X^H\right\}L^H$ of Y can be expressed as

$$R_Y = L\,R_X\,L^H, \tag{4.3.14}$$

and the covariance matrix $K_Y = \mathsf{E}\left\{(Y - m_Y)(Y - m_Y)^H\right\} = R_Y - m_Y m_Y^H = L\left(R_X - m_X m_X^H\right)L^H$ of Y can be expressed as

$$K_Y = L\,K_X L^H. \tag{4.3.15}$$

More generally, when $Y = LX + b$, we have $m_Y = Lm_X + b$, $R_Y = LR_X L^H + Lm_X b^H + b\,(Lm_X)^H + bb^H$, and $K_Y = L\,K_X L^H$. In essence, the results (4.3.13)–(4.3.15), shown in Fig. 4.14 as a visual representation, dictate that the mean vector, correlation matrix, and covariance matrix of $Y = LX$ can be obtained without first having to obtain the cdf, pdf, or pmf of Y, an observation similar to that on (3.3.9). ◇

Example 4.3.5 Assume $m_X = (1, 2)^T$ and $K_X = \begin{pmatrix} 2 & -1 \\ -1 & 1 \end{pmatrix}$ for $X = (X_1, X_2)^T$. When $L = \begin{pmatrix} 1 & 1 \\ -1 & 1 \end{pmatrix}$, obtain m_Y and K_Y for $Y = LX$.

Solution We easily get the mean $m_Y = L\,m_X = (3, 1)^T$ of $Y = LX$. Next, the covariance matrix of $Y = LX$ is $K_Y = L\,K_X L^H = \begin{pmatrix} 1 & 1 \\ -1 & 1 \end{pmatrix}\begin{pmatrix} 2 & -1 \\ -1 & 1 \end{pmatrix}\begin{pmatrix} 1 & -1 \\ 1 & 1 \end{pmatrix} = \begin{pmatrix} 1 & -1 \\ -1 & 5 \end{pmatrix}$. ◇

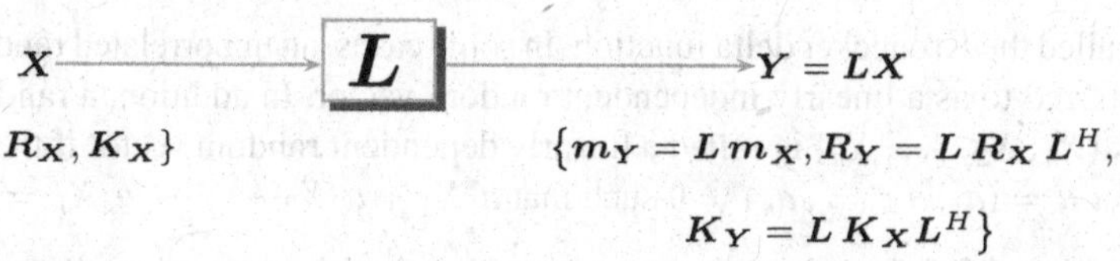

$$X \xrightarrow{\quad\boxed{\boldsymbol{L}}\quad} Y = LX$$

$$\{m_X, R_X, K_X\} \qquad \{m_Y = L m_X, R_Y = L R_X L^H,$$

$$K_Y = L K_X L^H\}$$

Fig. 4.14 The mean vector, correlation matrix, and covariance matrix of linear transformation

Theorem 4.3.3 *The correlation and covariance matrices are Hermitian.*[13]

Proof From $R_{X_i X_j} = \mathsf{E}\left\{X_i X_j^*\right\} = \left(\mathsf{E}\left\{X_j X_i^*\right\}\right)^* = R_{X_j X_i}^*$ for the correlation of X_i and X_j, the correlation matrix is Hermitian. Similarly, it is easy to see that the covariance matrix is also Hermitian by letting $Y_i = X_i - m_i$. ♠

Theorem 4.3.4 *The correlation and covariance matrices of any random vector are positive semi-definite.*

Proof Let $a = (a_1, a_2, \ldots, a_n)$ and $X = (X_1, X_2, \ldots, X_n)^T$. Then, the correlation matrix $R_X = \mathsf{E}\left\{XX^H\right\}$ is positive semi-definite because $\mathsf{E}\left\{|aX|^2\right\} = \mathsf{E}\left\{aXX^H a^H\right\} = a\mathsf{E}\left\{XX^H\right\} a^H \geq 0$. Letting $Y_i = X_i - m_i$, we can similarly show that the covariance matrix is positive semi-definite. ♠

Definition 4.3.5 (*uncorrelated random vector*) A random vector X is called an uncorrelated random vector if $\mathsf{E}\left\{X_i X_j^*\right\} = \mathsf{E}\{X_i\} \mathsf{E}\left\{X_j^*\right\}$ for all i and j such that $i \neq j$.

For an uncorrelated random vector X, we have the correlation matrix $R_X = \left[R_{X_i X_j}\right]$ with

$$R_{X_i X_j} = \begin{cases} \mathsf{E}\left\{|X_i|^2\right\}, & i = j, \\ \mathsf{E}\{X_i\} \mathsf{E}\left\{X_j^*\right\}, & i \neq j \end{cases} \tag{4.3.16}$$

and the covariance matrix $K_X = \left[K_{X_i X_j}\right]$ with $K_{X_i X_j} = \sigma_{X_i}^2 \delta_{ij}$, where

$$\delta_{ij} = \begin{cases} 1, & i = j, \\ 0, & i \neq j \end{cases} \tag{4.3.17}$$

and

$$\delta_k = \begin{cases} 1, & k = 0, \\ 0, & k \neq 0 \end{cases} \tag{4.3.18}$$

---

[13] A matrix A such that $A^H = A$ is called Hermitian.

are called the Kronecker delta function. In some cases, an uncorrelated random vector is referred to as a linearly independent random vector. In addition, a random vector $X = (X_1, X_2, \ldots, X_n)$ is called a linearly dependent random vector if there exists a vector $a = (a_1, a_2, \ldots, a_n) \neq 0$ such that $a_1 X_1 + a_2 X_2 + \cdots + a_n X_n = 0$.

Definition 4.3.6 (*random vectors uncorrelated with each other*) When we have $\mathsf{E}\left\{X_i Y_j^*\right\} = \mathsf{E}\{X_i\} \mathsf{E}\left\{Y_j^*\right\}$ for all i and j, the random vectors X and Y are called uncorrelated with each other.

Note that even when X and Y are uncorrelated with each other, each of X and Y may or may not be an uncorrelated random vector, and even when X and Y are both uncorrelated random vectors, X and Y may be correlated.

Theorem 4.3.5 (*McDonough and Whalen 1995*) *If the covariance matrix of a random vector is positive definite, then the random vector can be transformed into an uncorrelated random vector via a linear transformation.*

Proof The theorem can be proved by noting that, when an $n \times n$ matrix A is a normal[14] matrix, we can take n orthogonal unit vectors as the eigenvectors of A and that there exists a unitary[15] matrix P such that $P^{-1} A P = P^H A P$ is diagonal.

Let $\{\lambda_i\}_{i=1}^n$ be the eigenvalues of the positive definite covariance matrix K_X of X. Because a covariance matrix is Hermitian, $\{\lambda_i\}_{i=1}^n$ are all real. In addition, because K_X is positive definite, $\{\lambda_i\}_{i=1}^n$ are all larger than 0. Now, choose the eigenvectors corresponding to the eigenvalues $\{\lambda_i\}_{i=1}^n$ as

$$\{a_i\}_{i=1}^n = \left\{ (a_{i1}, a_{i2}, \ldots, a_{in})^T \right\}_{i=1}^n, \tag{4.3.19}$$

respectively, so that the eigenvectors are orthonormal, that is,

$$a_i^H a_j = \delta_{ij}. \tag{4.3.20}$$

Now, consider the unitary matrix

$$A = (a_1, a_2, \ldots, a_n)^H$$
$$= \left[a_{ij}^* \right] \tag{4.3.21}$$

composed of the eigenvectors (4.3.19) of K_X. Because $K_X a_i = \lambda_i a_i$, and therefore $K_X (a_1, a_2, \ldots, a_n) = (\lambda_1 a_1, \lambda_2 a_2, \ldots, \lambda_n a_n)$, the covariance matrix $K_Y = A K_X A^H$ of $Y = (Y_1, Y_2, \ldots, Y_n)^T = A X$ can be obtained as

---

[14] A matrix A such that $AA^H = A^H A$ is normal.

[15] A matrix A is unitary if $A^H = A^{-1}$ or if $A^H A = AA^H = I$. In the real space, a unitary matrix is referred to as an orthogonal matrix. A Hermitian matrix is always a normal matrix and a unitary matrix is always a normal matrix, but the converses are not necessarily true: for example, $\begin{pmatrix} 1 & -1 \\ 1 & 1 \end{pmatrix}$ is normal, but is neither Hermitian nor unitary.

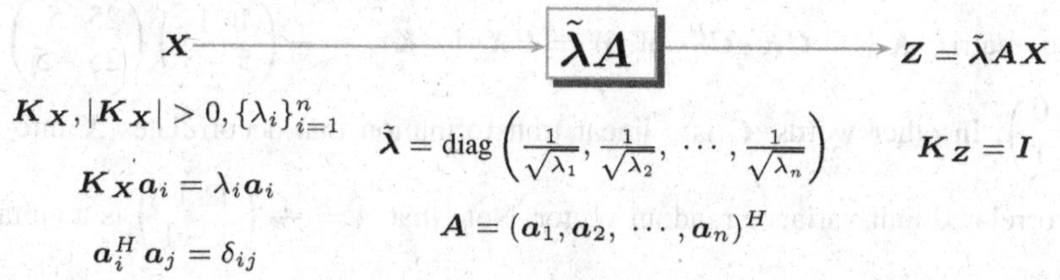

$$X \longrightarrow \boxed{\tilde{\lambda}A} \longrightarrow Z = \tilde{\lambda}AX$$

$K_X, |K_X| > 0, \{\lambda_i\}_{i=1}^{n}$

$\tilde{\lambda} = \text{diag}\left(\frac{1}{\sqrt{\lambda_1}}, \frac{1}{\sqrt{\lambda_2}}, \cdots, \frac{1}{\sqrt{\lambda_n}}\right)$

$K_Z = I$

$K_X a_i = \lambda_i a_i$

$a_i^H a_j = \delta_{ij}$

$A = (a_1, a_2, \cdots, a_n)^H$

Fig. 4.15 Decorrelating into uncorrelated unit-variance random vectors

$$K_Y = (a_1, a_2, \ldots, a_n)^H (\lambda_1 a_1, \lambda_2 a_2, \ldots, \lambda_n a_n)$$
$$= \left[\lambda_i a_i^H a_j\right]$$
$$= \text{diag}(\lambda_1, \lambda_2, \ldots, \lambda_n) \tag{4.3.22}$$

from (4.3.15) using (4.3.20). In essence, $Y = AX$ is an uncorrelated random vector.

♠

Let us proceed one step further from Theorem 4.3.5. Recollecting that the eigen-values $\{\lambda_i\}_{i=1}^{n}$ of the covariance matrix K_X are all larger than 0, let

$$\tilde{\lambda} = \text{diag}\left(\frac{1}{\sqrt{\lambda_1}}, \frac{1}{\sqrt{\lambda_2}}, \ldots, \frac{1}{\sqrt{\lambda_n}}\right) \tag{4.3.23}$$

and consider the linear transformation

$$Z = \tilde{\lambda}Y \tag{4.3.24}$$

of Y. Then, the covariance matrix $K_Z = \tilde{\lambda}K_Y\tilde{\lambda}^H$ of Z is

$$K_Z = I. \tag{4.3.25}$$

In other words, $Z = \tilde{\lambda}Y = \tilde{\lambda}AX$ is a vector of uncorrelated unit-variance random variables. Figure 4.15 summarizes the procedure.

Example 4.3.6 Assume that the covariance matrix of $X = (X_1, X_2)^T$ is $K_X = \begin{pmatrix} 13 & 12 \\ 12 & 13 \end{pmatrix}$. Find a linear transformation that decorrelates X into an uncorrelated unit-variance random vector.

Solution From the characteristic equation $|\lambda I - K_X| = 0$, we get the two pairs $\left\{\lambda_1 = 25, a_1 = \frac{1}{\sqrt{2}}(1\ 1)^T\right\}$ and $\left\{\lambda_2 = 1, a_2 = \frac{1}{\sqrt{2}}(1\ -1)^T\right\}$ of eigenvalue and unit eigenvector of K_X. With the linear transformation $C = \tilde{\lambda}A = \frac{1}{\sqrt{2}}\begin{pmatrix} \frac{1}{\sqrt{25}} & 0 \\ 0 & \frac{1}{\sqrt{1}} \end{pmatrix}\begin{pmatrix} 1 & 1 \\ 1 & -1 \end{pmatrix} = \frac{1}{\sqrt{50}}\begin{pmatrix} 1 & 1 \\ 5 & -5 \end{pmatrix}$ constructed from the two pairs, the covariance matrix $K_W = CK_XC^H$ of $W = CX$ is $K_W = \frac{1}{50}\begin{pmatrix} 1 & 1 \\ 5 & -5 \end{pmatrix}\begin{pmatrix} 25 & 5 \\ 25 & -5 \end{pmatrix} =$ $\begin{pmatrix} 1 & 0 \\ 0 & 1 \end{pmatrix}$. In other words, C is a linear transformation that decorrelates X into an uncorrelated unit-variance random vector. Note that $A = \frac{1}{\sqrt{2}}\begin{pmatrix} 1 & 1 \\ 1 & -1 \end{pmatrix}$ is a unitary matrix.

Meanwhile, for $B = \frac{1}{5}\begin{pmatrix} -2 & 3 \\ 3 & -2 \end{pmatrix}$, the covariance matrix of $Y = BX$ is $K_Y =$ $BK_XB^H = \frac{1}{25}\begin{pmatrix} -2 & 3 \\ 3 & -2 \end{pmatrix}\begin{pmatrix} 10 & 15 \\ 15 & 10 \end{pmatrix} = \begin{pmatrix} 1 & 0 \\ 0 & 1 \end{pmatrix}$. In other words, like C, the transformation B also decorrelates X into an uncorrelated unit-variance random vector. In addition, from $CK_XC^H = I$ and $BK_XB^H = I$, we get $C^HC = B^HB = \frac{1}{25}\begin{pmatrix} 13 & -12 \\ -12 & 13 \end{pmatrix} = K_X^{-1}$.   ◇

Example 4.3.7 When $U = (U_1, U_2)^T$ has mean vector $(10\ \ 0)^T$ and covariance matrix $\begin{pmatrix} 4 & 1 \\ 1 & 1 \end{pmatrix}$, consider the linear transformation $V = (V_1, V_2)^T = LU = \begin{pmatrix} -2 & 5 \\ 1 & 1 \end{pmatrix}\begin{pmatrix} U_1 \\ U_2 \end{pmatrix} = \begin{pmatrix} -2U_1 + 5U_2 \\ U_1 + U_2 \end{pmatrix}$ of U. Then, the mean vector of V is $\mathsf{E}\{V\} = L\mathsf{E}\{U\} = \begin{pmatrix} -2 & 5 \\ 1 & 1 \end{pmatrix}\begin{pmatrix} 10 \\ 0 \end{pmatrix} = \begin{pmatrix} -20 \\ 10 \end{pmatrix}$. In addition, the covariance matrix of V is $K_V = LK_UL^H = \begin{pmatrix} 21 & 0 \\ 0 & 7 \end{pmatrix}$.   ◇

Example 4.3.6 implies that the decorrelating linear transformation is generally not unique.

4.3.3 Joint Characteristic Function and Joint Moment Generating Function

By extending the notion of the cf and mgf discussed in Sect. 3.3.4, we introduce and discuss the joint cf and joint mgf of multi-dimensional random vectors.

Definition 4.3.7 (joint cf) The function

$$\varphi_X(\omega) = \mathsf{E}\left\{\exp\left(j\omega^T X\right)\right\} \tag{4.3.26}$$

is the joint cf of $X = (X_1, X_2, \ldots, X_n)^T$, where $\omega = (\omega_1, \omega_2, \ldots, \omega_n)^T$.

The joint cf $\varphi_X(\omega)$ of X can be expressed as

$$\varphi_X(\omega) = \int_{-\infty}^{\infty}\int_{-\infty}^{\infty}\cdots\int_{-\infty}^{\infty} f_X(x)\exp\left(j\omega^T x\right)dx \tag{4.3.27}$$

when X is a continuous random vector, where $x = (x_1, x_2, \ldots, x_n)^T$ and $dx = dx_1 dx_2 \cdots dx_n$. Thus, the joint cf $\varphi_X(\omega)$ is the complex conjugate of the multi-dimensional Fourier transform $\mathfrak{F}\{f_X(x)\}$ of the joint pdf $f_X(x)$. Clearly, we can obtain the joint pdf as $f_X(x) = \frac{1}{(2\pi)^n} \int_{-\infty}^{\infty} \int_{-\infty}^{\infty} \cdots \int_{-\infty}^{\infty} \varphi_X(\omega) \exp\left(-j\omega^T x\right) d\omega$ by inverse transforming the joint cf.

Definition 4.3.8 (joint mgf) The function

$$M_X(t) = \mathsf{E}\left\{\exp\left(t^T X\right)\right\} \tag{4.3.28}$$

is the joint mgf of X, where $t = (t_1, t_2, \ldots, t_n)^T$.

The joint mgf $M_X(t)$ is the multi-dimensional Laplace transform $\mathfrak{L}\{f_X(x)\}$ of the joint pdf $f_X(x)$ with t replaced by $-t$. The joint pdf can be obtained from the inverse Laplace transform of the joint mgf.

The marginal cf and marginal mgf can be obtained from the joint cf and joint mgf, respectively. For example, the marginal cf $\varphi_{X_i}(\omega_i)$ of X_i can be obtained as

$$\varphi_{X_i}(\omega_i) = \varphi_X(\omega)\big|_{\omega_j = 0 \text{ for all } j \neq i} \tag{4.3.29}$$

from the joint cf $\varphi_X(\omega)$, and the marginal mgf $M_{X_i}(t_i)$ of X_i as

$$M_{X_i}(t_i) = M_X(t)\big|_{t_j = 0 \text{ for all } j \neq i} \tag{4.3.30}$$

from the joint mgf $M_X(t)$.

When X is an independent random vector, it is easy to see that the joint cf $\varphi_X(\omega)$ is the product

$$\varphi_X(\omega) = \prod_{i=1}^{n} \varphi_{X_i}(\omega_i) \tag{4.3.31}$$

of marginal cf's from Theorem 4.1.3. Because cf's and distributions are related by one-to-one correspondences as we discussed in Theorem 3.3.2, a random vector whose joint cf is the product of the marginal cf's is an independent random vector.

Example 4.3.8 For an independent random vector X, let $Y = \sum_{i=1}^{n} X_i$. Then, the cf $\varphi_Y(\omega) = \mathsf{E}\left\{e^{j\omega Y}\right\} = \mathsf{E}\left\{e^{j\omega X_1} e^{j\omega X_2} \cdots e^{j\omega X_n}\right\}$ of Y can be expressed as

$$\varphi_Y(\omega) = \prod_{i=1}^{n} \varphi_{X_i}(\omega). \tag{4.3.32}$$

By inverse transforming the cf (4.3.32), we can get the pdf $f_Y(y)$ of Y, which is the convolution

$$f_{X_1+X_2+\cdots+X_n} = f_{X_1} * f_{X_2} * \cdots * f_{X_n} \tag{4.3.33}$$

of the marginal pdf's. The result (4.2.15) is a special case of (4.3.33) with $n = 2$. ◇

Example 4.3.9 Show that $Y = \sum_{i=1}^{n} X_i \sim P\left(\sum_{i=1}^{n} \lambda_i\right)$ when $\{X_i \sim P(\lambda_i)\}_{i=1}^{n}$ are independent of each other.

Solution The cf for the distribution $P(\lambda_i)$ is $\varphi_{X_i}(\omega) = \exp\{\lambda_i(e^{j\omega} - 1)\}$. Thus, the cf $\varphi_Y(\omega) = \prod_{i=1}^{n} \exp\{\lambda_i(e^{j\omega} - 1)\}$ of $Y = \sum_{i=1}^{n} X_i$ can be expressed as

$$\varphi_Y(\omega) = \exp\left\{\left(\sum_{i=1}^{n} \lambda_i\right)(e^{j\omega} - 1)\right\} \tag{4.3.34}$$

from (4.3.32), confirming the desired result. ◇

In Sect. 6.2.1, we will discuss again the sum of a number of independent random variables. As we observed in (4.3.32), when X and Y are independent of each other, the cf of $X + Y$ is the product of the cf's of X and Y. On the other hand, the converse does not hold true: when the cf of $X + Y$ is the product of the cf's of X and Y, X and Y may or may not be independent of each other. Specifically, assume $X_1 \sim F_1$ and $X_2 \sim F_2$ are independent of each other, where F_i is the cdf of X_i for $i = 1, 2$. Then, $X_1 + X_2 \sim F_1 * F_2$ and, if X_1 and X_2 are absolutely continuous random variables with pdf f_1 and f_2, respectively, $X_1 + X_2 \sim f_1 * f_2$. Yet, even when X_1 and X_2 are not independent of each other, in some cases we have $X_1 + X_2 \sim f_1 * f_2$ (Romano and Siegel 1986; Stoyanov 2013; Wies and Hall 1993). such cases include the non-independent Cauchy random variables and that shown in the example below.

Example 4.3.10 Assume the joint pmf $p_{X,Y}(x, y) = \mathsf{P}(X = x, Y = y)$

$$\begin{array}{lll} p_{X,Y}(1, 1) = \frac{1}{9}, & p_{X,Y}(1, 2) = \frac{1}{18}, & p_{X,Y}(1, 3) = \frac{1}{6}, \\ p_{X,Y}(2, 1) = \frac{1}{6}, & p_{X,Y}(2, 2) = \frac{1}{9}, & p_{X,Y}(2, 3) = \frac{1}{18}, \\ p_{X,Y}(3, 1) = \frac{1}{18}, & p_{X,Y}(3, 2) = \frac{1}{6}, & p_{X,Y}(3, 3) = \frac{1}{9} \end{array} \tag{4.3.35}$$

of a discrete random vector (X, Y). Then, the pmf $p_X(x) = \mathsf{P}(X = x)$ of X is $p_X(1) = p_X(2) = p_X(3) = \frac{1}{3}$, and the pmf $p_Y(y) = \mathsf{P}(Y = y)$ of Y is $p_Y(1) = p_Y(2) = p_Y(3) = \frac{1}{3}$, implying that X and Y are not independent of each other. However, the mgf's of X and Y are both $M_X(t) = M_Y(t) = \frac{1}{3}(e^t + e^{2t} + e^{3t})$ and the mgf of $X + Y$ is $M_{X+Y}(t) = \frac{1}{9}(e^{2t} + 2e^{3t} + 3e^{4t} + 2e^{5t} + e^{6t}) = M_X(t)M_Y(t)$. ◇

We have observed that when X and Y are independent of each other and g and h are continuous functions, $g(X)$ and $h(Y)$ are independent of each other in Theorem 4.1.3. We now discuss whether $g(X)$ and $h(Y)$ are uncorrelated or not when X and Y are uncorrelated.

Theorem 4.3.6 *When X and Y are independent, $g(X)$ and $h(Y)$ are uncorrelated. However, when X and Y are uncorrelated but not independent, $g(X)$ and $h(Y)$ are not necessarily uncorrelated.*

Proof When X and Y are independent of each other, we have $\mathsf{E}\{g(X)h(Y)\} = \int_{-\infty}^{\infty}\int_{-\infty}^{\infty} g(x)h(y)\, f_{X,Y}(x, y)dxdy = \int_{-\infty}^{\infty} g(x)f_X(x)dx \int_{-\infty}^{\infty} h(y)f_Y(y)dy = \mathsf{E}\{g(X)\}\mathsf{E}\{h(Y)\}$ and thus $g(X)$ and $h(Y)$ are uncorrelated. Next, when X and Y are uncorrelated but are not independent of each other, assume that $g(X)$ and $h(Y)$ are uncorrelated. Then, we have $\mathsf{E}\left\{e^{j(\omega_1 X+\omega_2 Y)}\right\} = \mathsf{E}\left\{e^{j\omega_1 X}\right\}\mathsf{E}\left\{e^{j\omega_2 Y}\right\}$ from $\mathsf{E}\{g(X)h(Y)\} = \mathsf{E}\{g(X)\}\mathsf{E}\{h(Y)\}$ with $g(x) = e^{j\omega_1 x}$ and $h(y) = e^{j\omega_2 y}$. This result implies that the joint cf $\varphi_{X,Y}(\omega_1, \omega_2)$ of X and Y can be expressed as $\varphi_{X,Y}(\omega_1, \omega_2) = \mathsf{E}\left\{e^{j(\omega_1 X+\omega_2 Y)}\right\} = \mathsf{E}\left\{e^{j\omega_1 X}\right\}\mathsf{E}\left\{e^{j\omega_2 Y}\right\} = \varphi_X(\omega_1)\,\varphi_Y(\omega_2)$ in terms of the marginal cf's of X and Y, a contradiction that X and Y are not independent of each other. In short, we have $\mathsf{E}\{XY\} = \mathsf{E}\{X\}\mathsf{E}\{Y\} \not\rightarrow \mathsf{E}\{g(X)h(Y)\} = \mathsf{E}\{g(X)\}\mathsf{E}\{h(Y)\}$, i.e., when X and Y are uncorrelated but are not independent of each other, $g(X)$ and $h(Y)$ are not necessarily uncorrelated.   ♠

The joint moments of random vectors can be easily obtained by using the joint cf or joint mgf as shown in the following theorem:

Theorem 4.3.7 *The joint moment* $m_{k_1 k_2 \cdots k_n} = \mathsf{E}\left\{X_1^{k_1} X_2^{k_2} \cdots X_n^{k_n}\right\} = \int_{-\infty}^{\infty}\int_{-\infty}^{\infty}\cdots$ $\int_{-\infty}^{\infty} x_1^{k_1} x_2^{k_2} \cdots x_n^{k_n} f_X(x)dx$ *can be obtained as*

$$m_{k_1 k_2 \cdots k_n} = j^{-K} \left. \frac{\partial^K \varphi_X(\boldsymbol{\omega})}{\partial \omega_1^{k_1} \partial \omega_2^{k_2} \cdots \partial \omega_n^{k_n}}\right|_{\boldsymbol{\omega}=0} \tag{4.3.36}$$

from the joint cf $\varphi_X(\boldsymbol{\omega})$, *where* $K = \sum_{i=1}^{n} k_i$.

From the joint mgf $M_X(t)$, we can obtain the joint moment $m_{k_1 k_2 \cdots k_n}$ also as

$$m_{k_1 k_2 \cdots k_n} = \left. \frac{\partial^K M_X(t)}{\partial t_1^{k_1} \partial t_2^{k_2} \cdots \partial t_n^{k_n}}\right|_{t=0}. \tag{4.3.37}$$

As a special case of (4.3.37) for the two-dimensional random vector (X, Y), we have

$$m_{kr} = \left. \frac{\partial^{k+r} M_{X,Y}(t_1, t_2)}{\partial t_1^k \partial t_2^r}\right|_{(t_1, t_2)=(0,0)}, \tag{4.3.38}$$

where $M_{X,Y}(t_1, t_2) = \mathsf{E}\{\exp(t_1 X + t_2 Y)\}$ is the joint mgf of (X, Y).

Example 4.3.11 (Romano and Siegel 1986) When the two functions G and H are equal, we have $F * G = F * H$. The converse does not always hold true, i.e., $F * G = F * H$ does not necessarily imply $G = H$. Let us consider an example. Assume

$$P(X = x) = \begin{cases} \frac{1}{2}, & x = 0, \\ \frac{2}{\pi^2(2n-1)^2}, & x = \pm(2n-1)\pi, \quad n = 1, 2, \ldots, \\ 0, & \text{otherwise} \end{cases} \quad (4.3.39)$$

for a random variable X. Then, the cf $\varphi_X(t) = \frac{1}{2} + \sum_{n=1}^{\infty} \frac{4}{\pi^2(2n-1)^2} \cos\{(2n-1)\pi t\}$ of X is a train of triangular pulses with period 2 and

$$\varphi_X(t) = 1 - |t| \quad (4.3.40)$$

for $|t| \leq 1$. Meanwhile, when the distribution is

$$P(Y = x) = \begin{cases} \frac{4}{\pi^2(2n-1)^2}, & x = \pm\frac{(2n-1)\pi}{2}, \quad n = 1, 2, \ldots, \\ 0, & \text{otherwise} \end{cases} \quad (4.3.41)$$

for Y, the cf is also a train of triangular pulses with period 4 and

$$\varphi_Y(t) = 1 - |t| \quad (4.3.42)$$

for $|t| \leq 2$. It is easy to see that $\varphi_X(t) = \varphi_Y(t)$ for $|t| \leq 1$ and that $|\varphi_X(t)| = |\varphi_Y(t)|$ for all t from (4.3.40) and (4.3.42). Now, for a random variable Z with the pdf

$$f_Z(x) = \begin{cases} \frac{1-\cos x}{\pi x^2}, & x \neq 0, \\ \frac{1}{2\pi}, & x = 0, \end{cases} \quad (4.3.43)$$

we have the cf $\varphi_Z(t) = (1 - |t|)u(1 - |t|)$. Then, we have $\varphi_Z(t)\varphi_X(t) = \varphi_Z(t)\varphi_Y(t)$ and $F_Z(x) * F_X(x) = F_Z(x) * F_Y(x)$, but $F_X(x) \neq F_Y(x)$, where F_X, F_Y, and F_Z denote the cdf's of X, Y, and Z, respectively.   ◇

4.4 Conditional Distributions

In this section, we discuss conditional probability functions (Ross 2009) and conditional expected values mainly for bi-variate random vectors.

4.4.1 Conditional Probability Functions

We first extend the discussion on the conditional distribution explored in Sect. 3.4. When the event A is assumed, the conditional joint cdf[16] $F_{Z,W|A}(z, w) = P(Z \leq z, W \leq w| A)$ of Z and W is

$$F_{Z,W|A}(z, w) = \frac{P(Z \leq z, W \leq w, A)}{P(A)}. \tag{4.4.1}$$

The conditional joint pdf[17] can be obtained as

$$f_{Z,W|A}(z, w) = \frac{\partial^2}{\partial z \partial w} F_{Z,W|A}(z, w) \tag{4.4.2}$$

by differentiating the conditional joint cdf $F_{Z,W|A}(z, w)$ with respect to z and w.

Example 4.4.1 Obtain the conditional joint cdf $F_{X,Y|A}(x, y)$ and the conditional joint pdf $f_{X,Y|A}(x, y)$ under the condition $A = \{X \leq x\}$.

Solution Recollecting that $P(X \leq x, Y \leq y| X \leq x) = \frac{P(X \leq x, Y \leq y)}{P(X \leq x)}$, we get the conditional joint cdf $F_{X,Y|A}(x, y) = F_{X,Y|X \leq x}(x, y) = \frac{P(X \leq x, Y \leq y)}{P(X \leq x)}$ as

$$F_{X,Y|X \leq x}(x, y) = \frac{F_{X,Y}(x, y)}{F_X(x)}. \tag{4.4.3}$$

Differentiating the conditional joint cdf (4.4.3) with respect to x and y, we get the conditional joint pdf $f_{X,Y|A}(x, y)$ as

$$\begin{aligned} f_{X,Y|X \leq x}(x, y) &= \frac{\partial}{\partial x} \left\{ \frac{1}{F_X(x)} \frac{\partial}{\partial y} F_{X,Y}(x, y) \right\} \\ &= \frac{f_{X,Y}(x, y)}{F_X(x)} - \frac{f_X(x)}{F_X^2(x)} \frac{\partial}{\partial y} F_{X,Y}(x, y). \end{aligned} \tag{4.4.4}$$

By writing $F_{Y|X \leq x}(y)$ as $F_{Y|X \leq x}(y) = \frac{P(X \leq x, Y \leq y)}{P(X \leq x)}$, i.e.,

$$F_{Y|X \leq x}(y) = \frac{F_{X,Y}(x, y)}{F_X(x)}, \tag{4.4.5}$$

we get

$$F_{Y|X \leq x}(y) = F_{X,Y|X \leq x}(x, y) \tag{4.4.6}$$

from (4.4.3) and (4.4.5). ◇

---

[16] As in other cases, the conditional joint cdf is also referred to as the conditional cdf if it does not cause any ambiguity.

[17] The conditional joint pdf is also referred to as the conditional pdf if it does not cause any ambiguity.

We now discuss the conditional distribution when the condition is expressed in terms of random variables.

Definition 4.4.1 (*conditional cdf; conditional pdf; conditional pmf*) For a random vector (X, Y), $\mathsf{P}(Y \leq y | X = x)$ is called the conditional joint cdf, or simply the conditional cdf, of Y given $X = x$, and is written as $F_{X,Y|X=x}(x, y)$, $F_{Y|X=x}(y)$, or $F_{Y|X}(y|x)$. For a continuous random vector (X, Y), the derivative $\frac{\partial}{\partial y} F_{Y|X}(y|x)$, denoted by $f_{Y|X}(y|x)$, is called the conditional pdf of Y given $X = x$. For a discrete random vector (X, Y), $\mathsf{P}(Y = y | X = x)$ is called the conditional joint pmf or the conditional pmf of Y given $X = x$ and is written as $p_{X,Y|X=x}(x, y)$, $p_{Y|X=x}(y)$, or $p_{Y|X}(y|x)$.

The relationships among the conditional pdf $f_{Y|X}(y|x)$, joint pdf $f_{X,Y}(x, y)$, and marginal pdf $f_X(x)$ and those among the conditional pmf $p_{Y|X}(y|x)$, joint pmf $p_{X,Y}(x, y)$, and marginal pmf $p_X(x)$ are described in the following theorem:

Theorem 4.4.1 *The conditional pmf $p_{Y|X}(y|x)$ can be expressed as*

$$p_{Y|X}(y|x) = \frac{p_{X,Y}(x, y)}{p_X(x)} \tag{4.4.7}$$

when (X, Y) is a discrete random vector. Similarly, the conditional pdf $f_{Y|X}(y|x)$ can be expressed as

$$f_{Y|X}(y|x) = \frac{f_{X,Y}(x, y)}{f_X(x)} \tag{4.4.8}$$

when (X, Y) is a continuous random vector.

Proof For a discrete random vector, we easily get $p_{Y|X}(y|x) = \frac{\mathsf{P}(X=x,Y=y)}{\mathsf{P}(X=x)} = \frac{p_{X,Y}(x,y)}{p_X(x)}$. For a continuous random vector, we have $F_{Y|X}(y|x) = \mathsf{P}(Y \leq y | X = x) = \lim_{dx \to 0} \frac{\mathsf{P}(x-dx < X \leq x, Y \leq y)}{\mathsf{P}(x-dx < X \leq x)}$ as

$$F_{Y|X}(y|x) = \lim_{dx \to 0} \frac{\frac{F_{X,Y}(x,y) - F_{X,Y}(x-dx, y)}{dx}}{\frac{F_X(x) - F_X(x-dx)}{dx}}$$

$$= \frac{1}{f_X(x)} \frac{\partial}{\partial x} F_{X,Y}(x, y) \tag{4.4.9}$$

and consequently, $f_{Y|X}(y|x) = \frac{\partial}{\partial y} F_{Y|X}(y|x) = \frac{1}{f_X(x)} \frac{\partial^2}{\partial x \partial y} F_{X,Y}(x, y)$, which is the same as (4.4.8). ♠

We can similarly obtain the conditional pdf $f_{X|Y}(x|y) = \frac{\partial}{\partial x} F_{X|Y}(x|y)$ as

$$f_{X|Y}(x|y) = \frac{f_{X,Y}(x, y)}{f_Y(y)}, \tag{4.4.10}$$

which can also be obtained directly from (4.4.8) by replacing X, Y, x, and y with Y, X, y, and x, respectively. Note the similarity among (4.4.7), (4.4.8), and (2.4.1).

Example 4.4.2 (Ross 1976) Obtain the conditional pdf $f_{X|Y}(x|y)$ when the joint pdf of (X, Y) is $f_{X,Y}(x, y) = 6xy(2 - x - y)u(x)u(1 - x)u(y)u(1 - y)$.

Solution Employing (4.4.10) and noting that $f_Y(y) = \int_{-\infty}^{\infty} f_{X,Y}(x, y)dx = \int_0^1 6xy(2 - x - y)dx$, we have $f_{X|Y}(x|y) = \frac{6xy(2-x-y)}{\int_0^1 6xy(2-x-y)dx}$, i.e.,

$$f_{X|Y}(x|y) = \frac{6x(2 - x - y)}{4 - 3y} \tag{4.4.11}$$

for $0 < x < 1$ and $0 < y < 1$. ◇

Example 4.4.3 (Ross 1976) Obtain the conditional pdf $f_{X|Y}(x|y)$ when the joint pdf of (X, Y) is $f_{X,Y}(x, y) = 4y(x - y)e^{-(x+y)}u(y)u(x - y)$.

Solution We easily get $f_{X|Y}(x|y) = \frac{4y(x-y)e^{-(x+y)}}{\int_y^{\infty} 4y(x-y)e^{-(x+y)}dx}$, i.e.,

$$f_{X|Y}(x|y) = (x - y)e^{-(x-y)} \tag{4.4.12}$$

for $0 \le y \le x < \infty$ from (4.4.10). ◇

It is noteworthy that, unlike (4.4.7) or (4.4.8), we have

$$F_{Y|X}(y|x) \neq \frac{F_{X,Y}(x, y)}{F_X(x)} \tag{4.4.13}$$

from $F_{Y|X}(y|x) = \mathsf{P}(Y \le y|X = x)$ and $\frac{F_{X,Y}(x,y)}{F_X(x)} = \frac{\mathsf{P}(X \le x, Y \le y)}{\mathsf{P}(X \le x)} = \mathsf{P}(Y \le y|X \le x)$. Employing (4.4.8) with $F_{Y|X}(y|t) = \int_{-\infty}^{y} f_{Y|X}(s|t)ds$, we get $F_{X,Y}(x, y) = \int_{-\infty}^{x} F_{Y|X}(y|t)f_X(t)dt = \int_{-\infty}^{x} \int_{-\infty}^{y} f_{Y|X}(s|t)ds f_X(t)dt = \int_{-\infty}^{x} \int_{-\infty}^{y} f_{X,Y}(t, s)dsdt$ from (4.4.9). This result is the same as (4.1.10) or (4.1.21) in that the cdf can be obtained by integrating the pdf.

Theorem 4.4.2 *The pmf p_X of X can be expressed as*

$$p_X(x) = \sum_{y=-\infty}^{\infty} p_{X|Y}(x|y)p_Y(y) \tag{4.4.14}$$

for a discrete random vector (X, Y), and the pdf f_X of X can be expressed as

$$f_X(x) = \int_{-\infty}^{\infty} f_{X|Y}(x|y)f_Y(y)dy \qquad (4.4.15)$$

for a continuous random vector (X, Y).

Theorem 4.4.2 can be easily proved by noting that $p_X(x) = \sum\limits_{y=-\infty}^{\infty} p_{X,Y}(x, y)$, $p_{X,Y}(x, y) = p_{X|Y}(x|y)p_Y(y)$, $f_X(x) = \int_{-\infty}^{\infty} f_{X,Y}(x, y)dy$, and $f_{X,Y}(x, y) = f_{X|Y}(x|y)f_Y(y)$. We can obtain the following theorem based on (4.1.23), (4.4.8), (4.4.10), and (4.4.15):

Theorem 4.4.3 *We can rewrite $f_{Y|X}(y|x) = \frac{f_{X,Y}(x,y)}{f_X(x)}$ as*

$$f_{Y|X}(y|x) = \frac{f_{X|Y}(x|y)f_Y(y)}{\int_{-\infty}^{\infty} f_{X|Y}(x|y)f_Y(y)dy} \qquad (4.4.16)$$

for any x such that $f_X(x) > 0$ by noting that $f_{X,Y}(x, y) = f_{X|Y}(x|y)f_Y(y)$ and $f_X(x) = \int_{-\infty}^{\infty} f_{X|Y}(x|y)f_Y(y)dy$.

Similarly to (4.4.16), we have

$$p_{Y|X}(y|x) = \frac{p_{X|Y}(x|y)p_Y(y)}{\sum\limits_{y=-\infty}^{\infty} p_{X|Y}(x|y)p_Y(y)} \qquad (4.4.17)$$

for any x such that $p_X(x) > 0$ when (X, Y) is a discrete random vector.

When X and Y are independent of each other, we have

$$F_{X|Y}(x|y) = F_X(x) \qquad (4.4.18)$$

for every point y such that $F_Y(y) > 0$,

$$f_{X|Y}(x|y) = f_X(x) \qquad (4.4.19)$$

for every point y such that $f_Y(y) > 0$, $F_{Y|X}(y|x) = F_Y(y)$ for every point x such that $F_X(x) > 0$, and $f_{Y|X}(y|x) = f_Y(y)$ for every point x such that $f_X(x) > 0$ because $F_{X,Y}(x, y) = F_X(x)F_Y(y)$ and $f_{X,Y}(x, y) = f_X(x)f_Y(y)$.

Example 4.4.4 Assume the pmf

$$p_X(x) = \begin{cases} \frac{1}{6}, & x = 3; \quad \frac{1}{2}, & x = 4; \\ \frac{1}{3}, & x = 5 \end{cases} \qquad (4.4.20)$$

of X. Consider

$$Y = \begin{cases} 0, & \text{if } X = 4, \\ 1, & \text{if } X = 3 \text{ or } 5 \end{cases} \tag{4.4.21}$$

and $Z = X - Y$. Obtain the conditional pmf's $p_{X|Y}$, $p_{Y|X}$, $p_{X|Z}$, $p_{Z|X}$, $p_{Y|Z}$, and $p_{Z|Y}$, and the joint pmf's $p_{X,Y}$, $p_{Y,Z}$, and $p_{Z,X}$.

Solution We easily get $p_Y(y) = P(X = 4)$ for $y = 0$ and $p_Y(y) = P(X = 3 \text{ or } 5)$ for $y = 1$, i.e.,

$$p_Y(y) = \begin{cases} \frac{1}{2}, & y = 0, \\ \frac{1}{2}, & y = 1. \end{cases} \tag{4.4.22}$$

Next, because $Y = 1$ and $Z = X - Y = 2$ when $X = 3$, $Y = 0$ and $Z = X - Y = 4$ when $X = 4$, and $Y = 1$ and $Z = X - Y = 4$ when $X = 5$, we get $p_Z(z) = P(X = 3)$ for $z = 2$ and $p_Z(z) = P(X = 4 \text{ or } 5)$ for $z = 4$, i.e.,

$$p_Z(z) = \begin{cases} \frac{1}{6}, & z = 2, \\ \frac{5}{6}, & z = 4. \end{cases} \tag{4.4.23}$$

Next, because $Y = 1$ when $X = 3$, $Y = 0$ when $X = 4$, and $Y = 1$ when $X = 5$ from (4.4.21), we get

$$\begin{aligned} p_{Y|X}(0|3) = 0, \quad p_{Y|X}(0|4) = 1, \quad p_{Y|X}(0|5) = 0, \\ p_{Y|X}(1|3) = 1, \quad p_{Y|X}(1|4) = 0, \quad p_{Y|X}(1|5) = 1. \end{aligned} \tag{4.4.24}$$

Noting that $p_{X,Y}(x, y) = p_{Y|X}(y|x)p_X(x)$ from (4.4.17) and using (4.4.20) and (4.4.24), we get

$$\begin{aligned} p_{X,Y}(3, 0) = 0, \quad p_{X,Y}(4, 0) = \tfrac{3}{6}, \quad p_{X,Y}(5, 0) = 0, \\ p_{X,Y}(3, 1) = \tfrac{1}{6}, \quad p_{X,Y}(4, 1) = 0, \quad p_{X,Y}(5, 1) = \tfrac{1}{3}. \end{aligned} \tag{4.4.25}$$

Similarly, noting that $p_{X|Y}(x|y) = \frac{p_{X,Y}(x,y)}{p_Y(y)}$ from (4.4.17) and using (4.4.22) and (4.4.25), we get

$$\begin{aligned} p_{X|Y}(3|0) = 0, \quad p_{X|Y}(4|0) = 1, \quad p_{X|Y}(5|0) = 0, \\ p_{X|Y}(3|1) = \tfrac{1}{3}, \quad p_{X|Y}(4|1) = 0, \quad p_{X|Y}(5|1) = \tfrac{2}{3}. \end{aligned} \tag{4.4.26}$$

Meanwhile, $Y = 1$ and $Z = 2$ when $X = 3$, $Y = 0$ and $Z = 4$ when $X = 4$, and $Y = 1$ and $Z = 4$ when $X = 5$ from (4.4.21) and the definition of Z. Thus, we have

$$\begin{aligned} p_{Z|X}(2|3) = 1, \quad p_{Z|X}(2|4) = 0, \quad p_{Z|X}(2|5) = 0, \\ p_{Z|X}(4|3) = 0, \quad p_{Z|X}(4|4) = 1, \quad p_{Z|X}(4|5) = 1. \end{aligned} \tag{4.4.27}$$

Noting that $p_{X,Z}(x, z) = p_{Z|X}(z|x)p_X(x)$ from (4.4.17) and using (4.4.20) and (4.4.27), we get

$$p_{X,Z}(3,2) = \tfrac{1}{6}, \ p_{X,Z}(4,2) = 0, \ p_{X,Z}(5,2) = 0,$$
$$p_{X,Z}(3,4) = 0, \ p_{X,Z}(4,4) = \tfrac{3}{6}, \ p_{X,Z}(5,4) = \tfrac{1}{3}. \tag{4.4.28}$$

Similarly, noting that $p_{X|Z}(x|z) = \frac{p_{X,Z}(x,z)}{p_Z(z)}$ from (4.4.17) and using (4.4.23) and (4.4.28), we get

$$p_{X|Z}(3|2) = 1, \ p_{X|Z}(4|2) = 0, \ p_{X|Z}(5|2) = 0,$$
$$p_{X|Z}(3|4) = 0, \ p_{X|Z}(4|4) = \tfrac{3}{5}, \ p_{X|Z}(5|4) = \tfrac{2}{5}. \tag{4.4.29}$$

Finally, we have $p_{Z|Y}(2|1) = \mathsf{P}(X - Y = 2|Y = 1) = \mathsf{P}(X = 3|X = 3 \text{ or } 5) = \frac{\mathsf{P}(X=3)}{\mathsf{P}(X=3 \text{ or } 5)}$ as

$$p_{Z|Y}(2|1) = \frac{1}{3} \tag{4.4.30}$$

and $p_{Z|Y}(4|1) = \mathsf{P}(X - Y = 4|Y = 1) = \frac{\mathsf{P}(X=5)}{\mathsf{P}(X=3 \text{ or } 5)}$ as

$$p_{Z|Y}(4|1) = \frac{2}{3} \tag{4.4.31}$$

from $p_{Z|Y}(2|0) = \mathsf{P}(X - Y = 2|Y = 0) = \mathsf{P}(X = 2|X = 4) = 0, \ p_{Z|Y}(4|0) = \mathsf{P}(X - Y = 4|Y = 0) = \mathsf{P}(X = 4|X = 4) = 1,$ and $\{X = 3\} \cap \{X = 3 \text{ or } 5\} = \{X = 3\}$. Therefore, noting that $p_{Y,Z}(y, z) = p_{Z|Y}(z|y)p_Y(y)$ from (4.4.17) and using (4.4.22) and (4.4.31), we get

$$p_{Y,Z}(0, 2) = 0, \ p_{Y,Z}(0, 4) = \tfrac{1}{2}, \ p_{Y,Z}(1, 2) = \tfrac{1}{6}, \ p_{Y,Z}(1, 4) = \tfrac{1}{3}. \tag{4.4.32}$$

We also get

$$p_{Y|Z}(0|2) = 0, \ p_{Y|Z}(0|4) = \tfrac{3}{5}, \ p_{Y|Z}(1|2) = 1, \ p_{Y|Z}(1|4) = \tfrac{2}{5} \tag{4.4.33}$$

from $p_{Y|Z}(y|z) = \frac{p_{Y,Z}(y,z)}{p_Z(z)}$ by using (4.4.23) and (4.4.32).   ◇

Example 4.4.5 When the joint pdf of (X, Y) is

$$f_{X,Y}(x, y) = \frac{1}{16}u(2 - |x|)u(2 - |y|), \tag{4.4.34}$$

obtain the conditional joint cdf $F_{X,Y|A}$ and the conditional joint pdf $f_{X,Y|A}$ for $A = \{|X| \le 1, |Y| \le 1\}$.

Solution First, we have $\mathsf{P}(A) = \int_{-1}^{1}\int_{-1}^{1} f_{X,Y}(u, v)dudv = \frac{1}{16} \times 4 = \frac{1}{4}$. Next, for

$$\mathsf{P}(X \le x, Y \le y, A) = \iint\limits_{\{|u|\le 1, \ |v|\le 1, \ u\le x, \ v\le y\}} f_{X,Y}(u, v)dudv, \tag{4.4.35}$$

Fig. 4.16 The region of integration to obtain the conditional joint cdf $F_{X,Y|A}(x, y|A)$ under the condition $A = \{|X| \le 1, |Y| \le 1\}$ when $f_{X,Y}(x, y) = \frac{1}{16}u(2 - |x|)u(2 - |y|)$

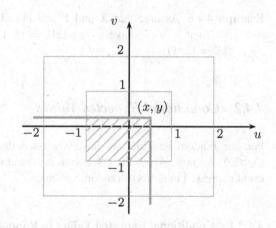

we get the following results by referring to Fig. 4.16: First, $P(X \le x, Y \le y, A) = P(A) = \frac{1}{4}$ when $x \ge 1$ and $y \ge 1$ and $P(X \le x, Y \le y, A) = P(\emptyset) = 0$ when $x \le -1$ or $y \le -1$. In addition, $P(X \le x, Y \le y, A) = \iint\limits_{\{|u| \le 1, \, -1 \le v \le y\}} f_{X,Y}(u, v)dudv = \frac{1}{16} \times 2(y + 1) = \frac{1}{8}(y + 1)$ because $\{|u| \le 1, \, |v| \le 1, \, u \le x, \, v \le y\} = \{|u| \le 1, \, -1 \le v \le y\}$ when $x \ge 1$ and $-1 \le y \le 1$. We similarly get $P(X \le x, Y \le y, A) = \frac{1}{8}(x + 1)$ when $-1 \le x \le 1$ and $y \ge 1$, and $P(X \le x, Y \le y, A) = \frac{1}{16}(x + 1)(y + 1)$ when $-1 \le x \le 1$ and $-1 \le y \le 1$. Taking these results into account, we get the conditional joint cdf $F_{X,Y|A}(x, y) = \frac{P(X \le x, Y \le y, A)}{P(A)}$ as

$$F_{X,Y|A}(x, y) = \begin{cases} 1, & x \ge 1, \ y \ge 1, \\ \frac{1}{2}(y + 1), & x \ge 1, \ |y| \le 1, \\ \frac{1}{4}(x + 1)(y + 1), & |x| \le 1, \ |y| \le 1, \\ \frac{1}{2}(x + 1), & |x| \le 1, \ y \ge 1, \\ 0, & x \le -1 \text{ or } y \le -1. \end{cases} \quad (4.4.36)$$

We subsequently get the conditional joint pdf

$$f_{X,Y|A}(x, y) = \frac{1}{4}u(1 - |x|)u(1 - |y|) \quad (4.4.37)$$

by differentiating $F_{X,Y|A}(x, y)$. $\diamond$

Let $Y = g(X)$ and $g^{-1}(\cdot)$ be the inverse of the function $g(\cdot)$. We then have

$$f_{Z|Y}(z|Y = y) = f_{Z|X}\left(z|X = g^{-1}(y)\right), \quad (4.4.38)$$

which implies that, when the relationship between X and Y can be expressed via an invertible function, conditioning on $X = g^{-1}(y)$ is equivalent to conditioning on $Y = y$.

Example 4.4.6 Assume that X and Y are related as $Y = g(X) = (X_1 + X_2, X_1 - X_2)$. Then, we have $f_{Z|Y}(z|Y = (3, 1)) = f_{Z|X}(z|X = g^{-1}(3, 1)) = f_{Z|X}(z|X = (2, 1))$. $\diamond$

4.4.2 Conditional Expected Values

For one random variable, we have discussed the conditional expected value in (3.4.30). We now extend the discussion into random vectors with the conditioning event expressed in terms of random variables.

4.4.2.1 Conditional Expected Values in Random Vectors

Let $B = \{X = x\}$ in the conditional expected value $E\{Y|B\} = \int_{-\infty}^{\infty} y f_{Y|B}(y) dy$ shown in (3.4.30). Then, the conditional expected value $m_{Y|X} = E\{Y|X = x\}$ of Y is

$$m_{Y|X} = \int_{-\infty}^{\infty} y f_{Y|X}(y|x) dy \tag{4.4.39}$$

when $X = x$.

Example 4.4.7 Obtain the conditional expected value $E\{X|Y = y\}$ in Example 4.4.2.

Solution We easily get $E\{X|Y = y\} = \int_0^1 \frac{6x^2(2-x-y)}{4-3y} dx = \frac{1}{4-3y}\{2(2-y)x^3 - \frac{3}{2}x^4\}\big|_{x=0}^1 = \frac{5-4y}{8-6y}$ for $0 < y < 1$. $\diamond$

The conditional expected value $E\{Y|X\}$ is a function of X and is thus a random variable with its value $E\{Y|X = x\}$ when $X = x$.

Theorem 4.4.4 *The expected value* $E\{Y\}$ *can be obtained as*

$$E\{E\{Y|X\}\} = E\{Y\}$$

$$= \begin{cases} \int_{-\infty}^{\infty} E\{Y|X = x\} f_X(x) dx, \\ \quad X \text{ is a continuous random variable}, \\ \sum_{x=-\infty}^{\infty} E\{Y|X = x\} p_X(x), \\ \quad X \text{ is a discrete random variable} \end{cases} \tag{4.4.40}$$

from the conditional expected value $E\{Y|X\}$.

Proof Considering only the case of a continuous random vector, (4.4.40) can be shown easily as $\mathsf{E}\{\mathsf{E}\{Y|X\}\} = \int_{-\infty}^{\infty} \mathsf{E}\{Y|X = x\} f_X(x)dx = \int_{-\infty}^{\infty} \left\{ \int_{-\infty}^{\infty} y f_{Y|X}(y|x)dy \right\} f_X(x)dx = \int_{-\infty}^{\infty} y \int_{-\infty}^{\infty} f_{X,Y}(x, y)dxdy = \int_{-\infty}^{\infty} y f_Y(y) dy = \mathsf{E}\{Y\}$. ♠

4.4.2.2 Conditional Expected Values for Functions of Random Vectors

The function $g(X, Y)$ is a function of two random vectors X and Y while $g(x, Y)$ is a function of a vector x and a random vector Y: in other words, $g(x, Y)$ and $g(X, Y)$ are different from each other. In addition, under the condition $X = x$, the conditional mean of $g(X, Y)$ is $\mathsf{E}\{g(X, Y)|X = x\} = \int_{\text{all } y} g(x, y) f_{Y|X}(y|x) dy$, which is the same as the conditional mean $\mathsf{E}\{g(x, Y)|X = x\} = \int_{\text{all } y} g(x, y) f_{Y|X}(y|x) dy$ of $g(x, Y)$. In other words,

$$\mathsf{E}\{g(X, Y)|X = x\} = \mathsf{E}\{g(x, Y)|X = x\}$$
$$= \int_{\text{all } y} g(x, y) f_{Y|X}(y|x) dy. \qquad (4.4.41)$$

Furthermore, for the expected value of the random vector $\mathsf{E}\{g(X, Y)|X\}$, we have

$$\mathsf{E}\{g(X, Y)\} = \mathsf{E}\{\mathsf{E}\{g(X, Y)|X\}\}$$
$$= \mathsf{E}\{\mathsf{E}\{g(X, Y)|Y\}\} \qquad (4.4.42)$$

from $\mathsf{E}\{\mathsf{E}\{g(X, Y)|X\}\} = \int_{\text{all } x} \left\{ \int_{\text{all } y} g(x, y) f_{Y|X}(y|x) dy \right\} f_X(x)dx = \int_{\text{all } y} \left\{ \int_{\text{all } x} g(x, y) f_{X,Y}(x, y)dx \right\} dy = \int_{\text{all } y} \left\{ \int_{\text{all } x} g(x, y) f_{X|Y}(x|y) dx \right\} f_Y(y)dy$. The result (4.4.42) can be obtained also from Theorem 4.4.4. When X and Y are both one-dimensional, if we let $g(X, Y) = (Y - m_{Y|X})^2$, we get the expected value

$$\mathsf{E}\{g(X, Y)|X = x\} = \mathsf{E}\left\{ (Y - m_{Y|X})^2 \Big| X = x \right\}, \qquad (4.4.43)$$

which is called the conditional variance of Y when $X = x$.

Example 4.4.8 (Ross 1976) Obtain the conditional expected value $\mathsf{E}\left\{ \exp\left(\frac{X}{2}\right) \Big| Y = 1 \right\}$ when the joint pdf of (X, Y) is $f_{X,Y}(x, y) = \frac{y}{2} e^{-xy} u(x)u(y)u(2 - y)$.

Solution From (4.4.10), we have $f_{X|Y}(x|1) = \frac{f_{X,Y}(x,1)}{f_Y(1)} = \frac{\frac{1}{2}e^{-x}}{\int_0^\infty \frac{1}{2}e^{-x}dx} = e^{-x}$ for $0 < x < \infty$. Thus, from (4.4.41), we get $\mathsf{E}\left\{ \exp\left(\frac{X}{2}\right) \Big| Y = 1 \right\} = \int_0^\infty e^{\frac{x}{2}} e^{-x}dx = 2$. ◇

When $g(X, Y) = g_1(X)g_2(Y)$, from (4.4.41) we get

$$\mathsf{E}\left\{ g_1(X)g_2(Y) \Big| X = x \right\} = g_1(x)\mathsf{E}\left\{ g_2(Y) \Big| X = x \right\}, \qquad (4.4.44)$$

which is called the factorization property (Gardner 1990). The factorization property implies that the random vector $g_1(X)$ under the condition $X = x$, or equivalently $g_1(x)$, is not probabilistic.

4.4.3   Evaluation of Expected Values via Conditioning

As we have observed in Sects. 2.4 and 3.4.3, we can obtain the probability and expected value more easily by first obtaining the conditional probability and conditional expected value with appropriate conditioning. Let us now discuss how we can obtain expected values for random vectors by first obtaining the conditional expected value with appropriate conditioning on random vectors.

Example 4.4.9 Consider the group $\{1, 1, 2, 2, \ldots, n, n\}$ of n pairs of numbers. When we randomly delete m numbers in the group, obtain the expected number of pairs remaining. For $n = 20$ and $m = 10$, obtain the value of the expected number.

Solution Denote by J_m the number of pairs remaining after m numbers have been deleted. After m numbers have been deleted, we have $2n - m - 2J_m$ non-paired numbers. When we delete one more number, the number of pairs is $J_m - 1$ with probability $\frac{2J_m}{2n-m}$ or J_m with probability $1 - \frac{2J_m}{2n-m}$. Based on this observation, we have

$$\mathsf{E}\{J_{m+1}|J_m\} = (J_m - 1)\frac{2J_m}{2n - m} + J_m\left(1 - \frac{2J_m}{2n - m}\right)$$

$$= J_m\frac{2n - m - 2}{2n - m}. \tag{4.4.45}$$

Noting that $\mathsf{E}\{\mathsf{E}\{J_{m+1}|J_m\}\} = \mathsf{E}\{J_{m+1}\}$ and $\mathsf{E}\{J_0\} = n$, we get $\mathsf{E}\{J_{m+1}\} = \mathsf{E}\{J_m\}\frac{2n-m-2}{2n-m} = \mathsf{E}\{J_{m-1}\}\frac{2n-m-2}{2n-m}\frac{2n-m-1}{2n-m+1} = \cdots = \mathsf{E}\{J_0\}\frac{2n-m-2}{2n-m}\frac{2n-m-1}{2n-m+1}\cdots\frac{2n-2}{2n} = \frac{(2n-m-2)(2n-m-1)}{2(2n-1)}$, i.e.,

$$\mathsf{E}\{J_m\} = \frac{(2n - m - 1)(2n - m)}{2(2n - 1)}$$

$$= n\frac{{}_{2n-2}C_m}{{}_{2n}C_m}. \tag{4.4.46}$$

For $n = 20$ and $m = 10$, the value is $\mathsf{E}\{J_{10}\} = \frac{29 \times 30}{2 \times 39} \approx 11.15$. ◇

Example 4.4.10 (Ross 1996) We toss a coin repeatedly until *head* appears r times consecutively. When the probability of a *head* is p, obtain the expected value of repetitions.

Solution Denote by C_k the number of repetitions until the first appearance of k consecutive *heads*. Then,

$$C_{k+1} = \begin{cases} C_k + 1, & C_{k+1}\text{-st outcome is } head, \\ C_{k+1} + C_k + 1, & C_{k+1}\text{-st outcome is } tail. \end{cases} \tag{4.4.47}$$

Let $\alpha_k = \mathsf{E}\{C_k\}$ for convenience. Then, using (4.4.47) in

$$\mathsf{E}\{C_{k+1}\} = \mathsf{E}\{C_{k+1}|\, C_{k+1}\text{-st outcome is } head\}\,\mathsf{P}(head)$$
$$+\mathsf{E}\{C_{k+1}|\, C_{k+1}\text{-st outcome is } tail\}\,\mathsf{P}(tail), \tag{4.4.48}$$

we get

$$\alpha_{k+1} = (\alpha_k + 1)\,p + (\alpha_{k+1} + \alpha_k + 1)\,(1 - p). \tag{4.4.49}$$

Now, solving (4.4.49), we get $\alpha_{k+1} = \frac{1}{p}\alpha_k + \frac{1}{p} = \frac{1}{p^2}\alpha_{k-1} + \frac{1}{p} + \frac{1}{p^2} = \cdots = \frac{1}{p^k}\alpha_1 + \frac{1}{p} + \frac{1}{p^2} + \cdots + \frac{1}{p^k} = \sum_{i=1}^{k+1} \frac{1}{p^i}$ because $\alpha_1 = \mathsf{E}\{C_1\} = 1 \times p + 2(1 - p)p + 3(1 - p)^2 p + \cdots = \frac{1}{p}$. In other words,

$$\alpha_k = \sum_{i=1}^{k} \frac{1}{p^i}. \tag{4.4.50}$$

A generalization of this problem, finding the mean time for a pattern, is discussed in Appendix 4.2. ◇

4.5 Impulse Functions and Random Vectors

As we have observed in Examples 2.5.23 and 3.1.34, the unit step and impulse functions are quite useful in representing the cdf and pdf of discrete and hybrid random variables. In addition, the unit step and impulse functions can be used for obtaining joint cdf's and joint pdf's expressed in several formulas depending on the condition.

Example 4.5.1 Obtain the joint cdf $F_{X,X+a}$ and joint pdf $f_{X,X+a}$ of X and $Y = X + a$ for a random variable X with pdf f_X and cdf F_X, where a is a constant.

Solution The joint cdf $F_{X,X+a}(x, y) = \mathsf{P}(X \leq x, X \leq y - a)$ of X and $Y = X + a$ can be obtained as $F_{X,X+a}(x, y) = \mathsf{P}(X \leq x)$ for $x \leq y - a$ and $F_{X,X+a}(x, y) = \mathsf{P}(X \leq y - a)$ for $x \geq y - a$, i.e.,

$$F_{X,X+a}(x, y) = F_X(\min(x, y - a))$$
$$= F_X(x)u(y - x - a) + F_X(y - a)u(x - y + a), \quad (4.5.1)$$

where it is assumed that $u(0) = \frac{1}{2}$. If we differentiate (4.5.1), we get the joint pdf $f_{X,X+a}(x, y) = \frac{\partial^2}{\partial x \partial y} F_{X,Y}(x, y) = \frac{\partial}{\partial y} \{f_X(x)u(y - x - a) - F_X(x)\delta(y - x - a) + F_X(y - a)\delta(x - y + a)\}$ of X and $Y = X + a$ as[18]

$$f_{X,X+a}(x, y) = f_X(x)\delta(y - x - a) \quad (4.5.2)$$

by noting that $\delta(t) = \delta(-t)$ as shown in (1.4.36) and that $F_X(x)\delta(y - x - a) = F_X(y - a)\delta(x - y + a)$ from $F_X(t)\delta(t - a) = F_X(a)\delta(t - a)$ as observed in (1.4.42). The result (4.5.2) can be written also as $f_{X,X+a}(x, y) = f_X(y - a)\delta(y - x - a)$. Another derivation of (4.5.2) based on $F_{X,X+a}(x, y) = F_X(\min(x, y - a))$ in (4.5.1) is discussed in Exercise 4.72.   ◇

Example 4.5.2 Obtain the joint cdf $F_{X,cX}$ and the joint pdf $f_{X,cX}$ of X and $Y = cX$ for a continuous random variable X with pdf f_X and cdf F_X, where c is a constant.

Solution For $c > 0$, the joint cdf $F_{X,cX}(x, y) = P(X \leq x, cX \leq y)$ of X and $Y = cX$ can be obtained as

$$F_{X,cX}(x, y) = \begin{cases} P(X \leq x), & x \leq \frac{y}{c}, \\ P\left(X \leq \frac{y}{c}\right), & x \geq \frac{y}{c} \end{cases}$$
$$= F_X(x)u\left(\frac{y}{c} - x\right) + F_X\left(\frac{y}{c}\right)u\left(x - \frac{y}{c}\right). \quad (4.5.3)$$

For $c < 0$, the joint cdf is $F_{X,cX}(x, y) = P\left(X \leq x, X \geq \frac{y}{c}\right)$, i.e.,

$$F_{X,cX}(x, y) = \begin{cases} 0, & x \leq \frac{y}{c}, \\ P\left(\frac{y}{c} \leq X \leq x\right), & x > \frac{y}{c} \end{cases}$$
$$= \left\{F_X(x) - F_X\left(\frac{y}{c}\right)\right\}u\left(x - \frac{y}{c}\right). \quad (4.5.4)$$

For $c = 0$, we have the joint cdf $F_{X,cX}(x, y) = P(X \leq x, cX \leq y)$ as

$$F_{X,cX}(x, y) = \begin{cases} 0, & y < 0, \\ P(X \leq x), & y \geq 0 \end{cases}$$
$$= F_X(x)u(y) \quad (4.5.5)$$

with $u(0) = 1$. Collecting (4.5.3)–(4.5.5), we eventually have

---

[18] Here, $\int_{-\infty}^{\infty} \int_{-\infty}^{\infty} f_X(x)\delta(y - x - a)dy\,dx = \int_{-\infty}^{\infty} f_X(x)dx = 1$.

$$F_{X,cX}(x, y) = \begin{cases} \{F_X(x) - F_X\left(\frac{y}{c}\right)\} u\left(x - \frac{y}{c}\right), & c < 0, \\ F_X(x)u(y), & c = 0, \\ F_X(x)u\left(\frac{y}{c} - x\right) & \\ \quad + F_X\left(\frac{y}{c}\right) u\left(x - \frac{y}{c}\right), & c > 0. \end{cases} \tag{4.5.6}$$

We now obtain the joint pdf of X and $Y = cX$ by differentiating (4.5.6). First, recollect that $F_X(x)\delta\left(x - \frac{y}{c}\right) = F_X\left(\frac{y}{c}\right)\delta\left(x - \frac{y}{c}\right)$ from $\delta(t) = \delta(-t)$ and $F_X(t)\delta(t - a) = F_X(a)\delta(t - a)$. Then, for $c < 0$, the joint pdf $f_{X,cX}(x, y) = \frac{\partial^2}{\partial x \partial y} F_{X,cX}(x, y)$ of X and $Y = cX$ can be obtained as

$$f_{X,cX}(x, y) = \frac{\partial}{\partial y}\left\{f_X(x)u\left(x - \frac{y}{c}\right) + \left\{F_X(x) - F_X\left(\frac{y}{c}\right)\right\}\delta\left(x - \frac{y}{c}\right)\right\}$$

$$= -\frac{1}{c}f_X(x)\delta\left(x - \frac{y}{c}\right). \tag{4.5.7}$$

Similarly, we get $f_{X,cX}(x, y) = \frac{\partial}{\partial y}\left\{f_X(x)u\left(\frac{y}{c} - x\right) - F_X(x)\delta\left(\frac{y}{c} - x\right) + F_X\left(\frac{y}{c}\right)\delta\left(x - \frac{y}{c}\right)\right\}$, i.e.,

$$f_{X,cX}(x, y) = \frac{1}{c}f_X(x)\delta\left(\frac{y}{c} - x\right) \tag{4.5.8}$$

for $c > 0$, and $f_{X,cX}(x, y) = \frac{\partial^2}{\partial x \partial y} F_X(x)u(y) = f_X(x)\delta(y)$ for $c = 0$: in short,

$$f_{X,cX}(x, y) = \begin{cases} f_X(x)\delta(y), & c = 0, \\ \frac{1}{|c|}f_X(x)\delta\left(x - \frac{y}{c}\right), & c \neq 0. \end{cases} \tag{4.5.9}$$

Note that $\int_{-\infty}^{\infty}\int_{-\infty}^{\infty} f_{X,cX}(x, y)dydx = \int_{-\infty}^{\infty} f_X(x)dx \int_{-\infty}^{\infty} \delta(y)dy = 1$ for $c = 0$. For $c \neq 0$, we have $\int_{-\infty}^{\infty}\int_{-\infty}^{\infty} f_{X,cX}(x, y)dydx = \frac{1}{|c|}\int_{-\infty}^{\infty} f_X(x) \int_{-\infty}^{\infty} \delta\left(x - \frac{y}{c}\right) dydx$. From $\int_{-\infty}^{\infty} \delta\left(x - \frac{y}{c}\right) dy = c \int_{\infty}^{-\infty} \delta(x - t)dt = -c$ for $c < 0$, and $\int_{-\infty}^{\infty} \delta\left(x - \frac{y}{c}\right) dy = c \int_{-\infty}^{\infty} \delta(x - t)dt = c$ for $c > 0$, we have $\int_{-\infty}^{\infty} \delta\left(x - \frac{y}{c}\right) dy = |c|$. Therefore, $\int_{-\infty}^{\infty}\int_{-\infty}^{\infty} f_{X,cX}(x, y)dydx = 1$. ◇

Letting $a = 0$ in Example 4.5.1, $c = 1$ in Example 4.5.2, or $a = 0$ and $c = 1$ in Exercise 4.68, we get

$$f_{X,X}(x, y) = f_X(x)\delta(x - y) \tag{4.5.10}$$

or $f_{X,X}(x, y) = f_X(x)\delta(y - x) = f_X(y)\delta(x - y) = f_X(y)\delta(y - x)$. Let us now consider the joint distribution of a random variable and its absolute value.

Example 4.5.3 Obtain the joint cdf $F_{X,|X|}$ and the joint pdf $f_{X,|X|}$ of X and $Y = |X|$ for a continuous random variable X with pdf f_X and cdf F_X.

Solution First, the joint cdf $F_{X,|X|}(x, y) = P(X \le x, |X| \le y)$ of X and $Y = |X|$ can be obtained as (Bae et al. 2006)

$$F_{X,|X|}(x, y) = \begin{cases} P(-y \le X \le x), & -y < x < y, \ y > 0, \\ P(-y \le X \le y), & x > y, \ y > 0, \\ 0, & \text{otherwise} \end{cases}$$

$$= u(y + x)u(y - x)G_1(x, y) + u(y)u(x - y)G_1(y, y),$$

(4.5.11)

where

$$G_1(x, y) = F_X(x) - F_X(-y) \tag{4.5.12}$$

satisfies $G_1(x, -x) = 0$. We have used $u(y + x)u(y - x)u(y) = u(y - x)u(y + x)$ in obtaining (4.5.11). Note that $F_{X,|X|}(x, y) = 0$ for $y < 0$ because $u(x + y)u(y - x) = 0$ from $(x + y)(y - x) = y^2 - x^2 < 0$.

Noting that

$$\delta(y - a)f_X(y) = \delta(y - a)f_X(a), \tag{4.5.13}$$

$G_1(x, -x) = 0$, $\delta(x - y) = \delta(y - x)$, and $u(\alpha x) = u(x)$ for $\alpha > 0$, we get

$$\frac{\partial}{\partial x}F_{X,|X|}(x, y) = \delta(x + y)u(y - x)G_1(x, y) - u(y + x)\delta(y - x)G_1(x, y)$$

$$+ u(x + y)u(y - x)f_X(x) + u(y)\delta(x - y)G_1(y, y)$$

$$= \delta(x + y)u(-2x)G_1(x, -x) - u(2x)\delta(y - x)G_1(x, x)$$

$$+ u(x + y)u(y - x)f_X(x) + u(x)\delta(x - y)G_1(x, x)$$

$$= u(x + y)u(y - x)f_X(x) \tag{4.5.14}$$

by differentiating (4.5.11) with respect to x. Consequently, the joint pdf $f_{X,|X|}(x, y) = \frac{\partial^2}{\partial x \partial y}F_{X,|X|}(x, y)$ of X and $Y = |X|$ is

$$f_{X,|X|}(x, y) = \{\delta(x + y)u(y - x) + u(x + y)\delta(y - x)\}f_X(x)$$

$$= \{\delta(x + y)u(-x) + \delta(x - y)u(x)\}f_X(x)$$

$$= \{\delta(x + y)f_X(-y) + \delta(x - y)f_X(y)\}u(y). \tag{4.5.15}$$

Obtaining the pdf $f_{|X|}(y) = \int_{-\infty}^{\infty} f_{X,|X|}(x, y)dx$ of $|X|$ from (4.5.15), we get

$$f_{|X|}(y) = \int_{-\infty}^{\infty} \{\delta(x + y)f_X(-y) + \delta(x - y)f_X(y)\}u(y)dx$$

$$= \{f_X(-y) + f_X(y)\}u(y), \tag{4.5.16}$$

which is equivalent to that obtained in Example 3.2.7. From $u(x) + u(-x) = 1$, we also have $\int_{-\infty}^{\infty} f_{X,|X|}(x, y)dy = f_X(x)u(-x) + f_X(x)u(x) = f_X(x)$.   ◇

Example 4.5.4 Using (4.5.16), it is easy to see that $Y = |X| \sim U(0, 1)$ when $X \sim U(-1, 1)$, $X \sim U(-1, 0)$, or $X \sim U[0, 1)$.   ◇

When the Jacobian $J(g(x))$ is 0, Theorem 4.2.1 is not applicable as mentioned in the paragraph just before Sect. 4.2.1.2. We have $J(g(x)) = 0$ if a function $g = (g_1, g_2, \ldots, g_n)$ in the n-dimensional space is in the form, for example, of

$$g_j(x) = c_j g_1(x) + a_j \qquad (4.5.17)$$

for $j = 2, 3, \ldots, n$, where $\{a_j\}_{j=2}^n$ and $\{c_j\}_{j=2}^n$ are all constants. Let us discuss how we can obtain the pdf of $(Y_1, Y_2) = g(X) = (g_1(X), g_2(X))$ in the two dimensional case, where

$$g_2(x) = c g_1(x) + a \qquad (4.5.18)$$

with a and c constants. First, we choose an appropriate auxiliary variable, for example, $Z = X_2$ or $Z = X_1$ when $g_1(X)$ is not or is, respectively, in the form of $dX_2 + b$. Then, using Theorem 4.2.1, we obtain the joint pdf $f_{Y_1, Z}$ of (Y_1, Z). We next obtain the pdf of Y_1 as

$$f_{Y_1}(x) = \int_{-\infty}^{\infty} f_{Y_1, Z}(x, v) dv \qquad (4.5.19)$$

from $f_{Y_1, Z}$. Subsequently, we get the joint pdf

$$f_{Y_1, Y_2}(x, y) = \begin{cases} f_{Y_1}(x)\delta(y - a), & c = 0, \\ \frac{1}{|c|} f_{Y_1}(x)\delta\left(x - \frac{y-a}{c}\right), & c \neq 0 \end{cases} \qquad (4.5.20)$$

of $(Y_1, Y_2) = (g_1(X), c g_1(X) + a)$ using (4.E.18).

Example 4.5.5 Obtain the joint pdf of $(Y_1, Y_2) = (X_1 - X_2, 2X_1 - 2X_2)$ and the joint pdf of $(Y_1, Y_3) = (X_1 - X_2, X_1 - X_2 + 2)$ when the joint pdf of $X = (X_1, X_2)$ is $f_X(x_1, x_2) = u(x_1) u(1 - x_1) u(x_2) u(1 - x_2)$.

Solution Because $g_1(X) = Y_1 = X_1 - X_2$ is not in the form of $dX_2 + b$, we choose the auxiliary variable $Z = X_2$. Let us then obtain the joint pdf of $(Y_1, Z) = (X_1 - X_2, X_2)$. Noting that $X_1 = Y_1 + Z$, $X_2 = Z$, and the Jacobian

$$J\left(\frac{\partial(x_1 - x_2, x_2)}{\partial(x_1, x_2)}\right) = \begin{vmatrix} 1 & -1 \\ 0 & 1 \end{vmatrix} = 1,$$ the joint pdf of (Y_1, Z) is

$$\begin{aligned} f_{Y_1, Z}(y, z) &= f_X(x_1, x_2)|_{x_1 = y + z, \, x_2 = z} \\ &= u(y + z)u(1 - y - z)u(z)u(1 - z). \end{aligned} \qquad (4.5.21)$$

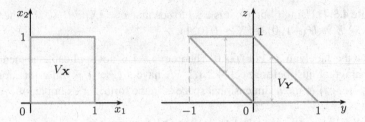

Fig. 4.17 The support V_Y of $f_{X_1-X_2,X_2}(y, z)$ when the support of $f_X(x_1, x_2)$ is V_X. The intervals of integrations in the cases $-1 < y < 0$ and $0 < y < 1$ are also represented as lines with two arrows

We can then obtain the pdf $f_{Y_1}(y) = \int_{-\infty}^{\infty} u(y+z)u(1 - y - z)u(z)u(1 - z)dz$ of $Y_1 = g_1(X) = X_1 - X_2$ as

$$
f_{Y_1}(y) = \begin{cases} 0, & |y| > 1, \\ \int_{-y}^{1} dz, & -1 < y < 0, \\ \int_{0}^{1-y} dz, & 0 < y < 1 \end{cases}
$$

$$
= \begin{cases} 0, & |y| > 1, \\ 1 + y, & -1 < y < 0, \\ 1 - y, & 0 < y < 1 \end{cases}
$$

$$
= (1 - |y|)u(1 - |y|) \tag{4.5.22}
$$

by integrating $f_{Y_1, Z}(y, z)$, for which Fig. 4.17 is useful in identifying the integration intervals.

Next, using (4.5.20), we get the joint pdf

$$
f_{Y_1, Y_2}(x, y) = \frac{1}{2}(1 - |x|)u(1 - |x|)\delta\left(x - \frac{y}{2}\right) \tag{4.5.23}
$$

of $(Y_1, Y_2) = (X_1 - X_2, 2X_1 - 2X_2)$ and the joint pdf

$$
f_{Y_1, Y_3}(x, y) = (1 - |x|)u(1 - |x|)\delta(x - y + 2) \tag{4.5.24}
$$

of $(Y_1, Y_3) = (X_1 - X_2, X_1 - X_2 + 2)$. Note that $\int_{-\infty}^{\infty} \delta\left(x - \frac{y}{2}\right) dy = \int_{\infty}^{-\infty} \delta(t)(-2dt) = 2\int_{-\infty}^{\infty} \delta(t)dt = 2$.   ◇

We have briefly discussed how we can obtain the joint pdf and joint cdf in some special cases by employing the unit step and impulse functions. This approach is also quite fruitful in dealing with the order statistics and rank statistics.

Appendices

Appendix 4.1 Multinomial Random Variables

Let us discuss in more detail the multinomial random variables introduced in Example 4.1.4.

Definition 4.A.1 (*multinomial distribution*) Assume n repetitions of an independent experiment of which the outcomes are a collection $\{A_i\}_{i=1}^{r}$ of disjoint events with probability $\{P(A_i) = p_i\}_{i=1}^{r}$, where $\sum_{i=1}^{r} p_i = 1$. Denote by X_i the number of occurrences of event A_i. Then, the joint distribution of $X = (X_1, X_2, \ldots, X_r)$ is called the multinomial distribution, and the joint pmf of X is

$$p_X(k_1, k_2, \ldots, k_r) = \frac{n!}{k_1! k_2! \cdots k_r!} p_1^{k_1} p_2^{k_2} \cdots p_r^{k_r} \qquad (4.A.1)$$

for $\{k_i \in \{0, 1, \ldots, n\}\}_{i=1}^{r}$ and $\sum_{i=1}^{r} k_i = n$.

The right-hand side of (4.A.1) is the coefficient of $\prod_{j=1}^{r} t_j^{k_j}$ in the multinomial expansion of $(p_1 t_1 + p_2 t_2 + \cdots + p_r t_r)^n$.

Example 4.A.1 In a repetition of rolling of a fair die ten times, let $\{X_i\}_{i=1}^{3}$ be the numbers of $A_1 = \{1\}$, $A_2 = \{$an even number$\}$, and $A_3 = \{3, 5\}$, respectively. Then, the joint pmf of $X = (X_1, X_2, X_3)$ is

$$p_X(k_1, k_2, k_3) = \frac{10!}{k_1! k_2! k_3!} \left(\frac{1}{6}\right)^{k_1} \left(\frac{1}{2}\right)^{k_2} \left(\frac{1}{3}\right)^{k_3} \qquad (4.A.2)$$

for $\{k_i \in \{0, 1, \ldots, 10\}\}_{i=1}^{3}$ such that $\sum_{i=1}^{3} k_i = 10$. Based on this pmf, the probability of the event $\{$three times of A_1, six times of $A_2\} = \{X_1 = 3, X_2 = 6, X_3 = 1\}$ can be obtained as $p_X(3, 6, 1) = \frac{10!}{3! 6! 1!} \left(\frac{1}{6}\right)^3 \left(\frac{1}{2}\right)^6 \left(\frac{1}{3}\right)^1 = \frac{35}{1728} \approx 2.025 \times 10^{-2}$. $\diamond$

Example 4.A.2 As in the binomial distribution, let us consider the approximation of the multinomial distribution in terms of the Poisson distribution. For $p_i \to 0$ and $np_i \to \lambda_i$ when $n \to \infty$, we have $\frac{n!}{k_r!} = \frac{n!}{\left(n - \sum_{i=1}^{r-1} k_i\right)!} =$

$$n(n-1)\cdots\left(n - \sum_{i=1}^{r-1} k_i + 1\right) \approx n^{\sum_{i=1}^{r-1} k_i}, \quad p_r = 1 - \sum_{i=1}^{r-1} p_i \approx \exp\left(-\sum_{i=1}^{r-1} p_i\right), \quad \text{and}$$

$$p_r^{k_r} = p_r^{n - \sum_{i=1}^{r-1} k_i} \approx \exp\left\{-\left(n - \sum_{i=1}^{r-1} k_i\right) \sum_{i=1}^{r-1} p_i\right\} \approx \exp\left(-n \sum_{i=1}^{r-1} p_i\right) = \exp(-\bar{\lambda}),$$

where $\bar{\lambda} = \sum_{i=1}^{r-1} \lambda_i$. Based on these results, we can show that $p_X(k_1, k_2, \ldots, k_r) = \frac{n!}{k_1!k_2!\cdots k_r!} p_1^{k_1} p_2^{k_2} \cdots p_r^{k_r} \rightarrow \frac{n^{k_1} n^{k_2} \cdots n^{k_{r-1}}}{k_1!k_2!\cdots k_{r-1}!} p_1^{k_1} p_2^{k_2} \cdots p_{r-1}^{k_{r-1}} \exp(-\bar{\lambda})$, i.e.,

$$p_X(k_1, k_2, \ldots, k_r) \rightarrow \prod_{i=1}^{r-1} e^{-\lambda_i} \frac{\lambda_i^{k_i}}{k_i!}. \tag{4.A.3}$$

The result (4.A.3) with $r = 2$ is clearly the same as (3.5.19) obtained in Theorem 3.5.2 for the binomial distribution. $\diamond$

Example 4.A.3 For $k_i = np_i + O(\sqrt{n})$ and $n \rightarrow \infty$, the multinomial pmf can be approximated as

$$p_X(k_1, k_2, \ldots, k_r) = \frac{n!}{k_1!k_2!\cdots k_r!} p_1^{k_1} p_2^{k_2} \cdots p_r^{k_r}$$

$$\approx \frac{1}{\sqrt{(2\pi n)^{r-1} p_1 p_2 \cdots p_r}} \exp\left\{ -\frac{1}{2} \sum_{i=1}^{r} \frac{(k_i - np_i)^2}{np_i} \right\}. \tag{4.A.4}$$

Consider the case of $r = 2$ in (4.A.4). Letting $k_1 = k$, $k_2 = n - k$, $p_1 = p$, and $p_2 = 1 - p_1 = q$, we get $p_X(k_1, k_2) = p_X(k, n - k)$ as

$$p_X(k_1, k_2) \approx \frac{1}{\sqrt{2\pi npq}} \exp\left[-\frac{1}{2} \left\{ \frac{(k - np)^2}{np} + \frac{(n - k - nq)^2}{nq} \right\} \right]$$

$$= \frac{1}{\sqrt{2\pi npq}} \exp\left\{ -\frac{1}{2} \frac{q(k - np)^2 + p(np - k)^2}{npq} \right\}$$

$$= \frac{1}{\sqrt{2\pi npq}} \exp\left\{ -\frac{(k - np)^2}{2npq} \right\}, \tag{4.A.5}$$

which is the same as (3.5.16) of Theorem 3.5.1 for the binomial distribution. $\diamond$

The multinomial distribution (Johnson and Kotz 1972) is a generalization of the binomial distribution, and the special case $r = 2$ of the multinomial distribution is the binomial distribution. For the multinomial random vector $X = (X_1, X_2, \ldots, X_r)$, the marginal distribution of X_i is a binomial distribution. In addition, assuming $n - \sum_{i=1}^{s} X_{a_i}$ as the $(s + 1)$-st random variable, the distribution of the subvector $X_s = (X_{a_1}, X_{a_2}, \ldots, X_{a_s})$ of X is also a multinomial distribution with the joint pmf

$$p_{X_s}(k_{a_1}, k_{a_2}, \ldots, k_{a_s}) = n! \frac{\left(1 - \sum_{i=1}^{s} p_{a_i} \right)^{n - \sum_{i=1}^{s} k_{a_i}}}{\left(n - \sum_{i=1}^{s} k_{a_i} \right)!} \prod_{i=1}^{s} \frac{p_{a_i}^{k_{a_i}}}{k_{a_i}!}. \tag{4.A.6}$$

Letting $s = 1$ in (4.A.6), we get the binomial pmf $p_X(k_a) = n! \frac{(1-p_a)^{n-k_a}}{(n-k_a)!} \frac{p_a^{k_a}}{k_a!}$.

In addition, when a subvector of X is given, the conditional joint distribution of the random vector of the remaining random variables is also a multinomial distribution, which depends not on the individual remaining random variables but on the sum of the remaining random variables. For example, assume $X = (X_1, X_2, X_3, X_4)$. Then, the joint distribution of (X_2, X_4) when (X_1, X_3) is given is a multinomial distribution, which depends not on X_1 and X_3 individually but on the sum $X_1 + X_3$.

Finally, when $X = (X_1, X_2, \ldots, X_r)$ has the pmf (4.A.1), it is known that

$$
\mathsf{E}\left\{ X_i | X_{b_1}, X_{b_2}, \ldots, X_{b_{r-1}} \right\} = \frac{p_i}{1 - \sum\limits_{j=1}^{r-1} p_{b_j}} \left(n - \sum_{j=1}^{r-1} X_{b_j} \right), \quad (4.A.7)
$$

where i is not equal to any of $\{b_j\}_{j=1}^{r-1}$. It is also known that we have the conditional expected value

$$
\mathsf{E}\left\{ X_i | X_j \right\} = \frac{(n - X_j) p_i}{1 - p_j}, \quad (4.A.8)
$$

the correlation coefficient

$$
\rho_{X_i, X_j} = -\sqrt{\frac{p_i p_j}{(1 - p_i)(1 - p_j)}}, \quad (4.A.9)
$$

and $\mathsf{Cov}(X_i, X_j) = -n p_i p_j$ for X_i and X_j with $i \neq j$.

Appendix 4.2 Mean Time to Pattern

Denote by X_k the outcome of the k-th trial of an experiment with the pmf

$$
p_{X_k}(j) = p_j \quad (4.A.10)
$$

for $j = 1, 2, \ldots$, where $\sum\limits_{j=1}^{\infty} p_j = 1$. The number of trials of the experiment until a pattern $M = (i_1, i_2, \ldots, i_n)$ is observed for the first time is called the time to pattern M, which is denoted by $T = T(M) = T(i_1, i_2, \ldots, i_n)$. For example, when the sequence of the outcomes is $(6, 4, 9, 5, 5, 9, 5, 7, 3, 2, \ldots)$, the time to pattern $(9, 5, 7)$ is $T(9, 5, 7) = 8$. Now, let us obtain (Nielsen 1973) the mean time $\mathsf{E}\{T(M)\}$ for the pattern M.

First, when M satisfies

$$(i_1, i_2, \ldots, i_k) = (i_{n-k+1}, i_{n-k+2}, \ldots, i_n) \tag{4.A.11}$$

for $n = 2, 3, \ldots$ and $k = 1, 2, \ldots, n-1$, the pattern M overlaps and $L_k = (i_1, i_2, \ldots, i_k)$ is an overlapping piece or a bifix of M. For instance, (A, B, C), (D, E, F, G), (S, S, P), (4, 4, 5), and (4, 1, 3, 3, 2) are non-overlapping patterns; and (A, B, G, A, B), (9, 9, 2, 4, 9, 9), (3, 4, 3), (5, 4, 5, 4, 5), and (5, 4, 5, 4, 5, 4) are overlapping patterns. Note that the length k of an overlapping piece can be longer than $\frac{n}{2}$ and that more than one overlapping pieces may exist in a pattern as in (5, 4, 5, 4, 5) and (5, 4, 5, 4, 5, 4). In addition, when the overlapping piece is of length k, the elements in the pattern are the same at every other $n - k - 1$: for instance, $M = (i_1, i_1, \ldots, i_1)$ when $k = n - 1$. A non-overlapping pattern can be regarded as an overlapping pattern with $k = n$.

(A) A Recursive Method

First, the mean time $\mathsf{E}\{T(i_1)\} = \sum_{k=1}^{\infty} k\mathsf{P}(T(i_1) = k) = \sum_{k=1}^{\infty} k\left(1 - p_{i_1}\right)^{k-1} p_{i_1}$ to pattern i_1 of length 1 is

$$\mathsf{E}\{T(i_1)\} = \frac{1}{p_{i_1}}. \tag{4.A.12}$$

When M has J overlapping pieces, let the lengths of the overlapping pieces be $K_0 < K_1 < \cdots < K_J < K_{J+1}$ with $K_0 = 0$ and $K_{J+1} = n$, and express M as $M = \left(i_1, i_2, \ldots, i_{K_1}, i_{K_1+1}, \ldots, i_{K_2}, i_{K_2+1}, \ldots, i_{K_J}, i_{K_J+1}, \ldots, i_{n-1}, i_n\right)$. If we write the overlapping pieces $\left\{L_{K_j} = \left(i_1, i_2, \ldots, i_{K_j}\right)\right\}_{j=1}^{J}$ as

$$
\begin{array}{ccccc}
 & & i_1 & i_2 & \cdots & i_{K_1} \\
i_1 & i_2 & \cdots i_{K_{2,1}+1} & i_{K_{2,1}+2} & \cdots & i_{K_2} \\
\vdots & \vdots & \vdots & \vdots & & \vdots \\
i_1\, i_2 & \cdots\, i_{K_{J,2}+1}\, i_{K_{J,2}+2} & \cdots\, i_{K_{J,1}+1} & i_{K_{J,1}+2} & \cdots & i_{K_J},
\end{array}
\tag{4.A.13}
$$

where $K_{\alpha,\beta} = K_\alpha - K_\beta$, then we have $i_m = i_{K_{b,a}+m}$ for $1 \le a \le b \le J$ and $m = 1, 2, \ldots, K_a - K_{a-1}$ because the values at the same column in (4.A.13) are all the same.

Denote by $T(A_1)$ the time to wait until the occurrence of $M_{+1} = (i_1, i_2, \ldots, i_n, i_{n+1})$ after the occurrence of M. Then, we have

$$\mathsf{E}\{T(M_{+1})\} = \mathsf{E}\{T(M)\} + \mathsf{E}\{T(A_1)\} \tag{4.A.14}$$

because $T(M_{+1}) = T(M) + T(A_1)$. Here, we can express $\mathsf{E}\{T(A_1)\}$ as

$$\mathsf{E}\{T(A_1)\} = \sum_{x=1}^{\infty} \mathsf{E}\{T(A_1)|X_{n+1} = x\}\,\mathsf{P}(X_{n+1} = x). \qquad (4.A.15)$$

Let us focus on the term $\mathsf{E}\{T(A_1)|X_{n+1} = x\}$ in (4.A.15). First, when $x = i_{K_j+1}$ for example, denote by $\tilde{L}_{K_j+1} = (i_1, i_2, \ldots, i_{K_j}, i_{K_j+1})$ the j-th overlapping piece with its immediate next element, and recollect (4.A.11). Then, we have

$$M_{+1} = (i_1, i_2, \ldots, i_{n-K_j}, i_{n-K_j+1}, i_{n-K_j+2}, \ldots, i_n, i_{K_j+1})$$

$$= (i_1, i_2, \ldots, i_{n-K_j}, i_1, i_2, \ldots, i_{K_j}, i_{K_j+1}), \qquad (4.A.16)$$

from which we can get $\mathsf{E}\left\{T(A_1)\,\middle|\,X_{n+1} = i_{K_j+1}\right\} = 1 + \mathsf{E}\left\{T(M_{+1})\,\middle|\,\tilde{L}_{K_j+1}\right\}$. We can similarly get

$$\mathsf{E}\{T(A_1)|X_{n+1} = x\} = \begin{cases} 1 + \mathsf{E}\left\{T(M_{+1})\,\middle|\,\tilde{L}_{K_0+1}\right\}, & x = i_{K_0+1}, \\ 1 + \mathsf{E}\left\{T(M_{+1})\,\middle|\,\tilde{L}_{K_1+1}\right\}, & x = i_{K_1+1}, \\ \quad\vdots & \quad\vdots \\ 1 + \mathsf{E}\left\{T(M_{+1})\,\middle|\,\tilde{L}_{K_J+1}\right\}, & x = i_{K_J+1}, \\ 1, & x = i_{n+1}, \\ 1 + \mathsf{E}\{T(M_{+1})\}, & \text{otherwise} \end{cases} \qquad (4.A.17)$$

when $i_1 = i_{K_0+1}, i_{K_1+1}, \ldots, i_{K_J+1}, i_{K_{J+1}+1} = i_{n+1}$ are all distinct. Here, recollecting

$$\mathsf{E}\left\{T(M_{+1})\,\middle|\,\tilde{L}_{K_j+1}\right\} = \mathsf{E}\{T(M_{+1})\} - \mathsf{E}\left\{T\left(\tilde{L}_{K_j+1}\right)\right\} \qquad (4.A.18)$$

from $\mathsf{E}\{T(M_{+1})\} = \mathsf{E}\left\{T(M_{+1})\,\middle|\,\tilde{L}_{K_j+1}\right\} + \mathsf{E}\left\{T\left(\tilde{L}_{K_j+1}\right)\right\}$, we get

$$\mathsf{E}\{T(M_{+1})\} = \mathsf{E}\{T(M)\}$$

$$+ \sum_{j=0}^{J} p_{i_{K_j+1}}\left[1 + \mathsf{E}\{T(M_{+1})\} - \mathsf{E}\left\{T\left(\tilde{L}_{K_j+1}\right)\right\}\right]$$

$$+ p_{i_{n+1}} \times 1 + \left(1 - p_{i_{n+1}} - \sum_{j=0}^{J} p_{i_{K_j+1}}\right)$$

$$\times \left[1 + \mathsf{E}\{T(M_{+1})\}\right] \qquad (4.A.19)$$

from (4.A.14), (4.A.15), and (4.A.17). We can rewrite (4.A.19) as

$$p_{i_{n+1}}\mathsf{E}\{T(M_{+1})\} = \mathsf{E}\{T(M)\} + 1 - \sum_{j=0}^{J} p_{i_{K_j+1}}\mathsf{E}\left\{T\left(\tilde{L}_{K_j+1}\right)\right\} \qquad (4.A.20)$$

after some steps.

Let us next consider the case in which some are the same among i_{K_0+1}, i_{K_1+1}, ..., i_{K_J+1}, and i_{n+1}. For example, assume $a < b$ and $i_{K_a+1} = i_{K_b+1}$. Then, for $x = i_{K_a+1} = i_{K_b+1}$ in (4.A.15) and (4.A.17), the line '$1 + \mathsf{E}\left\{T(M_{+1}) \middle| \tilde{L}_{K_a+1}\right\}, x = i_{K_a+1}$' corresponding to the K_a-th piece among the lines of (4.A.17) will disappear because the longest overlapping piece in the last part of M_{+1} is not $\tilde{L}_{K_a+1}$ but $\tilde{L}_{K_b+1}$. Based on this fact, if we follow steps similar to those leading to (4.A.19) and (4.A.20), we get

$$p_{i_{n+1}} \mathsf{E}\{T(M_{+1})\} = \mathsf{E}\{T(M)\} + 1 - \sum_j p_{i_{K_j+1}} \mathsf{E}\left\{T\left(\tilde{L}_{K_j+1}\right)\right\}, \quad (4.A.21)$$

where $\sum_j$ denotes the sum from $j = 0$ to J letting all $\mathsf{E}\left\{T\left(\tilde{L}_{K_a+1}\right)\right\}$ to 0 when $i_{K_a+1} = i_{K_b+1}$ for $0 \le a < b \le J + 1$. Note here that $\{K_j\}_{j=1}^J$ are the lengths of the overlapping pieces of M, not of M_{+1}. Note also that (4.A.20) is a special case of (4.A.21): in other words, (4.A.21) is always applicable.

In essence, starting from $\mathsf{E}\{T(i_1)\} = \frac{1}{p_{i_1}}$ shown in (4.A.12), we can successively obtain $\mathsf{E}\{T(i_1, i_2)\}$, $\mathsf{E}\{T(i_1, i_2, i_3)\}$, ..., $\mathsf{E}\{T(M)\}$ based on (4.A.21).

Example 4.A.4 For an i.i.d. random variables $\{X_k\}_{k=1}^\infty$ with the marginal pmf $p_{X_k}(j) = p_j$, obtain the mean time to $M = (5, 4, 5, 3)$.

Solution First, $\mathsf{E}\{T(5)\} = \frac{1}{p_5}$. When $(5, 4)$ is M_{+1}, because $J = 0$, $i_{K_0+1} = 5$, and $i_{n+1} = 4$, we get

$$\sum_j p_{i_{K_j+1}} \mathsf{E}\left\{T\left(\tilde{L}_{K_j+1}\right)\right\} = p_{i_1} \mathsf{E}\left\{T\left(\tilde{L}_1\right)\right\}, \quad (4.A.22)$$

i.e., $\mathsf{E}\{T(5, 4)\} = \frac{1}{p_4}\left[\mathsf{E}\{T(5)\} + 1 - p_5 \mathsf{E}\{T(5)\}\right] = \frac{1}{p_4 p_5}$. Next, when $(5, 4, 5)$ is M_{+1}, because $J = 0$ and $i_{K_0+1} = 5 = i_{n+1}$, we get $\sum_j p_{i_{K_j+1}} \mathsf{E}\left\{T\left(\tilde{L}_{K_j+1}\right)\right\} = 0$.

Thus, $\mathsf{E}\{T(5, 4, 5)\} = \frac{1}{p_5}\left[\mathsf{E}\{T(5, 4)\} + 1\right] = \frac{1}{p_4 p_5^2} + \frac{1}{p_5}$. Finally, when $(5, 4, 5, 3)$ is M_{+1}, because $J = 1$ and $K_1 = 1$, we have $i_{K_0+1} = i_1 = 5$, $i_{K_1+1} = i_2 = 4$, $i_{K_{J+1}+1} = i_4 = 3$, and

$$\sum_j p_{i_{K_j+1}} \mathsf{E}\left\{T\left(\tilde{L}_{K_j+1}\right)\right\} = p_5 \mathsf{E}\{T(5)\} + p_4 \mathsf{E}\{T(5, 4)\}, \quad (4.A.23)$$

i.e., $\mathsf{E}\{T(5, 4, 5, 3)\} = \frac{1}{p_3}\left[\mathsf{E}\{T(5, 4, 5)\} + 1 - p_5 \mathsf{E}\{T(5)\} - p_4 \mathsf{E}\{T(5, 4)\}\right] = \frac{1}{p_3 p_4 p_5^2}$.   ◇

(B) An Efficient Method

The result (4.A.21) is applicable always. However, as we have observed in Example 4.A.4, (4.A.21) possesses some inefficiency in the sense that we have to first obtain the expected values $\mathsf{E}\{(i_1)\}$, $\mathsf{E}\{(i_1, i_2)\}$, ..., $\mathsf{E}\{(i_1, i_2, \ldots, i_{n-1})\}$ before we can obtain the expected value $\mathsf{E}\{(i_1, i_2, \ldots, i_n)\}$. Let us now consider a more efficient method.

(B-1) Non-overlapping Patterns

When the pattern M is non-overlapping, we have

$$(i_1, i_2, \ldots, i_k) \neq (i_{n-k+1}, i_{n-k+2}, \ldots, i_n) \tag{4.A.24}$$

for every $k \in \{1, 2, \ldots, n-1\}$. Based on this observation, let us show that

$$\{T = j + n\} \rightleftarrows \{T > j, (X_{j+1}, X_{j+2}, \ldots, X_{j+n}) = M\}. \tag{4.A.25}$$

First, when $T = j + n$, the first occurrence of M is $(X_{j+1}, X_{j+2}, \ldots, X_{j+n})$, which implies that $T > j$ and

$$(X_{j+1}, X_{j+2}, \ldots, X_{j+n}) = M. \tag{4.A.26}$$

Next, let us show that $T = j + n$ when $T > j$ and (4.A.26) holds true. If $k \in \{1, 2, \ldots, n-1\}$ and $T = j + k$, then we have $X_{j+k} = i_n, X_{j+k-1} = i_{n-1}, \ldots, X_{j+1} = i_{n-k+1}$. This is a contradiction to $(X_{j+1}, X_{j+2}, \ldots, X_{j+n}) = (i_1, i_2, \ldots, i_k) \neq (i_{n-k+1}, i_{n-k+2}, \ldots, i_n)$ implied by (4.A.24) and (4.A.26). In short, for any value k in $\{1, 2, \ldots, n-1\}$, we have $T \neq j + k$ and thus $T \geq j + n$. Meanwhile, (4.A.26) implies $T \leq j + n$. Thus, we get $T = j + n$.

From (4.A.25), we have

$$\mathsf{P}(T = j + n) = \mathsf{P}(T > j, (X_{j+1}, X_{j+2}, \ldots, X_{j+n}) = M). \tag{4.A.27}$$

Here, the event $T > j$ is dependent only on $X_1, X_2, \ldots, X_j$ but not on $X_{j+1}, X_{j+2}, \ldots, X_{j+n}$, and thus

$$\mathsf{P}(T = j + n) = \mathsf{P}(T > j) \mathsf{P}\left((X_{j+1}, X_{j+2}, \ldots, X_{j+n}) = M\right)$$
$$= \mathsf{P}(T > j) \hat{p}, \tag{4.A.28}$$

where $\hat{p} = p_{i_1} p_{i_2} \cdots p_{i_n}$. Now, recollecting that $\sum_{j=0}^{\infty} \mathsf{P}(T = j + n) = 1$ and that

$$\sum_{j=0}^{\infty} \mathsf{P}(T > j) = \mathsf{P}(T > 0) + \mathsf{P}(T > 1) + \cdots = \left[\mathsf{P}(T = 1) + \mathsf{P}(T = 2) + \cdots\right] +$$

$$\left[P(T = 2) + P(T = 3) + \cdots \right] + \cdots = \sum_{j=0}^{\infty} j \, P(T = j), \text{ i.e.,}$$

$$\sum_{j=0}^{\infty} P(T > j) = \mathsf{E}\{T\}, \tag{4.A.29}$$

we get $\hat{p} \sum\limits_{j=0}^{\infty} P(T > j) = \hat{p} \mathsf{E}\{T\} = 1$ from (4.A.28). Thus, we have $\mathsf{E}\{T(M)\} = \frac{1}{\hat{p}}$, i.e.,

$$\mathsf{E}\{T(M)\} = \frac{1}{p_{i_1} p_{i_2} \cdots p_{i_n}}. \tag{4.A.30}$$

Example 4.A.5 For the pattern $M = (9, 5, 7)$, we have $\mathsf{E}\{T(9, 5, 7)\} = \frac{1}{p_5 p_7 p_9}$. Thus, to observe the pattern $(9, 5, 7)$, we have to wait on the average until the $\frac{1}{p_5 p_7 p_9}$-th repetition. In tossing a fair die, we need to repeat $\mathsf{E}\{T(3, 5)\} = \frac{1}{p_3 p_5} = 36$ times on the average to observe the pattern $(3, 5)$ for the first time. $\diamond$

(B-2) Overlapping Patterns

We next consider overlapping patterns. When M is an overlapping pattern, construct a non-overlapping pattern

$$M_x = (i_1, i_2, \ldots, i_n, x) \tag{4.A.31}$$

of length $n + 1$ by appropriately[19] choosing x as $x \notin \{i_1, i_2, \ldots, i_n\}$ or $x \notin \{i_{K_0+1}, i_{K_1+1}, \ldots, i_{K_J+1}\}$. Then, from (4.A.30), we have

$$\mathsf{E}\{T(M_x)\} = \frac{1}{p_x \hat{p}}. \tag{4.A.32}$$

When $x = i_{n+1}$, using (4.A.32) in (4.A.21), we get

$$\mathsf{E}\{T(M)\} = \frac{1}{\hat{p}} - 1 + \sum_j p_{i_{K_j+1}} \mathsf{E}\left\{ T\left(\tilde{L}_{K_j+1} \right) \right\} \tag{4.A.33}$$

by noting that M_x in (4.A.31) and M_{+1} in (4.A.21) are the same. Now, if we consider the case in which M is not an overlapping pattern, the last

---

[19] More generally, we can interpret 'appropriate x' as 'any x such that $(i_1, i_2, \ldots, i_n, x)$ is a non-overlapping pattern', and x can be chosen even if it is not a realization of any X_k. For example, when $p_1 + p_2 + p_3 = 1$, we could choose $x = 7$.

term of (4.A.33) becomes $\sum_j p_{i_{K_j+1}} \mathsf{E}\left\{T\left(\tilde{L}_{K_j+1}\right)\right\} = p_{i_{K_0+1}} \mathsf{E}\left\{T\left(\tilde{L}_{K_0+1}\right)\right\} = p_{i_1} \mathsf{E}\{T(i_1)\} = 1$. Consequently, (4.A.33) and (4.A.30) are the same. Thus, for any overlapping or non-overlapping pattern M, we can use (4.A.33) to obtain $\mathsf{E}\{T(M)\}$.

Example 4.A.6 In the pattern $(9, 5, 1, 9, 5)$, we have $J = 1$, $K_1 = 2$, and $\tilde{L}_{K_1+1} = (9, 5, 1)$. Thus, from (4.A.30) and (4.A.33), we get $\mathsf{E}\{T(9, 5, 1, 9, 5)\} = \frac{1}{p_1 p_5^2 p_9^2} - 1 + \left[p_9 \mathsf{E}\{T(9)\} + p_1 \mathsf{E}\{T(9, 5, 1)\}\right] = \frac{1}{p_1 p_5^2 p_9^2} + \frac{1}{p_5 p_9}$. Similarly, in the pattern $(9, 5, 9, 1, 9, 5, 9)$, we get $J = 2$, $K_1 = 1$, $K_2 = 3$, and $\tilde{L}_{K_1+1} = (9, 5)$ and $\tilde{L}_{K_2+1} = (9, 5, 9, 1)$. Therefore,

$$
\begin{aligned}
\mathsf{E}\{T(9, 5, 9, 1, 9, 5, 9)\} &= \frac{1}{p_1 p_5^2 p_9^4} - 1 + \left[p_9 \mathsf{E}\{T(9)\} + p_5 \mathsf{E}\{T(9, 5)\}\right. \\
&\quad \left. + p_1 \mathsf{E}\{T(9, 5, 9, 1)\}\right] \\
&= \frac{1}{p_1 p_5^2 p_9^4} + \frac{1}{p_5 p_9^2} + \frac{1}{p_9}
\end{aligned} \tag{4.A.34}
$$

from (4.A.30) and (4.A.33). ◇

Comparing Examples 4.A.4 and 4.A.6, it is easy to see that we can obtain $\mathsf{E}\{T(M)\}$ faster from (4.A.30) and (4.A.33) than from (4.A.21).

Theorem 4.A.1 *For a pattern* $M = (i_1, i_2, \ldots, i_n)$ *with* J *overlapping pieces, the mean time to* M *can be obtained as*

$$
\mathsf{E}\{T(M)\} = \sum_{j=1}^{J+1} \frac{1}{p_{i_1} p_{i_2} \cdots p_{i_{K_j}}}, \tag{4.A.35}
$$

where $K_1 < K_2 < \cdots < K_J$ *are the lengths of the overlapping pieces with* $K_{J+1} = n$.

Proof For convenience, let $\alpha_j = p_{i_{K_j+1}} \mathsf{E}\left\{T\left(\tilde{L}_{K_j+1}\right)\right\}$ and $\beta_j = \mathsf{E}\left\{T\left(L_{K_j}\right)\right\}$. Also let

$$
\epsilon_j = \begin{cases} 1, & \text{if } i_{K_j+1} \neq i_{K_m+1} \text{ for every value of} \\ & \quad m \in \{j+1, j+2, \ldots, J\}, \\ 0, & \text{otherwise} \end{cases} \tag{4.A.36}
$$

for $j = 0, 1, \ldots, J - 1$, and $\epsilon_J = 1$ by noting that the term with $j = J$ is always added in the sum in the right-hand side of (4.A.33). Then, we can rewrite (4.A.33) as

$$
\mathsf{E}\{T(M)\} = \frac{1}{p_{i_1} p_{i_2} \cdots p_{i_n}} - 1 + \sum_{j=0}^{J} \alpha_j \epsilon_j. \tag{4.A.37}
$$

Now, $\alpha_0 = p_{i_1} E\{T(i_1)\} = 1$ and $\alpha_j = \beta_j + 1 - \sum_{l=0}^{j-1} \alpha_l \epsilon_l$ for $j = 1, 2, \ldots, J$

from (4.A.21). Solving for $\{\alpha_j\}_{j=1}^{J}$, we get $\alpha_1 = \beta_1 + 1 - \epsilon_0$, $\alpha_2 = \beta_2 + 1 - (\epsilon_1 \alpha_1 + \epsilon_0 \alpha_0) = \beta_2 - \epsilon_1 \beta_1 + (1 - \epsilon_0)(1 - \epsilon_1)$, $\alpha_3 = \beta_3 + 1 - (\epsilon_2 \alpha_2 + \epsilon_1 \alpha_1 + \epsilon_0 \alpha_0) = \beta_3 - \epsilon_2 \beta_2 - \epsilon_1(1 - \epsilon_2)\beta_1 + (1 - \epsilon_0)(1 - \epsilon_1)(1 - \epsilon_2), \ldots$, and

$$
\begin{aligned}
\alpha_J = \beta_J &- \epsilon_{J-1}\beta_{J-1} - \epsilon_{J-2}(1 - \epsilon_{J-1})\beta_{J-2} - \cdots \\
&- \epsilon_1(1 - \epsilon_2)(1 - \epsilon_3) \cdots (1 - \epsilon_{J-1})\beta_1 \\
&+ (1 - \epsilon_0)(1 - \epsilon_1) \cdots (1 - \epsilon_{J-1}).
\end{aligned}
\tag{4.A.38}
$$

Therefore,

$$
\begin{aligned}
\sum_{j=0}^{J} \alpha_j \epsilon_j = \beta_J &+ (\epsilon_{J-1} - \epsilon_{J-1})\beta_{J-1} + \{\epsilon_{J-2} - \epsilon_{J-2}\epsilon_{J-1} \\
&- \epsilon_{J-2}(1 - \epsilon_{J-1})\}\beta_{J-2} + \cdots + \{\epsilon_1 - \epsilon_1\epsilon_2 - \epsilon_1(1 - \epsilon_2)\epsilon_3 \\
&- \cdots - \epsilon_1(1 - \epsilon_2)(1 - \epsilon_3) \cdots (1 - \epsilon_{J-1})\}\beta_1 + \{\epsilon_0 + (1 - \epsilon_0)\epsilon_1 \\
&+ (1 - \epsilon_0)(1 - \epsilon_1)\epsilon_2 + \cdots \\
&+ (1 - \epsilon_0)(1 - \epsilon_1) \cdots (1 - \epsilon_{J-1})\}.
\end{aligned}
\tag{4.A.39}
$$

In the right-hand side of (4.A.39), the second, third, $\ldots$, second last terms are all 0, and the last term is

$$
\begin{aligned}
\epsilon_0 &+ (1 - \epsilon_0)\epsilon_1 + (1 - \epsilon_0)(1 - \epsilon_1)\epsilon_2 + \cdots \\
&+ (1 - \epsilon_0)(1 - \epsilon_1) \cdots (1 - \epsilon_{J-3})\epsilon_{J-2} \\
&+ (1 - \epsilon_0)(1 - \epsilon_1) \cdots (1 - \epsilon_{J-2})\epsilon_{J-1} + (1 - \epsilon_0)(1 - \epsilon_1) \cdots (1 - \epsilon_{J-1}) \\
= \ \ \epsilon_0 &+ (1 - \epsilon_0)\epsilon_1 + (1 - \epsilon_0)(1 - \epsilon_1)\epsilon_2 + \cdots \\
&+ (1 - \epsilon_0)(1 - \epsilon_1) \cdots (1 - \epsilon_{J-3})\epsilon_{J-2} + (1 - \epsilon_0)(1 - \epsilon_1) \cdots (1 - \epsilon_{J-2}) \\
\vdots \\
= \ \ \epsilon_0 &+ (1 - \epsilon_0)\epsilon_1 + (1 - \epsilon_0)(1 - \epsilon_1) \\
= \ \ 1.
\end{aligned}
\tag{4.A.40}
$$

Thus, noting (4.A.40) and using (4.A.39) into (4.A.37), we get $E\{T(M)\} = \frac{1}{p_{i_1} p_{i_2} \cdots p_{i_n}} - 1 + \beta_J + 1$, i.e.,

$$
E\{T(M)\} = \frac{1}{p_{i_1} p_{i_2} \cdots p_{i_n}} + E\{T(L_{K_J})\}.
\tag{4.A.41}
$$

Next, if we obtain $E\{T(L_{K_J})\}$ after some steps similar to those for (4.A.41) by recollecting that the overlapping pieces of L_{K_J} are $L_{K_1}, L_{K_2}, \ldots, L_{K_{J-1}}$, we have

$\mathsf{E}\left\{T\left(L_{K_J}\right)\right\} = \frac{1}{p_{i_1} p_{i_2} \cdots p_{i_{K_J}}} + \mathsf{E}\left\{T\left(L_{K_{J-1}}\right)\right\}$. Repeating this procedure, and noting that L_1 is not an overlapping piece, we get (4.A.35) by using (4.A.30). ♠

Example 4.A.7 Using (4.A.35), it is easy to get $\mathsf{E}\{T(5, 4, 4, 5)\} = \frac{1+p_4^2 p_5}{p_4^2 p_5^2}$, $\mathsf{E}\{T(5, 4, 5, 4)\} = \frac{1+p_4 p_5}{p_4^2 p_5^2}$, $\mathsf{E}\{T(5, 4, 5, 4, 5)\} = \frac{1}{p_4^2 p_5^3} + \frac{1}{p_4 p_5^2} + \frac{1}{p_5}$, and $\mathsf{E}\{T(5, 4, 4, 5, 4, 4, 5)\} = \frac{1}{p_4^4 p_5^3} + \frac{1}{p_4^2 p_5^2} + \frac{1}{p_5}$. ◇

Example 4.A.8 Assume a coin with $\mathsf{P}(h) = p = 1 - \mathsf{P}(t)$, where h and t denote *head* and *tail*, respectively. Then, the expected numbers of tosses until the first occurrences of h, tht, $htht$, ht, hh, and $hthhthh$ are $\mathsf{E}\{T(h)\} = \frac{1}{p}$, $\mathsf{E}\{T(tht)\} = \frac{1}{pq^2} + \frac{1}{q}$, $\mathsf{E}\{T(htht)\} = \frac{1}{p^2 q^2} + \frac{1}{pq}$, $\mathsf{E}\{T(hthh)\} = \frac{1}{p^3 q} + \frac{1}{p}$, and $\mathsf{E}\{T(hthhthh)\} = \frac{1}{p^5 q^2} + \frac{1}{p^3 q} + \frac{1}{p}$, respectively, where $q = 1 - p$. ◇

Exercises

Exercise 4.1 Show that

$$f(x) = \frac{\mu e^{-\mu x} (\mu x)^{n-1}}{(n-1)!} u(x) \qquad (4.E.1)$$

is the pdf of the sum of n i.i.d. exponential random variables with rate μ.

Exercise 4.2 A box contains three red and two green balls. We choose a ball from the box, discard it, and choose another ball from the box. Let $X = 1$ and $X = 2$ when the first ball is red and green, respectively, and $Y = 4$ and $Y = 3$ when the second ball is red and green, respectively. Obtain the pmf p_X of X, pmf p_Y of Y, joint pmf $p_{X,Y}$ of X and Y, conditional pmf $p_{Y|X}$ of Y given X, conditional pmf $p_{X|Y}$ of X given Y, and pmf p_{X+Y} of $X + Y$.

Exercise 4.3 For two i.i.d. random variables X_1 and X_2 with marginal distribution $\mathsf{P}(1) = \mathsf{P}(-1) = 0.5$, let $X_3 = X_1 X_2$. Are X_1, X_2, and X_3 pairwise independent? Are they independent?

Exercise 4.4 When the joint pdf of a random vector (X, Y) is $f_{X,Y}(x, y) = a\{1 + xy(x^2 - y^2)\} u(1 - |x|) u(1 - |y|)$, determine the constant a. Are X and Y independent of each other? If not, obtain the correlation coefficient between X and Y.

Exercise 4.5 A box contains three red, six green, and five blue balls. A ball is chosen randomly from the box and then replaced to the box after the color is recorded. After six trials, let the numbers of red and blue be R and B, respectively. Obtain the conditional pmf $p_{R|B=3}$ of R when $B = 3$ and conditional mean $\mathsf{E}\{R|B = 1\}$ of R when $B = 1$.

Exercise 4.6 Two binomial random variables $X_1 \sim b(n_1, p)$ and $X_2 \sim b(n_2, p)$ are independent of each other. Show that, when $X_1 + X_2 = x$ is given, the conditional distribution of X_1 is a hypergeometric distribution.

Exercise 4.7 Show that $Z = \frac{X}{X+Y} \sim U(0, 1)$ for two i.i.d. exponential random variables X and Y.

Exercise 4.8 When the joint pdf of $X = (X_1, X_2)$ is

$$f_X(x_1, x_2) = u\left(x_1 + \frac{1}{2}\right) u\left(\frac{1}{2} - x_1\right) u\left(x_2 + \frac{1}{2}\right) u\left(\frac{1}{2} - x_2\right), \quad (4.E.2)$$

obtain the joint pdf f_Y of $Y = (Y_1, Y_2) = (X_1^2, X_1 + X_2)$. Based on the joint pdf f_Y, obtain the pdf f_{Y_1} of $Y_1 = X_1^2$ and pdf f_{Y_2} of $Y_2 = X_1 + X_2$.

Exercise 4.9 When the joint pdf of X_1 and X_2 is $f_{X_1,X_2}(x, y) = \frac{1}{4}u(1 - |x|)u(1 - |y|)$, obtain the cdf F_W and pdf f_W of $W = \sqrt{X_1^2 + X_2^2}$.

Exercise 4.10 Two random variables X and Y are independent of each other with the pdf's $f_X(x) = \lambda e^{-\lambda x} u(x)$ and $f_Y(y) = \mu e^{-\mu y} u(y)$, where $\lambda > 0$ and $\mu > 0$. When $W = \min(X, Y)$ and

$$V = \begin{cases} 1, & \text{if } X \leq Y, \\ 0, & \text{if } X > Y, \end{cases} \quad (4.E.3)$$

obtain the joint cdf of (W, V).

Exercise 4.11 Obtain the pdf of $U = X + Y + Z$ when the joint pdf of X, Y, and Z is $f_{X,Y,Z}(x, y, z) = \frac{6u(x)u(y)u(z)}{(1+x+y+z)^4}$.

Exercise 4.12 Consider the two joint pdf's (1) $f_X(x) = u(x_1)u(1 - x_1)u(x_2)u(1 - x_2)$ and (2) $f_X(x) = 2u(x_1)u(1 - x_2)u(x_2 - x_1)$ of $X = (X_1, X_2)$, where $x = (x_1, x_2)$. In each of the two cases, obtain the joint pdf f_Y of $Y = (Y_1, Y_2) = (X_1^2, X_1 + X_2)$, and then, obtain the pdf f_{Y_1} of $Y_1 = X_1^2$ and pdf f_{Y_2} of $Y_2 = X_1 + X_2$ based on f_Y.

Exercise 4.13 In each of the two cases of the joint pdf f_X described in Exercise 4.12, obtain the joint pdf f_Y of $Y = (Y_1, Y_2) = \left(\frac{1}{2}\left(X_1^2 + X_2\right), \frac{1}{2}\left(X_1^2 - X_2\right)\right)$, and then, obtain the pdf f_{Y_1} of Y_1 and pdf f_{Y_2} of Y_2 based on f_Y.

Exercise 4.14 Two random variables $X \sim G(\alpha_1, \beta)$ and $Y \sim G(\alpha_2, \beta)$ are independent of each other. Show that $Z = X + Y$ and $W = \frac{X}{Y}$ are independent of each other and obtain the pdf of Z and pdf of W.

Exercise 4.15 Denote the joint pdf of $X = (X_1, X_2)$ by f_X.

(1) Express the pdf of $Y_1 = \left(X_1^2 + X_2^2\right)^r$ in terms of f_X.

(2) When $f_X(x, y) = \frac{1}{\pi}u\left(1 - x^2 - y^2\right)$, show that the cdf F_W and pdf f_W of $W = \left(X_1^2 + X_2^2\right)^r$ are as follows:

1. $F_W(w) = u(w - 1)$ and $f_W(w) = \delta(w - 1)$ if $r = 0$.

2. $F_W(w) = \begin{cases} 0, & w \leq 0, \\ w^{\frac{1}{r}}, 0 \leq w \leq 1, & \text{and} \quad f_W(w) = \frac{1}{r}w^{\frac{1}{r}-1}u(w)u(1 - w) \text{ if } r > \\ 1, & w \geq 1 \end{cases}$

0.

3. $F_W(w) = \begin{cases} 0, & w < 1, \\ 1 - w^{\frac{1}{r}}, & w \geq 1 \end{cases}$ and $f_W(w) = -\frac{1}{r}w^{\frac{1}{r}-1}u(w - 1)$ if $r < 0$.

(3) Obtain F_W and f_W when $r = \frac{1}{2}$, 1, and -1 in (2).

Exercise 4.16 The marginal pdf of the three i.i.d. random variables X_1, X_2, and X_3 is $f(x) = u(x)u(1 - x)$.

(1) Obtain the joint pdf f_{Y_1, Y_2} of $(Y_1, Y_2) = (X_1 + X_2 + X_3, X_1 - X_3)$.
(2) Based on f_{Y_1, Y_2}, obtain the pdf f_{Y_2} of Y_2.
(3) Based on f_{Y_1, Y_2}, obtain the pdf f_{Y_1} of Y_1.

Exercise 4.17 Consider i.i.d. random variables X and Y with marginal pmf $p(x) = (1 - \alpha)\alpha^{x-1}\tilde{u}(x - 1)$, where $0 < \alpha < 1$.

(1) Obtain the pmf of $X + Y$ and pmf of $X - Y$.
(2) Obtain the joint pmf of $(X - Y, X)$ and joint pmf of $(X - Y, Y)$.
(3) Using the results in (2), obtain the pmf of X, pmf of Y, and pmf of $X - Y$.
(4) Obtain the joint pmf of $(X + Y, X - Y)$, and using the result, obtain the pmf of $X - Y$ and pmf of $X + Y$. Compare the results with those obtained in (1).

Exercise 4.18 Consider Exercise 2.30. Let R_n be the number of type O cells at $n + \frac{1}{2}$ minutes after the start of the culture. Obtain $E\{R_n\}$, the pmf $p_2(k)$ of R_2, and the probability η_0 that nothing will remain in the culture.

Exercise 4.19 Obtain the conditional expected value $E\{X|Y = y\}$ in Example 4.4.3.

Exercise 4.20 Consider an i.i.d. random vector $X = (X_1, X_2, X_3)$ with marginal pdf $f(x) = e^{-x}u(x)$. Obtain the joint pdf $f_Y(y_1, y_2, y_3)$ of $Y = (Y_1, Y_2, Y_3)$, where $Y_1 = X_1 + X_2 + X_3$, $Y_2 = \frac{X_1 + X_2}{X_1 + X_2 + X_3}$, and $Y_3 = \frac{X_1}{X_1 + X_2}$.

Exercise 4.21 Consider two i.i.d. random variables X_1 and X_2 with marginal pdf $f(x) = u(x)u(1 - x)$. Obtain the joint pdf of $Y = (Y_1, Y_2)$, pdf of Y_1, and pdf of Y_2 when $Y_1 = X_1 + X_2$ and $Y_2 = X_1 - X_2$.

Exercise 4.22 When $Y = (Y_1, Y_2)$ is obtained from rotating clockwise a point $X = (X_1, X_2)$ in the two dimensional plane by θ, express the pdf of Y in terms of the pdf f_X of X.

Exercise 4.23 Assume that the value of the joint pdf $f_{X,Y}(x, y)$ of X and Y is positive in a region containing $x^2 + y^2 < a^2$, where $a > 0$. Express the conditional joint cdf $F_{X,Y|A}$ and conditional joint pdf $f_{X,Y|A}$ in terms of $f_{X,Y}$ when $A = \{X^2 + Y^2 \leq a^2\}$.

Exercise 4.24 The joint pdf of (X, Y) is $f_{X,Y}(x, y) = \frac{1}{4}u(1 - |x|)u(1 - |y|)$. When $A = \{X^2 + Y^2 \leq a^2\}$ with $0 < a < 1$, obtain the conditional joint cdf $F_{X,Y|A}$ and conditional joint pdf $f_{X,Y|A}$. .

Exercise 4.25 Prove the following results:

(1) If X and Z are not orthogonal, then there exists a constant a for which Z and $X - aZ$ are orthogonal.
(2) It is possible that X and Y are uncorrelated even when X and Z are correlated and Y and Z are correlated.

Exercise 4.26 Prove the following results:

(1) If X and Y are independent of each other, then they are uncorrelated.
(2) If the pdf f_X of X is an even function, then X and X^2 are uncorrelated but are not independent of each other.

Exercise 4.27 Show that

$$E\{\max(X^2, Y^2)\} \leq 1 + \sqrt{1 - \rho^2}, \tag{4.E.4}$$

where ρ is the correlation coefficient between the random variables X and Y both with zero mean and unit variance.

Exercise 4.28 Consider a random vector $X = (X_1, X_2, X_3)^T$ with covariance matrix $K_X = \begin{pmatrix} 2 & 1 & 1 \\ 1 & 2 & 1 \\ 1 & 1 & 2 \end{pmatrix}$. Obtain a linear transformation making X into an uncorrelated random vector with unit variance.

Exercise 4.29 Obtain the pdf of Y when the joint pdf of (X, Y) is $f_{X,Y}(x, y) = \frac{1}{y}\exp\left(-y - \frac{x}{y}\right)u(x)u(y)$.

Exercise 4.30 When the joint pmf of (X, Y) is

$$p_{X,Y}(x, y) = \begin{cases} \frac{1}{2}, & (x, y) = (1, 1), \\ \frac{1}{8}, & (x, y) = (1, 2) \text{ or } (2, 2), \\ \frac{1}{4}, & (x, y) = (2, 1), \\ 0, & \text{otherwise}, \end{cases} \tag{4.E.5}$$

obtain the pmf of X and pmf of Y.

Exercise 4.31 For two i.i.d random variables X_1 and X_2 with marginal pmf $p(x) = e^{-\lambda}\frac{\lambda^x}{x!}\tilde{u}(x)$, where $\lambda > 0$, obtain the pmf of $M = \max(X_1, X_2)$ and pmf of $N = \min(X_1, X_2)$.

Exercise 4.32 For two i.i.d. random variables X and Y with marginal pdf $f(z) = u(z) - u(z - 1)$, obtain the pdf's of $W = 2X$, $U = -Y$, and $Z = W + U$.

Exercise 4.33 For three i.i.d. random variables X_1, X_2, and X_3 with marginal distribution $U\left[-\frac{1}{2}, \frac{1}{2}\right]$, obtain the pdf of $Y = X_1 + X_2 + X_3$ and $\mathsf{E}\left\{Y^4\right\}$.

Exercise 4.34 The random variables $\{X_i\}_{i=1}^n$ are independent of each other with pdf's $\{f_i\}_{i=1}^n$, respectively. Obtain the joint pdf of $\{Y_k\}_{k=1}^n$, where $Y_k = X_1 + X_2 + \cdots + X_k$ for $k = 1, 2, \ldots, n$.

Exercise 4.35 The joint pmf of X and Y is

$$p_{X,Y}(x, y) = \begin{cases} \frac{x+y}{32}, & x = 1, 2, \ y = 1, 2, 3, 4, \\ 0, & \text{otherwise.} \end{cases} \tag{4.E.6}$$

(1) Obtain the pmf of X and pmf of Y.
(2) Obtain $\mathsf{P}(X > Y)$, $\mathsf{P}(Y = 2X)$, $\mathsf{P}(X + Y = 3)$, and $\mathsf{P}(X \le 3 - Y)$.
(3) Discuss whether or not X and Y are independent of each other.

Exercise 4.36 For independent random variables X_1 and X_2 with pdf's $f_{X_1}(x) = u(x)u(1 - x)$ and $f_{X_2}(x) = e^{-x}u(x)$, obtain the pdf of $Y = X_1 + X_2$.

Exercise 4.37 Three Poisson random variables X_1, X_2, and X_3 with means 2, 1, and 4, respectively, are independent of each other.

(1) Obtain the mgf of $Y = X_1 + X_2 + X_3$.
(2) Obtain the distribution of Y.

Exercise 4.38 When the joint pdf of X, Y, and Z is $f_{X,Y,Z}(x, y, z) = k(x + y + z)u(x)u(y)u(z)u(1 - x)u(1 - y)u(1 - z)$, determine the constant k and obtain the conditional pdf $f_{Z|X,Y}(z|x, y)$.

Exercise 4.39 Consider a random variable with probability measure

$$\mathsf{P}(X = x) = \begin{cases} \frac{\lambda^x e^{-\lambda}}{x!}, & x = 0, 1, 2, \ldots, \\ 0, & \text{otherwise.} \end{cases} \tag{4.E.7}$$

Here, λ is a realization of a random variable Λ with pdf $f_\Lambda(v) = e^{-v}u(v)$. Obtain $\mathsf{E}\left\{e^{-\Lambda}\big| X = 1\right\}$.

Exercise 4.40 When U_1, U_2, and U_3 are independent of each other, obtain the joint pdf $f_{X,Y,Z}(x, y, z)$ of $X = U_1$, $Y = U_1 + U_2$, and $Z = U_1 + U_2 + U_3$ in terms of the pdf's of U_1, U_2, and U_3.

Exercise 4.41 Let (X, Y, Z) be the rectangular coordinate of a randomly chosen point in a sphere of radius 1 centered at the origin in the three dimensional space.

(1) Obtain the joint pdf $f_{X,Y}(x, y)$ and marginal pdf $f_X(x)$.
(2) Obtain the conditional joint pdf $f_{X,Y|Z}(x, y|z)$. Are X, Y, and Z independent of each other?

Exercise 4.42 Consider a random vector (X, Y) with joint pdf $f_{X,Y}(x, y) = c\, u\, (r - |x| - |y|)$, where c is a constant and $r > 0$.

(1) Express c in terms of r and obtain the pdf $f_X(x)$.
(2) Are X and Y independent of each other?
(3) Obtain the pdf of $Z = |X| + |Y|$.

Exercise 4.43 Assume X with cdf F_X and Y with cdf F_Y are independent of each other. Show that $P(X \geq Y) \geq \frac{1}{2}$ when $F_X(x) \leq F_Y(x)$ at every point x.

Exercise 4.44 The joint pdf of (X, Y) is $f_{X,Y}(x, y) = c\left(x^2 + y^2\right) u(x)u(y)u\, (1 - x^2 - y^2)$.

(1) Determine the constant c and obtain the pdf of X and pdf of Y. Are X and Y independent of each other?
(2) Obtain the joint pdf $f_{R,\Theta}$ of $R = \sqrt{X^2 + Y^2}$ and $\Theta = \tan^{-1}\frac{Y}{X}$.
(3) Obtain the pmf of the output $Q = q(R, \Theta)$ of polar quantizer, where

$$
q(r, \theta) = \begin{cases} k, & \text{if } 0 \leq r \leq \left(\frac{1}{2}\right)^{\frac{1}{4}}, \ \frac{\pi(k-1)}{8} \leq \theta \leq \frac{\pi k}{8}, \\ k+4, & \text{if } \left(\frac{1}{2}\right)^{\frac{1}{4}} \leq r \leq 1, \ \frac{\pi(k-1)}{8} \leq \theta \leq \frac{\pi k}{8} \end{cases} \tag{4.E.8}
$$

for $k = 1, 2, 3, 4$.

Exercise 4.45 Two types of batteries have the pdf $f(x) = 3\lambda x^2 \exp(-\lambda x^3)u(x)$ and $g(y) = 3\mu y^2 \exp(-\mu y^3)u(y)$, respectively, of lifetime with $\mu > 0$ and $\lambda > 0$. When the lifetimes of batteries are independent of each other, obtain the probability that the battery with pdf f of lifetime lasts longer than that with g, and obtain the value when $\lambda = \mu$.

Exercise 4.46 Two i.i.d. random variables X and Y have marginal pdf $f(x) = e^{-x}u(x)$.

(1) Obtain the pdf each of $U = X + Y$, $V = X - Y$, XY, $\frac{X}{Y}$, $Z = \frac{X}{X+Y}$, $\min(X, Y)$, $\max(X, Y)$, and $\frac{\min(X,Y)}{\max(X,Y)}$.
(2) Obtain the conditional pdf of V when $U = u$.
(3) Show that U and Z are independent of each other.

Exercise 4.47 Two Poisson random variables $X_1 \sim P(\lambda_1)$ and $X_2 \sim P(\lambda_2)$ are independent of each other.

(1) Show that $X_1 + X_2 \sim P(\lambda_1 + \lambda_2)$.
(2) Show that the conditional distribution of X_1 when $X_1 + X_2 = n$ is $b\left(n, \frac{\lambda_1}{\lambda_1+\lambda_2}\right)$.

Exercise 4.48 Consider Exercise 2.17.

(1) Obtain the mean and variance of the number M of matches.
(2) Assume that the students with matches will leave with their balls, and each of the remaining students will pick a ball again after their balls are mixed. Show that the mean of the number of repetitions until every student has a match is N.

Exercise 4.49 A particle moves back and forth between positions $0, 1, \ldots, n$. At any position, it moves to the previous or next position with probability $1 - p$ or p, respectively, after 1 second. At positions 0 and n, however, it moves only to the next position 1 and previous position $n - 1$, respectively. Obtain the expected value of the time for the particle to move from position 0 to position n.

Exercise 4.50 Let N be the number of tosses of a coin with probability p of *head* until we have two *head*'s in the last three tosses: we let $N = 2$ if the first two outcomes are both *head*'s. Obtain the expected value of N.

Exercise 4.51 Two people A_1 and A_2 with probabilities p_1 and p_2, respectively, of hit alternatingly fire at a target until the target has been hit two times consecutively.

(1) Obtain the mean number μ_i of total shots fired at the target when A_i starts the shooting for $i = 1, 2$.
(2) Obtain the mean number h_i of times the target has been hit when A_i starts the shooting for $i = 1, 2$.

Exercise 4.52 Consider Exercise 4.51, but now assume that the game ends when the target is hit twice (i.e., consecutiveness is unnecessary). When A_1 starts, obtain the probability α_1 that A_1 fires the last shot of the game and the probability α_2 that A_1 makes both hits.

Exercise 4.53 Assume i.i.d. random variables $X_1, X_2, \ldots$ with marginal distribution $U[0, 1)$. Let $g(x) = \mathsf{E}\{N\}$, where $N = \min\{n : X_n < X_{n-1}\}$ and $X_0 = x$. Obtain an integral equation for $g(x)$ conditional on X_1, and solve the equation.

Exercise 4.54 We repeat tossing a coin with probability p of *head*. Let X be the number of repetitions until *head* appears three times consecutively.

(1) Obtain a difference equation for $g(k) = \mathsf{P}(X = k)$.
(2) Obtain the generating function $G_X(s) = \mathsf{E}\{s^X\}$.
(3) Obtain $\mathsf{E}\{X\}$. (Hint. Use conditional expected value.)

Exercise 4.55 Obtain the conditional joint cdf $F_{X,Y|A}(x, y)$ and conditional joint pdf $f_{X,Y|A}(x, y)$ when $A = \{x_1 < X \le x_2\}$.

Exercise 4.56 For independent random variables X and Y, assume the pmf

$$p_X(x) = \begin{cases} \frac{1}{6}, x = 3; \quad \frac{1}{2}, x = 4; \\ \frac{1}{3}, x = 5 \end{cases} \tag{4.E.9}$$

of X and pmf

$$p_Y(y) = \begin{cases} \frac{1}{2}, y = 0, \\ \frac{1}{2}, y = 1 \end{cases} \tag{4.E.10}$$

of Y. Obtain the conditional pmf's $p_{X|Z}$, $p_{Z|X}$, $p_{Y|Z}$, and $p_{Z|Y}$ and the joint pmf's $p_{X,Y}$, $p_{Y,Z}$, and $p_{X,Z}$ when $Z = X - Y$.

Exercise 4.57 Two exponential random variables T_1 and T_2 with rate λ_1 and λ_2, respectively, are independent of each other. Let $U = \min(T_1, T_2)$, $V = \max(T_1, T_2)$, and I be the smaller index, i.e., the index I such that $T_I = U$.

(1) Obtain the expected values $\mathsf{E}\{U\}$, $\mathsf{E}\{V - U\}$, and $\mathsf{E}\{V\}$.
(2) Obtain $\mathsf{E}\{V\}$ using $V = T_1 + T_2 - U$.
(3) Obtain the joint pdf $f_{U,V-U,I}$ of $(U, V - U, I)$.
(4) Are U and $V - U$ independent of each other?

Exercise 4.58 Consider a bi-variate beta random vector (X, Y) with joint pdf

$$
f_{X,Y}(x, y) = \frac{\Gamma(p_1 + p_2 + p_3)}{\Gamma(p_1)\,\Gamma(p_2)\,\Gamma(p_3)} x^{p_1-1} y^{p_2-1} (1 - x - y)^{p_3-1}
$$
$$
\times u(x)u(y)u(1 - x - y), \tag{4.E.11}
$$

where p_1, p_2, and p_3 are positive numbers. Obtain the pdf f_X of X, pdf f_Y of Y, conditional pdf $f_{X|Y}$, and conditional pdf $f_{Y|X}$. In addition, obtain the conditional pdf $f_{\frac{Y}{1-X}|X}$ of $\frac{Y}{1-X}$ when X is given.

Exercise 4.59 Assuming the joint pdf $f_{X,Y}(x, y) = \frac{1}{16} u(2 - |x|) u(2 - |y|)$ of (X, Y), obtain the conditional joint cdf $F_{X,Y|B}$ and conditional joint pdf $f_{X,Y|B}$ when $B = \{|X| + |Y| \le 1\}$.

Exercise 4.60 Let the joint pdf of X and Y be $f_{X,Y}(x, y) = |xy| u(1 - |x|) u(1 - |y|)$. When $A = \{X^2 + Y^2 \le a^2\}$ with $0 < a < 1$, obtain the conditional joint cdf $F_{X,Y|A}$ and conditional joint pdf $f_{X,Y|A}$.

Exercise 4.61 For a random vector $X = (X_1, X_2, \ldots, X_n)$, show that

$$
\mathsf{E}\{X R^{-1} X^T\} = n, \tag{4.E.12}
$$

where R is the correlation matrix of X and R^{-1} is the inverse matrix of R.

Exercise 4.62 When the cf of (X, Y) is $\varphi_{X,Y}(t, s)$, show that the cf of $Z = aX + bY$ is $\varphi_{X,Y}(at, bt)$.

Exercise 4.63 The joint pdf of (X, Y) is

$$
f_{X,Y}(x, y) = \frac{n!}{(i - 1)!(k - i - 1)!(n - k)!} F^{i-1}(x)\{F(y) - F(x)\}^{k-i-1}
$$
$$
\times \{1 - F(y)\}^{n-k} f(x) f(y) u(y - x), \tag{4.E.13}
$$

where i, k, and n are natural numbers such that $1 \le i < k \le n$, F is the cdf of a random variable, and $f(t) = \frac{d}{dt} F(t)$. Obtain the pdf of X and pdf of Y.

Exercise 4.64 The number N of typographical errors in a book is a Poisson random variable with mean λ. Proofreaders A and B find a typographical error with probability p_1 and p_2, respectively. Let X_1, X_2, X_3, and X_4 be the numbers of typographical errors found by Proofreader A but not by Proofreader B, by Proofreader B but not by Proofreader A, by both proofreaders, and by neither proofreader, respectively. Assume that the event of a typographical error being found by a proofreader is independent of that by another proofreader.

(1) Obtain the joint pmf of X_1, X_2, X_3, and X_4.
(2) Show that

$$\frac{\mathsf{E}\{X_1\}}{\mathsf{E}\{X_3\}} = \frac{1 - p_2}{p_2}, \quad \frac{\mathsf{E}\{X_2\}}{\mathsf{E}\{X_3\}} = \frac{1 - p_1}{p_1}. \tag{4.E.14}$$

Now assume that the values of p_1, p_2, and λ are not available.
(3) Using X_i as the estimate of $\mathsf{E}\{X_i\}$ for $i = 1, 2, 3$, obtain the estimates of p_1, p_2, and λ.
(4) Obtain an estimate of X_4.

Exercise 4.65 Show that the correlation coefficient between X and $|X|$ is

$$\rho_{X|X|} = \frac{\int_{-\infty}^{\infty} |x| (x - m_X) f(x) dx}{\sqrt{\sigma_X^2} \sqrt{\sigma_X^2 + 4m_X^+ m_X^-}}, \tag{4.E.15}$$

where $m_X^\pm$, f, m_X, and σ_X^2 are the half means defined in (3.E.28), pdf, mean, and variance, respectively, of X. Obtain the value of $\rho_{X|X|}$ and compare it with what can be obtained intuitively in each of the following cases of the pdf $f_X(x)$ of X:

(1) $f_X(x)$ is an even function.
(2) $f_X(x) > 0$ only for $x \geq 0$.
(3) $f_X(x) > 0$ only for $x \leq 0$.

Exercise 4.66 For a random variable X with pdf $f_X(x) = u(x) - u(x - 1)$, obtain the joint pdf of X and $Y = 2X + 1$.

Exercise 4.67 Consider a random variable X and its magnitude $Y = |X|$. Show that the conditional pdf $f_{X|Y}$ can be expressed as

$$f_{X|Y}(x|y) = \frac{f_X(x)\delta(x + y)}{f_X(x) + f_X(-x)} u(-x) + \frac{f_X(x)\delta(x - y)}{f_X(x) + f_X(-x)} u(x) \tag{4.E.16}$$

for $y \in \{y \mid \{f_X(y) + f_X(-y)\} u(y) > 0\}$, where f_X is the pdf of X. Obtain the conditional pdf $f_{Y|X}(y|x)$. (Hint. Use (4.5.15).)

Exercise 4.68 Show that the joint cdf and joint pdf are

$$
F_{X,cX+a}(x, y) = \begin{cases} F_X(x)u\left(\frac{y-a}{c} - x\right) \\ \quad + F_X\left(\frac{y-a}{c}\right)u\left(x - \frac{y-a}{c}\right), & c > 0, \\ F_X(x)u(y-a), & c = 0, \\ \left\{F_X(x) - F_X\left(\frac{y-a}{c}\right)\right\}u\left(x - \frac{y-a}{c}\right), & c < 0 \end{cases} \quad (4.E.17)
$$

and

$$
f_{X,cX+a}(x, y) = \begin{cases} \frac{1}{|c|}f_X(x)\delta\left(\frac{y-a}{c} - x\right), & c \neq 0, \\ f_X(x)\delta(y-a), & c = 0, \end{cases} \quad (4.E.18)
$$

respectively, for X and $Y = cX + a$.

Exercise 4.69 Let f and F be the pdf and cdf, respectively, of a continuous random variable X, and let $Y = X^2$.

(1) Obtain the joint cdf $F_{X,Y}$.
(2) Obtain the joint pdf $f_{X,Y}$, and then confirm $\int_{-\infty}^{\infty} f_{X,Y}(x, y)dy = f(x)$ and

$$
\int_{-\infty}^{\infty} f_{X,Y}(x, y)dx = \frac{1}{2\sqrt{y}}\left\{f\left(\sqrt{y}\right) + f\left(-\sqrt{y}\right)\right\}u(y) \quad (4.E.19)
$$

by integration.
(3) Obtain the conditional pdf $f_{X|Y}$.

Exercise 4.70 Show that the pdf $f_{X,cX}$ shown in (4.5.9) satisfies $\int_{-\infty}^{\infty}\int_{-\infty}^{\infty} f_{X,cX}(x, y)dydx = 1$.

Exercise 4.71 Express the joint cdf and joint pdf of the input X and output $Y = Xu(X)$ of a half-wave rectifier in terms of the pdf f_X and cdf F_X of X.

Exercise 4.72 Obtain (4.5.2) from $F_{X,X+a}(x, y) = F_X(\min(x, y - a))$ shown in (4.5.1).

Exercise 4.73 Assume that the joint pdf of $X = (X_1, X_2)$ is

$$
f_X(x, y) = cxu(x)u(1 - x + y)u(1 - x - y). \quad (4.E.20)
$$

Determine c. Obtain and sketch the pdf's of X_1 and X_2.

Exercise 4.74 Consider a random vector $X = (X_1, X_2)$ with the pdf $f_X(x_1, x_2) = u(x_1)u(1 - x_1)u(x_2)u(1 - x_2)$.

(1) Obtain the joint pdf f_Y of $Y = (Y_1, Y_2) = \left(X_1 - X_2, X_1^2 - X_2^2\right)$.
(2) Obtain the pdf f_{Y_1} of Y_1 and pdf f_{Y_2} of Y_2 from f_Y.

(3) Compare the pdf f_{Y_2} of Y_2 with that we can obtain from

$$f_{aX^2}(y) = \frac{1}{2\sqrt{ay}} \left\{ f_X\left(\sqrt{\frac{y}{a}}\right) + f_X\left(-\sqrt{\frac{y}{a}}\right) \right\} u(y) \qquad \text{(4.E.21)}$$

for $a > 0$ shown in (3.2.35) and

$$f_{X_1-X_2}(y) = \int_{-\infty}^{\infty} f_{X_1,X_2}(y+y_2, y_2)\, dy_2 \qquad \text{(4.E.22)}$$

shown in (4.2.20).

References

M. Abramowitz, I.A. Stegun (eds.), *Handbook of Mathematical Functions* (Dover, New York, 1972)

J. Bae, H. Kwon, S.R. Park, J. Lee, I. Song, Explicit correlation coefficients among random variables, ranks, and magnitude ranks. IEEE Trans. Inform. Theory **52**(5), 2233–2240 (2006)

N. Balakrishnan, *Handbook of the Logistic Distribution* (Marcel Dekker, New York, 1992)

D.L. Burdick, A note on symmetric random variables. Ann. Math. Stat. **43**(6), 2039–2040 (1972)

W.B. Davenport Jr., *Probability and Random Processes* (McGraw-Hill, New York, 1970)

H.A. David, H.N. Nagaraja, *Order Statistics*, 3rd edn. (Wiley, New York, 2003)

A.P. Dawid, Some misleading arguments involving conditional independence. J. R. Stat. Soc. Ser. B (Methodological) **41**(2), 249–252 (1979)

W.A. Gardner, *Introduction to Random Processes with Applications to Signals and Systems*, 2nd edn. (McGraw-Hill, New York, 1990)

S. Geisser, N. Mantel, Pairwise independence of jointly dependent variables. Ann. Math. Stat. **33**(1), 290–291 (1962)

R.M. Gray, L.D. Davisson, *An Introduction to Statistical Signal Processing* (Cambridge University Press, Cambridge, 2010)

R.A. Horn, C.R. Johnson, *Matrix Analysis* (Cambridge University Press, Cambridge, 1985)

N.L. Johnson, S. Kotz, *Distributions in Statistics: Continuous Multivariate Distributions* (Wiley, New York, 1972)

S.A. Kassam, *Signal Detection in Non-Gaussian Noise* (Springer, New York, 1988)

S.M. Kendall, A. Stuart, *Advanced Theory of Statistics*, vol. II (Oxford University, New York, 1979)

A. Leon-Garcia, *Probability, Statistics, and Random Processes for Electrical Engineering*, 3rd edn. (Prentice Hall, New York, 2008)

K.V. Mardia, *Families of Bivariate Distributions* (Charles Griffin and Company, London, 1970)

R.N. McDonough, A.D. Whalen, *Detection of Signals in Noise*, 2nd edn. (Academic, New York, 1995)

P.T. Nielsen, On the expected duration of a search for a fixed pattern in random data. IEEE Trans. Inform. Theory **19**(5), 702–704 (1973)

A. Papoulis, S.U. Pillai, *Probability, Random Variables, and Stochastic Processes*, 4th edn. (McGraw-Hill, New York, 2002)

V.K. Rohatgi, A.KMd.E. Saleh, *An Introduction to Probability and Statistics*, 2nd edn. (Wiley, New York, 2001)

J.P. Romano, A.F. Siegel, *Counterexamples in Probability and Statistics* (Chapman and Hall, New York, 1986)

S.M. Ross, *A First Course in Probability* (Macmillan, New York, 1976)

S.M. Ross, *Stochastic Processes*, 2nd edn. (Wiley, New York, 1996)

S.M. Ross, *Introduction to Probability Models*, 10th edn. (Academic, Boston, 2009)

G. Samorodnitsky, M.S. Taqqu, *Non-Gaussian Random Processes: Stochastic Models with Infinite Variance* (Chapman and Hall, New York, 1994)

I. Song, J. Bae, S.Y. Kim, *Advanced Theory of Signal Detection* (Springer, Berlin, 2002)

J.M. Stoyanov, *Counterexamples in Probability*, 3rd edn. (Dover, New York, 2013)

A. Stuart and J. K. Ord, *Advanced Theory of Statistics: Vol. 1. Distribution Theory*, 5th edn. (Oxford University, New York, 1987)

J.B. Thomas, *Introduction to Probability* (Springer, New York, 1986)

Y.H. Wang, Dependent random variables with independent subsets. Am. Math. Mon. **86**(4), 290–292 (1979).

G.L. Wies, E.B. Hall, *Counterexamples in Probability and Real Analysis* (Oxford University, New York, 1993)

Chapter 5
Normal Random Vectors

In this chapter, we consider normal random vectors in the real space. We first describe the pdf and cf of normal random vectors, and then consider the special cases of bi-variate and tri-variate normal random vectors. Some key properties of normal random vectors are then discussed. The expected values of non-linear functions of normal random vectors are also investigated, during which an explicit closed form for joint moments is presented. Additional topics related to normal random vectors are then briefly described.

5.1 Probability Functions

Let us first describe the pdf and cf of normal random vectors (Davenport 1970; Kotz et al. 2000; Middleton 1960; Patel et al. 1976) in general. We then consider additional topics in the special cases of bi-variate and tri-variate normal random vectors.

5.1.1 Probability Density Function and Characteristic Function

Definition 5.1.1 (*normal random vector*) A vector $X = (X_1, X_2, \ldots, X_n)^T$ is called an n dimensional, n-variable, or n-variate normal random vector if it has the joint pdf

$$f_X(x) = \frac{1}{\sqrt{(2\pi)^n |K|}} \exp\left\{ -\frac{1}{2} (x - m)^T K^{-1} (x - m) \right\}, \qquad (5.1.1)$$

I. Song et al., *Probability and Random Variables: Theory and Applications*,
https://doi.org/10.1007/978-3-030-97679-8_5

where $\boldsymbol{m} = (m_1, m_2, \dots, m_n)^T$ and $\boldsymbol{K}$ is an $n \times n$ Hermitian matrix with the determinant $|\boldsymbol{K}| \geq 0$. The distribution is denoted by $\mathcal{N}(\boldsymbol{m}, \boldsymbol{K})$.

When $\boldsymbol{m} = (0, 0, \dots, 0)^T$ and all the diagonal elements of $\boldsymbol{K}$ are 1, the normal distribution is called a standard normal distribution. We will in most cases assume $|\boldsymbol{K}| > 0$: when $|\boldsymbol{K}| = 0$, the distribution is called a degenerate distribution and will be discussed briefly in Theorems 5.1.3, 5.1.4, and 5.1.5.

The distribution of a normal random vector is often called a jointly normal distribution and a normal random vector is also called jointly normal random variables. It should be noted that 'jointly normal random variables' and 'normal random variables' are strictly different. Specifically, the term 'jointly normal random variables' is a synonym for 'a normal random vector'. However, the term 'normal random variables' denotes several random variables with marginal normal distributions which may or may not be a normal random vector. In fact, all the components of a non-Gaussian random vector may be normal random variables in some cases as we shall see in Example 5.2.3 later, for instance.

Example 5.1.1 For a random vector $X \sim \mathcal{N}(\boldsymbol{m}, \boldsymbol{K})$, the mean vector, covariance matrix, and correlation matrix are $\boldsymbol{m}$, $\boldsymbol{K}$, and $\boldsymbol{R} = \boldsymbol{K} + \boldsymbol{m}\,\boldsymbol{m}^T$, respectively.   ◇

Theorem 5.1.1 *The joint cf of* $\boldsymbol{X} = (X_1, X_2, \dots, X_n)^T \sim \mathcal{N}(\boldsymbol{m}, \boldsymbol{K})$ *is*

$$\varphi_X(\omega) = \exp\left(j\boldsymbol{m}^T\omega - \frac{1}{2}\omega^T \boldsymbol{K}\omega\right), \tag{5.1.2}$$

where $\omega = (\omega_1, \omega_2, \dots, \omega_n)^T$.

Proof Letting $\alpha = \{(2\pi)^n |\boldsymbol{K}|\}^{-\frac{1}{2}}$ and $\boldsymbol{y} = (y_1, y_2, \dots, y_n)^T = \boldsymbol{x} - \boldsymbol{m}$, the cf $\varphi_X(\omega) = \mathsf{E}\left\{\exp\left(j\omega^T \boldsymbol{X}\right)\right\}$ of X can be calculated as

$$\varphi_X(\omega) = \int\limits_{\boldsymbol{x} \in \mathbb{R}^n} \alpha \exp\left\{-\frac{1}{2}(\boldsymbol{x} - \boldsymbol{m})^T \boldsymbol{K}^{-1}(\boldsymbol{x} - \boldsymbol{m})\right\} \exp\left(j\omega^T \boldsymbol{x}\right) d\boldsymbol{x}$$

$$= \alpha \exp\left(j\omega^T \boldsymbol{m}\right) \int\limits_{\boldsymbol{y} \in \mathbb{R}^n} \exp\left\{-\frac{1}{2}\left(\boldsymbol{y}^T \boldsymbol{K}^{-1}\boldsymbol{y} - 2j\omega^T \boldsymbol{y}\right)\right\} d\boldsymbol{y}. \tag{5.1.3}$$

Now, recollecting that $\omega^T \boldsymbol{y} = \left(\omega^T \boldsymbol{y}\right)^T = \boldsymbol{y}^T \omega$ because $\omega^T \boldsymbol{y}$ is scalar and that $\boldsymbol{K} = \boldsymbol{K}^T$, we have $\left(\boldsymbol{y} - j\boldsymbol{K}^T\omega\right)^T \boldsymbol{K}^{-1}\left(\boldsymbol{y} - j\boldsymbol{K}^T\omega\right) = \left(\boldsymbol{y}^T \boldsymbol{K}^{-1} - j\omega^T \boldsymbol{K}\,\boldsymbol{K}^{-1}\right)\left(\boldsymbol{y} - j\boldsymbol{K}^T\omega\right) = \boldsymbol{y}^T \boldsymbol{K}^{-1}\boldsymbol{y} - j\boldsymbol{y}^T \boldsymbol{K}^{-1}\boldsymbol{K}^T\omega - j\omega^T \boldsymbol{y} - \omega^T \boldsymbol{K}^T\omega$, i.e.,

$$\left(\boldsymbol{y} - j\boldsymbol{K}^T\omega\right)^T \boldsymbol{K}^{-1}\left(\boldsymbol{y} - j\boldsymbol{K}^T\omega\right) = \boldsymbol{y}^T \boldsymbol{K}^{-1}\boldsymbol{y} - 2j\omega^T \boldsymbol{y} - \omega^T \boldsymbol{K}\omega. \tag{5.1.4}$$

Thus, letting $z = (z_1, z_2, \ldots, z_n)^T = y - jK^T\omega$, recollecting that $\omega^T m = (\omega^T m)^T = m^T\omega$ because $\omega^T m$ is scalar, and using (5.1.4), we get

$$
\begin{aligned}
\varphi_X(\omega) &= \alpha \exp\left(j\omega^T m\right) \\
&\times \int_{y \in \mathbb{R}^n} \exp\left[-\frac{1}{2}\left\{\left(y - jK^T\omega\right)^T K^{-1}\left(y - jK^T\omega\right) + \omega^T K\omega\right\}\right] dy \\
&= \alpha \exp\left(j\omega^T m\right) \exp\left(-\frac{1}{2}\omega^T K\omega\right) \int_{z \in \mathbb{R}^n} \exp\left(-\frac{1}{2}z^T K^{-1} z\right) dz \\
&= \exp\left(jm^T\omega - \frac{1}{2}\omega^T K\omega\right)
\end{aligned}
\tag{5.1.5}
$$

from (5.1.3).   ♠

5.1.2 Bi-variate Normal Random Vectors

Let the covariance matrix of a bi-variate normal random vector $X = (X_1, X_2)$ be

$$
K_2 = \begin{bmatrix} \sigma_1^2 & \rho\sigma_1\sigma_2 \\ \rho\sigma_1\sigma_2 & \sigma_2^2 \end{bmatrix},
\tag{5.1.6}
$$

where ρ is the correlation coefficient between X_1 and X_2 with $|\rho| \le 1$. Then, we have the determinant $|K_2| = \sigma_1^2\sigma_2^2\left(1 - \rho^2\right)$ of K_2, the inverse

$$
K_2^{-1} = \frac{1}{\sigma_1^2\sigma_2^2\left(1 - \rho^2\right)} \begin{bmatrix} \sigma_2^2 & -\rho\sigma_1\sigma_2 \\ -\rho\sigma_1\sigma_2 & \sigma_1^2 \end{bmatrix}
\tag{5.1.7}
$$

of K_2, and the joint pdf

$$
\begin{aligned}
f_{X_1, X_2}(x, y) = \frac{1}{2\pi\sigma_1\sigma_2\sqrt{1 - \rho^2}} \exp\Bigg[-\frac{1}{2\left(1 - \rho^2\right)} \Bigg\{ \frac{(x - m_1)^2}{\sigma_1^2} \\
-2\rho\frac{(x - m_1)(y - m_2)}{\sigma_1\sigma_2} + \frac{(y - m_2)^2}{\sigma_2^2} \Bigg\} \Bigg]
\end{aligned}
\tag{5.1.8}
$$

of (X_1, X_2). The distribution $\mathcal{N}\left((m_1, m_2)^T, K_2\right)$ is often also denoted by $\mathcal{N}\left(m_1, m_2, \sigma_1^2, \sigma_2^2, \rho\right)$. In (5.1.7) and the joint pdf (5.1.8), it is assumed that $|\rho| < 1$: the pdf for $\rho \to \pm 1$ will be discussed later in Theorem 5.1.3.

The contour or isohypse of the bi-variate normal pdf (5.1.8) is an ellipse: specifically, referring to Exercise 5.42, the equation of the ellipse containing $100\alpha\%$ of the distribution can be expressed as $\frac{(x-m_1)^2}{\sigma_1^2} - 2\rho\frac{(x-m_1)(y-m_2)}{\sigma_1\sigma_2} + \frac{(y-m_2)^2}{\sigma_2^2} = -2\left(1-\rho^2\right)\ln(1-\alpha)$. As shown in Exercise 5.44, the major axis of the ellipse makes the angle

$$\theta = \frac{1}{2}\tan^{-1}\frac{2\rho\sigma_1\sigma_2}{\sigma_1^2 - \sigma_2^2} \tag{5.1.9}$$

with the positive x-axis. For $\rho > 0$, we have $0 < \theta < \frac{\pi}{4}, \theta = \frac{\pi}{4}$, and $\frac{\pi}{4} < \theta < \frac{\pi}{2}$ when $\sigma_1 > \sigma_2, \sigma_1 = \sigma_2$, and $\sigma_1 < \sigma_2$, respectively, as shown in Figs. 5.1, 5.2, and 5.3.

Denoting the standard bi-variate normal pdf by f_2, we have

$$f_2(0, y) = f_2(0, 0)\exp\left\{-\frac{y^2}{2\left(1-\rho^2\right)}\right\}, \tag{5.1.10}$$

where $f_2(0, 0) = \left(2\pi\sqrt{1-\rho^2}\right)^{-1}$.

Fig. 5.1 Contour of bi-variate normal pdf when $\rho > 0$ and $\sigma_1 > \sigma_2$, in which case $0 < \theta < \frac{\pi}{4}$

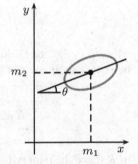

Fig. 5.2 Contour of bi-variate normal pdf when $\rho > 0$ and $\sigma_1 = \sigma_2$, in which case $\theta = \frac{\pi}{4}$

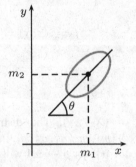

Fig. 5.3 Contour of
bi-variate normal pdf when
$\rho > 0$ and $\sigma_1 < \sigma_2$, in which
case $\frac{\pi}{4} < \theta < \frac{\pi}{2}$

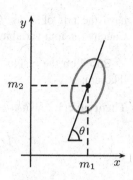

Example 5.1.2 By integrating the joint pdf (5.1.8) over y and x, it is easy to see that we have $X \sim \mathcal{N}\left(m_1, \sigma_1^2\right)$ and $Y \sim \mathcal{N}\left(m_2, \sigma_2^2\right)$, respectively, when $(X, Y) \sim \mathcal{N}\left(m_1, m_2, \sigma_1^2, \sigma_2^2, \rho\right)$. In other words, two jointly normal random variables are also individually normal, which is a special case of Theorem 5.2.1. $\diamond$

Example 5.1.3 For $(X, Y) \sim \mathcal{N}\left(m_1, m_2, \sigma_1^2, \sigma_2^2, \rho\right)$, X and Y are independent if $\rho = 0$. That is, two uncorrelated jointly normal random variables are independent, which will be generalized in Theorem 5.2.3. $\diamond$

Example 5.1.4 From Theorem 5.1.1, the joint cf is $\varphi_{X,Y}(u, v) = \exp\left\{j\left(m_1 u + m_2 v\right) - \frac{1}{2}\left(\sigma_1^2 u^2 + 2\rho\sigma_1\sigma_2 uv + \sigma_2^2 v^2\right)\right\}$ for $(X, Y) \sim \mathcal{N}\left(m_1, m_2, \sigma_1^2, \sigma_2^2, \rho\right)$. $\diamond$

Example 5.1.5 Assume that $X_1 \sim \mathcal{N}(0, 1)$ and $X_2 \sim \mathcal{N}(0, 1)$ are independent. Then, the pdf $f_Y(y) = \frac{1}{2\pi} \int_{-\infty}^{\infty} \exp\left\{-\frac{1}{2}x^2 - \frac{1}{2}(y - x)^2\right\} dx = \frac{1}{2\pi} \int_{-\infty}^{\infty} \exp\left\{-\left(x - \frac{y}{2}\right)^2 - \frac{1}{4}y^2\right\} dx = \frac{1}{2\sqrt{\pi}} \exp\left(-\frac{1}{4}y^2\right) \int_{-\infty}^{\infty} \frac{1}{\sqrt{2\pi}} \exp\left(-\frac{1}{2}v^2\right) dv$ of $Y = X_1 + X_2$ can eventually be obtained as

$$f_Y(y) = \frac{1}{2\sqrt{\pi}} \exp\left(-\frac{y^2}{4}\right) \tag{5.1.11}$$

using (4.2.37) and letting $v = \sqrt{2}\left(x - \frac{y}{2}\right)$. In other words, The sum of two independent, standard normal random variables is an $\mathcal{N}(0, 2)$ random variable. In general, when $X_1 \sim \mathcal{N}\left(m_1, \sigma_1^2\right)$, $X_2 \sim \mathcal{N}\left(m_2, \sigma_2^2\right)$, and X_1 and X_2 are independent of each other, we have $Y = X_1 + X_2 \sim \mathcal{N}\left(m_1 + m_2, \sigma_1^2 + \sigma_2^2\right)$. A further generalization of this result is expressed as Theorem 5.2.5 later. $\diamond$

Example 5.1.6 Obtain the pdf of $Z = \frac{X}{Y}$ when $X \sim \mathcal{N}(0, 1)$ and $Y \sim \mathcal{N}(0, 1)$ are independent of each other.

Solution Because X and Y are independent of each other, we get the pdf $f_Z(z) = \int_{-\infty}^{\infty} |y| f_Y(y) f_X(zy) dy = \int_{-\infty}^{\infty} |y| \frac{1}{2\pi} \exp\left\{-\frac{1}{2}\left(z^2 + 1\right) y^2\right\} dy = \frac{1}{\pi} \int_{0}^{\infty} y \exp\left\{-\frac{1}{2}\left(z^2 + 1\right) y^2\right\} dy$ of $Z = \frac{X}{Y}$ using (4.2.27). Next, letting $\frac{1}{2}\left(z^2 + 1\right) y^2 = t$, we have the pdf of Z as $f_Z(z) = \frac{1}{\pi}\frac{1}{z^2+1}\int_0^\infty e^{-t}dt = \frac{1}{\pi}\frac{1}{z^2+1}$. In other words, Z is a Cauchy random variable. $\diamond$

Based on the results in Examples 4.2.10 and 5.1.6, we can show the following theorem:

Theorem 5.1.2 *When* $(X, Y) \sim \mathcal{N}\left(0, 0, \sigma_X^2, \sigma_Y^2, \rho\right)$, *we have the pdf*

$$f_Z(z) = \frac{\sigma_X \sigma_Y \sqrt{1-\rho^2}}{\pi\left(\sigma_Y^2 z^2 - 2\rho\sigma_X\sigma_Y z + \sigma_X^2\right)} \tag{5.1.12}$$

and cdf

$$F_Z(z) = \frac{1}{2} + \frac{1}{\pi}\tan^{-1}\frac{\sigma_Y z - \rho\sigma_X}{\sigma_X\sqrt{1-\rho^2}} \tag{5.1.13}$$

of $Z = \frac{X}{Y}$.

Proof Let $\alpha = \left(2\pi\sigma_X\sigma_Y\sqrt{1-\rho^2}\right)^{-1}$ and $\beta = \frac{1}{2(1-\rho^2)}\left(\frac{z^2}{\sigma_X^2} - \frac{2\rho z}{\sigma_X\sigma_Y} + \frac{1}{\sigma_Y^2}\right)$. Using (4.2.27), we get $f_Z(z) = \int_{-\infty}^\infty |v| f_{X,Y}(zv, v)dv = \alpha \int_{-\infty}^\infty |v|\exp\left\{-\frac{1}{2(1-\rho^2)}\left(\frac{z^2 v^2}{\sigma_X^2} - \frac{2\rho z v^2}{\sigma_X\sigma_Y} + \frac{v^2}{\sigma_Y^2}\right)\right\}dv = 2\alpha\int_0^\infty v\exp\left\{-\frac{v^2}{2(1-\rho^2)}\left(\frac{z^2}{\sigma_X^2} - \frac{2\rho z}{\sigma_X\sigma_Y} + \frac{1}{\sigma_Y^2}\right)\right\}dv$, i.e.,

$$f_Z(z) = 2\alpha\int_0^\infty v\exp\left(-\beta v^2\right)dv. \tag{5.1.14}$$

Thus, noting that $\int_0^\infty v\exp\left(-\beta v^2\right)dv = -\frac{1}{2\beta}\exp\left(-\beta v^2\right)\Big|_0^\infty = \frac{1}{2\beta}$, we get the pdf of Z as $f_Z(z) = \frac{\alpha}{\beta} = \frac{\sigma_X\sigma_Y\sqrt{1-\rho^2}}{\pi}\left\{\sigma_Y^2\left(z - \frac{\rho\sigma_X}{\sigma_Y}\right)^2 + \sigma_X^2\left(1-\rho^2\right)\right\}^{-1}$, which is the same as (5.1.12).

Next, if we let $\tan\theta_z = \frac{\sigma_Y}{\sqrt{1-\rho^2}\,\sigma_X}\left(z - \frac{\rho\sigma_X}{\sigma_Y}\right)$ for convenience, the cdf of Z can be obtained as $F_Z(z) = \int_{-\infty}^z f_Z(t)dt = \frac{\sigma_X\sigma_Y\sqrt{1-\rho^2}}{\pi\sigma_X^2(1-\rho^2)}\int_{-\infty}^z b(t)dt = \frac{\sigma_X\sigma_Y\sqrt{1-\rho^2}}{\pi\sigma_X^2(1-\rho^2)}\frac{\sigma_X\sqrt{1-\rho^2}}{\sigma_Y}\int_{-\frac{\pi}{2}}^{\theta_z}d\theta = \frac{1}{\pi}\left(\theta_z + \frac{\pi}{2}\right)$, leading to (5.1.13), where $b(t) = \left\{1 + \frac{\sigma_Y^2}{\sigma_X^2(1-\rho^2)}\left(t - \frac{\rho\sigma_X}{\sigma_Y}\right)^2\right\}^{-1}$.   ♠

Theorem 5.1.3 *When* $\rho \to \pm 1$, *we have the limit*

$$\lim_{\rho\to\pm 1} f_{X_1,X_2}(x, y) = \frac{\exp\left\{-\frac{(y-m_2)^2}{2\sigma_2^2}\right\}}{\sqrt{2\pi}\sigma_1\sigma_2}\delta\left(\frac{x-m_1}{\sigma_1} - \xi\frac{y-m_2}{\sigma_2}\right) \tag{5.1.15}$$

of the bi-variate normal pdf (5.1.8), *where* $\xi = \mathrm{sgn}(\rho)$.

Proof We can rewrite $f_{X_1,X_2}(x, y)$ as

$$f_{X_1,X_2}(x, y) = \sqrt{\frac{\alpha}{\pi}} \frac{1}{\sqrt{2\pi}\sigma_1\sigma_2} \exp\left\{-\frac{(y - m_2)^2}{2\sigma_2^2}\right\}$$

$$\times \exp\left\{-\alpha\left(\frac{x - m_1}{\sigma_1} - \rho\frac{y - m_2}{\sigma_2}\right)^2\right\} \qquad (5.1.16)$$

by noting that $\frac{(x-m_1)^2}{\sigma_1^2} - 2\rho\frac{(x-m_1)(y-m_2)}{\sigma_1\sigma_2} + \frac{(y-m_2)^2}{\sigma_2^2} = \frac{(x-m_1)^2}{\sigma_1^2} - 2\rho\frac{(x-m_1)}{\sigma_1}\frac{(y-m_2)}{\sigma_2} + \rho^2\frac{(y-m_2)^2}{\sigma_2^2} + \frac{(y-m_2)^2}{\sigma_2^2} - \rho^2\frac{(y-m_2)^2}{\sigma_2^2}$, i.e.,

$$\frac{(x - m_1)^2}{\sigma_1^2} - 2\rho\frac{(x - m_1)(y - m_2)}{\sigma_1\sigma_2} + \frac{(y - m_2)^2}{\sigma_2^2}$$

$$= \left(\frac{x - m_1}{\sigma_1} - \rho\frac{y - m_2}{\sigma_2}\right)^2 + \left(1 - \rho^2\right)\left(\frac{y - m_2}{\sigma_2}\right)^2, \qquad (5.1.17)$$

where $\alpha = \left\{2\left(1 - \rho^2\right)\right\}^{-1}$. Now, noting that $\sqrt{\frac{\alpha}{\pi}}\exp\left(-\alpha x^2\right) \to \delta(x)$ for $\alpha \to \infty$ as shown in Example 1.4.6 and that $\alpha \to \infty$ for $\rho \to \pm 1$, we can obtain (5.1.15) from (5.1.16). ♠

Based on $f(x)\delta(x - b) = f(b)\delta(x - b)$ shown in (1.4.42) and the property $\delta(ax) = \frac{1}{|a|}\delta(x)$ shown in (1.4.49) of the impulse function, the degenerate pdf (5.1.15) can be expressed in various equivalent formulas. For instance, the term $\exp\left\{-\frac{(y-m_2)^2}{2\sigma_2^2}\right\}$ in (5.1.15) can be replaced with $\exp\left\{-\frac{(x-m_1)^2}{2\sigma_1^2}\right\}$ or $\exp\left\{-\xi\frac{(x-m_1)(y-m_2)}{2\sigma_1\sigma_2}\right\}$ and the term $\frac{1}{\sigma_1\sigma_2}\delta\left(\frac{x-m_1}{\sigma_1} - \xi\frac{y-m_2}{\sigma_2}\right)$ can be replaced with $\delta\left(\sigma_2(x - m_1) - \xi\sigma_1(y - m_2)\right)$.

5.1.3 Tri-variate Normal Random Vectors

For a standard tri-variate normal random vector (X_1, X_2, X_3), let us denote the covariance matrix as

$$K_3 = \begin{bmatrix} 1 & \rho_{12} & \rho_{31} \\ \rho_{12} & 1 & \rho_{23} \\ \rho_{31} & \rho_{23} & 1 \end{bmatrix} \qquad (5.1.18)$$

and the pdf as $f_3(x, y, z) = \frac{1}{\sqrt{8\pi^3|K_3|}}\exp\left\{-\frac{1}{2}(x\ y\ z)K_3^{-1}(x\ y\ z)^T\right\}$, i.e.,

$$f_3(x, y, z) = \frac{1}{\sqrt{8\pi^3 |K_3|}} \exp\left[-\frac{1}{2|K_3|}\{(1 - \rho_{23}^2) x^2 + (1 - \rho_{31}^2) y^2\right.$$

$$\left. + (1 - \rho_{12}^2) z^2 + 2c_{12}xy + 2c_{23}yz + 2c_{31}zx\}\right], \tag{5.1.19}$$

where $c_{ij} = \rho_{jk}\rho_{ki} - \rho_{ij}$. Then, we have $f_3(0, 0, 0) = (8\pi^3 |K_3|)^{-\frac{1}{2}}$,

$$\begin{aligned}|K_3| &= 1 - (\rho_{12}^2 + \rho_{23}^2 + \rho_{31}^2) + 2\rho_{12}\rho_{23}\rho_{31} \\ &= (1 - \rho_{jk}^2)(1 - \rho_{ki}^2) - c_{ij}^2 \\ &= \alpha_{ij,k}^2 (1 - \beta_{ij,k}^2), \end{aligned} \tag{5.1.20}$$

and

$$K_3^{-1} = \frac{1}{|K_3|} \begin{bmatrix} 1 - \rho_{23}^2 & c_{12} & c_{31} \\ c_{12} & 1 - \rho_{31}^2 & c_{23} \\ c_{31} & c_{23} & 1 - \rho_{12}^2 \end{bmatrix}, \tag{5.1.21}$$

where

$$\alpha_{ij,k} = \sqrt{(1 - \rho_{jk}^2)(1 - \rho_{ki}^2)} \tag{5.1.22}$$

and $\beta_{ij,k} = \dfrac{-c_{ij}}{\sqrt{(1 - \rho_{jk}^2)(1 - \rho_{ki}^2)}}$, i.e.,

$$\beta_{ij,k} = \frac{\rho_{ij} - \rho_{jk}\rho_{ki}}{\alpha_{ij,k}} \tag{5.1.23}$$

denotes the partial correlation coefficient between X_i and X_j when X_k is given.

Example 5.1.7 Note that we have $\rho_{ij} = \rho_{jk}\rho_{ki}$ and $|K_3| = \alpha_{ij,k}^2 = (1 - \rho_{jk}^2)$ $(1 - \rho_{ki}^2)$ when $\beta_{ij,k} = 0$. In addition, for $\rho_{ij} \to \pm 1$, we have $\rho_{jk} \to \text{sgn}(\rho_{ij})\rho_{ki}$ because $X_i \to \text{sgn}(\rho_{ij}) X_j$. Thus, when $\rho_{ij} \to \pm 1$, we have $\beta_{ij,k} \to \frac{1-\rho_{jk}^2}{1-\rho_{jk}^2} = 1$ and $|K_3| \to -(\rho_{jk} - \text{sgn}(\rho_{ij})\rho_{ki})^2 \to 0$.   ◇

Note also that $f_3(0, y, z) = f_3(0, 0, 0) \exp\left\{-\frac{1}{2}(0\ y\ z)K_3^{-1}(0\ y\ z)^T\right\} = f_3(0, 0, 0) \exp\left[-\frac{1}{2|K_3|}\{(1 - \rho_{31}^2) y^2 + 2c_{23}yz + (1 - \rho_{12}^2) z^2\}\right]$, i.e.,

$$f_3(0, y, z) = f_3(0, 0, 0) \exp\left\{-\frac{1}{2(1 - \beta_{23,1}^2)}\left(\frac{y^2}{1 - \rho_{12}^2} - \frac{2\beta_{23,1}yz}{\alpha_{23,1}}\right.\right.$$

$$\left.\left. + \frac{z^2}{1 - \rho_{31}^2}\right)\right\} \tag{5.1.24}$$

and

$$f_3(0, 0, z) = f_3(0, 0, 0) \exp\left\{-\frac{\left(1 - \rho_{12}^2\right)}{2\left|K_3\right|}z^2\right\}. \tag{5.1.25}$$

Example 5.1.8 Based on (5.1.24) and (5.1.25), we have $\int_{-\infty}^{\infty} h_1(z) f_3(0, 0, z) dz = \frac{1}{\sqrt{8\pi^3|K_3|}} \int_{-\infty}^{\infty} h_1(z) \exp\left(-\frac{z^2}{2\epsilon_{23,1}^2}\right) dz = \frac{\epsilon_{23,1}}{\sqrt{8\pi^3|K_3|}} \int_{-\infty}^{\infty} h_1(\epsilon_{23,1} w) \exp\left(-\frac{w^2}{2}\right) dw$, i.e.,

$$
\begin{aligned}
\int_{-\infty}^{\infty} h_1(z) f_3(0, 0, z) dz &= \frac{1}{\sqrt{8\pi^3\left(1 - \rho_{12}^2\right)}} \int_{-\infty}^{\infty} h_1\left(\frac{\sqrt{|K_3|}}{\sqrt{1 - \rho_{12}^2}}w\right) \\
&\quad \times \exp\left(-\frac{w^2}{2}\right) dw \\
&= \frac{1}{2\pi\sqrt{1 - \rho_{12}^2}} \mathsf{E}\left\{h_1\left(\frac{\sqrt{|K_3|}}{\sqrt{1 - \rho_{12}^2}}U\right)\right\} \tag{5.1.26}
\end{aligned}
$$

for a uni-variate function h_1, where $\epsilon_{ij,k}^2 = \left(1 - \beta_{ij,k}^2\right)\left(1 - \rho_{jk}^2\right)$ and $U \sim \mathcal{N}(0, 1)$. We also have

$$
\begin{aligned}
\int_{-\infty}^{\infty}\int_{-\infty}^{\infty} h_2(y, z) f_3(0, y, z) dy dz &= \frac{1}{\sqrt{8\pi^3|K_3|}} \int_{-\infty}^{\infty}\int_{-\infty}^{\infty} h_2(y, z) \\
&\quad \times \exp\left\{-\frac{1}{2\left(1 - \beta_{23,1}^2\right)}\left(\frac{y^2}{1 - \rho_{12}^2} - \frac{2\beta_{23,1}yz}{\alpha_{23,1}} + \frac{z^2}{1 - \rho_{31}^2}\right)\right\} dy dz \\
&= \frac{\alpha_{23,1}}{\sqrt{8\pi^3|K_3|}} \int_{-\infty}^{\infty}\int_{-\infty}^{\infty} h_2\left(\sqrt{1 - \rho_{12}^2}v, \sqrt{1 - \rho_{31}^2}w\right) \\
&\quad \times \exp\left\{-\frac{1}{2\left(1 - \beta_{23,1}^2\right)}\left(v^2 - 2\beta_{23,1}vw + w^2\right)\right\} dv dw \\
&= \frac{2\pi\alpha_{23,1}\sqrt{1 - \beta_{23,1}^2}}{\sqrt{8\pi^3|K_3|}} \int_{-\infty}^{\infty}\int_{-\infty}^{\infty} h_2\left(\sqrt{1 - \rho_{12}^2}v, \sqrt{1 - \rho_{31}^2}w\right) \\
&\quad \times f_2(v, w)|_{\rho=\beta_{23,1}} dv dw \\
&= \frac{1}{\sqrt{2\pi}} \mathsf{E}\left\{h_2\left(\sqrt{1 - \rho_{12}^2}V_1, \sqrt{1 - \rho_{31}^2}V_2\right)\right\} \tag{5.1.27}
\end{aligned}
$$

for a bi-variate function h_2, where $(V_1, V_2) \sim \mathcal{N}(0, 0, 1, 1, \beta_{23,1})$. The two results (5.1.26) and (5.1.27) are useful in obtaining the expected values of some non-linear functions. $\diamond$

Denote by $g_\rho(x, y, z)$ the standard tri-variate normal pdf $f_3(x, y, z)$ with the covariance matrix K_3 shown in (5.1.18) so that the correlation coefficients $\rho = (\rho_{12}, \rho_{23}, \rho_{31})$ are shown explicitly. Then, we have

$$g_\rho(-x, y, z) = g_\rho(x, y, z)\big|_{\langle 1 \rangle}, \tag{5.1.28}$$

$$g_\rho(x, -y, z) = g_\rho(x, y, z)\big|_{\langle 2 \rangle}, \tag{5.1.29}$$

$$g_\rho(x, y, -z) = g_\rho(x, y, z)\big|_{\langle 3 \rangle}, \tag{5.1.30}$$

and

$$\int_{-\infty}^{0} \int_{0}^{\infty} \int_{0}^{\infty} h(x, y, z) g_\rho(x, y, z) dx dy dz$$
$$= \int_{\infty}^{0} \int_{0}^{\infty} \int_{0}^{\infty} h(x, y, -t) g_\rho(x, y, -t) dx dy (-dt)$$
$$= \int_{0}^{\infty} \int_{0}^{\infty} \int_{0}^{\infty} h(x, y, -z) g_\rho(x, y, z) dx dy dz \bigg|_{\langle 3 \rangle} \tag{5.1.31}$$

for a tri-variate function $h(x, y, z)$. Here, $\langle k \rangle$ denotes the replacements of the correlation coefficients ρ_{jk} and ρ_{ki} with $-\rho_{jk}$ and $-\rho_{ki}$, respectively.

Example 5.1.9 We have $\beta_{ij,k} = \beta_{ji,k}$,

$$\beta_{ij,k}\big|_{\langle i \rangle} = -\beta_{ij,k}, \tag{5.1.32}$$

$$\beta_{ij,k}\big|_{\langle k \rangle} = \beta_{ij,k}, \tag{5.1.33}$$

$$\frac{\partial}{\partial \rho_{ij}} \sin^{-1} \beta_{ij,k} = \frac{1}{\sqrt{|K_3|}}, \tag{5.1.34}$$

and

$$\frac{\partial}{\partial \rho_{jk}} \sin^{-1} \beta_{ij,k} = -\frac{\rho_{ki} - \rho_{ij}\rho_{jk}}{\left(1 - \rho_{jk}^2\right) \sqrt{|K_3|}} \tag{5.1.35}$$

for the standard tri-variate normal distribution. $\diamond$

After steps similar to those used in obtaining (5.1.15), we can obtain the following theorem:

Theorem 5.1.4 *Letting $\xi_{ij} = \text{sgn}\left(\rho_{ij}\right)$, we have*

$$f_3(x, y, z) \ \rightarrow \ \frac{\exp\left(-\frac{1}{2}x^2\right)}{2\pi\sqrt{1 - \rho_{31}^2}} \exp\left[-\frac{\{z - \mu_1(x, y)\}^2}{2\left(1 - \rho_{31}^2\right)}\right] \delta\left(x - \xi_{12}y\right) \quad (5.1.36)$$

when $\rho_{12} \rightarrow \pm 1$, where $\mu_1(x, y) = \frac{1}{2}\xi_{12}\left(\rho_{23}x + \rho_{31}y\right)$. We subsequently have

$$f_3(x, y, z) \ \rightarrow \ \frac{\exp\left(-\frac{1}{2}x^2\right)}{\sqrt{2\pi}} \delta\left(x - \xi_{12}y\right) \delta\left(x - \xi_{31}z\right) \quad (5.1.37)$$

when $\rho_{12} \rightarrow \pm 1$ and $\rho_{31} \rightarrow \pm 1$.

Proof The proof is discussed in Exercise 5.41. ♠

In (5.1.36), we can replace ρ_{31}^2 with ρ_{23}^2 and $\exp\left(-\frac{1}{2}x^2\right)$ with $\exp\left(-\frac{1}{2}y^2\right)$. The 'mean' $\mu_1(x, y)$ of X_3, when (X_1, X_2, X_3) has the pdf (5.1.36), can be written also as $\mu_1(x, y) = \frac{1}{2}\xi_{12}\left(\xi_{12}\rho_{31}x + \rho_{31}y\right) = \frac{1}{2}\xi_{12}\rho_{31}\left(\xi_{12}x + y\right)$ because, due to the condition $|K_3| \geq 0$, the result $\lim_{\rho_{12}\rightarrow\pm 1} |K_3| = -\left(\rho_{23} \mp \rho_{31}\right)^2$ requires that $\rho_{23} \rightarrow \rho_{12}\rho_{31}$, i.e., $\rho_{23} = \xi_{12}\rho_{31}$ when $\rho_{12} \rightarrow \pm 1$. In addition, because of the function $\delta\left(x - \xi_{12}y\right)$, the mean can further be rewritten as $\mu_1(x, y) = \frac{1}{2}\xi_{12}\rho_{31}\left(\xi_{12}x + \xi_{12}x\right) = \rho_{31}x$ or as $\mu_1(x, y) = \rho_{23}y$. Similarly, (5.1.37) can be expressed in various equivalent formulas: for instance, $\exp\left(-\frac{1}{2}x^2\right)$ can be replaced with $\exp\left(-\frac{1}{2}y^2\right)$ or $\exp\left(-\frac{1}{2}z^2\right)$ and $\delta\left(x - \xi_{31}z\right)$ can be replaced with $\delta\left(z - \xi_{31}x\right)$ or $\delta\left(y - \xi_{23}z\right)$.

The result (5.1.37) in Theorem 5.1.4 can be generalized as follows:

Theorem 5.1.5 *For the pdf $f_X(x)$ shown in (5.1.1), we have*

$$f_X(x) \ \rightarrow \ \frac{\exp\left\{-\frac{(x_1 - m_1)^2}{2\sigma_1^2}\right\}}{\sqrt{2\pi\sigma_1^2}} \prod_{i=2}^{n} \frac{1}{\sigma_i} \delta\left(\frac{x_1 - m_1}{\sigma_1} - \xi_{1i}\frac{x_i - m_i}{\sigma_i}\right) \quad (5.1.38)$$

when $\rho_{1j} \rightarrow \pm 1$ for $j = 2, 3, \ldots, n$, where $\xi_{1j} = \text{sgn}\left(\rho_{1j}\right)$.

Note in Theorem 5.1.5 that, when $\rho_{1j} \rightarrow \pm 1$ for $j \in \{2, 3, \ldots, n\}$, the value of ρ_{ij} for $i \in \{2, 3, \ldots, n\}$ and $j \in \{2, 3, \ldots, n\}$ is determined as $\rho_{ij} \rightarrow \rho_{1i}\rho_{1j} = \xi_{1i}\xi_{1j}$. In the tri-variate case, for instance, when $\rho_{12} \rightarrow 1$ and $\rho_{31} \rightarrow 1$, we have $\rho_{23} \rightarrow 1$ from $\lim_{\rho_{12}, \rho_{13}\rightarrow 1} |K_3| = -\left(1 - \rho_{23}\right)^2 \geq 0$.

5.2 Properties

In this section, we discuss the properties (Hamedani 1984; Horn and Johnson 1985; Melnick and Tenenbein 1982; Mihram 1969; Pierce and Dykstra 1969) of normal random vectors. Some of the properties we will discuss in this chapter are based on those described in Chap. 4. We will also present properties unique to normal random vectors.

5.2.1 Distributions of Subvectors and Conditional Distributions

For $X = (X_1, X_2, \ldots, X_n)^T \sim \mathcal{N}(m, K)$, let us partition the covariance matrix K and its inverse matrix K^{-1} between the s-th and $(s+1)$-st rows, and between the s-th and $(s+1)$-st columns also, as

$$K = \begin{pmatrix} K_{11} & K_{12} \\ K_{21} & K_{22} \end{pmatrix} \tag{5.2.1}$$

and

$$K^{-1} = \begin{pmatrix} \Psi_{11} & \Psi_{12} \\ \Psi_{21} & \Psi_{22} \end{pmatrix}. \tag{5.2.2}$$

Then, we have $K_{ii} = K_{ii}^T$ and $\Psi_{ii} = \Psi_{ii}^T$ for $i = 1, 2$, $K_{21} = K_{12}^T$, and $\Psi_{21} = \Psi_{12}^T$. We also have

$$\Psi_{11} = K_{11}^{-1} + K_{11}^{-1} K_{12} \xi^{-1} K_{21} K_{11}^{-1}, \tag{5.2.3}$$

$$\Psi_{12} = -K_{11}^{-1} K_{12} \xi^{-1}, \tag{5.2.4}$$

$$\Psi_{21} = -\xi^{-1} K_{21} K_{11}^{-1}, \tag{5.2.5}$$

and

$$\Psi_{22} = \xi^{-1}, \tag{5.2.6}$$

where $\xi = K_{22} - K_{21} K_{11}^{-1} K_{12}$.

Theorem 5.2.1 *Assume $X = (X_1, X_2, \ldots, X_n)^T \sim \mathcal{N}(m, K)$ and the partition of the covariance matrix K as described in (5.2.1). Then, for a subvector $X_{(2)} = (X_{s+1}, X_{s+2}, \ldots, X_n)^T$ of X, we have (Johnson and Kotz 1972)*

$$X_{(2)} \sim \mathcal{N}\left(m_{(2)}, K_{22}\right), \tag{5.2.7}$$

where $m_{(2)} = (m_{s+1}, m_{s+2}, \ldots, m_n)^T$. In other words, any subvector of a normal random vector is a normal random vector.

Theorem 5.2.1 also implies that every element of a normal random vector is a normal random variable, which we have already observed in Example 5.1.2. However, it should again be noted that the converse of Theorem 5.2.1 does not hold true as we can see in the example below.

Example 5.2.1 (Romano and Siegel 1986) Assume the joint pdf

$$f_{X,Y}(x, y) = \begin{cases} 2g(x, y), & xy \geq 0, \\ 0, & xy < 0 \end{cases} \tag{5.2.8}$$

of (X, Y), where $g(x, y) = \frac{1}{2\pi} \exp\left\{-\frac{1}{2}\left(x^2 + y^2\right)\right\}$. Then, we have the marginal pdf

$$f_X(x) = \begin{cases} \frac{1}{\pi} \exp\left(-\frac{x^2}{2}\right) \int_{-\infty}^{0} \exp\left(-\frac{y^2}{2}\right) dy, & x < 0, \\ \frac{1}{\pi} \exp\left(-\frac{x^2}{2}\right) \int_{0}^{\infty} \exp\left(-\frac{y^2}{2}\right) dy, & x \geq 0 \end{cases}$$

$$= \frac{1}{\sqrt{2\pi}} \exp\left(-\frac{x^2}{2}\right) \tag{5.2.9}$$

of X. We can similarly show that Y is also a normal random variable. In other words, although X and Y are both normal random variables, (X, Y) is not a normal random vector. ◇

Theorem 5.2.2 Assume $X = (X_1, X_2, \ldots, X_n)^T \sim \mathcal{N}(m, K)$ and the partition of the inverse K^{-1} of the covariance matrix K as described in (5.2.2). Then, we have the conditional distribution (Johnson and Kotz 1972)

$$\mathcal{N}\left(m_{(1)}^T - \left(x_{(2)} - m_{(2)}\right)^T \Psi_{21} \Psi_{11}^{-1}, \Psi_{11}^{-1}\right) \tag{5.2.10}$$

of $X_{(1)} = (X_1, X_2, \ldots, X_s)^T$ when $X_{(2)} = (X_{s+1}, X_{s+2}, \ldots, X_n)^T = x_{(2)} = (x_{s+1}, x_{s+2}, \ldots, x_n)^T$ is given, where $m_{(1)} = (m_1, m_2, \ldots, m_s)^T$ and $m_{(2)} = (m_{s+1}, m_{s+2}, \ldots, m_n)^T$.

Example 5.2.2 From the joint pdf (5.1.8) of $\mathcal{N}\left(m_1, m_2, \sigma_1^2, \sigma_2^2, \rho\right)$ and the pdf of $\mathcal{N}\left(m_2, \sigma_2^2\right)$, the conditional pdf $f_{X|Y}(x|y)$ for a normal random vector $(X, Y) \sim \mathcal{N}\left(m_1, m_2, \sigma_1^2, \sigma_2^2, \rho\right)$ can be obtained as

$$f_{X|Y}(x|y) = \frac{1}{\sqrt{2\pi\sigma_1^2\left(1 - \rho^2\right)}} \exp\left\{-\frac{\left(x - m_{X|Y=y}\right)^2}{2\sigma_1^2\left(1 - \rho^2\right)}\right\}, \tag{5.2.11}$$

where $m_{X|Y=y} = m_1 + \rho\frac{\sigma_1}{\sigma_2}(y - m_2)$. In short, the distribution of X given $Y = y$ is $\mathcal{N}\left(m_{X|Y=y}, \sigma_1^2\left(1 - \rho^2\right)\right)$ for $(X, Y) \sim \mathcal{N}\left(m_1, m_2, \sigma_1^2, \sigma_2^2, \rho\right)$. This result can be obtained also from (5.2.10) using $\boldsymbol{m}_{(1)}^T = m_1$, $\left(\boldsymbol{x}_{(2)} - \boldsymbol{m}_{(2)}\right)^T = y - m_2$, $\boldsymbol{\Psi}_{21} = -\frac{\rho}{\sigma_1\sigma_2(1-\rho^2)}$, and $\boldsymbol{\Psi}_{11}^{-1} = \sigma_1^2\left(1 - \rho^2\right)$. $\diamondsuit$

Theorem 5.2.3 *If a normal random vector is an uncorrelated random vector, then it is an independent random vector.*

In general, two uncorrelated random variables are not necessarily independent as we have discussed in Chap. 4. Theorem 5.2.3 tells us that two jointly normal random variables are independent of each other if they are uncorrelated, which can promptly be confirmed because we have $f_{X,Y}(x, y) = f_X(x)f_Y(y)$ when $\rho = 0$ in the two-dimensional normal pdf (5.1.8). In Theorem 5.2.3, the key point is that the two random variables are jointly normal (Wies and Hall 1993): in other words, if two normal random variables are not jointly normal but are only marginally normal, they may or may not be independent when they are uncorrelated.

Example 5.2.3 (Stoyanov 2013) Let $\phi_1(x, y)$ and $\phi_2(x, y)$ be two standard bivariate normal pdf's with correlation coefficients ρ_1 and ρ_2, respectively. Assume that the random vector (X, Y) has the joint pdf

$$f_{X,Y}(x, y) = c_1\phi_1(x, y) + c_2\phi_2(x, y), \qquad (5.2.12)$$

where $c_1 > 0, c_2 > 0$, and $c_1 + c_2 = 1$. Then, when $\rho_1 \neq \rho_2$, $f_{X,Y}$ is not a normal pdf and, therefore, (X, Y) is not a normal random vector. Now, we have $X \sim \mathcal{N}(0, 1)$, $Y \sim \mathcal{N}(0, 1)$, and the correlation coefficient between X and Y is $\rho_{XY} = c_1\rho_1 + c_2\rho_2$. If we choose $c_1 = \frac{\rho_2}{\rho_2-\rho_1}$ and $c_2 = \frac{\rho_1}{\rho_1-\rho_2}$ for $\rho_1\rho_2 < 0$, then $c_1 > 0, c_2 > 0, c_1 + c_2 = 1$, and $\rho_{XY} = 0$. In short, although X and Y are both normal and uncorrelated with each other, they are not independent of each other because (X, Y) is not a normal random vector. $\diamondsuit$

Based on Theorem 5.2.3, we can show the following theorem:

Theorem 5.2.4 *For a normal random vector $\boldsymbol{X} = (X_1, X_2, \ldots, X_n)$, consider non-overlapping k subvectors $\boldsymbol{X}_1 = \left(X_{i_1}, X_{i_2}, \ldots, X_{i_{n_1}}\right)$, $\boldsymbol{X}_2 = \left(X_{j_1}, X_{j_2},\right.$ $\ldots, X_{j_{n_2}}\right)$, $\ldots$, $\boldsymbol{X}_k = \left(X_{l_1}, X_{l_2}, \ldots, X_{l_{n_k}}\right)$, where $\sum_{j=1}^{k} n_j = n$. If $\rho_{ij} = 0$ for every choice of $i \in S_a$ and $j \in S_b$, with $a \in \{1, 2, \ldots, k\}$ and $b \in \{1, 2, \ldots, k\}$, then $\boldsymbol{X}_1$, $\boldsymbol{X}_2$, $\ldots$, $\boldsymbol{X}_k$ are independent of each other, where $S_1 = \{i_1, i_2, \ldots, i_{n_1}\}$, $S_2 = \{j_1, j_2, \ldots, j_{n_2}\}$, $\ldots$, and $S_k = \{l_1, l_2, \ldots, l_{n_k}\}$.*

Example 5.2.4 For a normal random vector $\boldsymbol{X} = (X_1, X_2, \ldots, X_5)$ with the covariance matrix

$$K = \begin{pmatrix} 1 & 0 & \rho_{13} & \rho_{14} & 0 \\ 0 & 1 & 0 & 0 & \rho_{25} \\ \rho_{31} & 0 & 1 & \rho_{34} & 0 \\ \rho_{41} & 0 & \rho_{43} & 1 & 0 \\ 0 & \rho_{52} & 0 & 0 & 1 \end{pmatrix}, \tag{5.2.13}$$

the subvectors $X_1 = (X_1, X_3, X_4)$ and $X_2 = (X_2, X_5)$ are independent of each other.

$\diamond$

5.2.2 Linear Transformations

Let us first consider a generalization of the result obtained in Example 5.1.5 that the sum of two jointly normal random variables is a normal random variable.

Theorem 5.2.5 *When the random variables $\left\{ X_i \sim \mathcal{N}\left(m_i, \sigma_i^2\right) \right\}_{i=1}^{n}$ are independent of each other, we have*

$$\sum_{i=1}^{n} X_i \sim \mathcal{N}\left(\sum_{i=1}^{n} m_i, \sum_{i=1}^{n} \sigma_i^2 \right). \tag{5.2.14}$$

Proof Because the cf of X_i is $\varphi_{X_i}(\omega) = \exp\left(jm_i\omega - \frac{1}{2}\sigma_i^2\omega^2 \right)$, we can obtain the cf $\varphi_Y(\omega) = \prod_{i=1}^{n} \exp\left(jm_i\omega - \frac{1}{2}\sigma_i^2\omega^2 \right) = \exp\left\{ \sum_{i=1}^{n} \left(jm_i\omega - \frac{1}{2}\sigma_i^2\omega^2 \right) \right\}$ of $Y = \sum_{i=1}^{n} X_i$ as

$$\varphi_Y(\omega) = \exp\left\{ j\left(\sum_{i=1}^{n} m_i \right)\omega - \frac{1}{2}\left(\sum_{i=1}^{n} \sigma_i^2 \right)\omega^2 \right\} \tag{5.2.15}$$

using (4.3.32). This result implies (5.2.14). ♠

Generalizing Theorem 5.2.5 further, we have the following theorem that a linear transformation of a normal random vector is also a normal random vector:

Theorem 5.2.6 *When $X = (X_1, X_2, \ldots, X_n)^T \sim \mathcal{N}(m, K)$, we have $LX \sim \mathcal{N}\left(Lm, LKL^T \right)$ when L is an $n \times n$ matrix such that $|L| \neq 0$.*

Proof First, we have $X = L^{-1}Y$ because $|L| \neq 0$, and the Jacobian of the inverse transformation $x = g^{-1}(y) = L^{-1}y$ is $\left| \frac{\partial}{\partial y} g^{-1}(y) \right| = \left| L^{-1} \right| = \frac{1}{|L|}$. Thus, we have the pdf $f_Y(y) = \frac{1}{|L|} f_X(x)\Big|_{x=L^{-1}y}$ of Y as

$$f_Y(y) = \frac{\exp\left\{-\frac{1}{2}\left(L^{-1}y - m\right)^T K^{-1}\left(L^{-1}y - m\right)\right\}}{|L|\sqrt{(2\pi)^n|K|}} \tag{5.2.16}$$

from Theorem 4.2.1. Now, note that $\left(L^T\right)^{-1} = \left(L^{-1}\right)^T$, $\left|L^T\right| = |L|$, and $\left(L^{-1}y - m\right)^T K^{-1}\left(L^{-1}y - m\right) = (y - Lm)^T \left(L^{-1}\right)^T K^{-1}L^{-1}(y - Lm)$. In addition, letting $H = LKL^T$, we have $H^{-1} = \left(L^T\right)^{-1}K^{-1}L^{-1} = \left(L^{-1}\right)^T K^{-1}L^{-1}$ and $|H| = \left|LKL^T\right| = |L|^2|K|$. Then, we can rewrite (5.2.16) as

$$f_Y(y) = \frac{1}{\sqrt{(2\pi)^n|H|}}\exp\left\{-\frac{1}{2}(y - Lm)^T H^{-1}(y - Lm)\right\}, \tag{5.2.17}$$

which implies $LX \sim \mathcal{N}\left(Lm, LKL^T\right)$ when $X \sim \mathcal{N}(m, K)$.   ♠

Theorem 5.2.6 is a combined generalization of the facts that the sum of two jointly normal random variables is a normal random variable, as described in Example 5.1.5, and that the sum of a number of independent normal random variables is a normal random variable, as shown in Theorem 5.2.5.

Example 5.2.5 For $(X, Y) \sim \mathcal{N}(10, 0, 4, 1, 0.5)$, find the numbers a and b so that $Z = aX + bY$ and $W = X + Y$ are uncorrelated.

Solution Clearly, $\mathsf{E}\{ZW\} - \mathsf{E}\{Z\}\mathsf{E}\{W\} = \mathsf{E}\left\{aX^2 + bY^2 + (a + b)XY\right\} - 100a = 5a + 2b$ because $\mathsf{E}\{Z\} = 10a$ and $\mathsf{E}\{W\} = 10$. Thus, for any pair of two real numbers a and b such that $5a + 2b = 0$, the two random variables $Z = aX + bY$ and $W = X + Y$ will be uncorrelated.   ◇

Example 5.2.6 For a random vector $(X, Y) \sim \mathcal{N}(10, 0, 4, 1, 0.5)$, obtain the joint distribution of $Z = X + Y$ and $W = X - Y$.

Solution We first note that Z and W are jointly normal from Theorem 5.2.6. We thus only need to obtain $\mathsf{E}\{Z\}$, $\mathsf{E}\{W\}$, $\mathsf{Var}\{Z\}$, $\mathsf{Var}\{W\}$, and ρ_{ZW}. We first have $\mathsf{E}\{Z\} = 10$ and $\mathsf{E}\{W\} = 10$ from $\mathsf{E}\{X \pm Y\} = \mathsf{E}\{X\} \pm \mathsf{E}\{Y\}$. Next, we have the variance $\sigma_Z^2 = \mathsf{E}\left\{(X + Y - 10)^2\right\} = \mathsf{E}\left[\{(X - 10) + Y\}^2\right]$ of Z as

$$\sigma_Z^2 = \sigma_X^2 + 2\mathsf{E}\{XY - 10Y\} + \sigma_Y^2$$
$$= 7 \tag{5.2.18}$$

from $\mathsf{E}\{XY\} - \mathsf{E}\{X\}\mathsf{E}\{Y\} = \rho\sigma_X\sigma_Y = 1$ and, similarly, the variance $\sigma_W^2 = 3$ of W. In addition, we also get $\mathsf{E}\{ZW\} = \mathsf{E}\left\{X^2 - Y^2\right\} = m_X^2 + \sigma_X^2 - \sigma_Y^2 = 103$ and, consequently, the correlation coefficient $\rho_{ZW} = \frac{103 - 100}{\sqrt{7}\sqrt{3}} = \sqrt{\frac{3}{7}}$ between Z and W. Thus, we have $(Z, W) \sim \mathcal{N}\left(10, 10, 7, 3, \sqrt{\frac{3}{7}}\right)$. In passing, the joint pdf of (Z, W) is $f_{Z,W}(x, y) = \frac{1}{2\pi\sqrt{7}\sqrt{3}\sqrt{1-\frac{3}{7}}} \exp\left[-\frac{1}{2(1-\frac{3}{7})}\left\{\frac{(x-10)^2}{7} - 2\sqrt{\frac{3}{7}}\frac{x-10}{\sqrt{7}}\right.\right.$ $\frac{y-10}{\sqrt{3}} + \frac{(y-10)^2}{3}\bigg\}\bigg]$, i.e..

$$f_{Z,W}(x, y) = \frac{\sqrt{3}}{12\pi} \exp\left\{-\frac{1}{24}\left(3x^2 - 6xy + 7y^2 - 80y + 400\right)\right\}. \quad (5.2.19)$$

The distribution of (Z, W) can also be obtained from Theorem 5.2.6 more directly as follows: because $V = \begin{pmatrix} Z \\ W \end{pmatrix} = \begin{pmatrix} 1 & 1 \\ 1 & -1 \end{pmatrix}\begin{pmatrix} X \\ Y \end{pmatrix} = L\begin{pmatrix} X \\ Y \end{pmatrix}$, we have the mean vector $\mathsf{E}\{V\} = L\mathsf{E}\left\{\begin{pmatrix} X \\ Y \end{pmatrix}\right\} = \begin{pmatrix} 1 & 1 \\ 1 & -1 \end{pmatrix}\begin{pmatrix} 10 \\ 0 \end{pmatrix} = \begin{pmatrix} 10 \\ 10 \end{pmatrix}$ and the covariance matrix $K_V = LKL^T = \begin{pmatrix} 1 & 1 \\ 1 & -1 \end{pmatrix}\begin{pmatrix} 4 & 1 \\ 1 & 1 \end{pmatrix}\begin{pmatrix} 1 & 1 \\ 1 & -1 \end{pmatrix} = \begin{pmatrix} 7 & 3 \\ 3 & 3 \end{pmatrix}$ of V. Thus, $(Z, W) \sim$ $\mathcal{N}\left(\begin{pmatrix} 10 \\ 10 \end{pmatrix}, \begin{pmatrix} 7 & 3 \\ 3 & 3 \end{pmatrix}\right) = \mathcal{N}\left(10, 10, 7, 3, \sqrt{\frac{3}{7}}\right)$. $\diamond$

From Theorem 5.2.6, the linear combination $\sum_{i=1}^{n} a_i X_i$ of the components of a normal random vector $X = (X_1, X_2, \ldots, X_n)$ is a normal random variable. Let us again emphasize that, while Theorem 5.2.6 tells us that a linear transformation of jointly normal random variables produces jointly normal random variables, a linear transformation of random variables which are normal only marginally but not jointly is not guaranteed to produce normal random variables (Wies and Hall 1993).

As we can see in Examples 5.2.7–5.2.9 below, when $\{X_i\}_{i=1}^{n}$ are all normal random variables but $X = (X_1, X_2, \ldots, X_n)$ is not a normal random vector, (A) the normal random variables $\{X_i\}_{i=1}^{n}$ are generally not independent even if they are uncorrelated, (B) the linear combination of $\{X_i\}_{i=1}^{n}$ may or may not be a normal random variable, and (C) the linear transformation of X is not a normal random vector.

Example 5.2.7 (Romano and Siegel 1986) Let $X \sim \mathcal{N}(0, 1)$ and H be the outcome from a toss of a fair coin. Then, we have $Y \sim \mathcal{N}(0, 1)$ for the random variable

$$Y = \begin{cases} X, & H = head, \\ -X, & H = tail. \end{cases} \quad (5.2.20)$$

Now, because $\mathsf{E}\{X\} = 0$, $\mathsf{E}\{Y\} = 0$, and $\mathsf{E}\{XY\} = \mathsf{E}\{\mathsf{E}\{XY|H\}\} = \frac{1}{2}\left[\mathsf{E}\left\{X^2\right\} + \mathsf{E}\left\{-X^2\right\}\right] = 0$, the random variables X and Y are uncorrelated. However, X and Y are not independent because, for instance, $\mathsf{P}(|X| > 1)\mathsf{P}(|Y| < 1) > 0$ while $\mathsf{P}(|X| > 1, |Y| < 1) = 0$. In addition, $X + Y$ is not normal. In other words, even when X and Y are both normal random variables, $X + Y$ could be non-normal if (X, Y) is not a normal random vector. $\diamond$

Example 5.2.8 (Romano and Siegel 1986) Let $X \sim \mathcal{N}(0, 1)$ and

$$Y = \begin{cases} X, & |X| \leq \alpha, \\ -X, & |X| > \alpha \end{cases} \tag{5.2.21}$$

for a positive number α. Then, X and Y are not independent. In addition, Y is also a standard normal random variable because, for any set B such that $B \in \mathcal{B}(\mathbb{R})$, we have

$$\begin{aligned} \mathsf{P}(Y \in B) &= \mathsf{P}(Y \in B| |X| \leq \alpha)\mathsf{P}(|X| \leq \alpha) \\ &\quad + \mathsf{P}(Y \in B| |X| > \alpha)\mathsf{P}(|X| > \alpha) \\ &= \mathsf{P}(X \in B| |X| \leq \alpha)\mathsf{P}(|X| \leq \alpha) + \mathsf{P}(-X \in B| |X| > \alpha)\mathsf{P}(|X| > \alpha) \\ &= \mathsf{P}(X \in B| |X| \leq \alpha)\mathsf{P}(|X| \leq \alpha) + \mathsf{P}(X \in B| |X| > \alpha)\mathsf{P}(|X| > \alpha) \\ &= \mathsf{P}(X \in B). \end{aligned} \tag{5.2.22}$$

Now, the correlation coefficient $\rho_{XY} = \mathsf{E}\{XY\} = 2\int_0^\alpha x^2 \phi(x) dx - 2\int_\alpha^\infty x^2 \phi(x) dx$ between X and Y can be obtained as

$$\rho_{XY} = 4\int_0^\alpha x^2 \phi(x)\, dx - 1, \tag{5.2.23}$$

where ϕ denotes the standard normal pdf. Letting $g(\alpha) = \int_0^\alpha x^2 \phi(x) dx$, we can find a positive number α_0 such that[1] $g(\alpha_0) = \frac{1}{4}$ because $g(0) = 0$, $g(\infty) = \frac{1}{2}$, and g is a continuous function. Therefore, when $\alpha = \alpha_0$, X and Y are uncorrelated from (5.2.23). Meanwhile, because

$$X + Y = \begin{cases} 2X, & |X| \leq \alpha, \\ 0, & |X| > \alpha, \end{cases} \tag{5.2.24}$$

$X + Y$ is not normal. ◇

Example 5.2.9 (Stoyanov 2013) When $X = (X, Y)$ is a normal random vector, the random variables X, Y, and $X + Y$ are all normal. Yet, the converse is not necessarily true. We now consider an example. Let the joint pdf of $X = (X, Y)$ be

$$\begin{aligned} f_X(x, y) &= \frac{1}{2\pi} \exp\left\{-\frac{1}{2}\left(x^2 + y^2\right)\right\} \\ &\quad \times \left[1 + xy\left(x^2 - y^2\right)\exp\left\{-\frac{1}{2}\left(x^2 + y^2 + 2\epsilon\right)\right\}\right], \end{aligned} \tag{5.2.25}$$

where $\epsilon > 0$. Let us also note that

---

[1] Here, $\alpha_0 \approx 1.54$.

$$\left| xy \left(x^2 - y^2 \right) \exp \left\{ -\frac{1}{2} \left(x^2 + y^2 + 2\epsilon \right) \right\} \right| \leq 1 \qquad (5.2.26)$$

when $\epsilon \geq -2 + \ln 4 \approx -0.6137$ because $-4e^{-2} \leq xy \left(x^2 - y^2 \right) \exp \left\{ -\frac{1}{2} \left(x^2 + y^2 \right) \right\} \leq 4e^{-2}$. Then, the joint cf of X can be obtained as

$$\varphi_X(s, t) = \exp \left(-\frac{s^2 + t^2}{2} \right) + \frac{st \left(s^2 - t^2 \right)}{32} \exp \left(-\epsilon - \frac{s^2 + t^2}{4} \right), \qquad (5.2.27)$$

from which we can make the following observations:

(A) We have $X \sim \mathcal{N}(0, 1)$ and $Y \sim \mathcal{N}(0, 1)$ because $\varphi_X(t) = \varphi_X(t, 0) = \exp \left(-\frac{1}{2} t^2 \right)$ and $\varphi_Y(t) = \varphi_X(0, t) = \exp \left(-\frac{1}{2} t^2 \right)$.

(B) We have[2] $X + Y \sim \mathcal{N}(0, 2)$ because $\varphi_{X+Y}(t) = \varphi_X(t, t) = \exp \left(-t^2 \right)$.

(C) We have $X - Y \sim \mathcal{N}(0, 2)$ because $\varphi_{X-Y}(t) = \varphi_X(t, -t) = \exp \left(-t^2 \right)$.

(D) We have $\mathsf{E}\{XY\} = \left. \frac{\partial^2}{\partial t \partial s} \varphi_X(s, t) \right|_{(s,t)=(0,0)} = 0$ and $\mathsf{E}\{X\} = \mathsf{E}\{Y\} = 0$. Therefore, X and Y are uncorrelated.

(E) As it is clear from (5.2.25) or (5.2.27), the random vector (X, Y) is not a normal random vector. In other words, although X, Y, and $X + Y$ are all normal random variables, (X, Y) is not a normal random vector. $\diamond$

Theorem 5.2.7 *A normal random vector with a positive definite covariance matrix can be linearly transformed into an independent standard normal random vector.*

Proof Theorem 5.2.7 can be proved from Theorems 4.3.5, 5.2.3, and 5.2.6, or from (4.3.24) and (4.3.25). Specifically, when $X \sim \mathcal{N}(m, K)$ with $|K| > 0$, the eigenvalues of K are $\{\lambda_i\}_{i=1}^n$, and the eigenvector corresponding to λ_i is a_i, assume the matrix A and $\tilde{\lambda} = \mathrm{diag} \left(\frac{1}{\sqrt{\lambda_1}}, \frac{1}{\sqrt{\lambda_2}}, \ldots, \frac{1}{\sqrt{\lambda_n}} \right)$ considered in (4.3.21) and (4.3.23), respectively. Then, the mean vector of

$$Y = \tilde{\lambda} A (X - m) \qquad (5.2.28)$$

is $\mathsf{E}\{Y\} = \tilde{\lambda} A \mathsf{E}\{X - m\} = 0$. In addition, as we can see from (4.3.25), the covariance matrix of Y is $K_Y = I$. Therefore, using Theorem 5.2.6, we get $Y \sim \mathcal{N}(0, I)$. In other words, $Y = \tilde{\lambda} A (X - m)$ is a vector of independent standard normal random variables.   ♠

Example 5.2.10 Transform the random vector $U = (U_1, U_2)^T \sim \mathcal{N}(m, K)$ into an independent standard normal random vector when $m = (10 \ 0)^T$ and $K = \begin{pmatrix} 2 & -1 \\ -1 & 2 \end{pmatrix}$.

---

[2] Here, as we have observed in Exercise 4.62, when the joint cf of $X = (X, Y)$ is $\varphi_X(t, s)$, the cf of $Z = aX + bY$ is $\varphi_Z(t) = \varphi_X(at, bt)$.

Solution The eigenvalues and corresponding eigenvectors of the covariance matrix K of U are $\left\{\lambda_1 = 3, a_1 = \frac{1}{\sqrt{2}}(1 \ -1)^T\right\}$ and $\left\{\lambda_2 = 1, a_2 = \frac{1}{\sqrt{2}}(1 \ 1)^T\right\}$. Thus, for the linear transformation $L = \frac{1}{\sqrt{2}}\begin{pmatrix} \frac{1}{\sqrt{3}} & 0 \\ 0 & 1 \end{pmatrix}\begin{pmatrix} 1 & -1 \\ 1 & 1 \end{pmatrix}$, i.e.,

$$L = \begin{pmatrix} \frac{1}{\sqrt{6}} & -\frac{1}{\sqrt{6}} \\ \frac{1}{\sqrt{2}} & \frac{1}{\sqrt{2}} \end{pmatrix},$$
(5.2.29)

the random vector $V = L\left(U - (10 \ 0)^T\right)$ will be a vector of independent standard normal random variables: the covariance matrix of V is $K_V = L K_U L^H = \begin{pmatrix} \frac{1}{\sqrt{6}} & -\frac{1}{\sqrt{6}} \\ \frac{1}{\sqrt{2}} & \frac{1}{\sqrt{2}} \end{pmatrix}\begin{pmatrix} 2 & -1 \\ -1 & 2 \end{pmatrix}\begin{pmatrix} \frac{1}{\sqrt{6}} & \frac{1}{\sqrt{2}} \\ -\frac{1}{\sqrt{6}} & \frac{1}{\sqrt{2}} \end{pmatrix} = \begin{pmatrix} 1 & 0 \\ 0 & 1 \end{pmatrix}$. $\diamond$

In passing, the following theorem is noted without a proof:

Theorem 5.2.8 *If the linear combination $a^T X$ is a normal random variable for every vector $a = (a_1, a_2, \ldots, a_n)^T$, then the random vector X is a normal random vector and the converse is also true.*

5.3   Expected Values of Nonlinear Functions

In this section, expected values of some non-linear functions and joint moments (Bär and Dittrich 1971; Baum 1957; Brown 1957; Hajek 1969; Haldane 1942; Holmquist 1988; Kan 2008; Nabeya 1952; Song and Lee 2015; Song et al. 2020; Triantafyllopoulos 2003; Withers 1985) of normal random vectors are investigated. We first consider a few simple examples based on the cf and mgf.

5.3.1   Examples of Joint Moments

Let us start with some examples for obtaining joint moments of normal random vectors via cf and mgf.

Example 5.3.1 For the joint central moment $\mu_{ij} = \mathsf{E}\left\{(X - m_X)^i (Y - m_Y)^j\right\}$ of a random vector (X, Y), we have observed that $\mu_{00} = 1$, $\mu_{01} = \mu_{10} = 0$, $\mu_{20} = \sigma_1^2$, $\mu_{02} = \sigma_2^2$, and $\mu_{11} = \rho\sigma_1\sigma_2$ in Sect. 4.3.2.1. In addition, it is easy to see that $\mu_{30} = \mu_{03} = 0$ and $\mu_{40} = 3\sigma_1^4$ when $(X, Y) \sim \mathcal{N}\left(0, 0, \sigma_1^2, \sigma_2^2, \rho\right)$ from (3.3.31). Now, based on the moment theorem, show that $\mu_{31} = 3\rho\sigma_1^3\sigma_2$, $\mu_{22} = \left(1 + 2\rho^2\right)\sigma_1^2\sigma_2^2$, and $\mu_{41} = \mu_{32} = 0$ when $(X, Y) \sim \mathcal{N}\left(0, 0, \sigma_1^2, \sigma_2^2, \rho\right)$.

Solution For convenience, let $C = \mathsf{E}\{XY\} = \rho\sigma_1\sigma_2$, $A = \frac{1}{2}\left(\sigma_1^2 s_1^2 + 2Cs_1 s_2 + \sigma_2^2 s_2^2\right)$, and $A^{(ij)} = \frac{\partial^{i+j}}{\partial s_1^i \partial s_2^j} A$. Then, we easily have $A^{(10)}\big|_{s_1=0,s_2=0} = \left[\sigma_1^2 s_1 + Cs_2\right]_{s_1=0,s_2=0} = 0$, $A^{(01)}\big|_{s_1=0,s_2=0} = \left[\sigma_2^2 s_2 + Cs_2\right]_{s_1=0,s_2=0} = 0$, $A^{(20)} = \sigma_1^2$, $A^{(11)} = C$, $A^{(02)} = \sigma_2^2$, and $A^{(ij)} = 0$ for $i + j \geq 3$. Denoting the joint mgf of (X, Y) by $M = M(s_1, s_2) = \exp(A)$ and employing the notation $M^{(ij)} = \frac{\partial^{i+j} M}{\partial s_1^i \partial s_2^j}$, we get $M^{(10)} = M A^{(10)}$, $M^{(20)} = M\left\{A^{(20)} + \left(A^{(10)}\right)^2\right\}$,

$$
\begin{aligned}
M^{(21)} &= M\left\{A^{(21)} + A^{(20)}A^{(01)} + 2A^{(11)}A^{(10)} + \left(A^{(10)}\right)^2 A^{(01)}\right\} \\
&= M\left\{A^{(20)}A^{(01)} + 2A^{(11)}A^{(10)} + \left(A^{(10)}\right)^2 A^{(01)}\right\},
\end{aligned}
\tag{5.3.1}
$$

$$
\begin{aligned}
M^{(31)} &= M\left\{3A^{(20)}\left(A^{(11)} + A^{(10)}A^{(01)}\right)\right. \\
&\quad \left.+ \left(A^{(10)}\right)^2\left(3A^{(11)} + A^{(10)}A^{(01)}\right)\right\},
\end{aligned}
\tag{5.3.2}
$$

$$
\begin{aligned}
M^{(22)} &= M\left\{A^{(20)}A^{(02)} + 2\left(A^{(11)}\right)^2 + 4A^{(11)}A^{(10)}A^{(01)} + A^{(20)}\left(A^{(01)}\right)^2\right. \\
&\quad \left.+ \left(A^{(10)}\right)^2 A^{(02)} + \left(A^{(10)}\right)^2\left(A^{(01)}\right)^2\right\},
\end{aligned}
\tag{5.3.3}
$$

$$
M^{(41)} = B_{41} M,
\tag{5.3.4}
$$

and

$$
M^{(32)} = B_{32} M.
\tag{5.3.5}
$$

Here,

$$
\begin{aligned}
B_{41} &= 3\left(A^{(20)}\right)^2 A^{(01)} + 12A^{(20)}A^{(11)}A^{(10)} + 6A^{(20)}\left(A^{(10)}\right)^2 A^{(01)} \\
&\quad + 4A^{(11)}\left(A^{(10)}\right)^3 + \left(A^{(10)}\right)^3 A^{(01)}
\end{aligned}
\tag{5.3.6}
$$

and

$$
\begin{aligned}
B_{32} &= 6A^{(20)}A^{(11)}A^{(01)} + 3A^{(20)}A^{(02)}A^{(10)} + 6A^{(11)}\left(A^{(10)}\right)^2 A^{(01)} \\
&\quad + 6\left(A^{(11)}\right)^2 A^{(10)} + A^{(02)}\left(A^{(10)}\right)^3 + 3A^{(20)}A^{(10)}\left(A^{(01)}\right)^2 \\
&\quad + \left(A^{(10)}\right)^3\left(A^{(01)}\right)^2.
\end{aligned}
\tag{5.3.7}
$$

Recollecting that $M(0, 0) = 1$, we have $\mu_{31} = 3\rho\sigma_1^3\sigma_2$ from[3] (5.3.2), $\mu_{22} = (1 + 2\rho^2)\sigma_1^2\sigma_2^2$ from[4] (5.3.3), $\mu_{41} = M^{(41)}\big|_{s_1=0,s_2=0} = 0$ from (5.3.4) and (5.3.6), and $\mu_{32} = M^{(32)}\big|_{s_1=0,s_2=0} = 0$ from (5.3.5) and (5.3.7). $\diamond$

In Exercise 5.15, it is shown that

$$E\{X_1^2 X_2^2 X_3^2\} = 1 + 2\left(\rho_{12}^2 + \rho_{23}^2 + \rho_{31}^2\right) + 8\rho_{12}\rho_{23}\rho_{31} \tag{5.3.8}$$

for $(X_1, X_2, X_3) \sim \mathcal{N}(\mathbf{0}, \mathbf{K}_3)$. Similarly, it is shown in Exercise 5.18 that

$$E\{X_1 X_2 X_3\} = m_1 E\{X_2 X_3\} + m_2 E\{X_3 X_1\} + m_3 E\{X_1 X_2\}$$
$$-2m_1 m_2 m_3 \tag{5.3.9}$$

for a general tri-variate normal random vector (X_1, X_2, X_3) and that

$$E\{X_1 X_2 X_3 X_4\} = E\{X_1 X_2\} E\{X_3 X_4\} + E\{X_1 X_3\} E\{X_2 X_4\}$$
$$+ E\{X_1 X_4\} E\{X_2 X_3\} - 2m_1 m_2 m_3 m_4 \tag{5.3.10}$$

for a general quadri-variate normal random vector (X_1, X_2, X_3, X_4). The results (5.3.8)–(5.3.10) can also be obtained via the general formula (5.3.51).

5.3.2 Price's Theorem

We now discuss a theorem that is quite useful in evaluating the expected values of various non-linear functions such as the power functions, sign functions, and absolute values of normal random vectors.

Denoting the covariance between X_i and X_j by

$$\tilde{\rho}_{ij} = R_{ij} - m_i m_j, \tag{5.3.11}$$

where $R_{ij} = E\{X_i X_j\}$ and $m_i = E\{X_i\}$, the correlation coefficient ρ_{ij} between X_i and X_j and variance σ_i^2 of X_i can be expressed as $\rho_{ij} = \dfrac{\tilde{\rho}_{ij}}{\sqrt{\tilde{\rho}_{ii}\tilde{\rho}_{jj}}}$ and $\sigma_i^2 = \tilde{\rho}_{ii}$, respectively.

Theorem 5.3.1 *Let* $\mathbf{K} = \left[\tilde{\rho}_{rs}\right]$ *be the covariance matrix of an n-variate normal random vector* $\mathbf{X}$. *When* $\{g_i(\cdot)\}_{i=1}^n$ *are all memoryless functions, we have*

---

[3] More specifically, we have $\mu_{31} = M^{(31)}\big|_{s_1=0,s_2=0} = \{3MA^{(20)}A^{(11)}\}\big|_{s_1=0,s_2=0} = 3\sigma_1^2 C = 3\rho\sigma_1^3\sigma_2$.

[4] More specifically, we have $\mu_{22} = M^{(22)}\big|_{s_1=0,s_2=0} = \{MA^{(20)}A^{(02)} + 2M(A^{(11)})^2\}\big|_{s_1=0,s_2=0} = \sigma_1^2\sigma_2^2 + 2C^2 = (1 + 2\rho^2)\sigma_1^2\sigma_2^2$.

$$\frac{\partial^{\gamma_1} \mathsf{E}\left\{\prod_{i=1}^{n} g_i(X_i)\right\}}{\partial \tilde{\rho}_{r_1 s_1}^{k_1} \partial \tilde{\rho}_{r_2 s_2}^{k_2} \cdots \partial \tilde{\rho}_{r_N s_N}^{k_N}} = \frac{1}{2^{\gamma_2}} \mathsf{E}\left\{\prod_{i=1}^{n} g_i^{(\gamma_3)}(X_i)\right\}, \tag{5.3.12}$$

where $\gamma_1 = \sum_{j=1}^{N} k_j$, $\gamma_2 = \sum_{j=1}^{N} k_j \delta_{r_j s_j}$, and $\gamma_3 = \sum_{j=1}^{N} \epsilon_{ij} k_j$. Here, δ_{ij} is the Kronecker delta function defined as (4.3.17), $N \in \{1, 2, \ldots, \frac{1}{2}n(n+1)\}$, $g_i^{(k)}(x) = \frac{d^k}{dx^k} g_i(x)$, and $r_j \in \{1, 2, \ldots, n\}$ and $s_j \in \{1, 2, \ldots, n\}$ for $j = 1, 2, \ldots, N$. In addition, $\epsilon_{ij} = \delta_{ir_j} + \delta_{is_j} \in \{0, 1, 2\}$ denotes how many of r_j and s_j are equal to i for $i = 1, 2, \ldots, n$ and $j = 1, 2, \ldots, N$ and satisfies $\sum_{i=1}^{n} \epsilon_{ij} = 2$.

Based on Theorem 5.3.1, referred to as Price's theorem (Price 1958), let us describe how we can obtain the joint moments and the expected values of non-linear functions of normal random vectors in the three cases of $n = 1, 2$, and 3.

5.3.2.1 Uni-Variate Normal Random Vectors

When $n = 1$ in Theorem 5.3.1, we have $N = 1$, $r_1 = 1$, $s_1 = 1$; and $\epsilon_{11} = 2$. Let $k = k_1 \delta_{11}$, and use m for $m_1 = \mathsf{E}\{X_1\}$ and $\tilde{\rho}$ for $\tilde{\rho}_{11} = \sigma^2 = \mathsf{Var}\{X_1\}$ by deleting the subscripts for brevity. We can then express (5.3.12) as

$$\frac{\partial^k}{\partial \tilde{\rho}^k} \mathsf{E}\{g(X)\} = \left(\frac{1}{2}\right)^k \mathsf{E}\left\{g^{(2k)}(X)\right\}. \tag{5.3.13}$$

Meanwhile, for the pdf $f_X(x)$ of X, we have

$$\lim_{\tilde{\rho} \to 0} f_X(x) = \delta(x - m). \tag{5.3.14}$$

Based on (5.3.13) and (5.3.14), let us obtain the expected value $\mathsf{E}\{g(X)\}$ for a normal random variable X.

Example 5.3.2 For a normal random variable $X \sim \mathcal{N}(m, \sigma^2)$, obtain $\tilde{\Upsilon} = \mathsf{E}\{X^3\}$.

Solution Letting $g(x) = x^3$, we have $g^{(2)}(x) = 6x$. Thus, we get $\frac{\partial}{\partial \tilde{\rho}} \tilde{\Upsilon} = \frac{1}{2} \mathsf{E}\{g^{(2)}(X)\} = 3\mathsf{E}\{X\} = 3m$, i.e., $\tilde{\Upsilon} = 3m\tilde{\rho} + c$ from (5.3.13) with $k = 1$, where c is the integration constant. Subsequently, we have

$$\mathsf{E}\{X^3\} = 3m\sigma^2 + m^3 \tag{5.3.15}$$

because $c = m^3$ from $\tilde{\Upsilon} \to \int_{-\infty}^{\infty} x^3 \delta(x - m) dx = m^3$ recollecting (5.3.14). $\diamond$

We now derive a general formula for the moment $E\{X^a\}$. Let us use an underline as

$$\underline{n} = \begin{cases} \frac{n-1}{2}, & n \text{ is odd}, \\ \frac{n}{2}, & n \text{ is even} \end{cases} \tag{5.3.16}$$

to denote the quotient $\lfloor \frac{n}{2} \rfloor$ of a non-negative integer n when divided by 2.

Theorem 5.3.2 *For $X \sim \mathcal{N}(m, \tilde{\rho})$, we have*

$$E\{X^a\} = \sum_{j=0}^{\underline{a}} \frac{a!}{2^j j! (a - 2j)!} \tilde{\rho}^j m^{a-2j} \tag{5.3.17}$$

for $a = 0, 1, \ldots$.

Proof The proof is left as an exercise, Exercise 5.27.   ♠

Example 5.3.3 Using (5.3.17), we have

$$E\{X^4\} = 3\tilde{\rho}^2 + 6m^2\tilde{\rho} + m^4 \tag{5.3.18}$$

for $X \sim \mathcal{N}(m, \tilde{\rho})$. When $m = 0$, (5.3.18) is the same as (3.3.32).   ◇

5.3.2.2 Bi-variate Normal Random Vectors

When $n = 2$, specific simpler expressions of (5.3.12) for all possible pairs (n, N) are shown in Table 5.1. Let us consider the expected value $E\{g_1(X_1) g_2(X_2)\}$ for a normal random vector $X = (X_1, X_2)$ with mean vector $m = (m_1, m_2)$ and covariance matrix $K = \begin{bmatrix} \tilde{\rho}_{11} & \tilde{\rho}_{12} \\ \tilde{\rho}_{12} & \tilde{\rho}_{22} \end{bmatrix}$ assuming $n = 2, N = 1, r_1 = 1$, and $s_1 = 2$ in Theorem 5.3.1. Because $\epsilon_{11} = 1$ and $\epsilon_{21} = 1$, we can rewrite (5.3.12) as

$$\frac{\partial^k}{\partial \tilde{\rho}_{12}^k} E\{g_1(X_1) g_2(X_2)\} = E\left\{ g_1^{(k)}(X_1) g_2^{(k)}(X_2) \right\}. \tag{5.3.19}$$

First, find a value k for which the right-hand side $E\left\{ g_1^{(k)}(X_1) g_2^{(k)}(X_2) \right\}$ of (5.3.19) is simple to evaluate, and then obtain the expected value. Next, integrate the expected value with respect to $\tilde{\rho}_{12}$ to obtain $E\{g_1(X_1) g_2(X_2)\}$. Note that, when $\tilde{\rho}_{12} = 0$, we have $\rho_{12} = \frac{\tilde{\rho}_{12}}{\sigma_1 \sigma_2} = 0$ and therefore X_1 and X_2 are independent of each other from Theorem 5.2.3: this implies, from Theorem 4.3.6, that

$$E\left\{ g_1^{(k)}(X_1) g_2^{(l)}(X_2) \right\}\Big|_{\tilde{\rho}_{12}=0} = E\left\{ g_1^{(k)}(X_1) \right\} E\left\{ g_2^{(l)}(X_2) \right\} \tag{5.3.20}$$

Table 5.1 Specific formulas of Price's theorem for all possible pairs (n, N) when $n = 2$

(n, N)	$\left\{(r_j, s_j)\right\}_{j=1}^{N}, \left\{\delta_{r_j s_j}\right\}_{j=1}^{N}, \left\{\{\epsilon_{ij}\}_{i=1}^{n}\right\}_{j=1}^{N}$:
	Specific formula of (5.3.12)
$(2, 1)$	$(r_1, s_1) = (1, 1), \delta_{r_1 s_1} = 1, \epsilon_{11} = 2, \epsilon_{21} = 0$:
	$\dfrac{\partial^k}{\partial \tilde{\rho}_{11}^k} \mathsf{E}\{g_1(X_1) g_2(X_2)\} = \left(\dfrac{1}{2}\right)^k \mathsf{E}\left\{g_1^{(2k)}(X_1) g_2(X_2)\right\}$
$(2, 1)$	$(r_1, s_1) = (1, 2), \delta_{r_1 s_1} = 0, \epsilon_{11} = 1, \epsilon_{21} = 1$:
	$\dfrac{\partial^k}{\partial \tilde{\rho}_{12}^k} \mathsf{E}\{g_1(X_1) g_2(X_2)\} = \mathsf{E}\left\{g_1^{(k)}(X_1) g_2^{(k)}(X_2)\right\}$
$(2, 2)$	$(r_1, s_1) = (1, 1), (r_2, s_2) = (1, 2), \delta_{r_1 s_1} = 1, \delta_{r_2 s_2} = 0,$
	$\epsilon_{11} = 2, \epsilon_{21} = 0, \epsilon_{12} = 1, \epsilon_{22} = 1$:
	$\dfrac{\partial^{k_1 + k_2}}{\partial \tilde{\rho}_{11}^{k_1} \partial \tilde{\rho}_{12}^{k_2}} \mathsf{E}\{g_1(X_1) g_2(X_2)\} =$
	$\left(\dfrac{1}{2}\right)^{k_1} \mathsf{E}\left\{g_1^{(2k_1 + k_2)}(X_1) g_2^{(k_2)}(X_2)\right\}$
$(2, 2)$	$(r_1, s_1) = (1, 1), (r_2, s_2) = (2, 2), \delta_{r_1 s_1} = 1, \delta_{r_2 s_2} = 1,$
	$\epsilon_{11} = 2, \epsilon_{21} = 0, \epsilon_{12} = 0, \epsilon_{22} = 2$:
	$\dfrac{\partial^{k_1 + k_2}}{\partial \tilde{\rho}_{11}^{k_1} \partial \tilde{\rho}_{22}^{k_2}} \mathsf{E}\{g_1(X_1) g_2(X_2)\} =$
	$\left(\dfrac{1}{2}\right)^{k_1 + k_2} \mathsf{E}\left\{g_1^{(2k_1)}(X_1) g_2^{(2k_2)}(X_2)\right\}$
$(2, 3)$	$(r_1, s_1) = (1, 1), (r_2, s_2) = (1, 2), (r_3, s_3) = (2, 2),$
	$\delta_{r_1 s_1} = 1, \delta_{r_2 s_2} = 0, \delta_{r_3 s_3} = 1,$
	$\epsilon_{11} = 2, \epsilon_{21} = 0, \epsilon_{12} = 1, \epsilon_{22} = 1, \epsilon_{13} = 0, \epsilon_{23} = 2$:
	$\dfrac{\partial^{k_1 + k_2 + k_3}}{\partial \tilde{\rho}_{11}^{k_1} \partial \tilde{\rho}_{12}^{k_2} \partial \tilde{\rho}_{22}^{k_3}} \mathsf{E}\{g_1(X_1) g_2(X_2)\} =$
	$\left(\dfrac{1}{2}\right)^{k_1 + k_3} \mathsf{E}\left\{g_1^{(2k_1 + k_2)}(X_1) g_2^{(k_2 + 2k_3)}(X_2)\right\}$

for $k, l = 0, 1, \ldots$, which can be used to determine integration constants. In short, when we can easily evaluate the expected value of the product of the derivatives of g_1 and g_2 (e.g., when we have a constant or an impulse function after a few times of differentiations of g_1 and/or g_2), Theorem 5.3.1 is quite useful in obtaining $\mathsf{E}\{g_1(X_1) g_2(X_2)\}$.

Example 5.3.4 For a normal random vector $X = (X_1, X_2)$ with mean vector $m = (m_1, m_2)$ and covariance matrix $K = \begin{bmatrix} \tilde{\rho}_{11} & \tilde{\rho}_{12} \\ \tilde{\rho}_{12} & \tilde{\rho}_{22} \end{bmatrix}$, obtain $\tilde{\gamma} = \mathsf{E}\{X_1 X_2^2\}$.

Solution With $k = 1$, $g_1(x) = x$, and $g_2(x) = x^2$ in (5.3.19), we get $\frac{d\tilde{\gamma}}{d\tilde{\rho}_{12}} = \mathsf{E}\left\{g_1^{(1)}(X_1) g_2^{(1)}(X_2)\right\} = \mathsf{E}\{2X_2\} = 2m_2$, i.e., $\tilde{\gamma} = 2m_2\tilde{\rho}_{12} + c$. Recollecting (5.3.20), we have $c = \tilde{\gamma}\Big|_{\tilde{\rho}_{12}=0} = m_1\left(\tilde{\rho}_{22} + m_2^2\right)$. Thus, we finally have

$$\mathsf{E}\left\{X_1 X_2^2\right\} = 2m_2\tilde{\rho}_{12} + m_1\left(\tilde{\rho}_{22} + m_2^2\right). \tag{5.3.21}$$

The result (5.3.21) is the same as the result $E\{WZ^2\} = 2m_2\rho\sigma_1\sigma_2 + m_1(\sigma_2^2 + m_2^2)$ for a random vector $(W, Z) = (\sigma_1 X + m_1, \sigma_2 Y + m_2)$ which we would obtain after some steps based on $E\{XY^2\} = 0$ for $(X, Y) \sim \mathcal{N}(0, 0, 1, 1, \rho)$. In addition, when $X_1 = X_2 = X$, (5.3.21) is the same as (5.3.15).   ◇

A general formula for the joint moment $E\{X_1^a X_2^b\}$ is shown in the theorem below.

Theorem 5.3.3 *The joint moment* $E\{X_1^a X_2^b\}$ *can be expressed as*

$$E\{X_1^a X_2^b\} = \sum_{j=0}^{\min(a,b)} \sum_{p=0}^{a-j} \sum_{q=0}^{b-j} \frac{a!b!\, \tilde{\rho}_{12}^j \tilde{\rho}_{11}^p \tilde{\rho}_{22}^q\, m_1^{a-j-2p} m_2^{b-j-2q}}{2^{p+q} j!p!q!(a-j-2p)!(b-j-2q)!} \quad (5.3.22)$$

for $(X_1, X_2) \sim \mathcal{N}\left(m_1, m_2, \tilde{\rho}_{11}, \tilde{\rho}_{22}, \frac{\tilde{\rho}_{12}}{\sqrt{\tilde{\rho}_1 \tilde{\rho}_{22}}}\right)$, *where* $a, b = 0, 1, \ldots$.

Proof A proof is provided in Appendix 5.1.   ♠

Example 5.3.5 We can obtain $E\{X_1^2 X_2^3\} = \sum_{j=0}^{2} \sum_{p=0}^{2-j} \sum_{q=0}^{3-j} \frac{12\tilde{\rho}_{12}^j \tilde{\rho}_{11}^p \tilde{\rho}_{22}^q}{2^{p+q} j!p!q!} \frac{m_1^{2-j-2p}}{(2-j-2p)!}$ $\frac{m_2^{3-j-2q}}{(3-j-2q)!}$, i.e.,

$$E\{X_1^2 X_2^3\} = m_1^2 m_2^3 + 3\tilde{\rho}_{22} m_1^2 m_2 + \tilde{\rho}_{11} m_2^3 + 6\tilde{\rho}_{12} m_1 m_2^2$$
$$+ 3\tilde{\rho}_{11}\tilde{\rho}_{22} m_2 + 6\tilde{\rho}_{12}\tilde{\rho}_{22} m_1 + 6\tilde{\rho}_{12}^2 m_2 \quad (5.3.23)$$

from (5.3.22).   ◇

Theorem 5.3.4 *For* $(X_1, X_2) \sim \mathcal{N}(m_1, m_2, \sigma_1^2, \sigma_2^2, \rho)$, *the joint central moment* $\mu_{ab} = E\{(X_1 - m_1)^a (X_2 - m_2)^b\}$ *can be obtained as (Johnson and Kotz 1972; Mills 2001; Patel and Read 1996)*

$$\mu_{ab} = \begin{cases} 0, & a+b \text{ is odd}, \\ \frac{a!b!}{2^{g+h+\xi}} \sum_{j=0}^{t} \frac{(2\rho\sigma_1\sigma_2)^{2j+\xi}}{(g-j)!(h-j)!(2j+\xi)!}, & a+b \text{ is even} \end{cases} \quad (5.3.24)$$

for $a, b = 0, 1, \ldots$ *and satisfies the recursion*

$$\mu_{ab} = (a+b-1)\rho\sigma_1\sigma_2\mu_{a-1,b-1}$$
$$+ (a-1)(b-1)(1-\rho^2)\sigma_1^2\sigma_2^2\mu_{a-2,b-2}, \quad (5.3.25)$$

where g *and* h *are the quotients of* a *and* b, *respectively, when divided by 2;* ξ *is the residue when* a *or* b *is divided by 2; and* $t = \min(g, h)$.

Example 5.3.6 When $a = 2g$, $b = 2h$, and $m_1 = m_2 = 0$, all the terms except for those satisfying $a - j - 2p = 0$ and $b - j - 2q = 0$ will be zero in (5.3.22), and thus we have

$$\mathsf{E}\{X_1^a X_2^b\} = \sum_{j=0,2,\dots}^{\min(a,b)} \frac{a!b!}{2^{g+h-j} j! \left(\frac{a-j}{2}\right)! \left(\frac{b-j}{2}\right)! \, 0! 0!} \tilde{\rho}_{12}^j m_1^0 m_2^0$$

$$= \sum_{j=0}^{\min(g,h)} \frac{a!b!}{2^{g+h}(2j)!(g-j)!(h-j)!} (2\tilde{\rho}_{12})^{2j}, \qquad (5.3.26)$$

which is the same as the second line in the right-hand side of (5.3.24). Similarly, when $a = 2g + 1$, $b = 2h + 1$, and $m_1 = m_2 = 0$, the result (5.3.22) is the same as the second line in the right-hand side of (5.3.24). $\diamond$

Example 5.3.7 (Gardner 1990) Obtain $\tilde{\Upsilon} = \mathsf{E}\{\mathrm{sgn}(X_1)\,\mathrm{sgn}(X_2)\}$ for $X = (X_1, X_2) \sim \mathcal{N}\left(0, 0, \sigma_1^2, \sigma_2^2, \rho\right)$.

Solution First, note that $\frac{d}{dx}g(x) = \frac{d}{dx}\mathrm{sgn}(x) = 2\delta(x)$ and that $\mathsf{E}\{\delta(X_1)\delta(X_2)\} = f(0,0)$, where f denotes the pdf of $\mathcal{N}\left(0, 0, \sigma_1^2, \sigma_2^2, \rho\right)$. Letting $k = 1$ in (5.3.19), we have $\frac{d\tilde{\Upsilon}}{d\tilde{\rho}} = \mathsf{E}\left\{g_1^{(1)}(X_1) g_2^{(1)}(X_2)\right\} = 4f(0,0) = \frac{2}{\pi\sigma_1\sigma_2\sqrt{1-\rho^2}}$, i.e.,

$$\frac{d\tilde{\Upsilon}}{d\rho} = \frac{2}{\pi}\frac{1}{\sqrt{1-\rho^2}} \qquad (5.3.27)$$

because $\tilde{\rho} = \rho\sigma_1\sigma_2$. Integrating this result, we get[5]

$$\mathsf{E}\{\mathrm{sgn}(X_1)\,\mathrm{sgn}(X_2)\} = \frac{2}{\pi}\sin^{-1}\rho + c. \qquad (5.3.28)$$

Subsequently, because $\mathsf{E}\{\mathrm{sgn}(X_1)\,\mathrm{sgn}(X_2)\}|_{\rho=0} = \mathsf{E}\{\mathrm{sgn}(X_1)\}\mathsf{E}\{\mathrm{sgn}(X_2)\} = 0$ from (5.3.20), we finally have

$$\mathsf{E}\{\mathrm{sgn}(X_1)\,\mathrm{sgn}(X_2)\} = \frac{2}{\pi}\sin^{-1}\rho. \qquad (5.3.29)$$

The result (5.3.29) implies that, when $X_1 = X_2$, we have $\mathsf{E}\{\mathrm{sgn}(X_1)\,\mathrm{sgn}(X_2)\} = \mathsf{E}\{1\} = 1$. Table 5.2 provides the expected values for some non-linear functions of $(X_1, X_2) \sim \mathcal{N}(0, 0, 1, 1, \rho)$. $\diamond$

---

[5] Here, the range of $\sin^{-1} x$ is set as $\left[-\frac{\pi}{2}, \frac{\pi}{2}\right]$.

Table 5.2 Expected value $E\{g_1(X_1)g_2(X_2)\}$ for some non-linear functions g_1 and g_2 of $(X_1, X_2) \sim \mathcal{N}(0, 0, 1, 1, \rho)$

		$g_1(X_1)$			
		X_1	$\lvert X_1 \rvert$	$\mathrm{sgn}(X_1)$	$\delta(X_1)$
$g_2(X_2)$	X_2	ρ	0	$\sqrt{\frac{2}{\pi}}\rho$	0
	$\lvert X_2 \rvert$	0	$\frac{2}{\pi}\left(\rho\sin^{-1}\rho + \sqrt{1-\rho^2}\right)$	0	$\frac{1}{\pi}\sqrt{1-\rho^2}$
	$\mathrm{sgn}(X_2)$	$\sqrt{\frac{2}{\pi}}\rho$	0	$\frac{2}{\pi}\sin^{-1}\rho$	0
	$\delta(X_2)$	0	$\frac{1}{\pi}\sqrt{1-\rho^2}$	0	$\frac{1}{2\pi\sqrt{1-\rho^2}}$

Note. $\frac{d}{d\rho}\left(\rho\sin^{-1}\rho + \sqrt{1-\rho^2}\right) = \sin^{-1}\rho.\ \frac{d}{d\rho}\sin^{-1}\rho = \frac{1}{\sqrt{1-\rho^2}}.$
$E\{\lvert X_i \rvert\} = \sqrt{\frac{2}{\pi}}.\ E\{\delta(X_i)\} = \frac{1}{\sqrt{2\pi}}.$

Denoting the pdf of a standard bi-variate normal random vector (X_1, X_2) by $f_\rho(x, y)$, we have $f_\rho(-x, -y) = f_\rho(x, y)$ and $f_\rho(-x, y) = f_\rho(x, -y) = f_{-\rho}(x, y)$. Then, it is known (Kamat 1958) that the partial moment $[r, s] = \int_0^\infty \int_0^\infty x^r y^s f_\rho(x, y)\,dx\,dy$ is

$$
[r, s] = \begin{cases}
\frac{1}{4}\sqrt{\frac{2}{\pi}}(1+\rho), & r = 1, s = 0, \\
\frac{1}{2\pi}\left\{\rho\left(\frac{\pi}{2} + \sin^{-1}\rho\right) + \sqrt{1-\rho^2}\right\}, & r = 1, s = 1, \\
\frac{1}{2\pi}\left\{3\rho\left(\frac{\pi}{2} + \sin^{-1}\rho\right) \right. \\
\left. + \left(2+\rho^2\right)\sqrt{1-\rho^2}\right\}, & r = 3, s = 1
\end{cases}
\tag{5.3.30}
$$

and the absolute moment $\nu_{rs} = E\left\{\lvert X_1^r X_2^s \rvert\right\}$ is

$$
\nu_{rs} = \frac{2^{\frac{r+s}{2}}}{\pi}\Gamma\left(\frac{r+1}{2}\right)\Gamma\left(\frac{s+1}{2}\right){}_2F_1\left(-\frac{1}{2}r, -\frac{1}{2}s; \frac{1}{2}; \rho^2\right). \tag{5.3.31}
$$

Here, ${}_2F_1(\alpha, \beta; \gamma; z)$ denotes the hypergeometric function introduced in (1.A.24). Based on (5.3.30) and (5.3.31), we can obtain $\nu_{11} = \frac{2}{\pi}\left(\sqrt{1-\rho^2} + \rho\sin^{-1}\rho\right)$, $\nu_{12} = \nu_{21} = \sqrt{\frac{2}{\pi}}\left(1+\rho^2\right)$, $\nu_{13} = \nu_{31} = \frac{2}{\pi}\left\{\left(2+\rho^2\right)\sqrt{1-\rho^2} + 3\rho\sin^{-1}\rho\right\}$, $\nu_{22} = 1 + 2\rho^2$, $\nu_{14} = \nu_{41} = \sqrt{\frac{2}{\pi}}\left(3 + 6\rho^2 - \rho^4\right)$, $\nu_{23} = \sqrt{\frac{8}{\pi}}\left(1 + 3\rho^2\right)$, $\nu_{15} = \nu_{51} = \frac{2}{\pi}\left\{\left(8 + 9\rho^2 - 2\rho^4\right)\sqrt{1-\rho^2} + 15\rho\sin^{-1}\rho\right\}$, $\nu_{42} = \nu_{24} = 3\left(1 + 4\rho^2\right)$, and $\nu_{33} = \frac{2}{\pi}\left\{\left(4 + 11\rho^2\right)\sqrt{1-\rho^2} + 3\left(3 + 2\rho^2\right)\rho\sin^{-1}\rho\right\}$ (Johnson and Kotz 1972).

5.3.2.3 Tri-variate Normal Random Vectors

Let us briefly discuss the case $n = 3$ in Theorem 5.3.1. Letting $N = 3, r_1 = 1, s_1 = 2$, $r_2 = 2, s_2 = 3, r_3 = 3$, and $s_3 = 1$, we have $\epsilon_{11} = 1, \epsilon_{12} = 0, \epsilon_{13} = 1, \epsilon_{21} = 1, \epsilon_{22} = 1, \epsilon_{23} = 0, \epsilon_{31} = 0, \epsilon_{32} = 1$, and $\epsilon_{33} = 1$. Then, for $\tilde{\Upsilon} = \mathsf{E}\{g_1(X_1) g_2(X_2) g_3(X_3)\}$, we can rewrite (5.3.12) as

$$
\frac{\partial^{k_1+k_2+k_3}}{\partial \tilde{\rho}_{12}^{k_1} \partial \tilde{\rho}_{23}^{k_2} \partial \tilde{\rho}_{31}^{k_3}} \tilde{\Upsilon} = \left(\frac{1}{2}\right)^{\sum_{j=1}^{3} k_j \delta_{r_j s_j}} \mathsf{E}\left\{g_1^{(\epsilon_{11}k_1+\epsilon_{12}k_2+\epsilon_{13}k_3)}(X_1)\right.
$$
$$
\left. \times g_2^{(\epsilon_{21}k_1+\epsilon_{22}k_2+\epsilon_{23}k_3)}(X_2)\, g_3^{(\epsilon_{31}k_1+\epsilon_{32}k_2+\epsilon_{33}k_3)}(X_3)\right\}
$$
$$
= \mathsf{E}\left\{g_1^{(k_1+k_3)}(X_1)\, g_2^{(k_1+k_2)}(X_2)\, g_3^{(k_2+k_3)}(X_3)\right\}. \tag{5.3.32}
$$

In addition, similarly to (5.3.20), we have

$$
\mathsf{E}\left\{g_1^{(j)}(X_1)\, g_2^{(k)}(X_2)\, g_3^{(l)}(X_3)\right\}\Big|_{\tilde{\rho}_{12}=\tilde{\rho}_{23}=\tilde{\rho}_{31}=0}
$$
$$
= \mathsf{E}\left\{g_1^{(j)}(X_1)\right\} \mathsf{E}\left\{g_2^{(k)}(X_2)\right\} \mathsf{E}\left\{g_3^{(l)}(X_3)\right\} \tag{5.3.33}
$$

and

$$
\mathsf{E}\left\{g_1^{(j)}(X_1)\, g_2^{(k)}(X_2)\, g_3^{(l)}(X_3)\right\}\Big|_{\tilde{\rho}_{31}=\tilde{\rho}_{12}=0}
$$
$$
= \mathsf{E}\left\{g_1^{(j)}(X_1)\right\} \mathsf{E}\left\{g_2^{(k)}(X_2)\, g_3^{(l)}(X_3)\right\} \tag{5.3.34}
$$

for $j, k, l = 0, 1, \ldots$. For $\tilde{\rho}_{12} = \tilde{\rho}_{23} = 0$ and $\tilde{\rho}_{23} = \tilde{\rho}_{31} = 0$ as well, we can obtain formulas similar to (5.3.34). These formulas can all be used to determine $\mathsf{E}\{g_1(X_1) g_2(X_2) g_3(X_3)\}$.

Example 5.3.8 Obtain $\tilde{\Upsilon} = \mathsf{E}\{X_1 X_2 X_3\}$ for $X = (X_1, X_2, X_3) \sim \mathcal{N}(m, K)$ with $m = (m_1, m_2, m_3)$ and $K = [\tilde{\rho}_{ij}]$.

Solution From (5.3.32), we have $\frac{\partial}{\partial \tilde{\rho}_{12}} \tilde{\Upsilon} = \mathsf{E}\{X_3\} = m_3, \frac{\partial}{\partial \tilde{\rho}_{23}} \tilde{\Upsilon} = m_1$, and $\frac{\partial}{\partial \tilde{\rho}_{31}} \tilde{\Upsilon} = m_2$. Thus, $\tilde{\Upsilon} = m_3 \tilde{\rho}_{12} + m_1 \tilde{\rho}_{23} + m_2 \tilde{\rho}_{31} + c$. Now, when $\tilde{\rho}_{12} = \tilde{\rho}_{23} = \tilde{\rho}_{31} = 0$, we have $\tilde{\Upsilon} = c = \mathsf{E}\{X_1\}\mathsf{E}\{X_2\}\mathsf{E}\{X_3\} = m_1 m_2 m_3$ as we can see from (5.3.33). Thus, we have

$$
\mathsf{E}\{X_1 X_2 X_3\} = m_1 \tilde{\rho}_{23} + m_2 \tilde{\rho}_{31} + m_3 \tilde{\rho}_{12} + m_1 m_2 m_3. \tag{5.3.35}
$$

The result (5.3.35) is the same as (5.3.9), as (5.3.21) when $X_2 = X_3$, and as (5.3.15) when $X_1 = X_2 = X_3 = X$. For a zero-mean tri-variate normal random vector, (5.3.35) implies $\mathsf{E}\{X_1 X_2 X_3\} = 0$. $\diamond$

Consider a standard tri-variate normal random vector $X \sim \mathcal{N}(0, K_3)$ and its pdf $f_3(x, y, z)$ as described in Sect. 5.1.3. For the partial moment

$$[r, s, t] = \int_0^\infty \int_0^\infty \int_0^\infty x^r y^s z^t f_3(x, y, z) dx dy dz \qquad (5.3.36)$$

and absolute moment

$$\nu_{rst} = \mathsf{E}\left\{ |X_1^r X_2^s X_3^t| \right\}, \qquad (5.3.37)$$

it is known that we have[6]

$$[1, 0, 0] = \frac{1}{\sqrt{8\pi^3}} \left\{ \frac{\pi}{2} + \sin^{-1}\beta_{23,1} + \rho_{12}\left(\frac{\pi}{2} + \sin^{-1}\beta_{31,2}\right) \right.$$
$$\left. + \rho_{31}\left(\frac{\pi}{2} + \sin^{-1}\beta_{12,3}\right) \right\}, \qquad (5.3.38)$$

$$[2, 0, 0] = \frac{1}{4\pi} \left\{ \frac{\pi}{2} + \sum^c \sin^{-1}\rho_{ij} + \rho_{12}\sqrt{1 - \rho_{12}^2} + \rho_{31}\sqrt{1 - \rho_{31}^2} \right.$$
$$\left. + (2\rho_{31}\rho_{12} - \rho_{23})\sqrt{1 - \rho_{23}^2} + \frac{|K_3|\rho_{23}}{\sqrt{1 - \rho_{23}^2}} \right\}, \qquad (5.3.39)$$

$$[1, 1, 0] = \frac{1}{4\pi} \left\{ \rho_{12}\left(\frac{\pi}{2} + \sum^c \sin^{-1}\rho_{ij}\right) + \sqrt{1 - \rho_{12}^2} \right.$$
$$\left. + \rho_{23}\sqrt{1 - \rho_{31}^2} + \rho_{31}\sqrt{1 - \rho_{23}^2} \right\}, \qquad (5.3.40)$$

$$[1, 1, 1] = \frac{1}{\sqrt{8\pi^3}} \left\{ \sqrt{|K_3|} + \sum^c (\rho_{ij} + \rho_{jk}\rho_{ki})\left(\frac{\pi}{2} + \sin^{-1}\beta_{ij,k}\right) \right\}, \quad (5.3.41)$$

$$\nu_{211} = \frac{2}{\pi} \left\{ (\rho_{23} + 2\rho_{12}\rho_{31})\sin^{-1}\rho_{23} + (1 + \rho_{12}^2 + \rho_{31}^2)\sqrt{1 - \rho_{23}^2} \right\}, \quad (5.3.42)$$

and

---

[6] The last term $|K_3|\rho_{23}\sqrt{1 - \rho_{23}^2}$ of $[2, 0, 0]$ and the last two terms $\rho_{23}\sqrt{1 - \rho_{23}^2} + \rho_{31}\sqrt{1 - \rho_{31}^2}$ of $[1, 1, 0]$ given in (Johnson and Kotz 1972, Kamat 1958) some references should be corrected into $\frac{|K_3|\rho_{23}}{\sqrt{1 - \rho_{23}^2}}$ as in (5.3.39) and $\rho_{23}\sqrt{1 - \rho_{31}^2} + \rho_{31}\sqrt{1 - \rho_{23}^2}$ as in (5.3.40), respectively.

$$\nu_{221} = \sqrt{\frac{2}{\pi}} \left(1 + 2\rho_{12}^2 + \rho_{23}^2 + \rho_{31}^2 + 4\rho_{12}\rho_{23}\rho_{31} - \rho_{23}^2\rho_{31}^2 \right). \quad (5.3.43)$$

In (5.3.40) and (5.3.41), the symbol $\sum^c$ denotes the cyclic sum: for example, we have

$$\sum^c \sin^{-1} \rho_{ij} = \sin^{-1} \rho_{12} + \sin^{-1} \rho_{23} + \sin^{-1} \rho_{31}. \quad (5.3.44)$$

We will have $\binom{2+3-1}{3} = 4$, $\binom{3+3-1}{3} = 10$, $\binom{4+3-1}{3} = 20$ different[7] cases, respectively, of the expected value $\mathsf{E}\{g_1(X_1) g_2(X_2) g_3(X_3)\}$ for two, three, and four options as the function g_i. For a standard tri-variate normal random vector $X = (X_1, X_2, X_3)$, consider four functions $\{x, |x|, \operatorname{sgn}(x), \delta(x)\}$ of $g_i(x)$. Among the 20 expected values, due to the symmetry of the standard normal distribution, the four expected values $\mathsf{E}\{X_1 X_2 \operatorname{sgn}(X_3)\}$, $\mathsf{E}\{X_1 \operatorname{sgn}(X_2) \operatorname{sgn}(X_3)\}$, $\mathsf{E}\{\operatorname{sgn}(X_1) \operatorname{sgn}(X_2) \operatorname{sgn}(X_3)\}$, $\mathsf{E}\{X_1 X_2 X_3\}$ of products of three odd functions and the six expected values $\mathsf{E}\{X_1 \delta(X_2) \delta(X_3)\}$, $\mathsf{E}\{\operatorname{sgn}(X_1) |X_2| |X_3|\}$, $\mathsf{E}\{\operatorname{sgn}(X_1) |X_2| \delta(X_3)\}$, $\mathsf{E}\{\operatorname{sgn}(X_1) \delta(X_2) \delta(X_3)\}$, $\mathsf{E}\{X_1 |X_2| |X_3|\}$, $\mathsf{E}\{X_1 |X_2| \delta(X_3)\}$ of products of two odd functions and one even function are zero.

In addition, we easily get $\mathsf{E}\{\delta(X_1) \delta(X_2) \delta(X_3)\} = f_3(0, 0, 0)$, i.e.,

$$\mathsf{E}\{\delta(X_1) \delta(X_2) \delta(X_3)\} = \frac{1}{\sqrt{8\pi^3 |K_3|}} \quad (5.3.45)$$

based on (5.1.19), and the nine remaining expected values are considered in Exercises 5.21–5.23. The results of these ten expected values are summarized in Table 5.3. Meanwhile, some results shown in Table 5.3 can be verified using (5.3.38)–(5.3.43): for instance, we can reconfirm $\mathsf{E}\{X_1 |X_2| \operatorname{sgn}(X_3)\} = \frac{1}{2}\frac{\partial \nu_{211}}{\partial \rho_{31}} = \frac{2}{\pi}$ $\left(\rho_{12} \sin^{-1} \rho_{23} + \rho_{31}\sqrt{1 - \rho_{23}^2}\right)$ via ν_{211} shown in (5.3.42) and $\mathsf{E}\{X_1 X_2 |X_3|\}$ $= \frac{1}{4}\frac{\partial}{\partial \rho_{12}}\nu_{221} = \sqrt{\frac{2}{\pi}} (\rho_{12} + \rho_{23}\rho_{31})$ via ν_{221} shown in (5.3.43).

5.3.3 General Formula for Joint Moments

For a natural number n, let

$$a = \{a_1, a_2, \ldots, a_n\} \quad (5.3.46)$$

with $a_i \in \{0, 1, \ldots\}$ a set of non-negative integers and let

---

[7] This number is considered in (1.E.24).

Table 5.3 Expected values $\mathsf{E}\left\{g_1\left(X_1\right)g_2\left(X_2\right)g_3\left(X_3\right)\right\}$ of some products for a standard tri-variate normal random vector (X_1, X_2, X_3)

$\mathsf{E}\left\{\delta\left(X_1\right)\delta\left(X_2\right)\delta\left(X_3\right)\right\}$	$= \dfrac{1}{\sqrt{8\pi^3\lvert K_3\rvert}}$
$\mathsf{E}\left\{X_1 X_2\lvert X_3\rvert\right\}$	$= \sqrt{\dfrac{2}{\pi}}\left(\rho_{12}+\rho_{23}\rho_{31}\right)$
$\mathsf{E}\left\{\delta\left(X_1\right)\delta\left(X_2\right)\lvert X_3\rvert\right\}$	$= \dfrac{1}{\sqrt{2\pi^3}\left(1-\rho_{12}^2\right)}\sqrt{\lvert K_3\rvert}$
$\mathsf{E}\left\{\delta\left(X_1\right)X_2 X_3\right\}$	$= \dfrac{1}{\sqrt{2\pi}}\left(\rho_{23}-\rho_{31}\rho_{12}\right)$
$\mathsf{E}\left\{\delta\left(X_1\right)\operatorname{sgn}\left(X_2\right)X_3\right\}$	$= \dfrac{1}{\pi\sqrt{1-\rho_{12}^2}}\left(\rho_{23}-\rho_{31}\rho_{12}\right)$
$\mathsf{E}\left\{\delta\left(X_1\right)\lvert X_2\rvert\lvert X_3\rvert\right\}$	$= \sqrt{\dfrac{2}{\pi^3}}\left\{\left(\rho_{23}-\rho_{31}\rho_{12}\right)\sin^{-1}\beta_{23,1}+\sqrt{\lvert K_3\rvert}\right\}$
$\mathsf{E}\left\{\delta\left(X_1\right)\operatorname{sgn}\left(X_2\right)\operatorname{sgn}\left(X_3\right)\right\}$	$= \sqrt{\dfrac{2}{\pi^3}}\sin^{-1}\beta_{23,1}$
$\mathsf{E}\left\{X_1\lvert X_2\rvert\operatorname{sgn}\left(X_3\right)\right\}$	$= \dfrac{2}{\pi}\left(\rho_{12}\sin^{-1}\rho_{23}+\rho_{31}\sqrt{1-\rho_{23}^2}\right)$
$\mathsf{E}\left\{\operatorname{sgn}\left(X_1\right)\operatorname{sgn}\left(X_2\right)\lvert X_3\rvert\right\}$	$= \sqrt{\dfrac{8}{\pi^3}}\left(\sin^{-1}\beta_{12,3}+\rho_{23}\sin^{-1}\beta_{31,2}+\rho_{31}\sin^{-1}\beta_{23,1}\right)$
$\mathsf{E}\left\{\lvert X_1 X_2 X_3\rvert\right\}$	$= \sqrt{\dfrac{8}{\pi^3}}\left\{\sqrt{\lvert K_3\rvert}+\overset{c}{\sum}\left(\rho_{ij}+\rho_{jk}\rho_{ki}\right)\sin^{-1}\beta_{ij,k}\right\}$

$$l = \left\{l_{11}, l_{12}, \ldots, l_{1n}, l_{22}, l_{23}, \ldots, l_{2n}, \ldots, l_{n-1,n-1}, l_{n-1,n}, l_{nn}\right\} \qquad (5.3.47)$$

with $l_{ij} \in \{\ldots, -1, 0, 1, \ldots\}$ a set of integers. Given a and l, define the collection

$$S_a = \left\{l : \left\{\{l_{ij}\geq 0\}_{j=i}^{n}\right\}_{i=1}^{n},\ \{L_{a,k}\geq 0\}_{k=1}^{n}\right\} \qquad (5.3.48)$$

of l, where

$$L_{a,k} = a_k - l_{kk} - \sum_{j=1}^{n} l_{jk} \qquad (5.3.49)$$

for $k = 1, 2, \ldots, n$ and

$$l_{ji} = l_{ij} \qquad (5.3.50)$$

for $j > i$. A general formula for the joint moments of normal random vectors can now be obtained as shown in the following theorem:

Theorem 5.3.5 *For $X \sim \mathcal{N}(m, K)$ with $m = (m_1, m_2, \ldots, m_n)$ and $K = \left[\tilde{\rho}_{ij}\right]$, we have*

$$\mathsf{E}\left\{\prod_{k=1}^{n} X_k^{a_k}\right\} = \sum_{l\in S_a} d_{a,l}\left(\prod_{i=1}^{n}\prod_{j=i}^{n}\tilde{\rho}_{ij}^{l_{ij}}\right)\left(\prod_{j=1}^{n} m_j^{L_{a,j}}\right), \qquad (5.3.51)$$

where

$$d_{a,l} = 2^{-M_l} \left(\prod_{k=1}^{n} a_k!\right) \left(\prod_{i=1}^{n} \prod_{j=i}^{n} l_{ij}!\right)^{-1} \left(\prod_{j=1}^{n} L_{a,j}!\right)^{-1} \quad (5.3.52)$$

with $M_l = \sum_{i=1}^{n} l_{ii}$.

Proof The proof is shown in Appendix 5.1. ♠

Note that, when any of the $\frac{1}{2}n(n+1)$ elements of l or any of the n elements of $\{L_{a,j}\}_{j=1}^{n}$ is a negative integer, we have $\left(\prod_{i=1}^{n} \prod_{j=i}^{n} l_{ij}!\right)^{-1} \left(\prod_{j=1}^{n} L_{a,j}!\right)^{-1} = 0$ because $(-k)! \to \pm\infty$ for $k = 1, 2, \ldots$. Therefore, the collection S_a in $\sum_{l \in S_a}$ of (5.3.51) can be replaced with the collection of all sets of $\frac{1}{2}n(n+1)$ integers. Details for obtaining $\mathsf{E}\{X_1 X_2 X_3^2\}$ based on Theorem 5.3.5 as an example are shown in Table 5.4 in the case of $a = \{1, 1, 2\}$.

Example 5.3.9 Based on Theorem 5.3.5, we easily get

$$\mathsf{E}\{X_1 X_2 X_3\} = m_1 m_2 m_3 + \tilde{\rho}_{12} m_3 + \tilde{\rho}_{23} m_1 + \tilde{\rho}_{31} m_2 \quad (5.3.53)$$

and

$$\begin{aligned} \mathsf{E}\{X_1 X_2 X_3 X_4\} &= m_1 m_2 m_3 m_4 + m_1 m_2 \tilde{\rho}_{34} + m_1 m_3 \tilde{\rho}_{24} + m_1 m_4 \tilde{\rho}_{23} \\ &\quad + m_2 m_3 \tilde{\rho}_{14} + m_2 m_4 \tilde{\rho}_{13} + m_3 m_4 \tilde{\rho}_{12} \\ &\quad + \tilde{\rho}_{12} \tilde{\rho}_{34} + \tilde{\rho}_{13} \tilde{\rho}_{24} + \tilde{\rho}_{14} \tilde{\rho}_{23}. \end{aligned} \quad (5.3.54)$$

Table 5.4 Element sets $l = \{l_{11}, l_{12}, l_{13}, l_{22}, l_{23}, l_{33}\}$ of S_a, $\{L_{a,j}\}_{j=1}^{3}$, coefficient $d_{a,l}$, and the terms in $\mathsf{E}\{X_1 X_2 X_3^2\}$ for each of the seven element sets when $a = \{1, 1, 2\}$

	$\{l_{11}, l_{12}, l_{13}, l_{22}, l_{23}, l_{33}\}$	$\{L_{a,1}, L_{a,2}, L_{a,3}\}$	$d_{a,l}$	Terms
1	$\{0, 0, 0, 0, 0, 0\}$	$\{1, 1, 2\}$	1	$m_1 m_2 m_3^2$
2	$\{0, 0, 0, 0, 1, 0\}$	$\{1, 0, 1\}$	2	$2\tilde{\rho}_{23} m_1 m_3$
3	$\{0, 0, 1, 0, 0, 0\}$	$\{0, 1, 1\}$	2	$2\tilde{\rho}_{13} m_2 m_3$
4	$\{0, 0, 1, 0, 1, 0\}$	$\{0, 0, 0\}$	2	$2\tilde{\rho}_{13} \tilde{\rho}_{23}$
5	$\{0, 1, 0, 0, 0, 0\}$	$\{0, 0, 2\}$	1	$\tilde{\rho}_{12} m_3^2$
6	$\{0, 0, 0, 0, 0, 1\}$	$\{1, 1, 0\}$	1	$\tilde{\rho}_{33} m_1 m_2$
7	$\{0, 1, 0, 0, 0, 1\}$	$\{0, 0, 0\}$	1	$\tilde{\rho}_{12} \tilde{\rho}_{33}$

Note that (5.3.53) is the same as (5.3.9) and (5.3.35), and (5.3.54) is the same as (5.3.10).   ◇

When the mean vector is **0** in Theorem 5.3.5, we have

$$\mathsf{E}\left\{\prod_{k=1}^{n} X_k^{a_k}\right\} = \left(\prod_{k=1}^{n} a_k!\right) \sum_{l \in T_a} 2^{-M_l} \left(\prod_{i=1}^{n} \prod_{j=i}^{n} \frac{\tilde{\rho}_{ij}^{l_{ij}}}{l_{ij}!}\right), \qquad (5.3.55)$$

where T_a denotes the collection of l such that $l_{11} + \sum_{k=1}^{n} l_{k1} = a_1$, $l_{22} + \sum_{k=1}^{n} l_{k2} = a_2$, $\ldots, l_{nn} + \sum_{k=1}^{n} l_{kn} = a_n$, and $\left\{\{l_{ij} \geq 0\}_{j=i}^{n}\right\}_{i=1}^{n}$. In other words, T_a is the same as S_a with $L_{a,k} \geq 0$ replaced by $L_{a,k} = 0$ in (5.3.48).

Theorem 5.3.6 *We have* $\mathsf{E}\left\{X_1^{a_1} X_2^{a_2} \cdots X_n^{a_n}\right\} = 0$ *for* $X \sim \mathcal{N}(\mathbf{0}, K)$ *when* $\sum_{k=1}^{n} a_k$ *is an odd number.*

Proof Adding $l_{kk} + \sum_{j=1}^{n} l_{kj} = a_k$ for $k = 1, 2, \ldots, n$, we have $\sum_{k=1}^{n} a_k = M_l + \sum_{i=1}^{n} \sum_{j=1}^{n} l_{ij} = 2M_l + 2\sum_{i=1}^{n-1} \sum_{j=i+1}^{n} l_{ij}$, which is an even number. Thus, when $\sum_{k=1}^{n} a_k$ is an odd number, the collection T_a is a null set and $\mathsf{E}\left\{X_1^{a_1} X_2^{a_2} \ldots X_n^{a_n}\right\} = 0$.   ♠

Example 5.3.10 For a zero-mean n-variate normal random vector, assume $a = \mathbf{1} = \{1, 1, \ldots, 1\}$. When n is an odd number, $\mathsf{E}\{X_1 X_2 \cdots X_n\} = 0$ from Theorem 5.3.6. Next, assume n is even. Over a non-negative integer region, if $l_{kk} = 0$ for $k = 1, 2, \ldots, n$ and one of $\{l_{1k}, l_{2k}, \ldots, l_{nk}\} - \{l_{kk}\}$ is 1 and all the others are 0, we have $l_{kk} + \sum_{i=1}^{n} l_{ik} = 1$. Now, if $l \in T_a$, because $d_{a,l} = d_{1,l} = \frac{(1!)^n}{2^0 \left(\prod_{j=1}^{n} 1!\right)(0!)^n} = 1$, we have (Isserlis 1918)

$$\mathsf{E}\{X_1 X_2 \cdots X_n\} = \sum_{l \in T_a} \prod_{i=1}^{n} \prod_{j=i}^{n} \tilde{\rho}_{ij}^{l_{ij}}. \qquad (5.3.56)$$

Next, assigning 0, 1, and 0 to l_{kk}, one of $\{l_{1k}, l_{2k}, \ldots, l_{nk}\} - \{l_{kk}\}$, and all the others of $\{l_{1k}, l_{2k}, \ldots, l_{nk}\} - \{l_{kk}\}$, respectively, for $k = 1, 2, \ldots, n$ is the same as assigning 1 to each pair after dividing $\{1, 2, \ldots, n\}$ into $\underline{n}$ pairs of two numbers. Here, a pair (j, k) represents the subscript of l_{jk}. Now, recollecting that there are $\underline{n}!$ possibilities for the same choice with a different order, the number of ways to divide $\{1, 2, \ldots, n\}$ into $\underline{n}$ pairs of two numbers is $\binom{n}{2}\binom{n-2}{2} \cdots \binom{2}{2} (\underline{n}!)^{-1} = \frac{n!}{2^{\underline{n}}\underline{n}!}$. In short, the number of elements in T_a, i.e., the number of non-zero terms on the right-hand side of (5.3.56), is $\frac{n!}{2^{\underline{n}}\underline{n}!} = (2\underline{n} - 1)!!$.   ◇

5.4 Distributions of Statistics

Often, the terms sample and random sample (Abramowitz and Stegun 1972; Gradshteyn and Ryzhik 1980) are used to denote an i.i.d random vector, especially in statistics. A function of a sample is called a statistic. With a sample[8] $X = (X_1, X_2, \ldots, X_n)$ of size n, the mean $\mathsf{E}\{X_i\}$ and the variance $\mathsf{Var}(X_i)$ of the component random variable X_i are called the population mean and population variance, respectively. Unless stated otherwise, we assume that population mean and population variance are m and σ^2, respectively, for the samples considered in this section. We also denote by $\mathsf{E}\{(X_i - m)^k\} = \mu_k$ the k-th population central moment of X_i for $k = 0, 1, \ldots$.

5.4.1 Sample Mean and Sample Variance

Definition 5.4.1 (*sample mean*) The statistic

$$\overline{X}_n = \frac{1}{n} \sum_{i=1}^{n} X_i \qquad (5.4.1)$$

for a sample $X = (X_1, X_2, \ldots, X_n)$ is called the sample mean of X.

Theorem 5.4.1 *We have the expected value*

$$\mathsf{E}\{\overline{X}_n\} = m \qquad (5.4.2)$$

and variance

$$\mathsf{Var}\{\overline{X}_n\} = \frac{\sigma^2}{n} \qquad (5.4.3)$$

for the sample mean $\overline{X}_n$.

Proof First, $\mathsf{E}\{\overline{X}_n\} = \frac{1}{n} \sum_{i=1}^{n} \mathsf{E}\{X_i\} = m$. Using this result and the fact that $\mathsf{E}\{X_i X_j\} = \mathsf{E}\{X_i\}\mathsf{E}\{X_j\} = m^2$ for $i \neq j$ and $\mathsf{E}\{X_i X_j\} = \mathsf{E}\{X_i^2\} = \sigma^2 + m^2$ for $i = j$, we have (5.4.3) from $\mathsf{Var}\{\overline{X}_n\} = \mathsf{E}\{\overline{X}_n^2\} - \mathsf{E}^2\{\overline{X}_n\} = \mathsf{E}\left\{\frac{1}{n^2} \sum_{i=1}^{n} X_i \sum_{j=1}^{n} X_j\right\} - m^2 =$

---

[8] In several fields including engineering, the term sample is often used to denote an element X_i of $X = (X_1, X_2, \ldots, X_n)$.

$$\frac{1}{n^2}\left[\sum_{i=1}^{n}\mathsf{E}\left\{X_i^2\right\}+\sum_{i=1}^{n}\sum_{\substack{j=1 \\ i\neq j}}^{n}\mathsf{E}\left\{X_iX_j\right\}\right]-m^2=\frac{1}{n^2}\left\{n\left(\sigma^2+m^2\right)+n(n-1)m^2\right\}-m^2.$$

♠

Example 5.4.1 (Rohatgi and Saleh 2001) Obtain the third central moment $\mu_3\left(\overline{X}_n\right)$ of the sample mean $\overline{X}_n$.

Solution The third central moment $\mu_3\left(\overline{X}_n\right)=\mathsf{E}\left\{\left(\overline{X}_n-m\right)^3\right\}=\frac{1}{n^3}\mathsf{E}\left[\left\{\sum_{i=1}^{n}\right.\right.$

$\left.\left.(X_i-m)\right\}^3\right]$ of $X=(X_1,X_2,\ldots,X_n)$ can be expressed as

$$\mu_3\left(\overline{X}_n\right)=\frac{1}{n^3}\sum_{i=1}^{n}\mathsf{E}\left\{(X_i-m)^3\right\}+\frac{1}{n^3}\sum_{i=1}^{n}\sum_{\substack{j=1 \\ i\neq j}}^{n}\mathsf{E}\left\{(X_i-m)^2\left(X_j-m\right)\right\}$$

$$+\frac{1}{n^3}\sum_{i=1}^{n}\sum_{\substack{j=1 \\ i\neq j,\,j\neq k,\,k\neq i}}^{n}\sum_{k=1}^{n}\mathsf{E}\left\{(X_i-m)\left(X_j-m\right)(X_k-m)\right\}.\qquad(5.4.4)$$

Now, noting that $\mathsf{E}\left\{X_i-m\right\}=0$ and that X_i and X_j are independent of each other for $i\neq j$, we have $\mathsf{E}\left\{(X_i-m)^2\left(X_j-m\right)\right\}=\mathsf{E}\left\{(X_i-m)^2\right\}\mathsf{E}\{X_j-m\}=0$ for $i\neq j$ and $\mathsf{E}\left\{(X_i-m)\left(X_j-m\right)(X_k-m)\right\}=\mathsf{E}\{X_i-m\}$ $\mathsf{E}\{X_j-m\}\mathsf{E}\{X_k-m\}=0$ for $i\neq j$, $j\neq k$, and $k\neq i$. Thus, we have

$$\mu_3\left(\overline{X}_n\right)=\frac{\mu_3}{n^2}\qquad(5.4.5)$$

from $\mu_3\left(\overline{X}_n\right)=\frac{1}{n^3}\sum_{i=1}^{n}\mathsf{E}\left\{(X_i-m)^3\right\}$.   ◇

Definition 5.4.2 (*sample variance*) The statistic

$$W_n=\frac{1}{n-1}\sum_{i=1}^{n}\left(X_i-\overline{X}_n\right)^2\qquad(5.4.6)$$

is called the sample variance of $X=(X_1,X_2,\ldots,X_n)$.

Theorem 5.4.2 *We have*

$$\mathsf{E}\{W_n\}=\sigma^2,\qquad(5.4.7)$$

i.e., the expected value of sample variance is equal to the population variance.

Proof Let $Y_i = X_i - m$. Then, we have $\mathsf{E}\{Y_i\} = 0$, $\mathsf{E}\{Y_i^2\} = \sigma^2 = \mu_2$, and $\mathsf{E}\{Y_i^4\} = \mu_4$. Next, letting $\overline{Y} = \frac{1}{n}\sum_{i=1}^{n} Y_i = \frac{1}{n}\sum_{i=1}^{n} (X_i - m)$, we have

$$\sum_{i=1}^{n} (X_i - \overline{X})^2 = \sum_{i=1}^{n} (Y_i - \overline{Y})^2 \tag{5.4.8}$$

from $\sum_{i=1}^{n} (X_i - \overline{X})^2 = \sum_{i=1}^{n} \{X_i - m - (\overline{X} - m)\}^2 = \sum_{i=1}^{n} \left\{Y_i - \frac{1}{n}\sum_{k=1}^{n} (X_k - m)\right\}^2$. In addition, because $\sum_{i=1}^{n} (Y_i - \overline{Y})^2 = \sum_{i=1}^{n} \left(Y_i^2 - 2\overline{Y}Y_i + \overline{Y}^2\right) = \sum_{i=1}^{n} Y_i^2 - 2\overline{Y}\sum_{i=1}^{n} Y_i + n\overline{Y}^2 = \sum_{i=1}^{n} Y_i^2 - n\overline{Y}^2$, we have

$$\sum_{i=1}^{n} (X_i - \overline{X})^2 = \sum_{i=1}^{n} Y_i^2 - n\overline{Y}^2. \tag{5.4.9}$$

Therefore, we have $\mathsf{E}\left\{\sum_{i=1}^{n} (X_i - \overline{X})^2\right\} = \sum_{i=1}^{n} \mathsf{E}\{Y_i^2\} - n\mathsf{E}\left\{\overline{Y}^2\right\} = n\sigma^2 - \frac{n}{n^2}\mathsf{E}\left\{\sum_{i=1}^{n}\sum_{j=1}^{n} Y_iY_j\right\} = n\sigma^2 - \frac{1}{n}\left[n\mathsf{E}\{Y_i^2\} + 0\right] = (n-1)\sigma^2$ and $\mathsf{E}\{W_n\} = \mathsf{E}\left\{\frac{1}{n-1}\sum_{i=1}^{n} (X_i - \overline{X})^2\right\} = \sigma^2$. ♠

Note that, due to the factor $n - 1$ instead of n in the denominator of (5.4.6), the expected value of sample variance is equal to the population variance as shown in (5.4.7).

Theorem 5.4.3 (Rohatgi and Saleh 2001) *We have the variance*

$$\mathsf{Var}\{W_n\} = \frac{\mu_4}{n} - \frac{(n-3)\mu_2^2}{n(n-1)} \tag{5.4.10}$$

of the sample variance W_n.

Proof Letting $Y_i = X_i - m$, we have $\mathsf{E}\left\{\left(\sum_{i=1}^{n} Y_i^2 - n\overline{Y}^2\right)^2\right\} = \mathsf{E}\left\{\sum_{i=1}^{n}\sum_{j=1}^{n} Y_i^2 Y_j^2 - 2n\overline{Y}^2\sum_{i=1}^{n} Y_i^2 + n^2\overline{Y}^4\right\}$, i.e.,

$$\mathsf{E}\left\{\left(\sum_{i=1}^{n}Y_i^2 - n\overline{Y}^2\right)^2\right\} = \mathsf{E}\left\{\sum_{i=1}^{n}\sum_{j=1}^{n}Y_i^2 Y_j^2 - 2n\overline{Y}^2\sum_{i=1}^{n}Y_i^2 + n^2\overline{Y}^4\right\}$$

$$= n\mu_4 + n(n-1)\mu_2^2 - 2n\mathsf{E}\left\{\overline{Y}^2\sum_{i=1}^{n}Y_i^2\right\} + n^2\mathsf{E}\left\{\overline{Y}^4\right\}. \qquad (5.4.11)$$

In (5.4.11), $\mathsf{E}\left\{\overline{Y}^2\sum_{i=1}^{n}Y_i^2\right\} = \frac{1}{n^2}\mathsf{E}\left\{\sum_{i=1}^{n}\sum_{j=1}^{n}\sum_{k=1}^{n}Y_i^2 Y_j Y_k\right\} = \frac{1}{n^2}\mathsf{E}\left\{\sum_{i=1}^{n}Y_i^4 + \sum_{\substack{i=1 \\ i\neq j}}^{n}\sum_{j=1}^{n}Y_i^2 Y_j^2\right\}$ can

be evaluated as

$$\mathsf{E}\left\{\overline{Y}^2\sum_{i=1}^{n}Y_i^2\right\} = \frac{1}{n}\left\{\mu_4 + (n-1)\mu_2^2\right\} \qquad (5.4.12)$$

and $\mathsf{E}\left\{\overline{Y}^4\right\} = \frac{1}{n^4}\mathsf{E}\left\{\sum_{i=1}^{n}\sum_{j=1}^{n}\sum_{k=1}^{n}\sum_{l=1}^{n}Y_i Y_j Y_k Y_l\right\} = \frac{1}{n^4}\mathsf{E}\left\{\sum_{i=1}^{n}Y_i^4 + 3\sum_{\substack{i=1 \\ i\neq j}}^{n}\sum_{j=1}^{n}Y_i^2 Y_j^2\right\}$ can

be obtained as[9]

$$\mathsf{E}\left\{\overline{Y}^4\right\} = \frac{1}{n^3}\left\{\mu_4 + 3(n-1)\mu_2^2\right\}. \qquad (5.4.13)$$

Next, recollecting (5.4.9), (5.4.12), and (5.4.13), if we rewrite (5.4.11), we have $\mathsf{E}\left[\left\{\sum_{i=1}^{n}\left(X_i - \overline{X}\right)^2\right\}^2\right] = \mathsf{E}\left\{\left(\sum_{i=1}^{n}Y_i^2 - n\overline{Y}^2\right)^2\right\} = n\mu_4 + n(n-1)\mu_2^2 - \frac{2n}{n}\left\{\mu_4 + (n-1)\mu_2^2\right\} + \frac{n^2}{n^3}\left\{\mu_4 + 3(n-1)\mu_2^2\right\}$, i.e.,

$$\mathsf{E}\left[\left\{\sum_{i=1}^{n}\left(X_i - \overline{X}\right)^2\right\}^2\right] = \frac{(n-1)^2}{n}\mu_4 + \frac{(n-1)\left(n^2 - 2n + 3\right)}{n}\mu_2^2. \qquad (5.4.14)$$

We get (5.4.10) from $\mathsf{Var}\left\{W_n\right\} = \frac{1}{(n-1)^2}\mathsf{E}\left[\left\{\sum_{i=1}^{n}\left(X_i - \overline{X}\right)^2\right\}^2\right] - \mu_2^2$ using (5.4.7) and (5.4.14).   ♠

Theorem 5.4.4 *We have*

$$\overline{X}_n \sim \mathcal{N}\left(\mu, \frac{\sigma^2}{n}\right) \qquad (5.4.15)$$

---

[9] In this formula, the factor 3 results from the three distinct cases of $i = j \neq k = l$, $i = k \neq j = l$, and $i = l \neq k = j$.

and consequently

$$\frac{\sqrt{n}}{\sigma}\left(\overline{X}_n - \mu\right) \sim \mathcal{N}(0, 1) \qquad (5.4.16)$$

for a sample $X = (X_1, X_2, \ldots, X_n)$ *from* $\mathcal{N}\left(\mu, \sigma^2\right).$

Proof Recollecting the mgf $M(t) = \exp\left(\mu t + \frac{1}{2}\sigma^2 t^2\right)$ of $\mathcal{N}\left(\mu, \sigma^2\right)$ and using (5.E.23), the mgf of $\overline{X}_n$ can be obtained as $M_{\overline{X}_n}(t) = \left\{M\left(\frac{t}{n}\right)\right\}^n = \exp\left\{\mu t + \frac{1}{2}\left(\frac{\sigma}{\sqrt{n}}\right)^2 t^2\right\}.$ Thus, we have (5.4.15), and (5.4.16) follows. ♠

Theorem 5.4.4 can also be shown from Theorem 5.2.5. More generally, we have $\overline{X}_n \sim \mathcal{N}\left(\overline{\mu}, \frac{\overline{\sigma^2}}{n}\right)$ and $\frac{\sqrt{n}}{\sqrt{\overline{\sigma^2}}}\left(\overline{X}_n - \overline{\mu}\right) \sim \mathcal{N}(0, 1)$ when $\left\{X_i \sim \mathcal{N}\left(\mu_i, \sigma_i^2\right)\right\}_{i=1}^n$ are independent of each other, where $\overline{\mu} = \frac{1}{n}\sum\limits_{i=1}^n \mu_i$ and $\overline{\sigma^2} = \frac{1}{n}\sum\limits_{i=1}^n \sigma_i^2.$

Theorem 5.4.5 *The sample mean and sample variance of a normal sample are independent of each other.*

Proof We first show that $A = \overline{X}_n$ and $B = (V_1, V_2, \ldots, V_n) = \left(X_1 - \overline{X}_n, X_2 - \overline{X}_n, \ldots, X_n - \overline{X}_n\right)$ are independent of each other. Letting $t' = (t, t_1, t_2, \ldots, t_n)$, the joint mgf $M_{A,B}\left(t'\right) = \mathsf{E}\left\{\exp\left(t\overline{X}_n + t_1 V_1 + t_2 V_2 + \cdots + t_n V_n\right)\right\} = \mathsf{E}\left[\exp\left\{t\overline{X}_n + \sum\limits_{i=1}^n t_i\left(X_i - \overline{X}_n\right)\right\}\right]$ of A and B is obtained as

$$M_{A,B}\left(t'\right) = \mathsf{E}\left[\exp\left\{\left(\sum_{i=1}^n t_i X_i\right) - \left\{\left(\sum_{i=1}^n t_i\right) - t\right\}\overline{X}_n\right\}\right]$$

$$= \mathsf{E}\left[\exp\left\{\sum_{i=1}^n \frac{1}{n}\left(nt_i + t - \sum_{j=1}^n t_j\right)X_i\right\}\right]. \qquad (5.4.17)$$

Letting $\overline{t} = \frac{1}{n}\sum\limits_{i=1}^n t_i$, the joint mgf (5.4.17) can be expressed as $M_{A,B}\left(t'\right) = \prod\limits_{i=1}^n \mathsf{E}\left[\exp\left\{X_i\left(\frac{t+nt_i-n\overline{t}}{n}\right)\right\}\right] = \prod\limits_{i=1}^n \exp\left\{\mu\left(\frac{t+nt_i-n\overline{t}}{n}\right) + \frac{\sigma^2}{2}\left(\frac{t+nt_i-n\overline{t}}{n}\right)^2\right\} = \prod\limits_{i=1}^n \exp\left[\mu\left(t_i - \overline{t}\right) + \frac{\sigma^2}{2n^2}\left\{2nt\left(t_i - \overline{t}\right) + n^2\left(t_i - \overline{t}\right)^2\right\}\right]\prod\limits_{i=1}^n \exp\left(\frac{\mu t}{n} + \frac{\sigma^2 t^2}{2n^2}\right) = \exp\left(\mu t + \frac{\sigma^2 t^2}{2n}\right)\prod\limits_{i=1}^n \exp\left\{\frac{\sigma^2}{2}\left(t_i - \overline{t}\right)^2\right\}\prod\limits_{i=1}^n \exp\left\{\left(\mu + \frac{\sigma^2 t}{n}\right)\left(t_i - \overline{t}\right)\right\}$, or as

$$M_{A,B}\left(t'\right) = \exp\left(\mu t + \frac{\sigma^2 t^2}{2n}\right) \exp\left\{\frac{\sigma^2}{2} \sum_{i=1}^{n} \left(t_i - \bar{t}\right)^2\right\}$$

$$\times \exp\left\{\left(\mu + \frac{\sigma^2 t}{n}\right) \sum_{i=1}^{n} \left(t_i - \bar{t}\right)\right\}. \tag{5.4.18}$$

Noting that $\sum_{i=1}^{n} \left(t_i - \bar{t}\right) = \sum_{i=1}^{n} t_i - n\bar{t} = 0$, we eventually have

$$M_{A,B}\left(t'\right) = \exp\left(\mu t + \frac{\sigma^2 t^2}{2n}\right) \exp\left\{\frac{\sigma^2}{2} \sum_{i=1}^{n} \left(t_i - \bar{t}\right)^2\right\}. \tag{5.4.19}$$

Meanwhile, because $A = \overline{X}_n \sim \mathcal{N}\left(\mu, \frac{\sigma^2}{n}\right)$ as we have observed in Theorem 5.4.4, the mgf of A is

$$M_A(t) = \exp\left(\mu t + \frac{\sigma^2}{2n} t^2\right). \tag{5.4.20}$$

Recollecting $\sum_{i=1}^{n} \left(t_i - \bar{t}\right) = 0$, the mgf of $\boldsymbol{B} = (V_1, V_2, \ldots, V_n)$ can be obtained

as $\quad M_B(t) = \mathsf{E}\left\{\exp\left(t_1 V_1 + t_2 V_2 + \cdots + t_n V_n\right)\right\} = \mathsf{E}\left[\exp\left\{\sum_{i=1}^{n} t_i \left(X_i - \overline{X}_n\right)\right\}\right]$

$= \mathsf{E}\left[\exp\left\{\left(\sum_{i=1}^{n} t_i X_i\right) - \bar{t} \sum_{i=1}^{n} X_i\right\}\right] = \prod_{i=1}^{n} \mathsf{E}\left[\exp\left\{X_i \left(t_i - \bar{t}\right)\right\}\right] = \prod_{i=1}^{n} \exp$

$\left\{\mu\left(t_i - \bar{t}\right) + \frac{\sigma^2}{2}\left(t_i - \bar{t}\right)^2\right\} = \exp\left\{\mu \sum_{i=1}^{n}\left(t_i - \bar{t}\right) + \frac{\sigma^2}{2} \sum_{i=1}^{n}\left(t_i - \bar{t}\right)^2\right\}$ or, equivalently,

as

$$M_B(t) = \exp\left\{\frac{\sigma^2}{2} \sum_{i=1}^{n}\left(t_i - \bar{t}\right)^2\right\}, \tag{5.4.21}$$

where $t = (t_1, t_2, \ldots, t_n)$. In short, from (5.4.19)–(5.4.21), the random variable $\overline{X}_n$ and the random vector $\boldsymbol{B} = (V_1, V_2, \ldots, V_n)$ are independent of each other. Consequently, recollecting Theorem 4.1.3, the random variables $\overline{X}_n$ and $W_n = \frac{1}{n-1} \sum_{i=1}^{n} \left(X_i - \overline{X}_n\right)^2$, a function of $\boldsymbol{B} = (V_1, V_2, \ldots, V_n)$, are independent of each other.   ♠

Theorem 5.4.5 is an important property of normal samples, and its converse is known also to hold true: if the sample mean and sample variance of a sample are independent of each other, then the sample is from a normal distribution. In the more general case of samples from symmetric marginal distributions, the sample mean and sample variance are known to be uncorrelated (Rohatgi and Saleh 2001).

5.4.2 Chi-Square Distribution

Let us now consider the distributions of some statistics of normal samples.

Definition 5.4.3 (*central chi-square pdf*) The pdf

$$f(r) = \frac{1}{2^{\frac{n}{2}} \Gamma\left(\frac{n}{2}\right)} r^{\frac{n}{2}-1} \exp\left(-\frac{r}{2}\right) u(r) \tag{5.4.22}$$

is called the (central) chi-square pdf with its distribution denoted by $\chi^2(n)$, where n is called the degree of freedom.

The central chi-square pdf (5.4.22), an example of which is shown in Fig. 5.4, is the same as the gamma pdf $f(r) = \frac{1}{\beta^{\alpha}\Gamma(\alpha)} r^{\alpha-1} \exp\left(-\frac{r}{\beta}\right) u(r)$ introduced in (2.5.31) with α and β replaced by $\frac{n}{2}$ and 2, respectively: in other words, $\chi^2(n) = G\left(\frac{n}{2}, 2\right)$.

Theorem 5.4.6 *The square of a standard normal random variable is a $\chi^2(1)$ random variable.*

Theorem 5.4.6 is proved in Example 3.2.20. Based on Theorem 5.4.6, we have $\frac{n}{\sigma^2}\left(\overline{X}_n - \mu\right)^2 \sim \chi^2(1)$ for a sample of size n from $\mathcal{N}\left(\mu, \sigma^2\right)$ and, more generally, $\frac{n}{\sigma^2}\left(\overline{X}_n - \overline{\mu}\right)^2 \sim \chi^2(1)$ when $\left\{X_i \sim \mathcal{N}\left(\mu_i, \sigma_i^2\right)\right\}_{i=1}^n$ are independent of each other.

Theorem 5.4.7 *We have the mgf*

$$M_Y(t) = (1 - 2t)^{-\frac{n}{2}}, \quad t < \frac{1}{2} \tag{5.4.23}$$

and moments

$$E\left\{Y^k\right\} = 2^k \frac{\Gamma\left(k + \frac{n}{2}\right)}{\Gamma\left(\frac{n}{2}\right)} \tag{5.4.24}$$

for a random variable $Y \sim \chi^2(n)$.

Fig. 5.4 A central chi-square pdf

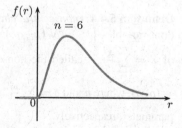

Proof Using (5.4.22), the mgf $M_Y(t) = \mathsf{E}\left\{e^{tY}\right\}$ can be obtained as

$$M_Y(t) = \int_0^\infty \frac{1}{2^{\frac{n}{2}} \Gamma\left(\frac{n}{2}\right)} x^{\frac{n}{2}-1} \exp\left(-\frac{1-2t}{2}x\right) dx. \qquad (5.4.25)$$

Now, letting $y = \frac{1-2t}{2}x$ for $t < \frac{1}{2}$, we have $x = \frac{2y}{1-2t}$ and $dx = \frac{2dy}{1-2t}$. Thus, recollecting (1.4.65), we obtain $M_Y(t) = \frac{1}{2^{\frac{n}{2}}\Gamma\left(\frac{n}{2}\right)} \int_0^\infty \left(\frac{2y}{1-2t}\right)^{\frac{n}{2}-1} e^{-y} \frac{2dy}{1-2t} = (1 - 2t)^{-\frac{n}{2}} \frac{1}{\Gamma\left(\frac{n}{2}\right)} \int_0^\infty y^{\frac{n}{2}-1} e^{-y} dy$, which results in (5.4.23). The moments of Y can easily be obtained as $\mathsf{E}\left\{Y^k\right\} = \frac{d^k}{dt^k} M_Y(t)\Big|_{t=0} = (-2)^k \left(-\frac{n}{2}\right)\left(-\frac{n}{2}-1\right)\cdots \left\{-\frac{n}{2} - (k-1)\right\} = 2^k \frac{n}{2}\left(\frac{n}{2}+1\right)\cdots\left(\frac{n}{2}+k-1\right)$, resulting in (5.4.24). ♠

Example 5.4.2 For $Y \sim \chi^2(n)$, we have the expected value $\mathsf{E}\{Y\} = n$ and variance $\mathrm{Var}(Y) = 2n$ from (5.4.24).   ◇

Example 5.4.3 (Rohatgi and Saleh 2001) For $X_n \sim \chi^2(n)$, obtain the limit distributions of $Y_n = \frac{X_n}{n^2}$ and $Z_n = \frac{X_n}{n}$.

Solution Recollecting that $M_{X_n}(t) = (1-2t)^{-\frac{n}{2}}$, we have $\lim_{n\to\infty} M_{Y_n}(t) = \lim_{n\to\infty} M_{X_n}\left(\frac{t}{n^2}\right) = \lim_{n\to\infty}\left(1 - \frac{2t}{n^2}\right)^{-\frac{n^2}{2t}\cdot\frac{t}{n}} = \lim_{n\to\infty} \exp\left(\frac{t}{n}\right) = 1$ and, consequently, $\mathsf{P}\left(Y_n = 0\right) \to 1$. Similarly, we get $\lim_{n\to\infty} M_{Z_n}(t) = \lim_{n\to\infty}\left(1 - \frac{2t}{n}\right)^{-\frac{n}{2t}\cdot t} = e^t$ and, consequently, $\mathsf{P}\left(Z_n = 1\right) \to 1$.   ◇

When $\left\{X_i \sim \chi^2(k_i)\right\}_{i=1}^n$ are independent of each other, the mgf of $S_n = \sum_{i=1}^n X_i$ can be obtained as $M_{S_n}(t) = \prod_{i=1}^n (1-2t)^{-\frac{k_i}{2}} = (1-2t)^{-\frac{1}{2}\sum_{i=1}^n k_i}$ based on the mgf shown in (5.4.23) of X_i. This result proves the following theorem:

Theorem 5.4.8 When $\left\{X_i \sim \chi^2(k_i)\right\}_{i=1}^n$ are independent of each other, we have $\sum_{i=1}^n X_i \sim \chi^2\left(\sum_{i=1}^n k_i\right)$. In addition, if $\left\{X_i \sim \mathcal{N}\left(\mu_i, \sigma_i^2\right)\right\}_{i=1}^n$ are independent of each other, then $\sum_{i=1}^n \left(\frac{X_i-\mu_i}{\sigma_i}\right)^2 \sim \chi^2(n)$.

Definition 5.4.4 (*non-central chi-square distribution*) For independent normal random variables $\left\{X_i \sim \mathcal{N}\left(\mu_i, \sigma^2\right)\right\}_{i=1}^n$ with an identical variance σ^2, the distribution of $Y = \sum_{i=1}^n \frac{X_i^2}{\sigma^2}$ is called the non-central chi-square distribution and is denoted by $\chi^2(n, \delta)$, where n and $\delta = \sum_{i=1}^n \frac{\mu_i^2}{\sigma^2}$ are called the degree of freedom and non-centrality parameter, respectively.

The pdf of $\chi^2(n, \delta)$ is

$$f(x) = \frac{1}{\sqrt{2^n \pi}} x^{\frac{n}{2}-1} \exp\left(-\frac{x+\delta}{2}\right) \sum_{j=0}^{\infty} \frac{(\delta x)^j \, \Gamma\left(j+\frac{1}{2}\right)}{(2j)! \, \Gamma\left(j+\frac{n}{2}\right)} u(x). \quad (5.4.26)$$

Recollecting $\Gamma\left(\frac{1}{2}\right) = \sqrt{\pi}$ shown in (1.4.83), it is easy to see that (5.4.26) with $\delta = 0$ is the same as (5.4.22): in other words, $\chi^2(n, 0)$ is $\chi^2(n)$. In Exercise 5.32, it is shown that

$$\mathsf{E}\{Y\} = n + \delta, \qquad\qquad\qquad (5.4.27)$$
$$\sigma_Y^2 = 2n + 4\delta, \qquad\qquad\qquad (5.4.28)$$

and

$$M_Y(t) = (1 - 2t)^{-\frac{n}{2}} \exp\left(\frac{\delta t}{1 - 2t}\right), \quad t < \frac{1}{2} \qquad (5.4.29)$$

are the mean, variance, and mgf, respectively, for $Y \sim \chi^2(n, \delta)$.

Theorem 5.4.9 If $\{X_i \sim \chi^2(k_i, \delta_i)\}_{i=1}^{n}$ are independent of each other, then $S_n = \sum_{i=1}^{n} X_i \sim \chi^2\left(\sum_{i=1}^{n} k_i, \sum_{i=1}^{n} \delta_i\right)$.

Proof From the mgf shown in (5.4.29) of X_i, the mgf of S_n can be obtained as

$$M_{S_n}(t) = \prod_{i=1}^{n}(1-2t)^{-\frac{k_i}{2}} \exp\left(\frac{t\delta_i}{1-2t}\right) = (1-2t)^{-\frac{1}{2}\sum_{i=1}^{n} k_i} \exp\left(\frac{t}{1-2t}\sum_{i=1}^{n} \delta_i\right).$$ In other

words, $\sum_{i=1}^{n} X_i \sim \chi^2\left(\sum_{i=1}^{n} k_i, \sum_{i=1}^{n} \delta_i\right)$. ♠

Theorem 5.4.8 is a special case of Theorem 5.4.9.

5.4.3 t Distribution

Definition 5.4.5 (*central t pdf*) The pdf

$$f(r) = \frac{\Gamma\left(\frac{n+1}{2}\right)}{\Gamma\left(\frac{n}{2}\right)\sqrt{n\pi}} \left(1 + \frac{r^2}{n}\right)^{-\frac{n+1}{2}} \qquad (5.4.30)$$

is called the central t pdf with the corresponding distribution denoted as $t(n)$, where the natural number n is called the degree of freedom.

Fig. 5.5 A central t pdf

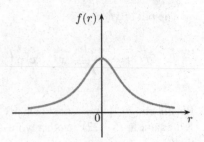

The central t pdf with the degree of freedom of 1 is a Cauchy pdf: in other words, $t(1) = C(0, 1)$. Figure 5.5 shows an example of the central t pdf.

Example 5.4.4 When $f(v) = \frac{5^3 a}{(5+v^2)^3}$ is a pdf, obtain the value of a.

Solution From $\int_{-\infty}^{\infty} f(v)dv = 5^3 a \int_{-\infty}^{\infty} \left(5 + v^2\right)^{-3} dv = 1$, we get $a = \frac{8}{3\sqrt{5}\pi}$.
Alternatively, comparing (5.4.30) and $f(v) = \frac{5^3 a}{(5+v^2)^3}$, we have $5^3 a = \frac{5^3 \Gamma(3)}{\Gamma(\frac{5}{2})\sqrt{5\pi}}$, i.e.,
$a = \frac{2}{\frac{3}{4}\sqrt{\pi}\sqrt{5\pi}} = \frac{8}{3\sqrt{5}\pi}$. $\diamond$

Example 5.4.5 Obtain the limit of the central t pdf (5.4.30) as $n \to \infty$.

Solution From (1.4.77), we have $\lim\limits_{n \to \infty} \frac{\Gamma\left(\frac{n+1}{2}\right)}{\sqrt{n\pi}\,\Gamma\left(\frac{n}{2}\right)} = \lim\limits_{n \to \infty} \sqrt{\frac{n}{2}} \frac{1}{\sqrt{n\pi}} = \frac{1}{\sqrt{2\pi}}$. In addition,
$\lim\limits_{n \to \infty} \left(1 + \frac{x^2}{n}\right)^{-\frac{n+1}{2}} = \lim\limits_{n \to \infty} \left(1 + \frac{x^2}{n}\right)^{\frac{n}{x^2}\left(-\frac{x^2}{2}\right)} = \exp\left(-\frac{x^2}{2}\right)$. In short, the limit of the
central t pdf for $n \to \infty$ is the standard normal pdf. $\diamond$

Theorem 5.4.10 (Rohatgi and Saleh 2001) *We have*

$$\sqrt{n}\frac{X}{\sqrt{Y}} \sim t(n) \qquad (5.4.31)$$

when $X \sim \mathcal{N}(0, 1)$ and $Y \sim \chi^2(n)$ are independent of each other.

Proof Let $T = \sqrt{n}\frac{X}{\sqrt{Y}}$ and $W = Y$. Then, $X = T\sqrt{\frac{W}{n}}$ and the Jacobian of the
inverse transformation $(X, Y) = g^{-1}(T, W) = \left(T\sqrt{\frac{W}{n}}, W\right)$ is $J\left(g^{-1}(t, w)\right) =$
$\left|\frac{\partial}{\partial(t,w)} g^{-1}(t, w)\right| = \begin{vmatrix} \sqrt{\frac{w}{n}} & \frac{1}{\sqrt{n}}\frac{t}{2\sqrt{w}} \\ 0 & 1 \end{vmatrix} = \sqrt{\frac{w}{n}}$. Thus, the joint pdf of (T, W) can be
obtained as $f_{T,W}(t, w) = \sqrt{\frac{w}{2\pi n}} \exp\left(-\frac{t^2 w}{2n}\right) \frac{w^{\frac{n}{2}-1}}{2^{\frac{n}{2}}\Gamma(\frac{n}{2})} \exp\left(-\frac{w}{2}\right) u(w)$, i.e.,

$$f_{T,W}(t, w) = \frac{w^{\frac{n-1}{2}}}{\sqrt{2\pi n}2^{\frac{n}{2}}\Gamma\left(\frac{n}{2}\right)} \exp\left\{-\frac{w}{2}\left(1 + \frac{t^2}{n}\right)\right\} u(w). \qquad (5.4.32)$$

Next, letting $\frac{w}{2}\left(1 + \frac{t^2}{n}\right) = v$, we have $\left(1 + \frac{t^2}{n}\right) dw = 2dv$. Thus, the pdf of T can be obtained as $f_T(t) = \int_{-\infty}^{\infty} f_{T,W}(t, w) dw = \frac{1}{\sqrt{\pi n}\,\Gamma(\frac{n}{2})} \left(1 + \frac{t^2}{n}\right)^{-\frac{n+1}{2}} \int_0^{\infty} v^{\frac{n-1}{2}} e^{-v} dv$, or as

$$f_T(t) = \frac{\Gamma\left(\frac{n+1}{2}\right)}{\Gamma\left(\frac{n}{2}\right)\sqrt{n\pi}} \left(1 + \frac{t^2}{n}\right)^{-\frac{n+1}{2}}, \tag{5.4.33}$$

confirming the theorem. ♠

The statistic $T = \frac{\sqrt{n}}{\sqrt{W_n}}\left(\overline{X}_n - \mu\right)$, a function of the sample mean $\overline{X}_n$ and the sample variance W_n, is called the t statistic and is widely used in statistics. Now, based on Theorem 5.4.10, let us consider a property of the t statistic from normal samples.

Theorem 5.4.11 *When X is a sample of size n from $X \sim \mathcal{N}\left(\mu, \sigma^2\right)$, we have*

$$\frac{n-1}{\sigma^2} W_n \sim \chi^2(n-1) \tag{5.4.34}$$

and

$$\sqrt{n}\frac{\overline{X}_n - \mu}{\sqrt{W_n}} \sim t(n-1) \tag{5.4.35}$$

for the sample mean $\overline{X}_n$ and the sample variance W_n of X.

Proof Recollecting that $X_i - \overline{X}_n = (X_i - \mu) - (\overline{X}_n - \mu)$ and $\sum_{i=1}^{n}\left\{(X_i - \mu)(\overline{X}_n - \mu)\right\} = n\left(\overline{X}_n - \mu\right)^2$, we can rewrite the sample variance as $W_n = \frac{1}{n-1}\sum_{i=1}^{n}(X_i - \overline{X}_n)^2 = \frac{1}{n-1}\sum_{i=1}^{n}\left\{(X_i - \mu)^2 - 2(X_i - \mu)(\overline{X}_n - \mu) + (\overline{X}_n - \mu)^2\right\} = \frac{1}{n-1}\left[\left\{\sum_{i=1}^{n}(X_i - \mu)^2\right\} - n(\overline{X}_n - \mu)^2\right]$. Thus, we have

$$\sum_{i=1}^{n}\left(\frac{X_i - \mu}{\sigma}\right)^2 = \frac{n-1}{\sigma^2} W_n + \frac{n}{\sigma^2}\left(\overline{X}_n - \mu\right)^2. \tag{5.4.36}$$

Now, we have

$$\frac{n}{\sigma^2}\left(\overline{X}_n - \mu\right)^2 \sim \chi^2(1) \tag{5.4.37}$$

as observed in Theorem 5.4.6 and

$$\sum_{i=1}^{n}\left(\frac{X_i - \mu}{\sigma}\right)^2 \sim \chi^2(n) \tag{5.4.38}$$

as observed in Theorem 5.4.8. Recollecting that $\frac{n-1}{\sigma^2}W_n$ and $\frac{n}{\sigma^2}\left(\overline{X}_n - \mu\right)^2$ are independent of each other as discussed in Theorem 5.4.5, the mgf of the statistic $\sum_{i=1}^{n}\left(\frac{X_i - \mu}{\sigma}\right)^2$ in (5.4.36) can be obtained as

$$\mathsf{E}\left[\exp\left\{t\sum_{i=1}^{n}\left(\frac{X_i - \mu}{\sigma}\right)^2\right\}\right] = \mathsf{E}\left[\exp\left\{t\frac{(n-1)W_n}{\sigma^2} + t\frac{n\left(\overline{X}_n - \mu\right)^2}{\sigma^2}\right\}\right]$$

$$= \mathsf{E}\left[\exp\left\{t\frac{(n-1)W_n}{\sigma^2}\right\}\right]\mathsf{E}\left[\exp\left\{t\frac{n\left(\overline{X}_n - \mu\right)^2}{\sigma^2}\right\}\right]. \tag{5.4.39}$$

Combining (5.4.37) and (5.4.38) into (5.4.39), we get

$$(1-2t)^{-\frac{n}{2}} = \mathsf{E}\left[\exp\left\{t\frac{(n-1)W_n}{\sigma^2}\right\}\right](1-2t)^{-\frac{1}{2}} \tag{5.4.40}$$

or $\mathsf{E}\left[\exp\left\{t\frac{(n-1)W_n}{\sigma^2}\right\}\right] = (1-2t)^{-\frac{n-1}{2}}$, which implies

$$\frac{n-1}{\sigma^2}W_n \sim \chi^2(n-1). \tag{5.4.41}$$

Next, the distribution of $\sqrt{n-1} \times \frac{\sqrt{n}(\overline{X}_n - \mu)}{\sigma}\left\{\frac{(n-1)W_n}{\sigma^2}\right\}^{-\frac{1}{2}} = \sqrt{n}\frac{\overline{X}_n - \mu}{\sqrt{W_n}}$ is $t(n-1)$ from Theorem 5.4.10. ♠

Example 5.4.6 For a sample from $\mathcal{N}\left(1, \sigma^2\right)$, it is easy to see that $T = \frac{\sqrt{7}(\overline{X}_7 - 1)}{\sqrt{W_7}} \sim t(6)$ from Theorem 5.4.11. ◇

Definition 5.4.6 (*non-central t distribution*) When the two random variables $X \sim \mathcal{N}\left(\mu, \sigma^2\right)$ and $\frac{Y}{\sigma^2} \sim \chi^2(n)$ are independent of each other, the distribution of

$$Z = \sqrt{n}\frac{X}{\sqrt{Y}} \tag{5.4.42}$$

is called the non-central t distribution with the degree of freedom of n and non-centrality parameter $\delta = \frac{\mu}{\sigma}$, and is denoted by $t(n, \delta)$.

The pdf of $t(n, \delta)$ is

$$f(x) = \frac{n^{\frac{n}{2}} \exp\left(-\frac{\delta^2}{2}\right)}{\sqrt{\pi}\left(n+x^2\right)^{\frac{n+1}{2}}} \sum_{j=0}^{\infty} \frac{\Gamma\left(\frac{n+j+1}{2}\right)}{\Gamma\left(\frac{n}{2}\right)} \left(\frac{\delta^j}{j!}\right) \left(\frac{2x^2}{n+x^2}\right)^{\frac{j}{2}}. \tag{5.4.43}$$

Comparing the pdf's (5.4.30) and (5.4.43), it is easy to see that the non-central t distribution $t(n, 0)$ is the same as the central t distribution $t(n)$. In Exercise 5.34, we obtain the mean

$$\mathsf{E}\{Z\} = \delta \frac{\Gamma\left(\frac{n-1}{2}\right)}{\Gamma\left(\frac{n}{2}\right)} \sqrt{\frac{n}{2}}, \quad n > 1 \tag{5.4.44}$$

and variance

$$\mathsf{Var}\{Z\} = \frac{n(1+\delta^2)}{n-2} - \frac{n\delta^2}{2} \left\{\frac{\Gamma\left(\frac{n-1}{2}\right)}{\Gamma\left(\frac{n}{2}\right)}\right\}^2, \quad n > 2 \tag{5.4.45}$$

of $Z \sim t(n, \delta)$.

Definition 5.4.7 (*bi-variate t distribution*) When the joint pdf of (X, Y) is

$$f_{X,Y}(x, y) = \frac{1}{2\pi\sigma_1\sigma_2\sqrt{1-\rho^2}} \left[1 + \frac{1}{n\left(1-\rho^2\right)} \left\{\left(\frac{x-\mu_1}{\sigma_1}\right)^2\right.\right.$$
$$\left.\left. -2\rho\frac{(x-\mu_1)(y-\mu_2)}{\sigma_1\sigma_2} + \left(\frac{y-\mu_2}{\sigma_2}\right)^2\right\}\right]^{-\frac{n+2}{2}}, \tag{5.4.46}$$

the random vector (X, Y) is called a bi-variate t random vector. The distribution, denoted by $t\left(\mu_1, \mu_2, \sigma_1^2, \sigma_2^2, \rho, n\right)$, of (X, Y) is called a bi-variate t distribution.

For $(X, Y) \sim t\left(\mu_1, \mu_2, \sigma_1^2, \sigma_2^2, \rho, n\right)$, we have the means $\mathsf{E}\{X\} = \mu_1$ and $\mathsf{E}\{Y\} = \mu_2$, variances $\mathsf{Var}\{X\} = \frac{n}{n-2}\sigma_1^2$ and $\mathsf{Var}\{Y\} = \frac{n}{n-2}\sigma_2^2$ for $n > 2$, and correlation coefficient ρ. The parameter n determines how fast the pdf (5.4.7) decays to 0 as $|x| \to \infty$ or as $|y| \to \infty$. When $n = 1$, the bi-variate t pdf is the same as the bi-variate Cauchy pdf, and the bi-variate t pdf converges to the bi-variate normal pdf as n gets larger. In addition, we have $\mathsf{E}\{X|Y\} = \rho s Y$, $\mathsf{E}\{XY\} = \frac{\rho s n}{n-2}$, and

$$\mathsf{E}\left\{X^2|Y\right\} = \frac{s^2}{(n-1)^2}\left[(1-\rho^2)n + \left\{1 + (n-2)\rho^2\right\}Y^2\right] \tag{5.4.47}$$

for $(X, Y) \sim t\left(0, 0, s^2, 1, \rho, n\right)$.

5.4.4   F Distribution

Definition 5.4.8 (*central F pdf*) The pdf

$$f(r) = \frac{\Gamma\left(\frac{m+n}{2}\right)}{\Gamma\left(\frac{m}{2}\right)\Gamma\left(\frac{n}{2}\right)} \frac{m}{n} \left(\frac{m}{n}r\right)^{\frac{m}{2}-1} \left(1 + \frac{m}{n}r\right)^{-\frac{m+n}{2}} u(r) \qquad (5.4.48)$$

is called the central F pdf with the degree of freedom of (m, n) and its distribution is denoted by $F(m, n)$.

The F distribution, together with the chi-square and t distributions, plays an important role in mathematical statistics. In Exercise 5.35, it is shown that the moment of $H \sim F(m, n)$ is

$$\mathsf{E}\{H^k\} = \left(\frac{n}{m}\right)^k \frac{\Gamma\left(\frac{m}{2}+k\right)\Gamma\left(\frac{n}{2}-k\right)}{\Gamma\left(\frac{m}{2}\right)\Gamma\left(\frac{n}{2}\right)} \qquad (5.4.49)$$

for $k = 1, 2, \ldots, \lceil\frac{n}{2}\rceil - 1$. Figure 5.6 shows the pdf of $F(4, 3)$.

Theorem 5.4.12 (Rohatgi and Saleh 2001) *We have*

$$\frac{nX}{mY} \sim F(m, n) \qquad (5.4.50)$$

when $X \sim \chi^2(m)$ *and* $Y \sim \chi^2(n)$ *are independent of each other.*

Proof Let $H = \frac{nX}{mY}$. Assuming the auxiliary variable $V = Y$, we have $X = \frac{m}{n}HV$ and $Y = V$. Because the Jacobian of the inverse transformation $(X, Y) = \boldsymbol{g}^{-1}(H, V) = \left(\frac{m}{n}HV, V\right)$ is $J\left(\boldsymbol{g}^{-1}(r, v)\right) = \left|\frac{\partial}{\partial(r,v)}\boldsymbol{g}^{-1}(r, v)\right| = \left|\begin{matrix} \frac{m}{n}v & 0 \\ \frac{m}{n}r & 1 \end{matrix}\right| = \frac{m}{n}v$, we have the joint pdf $f_{H,V}(r, v) = \frac{mv}{n}f_{X,Y}\left(\frac{m}{n}vr, v\right)$ of (H, V) as

$$f_{H,V}(r, v) = \frac{mv}{n}f_X\left(\frac{m}{n}vr\right)f_Y(v). \qquad (5.4.51)$$

Fig. 5.6 The pdf of $F(4, 3)$

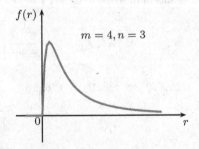

$f(r)$

$m = 4, n = 3$

r

Now, the marginal pdf $f_H(r) = \int_{-\infty}^{\infty} f_{H,V}(r,v)dv = \int_{-\infty}^{\infty} \frac{mv}{n} f_X\left(\frac{m}{n}vr\right) f_Y$

$(v)dv = \frac{mv}{n} \int_{-\infty}^{\infty} \frac{\left(\frac{1}{2}\right)^{\frac{m}{2}} \left(\frac{m}{n}vr\right)^{\frac{m}{2}-1}}{\Gamma\left(\frac{m}{2}\right)} \exp\left(-\frac{\frac{m}{n}vr}{2}\right) \frac{\left(\frac{1}{2}\right)^{\frac{n}{2}} v^{\frac{n}{2}-1}}{\Gamma\left(\frac{n}{2}\right)} \exp\left(-\frac{v}{2}\right) u(v)u\left(\frac{m}{n}vr\right) dv =$

$\frac{m}{n} \frac{\left(\frac{1}{2}\right)^{\frac{m}{2}} \left(\frac{1}{2}\right)^{\frac{n}{2}}}{\Gamma\left(\frac{m}{2}\right)\Gamma\left(\frac{n}{2}\right)} \left(\frac{m}{n}r\right)^{\frac{m}{2}-1} \int_0^{\infty} v^{\frac{m+n}{2}-1} \exp\left\{-\frac{v}{2}\left(\frac{m}{n}r+1\right)\right\} dv\, u(r)$ of H can be obtained

as[10]

$$f_H(r) = \frac{m}{n} \frac{1}{\Gamma\left(\frac{m}{2}\right)\Gamma\left(\frac{n}{2}\right)} \left(\frac{m}{n}r\right)^{\frac{m}{2}-1} \int_0^{\infty} \left(\frac{1}{2}\right)^{\frac{m+n}{2}} v^{\frac{m+n}{2}-1}$$
$$\times \exp\left\{-\frac{v}{2}\left(\frac{m}{n}r+1\right)\right\} dv\, u(r)$$
$$= \frac{\Gamma\left(\frac{m+n}{2}\right)}{\Gamma\left(\frac{m}{2}\right)\Gamma\left(\frac{n}{2}\right)} \frac{m}{n}\left(\frac{m}{n}r\right)^{\frac{m}{2}-1} \left(1+\frac{m}{n}r\right)^{-\frac{m+n}{2}} u(r) \qquad (5.4.52)$$

by noting that $f_X(x) = \left(\frac{1}{2}\right)^{\frac{m}{2}} \left\{\Gamma\left(\frac{m}{2}\right)\right\}^{-1} x^{\frac{m}{2}-1} e^{-\frac{x}{2}} u(x)$ and $f_Y(y) = \left(\frac{1}{2}\right)^{\frac{n}{2}} \left\{\Gamma\left(\frac{n}{2}\right)\right\}^{-1} y^{\frac{n}{2}-1} e^{-\frac{y}{2}} u(y)$. ♠

Example 5.4.7 Show that $\frac{1}{X} \sim F(n,m)$ when $X \sim F(m,n)$.

Solution If we obtain the pdf $f_Y(y) = \frac{1}{y^2} \frac{\left(\frac{m}{n}\right)\Gamma\left(\frac{m+n}{2}\right)}{\Gamma\left(\frac{m}{2}\right)\Gamma\left(\frac{n}{2}\right)} \left(1+\frac{m}{n}\frac{1}{y}\right)^{-\frac{m+n}{2}} \left(\frac{m}{n}\frac{1}{y}\right)^{\frac{m}{2}-1}$

$u\left(\frac{1}{y}\right) = \frac{1}{y^2} \frac{\left(\frac{m}{n}\right)\Gamma\left(\frac{m+n}{2}\right)}{\Gamma\left(\frac{m}{2}\right)\Gamma\left(\frac{n}{2}\right)} \left(\frac{ny}{m}\right)^{1-\frac{m}{2}} \left(\frac{ny}{ny+m}\right)^{\frac{m+n}{2}} u(y) = \frac{\Gamma\left(\frac{m+n}{2}\right)}{\Gamma\left(\frac{m}{2}\right)\Gamma\left(\frac{n}{2}\right)} \left(\frac{m}{n}\right)^{\frac{m}{2}} y^{-\frac{m}{2}-1} (ny)^{\frac{m+n}{2}}$

$(ny+m)^{-\frac{m+n}{2}} u(y)$ of $Y = \frac{1}{X}$ based on (3.2.27) and (5.4.48), we have

$f_Y(y) = \frac{\Gamma\left(\frac{m+n}{2}\right)}{\Gamma\left(\frac{m}{2}\right)\Gamma\left(\frac{n}{2}\right)} \left(\frac{m}{n}\right)^{\frac{m}{2}} y^{\frac{n}{2}-1} \left(y+\frac{m}{n}\right)^{-\frac{m+n}{2}} u(y)$, i.e.,

$$f_Y(y) = \frac{\Gamma\left(\frac{m+n}{2}\right)}{\Gamma\left(\frac{m}{2}\right)\Gamma\left(\frac{n}{2}\right)} \left(\frac{n}{m}\right)\left(\frac{n}{m}y\right)^{\frac{n}{2}-1} \left(1+\frac{n}{m}y\right)^{-\frac{m+n}{2}} u(y) \qquad (5.4.53)$$

by noting that $u\left(\frac{1}{y}\right) = u(y)$. ◇

Example 5.4.8 When $n \to \infty$, find the limit of the pdf of $F(m,n)$.

Solution Let us first rewrite the pdf $f_F(x)$ as

$$f_F(x) = \frac{1}{\Gamma\left(\frac{m}{2}\right)} \frac{1}{x} \underbrace{\frac{\Gamma\left(\frac{m+n}{2}\right)}{\Gamma\left(\frac{n}{2}\right)} \left(\frac{mx}{n}\right)^{\frac{m}{2}} \left(1+\frac{mx}{n}\right)^{-\frac{m+n}{2}}}_{A} u(x). \qquad (5.4.54)$$

---

[10] Here, note that $\left\{\Gamma\left(\frac{m+n}{2}\right)\right\}^{-1} \int_0^{\infty} \left(\frac{1}{2}\right)^{\frac{m+n}{2}} w^{\frac{m+n}{2}-1} e^{-\frac{w}{2}} dw = 1$ when we let $w = v\left(1+\frac{m}{n}r\right)$.

Using (1.4.77), we have $\lim_{n\to\infty} A = \lim_{n\to\infty} \left(\frac{n}{2}\right)^{\frac{m}{2}} \left(\frac{mx}{n}\right)^{\frac{m}{2}} = \left(\frac{mx}{2}\right)^{\frac{m}{2}}$ and

$$\lim_{n\to\infty} \left(1 + \frac{m}{n}x\right)^{-\frac{(m+n)}{2}} = \lim_{n\to\infty} \left(1 + \frac{m}{n}x\right)^{-\frac{n}{2}} \left(1 + \frac{m}{n}x\right)^{-\frac{m}{2}}$$
$$= \exp\left(-\frac{m}{2}x\right). \qquad (5.4.55)$$

Thus, letting $a = \frac{m}{2}$, we get

$$\lim_{n\to\infty} f_F(x) = \frac{a}{\Gamma(a)} (ax)^{a-1} \exp(-ax)u(x). \qquad (5.4.56)$$

In other words, when $n \to \infty$, $F(m, n) \to G\left(\frac{m}{2}, \frac{2}{m}\right)$, where $G(\alpha, \beta)$ denotes the gamma distribution described by the pdf (2.5.31). ◇

Example 5.4.9 In Example 5.4.8, we obtained the limit of the pdf of $F(m, n)$ when $n \to \infty$. Now, obtain the limit when $m \to \infty$. Then, based on the result, when $X \sim F(m, n)$ and $m \to \infty$, obtain the pdf of $\frac{1}{X}$.

Solution Rewrite the pdf $f_F(x)$ as

$$f_F(x) = \frac{1}{\Gamma\left(\frac{n}{2}\right)} \frac{1}{x} \underbrace{\frac{\Gamma\left(\frac{m+n}{2}\right)}{\Gamma\left(\frac{m}{2}\right)} \left(\frac{n}{n+mx}\right)^{\frac{n}{2}}}_{B} \underbrace{\left(\frac{mx}{n+mx}\right)^{\frac{m}{2}}}_{C} u(x). \qquad (5.4.57)$$

Then, if we let $b = \frac{n}{2}$, we get

$$\lim_{m\to\infty} f_F(x) = \frac{1}{b\Gamma(b)} \left(\frac{b}{x}\right)^{b+1} \exp\left(-\frac{b}{x}\right) u(x) \qquad (5.4.58)$$

noting that $\lim_{m\to\infty} B = \lim_{m\to\infty} \frac{\left(\frac{m}{2}\right)^{\frac{n}{2}} \Gamma\left(\frac{m}{2}\right)}{\Gamma\left(\frac{m}{2}\right)} \times \left(\frac{n}{mx}\right)^{\frac{n}{2}} = \left(\frac{n}{2x}\right)^{\frac{n}{2}}$ from (1.4.77) and that $\lim_{m\to\infty} C = \lim_{m\to\infty} \left(1 + \frac{n}{mx}\right)^{-\frac{n}{2}} = \exp\left(-\frac{n}{2x}\right)$. Figure 5.7 shows the pdf of $F(m, 10)$ for some values of m, and Fig. 5.8 shows three pdf's[11] of $F(m, n)$ for $m \to \infty$.

Next, for $m \to \infty$, the pdf $\lim_{m\to\infty} f_{\frac{1}{X}}(y) = \lim_{m\to\infty} \frac{1}{y^2} f_F\left(\frac{1}{y}\right) = \frac{1}{b\Gamma(b)} \frac{1}{y^2} (by)^{b+1}$ $\exp(-by)u\left(\frac{1}{y}\right)$ of $\frac{1}{X}$ can be obtained as

$$\lim_{m\to\infty} f_{\frac{1}{X}}(y) = \frac{b}{\Gamma(b)} (by)^{b-1} \exp(-by)u(y). \qquad (5.4.59)$$

---

[11] The maximum is at $x = \frac{a-1}{a} = \frac{m-2}{m}$ in Fig. 5.7 and at $x = \frac{b}{b+1} = \frac{n}{n+2}$ in Fig. 5.8.

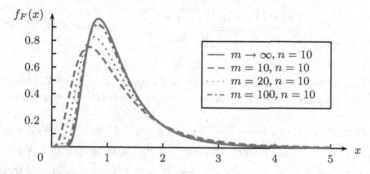

Fig. 5.7 The pdf $f_{F(m,10)}(x)$ for some values of m

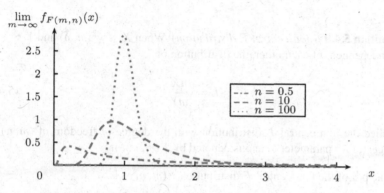

Fig. 5.8 The limit $\lim_{m \to \infty} f_{F(m,n)}(x)$ of the pdf of $F(m,n)$

In other words, $\frac{1}{X} \sim G\left(\frac{n}{2}, \frac{2}{n}\right)$ when $m \to \infty$. ◇

Because $\frac{1}{X} \sim F(n,m)$ when $X \sim F(m,n)$ as we have observed in (5.4.53) and $F(m,n) \to G\left(\frac{m}{2}, \frac{2}{m}\right)$ for $n \to \infty$ as we have observed in Example 5.4.8, we have $\frac{1}{X} \sim G\left(\frac{n}{2}, \frac{2}{n}\right)$ for $m \to \infty$ when $X \sim F(m,n)$: Example 5.4.9 shows this result directly. In addition, based on this result and (3.2.26), we have $\frac{n}{2X} \sim G\left(\frac{n}{2}, 1\right)$ for $m \to \infty$ when $X \sim F(m,n)$.

Theorem 5.4.13 (Rohatgi and Saleh 2001) *If* $X = (X_1, X_2, \ldots, X_m)$ *from* $\mathcal{N}\left(\mu_X, \sigma_X^2\right)$ *and* $Y = (Y_1, Y_2, \ldots, Y_n)$ *from* $\mathcal{N}\left(\mu_Y, \sigma_Y^2\right)$ *are independent of each other, then*

$$\frac{\sigma_Y^2}{\sigma_X^2} \frac{W_{X,m}}{W_{Y,m}} \sim F(m-1, n-1) \tag{5.4.60}$$

and

$$\frac{(\overline{X}_m - \mu_X) - (\overline{Y}_n - \mu_Y)}{\sqrt{\frac{(m-1)W_{X,m}}{\sigma_X^2} + \frac{(n-1)W_{Y,m}}{\sigma_Y^2}}} \sqrt{\frac{m+n-2}{\frac{\sigma_X^2}{m} + \frac{\sigma_Y^2}{n}}} \sim t(m+n-2), \qquad (5.4.61)$$

where $\overline{X}_m$ and $W_{X,m}$ are the sample mean and sample variance of X, respectively, and $\overline{Y}_n$ and $W_{Y,n}$ are the sample mean and sample variance of Y, respectively.

Proof From (5.4.41), we have $\frac{(m-1)}{\sigma_X^2} W_{X,m} \sim \chi^2(m-1)$ and $\frac{(n-1)}{\sigma_Y^2} W_{Y,n} \sim \chi^2(n-1)$. Thus, (5.4.60) follows from Theorem 5.4.12. Next, noting that $\overline{X}_m - \overline{Y}_n \sim \mathcal{N}\left(\mu_X - \mu_Y, \frac{\sigma_X^2}{m} + \frac{\sigma_Y^2}{n}\right)$, $\frac{(m-1)}{\sigma_X^2} W_{X,m} + \frac{(n-1)}{\sigma_Y^2} W_{Y,n} \sim \chi^2(m+n-2)$, and these two statistics are independent of each other, we easily get (5.4.61) from Theorem 5.4.10. ♠

Definition 5.4.9 (*non-central F distribution*) When $X \sim \chi^2(m, \delta)$ and $Y \sim \chi^2(n)$ are independent of each other, the distribution of

$$H = \frac{nX}{mY} \qquad (5.4.62)$$

is called the non-central F distribution with the degree of freedom of (m, n) and non-centrality parameter δ, and is denoted by $F(m, n, \delta)$.

The pdf of the non-central F distribution $F(m, n, \delta)$ is

$$f(x) = \frac{m^{\frac{m}{2}} n^{\frac{n}{2}} x^{\frac{m}{2}-1}}{\Gamma\left(\frac{n}{2}\right) \exp\left(\frac{\delta}{2}\right)} \sum_{j=0}^{\infty} \frac{\left(\frac{\delta m x}{2}\right)^j \Gamma\left(\frac{m+n+2j}{2}\right)}{j! \, \Gamma\left(\frac{m+2j}{2}\right) (mx+n)^{\frac{m+n+2j}{2}}} u(x). \qquad (5.4.63)$$

Here, the pdf (5.4.63) for $\delta = 0$ indicates that $F(m, n, 0)$ is the central F distribution $F(m, n)$. In Exercise 5.36, we obtain the mean

$$\mathsf{E}\{H\} = \frac{n(m + \delta)}{m(n - 2)} \qquad (5.4.64)$$

for $n = 3, 4, \ldots$ and variance

$$\mathsf{Var}\{H\} = \frac{2n^2 \left\{(m + \delta)^2 + (n - 2)(m + 2\delta)\right\}}{m^2 (n - 4)(n - 2)^2} \qquad (5.4.65)$$

for $n = 5, 6, \ldots$ of $F(m, n, \delta)$.

Appendices

Appendix 5.1 Proof of General Formula for Joint Moments

The general formula (5.3.51) for the joint moments of normal random vectors is proved via mathematical induction here.

First note that

$$\frac{\partial}{\partial \tilde{\rho}_{ij}} \mathsf{E}\left\{\prod_{k=1}^n X_k^{a_k}\right\} = a_i a_j \mathsf{E}\left\{ X_i^{a_i-1} X_j^{a_j-1} \prod_{\substack{k=1 \\ k \neq i,j}}^n X_k^{a_k}\right\} \tag{5.A.1}$$

for $i \neq j$ and

$$\frac{\partial}{\partial \tilde{\rho}_{ii}} \mathsf{E}\left\{\prod_{k=1}^n X_k^{a_k}\right\} = \frac{1}{2} a_i \left(a_i - 1\right) \mathsf{E}\left\{ X_i^{a_i-2} \prod_{\substack{k=1 \\ k \neq i}}^n X_k^{a_k}\right\} \tag{5.A.2}$$

for $i = j$ from (5.3.12).

(1) Let us show that (5.3.22) holds true. Express $\mathsf{E}\left\{X_1^a X_2^b\right\}$ as

$$\mathsf{E}\left\{X_1^a X_2^b\right\} = \sum_{j=0}^{\min(a,b)} \sum_{p=0}^{a-j} \sum_{q=0}^{b-j} d_{a,b,j,p,q}\, \tilde{\rho}_{12}^j \tilde{\rho}_{11}^p \tilde{\rho}_{22}^q m_1^{a-j-2p} m_2^{b-j-2q}. \tag{5.A.3}$$

Then, when $\tilde{\rho}_{12} = 0$, we have $\mathsf{E}\left\{X_1^a X_2^b\right\} = \mathsf{E}\left\{X_1^a\right\} \mathsf{E}\left\{X_2^b\right\}$, i.e.,

$$\sum_{p=0}^a \sum_{q=0}^b d_{a,b,0,p,q}\, \tilde{\rho}_{11}^p \tilde{\rho}_{22}^q m_1^{a-2p} m_2^{b-2q}$$

$$= \sum_{p=0}^a d_{a,p} \tilde{\rho}_{11}^p m_1^{a-2p} \sum_{q=0}^b d_{b,q} \tilde{\rho}_{22}^q m_2^{b-2q} \tag{5.A.4}$$

because the coefficient of $\tilde{\rho}_{11}^p m_1^{a-2p}$ is $d_{a,p} = \frac{a!}{2^p p!(a-2p)!}$ when $\mathsf{E}\left\{X_1^a\right\}$ is expanded as we can see from (5.3.17). Thus, we get

$$d_{a,b,0,p,q} = \frac{a!b!}{2^{p+q} p!q!(a-2p)!(b-2q)!}. \tag{5.A.5}$$

Next, from (5.A.1), we get

$$\frac{\partial}{\partial \tilde{\rho}_{12}} \mathsf{E}\left\{X_1^a X_2^b\right\} = ab \mathsf{E}\left\{X_1^{a-1} X_2^{b-1}\right\}. \tag{5.A.6}$$

The left- and right-hand sides of (5.A.6) can be expressed as

$$\frac{\partial}{\partial \tilde{\rho}_{12}} \mathsf{E}\left\{X_1^a X_2^b\right\} = \sum_{j=1}^{\min(a,b)} \sum_{p=0}^{a-j} \sum_{q=0}^{b-j} j \, d_{a,b,j,p,q} \, \tilde{\rho}_{12}^{j-1} \tilde{\rho}_{11}^p \tilde{\rho}_{22}^q$$

$$= \sum_{j=0}^{\min(a,b)-1} \sum_{p=0}^{a-1-j} \sum_{q=0}^{b-1-j} (j+1) d_{a,b,j+1,p,q} m_1^{a-1-j-2p} m_2^{b-1-j-2q} \tag{5.A.7}$$

and

$$\mathsf{E}\left\{X_1^{a-1} X_2^{b-1}\right\} = \sum_{j=0}^{\min(a-1,b-1)} \sum_{p=0}^{a-1-j} \sum_{q=0}^{b-1-j} d_{a-1,b-1,j,p,q} \, \tilde{\rho}_{12}^j \tilde{\rho}_{11}^p \tilde{\rho}_{22}^q$$

$$\times m_1^{a-1-j-2p} m_2^{b-1-j-2q}, \tag{5.A.8}$$

respectively, using (5.A.3). Taking into consideration that $\min(a, b) - 1 = \min(a - 1, b - 1)$, we get

$$d_{a,b,j+1,p,q} = \frac{ab}{j+1} d_{a-1,b-1,j,p,q} \tag{5.A.9}$$

from (5.A.6)–(5.A.8). Using (5.A.5) in (5.A.9) recursively, we obtain

$$d_{a,b,j+1,p,q} = \frac{a! b!}{2^{p+q} p! q! (j+1)! (a-j-1-2p)! (b-j-1-2q)!}, \tag{5.A.10}$$

which is equivalent to the coefficient shown in (5.3.22).

(2) We have so far shown that (5.3.51) holds true when $n = 2$, and that (5.3.51) holds true when $n = 1$ as shown in Exercise 5.27. Now, assume that (5.3.51) holds true when $n = m - 1$. Then, because

$$\mathsf{E}\left\{X_1^{a_1} X_2^{a_2} \cdots X_m^{a_m}\right\} = \mathsf{E}\left\{X_1^{a_1} X_2^{a_2} \cdots X_{m-1}^{a_{m-1}}\right\} \mathsf{E}\left\{X_m^{a_m}\right\} \tag{5.A.11}$$

when $\tilde{\rho}_{1m} = \tilde{\rho}_{2m} = \cdots = \tilde{\rho}_{m-1,m} = 0$, we get $\left[d_{a_2,l_2}\right]_{l_{1m}\mapsto 0, l_{2m}\mapsto 0, \ldots, l_{m-1,m}\mapsto 0} = d_{a_1,l_1} d_{a_m,l_{mm}}$, i.e.,

$$\left[d_{a_2,l_2}\right]_{l_{1m}\mapsto0,l_{2m}\mapsto0,\dots,l_{m-1,m}\mapsto0} = d_{a_1,l_1}\,d_{a_m,l_{mm}}$$

$$= \frac{\displaystyle\prod_{k=1}^{m} a_k!}{2^{M_{l_2}}\,\zeta_{m-1}\,(l)\,l_{mm}!\,(a_m - 2l_{mm})!\,\eta_{1,m-1}} \tag{5.A.12}$$

from (5.3.51) with $n = m - 1$ and (5.3.17), where $M_l = \sum_{i=1}^{n} l_{ii}$, $a_1 = \{a_i\}_{i=1}^{m-1}$, $a_2 = a_1 \cup \{a_m\}$, $l_1 = \left\{\{l_{ij}\}_{j=i}^{m-1}\right\}_{i=1}^{m-1}$, $l_2 = l_1 \cup \{l_{im}\}_{i=1}^{m}$, $\zeta_m\,(l) = \prod_{i=1}^{m}\prod_{j=i}^{m} l_{ij}!$, and $\eta_{k,m} = \prod_{j=1}^{m} L_{a_k,j}!$. Here, the symbol $\mapsto$ denotes a substitution: for example, $\alpha \mapsto \beta$ means the substitution of α with β.

Next, employing (5.A.1) with $(i, j) = (1, m), (2, m), \dots, (m - 1, m)$, we will get

$$\left[d_{a_2,l_2}\right]_{l_{im}\mapsto l_{im}+1} = \frac{a_i!\,a_m!\,\left[d_{a_2,l_2}\right]_{l_{im}\mapsto0,a_i\mapsto a_i-l_{im}-1,a_m\mapsto a_m-l_{im}-1}}{(a_i - l_{im} - 1)!\,(a_m - l_{im} - 1)!\,(l_{im} + 1)!} \tag{5.A.13}$$

for $i = 1, 2, \dots, m - 1$ in a fashion similar to that leading to (5.A.10) from (5.A.6). By changing $l_{im} + 1$ into l_{im} in (5.A.13), we have

$$d_{a_2,l_2} = \frac{a_i!\,a_m!\,\left[d_{a_2,l_2}\right]_{l_{im}\mapsto0,a_i\mapsto a_i-l_{im},a_m\mapsto a_m-l_{im}}}{(a_i - l_{im})!\,(a_m - l_{im})!\,l_{im}!} \tag{5.A.14}$$

for $i = 1, 2, \dots, m - 1$. Now, letting $l_{2m} = 0$ in (5.A.14) with $i = 1$, we get

$$\left[d_{a_2,l_2}\right]_{l_{2m}\mapsto0} = \frac{a_1!\,a_m!\,\left[d_{a_2,l_2}\right]_{l_{1m}\mapsto0,l_{2m}\mapsto0,a_1\mapsto a_1-l_{1m},a_m\mapsto a_m-l_{1m}}}{(a_1 - l_{1m})!\,(a_m - l_{1m})!\,l_{1m}!}. \tag{5.A.15}$$

Using (5.A.15) into (5.A.14) with $i = 2$, we obtain

$$d_{a_2,l_2} = \frac{a_2!\,a_m!}{(a_2 - l_{2m})!\,(a_m - l_{2m})!\,l_{2m}!} \times \frac{a_1!\,(a_m - l_{2m})!}{(a_1 - l_{1m})!\,(a_m - l_{1m} - l_{2m})!\,l_{1m}!}$$
$$\times \left[d_{a_2,l_2}\right]_{l_{1m}\mapsto0,l_{2m}\mapsto0,a_1\mapsto a_1-l_{1m},a_2\mapsto a_2-l_{2m},a_m\mapsto a_m-l_{1m}-l_{2m}}. \tag{5.A.16}$$

Subsequently, letting $l_{3m} = 0$ in (5.A.16), we get

$$\left[d_{a_2,l_2}\right]_{l_{3m}\mapsto0} = \frac{a_1!\,a_2!\,a_m!}{l_{1m}!\,l_{2m}!\,(a_1 - l_{1m})!\,(a_2 - l_{2m})!\,(a_m - l_{1m} - l_{2m})!}$$
$$\times \left[d_{a_2,l_2}\right]_{l_{km}\mapsto0 \text{ for } k=1,2,3,\,a_1\mapsto a_1-l_{1m},\,a_2\mapsto a_2-l_{2m},\,a_m\mapsto a_m-l_{1m}-l_{2m}}, \tag{5.A.17}$$

which can be employed into (5.A.14) with $i = 3$ to produce $d_{a_2, l_2} = $

$$\frac{a_3! a_m!}{(a_3 - l_{3m})! l_{3m}! (a_m - l_{3m})!} \frac{1}{l_{1m}! l_{2m}! (a_1 - l_{1m})! (a_2 - l_{2m})!} \frac{a_1! a_2! (a_m - l_{3m})!}{(a_m - l_{1m} - l_{2m} - l_{3m})!}$$

$$\left[d_{a_2, l_2}\right]_{l_{km} \mapsto 0, \, a_k \mapsto a_k - l_{km} \text{ for } k = 1,2,3; \, a_m \mapsto a_m - l_{1m} - l_{2m} - l_{3m}}, \text{ i.e.,}$$

$$d_{a_2, l_2} = \frac{\left(\prod_{k=1}^{3} a_k!\right) a_m!}{\left(\prod_{k=1}^{3} l_{km}!\right) \left\{\prod_{k=1}^{3} (a_k - l_{km})!\right\} \left(a_m - \sum_{k=1}^{3} l_{km}\right)!}$$

$$\times \left[d_{a_2, l_2}\right]_{l_{km} \mapsto 0, \, a_k \mapsto a_k - l_{km} \text{ for } k = 1,2,3; \, a_m \mapsto a_m - \sum_{k=1}^{3} l_{km}}. \tag{5.A.18}$$

If we repeat the steps above until we reach $i = m - 1$, using $\left[d_{a_2, l_2}\right]_{l_{m-1,m} \mapsto 0}$ obtained by letting $l_{m-1,m} = 0$ in (5.A.18) with $i = m - 2$ and recollecting (5.A.14) with $i = m - 1$, we will eventually get

$$d_{a_2, l_2} = \frac{\left(\prod_{k=1}^{m-1} a_k!\right) a_m!}{\left(\prod_{k=1}^{m-1} l_{km}!\right) \left\{\prod_{k=1}^{m-1} (a_k - l_{km})!\right\} \left(a_m - \sum_{k=1}^{m-1} l_{km}\right)!}$$

$$\times \left[d_{a_2, l_2}\right]_{l_{km} \mapsto 0, \, a_k \mapsto a_k - l_{km} \text{ for } k = 1,2,\ldots,m-1; \, a_m \mapsto a_m - \sum_{k=1}^{m-1} l_{km}}. \tag{5.A.19}$$

Finally, noting that $a_m - \sum_{k=1}^{m-1} l_{km} - 2 l_{mm} = a_m - l_{mm} - \sum_{k=1}^{m} l_{km} = L_{a_2, m}$ and

that $\left[L_{a_1, j}\right]_{a_k \mapsto a_k - l_{km} \text{ for } k = 1,2,\ldots,m-1} = a_j - l_{jm} - l_{jj} - \sum_{k=1}^{m-1} l_{kj} = a_j - l_{jj} - \sum_{k=1}^{m} l_{kj} = L_{a_2, j}$

for $j = 1, 2, \ldots, m - 1$, if we combine (5.A.12) and (5.A.19), we can get

$$d_{a_2, l_2} = \frac{\left(\prod_{k=1}^{m-1} a_k!\right) a_m!}{\left(\prod_{k=1}^{m-1} l_{km}!\right) \left\{\prod_{k=1}^{m-1} (a_k - l_{km})!\right\} \left(a_m - \sum_{k=1}^{m-1} l_{km}\right)!}$$

$$\times \frac{\left\{\prod_{k=1}^{m-1} (a_k - l_{km})!\right\} \left(a_m - \sum_{k=1}^{m-1} l_{km}\right)!}{2^{M_{l_2}} \zeta_{m-1}(l) \, l_{mm}! \left(a_m - \sum_{k=1}^{m-1} l_{km} - 2 l_{mm}\right)! D_{2,2}}$$

$$= \frac{\prod_{k=1}^{m} a_k!}{2^{M_{l_2}} \zeta_m(l) \, \eta_{2,m}(l)}, \tag{5.A.20}$$

which implies that (5.3.51) holds true also when $n = m$, where $D_{2,2} = \prod\limits_{j=1}^{m-1} \left([L_{a_1,j}]_{a_k \mapsto a_k - l_{km}} \text{ for } k=1,2,\ldots,m-1 \right)!$.

Appendix 5.2 Some Integral Formulas

For the quadratic function

$$Q(x) = \sum_{j=1}^{n} \sum_{i=1}^{n} a_{ij} x_i x_j \qquad (5.A.21)$$

of $x = (x_1, x_2, \ldots, x_n)$, consider

$$J_n = \int_0^\infty \int_0^\infty \cdots \int_0^\infty \exp\{-Q(x)\}\, dx, \qquad (5.A.22)$$

where $dx = dx_1 dx_2 \cdots dx_n$. When $n = 1$ with $Q(x) = a_{11} x_1^2$, we easily get

$$J_1 = \frac{\sqrt{\pi}}{2\sqrt{a_{11}}} \qquad (5.A.23)$$

for $a_{11} > 0$. When $n = 2$, assume $Q(x) = a_{11} x_1^2 + a_{22} x_2^2 + 2a_{12} x_1 x_2$, where $\Delta_2 = a_{11} a_{22} - a_{12}^2 > 0$. We then get

$$J_2 = \frac{1}{\sqrt{\Delta_2}} \left(\frac{\pi}{2} - \tan^{-1} \frac{a_{12}}{\sqrt{\Delta_2}} \right). \qquad (5.A.24)$$

In addition, when $n = 3$ assume $Q(x) = a_{11} x_1^2 + a_{22} x_2^2 + a_{33} x_3^2 + 2a_{12} x_1 x_2 + 2a_{23} x_2 x_3 + 2a_{31} x_3 x_1$, where $\Delta_3 = a_{11} a_{22} a_{33} - a_{11} a_{23}^2 - a_{22} a_{31}^2 - a_{33} a_{12}^2 + 2a_{12} a_{23} a_{31} > 0$ and $\{a_{ii} > 0\}_{i=1}^{3}$. Then, we will get

$$J_3 = \frac{\sqrt{\pi}}{4\sqrt{\Delta_3}} \left(\frac{\pi}{2} + \sum^c \tan^{-1} \frac{a_{ij} a_{ki} - a_{ii} a_{jk}}{\sqrt{a_{ii} \Delta_3}} \right) \qquad (5.A.25)$$

after some manipulations, where $\sum^c$ denotes the cyclic sum defined in (5.3.44).

Now, recollect the standard normal pdf

$$\phi(x) = \frac{1}{\sqrt{2\pi}} \exp\left(-\frac{x^2}{2}\right) \qquad (5.A.26)$$

and the standard normal cdf

$$\Phi(x) = \int_{-\infty}^{x} \phi(t)\, dt \tag{5.A.27}$$

defined in (3.5.2) and (3.5.3), respectively. Based on (5.A.23) or on $\int_{-\infty}^{\infty} \exp\left(-\alpha x^2\right) dx = \sqrt{\frac{\pi}{\alpha}}$ shown in (3.3.28), we get $\int_{-\infty}^{\infty} \phi^m(x) dx = \frac{1}{(2\pi)^{\frac{m}{2}}} \int_{-\infty}^{\infty} \exp\left(-\frac{mx^2}{2}\right) dx$, i.e.,

$$\int_{-\infty}^{\infty} \phi^m(x)\, dx = (2\pi)^{-\frac{m-1}{2}}\, m^{-\frac{1}{2}}. \tag{5.A.28}$$

For $n = 0, 1, \ldots$, consider

$$I_n(a) = 2\pi \int_{-\infty}^{\infty} \Phi^n(ax)\, \phi^2(x)\, dx$$
$$= \int_{-\infty}^{\infty} \Phi^n(ax)\, \exp\left(-x^2\right) dx. \tag{5.A.29}$$

Letting $n = 0$ in (5.A.29), we have

$$I_0(a) = \sqrt{\pi}, \tag{5.A.30}$$

and letting $a = 0$ in (5.A.29), we have

$$I_n(0) = \frac{\sqrt{\pi}}{2^n}. \tag{5.A.31}$$

Because $2m + 1$ is an odd number for $m = 0, 1, \ldots$, we have

$$\int_{-\infty}^{\infty} \left\{\Phi(ax) - \frac{1}{2}\right\}^{2m+1} \exp\left(-x^2\right) dx = 0, \tag{5.A.32}$$

which can subsequently be expressed as $\sum_{i=0}^{2m+1} \left(-\frac{1}{2}\right)^i {}_{2m+1}C_i I_{2m+1-i}(a) = 0$ from (5.A.29) and then $\left\{\Phi(ax) - \frac{1}{2}\right\}^{2m+1} = \sum_{i=0}^{2m+1} \left(-\frac{1}{2}\right)^i {}_{2m+1}C_i \Phi^{2m+1-i}(ax)$. This result in turn can be rewritten as

$$I_{2m+1}(a) = \sum_{i=1}^{2m+1} 2^{-i}(-1)^{i+1} {}_{2m+1}C_i \, I_{2m+1-i}(a) \tag{5.A.33}$$

for $m = 0, 1, \ldots$ after some steps. Thus, when $m = 0$, from (5.A.30) and (5.A.33), we get $I_1(a) = \frac{1}{2}I_0(a)$, i.e.,

$$I_1(a) = \frac{\sqrt{\pi}}{2}. \tag{5.A.34}$$

Similarly, when $m = 1$, from (5.A.30), (5.A.33), and (5.A.34), we get $I_3(a) = \frac{3}{2}I_2(a) - \frac{3}{4}I_1(a) + \frac{1}{8}I_0(a)$, i.e.,

$$I_3(a) = \frac{3}{2}I_2(a) - \frac{1}{4}I_0(a). \tag{5.A.35}$$

Next, recollecting that $\frac{d}{da}\Phi(ax) = x\phi(ax)$ and $\frac{d}{da}\Phi^2(ax) = 2x\Phi(ax)\phi(ax)$, if we differentiate $I_2(a)$ with respect to a using Leibnitz's rule (3.2.18), integrate by parts, and then use (3.3.29), we get $\frac{d}{da}I_2(a) = 2\pi \int_{-\infty}^{\infty} 2x\Phi(ax)\phi(ax)\phi^2(x) = \frac{2}{\sqrt{2\pi}}\int_{-\infty}^{\infty}\Phi(ax)x\exp\left(-\frac{2+a^2}{2}x^2\right)dx = \sqrt{\frac{2}{\pi}}\left[-\frac{\Phi(ax)}{2+a^2}\exp\left(-\frac{2+a^2}{2}x^2\right)\right]_{x=-\infty}^{\infty} + \sqrt{\frac{2}{\pi}}\frac{a}{2+a^2}\int_{-\infty}^{\infty}\phi(ax)\exp\left\{-\frac{(2+a^2)x^2}{2}\right\}dx = \frac{a}{\pi(2+a^2)}\int_{-\infty}^{\infty}\exp\left\{-\left(1+a^2\right)x^2\right\}dx$, i.e.,

$$\frac{d}{da}I_2(a) = \frac{a}{\pi(2+a^2)}\sqrt{\frac{\pi}{1+a^2}}. \tag{5.A.36}$$

Consequently, noting (5.A.31) and $\frac{d}{da}\tan^{-1}\sqrt{1+a^2} = \frac{a}{(2+a^2)\sqrt{1+a^2}}$ from $\frac{d}{dx}\tan^{-1}x = \frac{1}{1+x^2}$, we finally obtain

$$I_2(a) = \frac{1}{\sqrt{\pi}}\tan^{-1}\sqrt{1+a^2}, \tag{5.A.37}$$

and then,

$$I_3(a) = \frac{3}{2\sqrt{\pi}}\tan^{-1}\sqrt{1+a^2} - \frac{\sqrt{\pi}}{4} \tag{5.A.38}$$

from (5.A.35) and (5.A.37). The results $\{J_k\}_{k=1}^3$ and $\{I_k(a)\}_{k=0}^3$ we have derived so far, together with $\phi'(x) = -x\phi(x)$, are quite useful in obtaining the moments of order statistics of standard normal distribution for small values of n.

Appendix 5.3 Generalized Gaussian, Generalized Cauchy, and Stable Distributions

In many fields including signal processing, communication, and control, it is usually assumed that noise is a normal random variable. The rationale for this is as follows: the first reason is due to the central limit theorem, which will be discussed in Chap. 6. According to the central limit theorem, the sum of random variables will converge to a normal random variable under certain conditions and the sum can reasonably be approximated by a normal random variable even when the conditions are not satisfied perfectly. We have already observed such a case in Gaussian approximation of binomial distribution in Theorem 3.5.16. In essence, the mathematical model of Gaussian assumption on noise does not deviate much from reality. The second reason is that, if we assume that noise is Gaussian, many schemes of communications and signal processing can be obtained in a simple way and analysis of such schemes becomes relatively easy.

On the other hand, in some real environments, noise can be described only by non-Gaussian distributions. For example, it is reported that the low frequency noise in the atmosphere and noise environment in underwater acoustics can be modeled adequately only with non-Gaussian distributions. When noise is non-Gaussian, it would be necessary to adopt an adequate model other than the Gaussian model for the real environment in finding, for instance, signal processing techniques or communication schemes. Needless to say, in such an environment, we could still apply techniques obtained under Gaussian assumption on noise at the cost of some, and sometimes significant, loss and/or unpredictability in the performance.

Among the non-Gaussian distributions, impulsive distributions, also called long-tailed or heavy-tailed distributions also, constitute an important class. In general, when the tail of the pdf of a distribution is heavier (longer) than that of a normal distribution, the distribution is called an impulsive distribution. In impulsive distributions, noise of very large magnitude or absolute value (that is, values much larger or smaller than the median) can occur more frequently than that in the normal distribution.

Let us here discuss in a brief manner the generalized Gaussian distribution and the generalized Cauchy distribution. In addition, the class of stable distributions (Nikias and Shao 1995; Tsihrintzis and Nikias 1995), which has bases on the generalized central limit theorem, will also be introduced. In passing, the generalized central limit theorem, not covered in this book, allows us to consider the convergence of random variables of which the variance is not finite and, therefore, to which the central limit theorem cannot be applied.

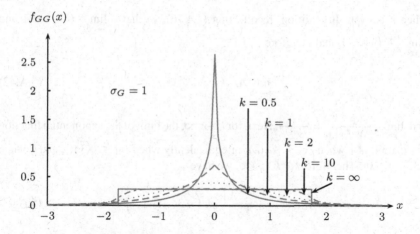

Fig. 5.9 The generalized normal pdf

(A) Generalized Gaussian Distribution

Definition 5.A.1 (*generalized normal distribution*) A distribution with the pdf

$$f_{GG}(x) = \frac{k}{2A_G(k)\Gamma\left(\frac{1}{k}\right)} \exp\left[-\left\{\frac{|x|}{A_G(k)}\right\}^k\right] \qquad (5.A.39)$$

is called a generalized normal or generalized Gaussian distribution, where $k > 0$ and $A_G(k) = \sqrt{\frac{\sigma_G^2 \Gamma\left(\frac{1}{k}\right)}{\Gamma\left(\frac{3}{k}\right)}}$.

As it is also clear in Fig. 5.9, the pdf of the generalized normal distribution is a unimodal even function, defined by two parameters. The two parameters are the variance σ_G^2 and the rate k of decay of the pdf.

The generalized normal pdf is usefully employed in representing many pdf's by adopting appropriate values of k. For example, when $k = 2$, the generalized normal pdf is a normal pdf. When $k < 2$, the generalized normal pdf is an impulsive pdf: specifically, when $k = 1$, the generalized normal pdf is the double exponential pdf

$$f_D(x) = \frac{1}{\sqrt{2}\sigma_G} \exp\left(-\frac{\sqrt{2}|x|}{\sigma_G}\right). \qquad (5.A.40)$$

The moment of a random variable X with the pdf $f_{GG}(x)$ in (5.A.39) is

$$\mathsf{E}\left\{X^r\right\} = \sigma_G^r \frac{\Gamma^{\frac{r}{2}-1}\left(\frac{1}{k}\right)\Gamma\left(\frac{r+1}{k}\right)}{\Gamma^{\frac{r}{2}}\left(\frac{3}{k}\right)} \qquad (5.A.41)$$

when r is even. In addition, recollecting (1.4.76), we have $\lim\limits_{k \to \infty} \frac{3}{k}\Gamma\left(\frac{3}{k}\right) = 1$ and $\lim\limits_{k \to \infty} \frac{1}{k}\Gamma\left(\frac{1}{k}\right) = 1$, and therefore

$$\lim_{k \to \infty} A_G(k) = \sqrt{3}\sigma_G \qquad (5.A.42)$$

and $\lim\limits_{k \to \infty} \frac{k}{2A_G(k)\Gamma\left(\frac{1}{k}\right)} = \frac{1}{2\sqrt{3}\sigma_G}$. Next, for $k \to \infty$, the limit of the exponential function in (5.A.39) is 1 when $|x| \le A_G(k)$, or equivalently when $|x| \le \sqrt{3}\sigma_G$, and 0 when $|x| > A_G(k)$. Therefore, for $k \to \infty$, we have

$$f_{GG}(x) \to \frac{1}{2\sqrt{3}\sigma_G} u\left(\sqrt{3}\sigma_G - |x|\right). \qquad (5.A.43)$$

In other words, for $k \to \infty$, the limit of the generalized normal pdf is a uniform pdf as shown in Fig. 5.9.

(B) Generalized Cauchy Distribution

Definition 5.A.2 (*generalized Cauchy distribution*) A distribution with the pdf

$$f_{GC}(x) = \frac{\tilde{B}_c(k, v)}{\tilde{D}_c^{v+\frac{1}{k}}(x)} \qquad (5.A.44)$$

is called the generalized Cauchy distribution and is denoted by $G_C(k, v)$. Here, $k > 0$, $v > 0$, $\tilde{B}_c(k, v) = \frac{k\Gamma\left(v+\frac{1}{k}\right)}{2v^{\frac{1}{k}} A_G(k)\Gamma(v)\Gamma\left(\frac{1}{k}\right)}$, and $\tilde{D}_c(x) = 1 + \frac{1}{v}\left\{\frac{|x|}{A_G(k)}\right\}^k$.

Figure 5.10 shows the generalized Cauchy pdf. When the parameter v is finite, the tail of the generalized Cauchy pdf shows not an exponential behavior, but an algebraic behavior. Specifically, when $|x|$ is large, the tail of the generalized Cauchy pdf $f_{GC}(x)$ decreases in proportion to $|x|^{-(kv+1)}$.

When $k = 2$ and $2v$ is an integer, the generalized Cauchy pdf is a t pdf, and when $k = 2$ and $v = \frac{1}{2}$, the generalized Cauchy pdf is a Cauchy pdf

$$f_C(x) = \frac{\sigma_G}{\pi\left(x^2 + \sigma_G^2\right)}. \qquad (5.A.45)$$

When the parameters σ_G^2 and k are fixed, we have $\lim\limits_{v \to \infty} \tilde{D}_c^{v+\frac{1}{k}}(x) = \lim\limits_{v \to \infty}$ $\left[\left[1 + \frac{1}{v}\left\{\frac{|x|}{A_G(k)}\right\}^k\right]^{v+\frac{1}{k}}\right]$, i.e.,

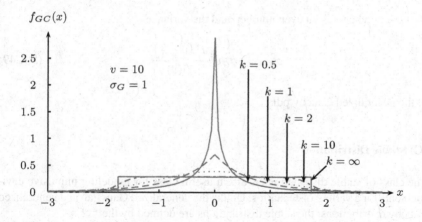

Fig. 5.10 The generalized Cauchy pdf

$$\lim_{v \to \infty} \tilde{D}_c^{v+\frac{1}{k}}(x) = \exp\left[\left\{\frac{|x|}{A_G(k)}\right\}^k\right]. \tag{5.A.46}$$

In addition, $\lim\limits_{v \to \infty} \tilde{B}_c(k, v) = \frac{k}{2A_G(k)\Gamma(\frac{1}{k})}\lim\limits_{v \to \infty}\frac{\Gamma(v+\frac{1}{k})}{v^{\frac{1}{k}}\Gamma(v)} = \frac{k}{2A_G(k)\Gamma(\frac{1}{k})}$ because $\lim\limits_{v \to \infty}\frac{\Gamma(v+\frac{1}{k})}{v^{\frac{1}{k}}\Gamma(v)} = 1$ from (1.4.77). Thus, for $v \to \infty$, the generalized Cauchy pdf converges to the generalized normal pdf. For example, when $k = 2$ and $v \to \infty$, the generalized Cauchy pdf is a normal pdf.

Next, using $\lim\limits_{p \to 0} p\Gamma(p) = \lim\limits_{p \to 0}\Gamma(p+1) = 1$ shown in (1.4.76) and $\lim\limits_{k \to \infty} A_G(k) = \sqrt{3}\sigma_G$ shown in (5.A.42), we get $\lim\limits_{k \to \infty}\tilde{B}_c(k, v) = \lim\limits_{k \to \infty}\frac{\Gamma(v)}{2A_G(k)\Gamma(v)\frac{1}{k}\Gamma(\frac{1}{k})} = \frac{1}{2\sqrt{3}\sigma_G}$ when v is fixed. In addition, $\lim\limits_{k \to \infty}\tilde{D}_c(x) = 1$ when $|x| < \sqrt{3}\sigma_G$ and $\lim\limits_{k \to \infty}\tilde{D}_c(x) = \infty$ when $|x| > \sqrt{3}\sigma_G$. In short, when v is fixed and $k \to \infty$, we have

$$f_{GC}(x) \to \frac{1}{2\sqrt{3}\sigma_G}u\left(\sqrt{3}\sigma_G - |x|\right), \tag{5.A.47}$$

i.e., the limit of the generalized Cauchy pdf is a uniform pdf as shown also in Fig. 5.10.

After some steps, we can obtain the r-th moment

$$\mathsf{E}\{X^r\} = v^{\frac{r}{k}}\sigma_G^r\frac{\Gamma\left(v - \frac{r}{k}\right)\Gamma\left(\frac{r+1}{k}\right)\Gamma^{\frac{r}{2}-1}\left(\frac{1}{k}\right)}{\Gamma(v)\Gamma^{\frac{r}{2}}\left(\frac{3}{k}\right)} \tag{5.A.48}$$

for $vk > r$ when r is an even number, and the variance

$$\sigma_{GC}^2 = \sigma_G^2 v^{\frac{2}{k}} \frac{\Gamma\left(v - \frac{2}{k}\right)}{\Gamma(v)} \tag{5.A.49}$$

of the generalized Cauchy pdf.

(C) Stable Distribution

The class of stable distributions is also a useful class for modeling impulsive environments for a variety of scenarios. Unlike the generalized Gaussian and generalized Cauchy distributions, the stable distributions are defined by their cf's.

Definition 5.A.3 *(stable distribution)* A distribution with the cf

$$\varphi(t) = \exp\left[jmt - \gamma|t|^\alpha \{1 + j\beta\mathrm{sgn}(t)\omega(t, \alpha)\}\right] \tag{5.A.50}$$

is called a stable distribution. Here, $0 < \alpha \le 2$, $|\beta| \le 1$, m is a real number, $\gamma > 0$, and

$$\omega(t, \alpha) = \begin{cases} \tan\frac{\alpha\pi}{2}, & \text{if } \alpha \ne 1, \\ \frac{2}{\pi}\log|t|, & \text{if } \alpha = 1. \end{cases} \tag{5.A.51}$$

In Definition 5.A.3, the numbers m, α, β, and γ are called the location parameter, characteristic exponent, symmetry parameter, and dispersion parameter, respectively. The location parameter m represents the mean when $1 < \alpha \le 2$ and the median when $0 < \alpha \le 1$. The characteristic exponent α represents the weight or length of the tail of the pdf, with a smaller value denoting a longer tail or a higher degree of impulsiveness. The symmetry parameter β determines the symmetry of the pdf with $\beta = 0$ resulting in a symmetric pdf. The dispersion parameter γ plays a role similar to the variance of a normal distribution. For instance, the stable distribution is a normal distribution and the variance is 2γ when $\alpha = 2$. When $\alpha = 1$ and $\beta = 0$, the stable distribution is a Cauchy distribution.

Definition 5.A.4 *(symmetric alpha-stable distribution)* When the symmetry parameter $\beta = 0$, the stable distribution is called the symmetric α-stable (SαS) distribution. When the location parameter $m = 0$ and the dispersion parameter $\gamma = 1$, the SαS distribution is called the standard SαS distribution.

By inverse transforming the cf

$$\varphi(t) = \exp\left(-\gamma|t|^\alpha\right) \tag{5.A.52}$$

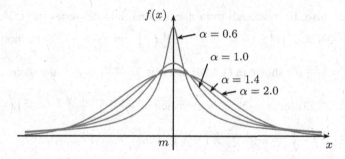

Fig. 5.11 The pdf of SαS distribution

of the SαS distribution with $m = 0$, we have the pdf

$$
f(x) = \begin{cases}
\frac{1}{\pi \gamma^{\frac{1}{\alpha}}} \sum_{k=1}^{\infty} \frac{(-1)^{k-1}}{k!} \Gamma(\alpha k + 1) \sin\left(\frac{k\alpha\pi}{2}\right) \left(\frac{|x|}{\gamma^{\frac{1}{\alpha}}}\right)^{-\alpha k - 1}, \\
\qquad\qquad \text{for } 0 < \alpha \le 1, \\
\frac{1}{\pi \alpha \gamma^{\frac{1}{\alpha}}} \sum_{k=0}^{\infty} \frac{(-1)^{k}}{(2k)!} \Gamma\left(\frac{2k+1}{\alpha}\right) \left(\frac{x}{\gamma^{\frac{1}{\alpha}}}\right)^{2k}, \\
\qquad\qquad \text{for } 1 \le \alpha \le 2.
\end{cases}
\tag{5.A.53}
$$

It is known that the pdf (5.A.53) can be expressed more explicitly in a closed form when $\alpha = 1$ and 2. Figure 5.11 shows pdf's of the SαS distributions.

Let us show that the two infinite series in (5.A.53) become the Cauchy pdf

$$
f(x) = \frac{\gamma}{\pi \left(x^2 + \gamma^2\right)}
\tag{5.A.54}
$$

when $\alpha = 1$, and that the second infinite series of (5.A.53) is the normal pdf

$$
f(x) = \frac{1}{2\sqrt{\pi \gamma}} \exp\left(-\frac{x^2}{4\gamma}\right)
\tag{5.A.55}
$$

when $\alpha = 2$. The first infinite series in (5.A.53) for $\alpha = 1$ can be expressed

as $\quad \frac{1}{\pi \gamma} \sum_{k=1}^{\infty} \frac{(-1)^{k-1}}{k!} \Gamma(k+1) \sin\left(\frac{k\pi}{2}\right) \left(\frac{|x|}{\gamma}\right)^{-k-1} = \frac{1}{\pi \gamma} \left\{ \left(\frac{\gamma}{|x|}\right)^2 - \left(\frac{\gamma}{|x|}\right)^4 + \left(\frac{\gamma}{|x|}\right)^6 \right.$

$\left. - \left(\frac{\gamma}{|x|}\right)^8 + \cdots \right\} = \frac{1}{\pi \gamma} \sum_{k=0}^{\infty} (-1)^k \left(\frac{\gamma^2}{x^2}\right)^{k+1}$, i.e.,

$$
\frac{1}{\pi \gamma} \sum_{k=1}^{\infty} \frac{(-1)^{k-1}}{k!} \Gamma(k+1) \sin\left(\frac{k\pi}{2}\right) \left(\frac{|x|}{\gamma}\right)^{-k-1} = \frac{\gamma}{\pi \left(x^2 + \gamma^2\right)},
\tag{5.A.56}
$$

which can also be obtained from the second infinite series of (5.A.53) as

$$\frac{1}{\pi\gamma}\sum_{k=0}^{\infty}\frac{(-1)^k}{(2k)!}\Gamma(2k+1)\left(\frac{x}{\gamma}\right)^{2k} = \frac{1}{\pi\gamma}\sum_{k=0}^{\infty}(-1)^k\left(\frac{x}{\gamma}\right)^{2k} = \frac{\gamma}{\pi(x^2+\gamma^2)}.$$ Next, noting that

$\Gamma\left(\frac{2k+1}{2}\right) = \frac{(2k)!}{2^{2k}k!}\sqrt{\pi}$ shown in (1.4.84) and that $\sum_{k=0}^{\infty}\frac{(-x)^k}{k!} = e^{-x}$, the second infinite

series of (5.A.53) for $\alpha = 2$ can be rewritten as $\frac{1}{2\pi\sqrt{\gamma}}\sum_{k=0}^{\infty}\frac{(-1)^k}{(2k)!}\Gamma\left(\frac{2k+1}{2}\right)\left(\frac{x}{\sqrt{\gamma}}\right)^{2k} =$

$\frac{1}{2\sqrt{\pi\gamma}}\sum_{k=0}^{\infty}\frac{(-1)^k}{2^{2k}k!}\left(\frac{x^2}{\gamma}\right)^{k} = \frac{1}{2\sqrt{\pi\gamma}}\sum_{k=0}^{\infty}\frac{1}{k!}\left(-\frac{x^2}{4\gamma}\right)^{k}$, i.e.,

$$\frac{1}{2\pi\sqrt{\gamma}}\sum_{k=0}^{\infty}\frac{(-1)^k}{(2k)!}\Gamma\left(\frac{2k+1}{2}\right)\left(\frac{x}{\sqrt{\gamma}}\right)^{2k} = \frac{1}{2\sqrt{\pi\gamma}}\exp\left(-\frac{x^2}{4\gamma}\right). \tag{5.A.57}$$

When $A \sim U\left(-\frac{\pi}{2}, \frac{\pi}{2}\right]$ and an exponential random variable B with mean 1 are independent of each other, it is known that

$$X = \frac{\sin(\alpha A)}{(\cos A)^{\frac{1}{\alpha}}}\left[\frac{\cos\{(1-\alpha)A\}}{B}\right]^{\frac{1-\alpha}{\alpha}} \tag{5.A.58}$$

is a standard SαS random variable. This result is useful when generating random numbers obeying the SαS distribution.

Definition 5.A.5 (*bi-variate isotropic SαS distribution*) When the joint pdf of a random vector (X, Y) can be expressed as

$$f_{X,Y}(x, y) = \frac{1}{4\pi^2}\int_{-\infty}^{\infty}\int_{-\infty}^{\infty}\exp\left\{-\gamma\left(\omega_1^2+\omega_2^2\right)^{\frac{\alpha}{2}}\right\}$$
$$\times \exp\left\{-j\left(x\omega_1 + y\omega_2\right)\right\}d\omega_1 d\omega_2, \tag{5.A.59}$$

the distribution of (X, Y) is called the bi-variate isotropic SαS distribution.

Expressing the pdf (5.A.59) of the bi-variate isotropic SαS distribution in infinite series, we have

$$f_{X,Y}(x, y) = \begin{cases} \frac{1}{\pi^2\gamma^{\frac{2}{\alpha}}}\sum_{k=1}^{\infty}\frac{1}{k!}2^{\alpha k}(-1)^{k-1}\Gamma^2\left(1+\frac{\alpha k}{2}\right) \\ \qquad \times \sin\left(\frac{k\alpha\pi}{2}\right)\left(\frac{\sqrt{x^2+y^2}}{\gamma^{\frac{1}{\alpha}}}\right)^{-\alpha k-2}, \\ \qquad\qquad \text{for } 0 < \alpha \leq 1, \\ \frac{1}{2\pi\alpha\gamma^{\frac{2}{\alpha}}}\sum_{k=0}^{\infty}\frac{1}{(k!)^2}\Gamma\left(\frac{2k+2}{\alpha}\right)\left(-\frac{x^2+y^2}{4\gamma^{\frac{2}{\alpha}}}\right)^{k}, \\ \qquad\qquad \text{for } 1 \leq \alpha \leq 2. \end{cases} \tag{5.A.60}$$

Example 5.A.1 Show that (5.A.60) represents a bi-variate Cauchy distribution and a bi-variate normal distribution for $\alpha = 1$ and $\alpha = 2$, respectively. In other words, show that the two infinite series of (5.A.60) become

$$f_{X,Y}(x, y) = \frac{\gamma}{2\pi \left(x^2 + y^2 + \gamma^2\right)^{\frac{3}{2}}} \tag{5.A.61}$$

when $\alpha = 1$ and that the second infinite series of (5.A.60) becomes

$$f_{X,Y}(x, y) = \frac{1}{4\pi\gamma} \exp\left(-\frac{x^2 + y^2}{4\gamma}\right) \tag{5.A.62}$$

when $\alpha = 2$.

Solution First, note that we have $\Gamma\left(\frac{2k+3}{2}\right) = \left(\frac{1}{2} + k\right) \Gamma\left(\frac{1}{2} + k\right) = \frac{(2k+1)!}{2^{2k+1}k!} \sqrt{\pi}$ from (1.4.75) and (1.4.84). Thus, recollecting that $(1 + x)^{-\frac{3}{2}} = \sum_{k=0}^{\infty} {}_{-\frac{3}{2}}C_k x^k$, i.e.,

$$(1 + x)^{-\frac{3}{2}} = \sum_{k=0}^{\infty} \frac{(-1)^k (2k + 1)!}{2^{2k}(k!)^2} x^k, \tag{5.A.63}$$

we get

$$\frac{1}{\pi^2\gamma^2} \sum_{k=1}^{\infty} \frac{2^k(-1)^{k-1}}{k!} \Gamma^2\left(\frac{k}{2} + 1\right) \sin\left(\frac{k\pi}{2}\right) \left(\frac{\sqrt{x^2 + y^2}}{\gamma}\right)^{-k-2}$$

$$= \frac{1}{\pi^2 (x^2 + y^2)} \left\{ \frac{2^1 \Gamma^2\left(\frac{3}{2}\right)}{1!} \left(\frac{\gamma}{\sqrt{x^2 + y^2}}\right) - \frac{2^3 \Gamma^2\left(\frac{5}{2}\right)}{3!} \left(\frac{\gamma}{\sqrt{x^2 + y^2}}\right)^3 \right.$$

$$\left. + \frac{2^5 \Gamma^2\left(\frac{7}{2}\right)}{5!} \left(\frac{\gamma}{\sqrt{x^2 + y^2}}\right)^5 - \frac{2^7 \Gamma^2\left(\frac{9}{2}\right)}{7!} \left(\frac{\gamma}{\sqrt{x^2 + y^2}}\right)^7 + \cdots \right\}$$

$$= \frac{1}{\pi^2 (x^2 + y^2)} \sum_{k=0}^{\infty} \frac{(-1)^k 2^{2k+1}}{(2k + 1)!} \Gamma^2\left(\frac{2k + 3}{2}\right) \left(\frac{\gamma}{\sqrt{x^2 + y^2}}\right)^{2k+1}$$

$$= \frac{1}{\pi (x^2 + y^2)} \sum_{k=0}^{\infty} \frac{(-1)^k (2k + 1)!}{2^{2k+1}(k!)^2} \left(\frac{\gamma}{\sqrt{x^2 + y^2}}\right)^{2k+1}$$

$$= \frac{1}{2\pi (x^2 + y^2)} \times \frac{\gamma}{\sqrt{x^2 + y^2}} \sum_{k=0}^{\infty} \frac{(-1)^k (2k + 1)!}{2^{2k}(k!)^2} \left(\frac{\gamma^2}{x^2 + y^2}\right)^k$$

$$= \frac{\gamma}{2\pi (x^2 + y^2)^{\frac{3}{2}}} \left(1 + \frac{\gamma^2}{x^2 + y^2}\right)^{-\frac{3}{2}}$$

$$= \frac{\gamma}{2\pi \left(x^2 + y^2 + \gamma^2\right)^{\frac{3}{2}}} \tag{5.A.64}$$

when $\alpha = 1$ from the first infinite series of (5.A.60). The result (5.A.64) can also be obtained as $\sum_{k=0}^{\infty} \frac{\Gamma(2k+2)}{2\pi\gamma^2(k!)^2}\left(-\frac{x^2+y^2}{4\gamma^2}\right)^k = \sum_{k=0}^{\infty} \frac{(2k+1)!(-1)^k}{2\pi\gamma^2(k!)^2 2^{2k}}\left(\frac{x^2+y^2}{\gamma^2}\right)^k = \frac{1}{2\pi\gamma^2}\left(1+\frac{x^2+y^2}{\gamma^2}\right)^{-\frac{3}{2}}$, i.e.,

$$\sum_{k=0}^{\infty} \frac{\Gamma(2k+2)}{2\pi\gamma^2(k!)^2}\left(-\frac{x^2+y^2}{4\gamma^2}\right)^k = \frac{\gamma}{2\pi\left(x^2+y^2+\gamma^2\right)^{\frac{3}{2}}} \qquad (5.A.65)$$

from the second infinite series of (5.A.60) using (5.A.63). Next, when $\alpha = 2$, from the second infinite series of (5.A.60), we get $\frac{1}{4\pi\gamma}\sum_{k=0}^{\infty} \frac{\Gamma(k+1)}{(k!)^2}\left(-\frac{x^2+y^2}{4\gamma}\right)^k = \frac{1}{4\pi\gamma}\sum_{k=0}^{\infty} \frac{1}{k!}\left(-\frac{x^2+y^2}{4\gamma}\right)^k$, which is the same as (5.A.62) because $\sum_{k=0}^{\infty} \frac{(-x)^k}{k!} = e^{-x}$. ◇

Exercises

Exercise 5.1 Assume a random vector (X, Y) with the joint pdf

$$f_{X,Y}(x, y) = \frac{1}{\pi\sqrt{3}} \exp\left\{-\frac{2}{3}\left(x^2+y^2\right)\right\} \cosh\left(\frac{2}{3}xy\right). \qquad (5.E.1)$$

(1) Show that $X \sim \mathcal{N}(0, 1)$ and $Y \sim \mathcal{N}(0, 1)$.
(2) Show that X and Y are uncorrelated.
(3) Is (X, Y) a bi-variate normal random vector?
(4) Are X and Y independent of each other?

Exercise 5.2 When $X_1 \sim \mathcal{N}(0, 1)$ and $X_2 \sim \mathcal{N}(0, 1)$ are independent of each other, obtain the conditional joint pdf of X_1 and X_2 given that $X_1^2 + X_2^2 < a^2$.

Exercise 5.3 Assume that $X_1 \sim \mathcal{N}(0, 1)$ and $X_2 \sim \mathcal{N}(0, 1)$ are independent of each other.

(1) Obtain the joint pdf of $U = \sqrt{X^2 + Y^2}$ and $V = \tan^{-1}\frac{Y}{X}$.
(2) Obtain the joint pdf of $U = \frac{1}{2}(X + Y)$ and $V = \frac{1}{2}(X - Y)^2$.

Exercise 5.4 Obtain the conditional pdf's $f_{Y|X}(y|x)$ and $f_{X|Y}(x|y)$ when $(X, Y) \sim \mathcal{N}(3, 4, 1, 2, 0.5)$.

Exercise 5.5 Obtain the correlation coefficient ρ_{ZW} between $Z = X_1 \cos\theta + X_2 \sin\theta$ and $W = X_2 \cos\theta - X_1 \sin\theta$, and show that

$$0 \leq \rho_{ZW}^2 \leq \left(\frac{\sigma_1^2 - \sigma_2^2}{\sigma_1^2 + \sigma_2^2}\right)^2 \tag{5.E.2}$$

when $X_1 \sim \mathcal{N}\left(\mu_1, \sigma_1^2\right)$ and $X_2 \sim \mathcal{N}\left(\mu_2, \sigma_2^2\right)$ are independent of each other.

Exercise 5.6 When the two normal random variables X and Y are independent of each other, show that $X + Y$ and $X - Y$ are independent of each other.

Exercise 5.7 Let us consider (5.2.1) and (5.2.2) when $n = 3$ and $s = 1$. Based on (5.1.18) and (5.1.21), show that $\left(\Psi_{22} - \Psi_{21}\Psi_{11}^{-1}\Psi_{12}\right)^{-1}$ is equal to $K_{22} = \begin{bmatrix} 1 & \rho_{23} \\ \rho_{23} & 1 \end{bmatrix}$.

Exercise 5.8 Consider the random variable

$$X = \begin{cases} Y, & \text{when } Z = +1, \\ -Y, & \text{when } Z = -1, \end{cases} \tag{5.E.3}$$

where Z is a binary random variable with pmf $p_Z(1) = p_Z(-1) = 0.5$ and $Y \sim \mathcal{N}(0, 1)$.

(1) Obtain the conditional cdf $F_{X|Y}(x|y)$.
(2) Obtain the cdf $F_X(x)$ of X and determine whether or not X is normal.
(3) Is the random vector (X, Y) normal?
(4) Obtain the conditional pdf $f_{X|Y}(x|y)$ and the joint pdf $f_{X,Y}(x, y)$.

Exercise 5.9 For a zero-mean normal random vector X with covariance matrix $\begin{pmatrix} 1 & \frac{1}{6} & \frac{1}{36} \\ \frac{1}{6} & 1 & \frac{1}{6} \\ \frac{1}{36} & \frac{1}{6} & 1 \end{pmatrix}$, find a linear transformation to decorrelate X.

Exercise 5.10 Let $X = (X, Y)$ denote the coordinate of a point in the two-dimensional plane and $C = (R, \Theta)$ be its polar coordinate. Specifically, as shown in Fig. 5.12, $R = \sqrt{X^2 + Y^2}$ is the distance from the origin to X, and $\Theta = \angle X$ is the angle between the positive x-axis and the line from the origin to X, where we assume $-\pi < \Theta \leq \pi$. Express the joint pdf of C in terms of the joint pdf of X. When X is an i.i.d. random vector with marginal distribution $\mathcal{N}\left(0, \sigma^2\right)$, prove or disprove that C is an independent random vector.

Exercise 5.11 For the limit pdf $\lim_{\rho \to \pm 1} f_{X_1, X_2}(x, y)$ shown in (5.1.15), show that $\int_{-\infty}^{\infty} \lim_{\rho \to \pm 1} f_{X_1, X_2}(x, y) dy = f_{X_1}(x)$ and $\int_{-\infty}^{\infty} \lim_{\rho \to \pm 1} f_{X_1, X_2}(x, y) dx = f_{X_2}(y)$.

Exercise 5.12 Consider a zero-mean normal random vector (X_1, X_2, X_3) with covariance matrix $\begin{pmatrix} 1 & 1 & 1 \\ 1 & 2 & 1 \\ 1 & 1 & 3 \end{pmatrix}$. Obtain the conditional distribution of X_3 when $X_1 = X_2 = 1$.

Fig. 5.12 Polar coordinate
$C = (R, \Theta) = (|X|, \angle X)$
for $X = (X, Y)$

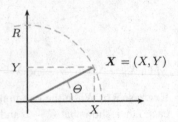

Exercise 5.13 Consider the linear transformation $(Z, W) = (aX + bY, cX + dY)$ of $(X, Y) \sim \mathcal{N}\left(m_X, m_Y, \sigma_X^2, \sigma_Y^2, \rho\right)$. When $ad - bc \neq 0$, find the requirement for $\{a, b, c, d\}$ for Z and W to be independent of each other.

Exercise 5.14 When $(X, Y) \sim \mathcal{N}\left(0, 0, \sigma_X^2, \sigma_Y^2, \rho\right)$, we have $\mathsf{E}\left\{X^2\right\} = \sigma_X^2$ and $\mathsf{E}\left\{X^4\right\} = 3\sigma_X^4$. Based on these two results and (4.4.44), obtain $\mathsf{E}\{XY\}$, $\mathsf{E}\left\{X^2Y\right\}$, $\mathsf{E}\left\{X^3Y\right\}$, and $\mathsf{E}\left\{X^2Y^2\right\}$. Compare the results with those you can obtain from (5.3.22) or (5.3.51).

Exercise 5.15 For a standard tri-variate normal random vector (X_1, X_2, X_3), denote the covariance by $\mathsf{E}\left\{X_i X_j\right\} - \mathsf{E}\{X_i\} \mathsf{E}\left\{X_j\right\} = \rho_{ij}$. Show that

$$\mathsf{E}\left\{X_1^2 X_2^2 X_3^2\right\} = 1 + 2\left(\rho_{12}^2 + \rho_{23}^2 + \rho_{31}^2\right) + 8\rho_{12}\rho_{23}\rho_{31} \qquad (5.E.4)$$

based on the moment theorem. Show (5.E.4) based on Taylor series of the cf.

Exercise 5.16 When $(Z, W) \sim \mathcal{N}\left(m_1, m_2, \sigma_1^2, \sigma_2^2, \rho\right)$, obtain $\mathsf{E}\left\{Z^2 W^2\right\}$.

Exercise 5.17 Denote the joint moment by $\mu_{ij} = \mathsf{E}\left\{X^i Y^j\right\}$ for a zero-mean random vector (X, Y). Based on the moment theorem, (5.3.22), (5.3.30), or (5.3.51), obtain μ_{51}, μ_{42}, and μ_{33} for a random vector $(X, Y) \sim \mathcal{N}\left(0, 0, \sigma_1^2, \sigma_2^2, \rho\right)$.

Exercise 5.18 Using the cf, prove (5.3.9) and (5.3.10).

Exercise 5.19 Denote the joint absolute moment by $\nu_{ij} = \mathsf{E}\left\{|X|^i |Y|^j\right\}$ for a zero-mean random vector (X, Y). By direct integration, show that $\nu_{11} = \frac{2}{\pi}\left(\sqrt{1 - \rho^2} + \rho \sin^{-1} \rho\right)$ and $\nu_{21} = \sqrt{\frac{2}{\pi}}\left(1 + \rho^2\right)$ for $(X, Y) \sim \mathcal{N}(0, 0, 1, 1, \rho)$. For $(X, Y) \sim \mathcal{N}\left(0, 0, \pi^2, 2, \frac{1}{\sqrt{2}}\right)$, calculate $\mathsf{E}\{|XY|\}$.

Exercise 5.20 Based on (5.3.30), obtain $\mathsf{E}\{|X_1|\}$, $\mathsf{E}\{|X_1 X_2|\}$, and $\mathsf{E}\left\{|X_1 X_2^3|\right\}$ for $X = (X_1, X_2) \sim \mathcal{N}(0, 0, 1, 1, \rho)$. Show

$$\rho_{|X_1||X_2|} = \frac{2}{\pi - 2}\left(\sqrt{1 - \rho^2} + \rho \sin^{-1} \rho - 1\right). \qquad (5.E.5)$$

Next, based on Price's theorem, show that

$$E\{X_1 u(X_1) X_2 u(X_2)\} = \frac{1}{2\pi}\left\{\left(\frac{\pi}{2} + \sin^{-1}\rho\right)\rho + \sqrt{1-\rho^2}\right\}, \quad (5.E.6)$$

which[12] implies that $E\{WY u(W)u(Y)\} = \frac{\sigma_W \sigma_Y}{2\pi}\left\{\rho\cos^{-1}(-\rho) + \sqrt{1-\rho^2}\right\}$ when $(W, Y) \sim \mathcal{N}\left(0, 0, \sigma_W^2, \sigma_Y^2, \rho\right)$. In addition, when $W = Y$, we can obtain $E\{W^2 u(W)\} = \frac{1}{2}\sigma_W^2$ with $\rho = 1$ and $\sigma_Y = \sigma_W$, which can be proved by a direct integration as $E\{W^2 u(W)\} = \int_0^\infty \frac{x^2}{\sqrt{2\pi\sigma_W^2}} \exp\left(-\frac{x^2}{2\sigma_W^2}\right) dx = \frac{1}{2}\int_{-\infty}^\infty \frac{x^2}{\sqrt{2\pi\sigma_W^2}} \exp\left(-\frac{x^2}{2\sigma_W^2}\right) dx = \frac{1}{2}\sigma_W^2$.

Exercise 5.21 Show

$$E\{X_1 X_2 |X_3|\} = \sqrt{\frac{2}{\pi}}\left(\rho_{12} + \rho_{23}\rho_{31}\right) \quad (5.E.7)$$

for a standard tri-variate normal random vector (X_1, X_2, X_3).

Exercise 5.22 Based on Price's theorem, (5.1.26), and (5.1.27), show that[13]

$$E\{\delta(X_1)\delta(X_2)|X_3|\} = \frac{\sqrt{|K_3|}}{\sqrt{2\pi^3}\left(1 - \rho_{12}^2\right)}, \quad (5.E.8)$$

$$E\{\delta(X_1) X_2 X_3\} = \frac{c_{23}}{\sqrt{2\pi}}, \quad (5.E.9)$$

$$E\{\delta(X_1)\,\mathrm{sgn}(X_2) X_3\} = \frac{c_{23}}{\pi\sqrt{1 - \rho_{12}^2}}, \quad (5.E.10)$$

$$E\{\delta(X_1) |X_2| |X_3|\} = \sqrt{\frac{2}{\pi^3}}\left\{\sqrt{|K_3|} + c_{23}\sin^{-1}\beta_{23,1}\right\}, \quad (5.E.11)$$

and

$$E\{\delta(X_1)\,\mathrm{sgn}(X_2)\,\mathrm{sgn}(X_3)\} = \sqrt{\frac{2}{\pi^3}}\sin^{-1}\beta_{23,1} \quad (5.E.12)$$

---

[12] Here, the range of the inverse cosine function $\cos^{-1}x$ is $[0, \pi]$, and $\cos\left(\sin^{-1}\rho\right) = \sqrt{1-\rho^2}$. Note that, letting $\frac{\pi}{2} + \sin^{-1}\rho = \theta$, we get $\cos\theta = \cos\left(\frac{\pi}{2} + \sin^{-1}\rho\right) = -\sin\left(\sin^{-1}\rho\right) = -\rho$ and, subsequently, $\left(\frac{\pi}{2} + \sin^{-1}\rho\right)\rho = \rho\cos^{-1}(-\rho)$. Thus, we have $\theta = \cos^{-1}(-\rho)$.

[13] Here, using $E\{\mathrm{sgn}(X_2)\,\mathrm{sgn}(X_3)\} = \frac{2}{\pi}\sin^{-1}\rho_{23}$ obtained in (5.3.29) and $E\{\delta(X_1)\} = \frac{1}{\sqrt{2\pi}}$, we can obtain $E\{\delta(X_1)\,\mathrm{sgn}(X_2)\,\mathrm{sgn}(X_3)\}|_{\rho_{31}=\rho_{12}=0} = E\{\delta(X_1)\} E\{\mathrm{sgn}(X_2)\,\mathrm{sgn}(X_3)\} = \sqrt{\frac{2}{\pi^3}}\sin^{-1}\rho_{23}$ from (5.E.12) when $\rho_{31} = \rho_{12} = 0$. This result is the same as $\sqrt{\frac{2}{\pi^3}}\sin^{-1}\beta_{23,1}\Big|_{\rho_{31}=\rho_{12}=0}$.

for a standard tri-variate normal random vector (X_1, X_2, X_3), where $c_{ij} = \rho_{jk}\rho_{ki} - \rho_{ij}$. Then, show that

$$E\{X_1 |X_2| \operatorname{sgn}(X_3)\} = \frac{2}{\pi}\left(\rho_{12}\sin^{-1}\rho_{23} + \rho_{31}\sqrt{1-\rho_{23}^2}\right) \tag{5.E.13}$$

based on Price's theorem and (5.E.10).

Exercise 5.23 Using (5.3.38)–(5.3.43), show that

$$E\{|X_1 X_2 X_3|\} = \sqrt{\frac{8}{\pi^3}}\left\{\sqrt{|\boldsymbol{K}_3|} + \sum^{c}\left(\rho_{ij} + \rho_{jk}\rho_{ki}\right)\kappa_{ijk}\right\}, \tag{5.E.14}$$

$$E\{\operatorname{sgn}(X_1)\operatorname{sgn}(X_2)|X_3|\} = \sqrt{\frac{8}{\pi^3}}\left(\kappa_{123} + \rho_{23}\kappa_{312} + \rho_{31}\kappa_{231}\right), \tag{5.E.15}$$

and

$$E\left\{X_1^2\operatorname{sgn}(X_2)\operatorname{sgn}(X_3)\right\} = \frac{2\left(2\rho_{31}\rho_{12} - \rho_{23}\rho_{12}^2 - \rho_{23}\rho_{31}^2\right)}{\pi\sqrt{1-\rho_{23}^2}}$$

$$+\frac{2}{\pi}\sin^{-1}\rho_{23} \tag{5.E.16}$$

for[14] a standard tri-variate normal random vector (X_1, X_2, X_3), where $\kappa_{ijk} = \sin^{-1}\beta_{ij,k}$. Confirm (5.E.7) and (5.E.13).

Exercise 5.24 Confirm (5.E.8)[15] and (5.E.12) based on (5.E.15). Based on (5.E.16), obtain $E\left\{X_1^2\delta(X_2)\delta(X_3)\right\}$ and confirm (5.E.10).

---

[14] We can easily get $E\{|X_1^2 X_2|\} = E\{|X_1 X_2 X_3|\}|_{X_3 \to X_1} = E\{|X_1 X_2 X_3|\}|_{\rho_{31}=1} = \sqrt{\frac{8}{\pi^3}}\{0 + \frac{\pi}{2}(1+\rho^2)\} = \sqrt{\frac{2}{\pi}}(1+\rho^2)$ with (5.E.14). Similarly, with (5.E.15), it is easy to get $E\{|X_3|\} = E\{\operatorname{sgn}(X_1)\operatorname{sgn}(X_2)|X_3|\}|_{X_2 \to X_1} = E\{\operatorname{sgn}(X_1)\operatorname{sgn}(X_2)|X_3|\}|_{\rho_{12}=1} = \sqrt{\frac{8}{\pi^3}}\left(\sin^{-1}\frac{1-\rho_{23}^2}{1-\rho_{23}^2} + 2\rho_{23}\sin^{-1}0\right) = \sqrt{\frac{2}{\pi}}$ and $E\{\operatorname{sgn}(X_1)X_2\} = E\{\operatorname{sgn}(X_1)\operatorname{sgn}(X_2)|X_3|\}|_{\rho_{23}=1} = \sqrt{\frac{8}{\pi^3}}\left(0 + 0 + \rho_{12}\sin^{-1}1\right) = \sqrt{\frac{2}{\pi}}\rho_{12}$. Next, when $|\rho_{23}| \to 1$, we have $\rho_{31} \to \operatorname{sgn}(\rho_{23})\rho_{12}$ because $X_3 \to X_2$, and thus $\lim_{\rho_{23}\to1}\left(2\rho_{31}\rho_{12} - \rho_{23}\rho_{12}^2 - \rho_{23}\rho_{31}^2\right) = 0$. Consequently, we get $\lim_{\rho_{23}\to1}\frac{2\rho_{31}\rho_{12} - \rho_{23}\rho_{12}^2 - \rho_{23}\rho_{31}^2}{\sqrt{1-\rho_{23}^2}} = \lim_{\rho_{23}\to1}\frac{-\rho_{12}^2-\rho_{31}^2}{\frac{-2\rho_{23}}{2\sqrt{1-\rho_{23}^2}}} = \lim_{\rho_{23}\to1}\frac{(\rho_{12}^2+\rho_{31}^2)\sqrt{1-\rho_{23}^2}}{\rho_{23}} = 0$ in (5.E.16) using L'Hospital's theorem.

[15] Based on this result, we have $\int\frac{\sqrt{|\boldsymbol{K}_3|}}{1-\rho_{12}^2}d\rho_{12} = \sin^{-1}\beta_{12,3} + \rho_{23}\sin^{-1}\beta_{31,2} + \rho_{31}\sin^{-1}\beta_{23,1} + h(\rho_{23}, \rho_{31})$ for a function h.

Exercise 5.25 Find the coefficient of the term $\tilde{\rho}_{12}^2 \tilde{\rho}_{22} \tilde{\rho}_{34}^4 m_1 m_4$ in the expansion of the joint moment $\mathsf{E}\left\{X_1^3 X_2^4 X_3^4 X_4^5\right\}$ for a quadri-variate normal random vector (X_1, X_2, X_3, X_4).

Exercise 5.26 Using the Price's theorem, confirm that

$$\mathsf{E}\{|X_1 X_2|\} = \frac{2}{\pi}\left(\sqrt{1-\rho^2} + \rho \sin^{-1}\rho\right) \tag{5.E.17}$$

for $(X_1, X_2) \sim \mathcal{N}(0, 0, 1, 1, \rho)$. The result (5.E.17) is obtained with other methods in Exercises 5.19 and 5.20. When $(X_1, X_2) \sim \mathcal{N}(0, 0, 1, 1, \rho)$ and $X_1 = X_2$, (5.E.17) can be written as $\mathsf{E}\left\{X_1^2\right\} = \frac{2}{\pi}\left(\rho \sin^{-1}\rho + \sqrt{1-\rho^2}\right)\sigma_1\sigma_2\Big|_{\rho=1, \sigma_2=\sigma_1}$ $= \sigma_1^2$, implying $\mathsf{E}\left\{X^2\right\} = \sigma^2$ when $X \sim \mathcal{N}\left(0, \sigma^2\right)$.

Exercise 5.27 Let us show some results related to the general formula (5.3.51) for the joint moments of normal random vectors.

(1) Confirm the coefficient

$$d_{a,j} = \frac{a!}{2^j j! (a - 2j)!} \tag{5.E.18}$$

in (5.3.17).
(2) Recollecting (5.3.46)–(5.3.48), show that

$$\mathsf{E}\left\{X_1^{a_1} X_2^{a_2} X_3^{a_3}\right\} = \sum_{l \in S_a} d_{a,l} \left(\prod_{i=1}^{3}\prod_{j=i}^{3} \tilde{\rho}_{ij}^{l_{ij}}\right)\left(\prod_{j=1}^{3} m_j^{L_{a,j}}\right) \tag{5.E.19}$$

for $\{a_i = 0, 1, \ldots\}_{i=1}^3$, where

$$d_{a,l} = \frac{a_1! a_2! a_3!}{2^{\sum_{j=1}^{3} l_{jj}}\left(\prod_{i=1}^{3}\prod_{j=i}^{3} l_{ij}!\right)\left(\prod_{j=1}^{3} L_{a,j}!\right)} \tag{5.E.20}$$

when $n = 3$.
(3) Show that (5.3.51) satisfies (5.A.1) and (5.A.2).

Exercise 5.28 When r is even, show the moment

$$\mathsf{E}\left\{X^r\right\} = \sigma_G^r \frac{\Gamma^{\frac{r}{2}-1}\left(\frac{1}{k}\right)\Gamma\left(\frac{r+1}{k}\right)}{\Gamma^{\frac{r}{2}}\left(\frac{3}{k}\right)} \tag{5.E.21}$$

for the generalized normal random variable X with the pdf (5.A.39).

Exercise 5.29 When $vk > r$ and r is even, show the r-th moment

$$\mathsf{E}\{X^r\} \;=\; v^{\frac{r}{k}}\sigma_G^r \frac{\Gamma\left(v-\frac{r}{k}\right)\Gamma\left(\frac{r+1}{k}\right)\Gamma^{\frac{r}{2}-1}\left(\frac{1}{k}\right)}{\Gamma(v)\Gamma^{\frac{r}{2}}\left(\frac{3}{k}\right)} \tag{5.E.22}$$

for the generalized Cauchy random variable X with the pdf (5.A.44).

Exercise 5.30 Obtain the pdf $f_X(x)$ of X from the joint pdf $f_{X,Y}(x,y) = \frac{\gamma}{2\pi}\left(x^2 + y^2 + \gamma^2\right)^{-\frac{3}{2}}$ shown in (5.A.61). Confirm that the pdf is the same as the pdf $f(r) = \frac{\alpha}{\pi}\left(r^2 + \alpha^2\right)^{-1}$ obtained by letting $\beta = 0$ in (2.5.28).

Exercise 5.31 Show that the mgf of the sample mean is

$$M_{\overline{X}_n}(t) = \left\{M\left(\frac{t}{n}\right)\right\}^n \tag{5.E.23}$$

for a sample $X = (X_1, X_2, \ldots, X_n)$ with marginal mgf $M(t)$.

Exercise 5.32 Obtain the mean and variance, and show the mgf

$$M_Y(t) \;=\; (1 - 2t)^{-\frac{n}{2}} \exp\left(\frac{\delta t}{1 - 2t}\right), \tag{5.E.24}$$

for $Y \sim \chi^2(n, \delta)$.

Exercise 5.33 Show the r-th moment

$$\mathsf{E}\{X^r\} = \begin{cases} \dfrac{k^{\frac{r}{2}}\Gamma\left(\frac{k-r}{2}\right)\Gamma\left(\frac{r+1}{2}\right)}{\sqrt{\pi}\Gamma\left(\frac{k}{2}\right)}, & \text{when } r < k \text{ and } r \text{ is even,} \\ 0, & \text{when } r < k \text{ and } r \text{ is odd} \end{cases} \tag{5.E.25}$$

of $X \sim t(k)$.

Exercise 5.34 Obtain the mean and variance of $Z \sim t(n, \delta)$.

Exercise 5.35 For $H \sim F(m, n)$, show that

$$\mathsf{E}\{H^k\} = \left(\frac{n}{m}\right)^k \frac{\Gamma\left(\frac{m}{2} + k\right)\Gamma\left(\frac{n}{2} - k\right)}{\Gamma\left(\frac{m}{2}\right)\Gamma\left(\frac{n}{2}\right)} \tag{5.E.26}$$

for $k = 1, 2, \ldots, \left\lceil\frac{n}{2}\right\rceil - 1$.

Exercise 5.36 Obtain the mean and variance of $H \sim F(m, n, \delta)$.

Exercise 5.37 For i.i.d. random variables X_1, X_2, X_3, and X_4 with marginal distribution $\mathcal{N}(0, 1)$, show that the pdf of $Y = X_1 X_2 + X_3 X_4$ is $f_Y(y) = \frac{1}{2}e^{-|y|}$.

Exercise 5.38 Show that the distribution of $Y = -2 \ln X$ is $\chi^2(2)$ when $X \sim U(0, 1)$. When $\{X_i \sim U(0, 1)\}_{i=1}^{k}$ are all independent of each other, show that $-2 \sum_{i=1}^{k} \ln X_i \sim \chi^2(2k)$.

Exercise 5.39 Prove that

$$\overline{X}_n = \overline{X}_{n-1} + \frac{1}{n}\left(X_n - \overline{X}_{n-1}\right) \tag{5.E.27}$$

for the sample mean $\overline{X}_n = \frac{1}{n}\sum_{i=1}^{n} X_i$ with $\overline{X}_0 = 0$.

Exercise 5.40 Let us denote the k-th central moment of X_i by $\mathsf{E}\left\{(X_i - m)^k\right\} = \mu_k$ for $k = 0, 1, \ldots$. Obtain the fourth central moment $\mu_4\left(\overline{X}_n\right)$ of the sample mean $\overline{X}_n$ for a sample $X = (X_1, X_2, \ldots, X_n)$.

Exercise 5.41 Prove Theorem 5.1.4 by taking the steps described below.

(1) Show that the pdf $f_3(x, y, z)$ shown in (5.1.19) can be written as

$$
f_3(x, y, z) = \frac{1}{\sqrt{8\pi^3 |K_3|}} \exp\left\{-\frac{(x + t_{12}y)^2}{2\left(1 - \rho_{12}^2\right)}\right\} \exp\left\{-\frac{\left(1 - t_{12}^2\right)y^2}{2\left(1 - \rho_{12}^2\right)}\right\}
$$
$$
\times \exp\left\{-\frac{1 - \rho_{12}^2}{2|K_3|}(z + b_{12})^2\right\}, \tag{5.E.28}
$$

where $t_{12} = \frac{q_{12}}{|K_3|}$ and $b_{12} = \frac{c_{23}y + c_{31}x}{1 - \rho_{12}^2}$ with $q_{12} = c_{12}\left(1 - \rho_{12}^2\right) - c_{23}c_{31}$ and $c_{ij} = \rho_{jk}\rho_{ki} - \rho_{ij}$.

(2) Show that $\lim_{\rho_{12} \to \pm 1} t_{12} = -\xi_{12}$ and

$$\lim_{\rho_{12} \to \pm 1} \frac{1 - \rho_{12}^2}{|K_3|} = \frac{1}{1 - \rho_{23}^2}. \tag{5.E.29}$$

Subsequently, using $\lim_{\alpha \to \infty} \sqrt{\frac{\alpha}{\pi}} \exp\left(-\alpha x^2\right) = \delta(x)$, show that

$$\lim_{\rho_{12} \to \pm 1} \frac{1}{\sqrt{8\pi^3 |K_3|}} \exp\left\{-\frac{(x + t_{12}y)^2}{2\left(1 - \rho_{12}^2\right)}\right\} = \frac{\delta(x - \xi_{12}y)}{2\pi\sqrt{1 - \rho_{23}^2}}, \tag{5.E.30}$$

where $\xi_{ij} = \text{sgn}\left(\rho_{ij}\right)$.

(3) Show that $\lim_{\rho_{12} \to \pm 1} \frac{1 - t_{12}^2}{1 - \rho_{12}^2} = 1$, which instantly yields

$$\lim_{\rho_{12}\to\pm1} \exp\left\{-\frac{\left(1-t_{12}^2\right)y^2}{2\left(1-\rho_{12}^2\right)}\right\} = \exp\left(-\frac{1}{2}y^2\right). \qquad (5.E.31)$$

(4) Using (5.E.29), show that

$$\lim_{\rho_{12}\to\pm1} \exp\left\{-\frac{1-\rho_{12}^2}{2\,|K_3|}\,(z+b_{12})^2\right\} = \exp\left\{-\frac{(z-\mu_1(x,y))^2}{2\left(1-\rho_{23}^2\right)}\right\},$$
$$(5.E.32)$$

where $\mu_1(x,y) = \frac{1}{2}\xi_{12}\,(\rho_{23}x + \rho_{31}y)$. Combining (5.E.30), (5.E.31), and (5.E.32) into (5.E.28), and noting that $\rho_{23} = \xi_{12}\rho_{31}$ when $\rho_{12} \to \pm1$ and that y can be replaced with $\xi_{12}x$ due to the function $\delta(x - \xi_{12}y)$, we get (5.1.36).

(5) Obtain (5.1.37) from (5.1.36).

Exercise 5.42 Assume (X, Y) has the standard bi-variate normal pdf ϕ_2.

(1) Obtain the pdf and cdf of $V = g(X, Y) = \frac{X^2 - 2\rho XY + Y^2}{2(1-\rho^2)}$.

(2) Note that $\phi_2(x, y) = c$ is equivalent to $x^2 - 2\rho xy + y^2 = c_1$, an ellipse, for positive constants c and c_1. Show that $c_1 = -2\left(1 - \rho^2\right)\ln(1 - \alpha)$ for the ellipse containing $100\alpha\%$ of the distribution of (X, Y).

Exercise 5.43 Consider $(X_1, X_2) \sim \mathcal{N}(0, 0, 1, 1, \rho)$ and $g(x) = 2\beta\int_0^{\frac{x}{\alpha}}\phi(z)dz$, i.e.,

$$g(x) = \beta\left\{2\Phi\left(\frac{x}{\alpha}\right) - 1\right\}, \qquad (5.E.33)$$

where $\alpha > 0$, $\beta > 0$, Φ is the standard normal cdf, and ϕ is the standard normal pdf. Obtain the correlation $R_Y = \mathsf{E}\{Y_1Y_2\}$ and correlation coefficient ρ_Y between $Y_1 = g(X_1)$ and $Y_2 = g(X_2)$. Obtain the values of ρ_Y when $\alpha^2 = 1$ and $\alpha^2 \to \infty$. Note that g is a smoothly increasing function from $-\beta$ to β. When $\alpha = 1$, we have $\beta\{2\Phi(X_i) - 1\} \sim U(-\beta, \beta)$ because $\Phi(X) \sim U(0, 1)$ when $X \sim \Phi$ from (3.2.50).

Exercise 5.44 In Figs. 5.1, 5.2 and 5.3, show that the angle θ between the major axis of the ellipse and the positive x-axis can be expressed as (5.1.9).

References

M. Abramowitz, I.A. Stegun (eds.), *Handbook of Mathematical Functions* (Dover, New York, 1972)

J. Bae, H. Kwon, S.R. Park, J. Lee, I. Song, Explicit correlation coefficients among random variables, ranks, and magnitude ranks. IEEE Trans. Inform. Theory **52**(5), 2233–2240 (2006)

W. Bär, F. Dittrich, Useful formula for moment computation of normal random variables with nonzero means. IEEE Trans. Automat. Control **16**(3), 263–265 (1971)

R.F. Baum, The correlation function of smoothly limited Gaussian noise. IRE Trans. Inform. Theory **3**(3), 193–197 (1957)

J.L. Brown Jr., On a cross-correlation property of stationary processes. IRE Trans. Inform. Theory **3**(1), 28–31 (1957)

W.B. Davenport Jr., *Probability and Random Processes* (McGraw-Hill, New York, 1970)

W.A. Gardner, *Introduction to Random Processes with Applications to Signals and Systems*, 2nd edn. (McGraw-Hill, New York, 1990)

I.S. Gradshteyn, I.M. Ryzhik, *Table of Integrals, Series, and Products* (Academic, New York, 1980)

J. Hajek, *Nonparametric Statistics* (Holden-Day, San Francisco, 1969)

J.B.S. Haldane, Moments of the distributions of powers and products of normal variates. Biometrika **32**(3/4), 226–242 (1942)

G,G. Hamedani, Nonnormality of linear combinations of normal random variables. Am. Stat. **38**(4), 295–296 (1984)

B. Holmquist, Moments and cumulants of the multivariate normal distribution. Stochastic Anal. Appl. **6**(3), 273–278 (1988)

R.A. Horn, C.R. Johnson, *Matrix Analysis* (Cambridge University Press, Cambridge, 1985)

L. Isserlis, On a formula for the product-moment coefficient of any order of a normal frequency distribution in any number of variables. Biometrika **12**(1/2), 134–139 (1918)

N.L. Johnson, S. Kotz, *Distributions in Statistics: Continuous Multivariate Distributions* (Wiley, New York, 1972)

A.R. Kamat, Incomplete moments of the trivariate normal distribution. Indian J. Stat. **20**(3/4), 321–322 (1958)

R. Kan, From moments of sum to moments of product. J. Multivariate Anal. **99**(3), 542–554 (2008)

S. Kotz, N. Balakrishnan, N.L. Johnson, *Continuous Multivariate Distributions*, 2nd edn. (Wiley, New York, 2000)

E.L. Melnick, A. Tenenbein, Misspecification of the normal distribution. Am. Stat. **36**(4), 372–373 (1982)

D. Middleton, *An Introduction to Statistical Communication Theory* (McGraw-Hill, New York, 1960)

G.A. Mihram, A cautionary note regarding invocation of the central limit theorem. Am. Stat. **23**(5), 38 (1969)

T.M. Mills, *Problems in Probability* (World Scientific, Singapore, 2001)

S. Nabeya, Absolute moments in 3-dimensional normal distribution. Ann. Inst. Stat. Math. **4**(1), 15–30 (1952)

C.L. Nikias, M. Shao, *Signal Processing with Alpha-Stable Distributions and Applications* (Wiley, New York, 1995)

J.K. Patel, C.H. Kapadia, D.B. Owen, *Handbook of Statistical Distributions* (Marcel Dekker, New York, 1976)

J.K. Patel, C.B. Read, *Handbook of the Normal Distribution*, 2nd edn. (Marcel Dekker, New York, 1996)

D.A. Pierce, R.L. Dykstra, Independence and the normal distribution. Am. Stat. **23**(4), 39 (1969)

R. Price, A useful theorem for nonlinear devices having Gaussian inputs. IRE Trans. Inform. Theory **4**(2), 69–72 (1958)

V.K. Rohatgi, A.K. Md. E. Saleh, *An Introduction to Probability and Statistics*, 2nd edn. (Wiley, New York, 2001)

J.P. Romano, A.F. Siegel, *Counterexamples in Probability and Statistics* (Chapman and Hall, New York, 1986)

I. Song and S. Lee, Explicit formulae for product moments of multivariate Gaussian random variables. Stat. Prob. Lett. **100**, 27–34 (2015)

I. Song, S. Lee, Y.H. Kim, S.R. Park, Explicit formulae and implication of the expected values of some nonlinear statistics of tri-variate Gaussian variables. J. Korean Stat. Soc. **49**(1), 117–138 (2020)

J.M. Stoyanov, *Counterexamples in Probability*, 3rd edn. (Dover, New York, 2013)

K. Triantafyllopoulos, On the central moments of the multidimensional Gaussian distribution. Math. Sci. **28**(2), 125–128 (2003)

G.A. Tsihrintzis, C.L. Nikias, Incoherent receiver in alpha-stable impulsive noise. IEEE Trans. Signal Process. **43**(9), 2225–2229 (1995)

G.L. Wies, E.B. Hall, *Counterexamples in Probability and Real Analysis* (Oxford University, New York, 1993)

C.S. Withers, The moments of the multivariate normal. Bull. Austral. Math. Soc. **32**(1), 103–107 (1985)

Chapter 6
Convergence of Random Variables

In this chapter, we discuss sequences of random variables and their convergence. The central limit theorem, one of the most important and widely-used results in many areas of the applications of random variables, will also be described.

6.1 Types of Convergence

In discussing the convergence of sequences (Grimmett and Stirzaker 1982; Thomas 1986) of random variables, we consider whether every or almost every sequence is convergent, and if convergent, whether the sequences converge to the same value or different values.

6.1.1 Almost Sure Convergence

Definition 6.1.1 (*sure convergence; almost sure convergence*) For every point ω of the sample space on which the random variable X_n is defined, if

$$\lim_{n \to \infty} X_n(\omega) = X(\omega), \tag{6.1.1}$$

then the sequence $\{X_n\}_{n=1}^{\infty}$ is called surely convergent to X, and if

$$P\left(\omega : \lim_{n \to \infty} X_n(\omega) = X(\omega)\right) = 1, \tag{6.1.2}$$

then the sequence $\{X_n\}_{n=1}^{\infty}$ is called almost surely convergent to X.

© The Author(s), under exclusive license to Springer Nature Switzerland AG 2022
I. Song et al., *Probability and Random Variables: Theory and Applications*,
https://doi.org/10.1007/978-3-030-97679-8_6

Sure convergence is also called always convergence, everywhere convergence, or certain convergence. When a sequence $\{X_n\}_{n=1}^\infty$ is surely convergent to X, it is denoted by $X_n \xrightarrow{c.} X$, $X_n \xrightarrow{e.} X$, or $X_n \xrightarrow{s.} X$. The sure convergence implies that all the sequences are convergent for all ω, yet the limit value of the convergence may depend on ω.

Almost sure convergence is synonymous with convergence with probability 1, almost always convergence, almost everywhere convergence, and almost certain convergence. When a sequence $\{X_n\}_{n=1}^\infty$ is almost surely convergent to X, it is denoted by $X_n \xrightarrow{a.c.} X$, $X_n \xrightarrow{a.e.} X$, $X_n \xrightarrow{a.s.} X$, or $X_n \xrightarrow{w.p.1} X$. For an almost surely convergent sequence $\{X_n(\omega)\}_{n=1}^\infty$, we have $\lim_{n\to\infty} X_n(\omega) = X(\omega)$ for any $\omega \in \tilde{\Omega}$ when $\mathsf{P}\left(\tilde{\Omega}\right) = 1$ and $\tilde{\Omega} \subseteq \Omega$. Although a sequence $\{X_n(\omega)\}_{n=1}^\infty$ for which $\omega \notin \tilde{\Omega}$ may or may not converge, the set of such ω has probability 0: in other words, $\mathsf{P}\left(\omega : \omega \notin \tilde{\Omega}, \omega \in \Omega\right) = 0$.

Example 6.1.1 Recollecting (1.5.17),

$$\mathsf{P}\,(\text{i.o.}\ |X_n| > \varepsilon) = 0 \tag{6.1.3}$$

for every positive number ε and

$$X_n \xrightarrow{a.s.} 0 \tag{6.1.4}$$

are the necessary and sufficient conditions of each other. $\diamond$

When a sequence $\{X_n\}_{i=1}^\infty$ of random variables is almost surely convergent, almost every random variable in the sequence will eventually be located within a range of 2ε for any number $\varepsilon > 0$: although some random variables may not converge, the probability of ω for such random variables will be 0. The strong law of large numbers, which we will consider later in this chapter, is an example of almost sure convergence.

Example 6.1.2 (Leon-Garcia 2008) For a randomly chosen point $\omega \in [0, 1]$, assume that $\mathsf{P}(\omega \in (a, b)) = b - a$ for $0 \le a \le b \le 1$. Now consider the five sequences of random variables $A_n(\omega) = \frac{\omega}{n}$, $B_n(\omega) = \omega\left(1 - \frac{1}{n}\right)$, $C_n(\omega) = \omega e^n$, $D_n(\omega) = \cos 2\pi n\omega$, and $H_n(\omega) = \exp\{-n(n\omega - 1)\}$. The sequence $\{A_n(\omega)\}_{n=1}^\infty$ converges always to 0 for any value of $\omega \in [0, 1]$, and thus it is surely convergent to 0. The sequence $\{B_n(\omega)\}_{n=1}^\infty$ converges to ω for any value of $\omega \in [0, 1]$, and thus it is surely convergent to ω with the limit distribution $U[0, 1]$. The sequence $\{C_n(\omega)\}_{n=1}^\infty$ converges to 0 when $\omega = 0$ and diverges when $\omega \in (0, 1]$: in other words, it is not convergent. The sequence $\{D_n(\omega)\}_{n=1}^\infty$ converges to 1 when $\omega \in \{0, 1\}$ and oscillates between -1 and 1 when $\omega \in (0, 1)$: in other words, it is not convergent. When $n \to \infty$, $H_n(0) = e^n \to \infty$ for $\omega = 0$ and $H_n(\omega) \to 0$ for $\omega \in (0, 1]$: in other words, $\{H_n(\omega)\}_{n=1}^\infty$ is not surely convergent. However, because $\mathsf{P}(\omega \in (0, 1]) = 1$, $\{H_n(\omega)\}_{n=1}^\infty$ converges almost surely to 0. $\diamond$

Example 6.1.3 (Stoyanov 2013) Consider a sequence $\{X_n\}_{n=1}^{\infty}$. When

$$\sum_{n=1}^{\infty} \mathsf{P}\left(|X_n| > \varepsilon\right) < \infty \tag{6.1.5}$$

for $\varepsilon > 0$, it is easy to see that $X_n \xrightarrow{a.s.} 0$ as $n \to \infty$ from the Borel-Cantelli lemma. In addition, even if we change the condition (6.1.5) into

$$\sum_{n=1}^{\infty} \mathsf{P}\left(|X_n| > \varepsilon_n\right) < \infty \tag{6.1.6}$$

for a sequence $\{\varepsilon_n\}_{n=1}^{\infty}$ such that $\varepsilon_n \downarrow 0$, we still have $X_n \xrightarrow{a.s.} 0$. Now, when ω is a randomly chosen point in $[0, 1]$, for a sequence $\{X_n\}_{n=1}^{\infty}$ with

$$X_n(\omega) = \begin{cases} 0, & 0 \le \omega \le 1 - \frac{1}{n}, \\ 1, & 1 - \frac{1}{n} < \omega \le 1, \end{cases} \tag{6.1.7}$$

we have $X_n \xrightarrow{a.s.} 0$ as $n \to \infty$. However, for any ε_n such that $\varepsilon_n \downarrow 0$, if we consider a sufficiently large n, we have $\mathsf{P}\left(|X_n| > \varepsilon_n\right) = \mathsf{P}\left(X_n = 1\right) = \frac{1}{n}$ and thus $\sum_{n=1}^{\infty} \mathsf{P}\left(|X_n| > \varepsilon_n\right) \to \infty$ for the sequence (6.1.7). In other words, (6.1.6) is a sufficient condition for the sequence to converge almost surely to 0, but not a necessary condition. $\diamond$

Theorem 6.1.1 (Rohatgi and Saleh 2001; Stoyanov 2013) *If $X_n \xrightarrow{a.s.} X$ for a sequence $\{X_n\}_{n=1}^{\infty}$, then*

$$\lim_{n \to \infty} \mathsf{P}\left(\sup_{m \ge n} |X_m - X| > \varepsilon\right) = 0 \tag{6.1.8}$$

holds true for every $\varepsilon > 0$, and the converse also holds true.

Proof If $X_n \xrightarrow{a.s.} X$, then we have $X_n - X \xrightarrow{a.s.} 0$. Thus, let us show that

$$X_n \xrightarrow{a.s.} 0 \tag{6.1.9}$$

and

$$\lim_{n \to \infty} \mathsf{P}\left(\sup_{m \ge n} |X_m| > \varepsilon\right) = 0 \tag{6.1.10}$$

are the necessary and sufficient conditions of each other.

Assume (6.1.9) holds true. Let $A_n(\varepsilon) = \left\{ \sup_{m \geq n} |X_m| > \varepsilon \right\}$, $C = \left\{ \lim_{n \to \infty} X_n = 0 \right\}$, and $B_n(\varepsilon) = C \cap A_n(\varepsilon)$. Then, $B_{n+1}(\varepsilon) \subseteq B_n(\varepsilon)$ and $\bigcap_{n=1}^{\infty} B_n(\varepsilon) = \emptyset$, and thus $\lim_{n \to \infty} P(B_n(\varepsilon)) = P\left(\bigcap_{n=1}^{\infty} B_n(\varepsilon) \right) = 0$. Recollecting $P(C) = 1$ and $P(C^c) = 0$, we get $P\left(C^c \cap A_n^c(\varepsilon) \right) \leq P(C^c) = 0$ because $C^c \cap A_n^c(\varepsilon) \subseteq C^c$. We also have $P(B_n(\varepsilon)) = P(C \cap A_n(\varepsilon)) = 1 - P\left(C^c \cup A_n^c(\varepsilon) \right) = 1 - P(C^c) - P\left(A_n^c(\varepsilon) \right) + P\left(C^c \cap A_n^c(\varepsilon) \right) = P(A_n(\varepsilon)) + P\left(C^c \cap A_n^c(\varepsilon) \right)$, i.e.,

$$P(B_n(\varepsilon)) = P(A_n(\varepsilon)). \tag{6.1.11}$$

Therefore, we have (6.1.10).

Next, assume that (6.1.10) holds true. Letting $D(\varepsilon) = \left\{ \limsup_{n \to \infty} |X_n| > \varepsilon > 0 \right\}$, we have $P(D(\varepsilon)) = 0$ because $D(\varepsilon) \subseteq A_n(\varepsilon)$ for $n = 1, 2, \ldots$. In addition, because $C^c = \left\{ \lim_{n \to \infty} X_n \neq 0 \right\} \subseteq \bigcup_{k=1}^{\infty} \left\{ \limsup_{n \to \infty} |X_n| > \frac{1}{k} \right\}$, we get $1 - P(C) \leq \sum_{k=1}^{\infty} P\left(D\left(\frac{1}{k} \right) \right) = 0$ and, consequently, (6.1.9). ♠

To show that a sequence of random variables is almost surely convergent, it is necessary that either the distribution of ω or the relationship between ω and the random variables are available, or that the random variables are sufficiently simple to show the convergence. Let us now consider a convergence weaker than almost sure convergence. For example, we may require most of the random variables in the sequence $\{X_n\}_{n=1}^{\infty}$ to be close to X in the sense that $E\left\{ (X_n - X)^2 \right\}$ is small enough. Such a convergence focuses on the time instances and is easier to show convergence or divergence than almost sure convergence because it does not require convergence of all the sequences.

6.1.2 Convergence in the Mean

Definition 6.1.2 (*convergence in the rth mean*) For a sequence $\{X_n\}_{n=1}^{\infty}$ and a random variable X, assume that the r-th absolute moments $\left\{ E\left\{ |X_n|^r \right\} \right\}_{n=1}^{\infty}$ and $E\left\{ |X|^r \right\}$ are all finite. If

$$\lim_{n \to \infty} E\left\{ |X_n - X|^r \right\} = 0, \tag{6.1.12}$$

then $\{X_n\}_{n=1}^{\infty}$ is called to converge to X in the r-th mean, and is denoted by $X_n \overset{r}{\to} X$ or $X_n \overset{L^r}{\longrightarrow} X$.

When $r = 2$, convergence in the r-th mean is called convergence in the mean square, and

$$\lim_{n \to \infty} \mathsf{E}\left\{|X_n - X|^2\right\} = 0 \tag{6.1.13}$$

is written also as

$$\underset{n \to \infty}{\text{l.i.m.}}\, X_n = X, \tag{6.1.14}$$

where l.i.m. is the acronym of 'limit in the mean'.

Example 6.1.4 (Rohatgi and Saleh 2001) Assume the distribution

$$\mathsf{P}\,(X_n = x) = \begin{cases} 1 - \frac{1}{n}, & x = 0, \\ \frac{1}{n}, & x = 1 \end{cases} \tag{6.1.15}$$

for a sequence $\{X_n\}_{n=1}^{\infty}$. Then, the sequence converges in the mean square to X such that $\mathsf{P}(X = 0) = 1$ because $\lim_{n \to \infty} \mathsf{E}\left\{|X_n|^2\right\} = \lim_{n \to \infty} \frac{1}{n} = 0$.

Example 6.1.5 (Leon-Garcia 2008) We have observed that the sequence $B_n(\omega) = \left(1 - \frac{1}{n}\right)\omega$ in Example 6.1.2 converges surely to ω. Now, because $\lim_{n \to \infty} \mathsf{E}\left[\{B_n(\omega) - \omega\}^2\right] = \lim_{n \to \infty} \mathsf{E}\left\{\left(\frac{\omega}{n}\right)^2\right\} = \lim_{n \to \infty} \frac{1}{3n^2} = 0$, the sequence $\{B_n(\omega)\}_{n=1}^{\infty}$ converges to ω also in the mean square.

Mean square convergence is easy to analyze and meaningful also in engineering applications because the quantity $\mathsf{E}\left\{|X_n - X|^2\right\}$ can be regarded as the power of an error. The Cauchy criterion shown in the following theorem allows us to see if a sequence converges in the mean square even when we do not know the limit X:

Theorem 6.1.2 *A necessary and sufficient condition for a sequence $\{X_n\}_{n=1}^{\infty}$ to converge in the mean square is*

$$\lim_{n,m \to \infty} \mathsf{E}\left\{|X_n - X_m|^2\right\} = 0. \tag{6.1.16}$$

Example 6.1.6 Consider the sequence $\{X_n\}_{n=1}^{\infty}$ discussed in Example 6.1.4. Then, we have $\lim_{n,m \to \infty} \mathsf{E}\left\{|X_n - X_m|^2\right\} = 0$ because $\mathsf{E}\left\{|X_n - X_m|^2\right\} = 1 \times \mathsf{P}\,(X_n = 0, X_m = 1) + 1 \times \mathsf{P}\,(X_n = 1, X_m = 0) = \frac{1}{n}\mathsf{P}\,(X_m = 0|\,X_n = 1) + \frac{1}{m}\mathsf{P}\,(X_n = 0|X_m = 1)$, i.e.,

$$\mathsf{E}\left\{|X_n - X_m|^2\right\} \leq \frac{1}{n} + \frac{1}{m}. \tag{6.1.17}$$

Therefore, $\{X_n\}_{n=1}^{\infty}$ converges in the mean square.   ◇

Example 6.1.7 We have $\lim\limits_{n,m\to\infty} \mathsf{E}\left\{|B_n - B_m|^2\right\} = \lim\limits_{n,m\to\infty} \mathsf{E}\left\{\left(\frac{1}{n} - \frac{1}{m}\right)^2 \omega^2\right\} =$ $\mathsf{E}\left\{\omega^2\right\} \lim\limits_{n,m\to\infty} \left(\frac{1}{n} - \frac{1}{m}\right)^2 = 0$ for the sequence $\{B_n\}_{n=1}^\infty$ in Example 6.1.5.

Mean square convergence implies that more and more sequences are close to the limit X as n becomes larger. However, unlike in almost sure convergence, the sequences close to X do not necessarily always stay close to X.

Example 6.1.8 (Leon-Garcia 2008) In Example 6.1.2, the sequence $H_n(\omega) = \exp\{-n(n\omega - 1)\}$ is shown to converge almost surely to 0. Now, because $\lim\limits_{n\to\infty} \mathsf{E}\left\{|H_n(\omega) - 0|^2\right\} = \lim\limits_{n\to\infty} e^{2n} \int_0^1 \exp\left(-2n^2\omega\right) d\omega = \lim\limits_{n\to\infty} \frac{e^{2n}}{2n^2}\left\{1 - \exp\left(-2n^2\right)\right\}$ $\to \infty$, the sequence $\{H_n(\omega)\}_{n=1}^\infty$ does not converge to 0 in the mean square.

6.1.3   Convergence in Probability and Convergence in Distribution

Definition 6.1.3 (*convergence in probability*) A sequence $\{X_n\}_{n=1}^\infty$ is said to converge stochastically, or converge in probability, to a random variable X if

$$\lim_{n\to\infty} \mathsf{P}\left(|X_n - X| > \varepsilon\right) = 0 \qquad (6.1.18)$$

for every $\varepsilon > 0$, and is denoted by $X_n \overset{p}{\to} X$.

Note that (6.1.18) implies that almost every sequence is within a range of 2ε at any given time but that the sequence is not required to stay in the range. However, (6.1.8) dictates that a sequence is required to stay within the range 2ε once it is inside the range. This can easily be confirmed by interpreting the meanings of $\{|X_n - X| > \varepsilon\}$ and $\left\{\sup\limits_{m\geq n} |X_m - X| > \varepsilon\right\}$.

Example 6.1.9 (Rohatgi and Saleh 2001) Assume the pmf

$$p_{X_n}(x) = \begin{cases} 1 - \frac{1}{n}, & x = 0, \\ \frac{1}{n}, & x = 1 \end{cases} \qquad (6.1.19)$$

for a sequence $\{X_n\}_{n=1}^\infty$. Then, because

$$\mathsf{P}\left(|X_n| > \varepsilon\right) = \begin{cases} 0, & \varepsilon \geq 1, \\ \frac{1}{n}, & 0 < \varepsilon < 1, \end{cases} \qquad (6.1.20)$$

we have $\lim\limits_{n\to\infty} \mathsf{P}\left(|X_n| > \varepsilon\right) = 0$ and thus $X_n \overset{p}{\to} 0$. $\diamond$

Example 6.1.10 Assume a sequence $\{X_n\}_{n=1}^{\infty}$ with the pmf

$$P(X_n = x) = \begin{cases} \frac{1}{2n}, & x = 3, 4, \\ 1 - \frac{1}{n}, & x = 5. \end{cases} \tag{6.1.21}$$

We then have $X_n \xrightarrow{p} 5$ because

$$P(|X_n - 5| > \varepsilon) = \begin{cases} 0, & \varepsilon \geq 2, \\ \frac{1}{2n}, & 1 \leq \varepsilon < 2, \\ \frac{1}{n}, & 0 < \varepsilon < 1, \end{cases} \tag{6.1.22}$$

and thus $\lim_{n \to \infty} P(|X_n - 5| > \varepsilon) = 0$. ◇

Theorem 6.1.3 (Rohatgi and Saleh 2001) *If a sequence $\{X_n\}_{n=1}^{\infty}$ converges to X in probability and g is a continuous function, then $\{g(X_n)\}$ converges to $g(X)$ in probability.*

Theorem 6.1.3 requires g to be a continuous function: note that the theorem may not hold true if g is not a continuous function.

Example 6.1.11 (Stoyanov 2013) Assume $X_n \sim \mathcal{N}\left(0, \frac{\sigma^2}{n}\right)$ and consider the unit step function $u(x)$ with $u(0) = 0$. Then, $\{X_n\}_{n=1}^{\infty}$ converges in probability to a random variable X which is almost surely 0 and $u(X)$ is a random variable which is almost surely 0. However, because $u(X_n) = 0$ and 1 each with probability $\frac{1}{2}$, we have $u(X_n) \xrightarrow{p} u(X)$. ◇

Definition 6.1.4 (*convergence in distribution*) If the cdf F_n of X_n satisfies

$$\lim_{n \to \infty} F_n(x) = F(x) \tag{6.1.23}$$

for all points at which the cdf F of X is continuous, then the sequence $\{X_n\}_{n=1}^{\infty}$ is said to converge weakly, in law, or in distribution to X, and is written as $X_n \xrightarrow{d} X$ or $X_n \xrightarrow{l} X$.

Example 6.1.12 For the cdf $F_n(x) = \int_{-\infty}^{x} \frac{\sqrt{n}}{\sigma\sqrt{2\pi}} \exp\left(-\frac{nt^2}{2\sigma^2}\right) dt$ of X_n, we have

$$\lim_{n \to \infty} F_n(x) = \begin{cases} 0, & x < 0, \\ \frac{1}{2}, & x = 0, \\ 1, & x > 0. \end{cases} \tag{6.1.24}$$

Thus, $\{X_n\}_{n=1}^{\infty}$ converges weakly to a random variable X with the cdf

$$F(x) = \begin{cases} 0, & x < 0, \\ 1, & x \geq 0. \end{cases} \tag{6.1.25}$$

Note that although $\lim_{n\to\infty} F_n(0) \neq F(0)$, the convergence in distribution does not require the convergence at discontinuity points of the cdf: in short, the convergence of $\{F_n(x)\}_{n=1}^{\infty}$ at the discontinuity point $x = 0$ of $F(x)$ is not a prerequisite for the convergence in distribution.   ◇

6.1.4 Relations Among Various Types of Convergence

We now discuss the relations among various types of convergence discussed in previous sections. First, let

$$\mathscr{A} = \{\text{collection of sequences almost surely convergent}\},$$
$$\mathscr{D} = \{\text{collection of sequences convergent in distribution}\},$$
$$\mathscr{M}_s = \{\text{collection of sequences convergent in the } s\text{-th mean}\},$$
$$\mathscr{M}_t = \{\text{collection of sequences convergent in the } t\text{-th mean}\},$$
$$\mathscr{P} = \{\text{collection of sequences convergent in probability}\},$$

and $t > s > 0$. Then, we have (Rohatgi and Saleh 2001)

$$\mathscr{D} \supset \mathscr{P} \supset \mathscr{M}_s \supset \mathscr{M}_t \tag{6.1.26}$$

and

$$\mathscr{P} \supset \mathscr{A}. \tag{6.1.27}$$

In addition, neither $\mathscr{A}$ and $\mathscr{M}_s$ nor $\mathscr{A}$ and $\mathscr{M}_t$ include each other.

Example 6.1.13 (Stoyanov 2013) Assume that the pdf of X is symmetric and let $X_n = -X$. Then, because[1]

$$X_n \overset{d}{=} X, \tag{6.1.28}$$

we have $X_n \overset{d}{\to} X$. However, because $\mathsf{P}\left(|X_n - X| > \varepsilon\right) = \mathsf{P}\left(|X| > \frac{\varepsilon}{2}\right) \nrightarrow 0$ when $n \to \infty$, we have $X_n \overset{p}{\nrightarrow} X$.   ◇

Example 6.1.14 For the sample space $S = \{\omega_1, \omega_2, \omega_3, \omega_4\}$, assume the event space 2^S and the uniform probability measure. Define $\{X_n\}_{n=1}^{\infty}$ by

$$X_n(\omega) = \begin{cases} 0, & \omega = \omega_3 \text{ or } \omega_4, \\ 1, & \omega = \omega_1 \text{ or } \omega_2. \end{cases} \tag{6.1.29}$$

---

[1] Here, $\overset{d}{=}$ means 'equal in distribution' as introduced in Example 3.5.18.

Also let

$$X(\omega) = \begin{cases} 0, & \omega = \omega_1 \, \text{or} \, \omega_2, \\ 1, & \omega = \omega_3 \, \text{or} \, \omega_4. \end{cases} \tag{6.1.30}$$

Then, the cdf's of X_n and X are both

$$F(x) = \begin{cases} 0, & x < 0, \\ \frac{1}{2}, & 0 \le x < 1, \\ 1, & x \ge 1. \end{cases} \tag{6.1.31}$$

In other words, $X_n \overset{d}{\to} X$. Meanwhile, because $|X_n(\omega) - X(\omega)| = 1$ for $\omega \in S$ and $n \ge 1$, we have $\mathsf{P}(|X_n - X| > \varepsilon) \not\to 0$ for $n \to \infty$. Thus, X_n does not converge to X in probability.

Example 6.1.15 (Stoyanov 2013) Assume that $\mathsf{P}(X_n = 1) = \frac{1}{n}$ and $\mathsf{P}(X_n = 0) = 1 - \frac{1}{n}$ for a sequence $\{X_n\}_{n=1}^{\infty}$ of independent random variables. Then, we have $X_n \overset{p}{\to} 0$ because $\mathsf{P}(|X_n - 0| > \varepsilon) = \mathsf{P}(X_n = 1) = \frac{1}{n} \to 0$ when $n \to \infty$ for any $\varepsilon \in (0, 1)$. Next, let $A_n(\varepsilon) = \{|X_n - 0| > \varepsilon\}$ and $B_m(\varepsilon) = \overset{\infty}{\underset{n=m}{\cup}} A_n(\varepsilon)$. Then, we have

$$\mathsf{P}(B_m(\varepsilon)) = 1 - \lim_{M \to \infty} \mathsf{P}(X_n = 0, \text{ for all } n \in [m, M]). \tag{6.1.32}$$

Noting that $\prod_{k=m}^{\infty} \left(1 - \frac{1}{k}\right) = 0$ for any natural number m, we get

$$\mathsf{P}(B_m(\varepsilon)) = 1 - \lim_{M \to \infty} \left(1 - \frac{1}{m}\right)\left(1 - \frac{1}{m+1}\right)\cdots\left(1 - \frac{1}{M}\right)$$
$$= 1 \tag{6.1.33}$$

because $\{X_n\}_{n=1}^{\infty}$ is an independent sequence. Thus, $\lim_{m \to \infty} \mathsf{P}(B_m(\varepsilon)) \ne 0$ and, from Theorem 6.1.1, $X_n \overset{a.s.}{\not\to} 0$. $\diamond$

Example 6.1.16 (Rohatgi and Saleh 2001; Stoyanov 2013) Based on the inequality $|x|^s \le 1 + |x|^r$ for $0 < s < r$, or on the Lyapunov inequality $\left[\mathsf{E}\{|X|^s\}\right]^{\frac{1}{s}} \le \left[\mathsf{E}\{|X|^r\}\right]^{\frac{1}{r}}$ for $0 < s < r$ shown in (6.A.21), we can easily show that

$$X_n \overset{L^r}{\longrightarrow} X \Rightarrow X_n \overset{L^s}{\longrightarrow} X \tag{6.1.34}$$

for $0 < s < r$. Next, if the distribution of the sequence $\{X_n\}_{n=1}^{\infty}$ is

$$\mathsf{P}(X_n = x) = \begin{cases} n^{-\frac{r+s}{2}}, & x = n, \\ 1 - n^{-\frac{r+s}{2}}, & x = 0 \end{cases} \tag{6.1.35}$$

for $0 < s < r$, then $X_n \xrightarrow{L^r} 0$ because $E\{X_n^s\} = n^{\frac{s-r}{2}} \to 0$ for $n \to \infty$: however, because $E\{X_n^r\} = n^{\frac{r-s}{2}} \to \infty$, we have $X_n \xrightarrow{L^s} 0$. In short,

$$X_n \xrightarrow{L^s} 0 \nRightarrow X_n \xrightarrow{L^r} 0 \tag{6.1.36}$$

for $r > s$. $\diamond$

Example 6.1.17 (Stoyanov 2013) If

$$P(X_n = x) = \begin{cases} \frac{1}{n}, & x = e^n, \\ 1 - \frac{1}{n}, & x = 0 \end{cases} \tag{6.1.37}$$

for a sequence $\{X_n\}_{n=1}^\infty$, then $X_n \xrightarrow{p} 0$ because $P(|X_n| < \varepsilon) = P(X_n = 0) = 1 - \frac{1}{n} \to 1$ for $\varepsilon > 0$ when $n \to \infty$. However, we have $X_n \xrightarrow{L^r} 0$ because $E\{X_n^r\} = \frac{e^{rn}}{n} \to \infty$.

Example 6.1.18 (Rohatgi and Saleh 2001) Note first that a natural number n can be expressed uniquely as

$$n = 2^k + m \tag{6.1.38}$$

with integers $k \in \{0, 1, \ldots\}$ and $m \in \{0, 1, \ldots, 2^k - 1\}$. Define a sequence $\{X_n\}_{n=1}^\infty$ by

$$X_n(\omega) = \begin{cases} 2^k, & \frac{m}{2^k} \le \omega < \frac{m+1}{2^k}, \\ 0, & \text{otherwise} \end{cases} \tag{6.1.39}$$

for $n = 1, 2, \ldots$ on $\Omega = [0, 1]$. Assume the pmf

$$P(X_n = x) = \begin{cases} \frac{1}{2^k}, & x = 2^k, \\ 1 - \frac{1}{2^k}, & x = 0. \end{cases} \tag{6.1.40}$$

Then, $\lim_{n\to\infty} X_n(\omega)$ does not exist for any choice $\omega \in \Omega$ and, therefore, the sequence does not converge almost surely. However, because

$$P(|X_n| > \varepsilon) = P(X_n > \varepsilon)$$
$$= \begin{cases} 0, & \varepsilon \ge 2^k, \\ \frac{1}{2^k}, & 0 < \varepsilon < 2^k, \end{cases} \tag{6.1.41}$$

we have $\lim_{n\to\infty} P(|X_n| > \varepsilon) = 0$, and thus $X_n \xrightarrow{p} 0$. $\diamond$

Example 6.1.19 (Stoyanov 2013) Consider a sequence $\{X_n\}_{n=1}^{\infty}$ with

$$P(X_n = x) = \begin{cases} 1 - \frac{1}{n^{\alpha}}, & x = 0, \\ \frac{1}{2n^{\alpha}}, & x = \pm n, \end{cases} \tag{6.1.42}$$

where $\alpha > 0$. Then, $\sum_{n=1}^{\infty} E\left\{|X_n|^{\frac{1}{2}}\right\} < \infty$ when $\alpha > \frac{3}{2}$ because $E\left\{|X_n|^{\frac{1}{2}}\right\} = n^{-\alpha+\frac{1}{2}}$. Letting $\alpha = \varepsilon$ and $X = \sqrt{\alpha}\sqrt{|X_n|}$ in the Markov inequality $P(X \geq \alpha) \leq \frac{E\{X\}}{\alpha}$ introduced in (6.A.15), we have $P(|X_n| > \varepsilon) \leq \varepsilon^{-\frac{1}{2}} E\left\{|X|^{\frac{1}{2}}\right\}$, and thus $\sum_{n=1}^{\infty} P(|X_n| > \varepsilon) < \infty$ when $\varepsilon > 0$. Now, employing Borel-Cantelli lemma as in Example 6.1.3, we have $X_n \xrightarrow{a.s.} 0$. Meanwhile, $X_n \xrightarrow{L^2} 0$ when $\alpha \leq 2$ because $E\left\{|X_n|^2\right\} = \frac{1}{n^{\alpha-2}}$. In essence, we have $X_n \xrightarrow{a.s.} 0$, yet $X_n \xrightarrow{L^2} 0$ for $\alpha \in \left(\frac{3}{2}, 2\right]$. ◇

Example 6.1.20 (Stoyanov 2013) When $X_n \xrightarrow{d} X$ and $Y_n \xrightarrow{d} Y$, we have $X_n + Y_n \xrightarrow{d} X + Y$ if the sequences $\{X_n\}_{n=1}^{\infty}$ and $\{Y_n\}_{n=1}^{\infty}$ are independent of each other. On the other hand, if the two sequences are not independent of each other, we may have different results. For example, assume $X_n \xrightarrow{d} X \sim \mathcal{N}(0, 1)$ and let $Y_n = \alpha X_n$. Then, we have

$$Y_n \xrightarrow{d} Y \sim \mathcal{N}\left(0, \alpha^2\right), \tag{6.1.43}$$

and the distribution of $X_n + Y_n = (1 + \alpha)X_n$ converges to $\mathcal{N}\left(0, (1+\alpha)^2\right)$. However, because $X \sim \mathcal{N}(0, 1)$ and $Y \sim \mathcal{N}\left(0, \alpha^2\right)$, the distribution of $X + Y$ is not necessarily $\mathcal{N}\left(0, (1+\alpha)^2\right)$. In other words, if the sequences $\{X_n\}_{n=1}^{\infty}$ and $\{Y_n\}_{n=1}^{\infty}$ are not independent of each other, it is possible that $X_n + Y_n \xrightarrow{d} X + Y$ even when $X_n \xrightarrow{d} X$ and $Y_n \xrightarrow{d} Y$. ◇

6.2 Laws of Large Numbers and Central Limit Theorem

In this section, we will consider the sum of random variables and its convergence. We will then introduce the central limit theorem (Davenport 1970; Doob 1949; Mihram 1969), one of the most useful and special cases of convergence.

6.2.1 Sum of Random Variables and Its Distribution

The sum of random variables is one of the key ingredients in understanding and applying the properties of convergence and limits. We have discussed the properties of the sum of random variables in Chap. 4. Specifically, the sum of two random variables as well as the cf and distribution of the sum of a number of random variables are discussed in Examples 4.2.4, 4.2.13, and 4.3.8. We now consider the sum of a number of random variables more generally.

Theorem 6.2.1 *The expected value and variance of the sum* $S_n = \sum_{i=1}^{n} X_i$ *of the random variables* $\{X_i\}_{i=1}^{n}$ *are*

$$E\{S_n\} = \sum_{i=1}^{n} E\{X_i\} \tag{6.2.1}$$

and

$$\text{Var}\{S_n\} = \sum_{i=1}^{n} \text{Var}\{X_i\} + 2 \sum_{i=1}^{n-1} \sum_{j=i+1}^{n} \text{Cov}\left(X_i, X_j\right), \tag{6.2.2}$$

respectively.

Proof First, it is easy to see that $E\{S_n\} = E\left\{\sum_{i=1}^{n} X_i\right\} = \sum_{i=1}^{n} E\{X_i\}$. Next, we have $\text{Var}\left\{\sum_{i=1}^{n} a_i X_i\right\} = E\left\{\left(\sum_{i=1}^{n} a_i \left[X_i - E\{X_i\}\right]\right)^2\right\}$, i.e.,

$$\text{Var}\left\{\sum_{i=1}^{n} a_i X_i\right\} = E\left\{\sum_{i=1}^{n} \sum_{j=1}^{n} a_i a_j \left(X_i - E\{X_i\}\right)\left(X_j - E\{X_j\}\right)\right\}$$

$$= \sum_{i=1}^{n} a_i^2 \text{Var}\{X_i\} + 2 \sum_{i=1}^{n-1} \sum_{j=i+1}^{n} a_i a_j \text{Cov}\left(X_i, X_j\right) \tag{6.2.3}$$

when $\{a_i\}_{i=1}^{n}$ are constants. Letting $\{a_i = 1\}_{i=1}^{n}$, we get (6.2.2). ♠

Theorem 6.2.2 *The variance of the sum* $S_n = \sum_{i=1}^{n} X_i$ *can be expressed as*

$$\text{Var}\{S_n\} = \sum_{i=1}^{n} \text{Var}\{X_i\} \tag{6.2.4}$$

when the random variables $\{X_i\}_{i=1}^{n}$ *are uncorrelated.*

Theorem 6.2.1 says that the expected value of the sum of random variables is the sum of the expected values of the random variables. In addition, the variance of the sum of random variables is obtained by adding the sum of the covariances between two distinct random variables to the sum of the variances of the random variables. Theorem 6.2.2 dictates that the variance of the sum of uncorrelated random variables is simply the sum of the variances of the random variables.

Example 6.2.1 (Yates and Goodman 1999) Assume that the joint moments are $E\{X_iX_j\} = \rho^{|i-j|}$ for zero-mean random variables $\{X_i\}_{i=1}^n$. Obtain the expected value and variance of $Y_i = X_{i-2} + X_{i-1} + X_i$ for $i = 3, 4, \ldots, n$.

Solution Using (6.2.1), we have $E\{Y_i\} = E\{X_{i-2}\} + E\{X_{i-1}\} + E\{X_i\} = 0$. Next, from (6.2.2), we get

$$
\begin{aligned}
\text{Var}\{Y_i\} &= \text{Var}\{X_{i-2}\} + \text{Var}\{X_{i-1}\} + \text{Var}\{X_i\} \\
&\quad + 2\{\text{Cov}(X_{i-2}, X_{i-1}) + \text{Cov}(X_{i-1}, X_i) + \text{Cov}(X_{i-2}, X_i)\} \\
&= 3\rho^0 + 2\rho^1 + 2\rho^1 + 2\rho^2 \\
&= 3 + 4\rho + 2\rho^2
\end{aligned}
\tag{6.2.5}
$$

because $\text{Var}\{X_i\} = \rho^0 = 1$.   ◇

Example 6.2.2 In a meeting of a group of n people, each person attends with a gift. The name tags of the n people are put in a box, from which each person randomly picks one name tag: each person gets the gift brought by the person on the name tag. Let G_n be the number of people who receive their own gifts back. Obtain the expected value and variance of G_n.

Solution Let us define

$$
X_i = \begin{cases} 1, & \text{when person } i \text{ picks her/his own name tag,} \\ 0, & \text{otherwise.} \end{cases}
\tag{6.2.6}
$$

Then,

$$
G_n = \sum_{i=1}^n X_i.
\tag{6.2.7}
$$

For any person, the probability of picking her/his own name tag is $\frac{1}{n}$. Thus, $E\{X_i\} = 1 \times \frac{1}{n} + 0 \times \frac{n-1}{n} = \frac{1}{n}$ and $\text{Var}\{X_i\} = E\{X_i^2\} - E^2\{X_i\} = \frac{1}{n} - \frac{1}{n^2}$. In addition, because $P(X_iX_j = 1) = \frac{1}{n(n-1)}$ and $P(X_iX_j = 0) = 1 - \frac{1}{n(n-1)}$ for $i \neq j$, we have $\text{Cov}(X_i, X_j) = E\{X_iX_j\} - E\{X_i\}E\{X_j\} = \frac{1}{n(n-1)} - \frac{1}{n^2}$, i.e.,

$$
\text{Cov}(X_i, X_j) = \frac{1}{n^2(n-1)}.
\tag{6.2.8}
$$

Therefore, $\mathsf{E}\{G_n\} = \sum_{i=1}^{n} \frac{1}{n} = 1$ and $\mathsf{Var}\{G_n\} = n\mathsf{Var}\{X_i\} + n(n-1)\mathsf{Cov}(X_i,$
$X_j) = 1$. In short, for any number n of the group, one person will get her/his own
gift back on average. $\diamond$

Theorem 6.2.3 *For independent random variables* $\{X_i\}_{i=1}^{n}$, *let the cf and mgf of* X_i
be $\varphi_{X_i}(\omega)$ *and* $M_{X_i}(t)$, *respectively. Then, we have*

$$\varphi_{S_n}(\omega) = \prod_{i=1}^{n} \varphi_{X_i}(\omega) \tag{6.2.9}$$

and

$$M_{S_n}(t) = \prod_{i=1}^{n} M_{X_i}(t) \tag{6.2.10}$$

as the cf and mgf, respectively, of $S_n = \sum_{i=1}^{n} X_i$.

Proof Noting that $\{X_i\}_{i=1}^{n}$ are independent of each other, we can easily obtain the
cf $\varphi_{S_n}(\omega) = \mathsf{E}\{e^{j\omega S_n}\} = \mathsf{E}\{e^{j\omega(X_1 + X_2 + \cdots + X_n)}\} = \prod_{i=1}^{n} \mathsf{E}\{e^{j\omega X_i}\}$, i.e.,

$$\varphi_{S_n}(\omega) = \prod_{i=1}^{n} \varphi_{X_i}(\omega) \tag{6.2.11}$$

as in (4.3.32). We can show (6.2.10) similarly.   ♠

Theorem 6.2.4 *For i.i.d. random variables* $\{X_i\}_{i=1}^{n}$, *let the cf and mgf of* X_i *be*
$\varphi_X(\omega)$ *and* $M_X(t)$, *respectively. Then, we have the cf*

$$\varphi_{S_n}(\omega) = \varphi_X^n(\omega) \tag{6.2.12}$$

and the mgf

$$M_{S_n}(t) = M_X^n(t) \tag{6.2.13}$$

of $S_n = \sum_{i=1}^{n} X_i$.

Proof Noting that the random variables $\{X_i\}_{i=1}^{n}$ are all of the same distribution,
Theorem 6.2.4 follows directly from Theorem 6.2.3.   ♠

Example 6.2.3 When $\{X_i\}_{i=1}^{n}$ are i.i.d. with marginal distribution $b(1, p)$, obtain
the distribution of $S_n = \sum_{i=1}^{n} X_i$.

Solution The mgf of X_i is $M_X(t) = 1 - p + pe^t$ as shown in (3.A.47). Therefore, the mgf of S_n is $M_{S_n}(t) = (1 - p + pe^t)^n$, implying that $S_n \sim b(n, p)$. ◇

Example 6.2.4 When $\{X_i\}_{i=1}^n$ are independent of each other with $X_i \sim b(k_i, p)$, obtain the distribution of $S_n = \sum_{i=1}^n X_i$.

Solution The mgf of X_i is $M_{X_i}(t) = (1 - p + pe^t)^{k_i}$ as shown in (3.A.49). Thus, the mgf of S_n is $M_{S_n}(t) = \prod_{i=1}^n (1 - p + pe^t)^{k_i}$, i.e.,

$$M_{S_n}(t) = (1 - p + pe^t)^{\sum_{i=1}^n k_i}. \tag{6.2.14}$$

This result implies $S_n \sim b\left(\sum_{i=1}^n k_i, p\right)$. ◇

Example 6.2.5 We have shown that $S_n = \sum_{i=1}^n X_i \sim \mathcal{N}\left(\sum_{i=1}^n m_i, \sum_{i=1}^n \sigma_i^2\right)$ when $\{X_i \sim \mathcal{N}(m_i, \sigma_i^2)\}_{i=1}^n$ are independent of each other in Theorem 5.2.5.

Example 6.2.6 Show that $S_n = \sum_{i=1}^n X_i \sim G\left(n, \frac{1}{\lambda}\right)$ when $\{X_i\}_{i=1}^n$ are i.i.d. with marginal exponential distribution of parameter λ.

Solution The mgf of X_i is $M_X(t) = \frac{\lambda}{\lambda - t}$ as shown in (3.A.67). Thus, the mgf of S_n is $M_{S_n}(t) = \left(\frac{\lambda}{\lambda - t}\right)^n$ and, therefore, $S_n \sim G\left(n, \frac{1}{\lambda}\right)$. ◇

Definition 6.2.1 (*random sum*) Assume that the support of the pmf of a random variable N is a subset of $\{0, 1, \ldots\}$ and that the random variables $\{X_1, X_2, \ldots, X_N\}$ are independent of N. The random variable

$$S_N = \sum_{i=1}^N X_i \tag{6.2.15}$$

is called the random sum or variable sum, where we assume $S_0 = 0$.

The mgf of the random sum S_N can be expressed as $M_{S_N}(t) = \mathsf{E}\{e^{tS_N}\} = \mathsf{E}[\mathsf{E}\{e^{tS_N} | N\}] = \sum_{n=0}^\infty \mathsf{E}\{e^{tS_N} | N = n\} p_N(n) = \sum_{n=0}^\infty \mathsf{E}\{e^{tS_n}\} p_N(n)$, i.e.,

$$M_{S_N}(t) = \sum_{n=0}^\infty M_{S_n}(t) p_N(n), \tag{6.2.16}$$

where $p_N(n)$ is the pmf of N and $M_{S_n}(t)$ is the mgf of $S_n = \sum_{i=1}^n X_i$.

Theorem 6.2.5 *When the random variables $\{X_i\}_{i=1}^N$ are i.i.d. with marginal mgf $M_X(t)$, the mgf of the random sum $S_N = \sum\limits_{i=1}^N X_i$ can be obtained as*

$$M_{S_N}(t) = M_N \left(\ln M_X(t) \right), \qquad (6.2.17)$$

where $M_N(t)$ is the mgf of N.

Proof Applying Theorem 6.2.4 in (6.2.16), we get $M_{S_N}(t) = \sum\limits_{n=0}^{\infty} M_X^n(t) p_N(n) = \sum\limits_{n=0}^{\infty} e^{n \ln M_X(t)} p_N(n) = \mathsf{E}\left\{ e^{N \ln M_X(t)} \right\}$, i.e.,

$$M_{S_N}(t) = M_N \left(\ln M_X(t) \right), \qquad (6.2.18)$$

where $p_N(n)$ is the pmf of N.   ♠

Meanwhile, if we write $\tilde{M}_N(z)$ as $\mathsf{E}\left\{ z^N \right\}$, the mgf (6.2.17) can be written as $M_{S_N}(t) = \mathsf{E}\left\{ e^{tS_N} \right\} = \mathsf{E}\left\{ \mathsf{E}\left\{ e^{tS_N} \,\middle|\, N \right\} \right\} = \sum\limits_{n=0}^{\infty} \mathsf{E}\left\{ e^{tS_N} \,\middle|\, N = n \right\} \mathsf{P}(N = n) = \sum\limits_{n=0}^{\infty}$ $\mathsf{E}\left\{ e^{tX_1 + tX_2 + \cdots + tX_n} \right\} \mathsf{P}(N = n) = \sum\limits_{n=0}^{\infty} \mathsf{E}\left\{ e^{tX_1} \right\} \mathsf{E}\left\{ e^{tX_2} \right\} \cdots \mathsf{E}\left\{ e^{tX_n} \right\} \mathsf{P}(N = n) = \sum\limits_{n=0}^{\infty} M_X^n(t) \mathsf{P}(N = n)$, i.e.,

$$M_{S_N}(t) = \tilde{M}_N \left(M_X(t) \right) \qquad (6.2.19)$$

using $\tilde{M}_N(g(t)) = \mathsf{E}\left\{ g^N(t) \right\} = \sum\limits_{n=0}^{\infty} g^n(t) \mathsf{P}(N = n)$.

Example 6.2.7 Assume that i.i.d. exponential random variables $\{X_n\}_{n=1}^{\infty}$ with mean $\frac{1}{\lambda}$ are independent of a geometric random variable N with pmf $p_N(k) = (1 - \alpha)^{k-1} \alpha$ for $k \in \{1, 2, \ldots\}$. Obtain the distribution of the random sum $S_N = \sum\limits_{i=1}^N X_i$.

Solution The mgf's of N and X_i are $M_N(t) = \frac{\alpha e^t}{1 - (1 - \alpha)e^t}$ and $M_X(t) = \frac{\lambda}{\lambda - t}$, respectively. Thus, the mgf of S_N is $M_{S_N}(t) = \frac{\alpha \exp\left(\ln \frac{\lambda}{\lambda - t} \right)}{1 - (1 - \alpha) \exp\left(\ln \frac{\lambda}{\lambda - t} \right)}$, i.e.,

$$M_{S_N}(t) = \frac{\alpha \lambda}{\alpha \lambda - t}. \qquad (6.2.20)$$

Therefore, S_N is an exponential random variable with mean $\frac{1}{\alpha \lambda}$. This result is also in agreement with the intuitive interpretation that S_N is the sum of, on average, $\frac{1}{\alpha}$ variables of mean $\frac{1}{\lambda}$.   ◇

Theorem 6.2.6 *When the random variables $\{X_i\}$ are i.i.d., we have the expected value*

$$E\{S_N\} = E\{N\}E\{X\} \tag{6.2.21}$$

and the variance

$$\mathrm{Var}\{S_N\} = E\{N\}\mathrm{Var}\{X\} + \mathrm{Var}\{N\}E^2\{X\} \tag{6.2.22}$$

of the random sum $S_N = \sum_{i=1}^{N} X_i$.

Proof

(Method 1) From (6.2.17), we have $M'_{S_N}(t) = M'_N(\ln M_X(t)) \frac{M'_X(t)}{M_X(t)}$ and

$$M''_{S_N}(t) = M''_N(\ln M_X(t)) \left\{\frac{M'_X(t)}{M_X(t)}\right\}^2$$

$$+ M'_N(\ln M_X(t)) \frac{M_X(t)M''_X(t) - \{M'_X(t)\}^2}{M_X^2(t)}. \tag{6.2.23}$$

Now, recollecting $M_X(0) = 1$, we get $E\{S_N\} = M'_{S_N}(0) = M'_N(0)M'_X(0)$, i.e.,

$$E\{S_N\} = E\{N\}E\{X\} \tag{6.2.24}$$

and $E\{S_N^2\} = M''_{S_N}(0) = M''_N(0)\{M'_X(0)\}^2 + M'_N(0)\left[M''_X(0) - \{M'_X(0)\}^2\right]$, i.e.,

$$E\{S_N^2\} = E\{N^2\}E^2\{X\} + E\{N\}\mathrm{Var}\{X\}. \tag{6.2.25}$$

Combining (6.2.24) and (6.2.25), we have (6.2.22).

(Method 2) Because $E\{Y|N\} = \sum_{i=1}^{N} E\{X_i\} = NE\{X\}$ from (4.4.40) with $Y = S_N$, we get the expected value of Y as $E\{Y\} = E[E\{Y|N\}] = E[NE\{X\}]$, i.e.,

$$E\{Y\} = E\{N\}E\{X\}. \tag{6.2.26}$$

Similarly, recollecting that $Y^2 = \sum_{i=1}^{N} X_i^2 + \sum_{i=1}^{N}\sum_{\substack{j=1 \\ i \neq j}}^{N} X_i X_j$, the second moment $E\{Y^2\} = E[E\{Y^2|N\}]$ can be evaluated as

$$E\left\{Y^2\right\} = E\left[NE\left\{X^2\right\} + N(N-1)E^2\left\{X\right\}\right]$$
$$= E\left\{N\right\}\left[E\left\{X^2\right\} - E^2\left\{X\right\}\right] + E\left\{N^2\right\}E^2\left\{X\right\}$$
$$= E\left\{N\right\}\text{Var}\left\{X\right\} + E\left\{N^2\right\}E^2\left\{X\right\}. \tag{6.2.27}$$

From (6.2.26) and (6.2.27), we can obtain (6.2.22).   ♠

Example 6.2.8 Assume that i.i.d. random variables $\left\{X_n \sim \mathcal{N}\left(m, \sigma^2\right)\right\}_{n=1}^{\infty}$ are independent of $N \sim P(\lambda)$. Then, the random sum S_N has the expected value $E\left\{S_N\right\} = \lambda m$ and variance $\text{Var}\left\{S_N\right\} = \lambda\left(\sigma^2 + m^2\right)$.

Let us note two observations.

(1) When the random variable N is a constant n: because $E\{N\} = n$ and $\text{Var}\{N\} = 0$, we have [2] $E\left\{S_n\right\} = nE\{X\}$ and $\text{Var}\left\{S_n\right\} = n\text{Var}\{X\}$ from Theorem 6.2.6.
(2) When the random variable X_i is a constant x: because $E\{X\} = x$, $\text{Var}\{X\} = 0$, and $S_N = xN$, we have $E\left\{S_N\right\} = xE\{N\}$ and $\text{Var}\left\{S_N\right\} = x^2\text{Var}\{N\}$.

6.2.2 Laws of Large Numbers

We now consider the limit of a sequence $\{X_i\}_{i=1}^{n}$ by taking into account the sum $S_n = \sum_{i=1}^{n} X_i$ for $n \to \infty$.

6.2.2.1 Weak Law of Large Numbers

Definition 6.2.2 (*weak law of large numbers*) When we have

$$\frac{S_n - a_n}{b_n} \xrightarrow{p} 0 \tag{6.2.28}$$

for two sequences $\{a_n\}_{n=1}^{\infty}$ and $\{b_n > 0\}_{n=1}^{\infty}$ of real numbers such that $b_n \uparrow \infty$, the sequence $\{X_i\}_{i=1}^{\infty}$ is called to follow the weak law of large numbers.

In Definition 6.2.2, $\{a_n\}_{n=1}^{\infty}$ and $\{b_n\}_{n=1}^{\infty}$ are called the central constants and normalizing constants, respectively. Note that (6.2.28) can be expressed as

$$\lim_{n \to \infty} P\left(\left|\frac{S_n - a_n}{b_n}\right| \geq \varepsilon\right) = 0 \tag{6.2.29}$$

for every positive number ε.

---

[2] This result is the same as (6.2.1) and (6.2.4).

Theorem 6.2.7 (Rohatgi and Saleh 2001) *Assume a sequence of uncorrelated random variables* $\{X_i\}_{i=1}^{\infty}$ *with means* $\mathsf{E}\{X_i\} = m_i$ *and variances* $\mathsf{Var}\{X_i\} = \sigma_i^2$. *If*

$$\sum_{i=1}^{\infty} \sigma_i^2 \to \infty, \tag{6.2.30}$$

then

$$\left(\sum_{i=1}^{n} \sigma_i^2\right)^{-1} \left(S_n - \sum_{i=1}^{n} m_i\right) \xrightarrow{P} 0. \tag{6.2.31}$$

In other words, the sequence $\{X_i\}_{i=1}^{\infty}$ *satisfies the weak law of large numbers with* $a_n = \sum_{i=1}^{n} m_i$ *and* $b_n = \sum_{i=1}^{n} \sigma_i^2$.

Proof Employing the Chebyshev inequality $\mathsf{P}(|Y - \mathsf{E}\{Y\}| \geq \varepsilon) \leq \frac{\mathsf{Var}\{Y\}}{\varepsilon^2}$ introduced in (6.A.16), we have

$$\mathsf{P}\left(\left|S_n - \sum_{i=1}^{n} m_i\right| > \varepsilon \sum_{i=1}^{n} \sigma_i^2\right) \leq \left(\varepsilon \sum_{i=1}^{n} \sigma_i^2\right)^{-2} \mathsf{E}\left[\left\{\sum_{i=1}^{n} (X_i - m_i)\right\}^2\right]$$

$$= \left(\varepsilon^2 \sum_{i=1}^{n} \sigma_i^2\right)^{-1}. \tag{6.2.32}$$

In short, $\mathsf{P}\left(\left|S_n - \sum_{i=1}^{n} m_i\right| > \varepsilon \sum_{i=1}^{n} \sigma_i^2\right) \to 0$ when $n \to \infty$. ♠

Example 6.2.9 (Rohatgi and Saleh 2001) If an uncorrelated sequence $\{X_i\}_{i=1}^{\infty}$ with mean $\mathsf{E}\{X_i\} = m_i$ and variance $\mathsf{Var}\{X_i\} = \sigma_i^2$ satisfies

$$\lim_{n \to \infty} \frac{1}{n^2} \sum_{i=1}^{n} \sigma_i^2 = 0, \tag{6.2.33}$$

then $\frac{1}{n}\left(S_n - \sum_{i=1}^{n} m_i\right) \xrightarrow{P} 0$. This result, called the Markov theorem, can be easily shown with the steps similar to those in the proof of Theorem 6.2.7. Here, (6.2.33) is called the Markov condition. ◇

Example 6.2.10 (Rohatgi and Saleh 2001) Assume an uncorrelated sequence $\{X_i\}_{i=1}^{\infty}$ with identical distribution, mean $\mathsf{E}\{X_i\} = m$, and variance $\mathsf{Var}\{X_i\} = \sigma^2$. Then, because $\sum_{i=1}^{\infty} \sigma^2 \to \infty$, we have $\frac{1}{\sigma^2}\left(\frac{S_n}{n} - m\right) \xrightarrow{P} 0$ from Theorem 6.2.7. Here, $a_n = nm$ and $b_n = n\sigma^2$.

From now on, we assume $b_n = n$ in discussing the weak law of large numbers unless specified otherwise.

Example 6.2.11 (Rohatgi and Saleh 2001) For an i.i.d. sequence of random variables with distribution $b(1, p)$, noting that the mean is p and the variance is $p(1 - p)$, we have $\frac{S_n}{n} \overset{p}{\to} p$ from Theorem 6.2.7 and Example 6.2.9.   ◇

Example 6.2.12 (Rohatgi and Saleh 2001) For a sequence of i.i.d. random variables with marginal distribution $C(1, 0)$, we have $\frac{S_n}{n} \sim C(1, 0)$ as discussed in Exercise 6.3. In other words, because $\frac{S_n}{n}$ does not converge to 0 in probability, the weak law of large numbers does not hold for sequences of i.i.d. Cauchy random variables.

Example 6.2.13 (Rohatgi and Saleh 2001; Stoyanov 2013) For an i.i.d. sequence $\{X_i\}_{i=1}^{\infty}$, if the absolute mean $\mathsf{E}\{|X_i|\}$ is finite, then $\frac{S_n}{n} \overset{p}{\to} \mathsf{E}\{X_1\}$ when $n \to \infty$ from Theorem 6.2.7 and Example 6.2.9. This result is called Khintchine's theorem.

6.2.2.2 Strong Law of Large Numbers

Definition 6.2.3 (*strong law of large numbers*) When we have

$$\frac{S_n - a_n}{b_n} \overset{a.s.}{\to} 0 \qquad\qquad (6.2.34)$$

for two sequences $\{a_n\}_{n=1}^{\infty}$ and $\{b_n > 0\}_{n=1}^{\infty}$ of real numbers such that $b_n \uparrow \infty$, the sequence $\{X_i\}_{i=1}^{\infty}$ is called to follow the strong law of large numbers.

Note that (6.2.34) implies

$$\mathsf{P}\left(\lim_{n \to \infty} \frac{S_n - a_n}{b_n} = 0\right) = 1. \qquad\qquad (6.2.35)$$

A sequence of random variables that follows the strong law of large numbers also follows the weak law of large numbers because almost sure convergence implies convergence in probability. As in the discussion of the weak law of large numbers, we often assume the normalizing constant $b_n = n$ also for the strong law of large numbers. We now consider sufficient conditions for a sequence $\{X_i\}_{i=1}^{\infty}$ to follow the strong law of large numbers when $b_n = n$.

Theorem 6.2.8 (Rohatgi and Saleh 2001) *The sum* $\sum\limits_{i=1}^{\infty} (X_i - \mu_i)$ *converges almost surely to 0 if*

$$\sum_{i=1}^{\infty} \sigma_i^2 < \infty \qquad\qquad (6.2.36)$$

for a sequence $\{X_i\}_{i=1}^{\infty}$ *with means* $\{\mu_i\}_{i=1}^{\infty}$ *and variances* $\{\sigma_i^2\}_{i=1}^{\infty}$.

Theorem 6.2.9 (Rohatgi and Saleh 2001) *If $\sum_{i=1}^{n} x_i$ converges for a sequence $\{x_n\}_{n=1}^{\infty}$,*

then $\lim_{n \to \infty} \frac{1}{b_n} \sum_{i=1}^{n} b_i x_i = 0$ for $\{b_n\}_{n=1}^{\infty}$ such that $b_n \uparrow \infty$. This result is called the Kronecker lemma.

Example 6.2.14 (Rohatgi and Saleh 2001) Let the means and variances be $\{\mu_i\}_{i=1}^{\infty}$ and $\{\sigma_i^2\}_{i=1}^{\infty}$, respectively, for independent random variables $\{X_i\}_{i=1}^{\infty}$. Then, we can easily show that

$$\frac{1}{b_n} \left[S_n - \mathsf{E}\{S_n\} \right] \xrightarrow{a.s.} 0 \tag{6.2.37}$$

from Theorems 6.2.8 and 6.2.9 when

$$\sum_{i=1}^{\infty} \frac{\sigma_i^2}{b_i^2} < \infty, \qquad b_i \uparrow \infty. \tag{6.2.38}$$

When $b_n = n$, (6.2.38) can be expressed as

$$\sum_{n=1}^{\infty} \frac{\sigma_n^2}{n^2} < \infty, \tag{6.2.39}$$

which is called the Kolmogorov condition. $\diamond$

Example 6.2.15 (Rohatgi and Saleh 2001) If the variances $\{\sigma_n^2\}_{n=1}^{\infty}$ of independent random variables are uniformly bounded, i.e., $\left|\sigma_n^2\right| \leq M$ for a finite number M, then

$$\frac{1}{n} \left[S_n - \mathsf{E}\{S_n\} \right] \xrightarrow{a.s.} 0 \tag{6.2.40}$$

from Kolmogorov condition because $\sum_{n=1}^{\infty} \frac{\sigma_n^2}{n^2} \leq \sum_{n=1}^{\infty} \frac{M}{n^2} < \infty$. $\diamond$

Example 6.2.16 (Rohatgi and Saleh 2001) Based on the result in Example 6.2.15, it is easy to see that Bernoulli trials with probability of success p satisfy the strong law of large numbers because the variance $p(1 - p)$ is no larger than $\frac{1}{4}$.

Note that the Markov condition (6.2.33) and the Kolmogorov condition (6.2.38) are sufficient conditions but are not necessary conditions.

Theorem 6.2.10 (Rohatgi and Saleh 2001) *If the fourth moment is finite for an i.i.d. sequence $\{X_i\}_{i=1}^{\infty}$ with mean $\mathsf{E}\{X_i\} = \mu$, then*

$$P\left(\lim_{n\to\infty} \frac{S_n}{n} = \mu\right) = 1. \qquad (6.2.41)$$

In other words, $\frac{S_n}{n}$ *converges almost surely to* μ.

Proof Let the variance of X_i be σ^2. Then, we have

$$E\left[\left\{\sum_{i=1}^{n}(X_i - \mu)\right\}^4\right] = E\left\{\sum_{i=1}^{n}(X_i - \mu)^4\right\}$$

$$+3E\left\{\sum_{i=1}^{n}(X_i - \mu)^2 \sum_{j=1, j\neq i}^{n}(X_j - \mu)^2\right\}$$

$$= nE\left\{(X_1 - \mu)^4\right\} + 3n(n-1)\sigma^4$$

$$\leq cn^2 \qquad (6.2.42)$$

for an appropriate constant c. From this result and the Bienayme-Chebyshev inequality (6.A.25), we get $P\left(\left|\sum_{i=1}^{n}(X_i - \mu)\right| > n\varepsilon\right) \leq \frac{1}{(n\varepsilon)^4}E\left[\left\{\sum_{i=1}^{n}(X_i - \mu)\right\}^4\right]$ $\leq \frac{cn^2}{(n\varepsilon)^4} = \frac{c'}{n^2}$ and, consequently, $\sum_{n=1}^{\infty} P\left(\left|\frac{S_n}{n} - \mu\right| > \varepsilon\right) < \infty$, where $c' = \frac{c}{\varepsilon^4}$. Therefore, letting $A_\varepsilon = \limsup_{n\to\infty}\left\{\left|\frac{S_n}{n} - \mu\right| > \varepsilon\right\}$, we get

$$P(A_\varepsilon) = 0 \qquad (6.2.43)$$

from the Borel-Cantelli lemma discussed in Theorem 2.A.3. Now, $\{A_\varepsilon\}$ is an increasing sequence of $\varepsilon \to 0$, and converges to $\left\{\omega: \lim_{n\to\infty}\left|\frac{S_n}{n} - \mu\right| > 0\right\}$ or, equivalently, to $\left\{\omega: \lim_{n\to\infty}\frac{S_n}{n} \neq \mu\right\}$: thus, we have $P\left(\lim_{n\to\infty}\frac{S_n}{n} \neq \mu\right) = P\left(\lim_{\varepsilon\to 0} A_\varepsilon\right) = \lim_{\varepsilon\to 0} P(A_\varepsilon) = 0$ from (6.2.43). Subsequently, we get (6.2.41). ♠

Example 6.2.17 (Rohatgi and Saleh 2001) Consider an i.i.d. sequence $\{X_i\}_{i=1}^{\infty}$ and a positive number B. If $P(|X_i| < B) = 1$ for every i, then $\frac{S_n}{n}$ converges almost surely to the mean $E\{X_i\}$ of X_i. This can be easily shown from Theorem 6.2.10 by noting that the fourth moment is finite when $P(|X_i| < B) = 1$. ◇

Theorem 6.2.11 (Rohatgi and Saleh 2001) *Consider an i.i.d. sequence* $\{X_i\}_{i=1}^{\infty}$ *with mean* μ. *If*

$$E\{|X_i|\} < \infty, \qquad (6.2.44)$$

then

$$\frac{S_n}{n} \xrightarrow{a.s.} \mu. \tag{6.2.45}$$

The converse also holds true.

Note that the conditions in Theorems 6.2.10 and 6.2.11 are on the fourth moment and absolute mean, respectively.

Example 6.2.18 (Stoyanov 2013) Consider an independent sequence $\{X_n\}_{n=2}^{\infty}$ with the pmf

$$p_{X_n}(x) = \begin{cases} 1 - \frac{1}{n \log n}, & x = 0, \\ \frac{1}{2n \log n}, & x = \pm n. \end{cases} \tag{6.2.46}$$

Letting $A_n = \{|X_n| \geq n\}$ for $n \geq 2$, we get $\sum_{n=2}^{\infty} P(A_n) \to \infty$ because $P(A_n)$ $= \frac{1}{n \log n}$. In other words, $\sum_{n=2}^{\infty} P(A_n)$ is divergent and $\{X_n\}_{n=2}^{\infty}$ are independent: therefore, the probability $P(|X_n| \geq n \text{ occurs i.o.}) = P(|\frac{X_n}{n}| \geq 1 \text{ occurs i.o.}) = P\left(\lim_{n \to \infty} \frac{S_n}{n} \neq 0\right)$ of $\{A_n \text{ occurs i.o.}\}$ is 1, i.e.,

$$P(|X_n| \geq n \text{ occurs i.o.}) = 1 \tag{6.2.47}$$

from the Borel-Cantelli lemma. In short, the sequence $\{X_n\}_{n=2}^{\infty}$ does not follow the strong law of large numbers. On the other hand, the sequence $\{X_n\}_{n=2}^{\infty}$, satisfying the Markov condition as

$$\frac{1}{n^2} \sum_{k=2}^{n} \text{Var}\{X_k\} \leq \frac{1}{n^2} \left(\frac{2}{\log 2} + \int_3^{n+1} \frac{x}{\log x} dx \right)$$

$$\leq \frac{2}{n^2 \log 2} + \frac{(n-2)(n+1)}{n^2 \log n}$$

$$\to 0 \tag{6.2.48}$$

from $\text{Var}\{X_k\} = \frac{k}{\log k}$, follows the weak law of large numbers.

6.2.3 Central Limit Theorem

Let us now discuss the central limit theorem (Feller 1970; Gardner 2010; Rohatgi and Saleh 2001), the basis for the wide-spread and most popular use of the normal distribution. Assume a sequence $\{X_n\}_{n=1}^{\infty}$ and the sum $S_n = \sum_{k=1}^{n} X_k$. Assume

$$\frac{1}{b_n}(S_n - a_n) \xrightarrow{l} Y \tag{6.2.49}$$

for appropriately chosen sequences $\{a_n\}_{n=1}^{\infty}$ and $\{b_n > 0\}_{n=1}^{\infty}$ of constants. It is known that the distribution of the limit random variable Y is always a stable distribution. For example, for an i.i.d. sequence $\{X_n\}_{n=1}^{\infty}$, we have $\frac{S_n}{\sqrt{n}} \sim \mathcal{N}(0,1)$ if $X_i \sim \mathcal{N}(0,1)$ and $\frac{S_n}{n} \sim C(1,0)$ if $X_i \sim C(1,0)$: the normal and Cauchy distributions are typical examples of the stable distribution. In this section, we discuss the conditions on which the limit random variable Y has a normal distribution.

Example 6.2.19 (Rohatgi and Saleh 2001)   Assume an i.i.d. sequence $\{X_i\}_{i=1}^{\infty}$ with marginal distribution $b(1, p)$. Letting $a_n = \mathsf{E}\{S_n\} = np$ and $b_n = \sqrt{\mathsf{Var}\{S_n\}} = \sqrt{np(1-p)}$, the mgf $M_n(t) = \mathsf{E}\left\{\exp\left(\frac{S_n - np}{\sqrt{np(1-p)}}t\right)\right\} = \prod_{i=1}^{n}\mathsf{E}\left\{\exp\left(\frac{X_i - p}{\sqrt{np(1-p)}}t\right)\right\}$ of $\frac{S_n - a_n}{b_n}$ can be obtained as

$$\begin{aligned}
M_n(t) &= \exp\left\{-\frac{npt}{\sqrt{np(1-p)}}\right\}\left\{(1-p) + p\,\exp\left(\frac{t}{\sqrt{np(1-p)}}\right)\right\}^n \\
&= \left\{(1-p)\exp\left(-\frac{pt}{\sqrt{np(1-p)}}\right) + p\,\exp\left(\frac{(1-p)t}{\sqrt{np(1-p)}}\right)\right\}^n \\
&= \left\{1 + \frac{t^2}{2n} + o\left(\frac{1}{n}\right)\right\}^n.
\end{aligned} \tag{6.2.50}$$

Thus, $M_n(t) \to \exp\left(\frac{t^2}{2}\right)$ when $n \to \infty$ and, subsequently,

$$\mathsf{P}\left(\frac{S_n - np}{\sqrt{np(1-p)}} \le x\right) \to \frac{1}{\sqrt{2\pi}}\int_{-\infty}^{x}\exp\left(-\frac{t^2}{2}\right)dt \tag{6.2.51}$$

because $\exp\left(\frac{t^2}{2}\right)$ is the mgf of $\mathcal{N}(0,1)$.   ◇

Theorem 6.2.12 (Rohatgi and Saleh 2001)   *For i.i.d. random variables $\{X_i\}_{i=1}^{\infty}$ with mean m and variance σ^2, we have*

$$\frac{S_n - nm}{\sqrt{n\sigma^2}} \xrightarrow{l} Z, \tag{6.2.52}$$

where $Z \sim \mathcal{N}(0,1)$.

Proof Letting $Y_i = X_i - m$, we have $\mathsf{E}\{Y_i\} = 0$ and $\mathsf{E}\{Y_i^2\} = \sigma^2$. Also let $V_i = \frac{Y_i}{\sqrt{n\sigma^2}}$. Denoting the pdf of Y_i by f_Y, the cf $\varphi_V(\omega) = \mathsf{E}\left\{\exp\left(\frac{j\omega Y_i}{\sqrt{n\sigma^2}}\right)\right\} = \int_{-\infty}^{\infty}\exp\left(\frac{j\omega y}{\sqrt{n\sigma^2}}\right)f_Y(y)dy = \int_{-\infty}^{\infty}\left\{1 + \frac{j\omega}{\sqrt{n\sigma^2}}y + \frac{1}{2}\left(\frac{j\omega}{\sqrt{n\sigma^2}}\right)^2 y^2 + \frac{1}{6}\left(\frac{j\omega}{\sqrt{n\sigma^2}}\right)^3 y^3 + \cdots\right\}f_Y(y)dy$ of V_i can be obtained as

$$\varphi_V(\omega) = 1 + \frac{j\omega}{\sqrt{n\sigma^2}}\mathsf{E}\{Y\} + \frac{1}{2}\left(\frac{j\omega}{\sqrt{n\sigma^2}}\right)^2 \mathsf{E}\{Y^2\} + \cdots$$

$$= 1 - \frac{\omega^2}{2n} + o\left(\frac{1}{n}\right). \tag{6.2.53}$$

Next, letting $Z_n = \frac{S_n - nm}{\sqrt{n\sigma^2}} = \sum_{i=1}^{n} V_i$ and denoting the cf of Z_n by φ_{Z_n}, we have

$$\lim_{n\to\infty} \varphi_{Z_n}(\omega) = \lim_{n\to\infty} \varphi_V^n(\omega) = \lim_{n\to\infty}\left\{1 - \frac{\omega^2}{2n} + o\left(\frac{1}{n}\right)\right\}^n, \text{ i.e.,}$$

$$\lim_{n\to\infty} \varphi_{Z_n}(\omega) = \exp\left(-\frac{\omega^2}{2}\right) \tag{6.2.54}$$

from (6.2.53) because $\{V_i\}_{i=1}^{\infty}$ are independent. In short, the distribution of $Z_n = \frac{S_n - nm}{\sqrt{n\sigma^2}}$ converges to $\mathcal{N}(0, 1)$ as $n \to \infty$. ◆

Theorem 6.2.12 is one of the many variants of the central limit theorem, and can be derived from the Lindeberg's central limit theorem introduced in Appendix 6.2.

Example 6.2.20 (Rohatgi and Saleh 2001) Assume an i.i.d. sequence $\{X_i\}_{i=1}^{\infty}$ with marginal pmf $p_{X_i}(k) = (1 - p)^k p\tilde{u}(k)$, where $0 < p < 1$ and $\tilde{u}(k)$ is the unit step function in discrete space defined in (1.4.17). We have $\mathsf{E}\{X_i\} = \frac{q}{p}$ and $\mathsf{Var}\{X_i\} = \frac{q}{p^2}$, where $q = 1 - p$. Thus, it follows from Theorem 6.2.12 that

$$\mathsf{P}\left(\frac{\sqrt{n}\left(p\frac{S_n}{n} - q\right)}{\sqrt{q}} \le x\right) \to \Phi(x) \tag{6.2.55}$$

for $x \in \mathbb{R}$ when $n \to \infty$.

The central limit theorem is useful in many cases: however, it should also be noted that there do exist cases in which the central limit theorem does not apply.

Example 6.2.21 (Stoyanov 2013) Assume an i.i.d. sequence $\{Y_k\}_{k=1}^{\infty}$ with $\mathsf{P}(Y_k = \pm 1) = \frac{1}{2}$, and let $X_k = \frac{\sqrt{15}}{4^k} Y_k$. Then, it is easy to see that $\mathsf{E}\{S_n\} = 0$ and $\mathsf{Var}\{S_n\} = 1 - 16^{-n}$. In other words, when n is sufficiently large, $\mathsf{Var}\{S_n\} \approx 1$. Meanwhile, because $\quad |S_n| = |X_1 + X_2 + \cdots + X_n| \ge |X_1| - (|X_2| + |X_3| + \cdots + |X_n|) = \frac{\sqrt{15}}{4} - \frac{\sqrt{15}}{12}\left(1 - \frac{1}{4^{n-1}}\right) \ge \frac{\sqrt{15}}{6} > \frac{1}{2}$, we have $\mathsf{P}\left(|S_n| \le \frac{1}{2}\right) = 0$. Thus, $\mathsf{P}(S_n)$ does not converge to the standard normal cdf $\Phi(x)$ at some point x. This fact implies that $\{X_i\}_{i=1}^{\infty}$ does not satisfy the central limit theorem: the reason is that the random variable X_1 is exceedingly large to virtually determine the distribution of S_n. ◇

The central limit theorem and laws of large numbers are satisfied for a wide range of sequences of random variables. As we have observed in Theorem 6.2.7 and Example 6.2.15, the laws of large numbers hold true for uniformly bounded independent sequences. As shown in Example 6.A.5 of Appendix 6.2, the central limit theorem holds true for an independent sequence even when the sum of variances diverges. Meanwhile, for an i.i.d. sequence $\{X_i\}_{i=1}^{\infty}$, noting that

$$P\left(\left|\frac{S_n}{n} - m\right| > \varepsilon\right) = P\left(\frac{|S_n - nm|}{\sigma\sqrt{n}} > \frac{\varepsilon}{\sigma}\sqrt{n}\right)$$
$$\approx 1 - P\left(|Z| \le \frac{\varepsilon}{\sigma}\sqrt{n}\right), \tag{6.2.56}$$

where $Z \sim \mathcal{N}(0, 1)$, we can obtain the laws of large numbers directly from the central limit theorem. In other words, the central limit theorem is stronger than the laws of large numbers: yet, in the laws of large numbers we are not concerned with the existence of the second moment. In some independent sequences for which the central limit theorem holds true, on the other hand, the weak law of large numbers does not hold true.

Example 6.2.22 (Feller 1970; Rohatgi and Saleh 2001)   Assume the pmf $P\left(X_k = k^\lambda\right) = P\left(X_k = -k^\lambda\right) = \frac{1}{2}$ for an independent sequence $\{X_k\}_{k=1}^{\infty}$, where $\lambda > 0$. Then, the mean and variance of X_k are $E\{X_k\} = 0$ and $Var\{X_k\} = k^{2\lambda}$, respectively. Now, letting $s_n^2 = \sum_{k=1}^{n} Var\{X_k\} = \sum_{k=1}^{n} k^{2\lambda}$, we have

$$s_n^2 \ge \frac{n^{2\lambda+1} - 1}{2\lambda + 1} \tag{6.2.57}$$

from $\sum_{k=1}^{n} k^{2\lambda} \ge \int_1^n x^{2\lambda} dx$. Here, we can assume $n > 1$ without loss of generality, and if we let $n > \frac{2\lambda+1}{\varepsilon^2} + 1$, we have $\varepsilon^2 > \frac{2\lambda+1}{n-1}$ and $\varepsilon^2 s_n^2 > \frac{2\lambda+1}{n-1} s_n^2 \ge \frac{2\lambda+1}{n-1} \frac{n^{2\lambda+1}-1}{2\lambda+1} = n^{2\lambda} + n^{2\lambda-1} + \cdots + 1 > n^{2\lambda}$. Therefore, for $n > \frac{2\lambda+1}{\varepsilon^2} + 1$, we have $|x_{kl}| > n^\lambda$ if $|x_{kl}| > \varepsilon s_n$. Noting in addition that $P\left(X_k = x\right)$ is non-zero only when $|x| \le n^\lambda$, we get

$$\frac{1}{s_n^2} \sum_{k=1}^{n} \sum_{|x_{kl}| > \varepsilon s_n} x_{kl}^2 p_{kl} = 0. \tag{6.2.58}$$

In short, the Lindeberg conditions[3] are satisfied and the central limit theorem holds true. Now, if we consider $s_n^2 = \sum_{k=1}^{n} k^{2\lambda} \le \int_0^n x^{2\lambda} dx$, i.e.,

$$s_n^2 \le \frac{n^{2\lambda+1}}{2\lambda + 1} \tag{6.2.59}$$

---

[3] Equations (6.A.5) and (6.A.6) in Appendix 6.2 are called the Lindeberg conditions.

and (6.2.57), we can write $s_n \approx \sqrt{\frac{n^{2\lambda+1}}{2\lambda+1}}$. Based on this, we have

$$P\left(a < \sqrt{\frac{2\lambda+1}{n^{2\lambda+1}}} S_n < b\right) \to \frac{1}{\sqrt{2\pi}} \int_a^b \exp\left(-\frac{t^2}{2}\right) dt \qquad (6.2.60)$$

from Theorem 6.A.1.

Next, let us discuss if the weak law of large numbers holds true. First, when $0 < \lambda < \frac{1}{2}$, it is easy to see that the weak law of large numbers holds true based on Example 6.2.9 because $\frac{s_n^2}{n^2} \le \frac{n^{2\lambda-1}}{2\lambda+1} \to 0$ from (6.2.59). When $\lambda \ge \frac{1}{2}$, however, rewriting (6.2.60), we get

$$P\left(\frac{an^{\lambda-\frac{1}{2}}}{\sqrt{2\lambda+1}} < \frac{S_n}{n} < \frac{bn^{\lambda-\frac{1}{2}}}{\sqrt{2\lambda+1}}\right) \to \frac{1}{\sqrt{2\pi}} \int_a^b \exp\left(-\frac{t^2}{2}\right) dt, \qquad (6.2.61)$$

which implies that $P\left(-\varepsilon < \frac{S_n}{n} < \varepsilon\right) \to 1$ is not always true when $\varepsilon > 0$. Thus, the weak law of large numbers does not hold true. $\diamondsuit$

Let us discuss one application of the central limit theorem. First, from Theorem 6.2.12, we get the following theorem:

Theorem 6.2.13 *For an i.i.d. sequence* $\{X_i\}_{i=1}^n$ *with mean* m *and variance* σ^2, *we have* $\sum_{k=1}^n X_k \sim \mathcal{N}\left(nm, n\sigma^2\right)$ *asymptotically.*

Example 6.2.23 For an i.i.d. sequence $\{X_i\}_{i=1}^n$ with marginal distribution $U(0, 1)$, compare the pdf of $S_n = \sum_{i=1}^n X_i$ with the pdf of the asymptotic distribution described in Theorem 6.2.13.

Solution From the pdf

$$f_{X_i}(x) = \begin{cases} 1, & x \in (0, 1), \\ 0, & \text{otherwise} \end{cases} \qquad (6.2.62)$$

of X_i, we get the pdf $f_{S_n}(x) = f_{X_1}(x) * f_{X_2}(x) * \cdots * f_{X_n}(x)$ of S_n. Meanwhile, from the mean $\frac{1}{2}$ and variance $\frac{1}{12}$ of X_i, the asymptotic distribution of S_n is $\mathcal{N}\left(\frac{n}{2}, \frac{n}{12}\right)$. Figure 6.1 shows the pdf and asymptotic pdf of S_n for $n = 1, 2, 3, 4$, which confirms that the two pdf's are closer when n is larger.

Example 6.2.24 Assume an i.i.d. sequence $\{X_i\}_{i=1}^n$ with marginal distribution $b\left(1, \frac{1}{2}\right)$. Compare the cdf of $S_n = \sum_{i=1}^n X_i$ and the cdf of the asymptotic distribution described in Theorem 6.2.13.

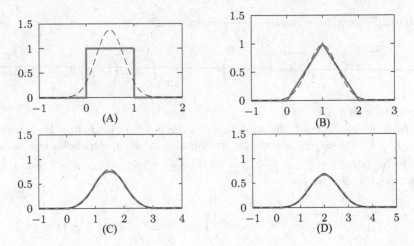

Fig. 6.1 The pdf (blue solid line) and asymptotic pdf (black dashed line) of S_n for an i.i.d. sequence with marginal distribution $U(0, 1)$: (A) $n = 1$, (B) $n = 2$, (C) $n = 3$, (D) $n = 4$

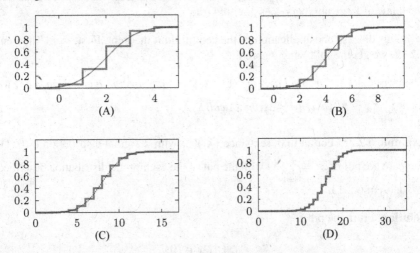

Fig. 6.2 The cdf (blue solid line) of S_n and approximate cdf (black dotted line) from the central limit theorem for an i.i.d. sequence with marginal distribution $b\left(1, \frac{1}{2}\right)$: (A) $n = 4$, (B) $n = 8$, (C) $n = 16$, (D) $n = 32$

Solution It is easy to see that $S_n \sim b\left(n, \frac{1}{2}\right)$ from Example 6.2.3. Noting that X_i has mean $\frac{1}{2}$ and variance $\frac{1}{4}$, the asymptotic distribution of S_n is $\mathcal{N}\left(\frac{n}{2}, \frac{n}{4}\right)$. Figure 6.2 shows the cdf and asymptotic cdf of S_n for $n = 4, 8, 16, 32$, which confirms that the two cdf's are closer when n is larger.

Theorem 6.2.14 *For an i.i.d. sequence* $\{X_i\}_{i=1}^n$ *with mean* m *and variance* σ^2, *we can approximate the distribution of* $\frac{S_n}{n} = \frac{1}{n}\sum_{i=1}^n X_i$ *as*

$$\mathcal{N}\left(m, \frac{\sigma^2}{n}\right) \tag{6.2.63}$$

and the cdf F_{S_n} of S_n as

$$F_{S_n}(x) \approx \Phi\left(\frac{x - nm}{\sqrt{n\sigma^2}}\right) \tag{6.2.64}$$

when n is sufficiently large.

Theorem 6.2.14 follows directly from Theorem 6.2.13.

Example 6.2.25 (Rohatgi and Saleh 2001) For an i.i.d. sequence $\{X_i\}_{i=1}^n$ with marginal distribution $b(1, p)$, the asymptotic distribution of S_n is $\mathcal{N}(np, np(1 - p))$. Therefore, based on

$$P\left(\frac{S_n - np}{\sqrt{np(1 - p)}} \le \frac{x - np}{\sqrt{np(1 - p)}}\right) \approx \Phi\left(\frac{x - np}{\sqrt{np(1 - p)}}\right), \tag{6.2.65}$$

we can approximately obtain $P(S_n \le x)$ when n is sufficiently large: practically, $n \ge 20$ is sufficient. When $n = 25$ and $p = \frac{1}{2}$, for example, $P(S_n \le 12) = P\left(\frac{S_n - 12.5}{2.5} \le \frac{-0.5}{2.5}\right) \approx P(Z \le -0.2) \approx 0.421$, where $Z \sim \mathcal{N}(0, 1)$. ◇

Note that taking the continuity correction[4] into account, a better approximation can be obtained. Specifically, for an integer random variable X, employing

$$P(x_1 \le X \le x_2) = P\left(x_1 - \frac{1}{2} < X < x_2 + \frac{1}{2}\right) \tag{6.2.66}$$

will provide us with a better approximation when x_1 and x_2 are integers.

Example 6.2.26 If we take the continuity correction into account in Example 6.2.25, then $P(S_n \le 12) = P(S_n < 12.5) \approx P(Z < 0) = 0.5$. This is the exact value of the probability that we have at most twelve *head* when a fair coin is tossed 25 times. ◇

Appendices

Appendix 6.1 Convergence of Probability Functions

For a sequence $\{X_n\}_{n=1}^\infty$, let F_n and M_n be the cdf and mgf, respectively, of X_n. We first note that, when $n \to \infty$, the sequence of cdf's does not always converge and, even when it does, the limit is not always a cdf.

---

[4] The continuity correction has been considered also in (3.5.18) of Chap. 3.

Example 6.A.1 Consider the cdf

$$F_n(x) = \begin{cases} 0, & x \leq 0, \\ \frac{nx}{n+1}, & 0 \leq x \leq 1 + \frac{1}{n}, \\ 1, & x \geq 1 + \frac{1}{n} \end{cases} \qquad (6.A.1)$$

of X_n. Then, the limit of the sequence $\{F_n\}_{n=1}^{\infty}$ is

$$\lim_{n \to \infty} F_n(x) = \begin{cases} 0, & x \leq 0, \\ x, & 0 \leq x \leq 1, \\ 1, & x \geq 1, \end{cases} \qquad (6.A.2)$$

which is a cdf.   ◇

Example 6.A.2 (Rohatgi and Saleh 2001) Consider the cdf $F_n(x) = u(x - n)$ of X_n. The limit of the sequence $\{F_n\}_{n=1}^{\infty}$ is $\lim_{n \to \infty} F_n(x) = 0$, which is not a cdf.

Example 6.A.3 (Rohatgi and Saleh 2001) Assume the pmf $\mathsf{P}(X_n = -n) = 1$ for a sequence $\{X_n\}_{n=1}^{\infty}$. Then, the mgf is $M_n(t) = e^{-nt}$ and its limit is $\lim_{n \to \infty} M_n(t) = M(t)$, where

$$M(t) = \begin{cases} 0, & t > 0, \\ 1, & t = 0, \\ \infty, & t < 0. \end{cases} \qquad (6.A.3)$$

The function $M(t)$ is not an mgf. In other words, the limit of a sequence of mgf's is not necessarily an mgf.   ◇

Example 6.A.4 Assume the pdf $f_n(x) = \frac{n}{\pi} \frac{1}{1+n^2 x^2}$ of X_n. Then, the cdf is $F_n(x) = \frac{n}{\pi} \int_{-\infty}^{x} \frac{dt}{1+n^2 t^2}$. We also have $\lim_{n \to \infty} f_n(x) = \delta(x)$ and $\lim_{n \to \infty} F_n(x) = u(x)$. These limits imply $\lim_{n \to \infty} \mathsf{P}(|X_n - 0| > \varepsilon) = \int_{-\infty}^{-\varepsilon} \delta(x)dx + \int_{\varepsilon}^{\infty} \delta(x)dx = 0$ and, consequently, $\{X_n\}_{n=1}^{\infty}$ converges to 0 in probability.   ◇

Appendix 6.2 The Lindeberg Central Limit Theorem

The central limit theorem can be expressed in a variety of ways. Among those varieties, the Lindeberg central limit theorem is one of the most general ones and does not require the random variables to have identical distribution.

Theorem 6.A.1 (Rohatgi and Saleh 2001) *For an independent sequence* $\{X_i\}_{i=1}^{\infty}$, *let the mean, variance, and cdf of* X_i *be* m_i, σ_i^2, *and* F_i, *respectively. Let*

$$s_n^2 = \sum_{i=1}^{n} \sigma_i^2. \qquad (6.A.4)$$

When the cdf F_i is absolutely continuous, assume that the pdf $f_i(x) = \frac{d}{dx} F_i(x)$ satisfies

$$\lim_{n \to \infty} \frac{1}{s_n^2} \sum_{i=1}^{n} \int_{|x-m_i|>\varepsilon s_n} (x - m_i)^2 f_i(x) dx = 0 \qquad (6.A.5)$$

for every value of $\varepsilon > 0$. When $\{X_i\}_{i=1}^{\infty}$ are discrete random variables, assume the pmf $p_i(x) = P(X_i = x)$ satisfies[5]

$$\lim_{n \to \infty} \frac{1}{s_n^2} \sum_{i=1}^{n} \sum_{|x_{il}-m_i|>\varepsilon s_n} (x_{il} - m_i)^2 p_i(x_{il}) = 0 \qquad (6.A.6)$$

for every value of $\varepsilon > 0$, where $\{x_{il}\}_{l=1}^{L_i}$ are the jump points of F_i with L_i the number of jumps of F_i. Then, the distribution of

$$\frac{1}{s_n} \left(\sum_{i=1}^{n} X_i - \sum_{i=1}^{n} m_i \right) \qquad (6.A.7)$$

converges to $\mathcal{N}(0, 1)$ as $n \to \infty$.

Example 6.A.5 (Rohatgi and Saleh 2001) Assume an independent sequence $\{X_k \sim U(-a_k, a_k)\}_{k=1}^{\infty}$. Then, $E\{X_k\} = 0$ and $Var\{X_k\} = \frac{1}{3} a_k^2$. Let $|a_k| < A$ and $s_n^2 = \sum_{k=1}^{n} Var\{X_k\} = \frac{1}{3} \sum_{k=1}^{n} a_k^2 \to \infty$ when $n \to \infty$. Then, from the Chebyshev inequality $P(|Y - E\{Y\}| \geq \varepsilon) \leq \frac{Var\{Y\}}{\varepsilon^2}$ discussed in (6.A.16), we get

$$\frac{1}{s_n^2} \sum_{k=1}^{n} \int_{|x|>\varepsilon s_n} x^2 F_k'(x) dx \leq \frac{A^2}{s_n^2} \sum_{k=1}^{n} \frac{Var\{X_k\}}{\varepsilon^2 s_n^2}$$

$$= \frac{A^2}{\varepsilon^2 s_n^2}$$

$$\to 0 \qquad (6.A.8)$$

as $n \to \infty$ because $\frac{1}{s_n^2} \sum_{k=1}^{n} \int_{|x|>\varepsilon s_n} x^2 F_k'(x) dx \leq \frac{1}{s_n^2} \sum_{k=1}^{n} \int_{|x|>\varepsilon s_n} A^2 \frac{1}{2a_k} dx = \frac{A^2}{s_n^2} \sum_{k=1}^{n} P(|X_k| > \varepsilon s_n)$.

Meanwhile, assume $\sum_{k=1}^{\infty} a_k^2 < \infty$, and let $s_n^2 \uparrow B^2$ for $n \to \infty$. Then, for a constant k, we can find ε_k such that $\varepsilon_k B < a_k$, and we have $\varepsilon_k s_n < \varepsilon_k B$. Thus, $P(|X_k| > \varepsilon_k s_n) \geq P(|X_k| > \varepsilon_k B) > 0$. Based on this result, for $n \geq k$, we get

---

[5] As mentioned in Example 6.2.22 already, (6.A.5) and (6.A.6) are called the Lindeberg condition.

$$\frac{1}{s_n^2} \sum_{j=1}^n \int_{|x|>\varepsilon_k s_n} x^2 F_j'(x)dx \geq \frac{s_n^2 \varepsilon_k^2}{s_n^2} \sum_{j=1}^n \int_{|x|>\varepsilon_k s_n} F_j'(x)dx$$

$$= \frac{s_n^2 \varepsilon_k^2}{s_n^2} \sum_{j=1}^n P\left(|X_j| > \varepsilon_k s_n\right)$$

$$\geq \varepsilon_k^2 P\left(|X_k| > \varepsilon_k s_n\right)$$

$$> 0, \tag{6.A.9}$$

implying that the Lindeberg condition is not satisfied. In essence, in a sequence of uniformly bounded independent random variables, a necessary and sufficient condition for the central limit theorem to hold true is $\sum_{k=1}^\infty \mathsf{Var}\{X_k\} \to \infty$.   ◇

Example 6.A.6 (Rohatgi and Saleh 2001)   Assume an independent sequence $\{X_k\}_{k=1}^\infty$. Let $\delta > 0$, $\alpha_k = \mathsf{E}\left\{|X_k|^{2+\delta}\right\} < \infty$, and $\sum_{j=1}^n \alpha_j = o\left(s_n^{2+\delta}\right)$. Then, the Lindeberg condition is satisfied and the central limit theorem holds true. This can be shown easily as

$$\frac{1}{s_n^2} \sum_{k=1}^n \int_{|x|>\varepsilon s_n} x^2 F_k'(x)dx \leq \frac{1}{s_n^2} \sum_{k=1}^n \int_{|x|>\varepsilon s_n} \frac{|x|^{2+\delta}}{\varepsilon^\delta s_n^\delta} F_k'(x)dx$$

$$\leq \frac{1}{\varepsilon^\delta s_n^{2+\delta}} \sum_{k=1}^n \int_{-\infty}^\infty |x|^{2+\delta} F_k'(x)dx$$

$$= \frac{1}{\varepsilon^\delta s_n^{2+\delta}} \sum_{k=1}^n \alpha_k$$

$$\to 0 \tag{6.A.10}$$

because $x^2 < \frac{|x|^{2+\delta}}{\varepsilon^\delta s_n^\delta}$ from $|x|^\delta x^2 > |\varepsilon s_n|^\delta x^2$ when $|x| > \varepsilon s_n$. We can similarly show that the central limit theorem holds true in discrete random variables.   ◇

The conditions (6.A.5) and (6.A.6) are the necessary conditions in the following sense: for a sequence $\{X_i\}_{i=1}^\infty$ of independent random variables, assume the variances $\{\sigma_i^2\}_{i=1}^\infty$ of $\{X_i\}_{i=1}^\infty$ are finite. If the pdf of X_i satisfies (6.A.5) or the pmf of X_i satisfies (6.A.6) for every value of $\varepsilon > 0$, then

$$\lim_{n\to\infty} \mathsf{P}\left(\frac{\overline{X}_n - \mathsf{E}\{\overline{X}_n\}}{\sqrt{\mathsf{Var}\{\overline{X}_n\}}} \leq x\right) = \Phi(x) \tag{6.A.11}$$

and

$$\lim_{n\to\infty} \mathsf{P}\left(\max_{1\le k\le n} |X_k - \mathsf{E}\{X_k\}| > n\varepsilon\sqrt{\mathrm{Var}\{\overline{X}_n\}}\right) = 0, \qquad (6.A.12)$$

and the converse also holds true, where $\overline{X}_n = \frac{1}{n}\sum_{i=1}^{n} X_i$ is the sample mean of $\{X_i\}_{i=1}^{n}$ defined in (5.4.1).

Appendix 6.3 Properties of Convergence

(A) Continuity of Expected Values

When the sequence $\{X_n\}_{n=1}^{\infty}$ converges to X, the sequence $\{\mathsf{E}\{X_n\}\}_{n=1}^{\infty}$ of expected values will also converge to the expected value $\mathsf{E}\{X\}$, which is called the continuity of expected values. The continuity of expected values (Gray and Davisson 2010) is a consequence of the continuity of probability discussed in Appendix 2.1.

(1) Monotonic convergence. If $0 \le X_n \le X_{n+1}$ for every integer n, then $\mathsf{E}\{X_n\} \to \mathsf{E}\{X\}$ as $n \to \infty$. In other words, $\mathsf{E}\left\{\lim_{n\to\infty} X_n\right\} = \lim_{n\to\infty} \mathsf{E}\{X_n\}$.

(2) Dominated convergence. If $|X_n| < Y$ for every integer n and $\mathsf{E}\{Y\} < \infty$, then $\mathsf{E}\{X_n\} \to \mathsf{E}\{X\}$ as $n \to \infty$.

(3) Bounded convergence. If there exists a constant c such that $|X_n| \le c$ for every integer n, then $\mathsf{E}\{X_n\} \to \mathsf{E}\{X\}$ as $n \to \infty$.

(B) Properties of Convergence

We list some properties among various types of convergence. Here, a and b are constants.

(1) If $X_n \xrightarrow{P} X$, then $X_n - X \xrightarrow{P} 0$, $aX_n \xrightarrow{P} aX$, and $X_n - X_m \xrightarrow{P} 0$ for $n, m \to \infty$.

(2) If $X_n \xrightarrow{P} X$ and $X_n \xrightarrow{P} Y$, then $\mathsf{P}(X = Y) = 1$.

(3) If $X_n \xrightarrow{P} a$, then $X_n^2 \xrightarrow{P} a^2$.

(4) If $X_n \xrightarrow{P} 1$, then $\frac{1}{X_n} \xrightarrow{P} 1$.

(5) If $X_n \xrightarrow{P} X$ and Y is a random variable, then $X_n Y \xrightarrow{P} XY$.

(6) If $X_n \xrightarrow{P} X$ and $Y_n \xrightarrow{P} Y$, then $X_n \pm Y_n \xrightarrow{P} X \pm Y$ and $X_n Y_n \xrightarrow{P} XY$.

(7) If $X_n \xrightarrow{P} a$ and $Y_n \xrightarrow{P} b \ne 0$, then $\frac{X_n}{Y_n} \xrightarrow{P} \frac{a}{b}$.

(8) If $X_n \xrightarrow{d} X$, then $X_n + a \xrightarrow{d} X + a$ and $bX_n \xrightarrow{d} bX$ for $b \ne 0$.

(9) If $X_n \xrightarrow{d} a$, then $X_n \xrightarrow{P} a$. Therefore, $X_n \xrightarrow{d} a \rightleftarrows X_n \xrightarrow{P} a$.

(10) If $|X_n - Y_n| \xrightarrow{P} 0$ and $Y_n \xrightarrow{d} Y$, then $X_n \xrightarrow{d} Y$. Based on this, it can be shown that $X_n \xrightarrow{d} X$ when $X_n \xrightarrow{P} X$.

(11) If $X_n \overset{d}{\to} X$ and $Y_n \overset{p}{\to} a$, then $X_n \pm Y_n \overset{d}{\to} X \pm a$, $X_n Y_n \overset{d}{\to} aX$ for $a \neq 0$,
     $X_n Y_n \overset{p}{\to} 0$ for $a = 0$, and $\frac{X_n}{Y_n} \overset{d}{\to} \frac{X}{a}$ for $a \neq 0$.

(12) If $X_n \overset{r=2}{\longrightarrow} X$, then $\lim_{n \to \infty} \mathsf{E}\{X_n\} = \mathsf{E}\{X\}$ and $\lim_{n \to \infty} \mathsf{E}\{X_n^2\} = \mathsf{E}\{X^2\}$.

(13) If $X_n \overset{L^r}{\to} X$, then $\lim_{n \to \infty} \mathsf{E}\{|X_n|^r\} = \mathsf{E}\{|X|^r\}$.

(14) If $X_1 > X_2 > \cdots > 0$ and $X_n \overset{p}{\to} 0$, then $X_n \overset{a.s.}{\longrightarrow} 0$.

(C) Convergence and Limits of Products

Consider the product

$$A_n = \prod_{k=1}^{n} a_k. \tag{6.A.13}$$

The infinite product $\prod_{k=1}^{\infty} a_k$ is called convergent to the limit A when $A_n \to A$ and $A \neq 0$ for $n \to \infty$; divergent to 0 when $A_n \to 0$; and divergent when A_n is not convergent to a non-zero value. The convergence of products is often related to the convergence of sums as shown below.

(1) When all the real numbers $\{a_k\}_{k=1}^{\infty}$ are positive, the convergence of $\prod_{k=1}^{\infty} a_k$ and

   that of $\sum_{k=1}^{\infty} \ln a_k$ are the necessary and sufficient conditions of each other.

(2) When all the real numbers $\{a_k\}_{k=1}^{\infty}$ are positive, the convergence of $\prod_{k=1}^{\infty} (1 + a_k)$

   and that of $\sum_{k=1}^{\infty} a_k$ are the necessary and sufficient conditions of each other.

(3) When all the real numbers $\{a_k\}_{k=1}^{\infty}$ are non-negative, the convergence of $\prod_{k=1}^{\infty} (1 -$

   $a_k)$ and that of $\sum_{k=1}^{\infty} a_k$ are the necessary and sufficient conditions of each other.

Appendix 6.4 Inequalities

In this appendix we introduce some useful inequalities (Beckenbach and Bellam 1965) in probability spaces.

(A) Inequalities for Random Variables

Theorem 6.A.2 (Rohatgi and Saleh 2001) *If a measurable function h is non-negative and $\mathsf{E}\{h(X)\}$ exists for a random variable X, then*

$$P(h(X) \geq \varepsilon) \leq \frac{E\{h(X)\}}{\varepsilon} \qquad (6.A.14)$$

for $\varepsilon > 0$, which is called the tail probability inequality.

Proof Assume X is a discrete random variable. Letting $P(X = x_k) = p_k$, we have $E\{h(X)\} = \left(\sum_A + \sum_{A^c}\right) h(x_k)\, p_k \geq \sum_A h(x_k)\, p_k$ when $A = \{k : h(x_k) \geq \epsilon\}$: this yields $E\{h(X)\} \geq \epsilon \sum_A p_k = \epsilon P(h(X) \geq \epsilon)$ and, subsequently, (6.A.14). ♠

Theorem 6.A.3 If X is a non-negative random variable, then[6]

$$P(X \geq \alpha) \leq \frac{E\{X\}}{\alpha} \qquad (6.A.15)$$

for $\alpha > 0$, which is called the Markov inequality.

The Markov inequality can be proved easily from (6.A.14) by letting $h(X) = |X|$ and $\varepsilon = \alpha$. We can show the Markov inequality also from $E\{X\} = \int_0^\infty x f_X(x)\, dx \geq \int_\alpha^\infty x f_X(x) dx \geq \alpha \int_\alpha^\infty f_X(x) dx = \alpha P(X \geq \alpha)$ by recollecting that a pdf is non-negative.

Theorem 6.A.4 The mean $E\{Y\}$ and variance $\mathrm{Var}\{Y\}$ of any random variable Y satisfy

$$P(|Y - E\{Y\}| \geq \varepsilon) \leq \frac{\mathrm{Var}\{Y\}}{\varepsilon^2} \qquad (6.A.16)$$

for any $\varepsilon > 0$, which is called the Chebyshev inequality.

Proof The random variable $X = (Y - E\{Y\})^2$ is non-negative. Thus, if we use (6.A.15), we get $P\left([Y - E\{Y\}]^2 \geq \varepsilon^2\right) \leq \frac{1}{\varepsilon^2} E\left\{[Y - E\{Y\}]^2\right\} = \frac{\mathrm{Var}\{Y\}}{\varepsilon^2}$. Now, noting that $P\left([Y - E\{Y\}]^2 \geq \varepsilon^2\right) = P(|Y - E\{Y\}| \geq \varepsilon)$, we get (6.A.16). ♠

Theorem 6.A.5 (Rohatgi and Saleh 2001) The absolute mean $E\{|X|\}$ of any random variable X satisfies

$$\sum_{n=1}^\infty P(|X| \geq n) \leq E\{|X|\} \leq 1 + \sum_{n=1}^\infty P(|X| \geq n), \qquad (6.A.17)$$

which is called the absolute mean inequality.

Proof Let the pdf of a continuous random variable X be f_X. Then, because $E\{|X|\} = \int_{-\infty}^\infty |x| f_X(x) dx = \sum_{k=0}^\infty \int_{k \leq |x| < k+1} |x| f_X(x) dx$, we have

---

[6] The inequality (6.A.15) holds true also when X is replaced by $|X|^r$ for $r > 0$.

$$\sum_{k=0}^{\infty} k P(k \le |X| < k+1) \le E\{|X|\}$$

$$\le \sum_{k=0}^{\infty} (k+1) P(k \le |X| < k+1). \qquad (6.A.18)$$

Now, employing $\sum_{k=0}^{\infty} k P(k \le |X| < k+1) = \sum_{n=1}^{\infty} \sum_{k=n}^{\infty} P(k \le |X| < k+1) = \sum_{n=1}^{\infty}$

$P(|X| \ge n)$ and $\sum_{k=0}^{\infty} (k+1) P(k \le |X| < k+1) = 1 + \sum_{k=0}^{\infty} k P(k \le |X| < k+1) =$

$1 + \sum_{n=1}^{\infty} P(|X| \ge n)$ in (6.A.18), we get (6.A.17). A similar procedure will show the result for discrete random variables.   ♠

Theorem 6.A.6 *If f is a convex[7] function, then*

$$E\{h(X)\} \ge h(E\{X\}), \qquad (6.A.19)$$

which is called the Jensen inequality.

Proof Let $m = E\{X\}$. Then, from the intermediate value theorem, we have

$$h(X) = h(m) + (X - m)h'(m) + \frac{1}{2}(X - m)^2 h''(\alpha) \qquad (6.A.20)$$

for $-\infty < \alpha < \infty$. Taking the expectation of the above equation, we get $E\{h(X)\} = h(m) + \frac{1}{2}h''(\alpha)\sigma_X^2$. Recollecting that $h''(\alpha) \ge 0$ and $\sigma_X^2 \ge 0$, we get $E\{h(X)\} \ge h(m) = h(E\{X\})$.   ♠

Theorem 6.A.7 (Rohatgi and Saleh 2001) *If the n-th absolute moment $E\{|X|^n\}$ is finite, then*

$$\left[E\{|X|^s\}\right]^{\frac{1}{s}} \le \left[E\{|X|^r\}\right]^{\frac{1}{r}} \qquad (6.A.21)$$

for $1 \le s < r \le n$, which is called the Lyapunov inequality.

Proof Consider the bi-variable formula

$$Q(u, v) = \int_{-\infty}^{\infty} \left(u|x|^{\frac{k-1}{2}} + v|x|^{\frac{k+1}{2}}\right)^2 f(x)\, dx, \qquad (6.A.22)$$

---

[7] A function h is called convex or concave up when $h(tx + (1-t)y) \le th(x) + (1-t)h(y)$ for every two points x and y and for every choice of $t \in [0, 1]$. A convex function is a continuous function with a non-decreasing derivative and is differentiable except at a countable number of points. In addition, the second order derivative of a convex function, if it exists, is non-negative.

where f is the pdf of X. Letting $\beta_n = E\{|X|^n\}$, (6.A.22) can be written as $Q(u, v) = (u\ v)\begin{pmatrix} \beta_{k-1} & \beta_k \\ \beta_k & \beta_{k+1} \end{pmatrix}(u\ v)^T$. Now, we have $\begin{vmatrix} \beta_{k-1} & \beta_k \\ \beta_k & \beta_{k+1} \end{vmatrix} \geq 0$, i.e., $\beta_k^{2k} \leq \beta_{k-1}^k \beta_{k+1}^k$ because $Q \geq 0$ for every choice of u and v. Therefore, we have

$$\beta_1^2 \leq \beta_0^1 \beta_2^1, \quad \beta_2^4 \leq \beta_1^2 \beta_3^2, \quad \cdots, \quad \beta_{n-1}^{2(n-1)} \leq \beta_{n-2}^{n-1} \beta_n^{n-1} \qquad (6.A.23)$$

with $\beta_0 = 1$. If we multiply the first $k - 1$ consecutive inequalities in (6.A.23), then we have $\beta_{k-1}^k \leq \beta_k^{k-1}$ for $k = 2, 3, \ldots, n$, from which we can easily get $\beta_1 \leq \beta_2^{\frac{1}{2}} \leq \beta_3^{\frac{1}{3}} \leq \cdots \leq \beta_n^{\frac{1}{n}}$. ♠

Theorem 6.A.8 Let $g(x)$ be a non-decreasing and non-negative function for $x \in (0, \infty)$. If $\frac{E\{g(|X|)\}}{g(\varepsilon)}$ is defined, then

$$P(|X| \geq \varepsilon) \leq \frac{E\{g(|X|)\}}{g(\varepsilon)} \qquad (6.A.24)$$

for $\varepsilon > 0$, which is called the generalized Bienayme-Chebyshev inequality.

Proof Let the cdf of X be $F(x)$. Then, we get $E\{g(|X|)\} \geq g(\varepsilon)\int_{|x|\geq\varepsilon} dF(x) = g(\varepsilon)P(|X| \geq \varepsilon)$ by recollecting $E\{g(|X|)\} = \int_{|x|<\varepsilon} g(|x|)dF(x) + \int_{|x|\geq\varepsilon} g(|x|) dF(x) \geq \int_{|x|\geq\varepsilon} g(|x|)dF(x)$. ♠

Letting $g(x) = x^r$ in the generalized Bienayme-Chebyshev inequality, we can easily get the Bienayme-Chebyshev inequality discussed below. In addition, the Chebyshev inequality discussed in Theorem 6.A.4 is a special case of the generalized Bienayme-Chebyshev inequality and of the Bienayme-Chebyshev inequality.

Theorem 6.A.9 When the r-th absolute moment $E\{|X|^r\}$ of X is finite, where $r > 0$, we have

$$P(|X| \geq \varepsilon) \leq \frac{E\{|X|^r\}}{\varepsilon^r} \qquad (6.A.25)$$

for $\varepsilon > 0$, which is called the Bienayme-Chebyshev inequality.

(B) Inequalities of Random Vectors

Theorem 6.A.10 (Rohatgi and Saleh 2001) For two random variables X and Y, we have

$$E^2\{XY\} \leq E\{X^2\}E\{Y^2\}, \qquad (6.A.26)$$

which is called the Cauchy-Schwarz inequality.

Proof First, note that $E\{|XY|\}$ exists when $E\{X^2\} < \infty$ and $E\{Y^2\} < \infty$ because $|ab| \leq \frac{a^2+b^2}{2}$ for real numbers a and b. Now, if $E\{X^2\} = 0$, then $P(X = 0) = 1$ and thus $E\{XY\} = 0$, implying that (6.A.26) holds true. Next when $E\{X^2\} > 0$, recollecting that $E\{(\alpha X + Y)^2\} = \alpha^2 E\{X^2\} + 2\alpha E\{XY\} + E\{Y^2\} \geq 0$ for any real number α, we have $\frac{E^2\{XY\}}{E\{X^2\}} - 2\frac{E^2\{XY\}}{E\{X^2\}} + E\{Y^2\} \geq 0$ by letting $\alpha = -\frac{E\{XY\}}{E\{X^2\}}$. This inequality is equivalent to (6.A.26). ♠

Theorem 6.A.11 (Rohatgi and Saleh 2001) *For zero-mean independent random variables* $\{X_i\}_{i=1}^n$ *with variances* $\{\sigma_i^2\}_{i=1}^n$, *let* $S_k = \sum_{j=1}^k X_j$. *Then,*

$$P\left(\max_{1 \leq k \leq n} |S_k| > \varepsilon\right) \leq \sum_{i=1}^n \frac{\sigma_i^2}{\varepsilon^2} \tag{6.A.27}$$

for $\varepsilon > 0$, *which is called the Kolmogorov inequality.*

Proof Let $A_0 = \Omega$, $A_k = \left\{\max_{1 \leq j \leq k} |S_j| \leq \varepsilon\right\}$ for $k = 1, 2, \ldots, n$, and $B_k = A_{k-1} \cap A_k^c = \{|S_1| \leq \varepsilon, |S_2| \leq \varepsilon, \ldots, |S_{k-1}| \leq \varepsilon\} \cap$ {at least one of $|S_1|, |S_2|, \ldots, |S_k|$ is larger than ε}, i.e.,

$$B_k = \{|S_1| \leq \varepsilon, |S_2| \leq \varepsilon, \ldots, |S_{k-1}| \leq \varepsilon, |S_k| > \varepsilon\}. \tag{6.A.28}$$

Then, $A_n^c = \bigcup_{k=1}^n B_k$ and $B_k \subseteq \{|S_{k-1}| \leq \varepsilon, |S_k| > \varepsilon\}$. Recollecting the indicator function $K_A(x)$ defined in (2.A.27), we get $E\left[\{S_n K_{B_k}(S_k)\}^2\right] = E\left[\{(S_n - S_k) K_{B_k}(S_k) + S_k K_{B_k}(S_k)\}^2\right]$, i.e.,

$$E\left[\{S_n K_{B_k}(S_k)\}^2\right] = E\{(S_n - S_k)^2 K_{B_k}(S_k)\} + E\left[\{S_k K_{B_k}(S_k)\}^2\right]$$
$$+ E\{2S_k(S_n - S_k) K_{B_k}(S_k)\}. \tag{6.A.29}$$

Noting that $S_n - S_k = X_{k+1} + X_{k+2} + \cdots + X_n$ and $S_k K_{B_k}(S_k)$ are independent of each other, that $E\{X_k\} = 0$, that $E\{K_{B_k}(S_k)\} = P(B_k)$, and that $|S_k| \geq \varepsilon$ under B_k, we have $E\left[\{S_n K_{B_k}(S_k)\}^2\right] = E\{(S_n - S_k)^2 K_{B_k}(S_k)\} + E\left[\{S_k K_{B_k}(S_k)\}^2\right] \geq E\left[\{S_k K_{B_k}(S_k)\}^2\right]$, i.e.,

$$E\left[\{S_n K_{B_k}(S_k)\}^2\right] \geq \varepsilon^2 P(B_k) \tag{6.A.30}$$

from (6.A.29). Subsequently, using $\sum_{k=1}^n E\left[\{S_n K_{B_k}(S_k)\}^2\right] = E\{S_n^2 K_{A_n^c}(S_n)\}$ $\leq E\{S_n^2\} = \sum_{k=1}^n \sigma_k^2$ and (6.A.30), we get $\sum_{k=1}^n \sigma_k^2 \geq \varepsilon^2 \sum_{k=1}^n P(B_k) = \varepsilon^2 P(A_n^c)$, which is the same as (6.A.27). ♠

Example 6.A.7 (Rohatgi and Saleh 2001) The Chebyshev inequality (6.A.16) with $E\{Y\} = 0$, i.e.,

$$P(|Y| > \varepsilon) \leq \frac{\text{Var}\{Y\}}{\varepsilon^2} \tag{6.A.31}$$

is the same as the Kolmogorov inequality (6.A.27) with $n = 1$.   ◇

Theorem 6.A.12 *Consider i.i.d. random variables* $\{X_i\}_{i=1}^n$ *with marginal mgf* $M(t) = E\{e^{tX_i}\}$. *Let* $Y_n = \sum_{i=1}^n X_i$ *and* $g(t) = \ln M(t)$. *If we let the solution to* $\alpha = ng'(t)$ *be* t_r *for a real number* α, *then*

$$P(Y_n \geq \alpha) \leq \exp\left[-n\left\{t_r g'(t_r) - g(t_r)\right\}\right], \quad t_r \geq 0 \tag{6.A.32}$$

and

$$P(Y_n \leq \alpha) \leq \exp\left[-n\left\{t_r g'(t_r) - g(t_r)\right\}\right], \quad t_r \leq 0. \tag{6.A.33}$$

The inequalities (6.A.32) and (6.A.33) are called the Chernoff bounds.

When $t_r = 0$, the right-hand sides of the two inequalities (6.A.32) and (6.A.33) are both 1 from $g(t_r) = \ln M(t_r) = \ln M(0) = 0$: in other words, the Chernoff bounds simply say that the probability is no larger than 1 when $t_r = 0$, and thus the Chernoff bounds are more useful when $t_r \neq 0$.

Example 6.A.8 (Thomas 1986) Let $X \sim \mathcal{N}(0, 1)$, $n = 1$, and $Y_1 = X$. From the mgf $M(t) = \exp\left(\frac{t^2}{2}\right)$, we get $g(t) = \ln M(t) = \frac{t^2}{2}$ and $g'(t) = t$. Thus, the solution to $\alpha = ng'(t) = t$ is $t_r = \alpha$. In other words, the Chernoff bounds can be written as

$$P(X \geq \alpha) \leq \exp\left(-\frac{\alpha^2}{2}\right), \quad \alpha \geq 0 \tag{6.A.34}$$

and

$$P(X \leq \alpha) \leq \exp\left(-\frac{\alpha^2}{2}\right), \quad \alpha \leq 0 \tag{6.A.35}$$

for $X \sim \mathcal{N}(0, 1)$.

◇

Example 6.A.9 For $X \sim P(\lambda)$, assume $n = 1$ and $Y_1 = X$. From the mgf $M(t) = \exp\{\lambda(e^t - 1)\}$, we get $g(t) = \ln M(t) = \lambda(e^t - 1)$ and $g'(t) = \lambda e^t$. Solving $\alpha = ng'(t) = \lambda e^t$, we get $t_r = \ln\left(\frac{\alpha}{\lambda}\right)$. Thus, $t_r > 0$ when $\alpha > \lambda$, $t_r = 0$ when $\alpha = \lambda$, and $t_r < 0$ when $\alpha < \lambda$. Therefore, we have

$$P(X \geq \alpha) \leq e^{-\lambda} \left(\frac{e\lambda}{\alpha} \right)^{\alpha}, \quad \alpha \geq \lambda \qquad (6.A.36)$$

and

$$P(X \leq \alpha) \leq e^{-\lambda} \left(\frac{e\lambda}{\alpha} \right)^{\alpha}, \quad \alpha \leq \lambda \qquad (6.A.37)$$

because $n \{ t_r g'(t_r) - g(t_r) \} = \ln \left(\frac{\alpha}{\lambda} \right)^{\alpha} - \alpha + \lambda$ from $g(t_r) = \lambda \left(\frac{\alpha}{\lambda} - 1 \right) = \alpha$
$- \lambda$ and $t_r g'(t_r) = \alpha \ln \left(\frac{\alpha}{\lambda} \right)$.

Theorem 6.A.13 *If p and q are both larger than 1 and $\frac{1}{p} + \frac{1}{q} = 1$, then*

$$E\{|XY|\} \leq E^{\frac{1}{p}} \{|X^p|\} E^{\frac{1}{q}} \{|Y^q|\}, \qquad (6.A.38)$$

which is called the Hölder inequality.

Theorem 6.A.14 *If $p > 1$, then*

$$E^{\frac{1}{p}} \{|X + Y|^p\} \leq E^{\frac{1}{p}} \{|X^p|\} + E^{\frac{1}{p}} \{|Y^p|\}, \qquad (6.A.39)$$

which is called the Minkowski inequality.

It is easy to see that the Minkowski inequality is a generalization of the triangle inequality $|a - b| \leq |a - c| + |c - b|$.

Exercises

Exercise 6.1 For the sample space $[0, 1]$, consider a sequence of random variables defined by

$$X_n(\omega) = \begin{cases} 1, & \omega \leq \frac{1}{n}, \\ 0, & \omega > \frac{1}{n} \end{cases} \qquad (6.E.1)$$

and let $X(\omega) = 0$ for $\omega \in [0, 1]$. Assume the probability measure $P(a \leq \omega \leq b) = b - a$, the Lebesgue measure mentioned following (2.5.24), for $0 \leq a \leq b \leq 1$. Discuss if $\{X_n(\omega)\}_{n=1}^{\infty}$ converges to $X(\omega)$ surely or almost surely.

Exercise 6.2 For the sample space $[0, 1]$, consider the sequence

$$X_n(\omega) = \begin{cases} 3, & 0 \leq \omega < \frac{1}{2n}, \\ 4, & 1 - \frac{1}{2n} < \omega \leq 1, \\ 5, & \frac{1}{2n} < \omega < 1 - \frac{1}{2n} \end{cases} \qquad (6.E.2)$$

and let $X(\omega) = 5$ for $\omega \in [0, 1]$. Assuming the probability measure $P(a \le \omega \le b) = b - a$ for $0 \le a \le b \le 1$, discuss if $\{X_n(\omega)\}_{n=1}^{\infty}$ converges to $X(\omega)$ surely or almost surely.

Exercise 6.3 When $\{X_i\}_{i=1}^{n}$ are independent random variables, obtain the distribution of $S_n = \sum_{i=1}^{n} X_i$ in each of the following five cases of the distribution of X_i.

(1) geometric distribution with parameter α,
(2) $NB(r_i, \alpha)$,
(3) $P(\lambda_i)$,
(4) $G(\alpha_i, \beta)$, and
(5) $C(\mu_i, \theta_i)$.

Exercise 6.4 To what does $\frac{S_n}{n}$ converge in Example 6.2.10?

Exercise 6.5 Let $Y = \frac{X-\lambda}{\sqrt{\lambda}}$ for a Poisson random variable $X \sim P(\lambda)$. Noting that the mgf of X is $M_X(t) = \exp\{\lambda(e^t - 1)\}$, show that Y converges to a standard normal random variable as $\lambda \to \infty$.

Exercise 6.6 For a sequence $\{X_n\}_{n=1}^{\infty}$ with the pmf

$$P(X_n = x) = \begin{cases} \frac{1}{n}, & x = 1, \\ 1 - \frac{1}{n}, & x = 0, \end{cases} \tag{6.E.3}$$

show that $X_n \overset{l}{\to} X$, where X has the distribution $P(X = 0) = 1$.

Exercise 6.7 Discuss if the weak law of large numbers holds true for a sequence of i.i.d. random variables with marginal pdf $f(x) = \frac{1+\alpha}{x^{2+\alpha}}u(x - 1)$, where $\alpha > 0$.

Exercise 6.8 Show that $S_n = \sum_{k=1}^{n} X_k$ converges to a Poisson random variable with distribution $P(np)$ when $n \to \infty$ for an i.i.d. sequence $\{X_n\}_{n=1}^{\infty}$ with marginal distribution $b(1, p)$.

Exercise 6.9 Discuss the central limit theorem for an i.i.d. sequence $\{X_i\}_{i=1}^{\infty}$ with marginal distribution $B(\alpha, \beta)$.

Exercise 6.10 An i.i.d. sequence $\{X_i\}_{i=1}^{n}$ has marginal distribution $P(\lambda)$. When n is large enough, we can approximate as $S_n = \sum_{k=1}^{n} X_k \sim \mathcal{N}(n\lambda, n\lambda)$. Using the continuity correction, obtain the probability $P(50 < S_n \le 80)$.

Exercise 6.11 Consider an i.i.d. Bernoulli sequence $\{X_i\}_{i=1}^{n}$ with $P(X_i = 1) = p$, a binomial random variable $M \sim b(n, p)$ which is independent of $\{X_i\}_{i=1}^{n}$, and $K = \sum_{i=1}^{n} X_i$. Note that K is the number of successes in n i.i.d. Bernoulli trials. Obtain the expected values of $U = \sum_{i=1}^{K} X_i$ and $V = \sum_{i=1}^{M} X_i$.

Exercise 6.12 The result of a game is independent of another game, and the probabilities of winning and losing are each $\frac{1}{2}$. Assume there is no tie. When a person wins, the person gets 2 points and then continues. On the other hand, if the person loses a round, the person gets 0 points and stops. Obtain the mgf, expected value, and variance of the score Y that the person may get from the games.

Exercise 6.13 Let P_n be the probability that we have more *head* than *tail* in a toss of n fair coins.

(1) Obtain P_3, P_4, and P_5.
(2) Obtain the limit $\lim_{n \to \infty} P_n$.

Exercise 6.14 For an i.i.d. sequence $\{X_n \sim \mathcal{N}(0, 1)\}_{n=1}^{\infty}$, let the cdf of $\overline{X}_n = \frac{1}{n} \sum_{i=1}^{n} X_i$ be F_n. Obtain $\lim_{n \to \infty} F_n(x)$ and discuss whether the limit is a cdf or not.

Exercise 6.15 Consider $X_{[1]} = \min (X_1, X_2, \ldots, X_n)$ for an i.i.d. sequence $\{X_n \sim U(0, \theta)\}_{n=1}^{\infty}$. Does $Y_n = n X_{[1]}$ converge in distribution? If yes, obtain the limit cdf.

Exercise 6.16 The marginal cdf F of an i.i.d. sequence $\{X_n\}_{n=1}^{\infty}$ is absolutely continuous. For the sequence $\{Y_n\}_{n=1}^{\infty} = \{n\{1 - F(M_n)\}\}_{n=1}^{\infty}$, obtain the limit $\lim_{n \to \infty} F_{Y_n}(y)$ of the cdf F_{Y_n} of Y_n, where $M_n = \max (X_1, X_2, \ldots, X_n)$.

Exercise 6.17 Is the sequence of cdf's

$$F_n(x) = \begin{cases} 0, & x < 0, \\ 1 - \frac{1}{n}, & 0 \le x < n, \\ 1, & x \ge n \end{cases} \tag{6.E.4}$$

convergent? If yes, obtain the limit.

Exercise 6.18 In the sequence $\{Y_i = X + W_i\}_{i=1}^{n}$, X and $\{W_i\}_{i=1}^{n}$ are independent of each other, and $\{W_i \sim \mathcal{N}(0, \sigma_i^2)\}_{i=1}^{n}$ is an i.i.d. sequence, where $\sigma_i^2 \le \sigma_{max}^2 < \infty$. We estimate X via

$$\hat{X}_n = \frac{1}{n} \sum_{i=1}^{n} Y_i \tag{6.E.5}$$

and let the error be $\varepsilon_n = \hat{X}_n - X$.

(1) Express the cf, mean, and variance of $\hat{X}_n$ in terms of those of X and $\{W_i\}_{i=1}^{n}$.
(2) Obtain the covariance $\mathsf{Cov}(Y_i, Y_j)$.
(3) Obtain the pdf $f_{\varepsilon_n}(\alpha)$ and the conditional pdf $f_{\hat{X}|X}(\alpha|\beta)$.
(4) Does $\hat{X}_n$ converge to X? If yes, what is the type of the convergence? If not, what is the reason?

Exercise 6.19 Assume an i.i.d. sequence $\{X_i\}_{i=1}^{\infty}$ with marginal pdf $f(x) = e^{-x+\theta}u(x-\theta)$. Show that

$$\min(X_1, X_2, \ldots, X_n) \xrightarrow{P} \theta \tag{6.E.6}$$

and that

$$Y \xrightarrow{P} 1 + \theta \tag{6.E.7}$$

for $Y = \frac{1}{n} \sum_{i=1}^{n} X_i$.

Exercise 6.20 Show that $\max(X_1, X_2, \ldots, X_n) \xrightarrow{P} \theta$ for an i.i.d. sequence $\{X_i\}_{i=1}^{\infty}$ with marginal distribution $U(0, \theta)$.

Exercise 6.21 Assume an i.i.d. sequence $\{X_n\}_{n=1}^{\infty}$ with marginal cdf

$$F(x) = \begin{cases} 0, & x \leq 0, \\ x, & 0 < x \leq 1, \\ 1, & x > 1. \end{cases} \tag{6.E.8}$$

Let $\{Y_n\}_{n=1}^{\infty}$ and $\{Z_n\}_{n=1}^{\infty}$ be defined by $Y_n = \max(X_1, X_2, \ldots, X_n)$ and $Z_n = n(1 - Y_n)$. Show that the sequence $\{Z_n\}_{n=1}^{\infty}$ converges in distribution to a random variable Z with cdf $F(z) = \left(1 - e^{-z}\right)u(z)$.

Exercise 6.22 For the sample space $\Omega = \{1, 2, \ldots\}$ and probability measure $P(n) = \frac{\alpha}{n^2}$, assume a sequence $\{X_n\}_{n=1}^{\infty}$ such that

$$X_n(\omega) = \begin{cases} n, & \omega = n, \\ 0, & \omega \neq n. \end{cases} \tag{6.E.9}$$

Show that, as $n \to \infty$, $\{X_n\}_{n=1}^{\infty}$ converges to $X = 0$ almost surely, but does not converge to $X = 0$ in the mean square, i.e., $E\left\{(X_n - 0)^2\right\} \nrightarrow 0$.

Exercise 6.23 The second moment of an i.i.d. sequence $\{X_i\}_{i=1}^{\infty}$ is finite. Show that $Y_n \xrightarrow{P} E\{X_1\}$ for $Y_n = \frac{2}{n(n+1)} \sum_{i=1}^{n} iX_i$.

Exercise 6.24 For a sequence $\{X_n\}_{n=1}^{\infty}$ with

$$P(X_n = x) = \begin{cases} 1 - \frac{1}{n}, & x = 0, \\ \frac{1}{n}, & x = 1, \end{cases} \tag{6.E.10}$$

we have $X_n \xrightarrow{r=2} 0$ because $\lim_{n\to\infty} E\left\{X_n^2\right\} = \lim_{n\to\infty} \frac{1}{n} = 0$. Show that the sequence $\{X_n\}_{n=1}^{\infty}$ does not converge almost surely.

Exercise 6.25 Consider a sequence $\{X_i\}_{i=1}^n$ with a finite common variance σ^2. When the correlation coefficient between X_i and X_j is negative for every $i \neq j$, show that the sequence $\{X_i\}_{i=1}^n$ follows the weak law of large numbers. (Hint. Assume $Y_n = \frac{1}{n} \sum_{k=1}^n (X_k - m_k)$ for a sequence $\{X_i\}_{i=1}^\infty$ with mean $\mathsf{E}\{X_i\} = m_i$. Then, it is known that a necessary and sufficient condition for $\{X_i\}_{i=1}^\infty$ to satisfy the weak law of large numbers is that

$$\mathsf{E}\left\{ \frac{Y_n^2}{1 + Y_n^2} \right\} \to 0 \tag{6.E.11}$$

as $n \to \infty$.)

Exercise 6.26 For an i.i.d. sequence $\{X_i\}_{i=1}^n$, let $\mathsf{E}\{X_i\} = \mu$, $\mathrm{Var}\{X_i\} = \sigma^2$, and $\mathsf{E}\{X_i^4\} < \infty$. Find the constants a_n and b_n such that $\frac{V_n - a_n}{b_n} \xrightarrow{l} Z$ for $V_n = \sum_{k=1}^n (X_k - \mu)^2$, where $Z \sim \mathcal{N}(0, 1)$.

Exercise 6.27 When the sequence $\{X_k\}_{k=1}^\infty$ with $\mathsf{P}(X_k = \pm k^\alpha) = \frac{1}{2}$ satisfies the strong law of large numbers, obtain the range of α.

Exercise 6.28 Assume a Cauchy random variable X with pdf $f_X(x) = \frac{a}{\pi(x^2 + a^2)}$.

(1) Show that the cf is

$$\varphi_X(w) = e^{-a|w|}. \tag{6.E.12}$$

(2) Show that the sample mean of n i.i.d. Cauchy random variables is a Cauchy random variable.

Exercise 6.29 Assume an i.i.d. sequence $\{X_i \sim P(0.02)\}_{i=1}^{100}$. For $S = \sum_{i=1}^{100} X_i$, obtain the value $\mathsf{P}(S \geq 3)$ using the central limit theorem and compare it with the exact value.

Exercise 6.30 Consider the sequence of cdf's

$$F_n(x) = \begin{cases} 0, & x \leq 0, \\ x\left\{ 1 - \frac{\sin(2n\pi x)}{2n\pi x} \right\}, & 0 < x \leq 1, \\ 1, & x \geq 1, \end{cases} \tag{6.E.13}$$

among which four are shown in Fig. 6.3. Obtain $\lim_{n \to \infty} F_n(x)$ and discuss if $\frac{d}{dx}\left\{ \lim_{n \to \infty} F_n(x) \right\}$ is the same as $\lim_{n \to \infty} \left\{ \frac{d}{dx} F_n(x) \right\}$.

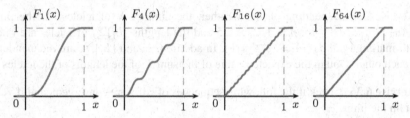

Fig. 6.3 Four cdf's $F_n(x) = x \left\{ 1 - \frac{\sin(2n\pi x)}{2n\pi x} \right\}$ for $n = 1, 4, 16,$ and 64 on $x \in [0, 1]$

Exercise 6.31 Assume an i.i.d. sequence $\{X_i \sim \chi^2(1)\}_{i=1}^{\infty}$. Then, we have $S_n \sim \chi^2(n)$, $\mathsf{E}\{S_n\} = n$, and $\mathsf{Var}\{S_n\} = 2n$. Thus, letting $Z_n = \frac{1}{\sqrt{2n}}(S_n - n) = \sqrt{\frac{n}{2}}\left(\frac{S_n}{n} - 1\right)$, the mgf $M_n(t) = \mathsf{E}\left\{e^{tZ_n}\right\} = \exp\left(-t\sqrt{\frac{n}{2}}\right)\left(1 - \frac{2t}{\sqrt{2n}}\right)^{-\frac{n}{2}}$ of Z_n can be obtained as

$$M_n(t) = \left\{ \exp\left(t\sqrt{\frac{2}{n}}\right) - t\sqrt{\frac{2}{n}}\exp\left(t\sqrt{\frac{2}{n}}\right) \right\}^{-\frac{n}{2}} \tag{6.E.14}$$

for $t < \sqrt{\frac{n}{2}}$. In addition, from Taylor approximation, we get

$$\exp\left(t\sqrt{\frac{2}{n}}\right) = 1 + t\sqrt{\frac{2}{n}} + \frac{t^2}{2}\left(\sqrt{\frac{2}{n}}\right)^2 + \frac{1}{6}\left(t\sqrt{\frac{2}{n}}\right)^3 \exp(\theta_n) \tag{6.E.15}$$

for $0 < \theta_n < t\sqrt{\frac{2}{n}}$. Show that $Z_n \xrightarrow{l} Z \sim \mathcal{N}(0, 1)$.

Exercise 6.32 In a soccer game, the number N of shootings of a player is a Poisson random variable with mean $\mu = 12$. The probability of a goal for a shooting is $\frac{1}{8}$ and is independent of N. Obtain the distribution, mean, and variance of the number of goals.

Exercise 6.33 For a facsimile (fax), the number W of pages sent is a geometric random variable with pmf $p_W(k) = \frac{3^{k-1}}{4^k}$ for $k \in \{1, 2, \ldots\}$ and mean $\frac{1}{\beta} = 4$. The amount B_i of information contained in the i-th page is a geometric random variable with pmf $p_B(k) = \frac{1}{10^5}\left(1 - 10^{-5}\right)^{k-1}$ for $k \in \{1, 2, \ldots\}$ with expected value $\frac{1}{\alpha} = 10^5$. Assuming that $\{B_i\}_{i=1}^{\infty}$ is an i.i.d. sequence and that W and $\{B_i\}_{i=1}^{\infty}$ are independent of each other, obtain the distribution of the total amount of information sent via this fax.

Exercise 6.34 Consider a sequence $\{X_i\}_{i=1}^{\infty}$ of i.i.d. exponential random variables with mean $\frac{1}{\lambda}$. A geometric random variable N is of mean $\frac{1}{p}$ and is independent of $\{X_i\}_{i=1}^{\infty}$. Obtain the expected value and variance of the random sum $S_N = \sum_{i=1}^{N} X_i$.

Exercise 6.35 Depending on the weather, the number N of icicles has the pmf $p_N(n) = \frac{1}{10} 2^{2-|3-n|}$ for $n = 1, 2, \ldots, 5$ and the lengths $\{L_i\}_{i=1}^{\infty}$ of icicles are i.i.d. with marginal pdf $f_L(v) = \lambda e^{-\lambda v} u(v)$. In addition, N and $\{L_i\}_{i=1}^{\infty}$ are independent of each other. Obtain the expected value of the sum T of the lengths of the icicles.

Exercise 6.36 Check if the following sequences of cdf's are convergent, and if yes, obtain the limit:

(1) sequence $\{F_n(x)\}_{n=1}^{\infty}$ with cdf

$$F_n(x) = \begin{cases} 0, & x < -n, \\ \frac{1}{2n}(x+n), & -n \le x < n, \\ 1, & x \ge n. \end{cases} \tag{6.E.16}$$

(2) sequence $\{F_n(x)\}_{n=1}^{\infty}$ such that $F_n(x) = F(x+n)$ for a continuous cdf $F(x)$.
(3) sequence $\{G_n(x)\}_{n=1}^{\infty}$ such that $G_n(x) = F\left(x + (-1)^n n\right)$ for a continuous cdf $F(x)$.

Exercise 6.37 For the sequence $\{X_n \sim G(n, \beta)\}_{n=1}^{\infty}$, obtain the limit distribution of $Y_n = \frac{X_n}{n^2}$.

References

E.F. Beckenbach, R. Bellam, *Inequalities* (Springer, Berlin, 1965)

W.B. Davenport Jr., *Probability and Random Processes* (McGraw-Hill, New York, 1970)

J.L. Doob, Heuristic approach to the Kolmogorov-Smirnov theorems. Ann. Math. Stat. **20**(3), 393–403 (1949)

W. Feller, *An Introduction to Probability Theory and Its Applications*, 3rd edn. revised printing (Wiley, New York, 1970)

W.A. Gardner, *Introduction to Random Processes with Applications to Signals and Systems*, 2nd edn. (McGraw-Hill, New York, 1990)

R.M. Gray, L.D. Davisson, *An Introduction to Statistical Signal Processing* (Cambridge University Press, Cambridge, 2010)

G. Grimmett, D. Stirzaker, *Probability and Random Processes* (Oxford University, London, 1982)

A. Leon-Garcia, *Probability, Statistics, and Random Processes for Electrical Engineering*, 3rd edn. (Prentice Hall, New York, 2008)

G.A. Mihram, A cautionary note regarding invocation of the central limit theorem. Am. Stat. **23**(5), 38 (1969)

V.K. Rohatgi, A.KMd.E. Saleh, *An Introduction to Probability and Statistics*, 2nd edn. (Wiley, New York, 2001)

J.M. Stoyanov, *Counterexamples in Probability*, 3rd edn. (Dover, New York, 2013)

J.B. Thomas, *Introduction to Probability* (Springer, New York, 1986)

R.D. Yates, D.J. Goodman, *Probability and Stochastic Processes* (Wiley, New York, 1999)

Answers to Selected Exercises

Chapter 1 Preliminaries

Exercise 1.3 $A - B = A \triangle (A \cap B)$. $A \cup B = A \triangle B \triangle (A \cap B)$.

Exercise 1.6 The set of polynomials with integer coefficients is countable.

Exercise 1.7 The set of algebraic numbers is countable.

Exercise 1.9 The collection of all non-overlapping open intervals with real end points is countable.

Exercise 1.10 (1) $f(n, m) = 2^n 3^m$. (2) $f(x) = \frac{1}{2} + \frac{1}{\pi} \arctan(x)$.

(3) Let $0.\alpha_1 \alpha_2 \cdots$ be the binary expression of the real number $\frac{1}{2} + \frac{1}{\pi} \arctan(x)$. Then, $\frac{1}{2} + \frac{1}{\pi} \arctan(x) \mapsto (\alpha_1, \alpha_2, \ldots)$, $\alpha_i \in \{0, 1\}$.

(4) infinite sequence $\epsilon_0, \epsilon_1, \ldots \mapsto$ ternary number $0. (2\epsilon_0)(2\epsilon_1) \cdots$, where $\epsilon_i \in \{0, 1\}$.

(5) sequence $n_0, n_1, \ldots$ of natural numbers $\mapsto$ sequence $n_0 + 1$ $0's$, 1, $n_0 + 1$ $0's$, 1, $\cdots$.

(6) Let $x_j = \pm \cdots \alpha_{-2}^{(j)} \alpha_{-1}^{(j)} . \alpha_1^{(j)} \alpha_2^{(j)} \cdots$ be the binary representation of a real number x_j. Let $\alpha_0^{(j)} = 1$ and 0 when $x_j > 0$ and < 0, respectively.

sequence $S = (x_0, x_1, \ldots) \mapsto$ sequence $\alpha_0^{(0)}, \alpha_{-1}^{(0)}, \alpha_0^{(1)}, \alpha_1^{(0)}, \alpha_{-2}^{(0)}, \alpha_{-1}^{(1)}, \alpha_0^{(2)}, \alpha_1^{(1)}, \alpha_2^{(0)}, \alpha_{-3}^{(0)}, \alpha_{-2}^{(1)}, \alpha_{-1}^{(2)}, \alpha_0^{(3)}, \alpha_1^{(2)}, \alpha_2^{(1)}, \alpha_3^{(0)}, \ldots$.

Exercise 1.11 (1) $f(n) = (k, l)$, where $n = 2^k \times 3^l \cdots$ is the factorization of n in prime factors.

(2) $f(n) = \frac{(-1)^k l}{m+1}$, where $n = 2^k \times 3^l \times 5^m \cdots$ is the factorization of n in prime factors.

(3) For an element $x = 0.\alpha_1 \alpha_2 \cdots$ of the Cantor set, let $f(x) = 0.\frac{\alpha_1}{2} \frac{\alpha_2}{2} \cdots$.

(4) a sequence $(\alpha_1, \alpha_2, \ldots)$ of 0 and $1 \mapsto$ the number $0.\alpha_1 \alpha_2 \cdots$.

Exercise 1.12 (1) When two intervals (a, b) and (c, d) are both finite, $f(x) = c + (d - c)\frac{x-a}{b-a}$.

When a is finite, $b = \infty$, and (c, d) is finite, $f(x) = c + \frac{2}{\pi}(d - c)\arctan(x - a)$. Similarly in other cases.

(2) $S_1 \mapsto S_2$, where S_2 is an infinite sequence of 0 and 1 obtained by replacing 1 with $(1, 0)$ and 2 with $(1, 1)$ in an infinite sequence $S_1 = (a_0, a_1, \ldots)$ of 0, 1, and 2.

© The Editor(s) (if applicable) and The Author(s), under exclusive license to Springer Nature Switzerland AG 2022
I. Song et al., *Probability and Random Variables: Theory and Applications*,
https://doi.org/10.1007/978-3-030-97679-8

(3) Denote a number $x \in [0, 1)$ by $x = 0.a_1a_2 \cdots$ in decimal system. Let us make consecutive 9's and the immediately following non-9 one digit into a group and each other digit into a group. For example, we have $x = 0.(1)(2)(97)(9996)(6)(5)$ $(99997)(93) \cdots$. Write the number as $x = 0. (x_1)(x_2)(x_3) \cdots$. Then, letting $y = 0. (x_1)(x_3)(x_5) \cdots$ and $z = 0. (x_2)(x_4)(x_6) \cdots$, $f(x) = (y, z)$ is the desired one-to-one correspondence.

Exercise 1.14 from (m, n) to k: $k = g(m, n) = m + \frac{1}{2}(m + n)(m + n + 1)$.

from k to (m, n): $m = k - \frac{1}{2}a(a + 1)$. $n = a - m$, where a is an integer such that $a(a + 1) \le 2k < (a + 1)(a + 2)$.

Exercise 1.15 The collection of intervals with rational end points in the space $\mathbb{R}$ of real numbers is countable.

Exercise 1.17 It is a rational number. It is a rational number. It is not always a rational number.

Exercise 1.18 (1) $\frac{a+b}{2}$. (2) $\frac{a+b}{2}$. (3) $a + \frac{b-a}{\sqrt{2}}$. (4) $c = \frac{2a+b}{3}$ and/or $d = \frac{a+2b}{3}$.

(5) Assume $a = 0.a_1a_2 \cdots$ and $b = 0.b_1b_2 \cdots$, where $a_i \in \{0, 1, \ldots, 9\}$ and $b_i \in \{0, 1, \ldots, 9\}$. Let $k = \arg\min_i \{b_i > a_i\}$ and $l = \arg\min_{i>k} \{b_i > 0\}$. Then, the number

$$c = 0.c_1c_2 \cdots c_k \cdots c_{l-1}c_l = 0.b_1b_2 \cdots b_k \cdots b_{l-1}(b_l - 1).$$

(6) The number c that can be found by the procedure, after replacing a with g, in (5), where $g = \frac{a+b}{2}$.

Exercise 1.19 Let $A = \{a_i\}_{i=0}^{\infty}$ and $a_i = 0.\alpha_1^{(i)}\alpha_2^{(i)}\alpha_3^{(i)} \cdots$. Assume the second player chooses $y_j = 4$ when $\alpha_{2j}^{(j)} \ne 4$ and $y_j = 6$ when $\alpha_{2j}^{(j)} = 4$. Then, for any sequence $x_1, x_2, \ldots$ of numbers the first player has chosen, the second player wins because the number $0.x_1y_1x_2y_2 \cdots$ is not the same as any number a_j and thus is not included in A.

Exercise 1.25 (1) $u(ax + b) = u\left(x + \frac{b}{a}\right)u(a) + u\left(-x - \frac{b}{a}\right)u(-a)$.

$$u(\sin x) = \sum_{n=-\infty}^{\infty} \{u(x - 2n\pi) - u(x - (2n + 1)\pi)\}.$$

$$u(e^x - \pi) = \begin{cases} 0, & x < \ln \pi, \\ 1, & x > \ln \pi \end{cases} = u(x - \ln \pi).$$

(2) $\int_{-\infty}^{x} u(t - y)dt = (x - y)u(x - y)$.

Exercise 1.27 $\delta'(x) \cos x = \delta'(x)$.

Exercise 1.28 $\int_{-2\pi}^{2\pi} e^{\pi x}\delta\left(x^2 - \pi^2\right) dx = \frac{\cosh \pi^2}{\pi}$. $\delta(\sin x) = \delta(x) + \delta(x - \pi)$.

Exercise 1.29 $\int_{-\infty}^{\infty} (\cos x + \sin x)\delta'\left(x^3 + x^2 + x\right) dx = 1$.

Exercise 1.31

(1) $\left\{\left[1 + \frac{1}{n}, 2\right)\right\}_{n=1}^{\infty} \to (1, 2)$. (2) $\left\{\left[1 + \frac{1}{n}, 2\right]\right\}_{n=1}^{\infty} \to (1, 2]$.

(3) $\left\{\left(1, 1 + \frac{1}{n}\right]\right\}_{n=1}^{\infty} \to (1, 1] = \emptyset$. (4) $\left\{\left[1, 1 + \frac{1}{n}\right]\right\}_{n=1}^{\infty} \to [1, 1] = \{1\}$.

(5) $\left\{\left[1 - \frac{1}{n}, 2\right)\right\}_{n=1}^{\infty} \to [1, 2)$. (6) $\left\{\left[1 - \frac{1}{n}, 2\right]\right\}_{n=1}^{\infty} \to [1, 2]$.

(7) $\left\{\left(1, 2 - \frac{1}{n}\right)\right\}_{n=1}^{\infty} \to (1, 2)$. (8) $\left\{\left[1, 2 - \frac{1}{n}\right)\right\}_{n=1}^{\infty} \to [1, 2)$.

Exercise 1.32 $\int_0^1 \lim_{n\to\infty} f_n(x)dx = 0$. $\lim_{n\to\infty} \int_0^1 f_n(x)dx = \frac{1}{2}$.

Exercise 1.33 $\int_0^1 \lim_{n\to\infty} f_n(x)dx = 0$. $\lim_{n\to\infty} \int_0^b f_n(x)dx = \infty$.

Exercise 1.34 The number of all possible arrangements with ten distinct red balls and ten distinct black balls $= 20! \approx 2.43 \times 10^{18}$.

Exercise 1.41 When $p > 0$, $_pC_0 - _pC_1 + _pC_2 - _pC_3 + \cdots = 0$.
When $p > 0$, $_pC_0 + _pC_1 + _pC_2 + _pC_3 + \cdots = 2^p$.
When $p > 0$, $\sum_{k=0}^{\infty} {}_pC_{2k+1} = \sum_{k=0}^{\infty} {}_pC_{2k} = 2^{p-1}$.

Exercise 1.42 $(1+z)^{\frac{1}{2}} = 1 + \frac{z}{2} - \frac{z^2}{8} + \frac{z^3}{16} - \cdots$

$$(1+z)^{-\frac{1}{2}} = \begin{cases} 1 - \frac{1}{2}z + \frac{3}{8}z^2 - \frac{5}{16}z^3 + \frac{35}{128}z^4 - \cdots, & |z| < 1, \\ z^{-\frac{1}{2}} - \frac{1}{2}z^{-\frac{3}{2}} + \frac{3}{8}z^{-\frac{5}{2}} - \frac{5}{16}z^{-\frac{7}{2}} + \frac{35}{128}z^{-\frac{9}{2}} - \cdots, & |z| > 1. \end{cases}$$

Chapter 2 Fundamentals of Probability

Exercise 2.1 $\mathcal{F}(\mathcal{C}) = \{\emptyset, \{a\}, \{b\}, \{a,b\}, \{c,d\}, \{b,c,d\}, \{a,c,d\}, S\}$.

Exercise 2.2 $\sigma(\mathcal{C}) = \{\emptyset, \{a\}, \{b\}, \{a,b\}, \{c,d\}, \{b,c,d\}, \{a,c,d\}, S\}$.

Exercise 2.3 (1) Denoting the lifetime of the battery by t, $S = \{t : 0 \le t < \infty\}$.
(2) $S = \{(n,m) : (0,0), (1,0), (1,1), (2,0), (2,1), (2,2)\}$.
(3) $S = \{(1, \text{red}), (2, \text{red}), (3, \text{green}), (4, \text{green}), (5, \text{blue})\}$.

Exercise 2.4 $P(AB^c + BA^c) = 0$ when $P(A) = P(B) = P(AB)$.

Exercise 2.5 $P(A \cup B) = \frac{1}{2}$. $P(A \cup C) = \frac{2}{3}$. $P(A \cup B \cup C) = 1$.

Exercise 2.6 (1) $C = A^c \cap B$.

Exercise 2.7 The probability that red balls and black balls are placed in an alternating fashion $= \frac{2 \times 10! \times 10!}{20!} \approx 1.08 \times 10^{-5}$.

Exercise 2.8 $P(\text{two nodes are disconnected}) = p^2(2 - p)$.

Exercise 2.12 Buying 50 tickets in one week brings us a higher probability of getting the winning ticket than buying one ticket over 50 weeks.
$\left(\frac{1}{2} \text{ versus } 1 - \left(\frac{99}{100}\right)^{50} \approx 0.395\right)$

Exercise 2.13 $\frac{1}{6}$.

Exercise 2.14 $\frac{1}{2}$.

Exercise 2.15 $P\left(C \cap (A - B)^c = \emptyset\right) = \left(\frac{5}{8}\right)^k$.

Exercise 2.16 $p_{n,A} = \left(-\frac{1}{4}\right)\left(-\frac{1}{3}\right)^{n-1} + \frac{1}{4}$. $p_{n,B} = \frac{1}{12}\left(-\frac{1}{3}\right)^{n-1} + \frac{1}{4}$.
$p_{10,A} = \frac{1}{4}\left\{1 - \left(-\frac{1}{3}\right)^9\right\} \approx 0.250013$. $p_{10,B} = \frac{1}{4}\left\{1 - \left(-\frac{1}{3}\right)^{10}\right\} \approx 0.249996$.

Exercise 2.17 (1), (2) probability of no match
$$= \begin{cases} 0, & N = 1, \\ \frac{1}{2!} - \frac{1}{3!} + \cdots + \frac{(-1)^N}{N!}, & N = 2, 3, \ldots. \end{cases}$$
(3) probability of k matches
$$= \begin{cases} \frac{1}{k!}\left\{\frac{1}{2!} - \frac{1}{3!} + \cdots + \frac{(-1)^{N-k}}{(N-k)!}\right\}, & k = 0, 1, \ldots, N-2, \\ 0, & k = N-1, \\ \frac{1}{k!}, & k = N. \end{cases}$$

Exercise 2.18 $\frac{3}{4}$.

Exercise 2.19 (1) $\alpha = p^2$. $P((k, m) : k \geq m) = \frac{1}{2-p}$.

(2) $P((k, m) : k + m = r) = p^2(1 - p)^{r-2}(r - 1)$.

(3) $P((k, m) : k$ is an odd number$) = \frac{1}{2-p}$.

Exercise 2.20 $P(A \cap B) = \frac{3}{10}$. $P(A|B) = \frac{3}{5}$. $P(B|A) = \frac{3}{7}$.

Exercise 2.21 Probability that only two will hit the target $= \frac{398}{1000}$.

Exercise 2.22 (4) $P(B^c | A) = 1 - p$ or $\frac{1-s-q+qr}{r}$. $P(B|A^c) = 1 - q$ or $\frac{s-rp}{1-r}$.

$P(A^c | B) = \frac{(1-q)(1-r)}{s}$ or $\frac{s-rp}{s}$. $P(A|B^c) = \frac{(1-p)r}{1-s}$ or $\frac{1-s-q+qr}{1-s}$.

(5) $S = \{$A defective element is identified to be defective, A defective element is identified to be functional, A functional element is identified to be defective, A functional element is identified to be functional$\}$.

Exercise 2.25 (1) A_i and A_j are independent if $P(A_i) = 0$ or $P(A_j) = 0$. A_i and A_j are not independent if $P(A_i) \neq 0$ and $P(A_j) \neq 0$.

(2) partition of $B = \{BA_1, BA_2, \ldots, BA_n\}$.

Exercise 2.26 $P(\text{red ball}) = \frac{2}{5} \times \frac{1}{2} + \frac{1}{5} \times \frac{1}{2} = \frac{3}{10}$.

Exercise 2.27 $P(\text{red ball}) = \frac{2n+1}{3n}$.

Exercise 2.28 $p_{n,k} = P(\text{ends in } k \text{ trials}) = \left(1 - \frac{n}{3^{n-1}}\right)^{k-1} \frac{n}{3^{n-1}}$.

Exercise 2.29 Probability of Candidate A leading always $= \frac{n-m}{n+m}$.

Exercise 2.30 $\beta_0 = \left(\frac{1}{4}\right)^{2^n-1}$.

Exercise 2.31 $P(A) = {}_nC_k \frac{2^{n-k}}{3^n}$.

Exercise 2.32 (1) $p_{10} = 1 - p_{11}$. $p_{01} = 1 - p_{00}$.

(2) $P(\text{error}) = (1 - p_{00})(1 - p) + (1 - p_{11})p$.

(3) $P(Y = 1) = pp_{11} + (1 - p)(1 - p_{00})$. $P(Y = 0) = p(1 - p_{11}) + (1 - p)p_{00}$.

(4) $P(X = j | Y = k) = \begin{cases} \frac{pp_{11}}{pp_{11}+(1-p)(1-p_{00})}, & j = 1, k = 1, \\ \frac{p(1-p_{11})}{p(1-p_{11})+(1-p)p_{00}}, & j = 1, k = 0, \\ \frac{(1-p)(1-p_{00})}{pp_{11}+(1-p)(1-p_{00})}, & j = 0, k = 1, \\ \frac{(1-p)p_{00}}{p(1-p_{11})+(1-p)p_{00}}, & j = 0, k = 0. \end{cases}$

(5) $P(Y = 1) = P(Y = 0) = \frac{1}{2}$.

$P(X = 1 | Y = 0) = P(X = 0 | Y = 1) = 1 - p_{11}$.

$P(X = 1 | Y = 1) = P(X = 0 | Y = 0) = p_{11}$.

Exercise 2.33 (1) $\alpha_{1,1} = \frac{m(m-1)}{n(n-1)}$. $\alpha_{0,1} = \frac{m(n-m)}{n(n-1)}$. $\alpha_{1,0} = \frac{m(n-m)}{n(n-1)}$. $\alpha_{0,0} = \frac{(n-m)(n-m-1)}{n(n-1)}$.

(2) $\tilde{\alpha}_{1,1} = \frac{m^2}{n^2}$. $\tilde{\alpha}_{0,1} = \frac{m(n-m)}{n^2}$. $\tilde{\alpha}_{1,0} = \frac{m(n-m)}{n^2}$. $\tilde{\alpha}_{0,0} = \frac{(n-m)^2}{n^2}$.

(3) $\beta_0 = \frac{(n-m)(n-m-1)}{n(n-1)}$. $\beta_1 = \frac{2m(n-m)}{n(n-1)}$. $\beta_2 = \frac{m(m-1)}{n(n-1)}$.

Exercise 2.34 (1) $P_2 = \frac{15}{280}$. $P_0 = \frac{143}{280}$. $P_1 = \frac{122}{280}$. (2) $P_3 = \frac{8}{15}$.

Exercise 2.35 (1) $P(r \text{ brown eye children}) = \frac{(1-p)(pb)^r}{(1-p+pb)^{r+1}}$.

(2) $P(r \text{ boys}) = \frac{2(1-p)p^r}{(2-p)^{r+1}}$.

(3) $P(\text{at least two boys} | \text{at least one boy}) = \frac{p}{2-p}$.

Exercise 2.37 $p_k = \frac{\left(\frac{q}{p}\right)^k - \left(\frac{q}{p}\right)^N}{1 - \left(\frac{q}{p}\right)^N}$.

Exercise 2.38 (1) $p_1 = \frac{3}{10}$. (2) $p_2 = \frac{7}{10}$. (3) P(15 red, 10 white|red flower) $= \frac{1}{2}$.

Exercise 2.39 $c = \frac{1}{5}$.

Exercise 2.41 (2) $3c_{11} = 2c_{12} + c_{13}$.

Exercise 2.42 (1) B_1 and R are independent of each other. B_1 and G are independent of each other.

(2) B_2 and R are not independent of each other. B_3 and G are not independent of each other.

Exercise 2.43 A_1, A_2, and A_3 are not mutually independent.

Exercise 2.44 A and C are not independent of each other.

Exercise 2.45 Probability of meeting $= \frac{60 \times 60 - 50 \times \frac{50}{2} \times 2}{60 \times 60} = \frac{11}{36} \approx 0.3056$.

Exercise 2.46 $p_1 = \frac{1}{2}$. $p_2 = \frac{1}{3}$.

Exercise 2.47 P (red ball| given condition) $= \frac{23}{40} = 0.575$.

Exercise 2.48 (2) P(one person wins eight times) $= {}_{10}C_8 \left(\frac{1}{4}\right)^8 \left(1 - \frac{1}{4}\right)^{10-8} = \frac{405}{4^{10}}$ $\approx 3.8624 \times 10^{-4}$.

P(one person wins at least eight times) $= \sum_{k=8}^{10} {}_{10}C_k \left(\frac{1}{4}\right)^k \left(1 - \frac{1}{4}\right)^{10-k} \approx 4.1580 \times 10^{-4}$.

Exercise 2.50 Probability that a person playing piano is a man $= \frac{1}{2}$.

Exercise 2.51 $\binom{50}{30,15,5} 0.5^{30} 0.3^{15} 0.2^5 = \frac{50!}{30!15!5!} \frac{1}{2^{30}} \frac{3^{15}}{10^{15}} \frac{1}{5^5} \approx 3.125 \times 10^{-4}$.

Chapter 3 Random Variables

Exercise 3.2 $F_{g(X)}(y) = \begin{cases} F_X(y+c), & y \geq 0, \\ F_X(y-c), & y < 0. \end{cases}$

Exercise 3.3 $F_Y(y) = \begin{cases} F_X(y-c), & y \geq c, \\ F_X(0) - P(X=0), & -c \leq y < c, \\ F_X(y+c), & y < -c. \end{cases}$

Exercise 3.4 Denoting the solutions to $y = a\sin(x + \theta)$ by $\{x_i\}_{i=1}^{\infty}$,

$f_Y(y) = \frac{1}{\sqrt{a^2 - y^2}} \sum_{i=1}^{\infty} f_X(x_i)$.

Exercise 3.5 For $G \neq 0$ and $B \neq 0$,

$p_X(k) = \begin{cases} {}_nC_k \left(\frac{B}{G+B}\right)^{n-k} \left(\frac{G}{G+B}\right)^k, & k = 0, 1, \ldots, n, \\ 0, & \text{otherwise.} \end{cases}$

For $G = 0$, $p_X(k) = \begin{cases} 1, & k = 0, \\ 0, & \text{otherwise.} \end{cases}$

For $B = 0$, $p_X(k) = \begin{cases} 1, & k = n, \\ 0, & \text{otherwise.} \end{cases}$

Exercise 3.6 pdf: $f_Y(y) = \begin{cases} \frac{1}{3\sqrt{y-1}}, & 1 < y \leq 2; \quad \frac{1}{6\sqrt{y-1}}, & 2 < y \leq 5; \\ 0, & \text{otherwise.} \end{cases}$

(In the pdf, the set $\{1 < y \le 2,\ 2 < y \le 5\}$ can be replaced with $\{1 < y \le 2,\ 2 < y < 5\}$, $\{1 < y < 2,\ 2 \le y \le 5\}$, $\{1 < y < 2,\ 2 \le y < 5\}$, $\{1 < y < 2,\ 2 < y \le 5\}$, or $\{1 < y < 2,\ 2 < y < 5\}$.)

cdf: $F_Y(y) = \begin{cases} 0, & y \le 1; & \frac{2}{3}\sqrt{y-1}, & 1 \le y \le 2; \\ \frac{1}{3} + \frac{1}{3}\sqrt{y-1}, & 2 \le y \le 5; & 1, & y \ge 5. \end{cases}$

Exercise 3.7 cdf $F_Y(y) = \begin{cases} 0, & y < -18; & \frac{1}{7}, & -18 \le y < -2; \\ \frac{3}{7}, & -2 \le y < 0; & \frac{4}{7}, & 0 \le y < 2; \\ \frac{6}{7}, & 2 \le y < 18; & 1, & y \ge 18. \end{cases}$

Exercise 3.8 $\text{E}\{X^{-1}\} = \frac{1}{n-2}, n = 3, 4, \ldots$.

Exercise 3.9 $f_Y(y) = F_X(y)\delta(y) + f_X(y)u(y) = F_X(0)\delta(y) + f_X(y)u(y)$, where F_X is the cdf of X.

Exercise 3.10 $f_Y(y) = \begin{cases} 0, & y \le 0, \\ \frac{\theta}{e\sqrt{y}} \cosh\left(\theta\sqrt{y}\right), & 0 < y \le \frac{1}{\theta^2}, \\ \frac{\theta}{2e\sqrt{y}} \exp\left(-\theta\sqrt{y}\right), & y > \frac{1}{\theta^2}. \end{cases}$

Exercise 3.11 $F_{X|b<X\le a}(x) = \frac{F_X(x)-F_X(b)}{F_X(a)-F_X(b)}$.

$f_{X|b<X\le a}(x) = \frac{f_X(x)}{F_X(a)-F_X(b)}u(x-b)u(a-x)$

Exercise 3.12 $\text{E}\{X|X > a\} = \frac{1+a}{2}$. $\text{Var}\{X|X > a\} = \frac{(1-a)^2}{12}$.

Exercise 3.13 $\text{P}(950 \le R \le 1050) = \frac{1}{2}$.

Exercise 3.14 Let X be the time to take to the location of the appointment with cdf F_X. Then, departing t^* minutes before the appointment time will incur the minimum cost, where $F_X(t^*) = \frac{k}{k+c}$.

Exercise 3.15 $\text{P}(X \le \alpha) = \frac{1}{3}u(\alpha) + \frac{2}{3}u(\alpha - \pi)$. $\text{P}(2 \le X < 4) = \frac{2}{3}$. $\text{P}(X \le 0) = \frac{1}{3}$.

Exercise 3.16 $\text{P}(U > 0) = \frac{1}{2}$. $\text{P}\left(|U| < \frac{1}{3}\right) = \frac{1}{3}$. $\text{P}\left(|U| \ge \frac{3}{4}\right) = \frac{1}{4}$. $\text{P}\left(\frac{1}{3} < U < \frac{1}{2}\right) = \frac{1}{12}$.

Exercise 3.17 $\text{P}(A) = \text{P}(A \cup B) = \frac{1}{32}$. $\text{P}(B) = \text{P}(A \cap B) = \frac{1}{1024}$. $\text{P}(B^c) = \frac{1023}{1024}$.

Exercise 3.18 (1) When $L_1 = \max(0, w_A + w_B - N)$ and $U_1 = \min(w_A, w_B)$, for $d = L_1, L_1 + 1, \ldots, U_1$, $\text{P}(D = d) = \frac{\binom{N}{d}\binom{N-d}{w_A-d}\binom{N-w_A}{w_B-d}}{\binom{N}{w_A}\binom{N}{w_B}}$.

(2) When $L_2 = \max(0, k - w_B)$, $U_2 = \min(w_A - d, k - d)$, $L_3 = L_1$ and $U_3 = \min(w_A + w_B, N)$, for $k = L_3, L_3 + 1, \ldots, U_3$

$\text{P}(K = k) = \frac{1}{\binom{N}{w_A}\binom{N}{w_B}} \sum_{d=L_1}^{U_1} \sum_{i=L_2}^{U_2} \binom{N}{d}\binom{N-d}{w_A-d}\binom{N-w_A}{w_B-d}\binom{w_A-d}{i}$
$\times \binom{w_B-d}{k-d-i} p^{w_B-k+2i} (1-p)^{k+w_A-2d-2i}$.

(3) $\text{P}(D = d) = \begin{cases} \frac{2}{3}, & d = 0, \\ \frac{1}{3}, & d = 1. \end{cases}$ $\text{P}(K = k) = \begin{cases} \frac{1}{6}, & k = 0, \\ \frac{2}{3}, & k = 1, \\ \frac{1}{6}, & k = 2. \end{cases}$

Exercise 3.19 (1) $c = 2$. (2) $\text{E}\{X\} = 2$.

Exercise 3.20 (1) $\text{P}(0 < X < 1) = \frac{1}{4}$. $\text{P}(1 \le X < 1.5) = \frac{3}{8}$.
(2) $\mu = \frac{19}{24}$. $\sigma^2 = \frac{191}{576}$.

Exercise 3.21 (1) $F_W(w) = \begin{cases} 0, & w < a; \quad \frac{w-a}{b-a}, \ a \le w < b; \\ 1, & w \ge b. \end{cases}$

(2) $W \sim U[a, b)$.

Exercise 3.22 $f_Y(y) = \frac{1}{(1-y)^2}\left\{u(y) - u\left(y - \frac{1}{2}\right)\right\}$.

Exercise 3.23 $f_Y(y) = \frac{1}{(1+y)^2} f_X\left(\frac{y}{1+y}\right)$.

When $X \sim U[0, 1)$, $f_Y(y) = \frac{1}{(1+y)^2} u(y)$.

Exercise 3.25 $f_Z(z) = u(z) - u(z - 1)$.

Exercise 3.26 $f_Y(t) = \frac{2}{(t+1)^2} u(t - 1)$. $f_Z(s) = \frac{2}{(1-s)^2}\{u(s + 1) - u(s)\}$.

Exercise 3.27 $f_Y(y) = (2 - 2y)u(y)u(1 - y)$.

Exercise 3.28 (1) $f_Y(y) = \frac{1}{\pi\sqrt{a^2 - y^2}} u\left(a - |y|\right)$.

(2) $f_Y(y) = \frac{2}{\pi\sqrt{1-y^2}} u(y)u(1 - y)$.

(3) With $0 \le \cos^{-1} y \le \pi$, $F_Y(y) = \begin{cases} 0, & y \le -1, \\ \frac{4}{3} - \frac{4}{3\pi}\cos^{-1} y, & -1 \le y \le 0, \\ 1 - \frac{2}{3\pi}\cos^{-1} y, & 0 \le y \le 1, \\ 1, & y \ge 1. \end{cases}$

Exercise 3.29 (1) $f_Y(y) = \frac{1}{1+y^2} \sum\limits_{i=1}^{\infty} f_X(x_i)$. (2) $f_Y(y) = \frac{1}{\pi(1+y^2)}$.

(3) $f_Y(y) = \frac{1}{\pi(1+y^2)}$.

Exercise 3.30 When $X \sim U[0, 1)$,
$Y = -\frac{1}{\lambda}\ln(1 - X) \sim F_Y(y) = \left(1 - e^{-\lambda y}\right)u(y)$.

Exercise 3.31 expected value: $E\{X\} = 3.5$. mode: $1, 2, \ldots$, or 6.
median: any real number in the interval $[3, 4]$.

Exercise 3.32 $c = 5 < \frac{101}{7} = b$.

Exercise 3.33 $E\{X\} = 0$. $\text{Var}\{X\} = \frac{2}{\lambda^2}$.

Exercise 3.34 $E\{X\} = \frac{\alpha}{\alpha+\beta}$. $\text{Var}\{X\} = \frac{\alpha\beta}{(\alpha+\beta)^2(\alpha+\beta+1)}$.

Exercise 3.36 $f_Y(y) = u(y + 2) - u(y + 1)$. $f_Z(z) = \frac{1}{2}\{u(z + 4) - u(z + 2)\}$.
$f_W(w) = \frac{1}{2}\{u(w + 3) - u(w + 1)\}$.

Exercise 3.37 $p_Y(k) = \frac{1}{4}$, $k = 3, 4, 5, 6$. $p_Z(k) = \frac{1}{4}$, $k = -1, 0, 1, 2$.
$p_W(r) = \frac{1}{4}$, $r = \pm\frac{1}{3}, 0, \frac{1}{5}$.

Exercise 3.39 (2) $E\left\{X_c^-\right\} = \begin{cases} c - \int_0^c F_X(x)dx, & c \ge 0, \\ c, & c < 0. \end{cases}$

(3) $E\left\{X_c^+\right\} = \begin{cases} \int_0^{\infty}\{1 - F_X(x)\}dx + \int_0^c F_X(x)dx, & c \ge 0, \\ \int_0^{\infty}\{1 - F_X(x)\}dx, & c < 0. \end{cases}$

Exercise 3.40 (1) $E\{X\} = \lambda\mu$. $\text{Var}\{X\} = (1 + \lambda)\lambda\mu$.

Exercise 3.41 $f_X(x) = 4\pi\rho x^2 \exp\left(-\frac{4}{3}\pi\rho x^3\right)u(x)$.

Exercise 3.42 $A = \frac{1}{16}$. $P(X \le 6) = \frac{7}{8}$.

Exercise 3.44 $E\{F(X)\} = \frac{1}{2}$.

Exercise 3.45 $M(t) = \frac{1}{t+1}$. $\varphi(\omega) = \frac{1}{1+j\omega}$. $m_1 = -1$. $m_2 = 2$. $m_3 = -6$. $m_4 = 24$.

Exercise 3.46 mgf $M(t) = \frac{2}{\pi}\int_0^{\frac{\pi}{2}} (\tan x)^{\frac{t}{\pi}} dx$.

Exercise 3.47 $\alpha = \frac{|\beta|}{2^{n-1} B\left(\frac{n}{2}, \frac{n}{2}\right)}$.

Exercise 3.48 $M_R(t) = 1 + \sqrt{2\pi}\sigma t \exp\left(\frac{\sigma^2 t^2}{2}\right) \Phi(\sigma t)$, where Φ is the standard normal cdf.

Exercise 3.51 $f_Y(y) = \begin{cases} 1 - \frac{y}{4}, & 0 \le y < 1; \quad \frac{1}{2} - \frac{y}{4}, \; 1 \le y < 2; \\ 0, & \text{otherwise.} \end{cases}$

Exercise 3.52 A cdf such that
$$\left\{ (\text{locaton of jump, height of jump}) \doteq \left(a + (b-a) \sum_{i=1}^{n} \left(\tfrac{1}{2}\right)^i, \left(\tfrac{1}{2}\right)^{n+1} \right) \right\}_{n=0}^{\infty}$$
and the interval between adjacent jumps are all the same.

Exercise 3.53 (1) $\left\{ a \ge 0, \; a + \frac{1}{3} \le b \le -3a + 1 \right\}$.

(2) $F_Y(x) = \begin{cases} 0, & x < 0; \quad \frac{1}{4}\sqrt{x}, & 0 \le x < 1; \\ \frac{1}{24}\left(11\sqrt{x} - 1\right), \; 1 \le x < 4; & \frac{1}{8}\left(\sqrt{x}+5\right), \; 4 \le x < 9; \\ 1, & x \ge 9. \end{cases}$

$P(Y = 1) = \frac{1}{6}$. $P(Y = 4) = 0$.

Exercise 3.54 $F_Y(x) = \begin{cases} 0, & x < 0; \quad \frac{5}{8}\sqrt{x}, \; 0 \le x < 1; \\ \frac{1}{8}\left(\sqrt{x}+4\right), \; 1 \le x < 16; & 1, \quad x \ge 16. \end{cases}$

Exercise 3.57 $F_Y(y) = \begin{cases} F_X(\alpha), & y < -2 \text{ or } y > 2, \\ F_X(-2) + p_X(1), & -2 \le y < 0, \\ F_X(-2) + p_X(0) + p_X(1), & 0 \le y < 2, \\ F_X(2), & y = 2, \end{cases}$

where α is the only real root of $y = x^3 - 3x$ when $y > 2$ or $y < -2$.

Exercise 3.58 $F_Y(x) = 0$ for $x < 0$, x for $0 \le x < 1$, and 1 for $x \ge 1$.

Exercise 3.59 (1) For $\alpha(\theta) = \frac{1}{2}$, $\varphi(\omega) = \exp\left(-\frac{\omega^2}{4}\right)$.

For $\alpha(\theta) = \cos^2\theta$, $\varphi(\omega) = \exp\left(-\frac{\omega^2}{4}\right) I_0\left(\frac{\omega^2}{4}\right)$.

(2) The normal pdf with mean 0 and variance $\frac{1}{2}$.

(4) $E\{X_1\} = 0$. $\text{Var}\{X_1\} = \frac{1}{2}$. $E\{X_2\} = 0$. $\text{Var}\{X_2\} = \frac{1}{2}$.

Exercise 3.63 $E\{X\} = \frac{\pi}{2}$. $E\left\{X^2\right\} = \frac{\pi^2}{2} - 2$.

Exercise 3.65 $\text{Var}\{Y\} = \sigma_X^2 + 4m_X^+ m_X^-$.

Exercise 3.67 pmf $p(k) = (1-\alpha)^k \alpha$ for $k \in \{0, 1, \ldots\}$: $E\{X\} = \frac{1-\alpha}{\alpha}$. $\sigma_X^2 = \frac{1-\alpha}{\alpha^2}$.

pmf $p(k) = (1-\alpha)^{k-1}\alpha$ for $k \in \{1, 2, \ldots\}$: $E\{X\} = \frac{1}{\alpha}$. $\sigma_X^2 = \frac{1-\alpha}{\alpha^2}$.

Exercise 3.68 $E\{X\} = \frac{\beta\gamma}{\alpha+\beta}$. $\text{Var}\{X\} = \frac{\alpha\beta\gamma(\alpha+\beta-\gamma)}{(\alpha+\beta)^2(\alpha+\beta-1)}$.

Exercise 3.70 For $t < \lambda$, $M_Y(t) = \frac{\lambda^n}{(\lambda-t)^n}$. $E\{Y\} = \frac{n}{\lambda}$. $\text{Var}\{Y\} = \frac{n}{\lambda^2}$.

Exercise 3.71 $F_Y(y) = \begin{cases} 1 - \left(2p^2 - 2p + 1\right)^{\lfloor y \rfloor}, & y \ge 1, \\ 0, & y < 1. \end{cases}$ $E\{Y\} = \frac{1}{2p(1-p)}$.

Exercise 3.74 In $b\left(10, \frac{1}{3}\right)$, $P_{10}(k)$ is the largest at $k = \left[\frac{11}{3}\right] = 3$.
In $b\left(11, \frac{1}{2}\right)$, $P_{11}(k)$ is the largest at $k = 5, 6$.

Exercise 3.75 (1) $P_{01} = (0.995)^{1000} + {}_{1000}C_1(0.005)^1(0.995)^{999} \approx 0.0401$.

approximate value with (3.5.17): $\Phi(2.2417) - \Phi(1.7933) \approx 0.0242$.

approximate value with (3.5.18): $\Phi(2.4658) - \Phi(1.5692) \approx 0.0515$.

approximate value with (3.5.19): $\left(\frac{5^0}{0!} + \frac{5^1}{1!}\right) e^{-5} \approx 0.0404$.

(2) $P_{456} = {}_{1000}C_4(0.005)^4(0.995)^{996} + {}_{1000}C_5(0.005)^5$

$\times(0.995)^{995} + {}_{1000}C_6(0.005)^6(0.995)^{994} \approx 0.4982.$

approximate value with (3.5.17): $2\{\Phi(0.4483) - 0.5\} \approx 0.3472.$

approximate value with (3.5.18): $2\{\Phi(0.6725) - 0.5\} \approx 0.4988.$

approximate value with (3.5.19): $\left(\frac{5^4}{4!} + \frac{5^5}{5!} + \frac{5^6}{6!}\right)e^{-5} \approx 0.4972.$

Exercise 3.77 (2) coefficient of variation $= \frac{1}{\sqrt{\lambda}}.$

(3) skewness $= \frac{1}{\sqrt{\lambda}}.$ kurtosis $= 3 + \frac{1}{\lambda}.$

Exercise 3.81 $f_Y(y) = \frac{y}{2}\left\{2u(y)u(1-y) + u(y-1)u\left(\sqrt{3}-y\right)\right\}.$

Chapter 4 Random Vectors

Exercise 4.2 $p_X(1) = \frac{3}{5}, p_X(2) = \frac{2}{5}. \qquad p_Y(4) = \frac{3}{5}, p_Y(3) = \frac{2}{5}.$

$p_{X,Y}(1, 4) = \frac{3}{10}, p_{X,Y}(1, 3) = \frac{3}{10}, p_{X,Y}(2, 4) = \frac{3}{10}, p_{X,Y}(2, 3) = \frac{1}{10}.$

$p_{Y|X}(4|1) = \frac{1}{2}, p_{Y|X}(3|1) = \frac{1}{2}, p_{Y|X}(4|2) = \frac{3}{4}, p_{Y|X}(3|2) = \frac{1}{4}.$

$p_{X|Y}(1|4) = \frac{1}{2}, p_{X|Y}(2|4) = \frac{1}{2}, p_{X|Y}(1|3) = \frac{3}{4}, p_{X|Y}(2|3) = \frac{1}{4}.$

$p_{X+Y}(4) = \frac{3}{10}, p_{X+Y}(5) = \frac{2}{5}, p_{X+Y}(6) = \frac{3}{10}.$

Exercise 4.3 pairwise independent. not mutually independent.

Exercise 4.4 $a = \frac{1}{4}.$ X and Y are not independent of each other. $\rho_{XY} = 0.$

Exercise 4.5 $p_{R|B=3}(0) = \frac{8}{27}, p_{R|B=3}(1) = \frac{4}{9}, p_{R|B=3}(2) = \frac{2}{9},$

$p_{R|B=3}(3) = \frac{1}{27}.$ $E\{R|B = 1\} = \frac{5}{3}.$

Exercise 4.8

$$f_Y(y_1, y_2) = \begin{cases} \frac{1}{2\sqrt{y_1}}, & 0 < y_1 < \frac{1}{4}, -\sqrt{y_1} + \frac{1}{2} < y_2 < \sqrt{y_1} + \frac{1}{2}, \\ \frac{1}{\sqrt{y_1}}, & 0 < y_1 < \frac{1}{4}, \sqrt{y_1} - \frac{1}{2} < y_2 < -\sqrt{y_1} + \frac{1}{2}, \\ \frac{1}{2\sqrt{y_1}}, & 0 < y_1 < \frac{1}{4}, -\sqrt{y_1} - \frac{1}{2} < y_2 < \sqrt{y_1} - \frac{1}{2}, \\ 0, & \text{otherwise.} \end{cases}$$

$f_{Y_1}(y) = \frac{1}{\sqrt{y}}u(y)u\left(\frac{1}{4} - y\right).$ $f_{Y_2}(y) = (1 - |y|)u(1 - |y|).$

Exercise 4.9 $F_W(w) = \begin{cases} 0, & w < 0, \\ \frac{\pi}{4}w^2, & 0 \le w < 1, \\ \left(\frac{\pi}{4} - \sin^{-1}\frac{\sqrt{w^2-1}}{w}\right)w^2 + \sqrt{w^2-1}, & 1 \le w < \sqrt{2}, \\ 1, & w \ge \sqrt{2}. \end{cases}$

$f_W(w) = \begin{cases} \frac{\pi}{2}w, & 0 \le w < 1, \\ 2\left(\frac{\pi}{4} - \sin^{-1}\frac{\sqrt{w^2-1}}{w}\right)w, & 1 \le w < \sqrt{2}, \\ 0, & \text{otherwise.} \end{cases}$

Exercise 4.10 $F_{W,V}(w, v) = \begin{cases} 1 - e^{-(\mu+\lambda)w}, & \text{if } w \ge 0, \ v \ge 1, \\ \frac{\mu}{\mu+\lambda}\left\{1 - e^{-(\mu+\lambda)w}\right\}, & \text{if } w \ge 0, \ 0 \le v < 1, \\ 0, & \text{otherwise.} \end{cases}$

Exercise 4.11 $f_U(v) = \frac{3v^2}{(1+v)^4}u(v).$

Exercise 4.12 (1) $f_Y(y_1, y_2) = \frac{u(y_1)}{2\sqrt{y_1}}u(1 - y_1)u(y_2 - \sqrt{y_1})u(1 - y_2 + \sqrt{y_1}).$

$f_{Y_1}(y) = \frac{1}{2\sqrt{y}}u(y)u(1-y)$. $f_{Y_2}(y) = \begin{cases} y, & 0 < y \le 1, \\ 2-y, & 1 < y \le 2, \\ 0, & \text{otherwise}. \end{cases}$

(2) $f_Y(y_1, y_2) = \frac{1}{\sqrt{y_1}}u(y_1)u(1 - y_2 + \sqrt{y_1})u(y_2 - 2\sqrt{y_1})$.

$f_{Y_1}(y_1) = \begin{cases} \frac{1}{\sqrt{y_1}} - 1, & 0 < y_1 \le 1, \\ 0, & \text{otherwise}. \end{cases}$ $\quad f_{Y_2}(y_2) = \begin{cases} y_2, & 0 < y_2 \le 1, \\ 2 - y_2, & 1 < y_2 \le 2, \\ 0, & \text{otherwise}. \end{cases}$

Exercise 4.13 (1) $f_Y(y_1, y_2) = \frac{1}{\sqrt{y_1+y_2}}u(y_1+y_2)u(1-y_1-y_2)u(y_1-y_2)u(1-y_1+y_2)$.

$f_{Y_1}(y_1) = \begin{cases} 2\sqrt{2y_1}, & 0 < y_1 \le \frac{1}{2}; \quad 2\left(1 - \sqrt{2y_1 - 1}\right), \frac{1}{2} < y_1 \le 1; \\ 0, & \text{otherwise}. \end{cases}$

$f_{Y_2}(y_2) = \begin{cases} 2\sqrt{2y_2 + 1}, & -\frac{1}{2} < y_2 \le 0; \quad 2\left(1 - \sqrt{2y_2}\right), 0 < y_2 \le \frac{1}{2}; \\ 0, & \text{otherwise}. \end{cases}$

(2) $f_Y(y_1, y_2) = \frac{2}{\sqrt{y_1+y_2}}u(y_1+y_2)u(1-y_1+y_2)u(y_1-y_2-\sqrt{y_1+y_2})$.

$f_{Y_1}(y_1) = \begin{cases} 2\left(\sqrt{8y_1 + 1} - 1\right), & 0 < y_1 \le \frac{1}{2}, \\ 2\left(\sqrt{8y_1 + 1} - 1\right) - 4\sqrt{2y_1 - 1}, & \frac{1}{2} < y_1 \le 1, \\ 0, & \text{otherwise}. \end{cases}$

$f_{Y_2}(y_2) = \begin{cases} 4\sqrt{2y_2 + 1}, & -\frac{1}{2} < y_2 \le -\frac{1}{8}, \\ 4\left(\sqrt{2y_2 + 1} - \sqrt{8y_2 + 1}\right), & -\frac{1}{8} < y_2 \le 0, \\ 0, & \text{otherwise}. \end{cases}$

Exercise 4.14 $f_Z(z) = \frac{z^{\alpha_1+\alpha_2-1}}{\beta^{\alpha_1+\alpha_2}\Gamma(\alpha_1+\alpha_2)}\exp\left(-\frac{z}{\beta}\right)u(z)$.

$f_W(w) = \frac{\Gamma(\alpha_1+\alpha_2)}{\Gamma(\alpha_1)\Gamma(\alpha_2)}\frac{w^{\alpha_1-1}}{(1+w)^{\alpha_1+\alpha_2}}u(w)$.

Exercise 4.15 (1) $f_{Y_1}(y_1) = \frac{1}{2r}u(y_1)\int_{-y_1^{\frac{1}{2r}}}^{y_1^{\frac{1}{2r}}} y_1^{-\frac{r-1}{r}}\left(y_1^{\frac{1}{r}} - y_2^2\right)^{-\frac{1}{2}}$

$\left\{f_X\left(\sqrt{y_1^{\frac{1}{r}} - y_2^2}, y_2\right) + f_X\left(-\sqrt{y_1^{\frac{1}{r}} - y_2^2}, y_2\right)\right\} dy_2$.

(3) For $r = \frac{1}{2}$, $F_W(w) = \begin{cases} 1, & w \ge 1, \\ w^2, & w \in [0, 1], \\ 0, & \text{otherwise}. \end{cases}$ $f_W(w) = \begin{cases} 2w, & w \in [0, 1], \\ 0, & \text{otherwise}. \end{cases}$

For $r = 1$, $F_W(w) = \begin{cases} 1, & w \ge 1, \\ w, & w \in [0, 1], \\ 0, & \text{otherwise}. \end{cases}$ $f_W(w) = \begin{cases} 1, & w \in [0, 1], \\ 0, & \text{otherwise}. \end{cases}$

For $r = -1$, $F_W(w) = \begin{cases} 0, & w < 1, \\ 1 - w^{-1}, & w \ge 1. \end{cases}$ $f_W(w) = \begin{cases} 0, & w < 1, \\ w^{-2}, & w > 1. \end{cases}$

Exercise 4.16 (1) $f_{Y_1, Y_2}(y_1, y_2)$

$= \begin{cases} \frac{1}{2}(y_1 - |y_2|), & (y_1, y_2) \in (1:3) \cup (2:3), \\ 1 - |y_2|, & (y_1, y_2) \in (1:2) \cup (2:1), \\ \frac{1}{2}(3 - y_1 - |y_2|), & (y_1, y_2) \in (3:2) \cup (3:1), \\ \frac{1}{2}, & (y_1, y_2) \in (3:3), \\ 0, & \text{otherwise}. \end{cases}$

(refer to Fig. A.1).

(2) $f_{Y_2}(y) = (1 - |y|)u(1 - |y|)$.

Fig. A.1 The regions for $f_{Y_1,Y_2}(y_1, y_2)$ in Exercise 4.16

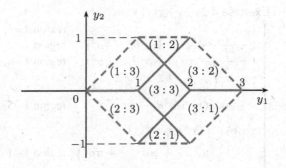

(3) $f_{Y_1}(y) = \begin{cases} \frac{1}{2}y^2, & 0 \le y \le 1, \\ -y^2 + 3y - \frac{3}{2}, & 1 \le y \le 2, \\ \frac{1}{2}(3-y)^2, & 2 \le y \le 3, \\ 0, & y \le 1, y \ge 3. \end{cases}$

Exercise 4.17 (1) $p_{X+Y}(v) = (v-1)(1-\alpha)^2 \alpha^{v-2} \tilde{u}(v-2)$.
$p_{X-Y}(w) = \frac{1-\alpha}{1+\alpha} \alpha^{|w|} \tilde{u}(|w|)$.
(2) $p_{X-Y,X}(w, x) = (1-\alpha)^2 \alpha^{2x-w-2} \tilde{u}(x-1) \tilde{u}(x-w-1)$.
$p_{X-Y,Y}(w, y) = (1-\alpha)^2 \alpha^{w+2y-2} \tilde{u}(w+y-1) \tilde{u}(y-1)$.
(3) $p_{X-Y}(w) = \frac{1-\alpha}{1+\alpha} \alpha^{|w|} \tilde{u}(|w|)$. $p_X(x) = (1-\alpha)\alpha^{x-1}\tilde{u}(x-1)$.
$p_Y(y) = (1-\alpha)\alpha^{y-1}\tilde{u}(y-1)$.
(4) $p_{X+Y,X-Y}(v, w) = (1-\alpha)^2 \alpha^{v-2} \tilde{u}\left(\frac{v+w}{2}-1\right) \tilde{u}\left(\frac{v-w}{2}-1\right)$.
$p_{X+Y}(v) = (v-1)(1-\alpha)^2 \alpha^{v-2} \tilde{u}(v-2)$. $p_{X-Y}(w) = \frac{1-\alpha}{1+\alpha} \alpha^{|w|} \tilde{u}(|w|)$.
the same as the results in (1).

Exercise 4.18 $E\{R_n\} = \left(\frac{7}{6}\right)^n$. $p_2(k) = \begin{cases} \frac{1}{64}, & k=4; & \frac{1}{12}, k=3; \\ \frac{83}{288}, & k=2; & \frac{17}{36}, k=1; \\ \frac{9}{64}, & k=0. \end{cases}$

$\eta_0 = \frac{1}{3}$.

Exercise 4.19 $E\{X|Y = y\} = 2 + y$ for $y \ge 0$.

Exercise 4.20 $f_Y(y) = y_1^2 y_2 e^{-y_1} u(y_1) u(y_2) u(1-y_2) u(y_3) u(1-y_3)$.

Exercise 4.21 $f_Y(y_1, y_2) = \frac{u(y_1+y_2)}{2} u(2-y_1-y_2)u(y_1-y_2)u(2-y_1+y_2)$.

$f_{Y_1}(y_1) = \begin{cases} y_1, & 0 < y_1 \le 1, \\ 2-y_1, & 1 \le y_1 < 2, \\ 0, & \text{otherwise.} \end{cases}$

$f_{Y_2}(y_2) = \begin{cases} 1+y_2, & -1 < y_2 \le 0, \\ 1-y_2, & 0 \le y_2 < 1, \\ 0, & \text{otherwise.} \end{cases}$

Exercise 4.22 $f_{Y_1,Y_2}(y_1, y_2) = f_{X_1,X_2}(y_1 \cos\theta - y_2 \sin\theta, y_1 \sin\theta + y_2 \cos\theta)$.

Exercise 4.24 $F_{X,Y|A}(x, y) =$

$$
\begin{cases}
1, & \text{region } 1\text{--}3, \\
1 - \frac{1}{\pi a^2}\left(-x\psi(x) + a^2\theta_x - \frac{\pi}{2}a^2\right), & \text{region } 1\text{--}2, \\
1 - \frac{1}{\pi a^2}\left(-y\psi(y) + a^2\theta_y - \frac{\pi}{2}a^2\right), & \text{region } 1\text{--}4, \\
1 - \frac{1}{\pi a^2}\left(a^2\cos^{-1}\frac{y}{a} - y\psi(y)\right. \\
\quad \left. -x\psi(x) + a^2\theta_x - \frac{\pi}{2}a^2\right), & \text{region } 1\text{--}5, \\
\frac{1}{\pi a^2}\left(xy - \frac{a^2}{2}\theta_y + \frac{y}{2}\psi(y)\right. \\
\quad \left. +\frac{x}{2}\psi(x) - \frac{a^2}{2}\cos^{-1}\frac{x}{a} + \pi a^2\right), & \text{region } 1\text{--}1 \text{ or } 2\text{--}1, \\
\frac{1}{\pi a^2}\left(xy - \frac{a^2}{2}\theta_x + \frac{x}{2}\psi(x)\right. \\
\quad \left. +\frac{y}{2}\psi(y) - \frac{a^2}{2}\cos^{-1}\frac{y}{a} + \pi a^2\right), & \text{region } 4\text{--}1 \text{ or } 1\text{--}1, \\
\frac{1}{\pi a^2}\left(xy - \frac{a^2}{2}\cos^{-1}\frac{y}{a} + \frac{y}{2}\psi(y)\right. \\
\quad \left. +\frac{x}{2}\psi(x) + \frac{a^2}{2}\theta_x\right), & \text{region } 2\text{--}1 \text{ or } 3\text{--}1, \\
\frac{1}{\pi a^2}\left(xy - \frac{a^2}{2}\cos^{-1}\frac{x}{a} + \frac{x}{2}\psi(x)\right. \\
\quad \left. +\frac{y}{2}\psi(y) + \frac{a^2}{2}\theta_y\right), & \text{region } 3\text{--}1 \text{ or } 4\text{--}1, \\
\frac{1}{\pi a^2}\left(x\psi(x) + a^2\theta_x - \frac{\pi}{2}a^2\right), & \text{region } 2\text{--}2, \\
\frac{1}{\pi a^2}\left(y\psi(y) + a^2\theta_y - \frac{\pi}{2}a^2\right), & \text{region } 4\text{--}2, \\
0, & \text{otherwise.}
\end{cases}
$$

Here, $\psi(t) = \sqrt{a^2 - t^2}$, $\theta_w = \cos^{-1}\frac{-\psi(w)}{a}$, and 'region' is shown in Fig. A.2.
$f_{X,Y|A}(x, y) = \frac{1}{\pi a^2} u\left(a^2 - x^2 - y^2\right)$.

Fig. A.2 The regions of $F_{X,Y|A}(u, v)$ in Exercise 4.24

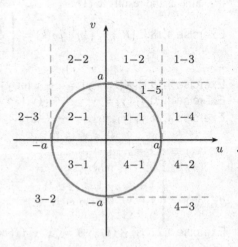

Exercise 4.28 A linear transformation transforming X into an uncorrelated random

vector: $A = \begin{pmatrix} \frac{1}{\sqrt{2}} & -\frac{1}{\sqrt{2}} & 0 \\ \frac{1}{\sqrt{6}} & \frac{1}{\sqrt{6}} & -\frac{2}{\sqrt{6}} \\ \frac{1}{\sqrt{3}} & \frac{1}{\sqrt{3}} & \frac{1}{\sqrt{3}} \end{pmatrix}$.

a linear transformation transforming X into an uncorrelated random vector with unit

variance: $\begin{pmatrix} \frac{1}{\sqrt{2}} & -\frac{1}{\sqrt{2}} & 0 \\ \frac{1}{\sqrt{6}} & \frac{1}{\sqrt{6}} & -\frac{2}{\sqrt{6}} \\ \frac{1}{2\sqrt{3}} & \frac{1}{2\sqrt{3}} & \frac{1}{2\sqrt{3}} \end{pmatrix}$.

Exercise 4.29 $f_Y(y) = \exp(-y)u(y)$.

Exercise 4.30 $p_X(1) = \frac{5}{8}$, $p_X(2) = \frac{3}{8}$. $p_Y(1) = \frac{3}{4}$, $p_Y(2) = \frac{1}{4}$.

Exercise 4.31 pmf of $M = \max(X_1, X_2)$:

$$P(M = m) = e^{-2\lambda}\frac{\lambda^m}{m!}\left(2\sum_{k=0}^{m}\frac{\lambda^k}{k!} - \frac{\lambda^m}{m!}\right)\tilde{u}(m).$$

pmf of $N = \min(X_1, X_2)$:

$$P(N = n) = e^{-2\lambda}\frac{\lambda^n}{n!}\left(2\sum_{k=n+1}^{\infty}\frac{\lambda^k}{k!} + \frac{\lambda^n}{n!}\right)\tilde{u}(n).$$

Exercise 4.32 $f_W(w) = \frac{1}{2}\{u(w) - u(w - 2)\}$. $f_U(v) = u(v + 1) - u(v)$.

$$f_Z(z) = \begin{cases} 0, & z \le -1 \text{ or } z > 2; & \frac{z+1}{2}, & -1 < z \le 0; \\ \frac{1}{2}, & 0 < z \le 1; & \frac{2-z}{2}, & 1 < z \le 2. \end{cases}$$

Exercise 4.33 $f_Y(t) = \begin{cases} \frac{1}{2}\left(t + \frac{3}{2}\right)^2, & -\frac{3}{2} \le t < -\frac{1}{2}, \\ \frac{3}{4} - t^2, & -\frac{1}{2} \le t < \frac{1}{2}, \\ \frac{1}{2}\left(t - \frac{3}{2}\right)^2, & \frac{1}{2} \le t < \frac{3}{2}, \\ 0, & t > \frac{3}{2} \text{ or } t < -\frac{3}{2}. \end{cases}$ $\quad E\{Y^4\} = \frac{13}{80}$.

Exercise 4.34 $f_Y(y) = f_{X_1}(y_1) f_{X_2}(y_2 - y_1) \cdots f_{X_n}(y_n - y_{n-1})$, where $Y = (Y_1, Y_2, \ldots, Y_n)$ and $y = (y_1, y_2, \ldots, y_n)$.

Exercise 4.35 (1) $p_X(x) = \frac{2x+5}{16}$, $x = 1, 2$. $p_Y(y) = \frac{3+2y}{32}$, $y = 1, 2, 3, 4$.

(2) $P(X > Y) = \frac{3}{32}$. $P(Y = 2X) = \frac{9}{32}$. $P(X + Y = 3) = \frac{3}{16}$.

$P(X \le 3 \div Y) = \frac{1}{4}$. (3) not independent.

Exercise 4.36 $f_Y(y) = \begin{cases} 0, & y \le 0; \quad 1 - e^{-y}, \ 0 < y < 1; \\ (e - 1)e^{-y}, & y \ge 1. \end{cases}$

Exercise 4.37 (1) $M_Y(t) = \exp\{7(e^t - 1)\}$. (2) Poisson distribution $P(7)$.

Exercise 4.38 $k = \frac{2}{3}$. $f_{Z|X,Y}(z|x, y) = \frac{x+y+z}{x+y+\frac{1}{2}}$, $0 \le x, y, z \le 1$.

Exercise 4.39 $E\{\exp(-\Lambda)|X = 1\} = \frac{4}{9}$.

Exercise 4.40 $f_{X,Y,Z}(x, y, z) = f_{U_1}(x) f_{U_2}(y - x) f_{U_3}(z - y)$.

Exercise 4.41 (1) $f_{X,Y}(x, y) = \frac{3}{2\pi}\sqrt{1 - x^2 - y^2}u(1 - x^2 - y^2)$.

$f_X(x) = \frac{3}{4}(1 - x^2)u(1 - |x|)$.

(2) $f_{X,Y|Z}(x, y|z) = \frac{1}{\pi(1-z^2)}u(1 - x^2 - y^2 - z^2)$.

not independent of each other.

Exercise 4.42 (1) $c = \frac{1}{2r^2}$. $f_X(x) = \begin{cases} \frac{x+r}{r^2}, & -r \le x \le 0, \\ \frac{r-x}{r^2}, & 0 \le x \le r. \end{cases}$

(2) not independent of each other. (3) $f_Z(z) = \frac{2z}{r^2}u(z)u(r - z)$.

Exercise 4.44 (1) $c = \frac{8}{\pi}$. $f_X(x) = f_Y(x) = \frac{8}{3\pi}\left(1 + 2x^2\right)\sqrt{1 - x^2}u(x)u(1 - x)$.
X and Y are not independent of each other.
(2) $f_{R,\theta}(r,\theta) = \frac{8}{\pi}r^3$, $0 \leq r < 1$, $0 \leq \theta < \frac{\pi}{2}$. (3) $p_Q(q) = \frac{1}{8}$, $q = 1, 2, \ldots, 8$.

Exercise 4.45 probability that the battery with pdf of lifetime f lasts longer than that with $g = \frac{\mu}{\lambda + \mu}$. When $\lambda = \mu$, the probability is $\frac{1}{2}$.

Exercise 4.46 (1) $f_U(x) = xe^{-x}u(x)$. $f_V(x) = \frac{1}{2}e^{-|x|}$.
$f_{XY}(g) = \int_0^\infty e^{-x}e^{-\frac{g}{x}}\frac{1}{x}dxu(g)$. $f_{\frac{X}{Y}}(w) = \frac{u(w)}{(1+w)^2}$.
For $Z = \frac{X}{X+Y} = \frac{W}{1+W}$, $f_Z(z) = u(z)u(1 - z)$.
$f_{\min(X,Y)}(z) = 2e^{-2z}u(z)$. $f_{\max(X,Y)}(z) = 2\left(1 - e^{-z}\right)e^{-z}u(z)$.
$f_{\frac{\min(X,Y)}{\max(X,Y)}}(x) = \frac{2}{(1+x)^2}u(x)u(1-x)$. (2) $f_{V|U}(x|y) = \frac{1}{2y}u(y)u(|y| - x)$.

Exercise 4.48 (1) $E\{M\} = 1$. $Var\{M\} = 1$.

Exercise 4.49 expected value $= \begin{cases} \frac{1}{2p-1}\left[n - \frac{1-p}{2p-1}\left\{1 - \left(\frac{1-p}{p}\right)^n\right\}\right], & p \neq \frac{1}{2}, \\ n^2, & p = \frac{1}{2}. \end{cases}$

Exercise 4.50 $E\{N\} = \frac{1+2p-p^2}{p^2(2-p)}$.

Exercise 4.51 (1) $\mu_1 = \frac{2-p_1p_2+p_1^2p_2}{2p_1p_2-p_1^2p_2-p_1p_2^2+p_1^2p_2^2}$. $\mu_2 = \frac{2-p_1p_2+p_1p_2^2}{2p_1p_2-p_1^2p_2-p_1p_2^2+p_1^2p_2^2}$.
(2) $h_1 = \frac{p_1+p_2-p_1^2p_2+p_1^2p_2^2}{2p_1p_2-p_1^2p_2-p_1p_2^2+p_1^2p_2^2}$. $h_2 = \frac{p_1+p_2-p_1p_2^2+p_1^2p_2^2}{2p_1p_2-p_1^2p_2-p_1p_2^2+p_1^2p_2^2}$.

Exercise 4.52 $\alpha_1 = \frac{p_1^2-2p_1^2p_2+p_1p_2}{\{1-(1-p_1)(1-p_2)\}^2}$. $\alpha_2 = \frac{p_1^2(1-p_2)}{\{1-(1-p_1)(1-p_2)\}^2}$.

Exercise 4.53 integral equation: $g(x) = 1 + \int_x^1 g(y)dy$. $g(x) = e^{1-x}$.

Exercise 4.54 (1) $g(k) = g(k-1)q + g(k-2)pq + g(k-3)p^2q + \delta_{k3}p^3$.
(2) $G_X(s) = \frac{p^3s^3}{1-qs-pqs^2-p^2qs^3}$. (3) $E\{X\} = \frac{1+p+p^2}{p^3}$.

Exercise 4.55 $F_{X,Y|A}(x,y) = \begin{cases} 0, & x < x_1, \\ \frac{F_{X,Y}(x,y)-F_{X,Y}(x_1,y)}{F_X(x_2)-F_X(x_1)}, & x_1 \leq x < x_2, \\ \frac{F_{X,Y}(x_2,y)-F_{X,Y}(x_1,y)}{F_X(x_2)-F_X(x_1)}, & x \geq x_2. \end{cases}$

$f_{X,Y|A}(x,y) = \frac{f_{X,Y}(x,y)}{F_X(x_2)-F_X(x_1)}u(x - x_1)u(x_2 - x)$.

Exercise 4.56 $p_{X,Y}(3,0) = \frac{1}{12}$, $p_{X,Y}(4,0) = \frac{3}{12}$, $p_{X,Y}(5,0) = \frac{2}{12}$,
$p_{X,Y}(3,1) = \frac{1}{12}$, $p_{X,Y}(4,1) = \frac{3}{12}$, $p_{X,Y}(5,1) = \frac{2}{12}$.
$p_{Z|X}(z|x) = \begin{cases} \frac{1}{2}, & x = 3,4,5; \quad z = x - 1, x, \\ 0, & \text{otherwise}. \end{cases}$

$p_{X,Z}(x,z) = \begin{cases} \frac{1}{12}, & x = 3, \quad z = 2,3, \\ \frac{3}{12}, & x = 4, \quad z = 3,4, \\ \frac{2}{12}, & x = 5, \quad z = 4,5, \\ 0, & \text{otherwise}. \end{cases}$

$p_{X|Z}(3|2) = 1$, $p_{X|Z}(3|3) = \frac{1}{4}$, $p_{X|Z}(3|4) = 0$, $p_{X|Z}(3|5) = 0$,
$p_{X|Z}(4|2) = 0$, $p_{X|Z}(4|3) = \frac{3}{4}$, $p_{X|Z}(4|4) = \frac{3}{5}$, $p_{X|Z}(4|5) = 0$,
$p_{X|Z}(5|2) = 0$, $p_{X|Z}(5|3) = 0$, $p_{X|Z}(5|4) = \frac{2}{5}$, $p_{X|Z}(5|5) = 1$.
$p_{Z|Y}(2|0) = 0$, $p_{Z|Y}(3|0) = \frac{1}{6}$, $p_{Z|Y}(4|0) = \frac{3}{6}$, $p_{Z|Y}(5|0) = \frac{1}{3}$,
$p_{Z|Y}(2|1) = \frac{1}{6}$, $p_{Z|Y}(3|1) = \frac{2}{6}$, $p_{Z|Y}(4|1) = \frac{1}{3}$, $p_{Z|Y}(5|1) = 0$.
$p_{Y,Z}(0,2) = 0$, $p_{Y,Z}(0,3) = \frac{1}{12}$, $p_{Y,Z}(0,4) = \frac{3}{12}$, $p_{Y,Z}(0,5) = \frac{2}{12}$,
$p_{Y,Z}(1,2) = \frac{1}{12}$, $p_{Y,Z}(1,3) = \frac{3}{12}$, $p_{Y,Z}(1,4) = \frac{2}{12}$, $p_{Y,Z}(1,5) = 0$.

$p_{Y|Z}(0|2) = 0$, $p_{Y|Z}(0|3) = \frac{1}{4}$, $p_{Y|Z}(0|4) = \frac{3}{5}$, $p_{Y|Z}(0|5) = 1$,
$p_{Y|Z}(1|2) = 1$, $p_{Y|Z}(1|3) = \frac{3}{4}$, $p_{Y|Z}(1|4) = \frac{2}{5}$, $p_{Y|Z}(1|5) = 0$.

Exercise 4.57 (1) $\mathsf{E}\{U\} = \frac{1}{\lambda_1 + \lambda_2}$. $\mathsf{E}\{V - U\} = \frac{1}{\lambda_1} - \frac{1}{\lambda_2} + \frac{2\lambda_1}{\lambda_2} \frac{1}{\lambda_1 + \lambda_2}$.

$\mathsf{E}\{V\} = \frac{1}{\lambda_1} + \frac{\lambda_1}{\lambda_2(\lambda_1 + \lambda_2)}$. (2) $\mathsf{E}\{V\} = \frac{1}{\lambda_1} + \frac{\lambda_1}{\lambda_2(\lambda_1 + \lambda_2)}$.

(3) $f_{U,V-U,I}(x, y, i) = \lambda_1 \lambda_2 e^{-(\lambda_1 + \lambda_2)x} \left\{ \delta(i - 1)e^{-\lambda_2 y} + \delta(i - 2)e^{-\lambda_1 y} \right\} \times u(x)u(y)$.

(4) independent.

Exercise 4.58 $f_X(x) = \frac{\Gamma(p_1 + p_2 + p_3)}{\Gamma(p_1)\Gamma(p_2 + p_3)} x^{p_1 - 1} (1 - x)^{p_2 + p_3 - 1} u(x)u(1 - x)$.

$f_Y(y) = \frac{\Gamma(p_1 + p_2 + p_3)}{\Gamma(p_2)\Gamma(p_1 + p_3)} y^{p_2 - 1} (1 - y)^{p_3 + p_1 - 1} u(y)u(1 - y)$.

$f_{X|Y}(x|y) = \frac{\Gamma(p_1 + p_3)}{\Gamma(p_1)\Gamma(p_3)} x^{p_1 - 1} (1 - x - y)^{p_3 - 1} (1 - y)^{1 - p_1 - p_3} u(x)u(y)u(1 - x - y)$.

$f_{Y|X}(y|x) = \frac{\Gamma(p_2 + p_3)}{\Gamma(p_2)\Gamma(p_3)} y^{p_2 - 1} (1 - x - y)^{p_3 - 1} (1 - x)^{1 - p_2 - p_3} u(x)u(y)u(1 - x - y)$.

$f_{\frac{Y}{1-X}, X}(z, x) = \frac{\Gamma(p_1 + p_2 + p_3)}{\Gamma(p_1)\Gamma(p_2)\Gamma(p_3)} z^{p_2 - 1} (1 - z)^{p_3 - 1} x^{p_1 - 1} (1 - x)^{p_2 + p_3 - 1} u(x)u(z)u(1 - x)u(1 - z)$.

$f_{\frac{Y}{1-X}|X}(z|x) = \frac{\Gamma(p_2 + p_3)}{\Gamma(p_2)\Gamma(p_3)} z^{p_2 - 1} (1 - z)^{p_3 - 1} u(x)u(z)u(1 - x)u(1 - z)$.

Exercise 4.59 $F_{X,Y|B}(x, y)$

$$= \begin{cases}
1, & x \geq 1, \ y \geq 1, \\
0, & x \leq -1, \ y \leq -1, \\
 & \text{or } y \leq -x - 1, \\
\frac{1}{2}(x + 1)^2, & -1 \leq x \leq 0, \ y \geq x + 1, \\
\frac{1}{2}(y + 1)^2, & -1 \leq y \leq 0, \ y \leq x - 1, \\
\frac{1}{2}\{2 - (1 - x)^2\}, & 0 \leq x \leq 1, \ y \geq 1, \\
\frac{1}{2}\{2 - (1 - y)^2\}, & x \geq 1, \ 0 \leq y \leq 1, \\
\frac{1}{2}\{2 - (1 - x)^2 - (1 - y)^2\}, & x \leq 1, \ y \leq 1, \ y \geq -x + 1, \\
\frac{1}{4}(x + y + 1)^2, & x \leq 0, \ y \leq 0, \ y \geq -x - 1, \\
\frac{1}{4}\{(x + y + 1)^2 - 2x^2\}, & x \geq 0, \ y \leq 0, \ y \geq x - 1, \\
\frac{1}{4}\{(x + y + 1)^2 - 2y^2\}, & x \leq 0, \ y \geq 0, \ y \leq x + 1, \\
\frac{1}{4}\{(x + y + 1)^2 & \\
\quad - 2(x^2 + y^2)\}, & x \geq 0, \ y \geq 0, \ y \leq -x + 1.
\end{cases}$$

$f_{X,Y|B}(x, y) = \frac{1}{2} u(1 - |x| - |y|)$.

Exercise 4.60 $F_{X,Y|A}(x, y)$

$$= \begin{cases}
1, & \text{region } 1\text{-}3, \\
1 - \frac{1}{2a^4}\psi^4(x), & \text{region } 1\text{-}2, \\
1 - \frac{1}{2a^4}\psi^4(y), & \text{region } 1\text{-}4, \\
1 - \frac{1}{2a^4}\{\psi^4(x) + \psi^4(y)\}, & \text{region } 1\text{-}5, \\
\frac{1}{4a^4}\{(a^2 + x^2 + y^2)^2 - 2(x^4 + y^4)\}, & \text{region } 1\text{-}1, \\
\frac{1}{4a^4}\{(\psi^2(x) + y^2)^2 - 2y^4\}, & \text{region } 2\text{-}1, \\
\frac{1}{2a^4}\psi^4(x), & \text{region } 2\text{-}2, \\
\frac{1}{4a^4}\{(\psi^2(y) + x^2)^2 - 2x^4\}, & \text{region } 4\text{-}1, \\
\frac{1}{2a^4}\psi^4(y), & \text{region } 4\text{-}2, \\
\frac{1}{4a^4}(\psi^2(x) - y^2)^2, & \text{region } 3\text{-}1, \\
0, & \text{otherwise,}
\end{cases}$$

where $\psi(t) = \sqrt{a^2 - t^2}$ (refer to Fig. A.2).

$f_{X,Y|A}(x, y) = \frac{2|xy|}{a^4} u(\psi^2(x) - y^2)$.

Exercise 4.63 $f_X(x) = \frac{n!}{(n-i)!(i-1)!} F^{i-1}(x)\{1 - F(x)\}^{n-i} f(x).$

$f_Y(y) = \frac{n!}{(n-k)!(k-1)!} F^{k-1}(y)\{1 - F(y)\}^{n-k} f(y).$

Exercise 4.64 (1) $p_{X_1,X_2,X_3,X_4}(x_1, x_2, x_3, x_4)$

$$= \begin{cases} \binom{N}{x_1,x_2,x_3,x_4} \{p_1(1-p_2)\}^{x_1} \{(1-p_1)p_2\}^{x_2} \\ \quad \times (p_1 p_2)^{x_3} \{(1-p_1)(1-p_2)\}^{x_4}, & \text{if } \sum_{i=1}^{4} x_i = N, \\ 0, & \text{otherwise.} \end{cases}$$

(3) $\hat{p}_1 = \frac{X_3}{X_2+X_3}. \; \hat{p}_2 = \frac{X_3}{X_1+X_3}. \; \hat{\lambda} = \frac{(X_2+X_3)(X_1+X_3)}{X_3}.$ (4) $\hat{X}_4 = \frac{X_1 X_2}{X_3}.$

Exercise 4.65 (1) $\rho_{X|X|} = 0.$ (2) $\rho_{X|X|} = 1.$ (3) $\rho_{X|X|} = -1.$

Exercise 4.66 $f_{X,2X+1}(x, y) = \frac{1}{2} \{u(x) - u(x-1)\} \delta \left(\frac{y-1}{2} - x \right).$

Exercise 4.67 For $x \in \{x | f_X(x) > 0\},$

$f_{Y|X}(y|x) = \{\delta(x+y) + \delta(x-y)\} u(y).$

Exercise 4.69 (1) $F_{X,Y}(x, y) = [\{F(x) - F(-\sqrt{y})\} u(x + \sqrt{y}) - \{F(x) - F(\sqrt{y})\}$

$u(x - \sqrt{y})] u(y) = \{F(\min(x, \sqrt{y})) - F(-\sqrt{y})\} u(y) u(x + \sqrt{y}).$

(2) $f_{X,Y}(x, y) = \frac{f(x)}{2\sqrt{y}} \{\delta(x + \sqrt{y}) + \delta(x - \sqrt{y})\} u(y)$

$= \frac{f(x)}{2|x|} \delta(\sqrt{y} - |x|) u(y).$

(3) $f_{X|Y}(x|y) = \frac{f(x)}{f(\sqrt{y})+f(-\sqrt{y})} \{\delta(x + \sqrt{y}) + \delta(x - \sqrt{y})\} u(y).$

Exercise 4.71 $F_{X,Y}(x, y) = \{F_X(x)u(y - x) + F_X(y)u(x - y)\} u(y).$

$f_{X,Y}(x, y) = f_X(x) \{u(y)\delta(y - x) + u(y - x)\delta(y)\}.$

Exercise 4.73 $f_{X_1}(t) = 6t(1 - t)u(t)u(1 - t). \; f_{X_2}(t) = \frac{3}{2}(1 - |t|)^2 u(1 - |t|)$

Exercise 4.74 (1) $f_Y(y_1, y_2) = \frac{1}{2|y_1|} u \left(\frac{y_2}{y_1} + y_1 \right) u \left(1 - \frac{1}{2} \left(\frac{y_2}{y_1} + y_1 \right) \right) u \left(\frac{y_2}{y_1} - y_1 \right)$

$u \left(1 - \frac{1}{2} \left(\frac{y_2}{y_1} - y_1 \right) \right)$

(2) $f_{Y_1}(y) = (1 - |y|)u(1 - |y|).$

$f_{Y_2}(y) = \frac{1}{2}u(1 - |y|) \ln \frac{\sqrt{|y|}}{1-\sqrt{1-|y|}} = \frac{1}{4}u(1 - |y|) \ln \frac{1+\sqrt{1-|y|}}{1-\sqrt{1-|y|}}.$

Chapter 5 Normal Random Vectors

Exercise 5.1 (3) The vector (X, Y) is not a bi-variate normal random vector.
(4) The random variables X and Y are not independent of each other.

Exercise 5.2 $f_{X_1,X_2|X_1^2+X_2^2<a^2}(x, y) = \frac{\exp\{-\frac{1}{2}(x^2+y^2)\}}{2\pi\{1-\exp(-\frac{a^2}{2})\}} u(a^2 - x^2 - y^2).$

Exercise 5.3 (1) $f_{U,V}(t, v) = \frac{t}{\pi} \exp \left(-\frac{t^2}{2} \right) u(t) \{u(v + \frac{\pi}{2}) - u(v - \frac{\pi}{2})\}.$

(2) $f_{U,V}(t, v) = \frac{1}{\pi\sqrt{2v}} \exp \left(-\frac{2t^2+v}{2} \right) u(v).$

Exercise 5.4 $f_{Y|X}(y|x) = \frac{1}{\sqrt{3\pi}} \exp \left\{ -\frac{1}{3} \left(y - 4 - \frac{x-3}{\sqrt{2}} \right)^2 \right\}.$

$f_{X|Y}(x|y) = \sqrt{\frac{2}{3\pi}} \exp \left\{ -\frac{2}{3} \left(x - 3 - \frac{y-4}{2\sqrt{2}} \right)^2 \right\}.$

Exercise 5.5 $\rho_{ZW} = \dfrac{(\sigma_2^2-\sigma_1^2)\cos\theta\sin\theta}{\sqrt{(\sigma_1^2+\sigma_2^2)^2\cos^2\theta\sin^2\theta+\sigma_1^2\sigma_2^2(\cos^2\theta-\sin^2\theta)^2}}$.

Exercise 5.8 (1) $F_{X|Y}(x|y) = \begin{cases} 0, & x < -|y|; \quad \frac{1}{2}, \quad -|y| \le x < |y|; \\ 1, & x > |y|. \end{cases}$

(2) $F_X(x) = \frac{1}{\sqrt{2\pi}}\int_{-\infty}^{x}\exp\left(-\frac{v^2}{2}\right)dv$. The random variable X is normal.

(3) The vector (X, Y) is not a normal random vector.

(4) $f_{X|Y}(x|y) = \frac{1}{2}\delta(x+y) + \frac{1}{2}\delta(x-y)$.

$f_{X,Y}(x, y) = \frac{1}{2}\{\delta(x+y) + \delta(x-y)\}\frac{1}{\sqrt{2\pi}}\exp\left(-\frac{y^2}{2}\right)$.

Exercise 5.9 $\begin{pmatrix} \frac{1}{\sqrt{2}} & 0 & -\frac{1}{\sqrt{2}} \\ \frac{3}{\sqrt{34}} & \frac{4}{\sqrt{34}} & \frac{3}{\sqrt{34}} \\ \frac{2}{\sqrt{17}} & -\frac{3}{\sqrt{17}} & \frac{2}{\sqrt{17}} \end{pmatrix}$

Exercise 5.10 $f_C(r, \theta) = r f_X(r\cos\theta, r\sin\theta)u(r)u(\pi-|\theta|)$. The random vector C is an independent random vector when X is an i.i.d. random vector with marginal distribution $\mathcal{N}(0, \sigma^2)$.

Exercise 5.12 The conditional distribution of X_3 when $X_1 = X_2 = 1$ is $\mathcal{N}(1, 2)$.

Exercise 5.13 $ac\sigma_X^2 + (bc+ad)\rho\sigma_X\sigma_Y + bd\sigma_Y^2 = 0$.

Exercise 5.14 $E\{XY\} = \rho\sigma_X\sigma_Y$. $E\{X^2Y\} = 0$. $E\{X^3Y\} = 3\rho\sigma_X^3\sigma_Y$.
$E\{X^2Y^2\} = (1+2\rho^2)\sigma_X^2\sigma_Y^2$.

Exercise 5.16
$E\{Z^2W^2\} = (1+2\rho^2)\sigma_1^2\sigma_2^2 + m_2^2\sigma_1^2 + m_1^2\sigma_2^2 + m_1^2m_2^2 + 4m_1m_2\rho\sigma_1\sigma_2$.

Exercise 5.17 $\mu_{51} = 15\rho\sigma_1^5\sigma_2$. $\mu_{42} = 3\rho(1+4\rho^2)\sigma_1^4\sigma_2^2$.
$\mu_{33} = 3\rho(3+2\rho^2)\sigma_1^3\sigma_2^3$.

Exercise 5.19 $E\{|XY|\} = 2 + \frac{\sqrt{2}}{6}\pi$.

Exercise 5.20 $E\{|X_1|\} = \sqrt{\frac{2}{\pi}}$. $E\{|X_1X_2|\} = \frac{2}{\pi}\left(\sqrt{1-\rho^2}+\rho\sin^{-1}\rho\right)$.

$E\{|X_1X_2^3|\} = \frac{2}{\pi}\left\{3\rho\sin^{-1}\rho + (2+\rho^2)\sqrt{1-\rho^2}\right\}$.

Exercise 5.25 $\frac{3!4!4!5!}{2!\times 2!4!} = 4320$.

Exercise 5.30 $f_X(x) = \frac{r}{\pi(x^2+r^2)}$.

Exercise 5.32 $E\{Y\} = n + \delta$. $\sigma_Y^2 = 2n + 4\delta$.

Exercise 5.34 $E\{Z\} = \delta\frac{\Gamma\left(\frac{n-1}{2}\right)}{\Gamma\left(\frac{n}{2}\right)}\sqrt{\frac{n}{2}}$ for $n > 1$.

$\mathrm{Var}\{Z\} = \frac{n(1+\delta^2)}{n-2} - \frac{n\delta^2}{2}\frac{\Gamma^2\left(\frac{n-1}{2}\right)}{\Gamma^2\left(\frac{n}{2}\right)}$ for $n > 2$.

Exercise 5.36 $E\{H\} = \frac{n(m+\delta)}{m(n-2)}$ for $n > 2$.

$\mathrm{Var}\{H\} = \frac{2n^2\{(m+\delta)^2+(n-2)(m+2\delta)\}}{m^2(n-4)(n-2)^2}$ for $n > 4$.

Exercise 5.40 $\mu_4(\overline{X}_n) = \frac{\mu_4}{n^3} + \frac{3(n-1)\mu_2^2}{n^3}$.

Exercise 5.42 (1) pdf: $f_V(v) = e^{-v}u(v)$. cdf: $F_V(v) = (1-e^{-v})u(v)$.

Exercise 5.43 $R_Y = \frac{2\beta^2}{\pi}\sin^{-1}\frac{\rho}{1+\alpha^2}$. $\rho_Y = \frac{\sin^{-1}\frac{\rho}{1+\alpha^2}}{\sin^{-1}\frac{1}{1+\alpha^2}}$.

$\rho_Y|_{\alpha^2=1} = \frac{6}{\pi}\sin^{-1}\frac{\rho}{2}$. $\lim_{\alpha^2\to 0}\rho_Y = \frac{2}{\pi}\sin^{-1}\rho$.

Chapter 6 Convergence of Random Variables

Exercise 6.1 $\{X_n(\omega)\}$ converges almost surely, but not surely, to $X(\omega)$.

Exercise 6.2 $\{X_n(\omega)\}$ converges almost surely, but not surely, to $X(\omega)$.

Exercise 6.3 (1) $S_n \sim NB(n, \alpha)$. (2) $S_n \sim NB\left(\sum_{i=1}^{n} r_i, \alpha\right)$.

(3) $S_n \sim P\left(\sum_{i=1}^{n} \lambda_i\right)$. (4) $S_n \sim G\left(\sum_{i=1}^{n} \alpha_i, \beta\right)$. (5) $S_n \sim C\left(\sum_{i=1}^{n} \mu_i, \sum_{i=1}^{n} \theta_i\right)$.

Exercise 6.4 $\frac{S_n}{n} \xrightarrow{P} m$.

Exercise 6.7 The weak law of large numbers holds true.

Exercise 6.10 $P(50 < S_n \leq 80) = P(50.5 < S_n < 80.5) \approx 0.9348$.

Exercise 6.11 $E\{U\} = p(1 - p) + np^2$. $E\{V\} = np^2$.

Exercise 6.12 mgf of Y $M_Y(t) = \frac{1}{2 - e^{2t}}$. expected value= 2. variance= 8.

Exercise 6.13 (1) $P_3 = P_5 = \frac{1}{2}$. $P_4 = \frac{5}{16}$. (2) $\lim_{n \to \infty} P_n = \frac{1}{2}$.

Exercise 6.14 $\lim_{n \to \infty} F_n(x) = u(x)$. It is a cdf.

Exercise 6.15 It is convergent. $F_{Y_n}(y) \to \left\{1 - \exp\left(-\frac{y}{\theta}\right)\right\} u(y)$.

Exercise 6.16 $\lim_{n \to \infty} F_{Y_n}(y) = u(y)$.

Exercise 6.17 It is convergent. $F_n(x) \to u(x)$.

Exercise 6.18 (1) $\varphi_{\hat{X}_n}(w) = \varphi_X(w) \prod_{i=1}^{n} \varphi_{W_i}\left(\frac{w}{n}\right)$. $E\left\{\hat{X}_n\right\} = E\{X\}$.

$\text{Var}\left\{\hat{X}_n\right\} = \text{Var}\{X\} + \frac{1}{n^2} \sum_{i=1}^{n} \sigma_i^2$.

(2) $\text{Cov}\left(Y_i, Y_j\right) = \text{Var}\{X\} + \sigma_i^2$ for $i = j$ and $\text{Var}\{X\}$ for $i \neq j$.

(3) Denoting $\sigma^2 = \frac{1}{n^2} \sum_{i=1}^{n} \sigma_i^2$, $f_{\varepsilon_n}(\alpha) = \frac{1}{\sqrt{2\pi\sigma^2}} \exp\left(-\frac{\alpha^2}{2\sigma^2}\right)$.

$f_{\hat{X}_n|X}(\alpha|\beta) = \frac{1}{\sqrt{2\pi\sigma^2}} \exp\left\{-\frac{(\alpha - \beta)^2}{2\sigma^2}\right\}$.

(4) $\hat{X}_n$ is mean square convergent to X.

Exercise 6.26 $a_n = n\sigma^2$. $b_n = n\mu_4$.

Exercise 6.27 $\alpha \leq 0$.

Exercise 6.29 Exact value: $P(S \geq 3) = 1 - 5e^{-2} \approx 0.3233$.

approximate value: $P(S \geq 3) = P\left(\frac{S-2}{\sqrt{2}} \geq \frac{1}{\sqrt{2}}\right) = P\left(Z \geq \frac{1}{\sqrt{2}}\right)$

≈ 0.2398. $\left(\text{or } P(S \geq 3) = P(S > 2.5) = P\left(Z > \frac{1}{2\sqrt{2}}\right) \approx 0.3618\right)$.

Exercise 6.30 $\lim_{n \to \infty} F_n(x) = F(x)$, where F is the cdf of $U[0, 1)$.

$\frac{d}{dx}\left\{\lim_{n \to \infty} F_n(x)\right\} \neq \lim_{n \to \infty}\left\{\frac{d}{dx} F_n(x)\right\}$.

Exercise 6.32 Distribution of points: Poisson with parameter μp.

mean: $\mu p = 1.5$. variance: $\mu p = 1.5$.

Exercise 6.33 Distribution of the total information sent via the fax: geometric distribution with expected value $\frac{1}{\alpha\beta} = 4 \times 10^5$.

Exercise 6.34 Expected value: $E\{S_N\} = \frac{1}{p\lambda}$. variance: $\text{Var}\{S_N\} = \frac{1}{p^2\lambda^2}$.

Exercise 6.35 Expected value of T: $\frac{3}{\lambda}$.

Exercise 6.36 (1) $\lim\limits_{n\to\infty} F_n(x) = \frac{1}{2}$. This limit is not a cdf.

(2) $F_n(x) \to 1$. This limit is not a cdf.

(3) Because $G_{2n}(x) \to 1$ and $G_{2n+1}(x) \to 0$, $\{G_n(x)\}_{n=1}^{\infty}$ is not convergent.

Exercise 6.37 $F_{Y_n}(y) \to u(y - \beta)$.

Index

Printed in the United States
by Baker & Taylor Publisher Services

Printed in the United States
by Baker & Taylor Publisher Services